World – Political

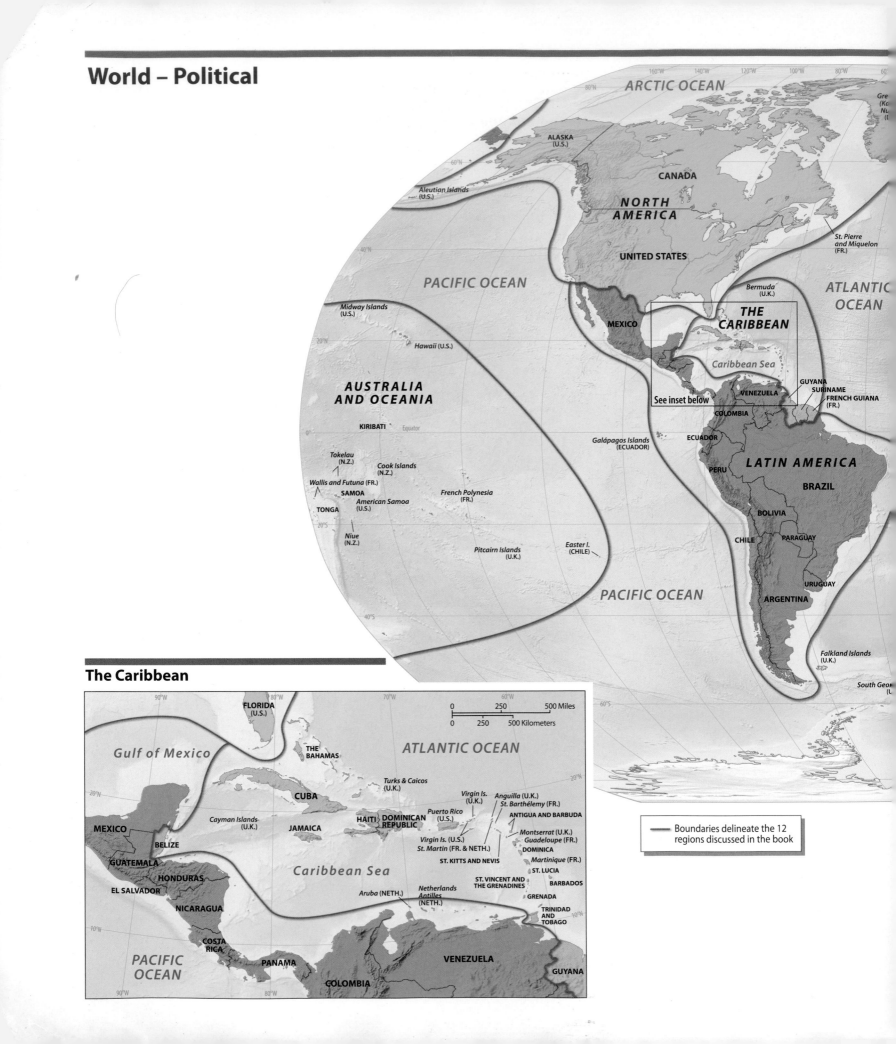

ARCTIC OCEAN

ALASKA (U.S.)

CANADA

NORTH AMERICA

UNITED STATES

Aleutian Islands (U.S.)

PACIFIC OCEAN

Midway Islands (U.S.)

Hawaii (U.S.)

MEXICO

THE CARIBBEAN

Bermuda (U.K.)

St. Pierre and Miquelon (FR.)

ATLANTIC OCEAN

Caribbean Sea

See inset below

GUYANA
SURINAME
FRENCH GUIANA (FR.)

VENEZUELA

COLOMBIA

AUSTRALIA AND OCEANIA

KIRIBATI

Equator

Tokelau (N.Z.)

Cook Islands (N.Z.)

Wallis and Futuna (FR.)

SAMOA

American Samoa (U.S.)

TONGA

Niue (N.Z.)

French Polynesia (FR.)

ECUADOR

LATIN AMERICA

PERU

BRAZIL

BOLIVIA

Galápagos Islands (ECUADOR)

CHILE

PARAGUAY

Pitcairn Islands (U.K.)

Easter I. (CHILE)

PACIFIC OCEAN

URUGUAY

ARGENTINA

Falkland Islands (U.K.)

South Geor (U

The Caribbean

FLORIDA (U.S.)

Gulf of Mexico

THE BAHAMAS

ATLANTIC OCEAN

0   250   500 Miles

0   250   500 Kilometers

Turks & Caicos (U.K.)

CUBA

Cayman Islands (U.K.)

HAITI

DOMINICAN REPUBLIC

Puerto Rico (U.S.)

Virgin Is. (U.K.)

Anguilla (U.K.)
St. Barthélemy (FR.)

ANTIGUA AND BARBUDA

Virgin Is. (U.S.)
St. Martin (FR. & NETH.)

Montserrat (U.K.)
Guadeloupe (FR.)

DOMINICA

Martinique (FR.)

MEXICO

JAMAICA

ST. KITTS AND NEVIS

ST. LUCIA

BELIZE

GUATEMALA

HONDURAS

EL SALVADOR

Caribbean Sea

ST. VINCENT AND THE GRENADINES

BARBADOS

NICARAGUA

Aruba (NETH.)

Netherlands Antilles (NETH.)

GRENADA

TRINIDAD AND TOBAGO

COSTA RICA

PACIFIC OCEAN

PANAMA

VENEZUELA

GUYANA

COLOMBIA

Boundaries delineate the 12 regions discussed in the book

ARCTIC OCEAN

THE RUSSIAN DOMAIN

RUSSIA

Svalbard
(NOR.)

See inset below

EUROPE

CENTRAL ASIA

KAZAKHSTAN

MONGOLIA

Kuril Is.
(RUS.)

GEORGIA

UZBEKISTAN

KYRGYZSTAN

ARMENIA

NORTH
KOREA

PACIFIC OCEAN

TURKEY

TURKMENISTAN

TAJIKISTAN

AZERBAIJAN

CHINA

EAST
ASIA

SOUTH
KOREA

TUNISIA

LEBANON

SYRIA

JAPAN

ISRAEL

JORDAN IRAQ

IRAN

AFGHANISTAN

MOROCCO

ALGERIA

LIBYA

GAZA STRIP

WEST BANK

KUWAIT

PAKISTAN

NEPAL

BHUTAN

NORTH AFRICA/
SOUTHWEST ASIA

EGYPT

BAHRAIN
QATAR

SAUDI
ARABIA

TAIWAN

ary Is.
SP.)

WESTERN SAHARA
(MOR.)

MAURITANIA

MALI

NIGER

CHAD

SUDAN

UNITED ARAB
EMIRATES

OMAN

ERITREA

YEMEN

SOUTH
ASIA

MYANMAR
(BURMA)

LAOS

Northern
Mariana Is.
(U.S.)

Wake Island
(U.S.)

BANGLADESH

SENEGAL

BURKINA
FASO

BENIN

NIGERIA

INDIA

THAILAND

VIETNAM

Guam
(U.S.)

GUINEA

GHANA

CENTRAL
AFRICAN
REP.

SOUTH
SUDAN

ETHIOPIA

CAMBODIA

PHILIPPINES

FEDERATED STATES
OF MICRONESIA

MARSHALL
ISLANDS

SIERRA
LEONE

SRI
LANKA

SOUTHEAST
ASIA

LIBERIA

CÔTE D'IVOIRE

TOGO

CAMEROON

SOMALIA

BRUNEI

PALAU

SÃO TOMÉ AND PRÍNCIPE

REP.
OF
THE
CONGO

GABON

DEM. REP.
OF THE
CONGO

UGANDA

KENYA

MALDIVES

MALAYSIA

EQUATORIAL
GUINEA

RWANDA

BURUNDI

SINGAPORE

NAURU

KIRIBATI

TANZANIA

SEYCHELLES

INDIAN
OCEAN

INDONESIA

PAPUA
NEW
GUINEA

SOLOMON
ISLANDS

TIMOR-LESTE

TUVALU

SUB-SAHARAN
AFRICA

COMOROS

Mayotte (FR.)

Cocos (Keeling) Islands
(AUS.)

Christmas Island
(AUS.)

St. Helena
(U.K.)

ANGOLA

ZAMBIA

MALAWI

MOZAMBIQUE

MADAGASCAR

MAURITIUS

VANUATU

AUSTRALIA
AND OCEANIA

FIJI

NAMIBIA

ZIMBABWE

Réunion
(FR.)

New
Caledonia
(FR.)

BOTSWANA

2,000 Miles

AUSTRALIA

ATLANTIC
OCEAN

SOUTH
AFRICA

SWAZILAND

LESOTHO

0   1,000

0   1,000   2,000 Kilometers

Norfolk
Island
(AUS.)

NEW
ZEALAND

Kerguelen Is.
(FR.)

Europe

20°W

0°

250   500 Miles

ICELAND

0   250   500 Kilometers

Faroe Is.
(DEN.)

SWEDEN

FINLAND

60°N

NORWAY

ESTONIA

RUSSIA

North
Sea

DENMARK

LATVIA

LITHUANIA

BELARUS

IRELAND

UNITED
KINGDOM

NETHERLANDS

RUSSIA

POLAND

BELGIUM

GERMANY

UKRAINE

ATLANTIC
OCEAN

Channel Islands
(U.K.)

LUXEMBOURG

CZECH REP.

SLOVAKIA

LIECHTENSTEIN

AUSTRIA

HUNGARY

MOLDOVA

FRANCE

SWITZERLAND

SLOVENIA

CROATIA

ROMANIA

MONACO

ITALY

SERBIA

SAN
MARINO

BOSNIA AND
HERZEGOVINA

BULGARIA

ANDORRA

Corsica
(FR.)

MONTENEGRO

KOSOVO

MACEDONIA

Black Sea

SPAIN

Balearic Is.
(SP.)

Sardinia
(IT.)

VATICAN
CITY

ALBANIA

PORTUGAL

40°N

GREECE

TURKEY

20°W

Gibraltar (U.K.)

Mediterranean

Sicily (IT.)

MALTA

Crete
(GR.)

CYPRUS

SYRIA

MOROCCO

ALGERIA

TUNISIA

20°E

LEBANON

GLOBALIZATION
AND DIVERSITY
GEOGRAPHY OF A CHANGING WORLD

FOURTH EDITION

LES ROWNTREE

University of California, Berkeley

MARTIN LEWIS

Stanford University

MARIE PRICE

George Washington University

WILLIAM WYCKOFF

Montana State University

PEARSON

Boston   Columbus   Indianapolis   New York   San Francisco   Upper Saddle River   Amsterdam
Cape Town   Dubai   London   Madrid   Milan   Munich   Paris   Montréal   Toronto   Delhi
Mexico City   São Paulo   Sydney   Hong Kong   Seoul   Singapore   Taipei   Tokyo

Geography Editor: Christian Botting
Senior Marketing Manager: Maureen McLaughlin
Project Editor: Anton Yakovlev
Director of Development: Jennifer Hart
Development Editor: David Chelton
Assistant Editor: Kristen Sanchez
Editorial Assistant: Bethany Sexton
Senior Marketing Assistant: Nicola Houston
Media Producer: Tod E. Regan
Managing Editor, Geosciences and Chemistry: Gina M. Cheselka
Production Project Manager: Connie M. Long
Full Service/Composition: Element LLC
Full Service Project Manager: Kelly Keeler, Element LLC
Illustrations: International Mapping
Image Lead: Maya Melenchuk
Photo Researcher: Stefanie Ramsey
Text Permissions Manager: Joanna Green
Text Permission Researcher: Varoon Deo-Singh
Design Manager: Laura Gardner
Interior and Cover Designer: Mark Scala, Symmetre Design Group LLC
Operations Specialist: Michael Penne
Cover Photo Credit: DreamPictures/age fotostock

Credits and acknowledgments borrowed from other sources and reproduced, with permission, in this textbook appear on the appropriate page within the text or on the credits page beginning on page CR-1.

Many of the designations used by manufacturers and sellers to distinguish their products are claimed as trademarks. Where those designations appear in this book, and the publisher was aware of a trademark claim, the designations have been printed in initial caps or all caps.

Library of Congress Cataloging-in-Publication Data

Globalization and diversity : geography of a changing world / Les Rowntree . . . [et al.]. — 4th ed.
    p. cm.
  Includes index.
  ISBN 978-0-321-82146-1
  1. Economic geography. 2. Globalization. 3. Cultural pluralism. I. Rowntree, Lester,
  HF1025.G59 2014
  330.9--dc23

                                            2012034151

5 6 7 8 9 10—V011—17 16 15 14

www.pearsonhighered.com

ISBN-10: 0-321-82146-7; ISBN-13: 978-0-321-82146-1 (Student Edition)
ISBN-10: 0-321-86219-8; ISBN-13: 978-0-321-86219-8 (Instructor's Review Copy)

BRIEF CONTENTS

About Our Sustainability Initiatives

Pearson recognizes the environmental challenges facing this planet, as well as acknowledges our responsibility in making a difference. This book is carefully crafted to minimize environmental impact. The binding, cover, and paper come from facilities that minimize waste, energy consumption, and the use of harmful chemicals. Pearson closes the loop by recycling every out-of-date text returned to our warehouse.

Along with developing and exploring digital solutions to our market's needs, Pearson has a strong commitment to achieving carbon-neutrality. As of 2009, Pearson became the first carbon- and climate-neutral publishing company. Since then, Pearson remains strongly committed to measuring, reducing, and offsetting our carbon footprint.

The future holds great promise for reducing our impact on Earth's environment, and Pearson is proud to be leading the way. We strive to publish the best books with the most up-to-date and accurate content, and to do so in ways that minimize our impact on Earth. To learn more about our initiatives, please visit **www.pearson.com/responsibility**.

PEARSON

CONTENTS

5 THE CARIBBEAN   139

6 SUB-SAHARAN AFRICA   171

7 SOUTHWEST ASIA AND NORTH AFRICA   215

PREFACE

Globalization and Diversity: Geography of a Changing World, Fourth Edition, is intended for college and university world regional geography classes. Our perspective is that globalization is the most fundamental reorganization of the world's environmental economic, cultural, and geopolitical structure since the Industrial Revolution, therefore our focus throughout is to discuss and analyze the geographic changes accompanying these changes.

As geographers and teachers, we think it essential for students to understand two interactive realities in the modern world. First, they need to appreciate and evaluate the changes in environmental, cultural, political, and economic systems resulting from globalization. Second, they need to deepen their understanding of the development and persistence of geographic diversity. These opposing and interactive forces form a theme running throughout our book and are reflected in the title, *Globalization and Diversity.*

NEW TO THE FOURTH EDITION

- **Working Toward Sustainability** features in all chapters describe sustainability projects throughout the world, emphasizing positive environmental and social initiatives and their results.
- **Chapter opening vignettes** with panoramic photographs provide a current real-world introduction to each region.
- Each chapter begins with a set of **Learning Objectives** to help students recognize the most important concepts in the chapter.
- **Review Questions** at the end of each major thematic section help students check their comprehension of the material as they read.
- Following an introduction in Chapter 1, **gender issues** are discussed in all regional chapters.
- **Migration maps and text material** highlight the importance of regional and global migration flows.
- All regional chapters now contain graphics showing how **population pyramids** at different scales help us better understand population dynamics.
- Updated **environmental issues maps** in all regional chapters have been standardized and brought up to date.
- Updated **population maps** in all regional chapters contain more effective categorizations of density and urban size.
- **Current information on climate change** is integrated in all of our chapters, including the policies and programs enacted by different countries and regions to address climate change, emission reduction, and energy usage. Material on these topics is introduced in Chapter 2 and subsequently presented through the regional chapters in dedicated climate change sections.
- **Current data and information** are integrated into all text, tables, and maps, based on the the lastest information from the

2010 U.S. Census, the World Bank, United Nations, and 2012 Population Reference Bureau Population Data.

- **Quick Response (QR) codes** at the end of each of chapter enable students to link smartphones and other mobile devices to the new Globalization and Diversity authors' blog, as well as to Martin Lewis' innovative and provocative *GeoCurrents* blog.
- The Fourth Edition is now supported by **MasteringGeography**™, the most widely used and effective online homework, tutorial, and assessment system for the geosciences. Assignable media and activities include MapMaster™ interactive maps, *Encounter World Regional Geography* Google Earth Explorations, geography videos, Thinking Spatially and Data Analysis Activities, end-of-chapter questions, reading quizzes, Test Bank questions, and select Geoscience Animations.

New and Updated in Chapter 1

- *Geography Matters.* New discussion of fundamental geographic concepts, including areal differentiation, regions, and the cultural landscape.
- *Geographer's Toolbox.* New discussion of latitude and longitude, map projections, scale, different kinds of maps, aerial photos remote sensing, and GIS.
- **Expanded, integrated treatment of globalization.** A revised presentation of globalization in different contexts that provides students with a critical framework for understanding its varied aspects.
- **Demographic transition revised.** Following the lead of professional demographers, a fifth stage has been added to the traditional demographic transition model to account for the current very low natural population rates in developed countries.
- *The Nation-State Revisited.* A critical view of the traditional nation-state concept that sets the scene for regional material on micro-regionalism, ethnic separatism, migrant enclaves, and multicultural nationalism.

New and Updated in Chapter 2

- *Geology: A Restless Earth.* A new section that includes explanation of plate tectonics and geologic hazards.
- *Climate Controls.* New content that discusses the effects of latitude, land-water interactions, global wind patterns, and topography on weather and climate.
- **Updated section on climate change and international activity to limit atmospheric emissions.** New material based on the latest CO_2 emission data and global discussions regarding post-Kyoto activity.
- *Water: A Scarce World Resource.* A new section that focuses on global water problems and solutions.

- **Revised discussion of bioregions.** A more detailed cartographic depiction of biomes and bioregions, complemented by a fuller discussion of the world's ecological diversity, as well as the issues faced in protecting those environments around the globe.

CHAPTER ORGANIZATION

We organized *Globalization and Diversity* to describe and explain the major world regions, including those of Asia, Europe, Africa, and the Americas. Our 12 regional chapters, however, depart somewhat from traditional world regional textbooks. Instead of filling these regional chapters with descriptions of individual countries, we develop five important themes as the structure for each regional chapter. We begin with *Environmental Geography,* describing the physical geography of each region, as well as current environmental issues. Then we assess the region's *Population and Settlement,* discussing its demography, migration, land use, and settlement (including cities). Third is a section on *Cultural Coherence and Diversity* that examines the geography of language and religion. It also explores current popular culture and cultural tensions resulting from the interplay of globalization and diversity. Following this is a thematic section on each region's *Geopolitical Framework* that examines the political geography of the region, including micro-regionalism, separatism, ethnic conflicts, global terrorism, and supranational organizations. We conclude each regional chapter with a section on *Economic and Social Development,* in which we analyze each region's economic framework; this section also examines social development, a topic that includes health, education, and gender issues.

The regional framework follows two substantive introductory chapters that provide the conceptual framework of human and physical geography necessary for understanding the contemporary globalized world. In the first chapter, we introduce the idea of globalization and ask students to ponder the costs and benefits of the globalization process, a critical perspective that is becoming increasingly necessary in the contentious global arena. Following this material, we examine the geographical foundation for each of the five thematic sections. This discussion draws heavily on the major concepts fundamental to an introductory university geography course. The second introductory chapter, "The Changing Global Environment," presents the themes and concepts of global physical geography, including landforms and geology, climate, climate change, hydrology, and biogeography.

CHAPTER FEATURES

Within each regional chapter, several unique features complement the thematic pedagogy of our approach:

- **Comparable maps.** Of the many maps in each regional chapter, eight are constructed on the same themes and with similar data so that readers can easily draw comparisons between different regions. Thus, almost every regional chapter has maps of physical geography, climate, environmental issues, population density, migration, language, religion, and geopolitical issues of the region.

- **Other maps.** Each regional chapter has additional maps illustrating such major themes as urban growth, ethnic tensions, social development, regional development, and linkages to the global economy.
- **Comparable regional data sets.** To facilitate comparison between regions, as well as to provide important insight into the characteristics of each region, each chapter contains two thematic tables. The first provides population data of all sorts, including population density, level of urbanization, total fertility rates, and proportions of the population under 15 and over 65 years of age, as well as net migration rates for each country within the region. The second table presents economic and social development data for each country, including GNI per capita, GDP growth, life expectancy, percentage of the population living on less than $2 per day, child mortality rates, and the UN gender equity index.
- **Working Toward Sustainability case studies.** This new feature in each regional chapter highlights examples of sustainability in multiple settings and at varied scales.
- **Exploring Global Connections case studies.** In each regional chapter, this feature explores how activities in different world regions are linked so that students understand that in a globalized world, regions are neither isolated nor discrete entities.

ACKNOWLEDGMENTS

We have many people to thank for their help in the conceptualization, writing, rewriting, and production of *Globalization and Diversity,* Fourth Edition. First, we'd like to thank the thousands of students in our world regional geography classes who have inspired us with their energy, engagement, and curiosity; challenged us with their critical insights; and demanded a textbook that better meets their need to understand the diverse people and places of our dynamic world.

Next, we are deeply indebted to many professional geographers and educators for their assistance, advice, inspiration, encouragement, and constructive criticism as we labored through the different stages of this book. Among the many who provided invaluable comments on various drafts and editions of *Globalization and Diversity,* or worked on supporting print or digital material, are:

Gillian Acheson, Southern Illinois University: Edwardsville
Iddi Adam, University of Wisconsin, Marshfield/Wood County
Dan Arreola, Arizona State University
Bernard BakamaNume, Texas A&M University
Brad Baltensperger, Michigan Technological University
Max Beavers, Samford University
Laurence Becker, Oregon State University
James Bell, University of Colorado
William H. Berentsen, University of Connecticut
Kevin Blake, Kansas State University
Karl Byrand, University of Wisconsin, Sheboygan
Michelle Calvarese, California State University, Fresno
Craig Campbell, Youngstown State University
Perry Carter, Texas Tech University
Elizabeth Chacko, George Washington University
Philip Chaney, Auburn University
David B. Cole, University of Northern Colorado
Malcolm Comeaux, Arizona State University

Jonathan C. Comer, Oklahoma State University
Catherine Cooper, George Washington University
Jeremy Crampton, George Mason University
Kevin Curtin, University of Texas at Dallas
James Curtis, California State University, Long Beach
Dydia DeLyser, Louisiana State University
Francis H. Dillon, George Mason University
Jason Dittmer, Georgia Southern University
Jerome Dobson, University of Kansas
Caroline Doherty, Northern Arizona University
Vernon Domingo, Bridgewater State College
Roy Doyon, Ball State University
Jane Ehemann, Shippensburg University
Doug Fuller, George Washington University
Gary Gaile, University of Colorado
Sherry Goddicksen, California State University, Fullerton
Reuel Hanks, Oklahoma State University
Kelsey Hanrahan, University of Kentucky
Steven Hoelscher, University of Texas, Austin
Lance Howard, Clemson University
Thomas Howard, Armstrong Atlantic State University
Peter J. Hugill, Texas A&M University
Eva Humbeck, Arizona State University
Ryan S. Kelly, University of Kentucky
Richard H. Kesel, Louisiana State University
Erin Kloster, Southern Illinois University, Edwardsville
Rob Kremer, Front Range Community College
Robert C. Larson, Indiana State University
Mathias Le Bosse, Kutztown University of Pennsylvania
Alan A. Lew, Northern Arizona University
Catherine Lockwood, Chadron State College
John F. Looney, Jr., University of Massachusetts, Boston
Max Lu, Kansas State University
Luke Marzen, Auburn University
Kent Matthewson, Louisiana State University
James Miller, Clemson University
Bob Mings, Arizona State University
Sherry D. Morea-Oakes, University of Colorado, Denver
Anne E. Mosher, Syracuse University
Tim Oakes, University of Colorado
Nancy Obermeyer, Indiana State University
Karl Offen, Oklahoma University
Kefa Otiso, Bowling Green State University
David Padgett, Tennessee State University
Joseph Palis, University of North Carolina, Chapel Hill
Jean Palmer-Moloney, Hartwick College
Bimal K. Paul, Kansas State University
Michael P. Peterson, University of Nebraska–Omaha
Richard Pillsbury, Georgia State University
Brandon Plewe, Brigham Young University
Patricia Price, Florida International University
Erik Prout, Texas A&M University
David Rain, United States Census Bureau
Donald Rallis, University of Mary Washington
Rhonda Reagan, Blinn College
Craig S. Revels, Portland State University
Scott M. Robeson, Indiana State University
Paul A. Rollinson, Southwest Missouri State University

Yda Schreuder, University of Delaware
Kay L. Scott, University of Central Florida
Patrick Shabram, South Plains College
J. Duncan Shaeffer, Arizona State University
Dimitrii Sidorov, California State University, Long Beach
Susan C. Slowey, Blinn College
Andrew Sluyter, Louisiana State University
Christa Ann Smith, Clemson University
Joseph Spinelli, Bowling Green State University
William Strong, University of Northern Alabama
Philip W. Suckling, University of Northern Iowa
Curtis Thomson, University of Idaho
Benjamin Timms, California Polytechnic State University
Suzanne Traub-Metlay, Front Range Community College
Jim Tyner, Kent State University
Nina Veregge, University of Colorado
Gerald R. Webster, University of Alabama
Mark Welford, Georgia Southern University
Scott White, Fort Lewis College
Nikki Williams, Texas State University
Keith Yearman, College of DuPage
Emily Young, University of Arizona
Bin Zhon, Southern Illinois University at Edwardsville
Henry J. Zintambila, Illinois State University
Sandra Zupan, University of Kentucky

In addition, we wish to thank the many publishing professionals who have been involved with this project, for it is a privilege working with you. We thank Paul F. Corey, President of Pearson's Science, Business and Technology division, for his early and continued support for this book project; Geography editor and good friend Christian Botting for his professional guidance, enduring patience, and high standards; Project Manager Anton Yakovlev for his daily attention to production matters and his diplomatic interaction with four demanding (and often cranky) authors; Development Editor David Chelton for his valuable editorial insights and cogent suggestions; Editorial Assistant Bethany Sexton for gracefully taking care of the many incidental tasks connected to this project; Senior Marketing Manager Maureen McLaughlin for her sales and promotion work; Production Project Manager Connie Long and *Element* Production Editor Kelly Keeler for somehow turning thousands of pages of manuscript into a finished book; and *International Mapping* Project Manager and Cartographer Kevin Lear for his outstanding work on our maps.

Last, the authors want to thank that special group of friends and family who were there when we needed you most—early in the morning and late at night; in foreign countries and familiar places; when we were on the verge of crying yet needing to laugh; for your love, patience, companionship, inspiration, solace, understanding, and enthusiasm: Elizabeth Chacko, Meg Conkey, Rob Crandall, Marie Dowd, Camille Fisher, Evan and Eleanor Lewis, Karen Wigen, and Linda, Tom, Katie Wyckoff, and Magdalena Cooper—Words cannot thank you enough.

Les Rowntree

Martin Lewis

Marie Price

William Wyckoff

THE TEACHING AND LEARNING PACKAGE

In addition to the text itself, the authors and publisher have been pleased to work with a number of talented people to produce an excellent instructional package.

FOR TEACHERS AND STUDENTS

MasteringGeography™ with Pearson eText. The **Mastering** platform is the most widely used and effective online homework, tutorial, and assessment system for the sciences. It delivers self-paced tutorials that provide individualized coaching, focus on course objectives, and are responsive to each student's progress. The Mastering system helps teachers maximize class time with customizable, easy-to-assign, and automatically graded assessments that motivate students to learn outside of class and arrive prepared for lecture. MasteringGeography offers:

- **Assignable activities** that include MapMaster™ interactive map activities, *Encounter World Regional Geography* Google Earth Explorations, video activities, Geoscience Animation activities, map projection activities, coaching activities on the toughest topics in geography, end-of-chapter questions and exercises, reading quizzes, Test Bank questions, and more.
- **A student Study Area** with MapMaster™ interactive maps, videos, Geoscience Animations, web links, glossary flashcards, "In the News" RSS feeds, chapter quizzes, an optional Pearson eText including versions for iPad and Android devices, and more.

Pearson eText gives students access to the text whenever and wherever they can access the Internet. The eText pages look exactly like the printed text and include powerful interactive and customization functions, including links to the multimedia.

Practicing Geography: Careers for Enhancing Society and the Environment by Association of American Geographers (0321811151).
This book examines career opportunities for geographers and geospatial professionals in the business, government, nonprofit, and education sectors. A diverse group of academic and industry professionals shares insights on career planning, networking, transitioning between employment sectors, and balancing work and home life. The book illustrates the value of geographic expertise and technologies through engaging profiles and case studies of geographers at work.

Teaching College Geography: A Practical Guide for Graduate Students and Early Career Faculty by Association of American Geographers (0136054471).
This two-part resource provides a starting point for becoming an effective geography teacher from the very first day of class. Part One addresses "nuts-and-bolts" teaching issues. Part Two explores being an effective teacher in the field, supporting critical thinking with GIS and mapping technologies, engaging learners in large geography classes, and promoting awareness of international perspectives and geographic issues.

Aspiring Academics: A Resource Book for Graduate Students and Early Career Faculty by Geographers Association of American Geographers (0136048919).
Drawing on several years of research, this set of essays is designed to help graduate students and early career faculty start their careers in geography and related social and environmental sciences. *Aspiring Academics* stresses the interdependence of teaching, research, and service—and the importance of achieving a healthy balance of professional and personal life—while doing faculty work. Each chapter provides accessible, forward-looking advice on topics that often cause the most stress in the first years of a college or university appointment.

Television for the Environment Earth Report Geography Videos on DVD (0321662989).
This three-DVD set helps students visualize how human decisions and behavior have affected the environment and how individuals are taking steps toward recovery. With topics ranging from the poor land management promoting the devastation of river systems in Central America to the struggles for electricity in China and Africa, these 13 videos from Television for the Environment's global *Earth Report* series recognize the efforts of individuals around the world to unite and protect the planet.

Television for the Environment Life World Regional Geography Videos on DVD (013159348X).
From the Television for the Environment's global *Life* series, this two-DVD set brings globalization and the developing world to the attention of any world regional geography course. These 10 full-length video programs highlight matters such as the growing number of homeless children in Russia, the lives of immigrants living in the United States and trying to aid family still living in their native countries, and the European conflict between commercial interests and environmental concerns.

Television for the Environment Life Human Geography Videos on DVD (0132416565).
This three-DVD set is designed to enhance any human geography course. These DVDs include 14 full-length video programs from Television for the Environment's global *Life* series, covering a wide array of issues affecting people and places in the contemporary world, including the serious health risks of pregnant women in Bangladesh, the social inequalities of the "untouchables" in the Hindu caste system, and Ghana's struggle to compete in a global market.

FOR TEACHERS

Instructor Resource Manual (Download) (0321862252).

The *Instructor Resource Manual,* created by Jim Tyner, follows the new organization of the main text. It includes a sample syllabus, chapter learning objectives, lecture outlines, a list of key terms, and answers to the textbook's Thinking Geographically and Review Questions. Discussion questions, classroom activities, and advice about how to integrate visual supplements (including web-based resources) are integrated throughout the chapter lecture outlines.

TestGen/Test Bank (Download) (0321861418).

TestGen is a computerized test generator that lets teachers view and edit *Test Bank* questions, transfer questions to tests, and print the test in a variety of customized formats. Authored by Iddi Adam, this *Test Bank* includes approximately 1,500 multiple-choice, true/false, and short answer/essay questions. Questions are correlated against the book's Learning Objectives, the revised U.S. National Geography Standards, chapter-specific learning outcomes, and Bloom's Taxonomy to help teachers to better map the assessments against both broad and specific teaching and learning objectives. The *Test Bank* is also available in Microsoft Word® and is importable into Blackboard.

Instructor Resource DVD (032186140X).

The *Instructor Resource DVD* provides teachers everything they need where they want it and helps make them more effective by saving them time and effort. All digital resources can be found in one well-organized, easy-to-access place. The IRC DVD includes:

- All textbook images as JPEGs, PDFs, and PowerPoint™ Presentations
- Pre-authored Lecture Outline PowerPoint™ Presentations, which outline the concepts of each chapter with embedded art and can be customized to fit instructors' lecture requirements
- CRS "Clicker" Questions in PowerPoint™ format, which correlate to the book's Learning Objectives, the U.S. National Geography Standards, chapter-specific learning outcomes, and Bloom's Taxonomy
- The TestGen software, *Test Bank* questions, and answers for both MACs and PCs
- Electronic files of the *Instructor Resource Manual* and *Test Bank*

This Instructor Resource content is also available completely online via the Instructor Resources section of MasteringGeography and **www.pearsonhighered.com/irc**.

FOR STUDENTS

Mapping Workbook (0321862201).

This workbook, which can be used in conjunction with either the main text or an atlas, features political and physical shaded relief base maps of every global region. These maps, along with the names of the regional key locations and physical features, are the basis for identification exercises. Conceptual exercises are included to further students' comprehension of the key points presented in the main text's chapters. An answer key is available to download from **www.pearsonhighered.com/irc**.

Goode's World Atlas, 22nd Edition (0321652002).

Goode's World Atlas has been the world's premiere educational atlas since 1923—and for good reason. It features over 250 pages of maps, from definitive physical and political maps to important thematic maps that illustrate the spatial aspects of many important topics. The 22nd Edition includes 160 pages of new, digitally produced reference maps, as well as new thematic maps on global climate change, sea-level rise, CO_2 emissions, polar ice fluctuations, deforestation, extreme weather events, infectious diseases, water resources, and energy production.

Pearson's Encounter Series provides rich, interactive explorations of geoscience concepts through GoogleEarth™ activities, covering a range of topics in regional, human, and physical geography. For those who do not use MasteringGeography, all chapter explorations are available in print workbooks, as well as in online quizzes at **www.mygeoscienceplace.com**, accommodating different classroom needs. Each exploration consists of a worksheet, online quizzes, and a corresponding Google Earth™ KMZ file.

- *Encounter World Regional Geography* Workbook and Website by Jess C. Porter (0321681754)
- *Encounter Human Geography* Workbook and Website by Jess C. Porter (0321682203)
- *Encounter Physical Geography* Workbook and Website by Jess C. Porter and Stephen O'Connell (0321672526)
- *Encounter Geosystems* Workbook and Website by Charlie Thomsen (0321636996)
- *Encounter Earth* by Steve Kluge (0321581296)

Dire Predictions: Understanding Global Warming by Michael Mann, Lee R. Kump (0136044352).

Dire Predictions is appropriate for any science or social science course in need of a basic understanding of the reports from the Intergovernmental Panel on Climate Change (IPCC). These periodic reports evaluate the risk of climate change brought on by humans. But the sheer volume of scientific data remains inscrutable to the general public, particularly to those who may still question the validity of climate change. In just over 200 pages, this practical text presents and expands upon the essential findings in a visually stunning and undeniably powerful way to the lay reader. Scientific findings that provide validity to the implications of climate change are presented in clear-cut graphic elements, striking images, and understandable analogies.

ABOUT THE AUTHORS

Les Rowntree is a Research Associate at the University of California, Berkeley, where he researches and writes about global and local environmental issues. This career change comes after more than three decades teaching both Geography and Environmental Studies at San Jose State University. As an environmental geographer, Dr. Rowntree's interests focus on international environmental issues, biodiversity conservation, and human-caused global change. He sees world regional geography as a way to engage and inform students by giving them the conceptual tools needed to critically assess a wide array of global issues.. Current projects include a natural history book on California's Coast Range, as well as creating and maintaining an assortment of web-based natural history, geography, and environmental blogs and Websites.

Martin Lewis is a Senior Lecturer in History at Stanford University, where he teaches courses on global geography. He has conducted extensive research on environmental geography in the Philippines and on the intellectual history of world geography. His publications include *Wagering the Land: Ritual, Capital, and Environmental Degradation in the Cordillera of Northern Luzon, 1900–1986* (1992), and, with Karen Wigen, *The Myth of Continents: A Critique of Metageography* (1997). Dr. Lewis has traveled extensively in East, South, and Southeast Asia. His current research focuses on the geography of languages. In April 2009, Dr. Lewis was recognized by *Time* magazine as one of American's most favorite lecturers.

Marie Price is a Professor of Geography and International Affairs at George Washington University. A Latin American specialist, Marie has conducted research in Belize, Mexico, Venezuela, Cuba, and Bolivia. She has also traveled widely throughout Latin America and Sub-Saharan Africa. Her studies have explored human migration, natural resource use, environmental conservation, and regional development. She is a non-resident fellow of the Migration Policy Institute, a nonpartisan think tank that focuses on immigration, and the Vice President of the American Geographical Society. Dr. Price brings to *Globalization and Diversity* a special interest in regions as dynamic spatial constructs that are shaped over time through both global and local forces. Her publications include the co-edited book *Migrants to the Metropolis: The Rise of Immigrant Gateway Cities* (2008, Syracuse University Press) and numerous academic articles and book chapters.

William Wyckoff is a geographer in the Department of Earth Sciences at Montana State University specializing in the cultural and historical geography of North America. He has written and co-edited several books on North American settlement geography, including *The Developer's Frontier: The Making of the Western New York Landscape* (1988), *The Mountainous West: Explorations in Historical Geography* (1995) (with Lary M. Dilsaver), *Creating Colorado: The Making of a Western American Landscape 1860–1940* (1999), and *On the Road Again: Montana's Changing Landscape* (2006). In 2003, he received Montana State's Cox Family Fund for Excellence Faculty Award for Teaching and Scholarship. A World Regional Geography instructor for 26 years, Dr. Wyckoff emphasizes in the classroom the connections between the everyday lives of his students and the larger global geographies that surround them and increasingly shape their future.

Conveying a strong sense of place and context, this contemporary approach to world regional geography helps students understand the unique connections among the world's diverse regions.

Globalization and Diversity

For most of the past 200 years, the landlocked region of Central Asia has been geopolitically dominated by countries located in other world regions and partially cut off from the main currents of global trade. Since the downfall of the Soviet Union in 1991, however, Central Asia has emerged as a key producer of globally traded resources and as a focus of international geopolitical rivalry.

ENVIRONMENTAL GEOGRAPHY
Intensive agriculture along the rivers that flow into the deserts of Central Asia has resulted in serious water shortages, leading to the drying up of many of the region's lakes and wetlands.

POPULATION AND SETTLEMENT
Pastoral nomadism, the traditional way of life across much of Central Asia, is gradually disappearing as people settle in towns and cities.

CULTURAL COHERENCE AND DIVERSITY
In much of eastern Central Asia, the growing Han Chinese population is sometimes seen as a threat to the long-term survival of the indigenous cultures of the Tibetan and Uyghur peoples.

GEOPOLITICAL FRAMEWORK
Afghanistan and its neighbors to the north are frontline states in the struggle between radical Islamic fundamentalism and secular governments.

ECONOMIC AND SOCIAL DEVELOPMENT
Despite its abundant resources, Central Asia remains a poor region, although much of it enjoys relatively high levels of social development.

➤ Soldiers in the U.S.-allied Afghan National Army walk through a field of opium poppies as they patrol in the Taliban stronghold of Panjwaii in Kandahar province in 2009.

326

CENTRAL ASIA 10

327

NEW! Chapter-opening vignettes with engaging photos provide students with a strong and immediate sense of the region.

EXPLORING GLOBAL CONNECTIONS

Social Media and Political Change: Lessons from the Arab Spring

As civil unrest spread with lightning speed through much of the Middle East in 2011, and as several regional conflicts continued thereafter, discussion focused on the role played by social media in spreading the conflicts and in popularizing the underlying reform movements (Figure 7.2.1). So was the Arab Spring a child of the Facebook generation? Well, yes and no. Ultimately, the chain of events producing dramatic protests in settings such as Tunisia, Egypt, Libya, Bahrain, Yemen, and Syria was probably not about the tweets, viral videos, and cell-phone connections that were sensationalized in the global media. Rather, it was more about how complex *local* geographies of people (from diverse ethnic, class, and tribal backgrounds) found themselves dealing with an equally complex assortment of country-specific political issues.

Still, technology and the Arab Spring were inextricably intertwined, particularly in three ways. First, cell phones, blogs, email, and tweets *facilitated the flow of information* that helped protesters plan events and coordinate strategies with allies. Second, local videos from smart phones and pinhole cameras *documented government abuse* and often provided (in settings such as Syria) the only proof of widespread state-supported violence. Third, the global diffusion of this information promoted the *internationalization of political*

discourse and made it easier to spread the word about local conflicts and to identify common threads among different protest movements. A related point is that about 60 percent of the region's population is under 30—precisely the group most inclined to use these technologies and often the people most frustrated by unresponsive governments that refuse to change their ways.

But is this technology inherently more democratic or liberating? Critics argue that such notions are merely naïve cyber-utopianism. Consider what has happened in settings such as Iran, Syria, and China. In these countries, repressive governments have used the Internet and other forms of electronic communication (such as blogs and tweets) to compile dossiers on critics of the government, spy on online chatter, and even identify and arrest suspected enemies. In Iran, thousands were arrested after the 2009 uprisings as secret police made use of information from the Internet and social media. Similarly, Syria's repressive Assad regime made widespread use of Facebook to follow its opponents during its civil war. The

FIGURE 7.2.1 Communicating from the Front Lines, Cairo, Egypt, 2011 A young Egyptian woman talks on a mobile phone in Cairo's Tahrir Square during demonstrations in February, 2011.

Chinese government now employs thousands of workers to add pro-government online chatter to various Internet outlets.

So what lessons can we learn from the interplay of the Arab Spring and the new world of electronic and social media? No doubt, in that moment in time, tweets and cell phones mattered greatly. But equally clear is the fact that protesters have no monopoly on their knowledge and use of these communications technologies. Bloggers beware: Repressive governments can be fast learners. Finally, as events continue to unfold in the region, it appears that at least for now, local rivalries and shifting political alliances in particular locales may trump any potential for more pan-regional democratic and reform-minded discourse.

Updated **Exploring Global Connections** case studies reinforce the book's popular globalization theme by illustrating the interesting and sometimes unexpected interconnections between regions. Examples include social media and the Arab Spring; S. Korean's investment in Central Asia; Burma's connections with China; China's recent heavy investment in Australia; and many others.

Socially conscious topics—sustainability, gender issues, global climate change—combine with geography fundamentals to engage and involve students on multiple levels.

NEW! Working Towards Sustainability features show diverse applications of how sustainability initiatives apply to people, groups, and settlements in different places and at different scales, emphasizing positive environmental and social initiatives and their outcomes. Examples include urban rooftops in U.S. cities, preservation of the Azraq Basin in Jordan, Germany's renewable energy program, sustainable development in China's Loess Plateau, and many others.

FIGURE 7.42 **Algerian Women** Women now make up more than 30 percent of Algeria's newly-elected National Assembly, a higher proportion of female representation than in many Western nations

Expanded coverage of gender issues, food, art, music, film, and sports brings these high-interest cultural topics to the forefront.

FIGURE 14.35 **Samoan Football Players** Even though rugby is the most popular sport in Oceania, there are a large number of Samoan football players in the United States, at both college and professional levels. Here, Joey Iosefa, running back for University of Hawaii moves the ball against San Jose State.

FIGURE 9.30 **Svetlana Loboda** Ukrainian singing star Svetlana Loboda was one of the top performers at Eurovision 2009, held in Moscow. In addition to a successful career in music and television, Loboda has championed the plight of battered women throughout the region.

WORKING TOWARD SUSTAINABILITY

Preserving Land and Life in Jordan's Azraq Basin

The Azraq Basin contains some of Jordan's largest supplies of groundwater (Figure 7.1.1). Associated surface springs and wetlands traditionally support both sedentary farmers (who depend on the basin for irrigation water for their desert grain and tree crops) and nomadic Bedouin populations (who use seasonal pastures to graze their camels, sheep, and goats). In addition, large flocks of migrating birds (including species from Scandinavia, Africa, and Siberia) use lakes in the area as stopovers on their multicontinental journeys (Figure 7.1.2).

Since 1980, however, unwelcome changes have threatened the long-term viability of the basin. Rapidly growing urban populations in nearby Amman and Zarqa have placed increased demands on regional aquifers. In rural areas, many farmers have shifted to more commercialized and intensive forms of irrigated agriculture. Large numbers of new wells have been dug, many of them illegal. Periodic regional droughts—potentially related to global warming—have also been on the rise, limiting surface recharge. The results have been dramatic and painful. Beginning in the 1990s, many springs and shallow wells have

FIGURE 7.1.2 **Azraq Wetlands, Jordan** This small freshwater resource is an invaluable part of the Azraq Wetlands Nature Reserve in Jordan.

seen lower flows or have dried up. The salinity of the remaining water resources has risen. Wetlands have shrunk greatly in size, and many migratory birds have disappeared.

In a coordinated response to the crisis, the Jordanian government established the Azraq Oasis Restoration Project, designed to study the problem and come up with sustainable solutions to reverse falling groundwater levels, preserve traditional farming and grazing lifestyles, and restore the ecological integrity of the critical wetlands environment. The project has also teamed up with the International Union for Conservation of Nature, the Arab Women Organization (focused on improving the lives of ordinary Jordanian women), and numerous stakeholders in the area. Project researchers have emphasized a *participatory approach*, in which they spend large amounts of time in local workshops and seminars and do extensive fieldwork partnering with residents. Although only about 5–10 percent of the surface wetlands have been temporarily restored (through increased groundwater pumping), project leaders hope that their efforts will slow the drilling of new wells, encourage farmers to rethink the mix of crops they grow (and adopt more efficient drip irrigation systems), and offer more examples of urban water consumption and recycling.

The fate of the Azraq Basin remains in doubt. Lower population growth and widespread local participation in water conservation efforts are crucial elements in restoring the basin and its groundwater to sustainable levels. Ultimately, the project's chances for long-term success may lie in a creative blending of tradition and innovation that combines local and global knowledge and in the process redefines water as a sustainable resource in a part of the world where it increasingly seems in short supply.

FIGURE 7.1.1 **Azraq Basin, Jordan** The map shows the size and centrality of Jordan's Azraq Basin. Note the proximity to Amman and Zarqa.

A stronger focus on geography fundamentals and tools of the trade includes expanded information in the introductory chapters on map reading and modern geospatial tools, and critical physical geography concepts such as plate tectonics and natural hazards.

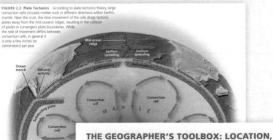

FIGURE 2.2 **Plate Tectonics** According to plate tectonics theory, large convection cells circulate molten rock in different directions within Earth's mantle. Near the crust, the slow movement of the cells drags tectonic plates away from the mid-oceanic ridges, resulting in the collision of plates in convergent plate boundaries. While the rate of movement differs between convection cells, in general it is only a few inches (or centimeters) per year.

THE GEOGRAPHER'S TOOLBOX: LOCATION, MAPS, REMOTE SENSING, AND GIS

Geographers use many different tools to represent the world in a convenient form for examination and analysis. You will need very different kinds of images and data if you're studying vegetation in Brazil or mining in Mongolia; population density in Tokyo or languages spoken in Europe; religions practiced in the Middle East or rainfall in southern India. Today's modern satellite and communications systems offer geographers an array of tools not imagined 50 years ago.

Structured to facilitate learning

Each of the regional chapters is organized into five thematic sections—Environmental Geography, Population and Settlement, Cultural Coherence and Diversity, Geopolitical Framework, and Economic and Social Development—to encourage cross-regional comparisons and highlight issues in today's globalized world. Each chapter also includes a new structured learning path to help students focus on important concepts and check their understanding.

NEW! Learning Objectives, which are presented in each chapter's opening pages, guide students in prioritizing the most important concepts in the chapter, and are connected to each chapter's Review Questions and MasteringGeography™ assessments.

LEARNING OBJECTIVES

After reading this chapter you should be able to:

- Identify the different components of globalization, including their controversial aspects, and list several ways in which globalization is changing world geographies.
- Describe the conceptual framework of world regional geography.
- Summarize the major tools used by geographers to study Earth's surface.
- Explain the concepts and metrics used to document changes in global population and settlement patterns.

- Describe the themes and concepts used to study the interaction between globalization and the world's cultural geographies.
- Explain how different aspects of globalization have interacted with global geopolitics from the colonial period to the present day.
- Identify the concepts and data important to documenting changes in the economic and social development of more and less developed countries.

NEW! Review Questions at the end of each thematic section help students check and apply their comprehension as they read each chapter.

REVIEW QUESTIONS

1. What is sectoral transformation, and how does it help explain economic change in North America?
2. Cite five types of location factors, and illustrate each with examples from your local economy.

REVIEW QUESTIONS

1. How did the colonization of Latin America by Iberia lead to the formation of the modern states of Latin America?
2. How are trade blocs reshaping the region's geopolitics?

End-of-chapter **Thinking Geographically** questions offer students opportunities to apply and synthesize their understanding and engage in collaborative/group activities.

Thinking Geographically

1. Select an economic, political, or cultural activity in your city or town, and discuss how it has been influenced by globalization.

2. Choose a specific country or region of the world and examine the benefits, and liabilities that globalization has posed for that country or region. Remember to look at different facets of globalization, such as the environment, cultural change and conflict, and the economic effects on different segments of the population.

3. Drawing on information in current newspapers and magazines, as well as TV and the Internet, apply the concepts of cultural imperialism, nationalism, and cultural syncretism to a region or place experiencing cultural tensions.

4. Select an African country with a colonial past. (a) Trace its pathway of decolonialization; (b) describe and analyze its contemporary relations with its former colonial overseer, being sensitive to the matter of whether a neocolonial relationship has been established.

5. Using the tables of social indicators in the regional chapters of this book, identify traits shared by countries that have a high percentage of female illiteracy. Based on your inquiry, what general conclusions do you reach?

NEW! Quick Response (QR) Codes at the end of each chapter enable students to link smartphones and other mobile devices from the book to blogs written by the authors of the book, providing easy and immediate access to current updates on topics covered in the chapters.

Scan to visit the author's blog for chapter updates.

http://gad4blog.wordpress.com/

Authors' Blogs

Scan now to access the authors' blogs for up-to-date information on Concepts of World Geography.

Scan to visit the GeoCurrents blog.

http://geocurrents.info/

The power of visualization

A pedagogically oriented cartography program, large-format photos, illustrations, and other visualizations of current data help students experience and understand the world's diverse regions.

The **cartography program features standardized maps for all regional chapters.**
Each regional chapter contains approximately 12 maps, with each region including many of the same thematic maps to help with cross-regional comparisons. This system allows students to compare and contrast concepts and data both within and between regions. Additionally, this system reinforces the global effect on local issues.

A highly graphical presentation includes a large-format map or photo on almost every page, with more than 50% of photos new to this edition.

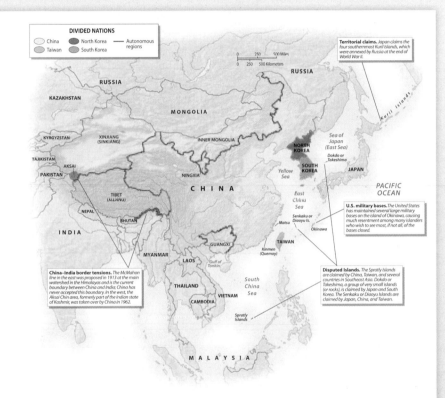

FIGURE 11.31 Geopolitical Issues in East Asia East Asia remains one of the world's geopolitical hot spots. Tensions are particularly high between capitalist, democratic South Korea and the isolated communist regime of North Korea, as well as between China and Taiwan. China has had several border disputes, one of which involves a group of small islands in the South China Sea. Japan and Russia have not been able to resolve their quarrel over the southern Kuril Islands.

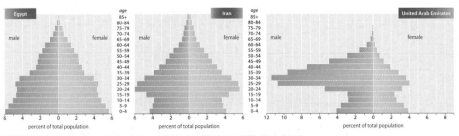

FIGURE 7.23 Population Pyramids: Egypt, Iran, and United Arab Emirates, 2012 Three distinctive demographic snapshots highlight regional diversity: Egypt's above-average growth rates differ sharply from those of Iran, where a focused campaign on family planning has reduced recent family sizes. Male immigrant laborers play a special role in skewing the pattern within the United Arab Emirates.

NEW! Paired population pyramids contrast subregions and settlements within each region, comparing either historic or forecasted change or current population trends.

The most current data throughout the book are pulled from the 2010 U.S. Census, 2012 Population Reference Bureau Data, and other important sources.

MasteringGeography™

www.masteringgeography.com

MasteringGeography delivers engaging, dynamic learning opportunities—focusing on course objectives and responsive to each student's progress—that are proven to help students absorb world regional course material and understand difficult geographic concepts.

Tools for improving geographic literacy and exploring Earth's dynamic landscape

MapMaster™ is a powerful interactive map tool that presents assignable layered thematic and place name interactive maps at world and regional scales for students to test their geographic literacy and spatial reasoning skills, and explore the modern geographer's tools.

MapMaster Layered Thematic Interactive Map Activities act as a mini-GIS tool, allowing students to layer various thematic maps to analyze spatial patterns and data at regional and global scales and answer multiple-choice and short-answer questions organized by region and theme.

NEW! MapMaster has been updated to include:

- 90 new map layers
- Zoom and annotation functionalities
- Current U.S. Census, United Nations, and Population Reference Bureau Data

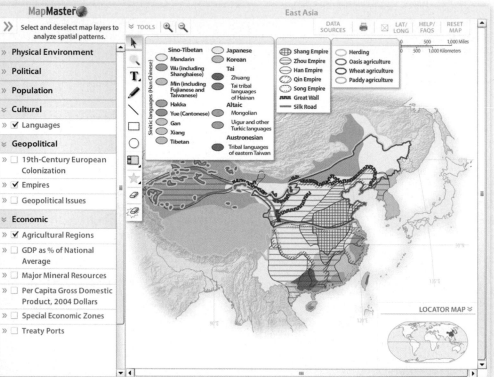

MapMaster Place Name Interactive Map Activities have students identify place names of political and physical features at regional and global scales, explore select recent country data from the CIA World Factbook, and answer associated assessment questions.

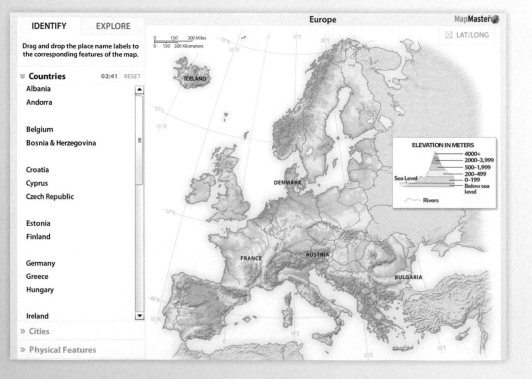

Help students develop spatial reasoning and a sense of place

Encounter Activities provide rich, interactive explorations of geoscience concepts through Google Earth™ activities, exploring a range of topics in world regional geography. Dynamic assessment includes questions related to core world regional geography concepts. All explorations include corresponding Google Earth KMZ media files, and questions include hints and specific wrong-answer feedback to coach students towards mastery of the concepts.

Geography videos provide students a sense of place and allow them to explore a range of locations and topics related to world regional and physical geography. Covering issues of economy, development, globalization, climate and climate change, culture, etc., there are 10 multiple choice questions for each video. These video activities allow teachers to test students' understanding and application of concepts, and offer hints and wrong-answer feedback.

Thinking Spatially and Data Analysis Activities help students master the toughest concepts and develop spatial reasoning and critical thinking skills by identifying and labeling features from maps, illustrations, graphs, and charts. Students then examine related data sets, answering multiple-choice and increasingly higher-order conceptual questions, which include hints and specific wrong-answer feedback.

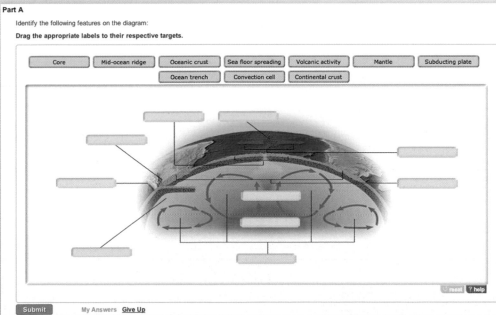

Student Study Resources in **MasteringGeography** include:

- MapMaster™ interactive maps
- Practice quizzes
- Geography videos
- Select Geoscience Animations
- "In the News" RSS feeds
- Glossary flashcards
- Optional Pearson eText and more

Callouts to MasteringGeography appear at the end of each chapter to direct students to extend their learning beyond the textbook.

MasteringGeography™

www.masteringgeography.com

With the Mastering gradebook and diagnostics, you'll be better informed about your students' progress than ever before. Mastering captures the step-by-step work of every student—including wrong answers submitted, hints requested, and time taken at every step of every problem—all providing unique insight into the most common misconceptions of your class.

Quickly monitor and display student results

The **Gradebook** records all scores for automatically graded assignments. Shades of red highlight struggling students and challenging assignments.

Diagnostics provide unique insight into class and student performance. With a single click, charts summarize the most difficult questions, vulnerable students, grade distribution, and score improvement over the duration of the course.

With a single click, **Individual Student Performance Data** provides **at-a-glance statistics** into each individual student's performance, including time spent on the question, number of hints opened, and number of wrong and correct answers submitted.

Easily measure student performance against your Learning Outcomes

Learning Outcomes

MasteringGeography provides quick and easy access to information on student performance against your learning outcomes and makes it easy to share those results.

- Quickly add your own learning outcomes, or use publisher-provided ones, to track student performance and report it to your administration.
- View class and individual student performance against specific learning outcomes.
- Effortlessly export results to a spreadsheet that you can further customize and/or share with your chair, dean, administrator, and/or accreditation board.

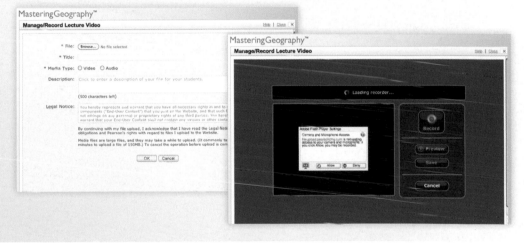

Easy to customize

Customize publisher-provided items or quickly add your own. MasteringGeography makes it easy to edit any questions or answers, import your own questions, and quickly add images, links, and files to further enhance the student experience.

Upload your own video and audio files from your hard drive to share with students, as well as record video from your computer's webcam directly into MasteringGeography—no plug-ins required. Students can download video and audio files to their local computer or launch them in Mastering to view the content.

Pearson eText gives students access to ***Globalization and Diversity: Geography of a Changing World,* Fourth Edition** whenever and wherever they can access the Internet. The eText pages look exactly like the printed text, and include powerful interactive and customization functions. Users can create notes, highlight text in different colors, create bookmarks, zoom, click hyperlinked words and phrases to view definitions, and view as a single page or as two pages. Pearson eText also links students to associated media files, enabling them to view an animation as they read the text, and offers a full-text search and the ability to save and export notes. The Pearson eText also includes embedded URLs in the chapter text with active links to the Internet.

NEW! The Pearson eText app is a great companion to Pearson's eText browser-based book reader. It allows existing subscribers who view their Pearson eText titles on a Mac or PC to additionally access their titles in a bookshelf on the iPad or an Android tablet either online or via download.

GLOBALIZATION AND DIVERSITY

GEOGRAPHY OF A CHANGING WORLD

FOURTH EDITION

Globalization and Diversity

Globalization affects all people and places throughout the world because it's everywhere, all the time. While some embrace the changes brought by globalization, others resist and push back, taking refuge in the traditional and familiar. As a result a complex world geography is produced consisting of new and old, containing intertwined and inseparable elements of both globalization and diversity.

CONVERGING CURRENTS OF GLOBALIZATION
One of the most important and controversial challenges facing the world in the 21st century is associated with globalization— the increasing interconnectedness of people and places through converging economic, political, and cultural activities.

GEOGRAPHY MATTERS
Geography is one of the fundamental sciences with its roots in the Greek word for "describing the Earth". As a result this discipline has been central to all cultures and civilizations as they seek to better understand the world.

THE GEOGRAPHER'S TOOLBOX
Geographers use many different tools in their study of the world— maps, aerial photos, satellite images, GPS, and digital geographical information systems to name just a few.

ENVIRONMENTAL GEOGRAPHY
Fundamental to world regional geography is an understanding of the physical world—geology, climate, vegetation—as well as analysis of environmental issues like global climate change.

POPULATION AND SETTLEMENT
The key issues are vastly different rates of natural growth, population migration, and urbanization.

CULTURAL COHERENCE AND DIVERSITY
Globalization affects people and cultures worldwide, often creating new cultural geographies, but also at times reinforcing traditional ways of life.

GEOPOLITICAL FRAMEWORK
Unquestionably, one of the characteristics of the last several decades has been the speed, scope, and character of geopolitical change in different regions of the world.

ECONOMIC AND SOCIAL DEVELOPMENT
Economic globalization is a driving force behind changing geographies in all parts of the world and understanding and evaluating these processes and changes is central to world regional geography.

➤ An icon of diversity and globalization, Hamburg has been a world trade center since medieval times. Today, Hamburg is not only Europe's second largest port (after Rotterdam) but also one of Europe's most ethnically diverse cities.

CONCEPTS OF WORLD GEOGRAPHY

Globalization affects the geography of people and places throughout the world, from small rural villages to large bustling cities. Pundits say globalization is like the weather: It's everywhere, all the time. It is a ubiquitous part of our lives and landscapes that is both beneficial and negative, depending on our needs and point of view. While some people in some places embrace the changes brought by globalization, others resist and push back, seeking refuge in traditional habits and places. As a result, the handmaiden of globalization is **diversity**: a tension between the global and the local. In Asian philosophy, *yin* and *yang* are polar opposites, yet what are seemingly contrary are actually interconnected and interdependent. Indeed, this is the case with the globalization and diversity that make up world regional geography.

CONVERGING CURRENTS OF GLOBALIZATION

One of the most important challenges facing the world in the 21st century is associated with **globalization**—the increasing interconnectedness of people and places through converging economic, political, and cultural activities. Once-distant regions and cultures are now increasingly linked through commerce, communications, and travel. Although earlier forms of globalization existed, especially during Europe's colonial period, the current degree of planetary integration is stronger than ever. In fact, many observers argue that contemporary globalization is the most fundamental reorganization of the world's socioeconomic structure since the Industrial Revolution.

Although economic activities may be the major force behind globalization, the consequences affect all aspects of land and life: Cultural patterns, political arrangements, environment, and social development are all undergoing profound change. Because natural resources are now global commodities, the planet's physical environment is also affected by globalization. Financial decisions made thousands of miles away now affect local ecosystems and habitats, often with far-reaching consequences for Earth's health and sustainability.

These immense and widespread global changes make understanding our contemporary world a challenging, yet necessary task. World regional geography is central to this task because of its integration of environmental, cultural, political, and economic themes and topics.

FIGURE 1.1 Global Communications The effects of globalization are everywhere, even in remote villages in developing countries. Here, in a small village in southwestern India, a rural family earns a few dollars a week by renting out viewing time on its globally linked television set.

Economic Globalization

Most scholars agree that the major component of globalization is the economic reorganization of the world. Although different forms of a world economy have existed for centuries, a well-integrated and truly global economy is primarily the product of the past several decades. The attributes of this system, while familiar, bear repeating:

- Global communication systems that link all regions and most people on the planet instantaneously (Figure 1.1)
- Transportation systems capable of moving goods quickly by air, sea, and land
- Transnational business strategies that have created global corporations more powerful than many sovereign nations
- New and more flexible forms of capital accumulation and international financial institutions that make 24-hour trading possible
- Global agreements that promote free trade

LEARNING OBJECTIVES

After reading this chapter you should be able to:

- Identify the different components of globalization, including their controversial aspects, and list several ways in which globalization is changing world geographies.
- Describe the conceptual framework of world regional geography.
- Summarize the major tools used by geographers to study Earth's surface.
- Explain the concepts and metrics used to document changes in global population and settlement patterns.

- Describe the themes and concepts used to study the interaction between globalization and the world's cultural geographies.
- Explain how different aspects of globalization have interacted with global geopolitics from the colonial period to the present day.
- Identify the concepts and data important to documenting changes in the economic and social development of more and less developed countries.

FIGURE 1.2 Global Shopping Malls   Once a fixture only of suburban North America, the shopping mall is now found throughout the world. This mall is in downtown Kunming, the capital city of Yunnan Province, China.

- Market economies and private enterprises that have replaced state-controlled economies and services
- An abundance of planetary goods and services that have arisen to fulfill consumer demand—real or imagined (Figure 1.2)
- Economic disparities between rich and poor regions and countries that drive people to migrate, both legally and illegally, in search of a better life
- An army of international workers, managers, and executives who give this powerful economic force a human dimension

As a result of this global reorganization, economic growth in some areas of the world has been unprecedented during recent decades; China is a good example. But not everyone has gained from economic globalization, nor have all world regions shared equally in the benefits. While globalization is often touted as universally beneficial through trickle-down economics, there is mounting evidence that this process is happening neither in all places nor for all peoples. Additionally, the global recession of 2008–2010 demonstrated that economic interconnectivity can also increase economic vulnerability, as illustrated by the precipitous decline in Hawaii's tourist trade as the economies of both Japan and the United States went flat at the same time.

Globalization and Changing Human Geographies

Economic changes also trigger cultural changes. The spread of a global consumer culture, for example, often accompanies globalization and frequently hurts local economies. It sometimes creates deep and serious social tensions between traditional cultures and new, external global culture. Global TV, movies, Facebook, Twitter, and videos implicitly promote Western values and culture that are then imitated by millions throughout the world.

Fast-food franchises are changing—some would say corrupting—traditional diets, with the explosive growth of McDonald's, Burger King, and KFC outlets in many of the world's cities. Although these changes may seem harmless to North Americans because of their familiarity, they are not just expressions of the deep cultural changes the world is experiencing through globalization; they are also generally unhealthy and environmentally destructive. The expansion of the cattle industry, for example, as a result of the new global demand for beef, is doing serious environmental damage to tropical rain forests.

Although the media give much attention to the rapid spread of Western consumer culture, nonmaterial culture is also becoming more dispersed and homogenized through globalization. Language is an obvious example—American tourists in far-flung places are often startled to hear locals speaking an English made up primarily of movie or TV clichés. However, far more than speech is involved, as social values also are dispersed globally. Changing expectations about human rights, the role of women in society, and the intervention of nongovernmental organizations are also expressions of globalization that may have far-reaching effects on cultural change.

It would be a mistake, however, to view cultural globalization as a one-way flow that spreads from the United States and Europe into the corners of the world. In actuality, when U.S. popular culture spreads abroad, it is typically melded with local cultural traditions in a process known as *hybridization*. The resulting cultural hybrids, such as hip hop and rap music or Asian food, can themselves resonate across the planet, adding yet another layer to globalization.

In addition, ideas and forms from the rest of the world are also having a great impact on U.S. culture. The growing internationalization of American food, the multiple languages spoken in the United States, or the spread of Japanese comic book culture among U.S. youngsters are all expressions of globalization within the United States (Figure 1.3).

Globalization also has a clear demographic dimension. Although international migration is not new, increasing numbers of people from all parts of the world are now crossing national boundaries, legally and illegally, temporarily and permanently (Figure 1.4). Migration from Latin America and Asia has drastically changed the demographic configuration of the United States, and migration from Africa and Asia has transformed western Europe. Countries such as Japan and South Korea that have long been perceived as ethnically homogeneous now have substantial immigrant populations. Even several relatively poor countries, such as Nigeria and the Ivory Coast, have large numbers of immigrants coming from even poorer countries, such as Burkina Faso and Mali. Although international migration is curtailed by the laws of every country—much more so, in fact, than the movement of goods or capital—it is still rapidly mounting, propelled by the uneven economic development associated with globalization.

FIGURE 1.3 **Global Culture in the United States** The multilingual welcome offered by a public library in Montgomery County, Maryland, not only illustrates the many different languages spoken by people in the suburbs of Washington, DC, but also reminds us that expressions of globalization are found throughout North America.

Finally, a significant criminal element also operates in contemporary globalization, including terrorism (discussed later in this chapter), drugs, pornography, slavery, and prostitution. Illegal narcotics, for example, are definitely a global commodity (Figure 1.5). Some of the most remote parts of the world, such as the mountains of northern Burma, are thoroughly integrated into the circuits of global exchange through the production of opium and therefore into the world heroin trade. Even many areas that do not directly produce drugs are involved in their global sale and transhipment. Nigerians often occupy prominent positions in the international drug trade, as do members of the Russian Mafia. Many Caribbean countries have seen their economies become reoriented to drug transhipments and the laundering of drug money. Prostitution, pornography, and gambling have also emerged as highly profitable global businesses. Over

FIGURE 1.4 **International Migration** Workers from southern India dig a hole to install a street sign in Dubai, United Arab Emirates. This Persian Gulf emirate is experiencing a massive construction boom as it shifts from an oil-based economy to an economy based on real estate, tourism, and international finance. As a result, temporary migrant workers from India and Pakistan constitute much of the labor force.

the past decades, for example, parts of eastern Europe have become major sources of both pornography and prostitution, finding a lucrative, but morally questionable niche in the new global economy.

Geopolitics and Globalization

Globalization also has important geopolitical components. To many, an essential dimension of globalization is that it is not restricted by territorial or national boundaries. For example, the creation of the United Nations (UN) following World War II was a step toward creating an international governmental structure in which all nations could find representation. The simultaneous emergence of the Soviet Union as a military and political superpower at that time led to a rigid division into Cold War blocs that slowed further geopolitical integration. However, with the peaceful end of the Cold War in the late 1980s and early 1990s, the former communist countries of eastern Europe and the Soviet Union were opened almost immediately to global trade and cultural exchange, which have changed those countries immensely (Figure 1.6).

Further, there is a strong argument that globalization—almost by definition—has weakened the political power of individual states by strengthening the power of regional economic and political organizations, such as the European Union and the World Trade Organization. In some world regions, a weakening of traditional state power has resulted in stronger local and separatist movements, as illustrated by the turmoil on Russia's southern borders or the plethora of separatist organizations in Europe (see "Exploring Global Connections: A Closer Look at Globalization").

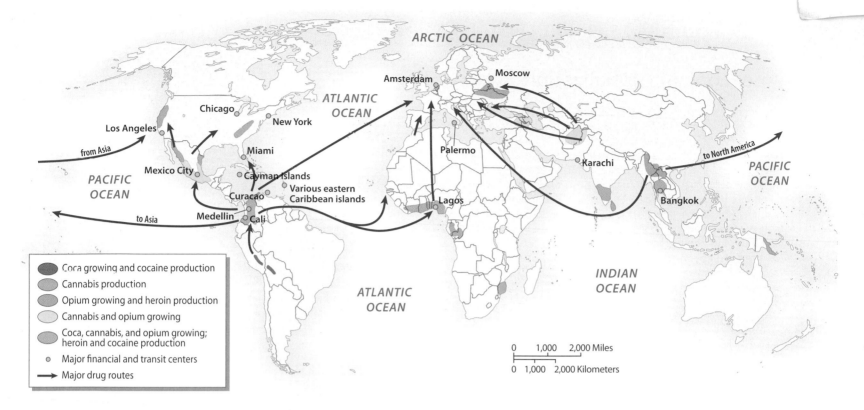

FIGURE 1.5 The Global Drug Trade The cultivation, processing, and transhipment of coca (cocaine), opium (heroin), and cannabis (marijuana) are global issues. The most important cultivation centers are Colombia, Mexico, Afghanistan, and northern Southeast Asia, and the major drug financing centers are located mostly in the Caribbean, the United States, and Europe. In addition, Nigeria and Russia play significant roles in the global transhipment of illegal drugs.

FIGURE 1.6 End of the Cold War The peaceful end of the Cold War in 1990 greatly facilitated global economic expansion and jumpstarted cultural and political globalization. In this photo, Germans celebrate the opening of the Berlin Wall that divided East and West Berlin from August 1961 to November 1989.

EXPLORING GLOBAL CONNECTIONS

A Closer Look at Globalization

Globalization comes in many shapes and forms as it connects far-flung people and places. While many of these interactions are expected and are common knowledge, such as the global reach of multinational corporations, other global connections are more surprising. Who would expect to find Australian firefighters dowsing California wildfires, or Russians investing in Thailand's coastal resorts, or Bronx hip-hop music becoming the favored voice of European youth?

Indeed, global connections are ubiquitous and often complex—so much so that an understanding of the many different shapes, forms, and scales of these interactions is a key component of the study of global geography. To complement that study, each chapter of this book contains an *Exploring Global Connections* sidebar, which presents an example drawn from a wide variety of topics. In Chapter 2, for example, a brief case study shows how the increasingly complicated linkages involving logging in Canada, Russia, and China are affecting the economic well-being of a small lumber-milling town in Washington State. Other case studies are presented throughout the book.

FIGURE 1.1.1 Global use of cell phones
Cell phones have revolutionized the way people communicate and interact in a globalized world for today phones are used for far more than simply talking. For example, rural farmers commonly use cell phones to check crop prices so they can take advantage of changing market conditions, while others pay bills and manage their money via cell phones. This woman is of the Swazi tribe in rural South Africa.

The Environment and Globalization

As we mentioned, the expansion of a globalized economy is creating and intensifying environmental problems throughout the world. Transnational firms, which do global business through international subsidiaries, disrupt local ecosystems with their incessant search for natural resources and manufacturing sites. Landscapes and resources previously used by only small groups of local peoples are now thought of as global commodities to be exploited and traded in the world marketplace. As a result, native peoples are often deprived of their traditional resource base and displaced into marginal environments.

On a larger scale, economic globalization is aggravating worldwide environmental problems such as global warming, air pollution, water pollution, and deforestation. Yet it is only through global cooperation, such as the UN treaties on biodiversity protection or global warming, that these problems can be addressed. These topics are discussed further in Chapter 2.

Controversy About Globalization

Globalization, especially in its economic aspect, is one of today's most contentious issues. Supporters believe that it results in a greater economic efficiency that will eventually result in rising prosperity for the entire world. In contrast, critics claim that globalization largely benefits those who are already prosperous, leaving most of the world poorer than before as the rich and powerful exploit the less fortunate.

Economic globalization is generally applauded by corporate leaders and economists, and it has substantial support among the leaders of both major political parties in the United States. Beyond North America, moderate and conservative politicians in most countries generally support free trade and other aspects of economic globalization. Opposition to economic globalization is widespread in the labor and environmental movements, as well as among many student groups worldwide. Hostility toward globalization is sometimes deeply felt, as massive protests at World Bank and World Trade Organization meetings have made obvious (Figure 1.7).

Pro-Globalization Arguments   Advocates of globalization argue that globalization is a logical and inevitable expression of contemporary international capitalism and that it benefits all nations and all peoples. Economic globalization can work wonders, they contend, by enhancing competition, allowing the flow of capital to poor areas, and encouraging the spread of beneficial new technologies and ideas. As countries reduce their barriers to trade, inefficient local industries are forced to become more efficient in order to compete with the new flood of imports, thereby enhancing overall national productivity.

FIGURE 1.7 Protests Against Globalization   Meetings of international groups such as the World Trade Organization (WTO) and International Monetary Fund (IMF) commonly draw large numbers of protesters against economic globalization. This group of protesters is at a recent meeting of the WTO meeting in Geneva, Switzerland.

Those that cannot adjust will most likely go out of business, making the global marketplace more efficient.

Every country and region of the world, moreover, ought to be able to concentrate on those activities for which it is best suited in the global economy. Enhancing such geographic specialization, the pro-globalizers argue, creates a more efficient world economy. Such economic restructuring is made increasingly possible by the free flow of capital to those areas that have the greatest opportunities. By making access to capital more readily available throughout the world, economists contend, globalization should eventually result in a certain global **economic convergence**, implying that the world's poorer countries will gradually catch up with the more advanced economies.

The American journalist and author Thomas Friedman, one of the most influential advocates of economic globalization, argues that the world has not only shrunk, but also become economically "flat," so that financial capital, goods, and services can flow freely from place to place. For example, the need to attract capital from abroad forces countries to adopt new economic policies. Friedman describes the great power of the global "electronic herd" of bond traders, currency speculators, and fund managers who either direct money to or withhold it from developing economies, resulting in economic winners and losers (Figure 1.8).

The pro-globalizers also strongly support the large multinational organizations that facilitate the flow of goods and capital across international boundaries. Three such organizations are particularly important: the World Bank, the International Monetary Fund (IMF), and the World Trade Organization (WTO). The primary function of the World Bank is to make loans to poor countries so that they can invest in infrastructure and build more modern economic foundations. The IMF is concerned with making short-term loans to countries that

FIGURE 1.8 The Electronic Herd One aspect of globalization is the rapid movement of capital within the global economic system, movement driven by bond traders, currency speculators, hedge funds, and similar institutions. Here traders work on the Eurodollars Futures floor of the Chicago Mercantile Exchange.

are in financial difficulty—those having trouble, for example, making interest payments on the loans that they had previously taken. The WTO, a much smaller organization than the other two, works to reduce trade barriers between countries to enhance economic globalization. It also tries to mediate between countries and trading blocs that are engaged in trade disputes (Figure 1.9).

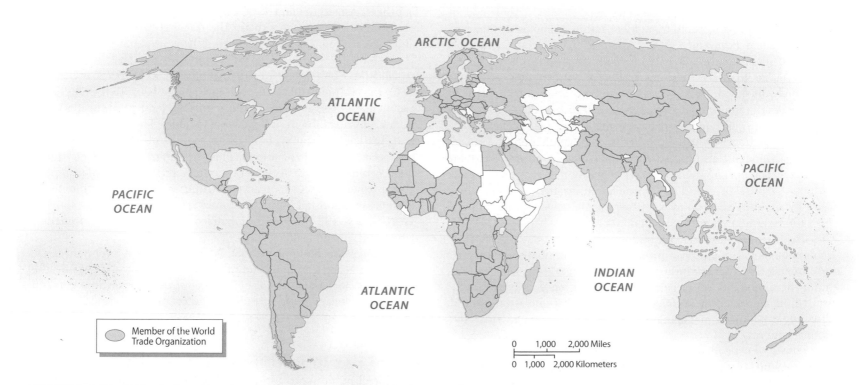

Member of the World Trade Organization

FIGURE 1.9 World Trade Organization One of the most powerful institutions of economic globalization is the World Trade Organization (WTO), which was created in 1995 to oversee trade agreements, encourage open markets, enforce trade rules, and settle disputes. The WTO currently consists of 154 member countries. In addition to these member countries, more than 30 states have "observer status," including Iran and Iraq.

To support their claims, pro-globalizers argue that countries that have been highly open to the global economy have generally had much more economic success than those that have isolated themselves by seeking self-sufficiency. The world's most isolated countries, Burma (Myanmar) and North Korea, have become economic disasters, with little growth and rampant poverty, whereas those that have opened themselves to global forces in the same period, such as Singapore and Thailand, have seen rapid growth and substantial reductions in poverty.

Critics of Globalization Virtually all of the claims of the pro-globalizers are strongly contradicted by the critics of globalization. Opponents often begin by arguing that globalization is not a "natural" process. Instead, it is the product of an explicit economic policy promoted by free-trade advocates, capitalist countries (mainly the United States, but also Japan and the countries of Europe), financial interests, international investors, and multinational firms.

Further, because the globalization of the world economy is creating greater inequity between rich and poor, the trickle-down model of developmental benefits for all people in all regions has yet to be validated. On a global scale, the richest 20 percent of the world's people consume 86 percent of the world's resources, whereas the poorest 80 percent use only 14 percent. The growing inequality of this age of globalization is apparent on both global and national scales. Globally, the wealthiest countries have grown much richer over the past two decades, while many of the poorest countries have lost ground. Nationally, even in developed countries such as the United States, the wealthiest 10 percent of the population have reaped almost all of the gains that globalization has offered, while the poorest 10 percent have seen their income decline in recent decades as wages have remained static or jobs have been lost to outsourcing.

Opponents also contend that globalization promotes free-market, export-oriented economies at the expense of localized, sustainable activities. World forests, for example, are increasingly cut for export timber, rather than serving local needs. As part of their economic structural adjustment package, the World Bank and the IMF often encourage developing countries to expand their resource exports so they have more hard currency to make payments on their foreign debts. This strategy, however, usually leads to overexploitation of local resources. Opponents also note that the IMF often requires developing countries to adopt programs of fiscal austerity that entail substantial reductions in public spending for education, health, and food subsidies. By adopting such policies, critics warn, poor countries will end up with even more impoverished populations than before.

Furthermore, anti-globalizers contend that the "free-market" economic model commonly promoted for developing countries is not the one that Western industrial countries used for their own economic development. In Germany, France, and even to some extent the United States, governments historically have played a strong role in directing investment, managing trade, and subsidizing chosen sectors of the economy.

Those who challenge globalization also worry that the entire system—with its instantaneous transfers of vast sums of money over nearly the entire world on a daily basis—is inherently unstable. The British author and noted critic of globalization John Gray, for example, argues that the same "electronic herd" that Thomas Friedman applauds is a dangerous force because it is susceptible to "stampedes." International managers of capital tend to panic when they think their

FIGURE 1.10 Economic Turmoil in Iceland Protesters burn an effigy of Iceland's prime minister during a demonstration against the government's handling of the 2009 economic crisis. Given the relatively small size of Iceland's economy, the collapse of its bubble economy was the largest downturn in percentage terms suffered by any country in history.

funds are at risk; when they do so, the entire intricately linked global financial system can quickly become destabilized, leading to a crisis of global proportions. The rapid downturn of the global economy in late 2008 seems to support that assertion.

Even when the "herd" spots opportunity, trouble may still ensue. As vast sums of money flow into a developing country, they may create a speculatively inflated **bubble economy** that cannot be sustained. Such a bubble economy emerged in Thailand and many other parts of Southeast Asia in the mid-1990s. Analysts have also used the concept of a bubble economy to explain the tragic collapse of the Icelandic and Irish economies in 2009 (Figure 1.10).

A Middle Position Not surprisingly, many experts argue that both the anti-globalization and the pro-globalization stances are exaggerated. Those in the middle ground tend to argue that economic globalization is indeed unavoidable. They further contend that, while globalization holds both promises and pitfalls, it can be managed, at both the national and the international levels, to reduce economic inequalities and protect the natural environment. These experts stress the need for strong, yet efficient national governments, supported by international institutions (such as the UN, World Bank, and IMF) and globalized networks of environmental, labor, and human rights groups.

Unquestionably, globalization is one of the most important issues of the day—and certainly one of the most complicated. While this book does not pretend to resolve the controversy, nor does it take a position, it does encourage readers to reflect on these critical points as they apply to different world regions.

Diversity in a Globalizing World

As globalization increases, many observers foresee a world far more uniform and homogeneous than today's. The optimists among them imagine a universal global culture uniting all humankind into a single community untroubled by war, ethnic strife, or resource shortage—a global utopia of sorts.

FIGURE 1.11 Local Cultures This market place in Ghana, Africa reminds us that although there are few places beyond the reach of globalization, nevertheless, many unique local landscapes, economies, and cultures still exist.

A more common view, however, is that the world is becoming blandly homogeneous as different places, peoples, and environments lose their distinctive character and become indistinguishable from their neighbors. Diversity may be difficult for a society to live with, but it also may be dangerous to live without. Nationality, ethnicity, cultural distinctiveness—all are the legitimate legacy of humanity. If this diversity is blurred, denied, or repressed through global homogenization, humanity loses one of its defining traits.

However, even if globalization is generating a certain degree of homogenization, the world is still a highly diverse place (Figure 1.11). You can still find marked differences in culture (language, religion, architecture, foods, and many other attributes of daily life), economy, and politics—as well as in the physical environment. Such diversity is so vast that it cannot readily be extinguished, even by the most powerful forces of globalization.

In fact, globalization often provokes a strong reaction on the part of local people, making them all the more determined to maintain what is distinctive about their way of life. Thus, globalization is understandable only if we also examine the diversity that continues to characterize the world and, perhaps most important, the tension between these two forces: the homogenization of globalization and the reaction against it in terms of protecting cultural and political diversity.

The politics of diversity also demand increasing attention as we try to understand worldwide tensions over terrorism, ethnic separateness, regional autonomy, and political independence. Groups of people throughout the world seek self-rule of territory they can call their own. Today most wars are fought *within* countries, not *between* them. As a result, our interest in geographic diversity takes many forms and goes far beyond simply celebrating traditional cultures and unique places. People have many ways of making a living throughout the world, and it is important to recognize this fact as the globalized economy becomes increasingly focused on mass-produced retail goods. Furthermore, a stark reality of today's economic landscape is unevenness: While some people and places prosper, others suffer from unrelenting poverty. This, unfortunately, also is a form of diversity amid globalization (Figure 1.12).

GEOGRAPHY MATTERS: ENVIRONMENTS, REGIONS, LANDSCAPES

Geography is one of the most fundamental sciences, a discipline awakened and informed by the long-standing human curiosity about our surroundings. The term *geography* has its roots in the Greek words for "describing the Earth," and as a result this discipline has been central to all cultures and civilizations as they explore the world, seeking natural resources, commercial trade, military advantage, and scientific knowledge about world environments. In some ways, geography can be compared to history: Historians describe and explain what has happened over time, whereas geographers describe and explain the world's spatial dimensions, how it differs from place to place.

Given the broad scope of geography, it is no surprise that geographers have different conceptual approaches to investigating the world. At the most basic level, geography can be broken into two complementary pursuits: *physical* and *human geography*. Physical geography examines climate, landforms, soils, vegetation, and hydrology. Human geography concentrates on the spatial analysis of economic, social, and cultural systems.

A physical geographer, for example, studying the Amazon Basin of Brazil, might be interested primarily in the ecological diversity of the tropical rain forest or the ways in which the destruction of that environment changes the local climate and hydrology. A human geographer, in contrast, would focus on the social and economic factors explaining the migration of settlers into the rain forest or the tensions and conflicts over resources between new migrants and indigenous peoples.

Another basic division is that between focusing on a specific topic or theme and analyzing a place or a region. The theme approach is referred to as *thematic* or *systematic geography*, while the regional approach is called *regional geography*. These two perspectives are complementary and by no means mutually exclusive. This textbook, for example, draws upon a regional scheme for its overall organization, dividing the globe into 12 separate world regions. It then presents each chapter thematically, examining the topics of environment, population and settlement, cultural differentiation, geopolitics, and economic development in a systematic way. In doing so, each chapter combines four kinds of geography: physical, human, thematic, and regional geography.

Areal Differentiation and Integration

As a spatial science, geography is charged with the study of Earth's surface. A central theme of that responsibility is describing and explaining the differences that distinguish one piece of the world from another. The geographical term for this is **areal differentiation** (*areal* means "pertaining to area"). Why is one part of Earth humid and lush,

FIGURE 1.12 The Landscape of Economic Diversity   The geography of diversity takes many expressions, one of which is economic unevenness, as depicted in this photo from New Delhi, India, where squatter settlements of the poor contrast with the hi-rise office buildings and apartment houses of the more affluent.

while another, just a few hundred kilometers away, is an arid desert (Figure 1.13)?

Geographers are also interested in the connections between different places and how they are linked. This concern is one of **areal integration**, or the study of how places interact with one another. An example is the analysis of how and why the economies of Singapore and the United States are closely intertwined, even though the two countries are situated in entirely different physical, cultural, and political environments. Questions of areal integration are becoming increasingly important because of the new global linkages inherent to globalization.

Global to Local   All systematic inquiry has a sense of scale, whatever the discipline. In biology, some scientists study the smaller units of cells, genes, or molecules, while others take a larger view, analyzing plants, animals, or whole ecosystems. Geographers also work at different scales. While one may concentrate on analysis of a local landscape—perhaps a single village in southern China—another might focus on the broader regional picture, examining all of southern China (Figure 1.14). Other geographers do research on a still larger global scale, perhaps studying emerging trade networks between southern India's center of information technology in Bangalore and North America's Silicon Valley or investigating how India's monsoon might be connected to and affected by the Pacific Ocean's El Niño. But even though geographers may be working at different scales, they never lose sight of the interactivity and connectivity among local, regional, and global scales. They will note the ways that the village in southern India might be linked to world trade patterns or how the late arrival of the monsoon could affect agriculture and food supplies in different parts of India.

Regions: Formal, Functional, and Vernacular

The human intellect seems driven to make sense of the universe by lumping phenomena together into categories that emphasize similarities. Biology has its taxa of living organisms, history marks off eras and periods of time, and geology classifies epochs of Earth history. Geography, too, organizes information about the world, by compressing it into units of spatial similarity called **regions**.

Sometimes the unifying threads of a region are physical, such as climate and vegetation, resulting in a regional designation like the *Sahara Desert* or the *Amazonian rain forest*. Other times the threads are more complex, combining economic and cultural traits, as in the use of the term *Corn Belt* for parts of the central United States. People commonly compress large amounts of information into stereotypes, and in a way geographic regions are just that—a spatial stereotype for a portion of Earth that has some special signature or characteristic that sets it apart from other places.

Geographers designate three types of regions: formal, functional, and vernacular (Figure 1.15). **Formal regions** take their name from the fact that these regions are defined by some aspect of physical form, such as a mountain range, valley, or climate. Cultural features can also be used to define formal regions. An example is the area where a certain language is spoken or a specific religion dominates. Many of the maps in this book denote formal regions. In contrast, a **functional region** is where a certain activity (or cluster of activities) takes place. The earlier example of America's Corn Belt fits this terminology because it forms a region where a specific economic activity dominates. The Bos–Wash (Boston to Washington, DC) megalopolis

FIGURE 1.13 Areal Differentiation In this satellite photo, the California–Mexico border is apparent because of the different sizes of agricultural fields. The larger red pattern in the upper left is the irrigated fields of the United States, while in the bottom right the much smaller and less irrigated fields of Mexico have a different landscape signature when seen from space. Explaining such differences is a central focus of geography.

FIGURE 1.14 Global to Local Just about all places in the world are connected in someway because of globalization, and because of that, global affairs affect local peoples and places. Here a farmer in rural China uses the village laptop to check world grain prices to determine if the time is right for selling their crop.

of the eastern United States is another example, as are newspaper circulation areas and the spatial dimension of a sports team's fan base (think of the line somewhere in the Midwest between Chicago and St. Louis that divides baseball and football fans of each city). Last, **vernacular regions** are defined solely in people's minds as spatial stereotypes that have no visible boundaries in the physical landscape. Examples abound: the South, the Midwest, southern California, New England, and so on.

The Cultural Landscape: Space into Place

Humans transform space into distinct places that are unique and heavily loaded with meaning and symbolism. This diverse fabric of *placefulness* is of great interest to geographers because it tells us much about the human condition throughout the world. Places can tell us how humans interact with nature and among themselves; where there are tensions and where there is peace; where people are rich and where they are poor.

A common tool for the analysis of place is the concept of the **cultural landscape**, which is, simply stated, the visible, material expression of human settlement, past and present. Thus, the cultural landscape is the tangible expression of the human habitat. It visually reflects the most basic human needs—shelter, food, and work. Additionally, the cultural landscape acts to bring people together (or keep them apart) because it is a marker of cultural values, attitudes, and symbols. As cultures vary greatly around the world, so do cultural landscapes (Figure 1.16).

Increasingly, however, we see the uniqueness of places being eroded by the homogeneous landscapes of globalization—shopping malls, fast-food outlets, business towers, theme parks, and industrial complexes. Understanding the forces behind the spread of these landscapes is important because they tell us much about the expansion of global economies and cultures. Although a modern shopping mall in Hanoi, Vietnam, may seem familiar to someone from North America, this new landscape represents yet another component of globalized world culture that has been implanted into a once remote and distinctive city.

REVIEW QUESTIONS

1. Explain the difference between areal differentiation and areal integration.

2. How do functional regions differ from formal and vernacular regions?

3. How is the concept of the cultural landscape related to areal differentiation?

FIGURE 1.15 Geographic Regions This map illustrates three different kinds of geographic regions: vernacular, formal, and functional. A vernacular region is an abstraction that has indistinct cognitive borders, as shown by the general outline of what the public considers to be Silicon Valley. Other vernacular regions would be "the Midwest," "the Deep South," and "the Pacific Northwest." In contrast, formal regions have distinct boundaries, such as that for the Santa Clara Valley, defined as the lowland between two bordering mountain ranges. A functional region is based on a certain activity or organizational structure, such as the civic government of San Jose, bounded by its legal city limits.

FIGURE 1.16 The Cultural Landscape Despite globalization, the world's landscapes still have great diversity, as shown by this village and its surrounding rice terraces on the island of Luzon, Philippines. Geographers use the cultural landscape concept to better understand how people interact with their environment.

THE GEOGRAPHER'S TOOLBOX: LOCATION, MAPS, REMOTE SENSING, AND GIS

Geographers use many different tools to represent the world in a convenient form for examination and analysis. You will need very different kinds of images and data if you're studying vegetation in Brazil or mining in Mongolia; population density in Tokyo or languages spoken in Europe; religions practiced in the Middle East or rainfall in southern India. Today's modern satellite and communications systems offer geographers an array of tools not imagined 50 years ago.

Latitude and Longitude

To navigate their way through their daily tasks, people generally use a mental map of *relative locations* that locate specific places in terms of their relationship to other landscape features. The shopping mall is near the highway, for example, or the college campus is near the river. In contrast, map makers use *absolute location*, often called a mathematical location, that draws upon a universally accepted coordinate system giving every place on Earth a specific numerical address based upon latitude and longitude. The absolute location for the Geography Department at the University of Oregon, for example, has the

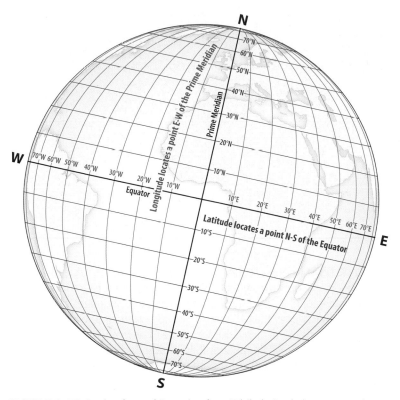

FIGURE 1.17 Latitude and Longitude   While latitude locates a point between the Equator and the poles and is designated as so many degrees north or south, longitude locates a point east or west of the Prime Meridian, which is located just east of London, England.

mathematical address of 44 degrees, 02 minutes, and 42.95 seconds north and 123 degrees, 04 minutes, and 41.29 seconds east. This is written 44° 02′ 42.95″ N and 123° 04′ 41.29″ E.

Lines of latitude, often called **parallels**, run east–west around the globe and are used to locate places north and south of the equator, which is 0 degrees latitude. In contrast, lines of longitude, referred to as **meridians**, run from the north pole, located at 90 degrees north latitude, to the south pole, located at 90 degrees south latitude. Longitude values locate places east or west of the **prime meridian**, located at 0 degrees longitude at the Royal Naval Observatory in Greenwich, England (just east of London) (Figure 1.17). The equator divides the globe into northern and southern hemispheres, whereas the prime meridian divides the world into eastern and western hemispheres; these latter two hemispheres meet at 180 degrees longitude in the western Pacific Ocean. The International Date Line, where each new solar day begins, lies along much of 180 degrees longitude, deviating where necessary to ensure that small Pacific island nations remain on the same calendar day.

Each degree of latitude measures 60 nautical miles or 69 land miles (111 km) and is made up of 60 minutes, each of which is 1 nautical mile (1.15 land miles). Each minute has 60 seconds of distance, each of which is approximately 100 feet (30.5 meters).

From the equator, parallels of latitude are used to mathematically define the tropics: the Tropic of Cancer at 23.5 degrees north and the Tropic of Capricorn at 23.5 degrees south. These lines of latitude denote where the Sun is directly overhead at noon on the solar solstices in June and December. Also, the Arctic and Antarctic

circles, at 66.5 degrees north and south latitude, respectively, mathematically define the polar regions. These regions experience 24 hours of daylight in the summer and 24 hours of darkness during the winter.

Global Positioning Systems (GPS)   Historically, precise measurements of latitude and longitude were determined by a complicated method of celestial navigation, based upon one's location relative to the Sun, Moon, planets, and stars. Today, though, absolute location on Earth (or in airplanes above Earth's surface) is achieved through satellite-based **global positioning systems (GPS)**. These systems use time signals sent from your location to a satellite and back to your receiver to calculate precise coordinates of latitude and longitude. These systems were first used by the U.S. military in the 1960s and then made available to the public in the later decades of the 20th century. Today GPS guides airplanes across the skies, ships across oceans, private autos on the roads, and hikers through wilderness areas, to name only a few of many uses. While most smartphones use locational systems based upon triangulation from cell phone towers, some smartphones are capable of true satellite-based GPS that is accurate to about 30 feet (9 meters).

Map Projections

Because the world is spherical, mapping the globe on a flat piece of paper creates inherent distortions in the latitudinal, or north–south, depiction of Earth's land and water areas. Cartographers (those who make maps) have tried to limit these distortions by using various **map projections**, which are defined simply as the different ways maps are projected onto a flat service. Historically, the Mercator projection was the projection of choice for maps used for oceanic exploration. However, just a brief look at the inflated Greenlandic and Russian landmasses shows its weakness in accurate depiction of high-latitude land areas (Figure 1.18). Over time, cartographers have created literally hundreds of different map projections in their attempts to find the best and most accurate way of mapping the world.

Without going into the details of this vexatious quest, in the last several decades cartographers have generally used the Robinson projection for their maps and atlases. In fact, several professional cartographic societies tried unsuccessfully in 1989 to actually ban projections such as the Mercator because of their spatial distortions. Like many other professional publications, in this book we use only the Robinson projection for our maps.

Map Scale

All maps must reduce the area being mapped to a smaller piece of paper. This reduction involves the use of **map scale**, or the mathematical ratio between the map and the surface area being mapped. Many maps note their scale as a ratio or fraction between a unit on the map and the same unit in the area being mapped, such as 1:63,360 or 1/63,360. This means that 1 inch on the map represents 63,360 inches on the land surface; thus, the scale is 1 inch equals 1 mile. Although 1:63,360 (1 inch equals 1 mile) is a convenient mapping scale to understand, the amount of surface area that can be mapped and fitted on a letter-sized sheet of paper is limited to about 20 square miles. At this scale, mapping 100 square miles would produce a bulky map 8 feet square. Therefore, the ratio must be changed to a larger number, such

(a)

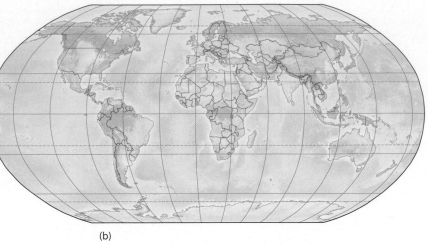

(b)

FIGURE 1.18 Map Projections Cartographers have long struggled with how best to accurately map the world given the inherent distortions when transferring features on a round globe to a flat piece of paper. Early map makers commonly used the Mercator Projection on the left, which while distorting features in the high latitudes worked fairly well for sea-going explorers. On the right is the Robinson Projection that was developed in the 1960s and is now the industry standard because it minimizes cartographic distortion.

as 1:316,800. This ratio means that 1 inch on the map now represents 5 miles (8 kilometers) of distance on land.

Based upon the **representative fraction**, which is the cartographic term for the ratio between the map and the area being mapped, maps are categorized as having either large or small scales (Figure 1.19). It may be easy to remember that large-scale maps make landscape features like rivers, roads, and cities *larger*, but because the features are larger, the maps must cover *smaller* areas. Conversely, small-scale maps cover *larger* areas, but to do so, these maps must make landscape features *smaller*. A bit harder to remember is that the larger the second number of the representative

fraction—the 63,360 in the fraction 1:63,360 or the 100,000 in the fraction 1:100,000, for example—the smaller the scale of the map.

Map scale is probably the easiest to interpret when it is simply portrayed in a **graphic or linear scale**, which visually depicts in a horizontal bar distance units such as feet, meters, miles, or kilometers. Most of the maps in this book are small-scale maps of large areas; thus, the graphic scale is in miles and kilometers. Distances between two points on the map can be calculated by making two tick marks on a piece of paper held next to the points and then measuring the distance between the two marks on the linear scale.

FIGURE 1.19 Small and Large Scale Maps A portion of Australia's east coast north of Sydney is mapped at two scales, one (on the left) at a small scale, and other (on the right) at a large scale. Note the differences in distance depicted on the linear scales of the two different maps. While there's more close up detail in the large scale map, it covers only a small portion of the area mapped at a small scale.

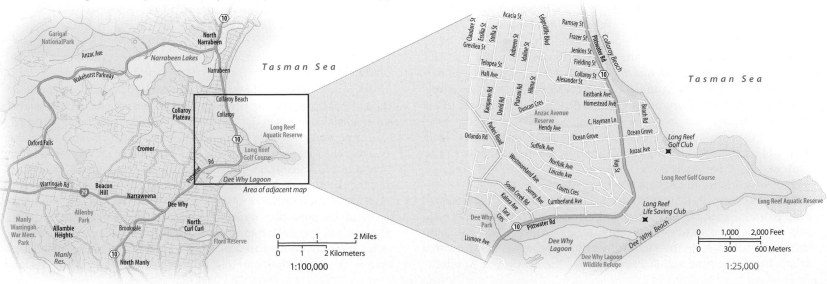

FIGURE 1.20 Choropleth Maps Two different cartographic techniques are shown in these maps. The population density of India is mapped using different categories of density, from sparsely populated to very high densities, depicted with increasing intensity of colors so that one sees immediately the gradients from low to high population density. In the second example, the climates of Sub-Saharan Africa, different climate categories are given different colors, with drier colors represented with sand-like tan, and wetter climates in darker colors.

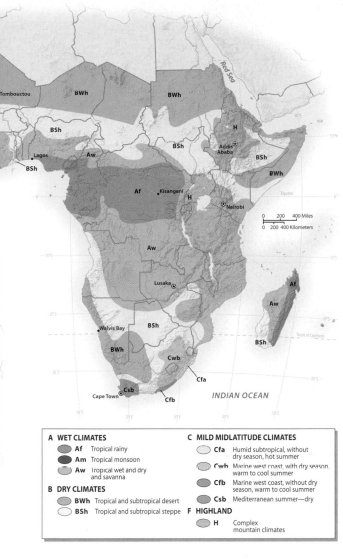

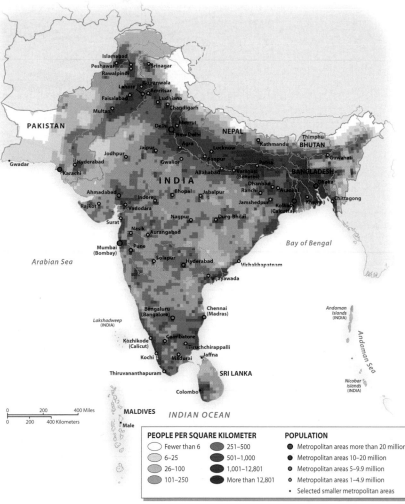

A WET CLIMATES		C MILD MIDLATITUDE CLIMATES	
Af	Tropical rainy	Cfa	Humid subtropical, without dry season, hot summer
Am	Tropical monsoon	Cwb	Marine west coast, with dry season, warm to cool summer
Aw	Tropical wet and dry and savanna	Cfb	Marine west coast, without dry season, warm to cool summer
B DRY CLIMATES		Csb	Mediterranean summer—dry
BWh	Tropical and subtropical desert	F HIGHLAND	
BSh	Tropical and subtropical steppe	H	Complex mountain climates

PEOPLE PER SQUARE KILOMETER		POPULATION	
Fewer than 6	251–500	Metropolitan areas more than 20 million	
6–25	501–1,000	Metropolitan areas 10–20 million	
26–100	1,001–12,801	Metropolitan areas 5–9.9 million	
101–250	More than 12,801	Metropolitan areas 1–4.9 million	
		Selected smaller metropolitan areas	

Map Patterns and Map Legends

Maps, quite obviously, come in a wide array of colors and patterns, which depict everything from the most basic representation of topographic and landscape features to complicated patterns of population, migration, economic conditions, and so forth. Whether the map is a simple *reference map* that shows the location of certain features or a *thematic map* that displays more complicated spatial phenomena, the map legend provides the details by explaining the different map patterns.

Many maps in this book are **choropleth maps**, which map different levels of intensity of data, such as per capita income or population density, placed within discrete spatial units, such as countries, cities, counties, or cultural regions (Figure 1.20). Along with choropleth data, many of our maps contain flow arrows that depict the movement of people or the flow of trade goods.

Aerial Photos and Remote Sensing

Although maps are a primary tool of geography, much can be learned about Earth's surface by deciphering patterns on aerial photographs taken from airplanes, balloons, or satellites. Originally, these photographs were only in black and white, but today color aerial photographs are common.

Even more information about Earth comes from electromagnetic images taken from aircraft or satellites referred to as **remote sensing** (Figure 1.21). This technology has many scientific applications, including monitoring the loss of rain forests, tracking the biological health of crops and woodlands, and even measuring changes in ocean surface temperatures. It is also central to national defense issues, such as monitoring troop movements or the building of missile sites in hostile countries. In simple terms, aerial photographs are merely photographs taken from balloons, airplanes, or satellites, whereas remote sensing gathers

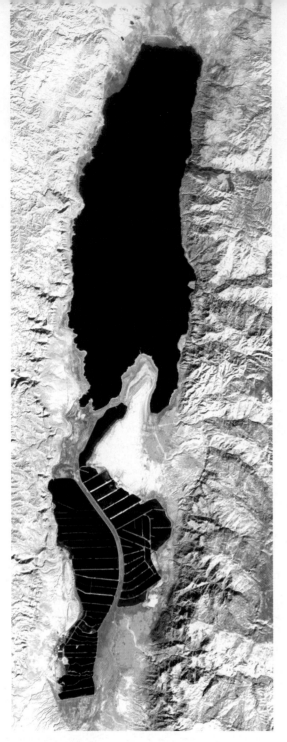

FIGURE 1.21 Remote Sensing of the Dead Sea This NASA satellite image of the Dead Sea, the lowest spot on Earth at 1300 feet (400 meters) below sea level, uses false color remote sensing to capture different elements of the environment. Black is deeper water, with light blue showing shallow waters. The green areas along the shoreline are irrigated crops, while the white areas are salt evaporation ponds.

electromagnetic data that then must be processed and interpreted by computer software to produce images of Earth's surface.

The Landsat satellite program launched by the United States in 1972 is a good example of both the technology and the uses of remote sensing. These satellites collect data simultaneously in four broad bands of electromagnetic energy, from visible through near-infrared

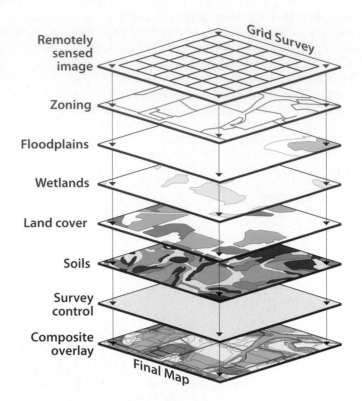

FIGURE 1.22 GIS Layers Geographic Information Systems (GIS) maps usually consist of many different layers of information that can be either viewed and analyzed separately or as a composite overlay. This illustration is of a typical environmental planning map where different physical features (such as wetlands and soils) are combined with zoning regulations.

wavelengths, that is reflected or emitted from Earth. Once these data are processed by computers, they display a range of images, as illustrated in Figure 1.22. The resolution on Earth's surface ranges from areas 260 feet (80 meters) square down to 98 feet (30 meters) square.

Landsat 4 and 5 pass from north to south over the equator at an altitude of 440 miles (700 kilometers) each day at about 10 AM, which allows detection of change to the environment on a continual basis. Of course, cloud cover often compromises the continuous coverage of many parts of the world.

Geographic Information Systems (GIS)

Vast amounts of computerized data from different sources, such as maps, aerial photos, remote sensing, and census tracts, are brought together in **geographic information systems (GIS)**. The resulting spatial databases are used to analyze a wide range of resource problems. Conceptually, GIS can be thought of as a computer system for producing a series of overlay maps showing spatial patterns and relationships (Figure 1.22). A GIS map, for example, might combine a conventional map with data on toxic waste sites, local geology, groundwater flow, and surface hydrology to determine the source of pollutants appearing in household water systems.

Although the earliest GIS dates back to the 1960s, it is only in the last several decades—with the advent of desktop computer systems and remote sensing data—that GIS has become absolutely central to geographic problem-solving. It has a central role in city planning, environmental science, earth science, and real-estate development, to name only a few of the many activities using these systems.

THEMES AND ISSUES IN WORLD REGIONAL GEOGRAPHY

Following two introductory chapters, this book adopts a regional perspective, grouping all of Earth's countries into a framework of 12 world regions (Figure 1.23). We begin with a region familiar to most of our readers—North America—and then move on to Latin America, the Caribbean, Africa, the Middle East, Europe, Russia, and the different regions of Asia, before concluding with Australia and Oceania. Each of the 12 regional chapters employs the same five part thematic structure—environmental geography, population and settlement, cultural coherence and diversity, geopolitical framework, and economic and social development. The concepts and data central to each theme are discussed in the following sections.

ENVIRONMENTAL GEOGRAPHY: THE CHANGING GLOBAL ENVIRONMENT

Chapter 2 provides background on world environmental geography, outlining the global environmental elements fundamental to human settlement—climate, geology, hydrology, and vegetation. In the regional chapters, the environmental geography sections explain the environmental issues relevant to each world region, topics such as climate change, sea-level rise, acid rain, tropical rain forest destruction, and wildlife conservation. These environmental issues sections are not simply a list of problems, but also cover plans and policies developed to resolve those issues (see "Working Toward Sustainability: A Concept with Many Meanings").

FIGURE 1.23 World Regions   These regions are the basis for the 12 regional chapters in this book. Countries or areas within countries that are treated in more than one chapter are designated on the map with a striped pattern. For example, western China is discussed in both Chapter 10, on Central Asia, and Chapter 11, on East Asia. Also, three countries on the South American continent are discussed as part of the Caribbean region because of their close cultural similarities to the island region.

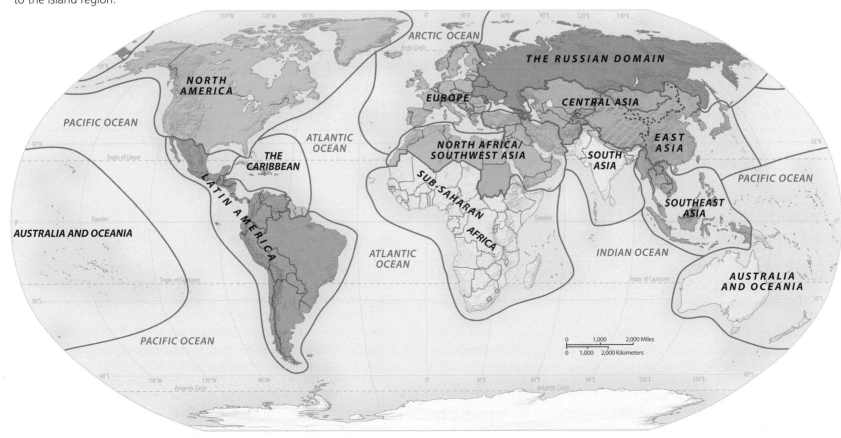

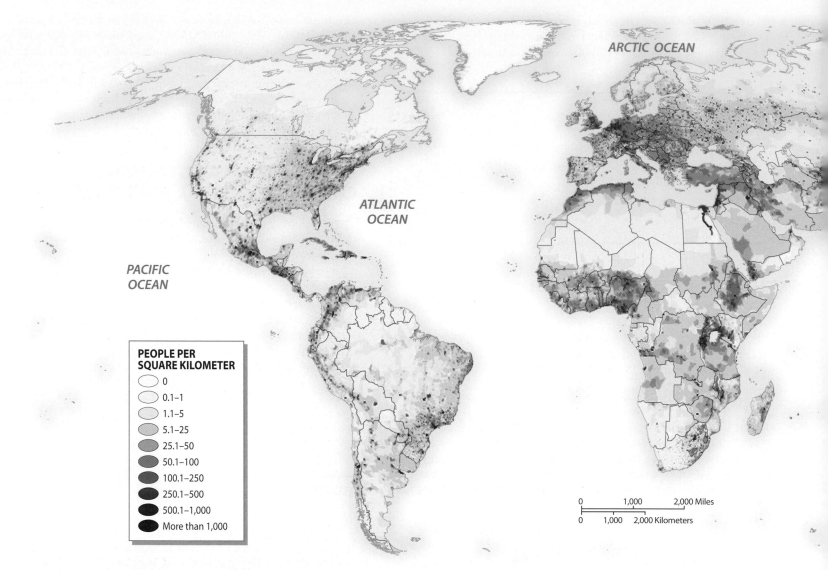

FIGURE 1.24 World Population This map emphasizes the different population densities in the areas of the world. East Asia stands out as the most populated region, with high densities in Japan, Korea, and eastern China. The second most populous region is South Asia, dominated by India, which is second only to China in total population. In North Africa and Southwest Asia, population clusters are often linked to the availability of water for irrigated agriculture, as is apparent with the population cluster along the Nile River. Higher population densities in Europe, North America, and other countries are usually associated with large cities, their extensive suburbs, and nearby economic activities.

POPULATION AND SETTLEMENT: PEOPLE ON THE LAND

Currently, Earth has more than 7 billion people, with demographers forecasting the addition of another half billion by 2025. Most of this increase will take place in Africa, Asia, and South America (Figure 1.24). Because of high rates of population growth in developing countries, difficult questions dominate discussions of many global issues. Can these countries absorb the rapidly increasing population and still achieve the necessary economic and social development to ensure some level of well-being and stability for their populations? What role, if any, should the developed countries of North America, Europe, and East Asia play in helping developing countries with their population problems?

Population is a complex and even contentious topic, but several points may help to focus the issues:

- Very different rates of population growth are found in different regions of the world. Some countries, such as India, are growing rapidly; others, such as Italy, have no natural growth at all, with all population growth coming from in-migration.

- The current rate of population growth is now half the peak rate experienced in the 1960s. At that time, talk of "the population bomb" and "population explosions" was common, warning that the world either was very close to or had already passed a sustainable carrying capacity for Earth. Still, even with slower growth, about 20 percent of the world's population is undernourished.

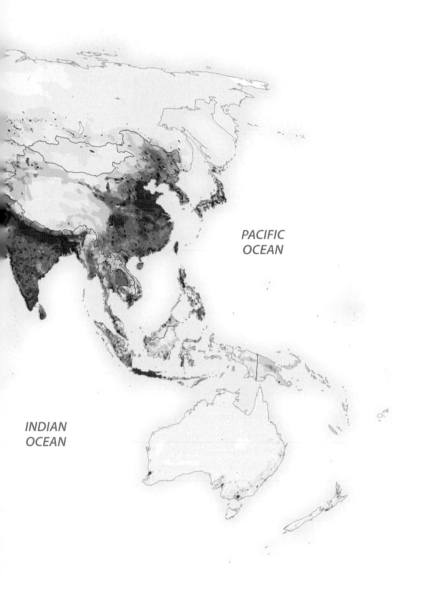

FIGURE 1.25 Family Planning Policies Many countries in the developing world have concluded that unrestrained population growth may keep them from realizing their development goals. Therefore, they have put family planning policies in place that promote birth control. This billboard is in China and encourages citizens to continue the state's policy of family planning.

- Population planning takes many forms, from the fairly rigid one- or two-child policies of China and other countries attempting to slow population growth, to the family-friendly policies of no-growth countries that would like to increase their natural birth rates (Figure 1.25).

- Not all attention should be focused on natural growth because migration is increasingly the root cause of population growth in the globalized world. Although most international migration is driven by a desire for a better life in the richer regions of the developed world, millions of migrants are refugees from civil strife, political persecution, and environmental disasters.

- The greatest migration in human history is going on now, as millions of people move from rural to urban environments. As of 2009, a landmark was reached when demographers estimated that more than half the world's population now lives in towns and cities. As with natural growth, the most rapid rates of urbanization are in the developing regions of Africa, South America, and Asia.

Population Growth and Change

Because of the central importance of population growth, each regional chapter in this book includes a table of population data for the countries in that region (Table 1.1). Although at first glance these statistics might seem daunting, this information is crucial to understanding the population geography of the regions.

Natural Population Increase A common starting point for measuring demographic change is the **rate of natural increase (RNI)**, which provides the annual growth rate for a country or region as a percentage. This statistic is produced by subtracting the number of deaths from the number of births in a given year. Important to remember is that population gains or losses through migration are not considered in the RNI.

Also, instead of using raw numbers for a large population, demographers divide the gross numbers of births and deaths by the total population, thereby producing a number per 1,000 of the population. This is referred to as the *crude birthrate* or the *crude death rate*. For example, in

A Concept with Many Meanings

The idea of "sustainability" seems to be everywhere, as we hear about sustainable cities, agriculture, forestry, businesses, corporations, lifestyles—even a sustainable world. The list seems endless at times, and since we see so many varied usages we can rightfully ask, whether the term has any substance? Or is sustainability overused to the point where it's merely a superficial buzzword for expressing general concern about the environment?

The dictionary tells us that *sustainable* has two main definitions. The first meaning is to endure and be able to maintain something at a certain level so that it lasts. The second refers to something that can be upheld or defended, such as a *sustainable idea* or *action*. Resource management has long used terms such as *sustained-yield forestry* to refer to timber practices where the amount of tree harvesting is attuned to the natural rate of forest growth—so that the resource is not overused, but is able to renew itself over time.

Moral and ethical dimensions were added to this traditional usage in 1987 when the UN World Commission on Environment and Development addressed the complicated relationship between economic development and environmental deterioration. The commission stated that "sustainable development is development that meets the needs of the present without compromising the ability of future generations to meet their own needs." This cautionary message expands the notion of sustainability from a narrow focus on managing a specific resource, such as trees or grass, to include the whole range of human "needs," present and future (Figure 1.2.1). Fossil fuels, for example, are finite, so we should not consume them greedily in the present without considering their availability for future generations. Similar cautions are given about the sustainable uses of all other resources—air, water, soil, genetic biodiversity, wildlife habitats, farming, and so on.

Achieving the sustainable use of a specific resource can be extremely difficult, however, because it requires knowing the total amount of the resource in question, as well as the current rate of consumption, and estimating the needs of future generations. These challenges have given rise to the new field of *sustainability science*, which emphasizes measuring and quantifying these factors. Because of these measurement difficulties, many researchers suggest that sustainability is better thought of as a process, rather than an achievable state. Regardless, even as a process, the concept of sustainability is still infused with strong ethical and moral dimensions in terms of constraint and stewardship.

FIGURE 1.2.1 Sustainable Energy Most of the world's energy comes from fossil fuels, mainly oil and coal, which, because those supplies are finite, is by definition unsustainable because at some point in time they will be completely used up and unavailable to produce energy. In contrast are sustainable energy sources such as solar, wind, biofuels, hydro, and, as shown in the photo, geothermal. Because of its abundant supplies of water and geothermal heat, Iceland promotes itself as only the country that runs completely on sustainable energy, with 25 percent of its energy coming from geothermal sources, and the rest from hydroelectricity. This, however, is not quite true since Iceland's large fishing fleet, which contributes considerably to the country's economy, is powered by fossil fuel.

In the following chapters, we explore the many different ways people are thinking about and working toward environmental and resource sustainability worldwide.

2011 the crude birthrate for the whole world was 20 per 1,000, with a crude death rate of 8 per 1,000. Thus, the natural growth rate was 12 per 1,000. Converting that figure to a percentage produces the RNI; therefore, the global RNI in 2011 was 1.2 percent per year.

Because birthrates vary greatly amongst peoples and cultures (and between countries and regions of the world), rates of natural increase also vary greatly. In Africa, for example, several countries have crude birthrates of more than 40 per 1,000 people. Because in these countries death rates are generally less than 11 per 1,000, the rates of natural increase are often greater than 2.9 percent per year, which are the highest population growth numbers found anywhere in the world.

Total Fertility Rate Although the crude birthrate gives some insight into current conditions in a country, demographers place more emphasis on the **total fertility rate (TFR)** to predict future growth. The TFR is an artificial and synthetic number that measures the fertility of a statistically fictitious, yet average group of women moving through their childbearing years. If women marry early and have many children over a long span of years, the TFR is a high number. Conversely, if data show that women marry late and have few children, the number is correspondingly low (Figure 1.26). Important to note is that any number less than 2.1 implies that a population has no natural growth because it takes a minimum of two children to replace their parents, with a fraction more to compensate for infant mortality. From population data collected in the past decade, the current TFR for the world is 2.4. While that is the average for the whole world, the variability among regions is striking. To illustrate, the current TFR for Africa is 4.7, whereas in no-growth Europe it is only 1.6.

Young and Old Populations One of the best indicators of the momentum (or its lack) for continued population growth is the youthfulness of a population, since this shows the proportion of a population about to enter the prime reproductive years. The common statistic for this measure is the percentage of a population under age 15. Currently, the global average is that 26 percent of the population is younger than age 15. However, in fast-growing Africa, that figure is 41 percent, with several African countries approaching 50 percent. This strongly suggests that rapid population growth in Africa will continue for at least another generation, despite the tragedy of the AIDS epidemic.

TABLE 1.1 POPULATION INDICATORS

Country	Population (millions) 2012	Population Density (per square kilometer)	Rate of Natural Increase (RNI)	Total Fertility Rate	Percent Urban	Percent <15	Percent >65	Net Migration (Rate per 1000) 2010–15[a]
China	1,350.4	141	0.5	1.5	51	16	9	−0.3
India	1,259.7	383	1.5	2.5	31	31	5	−0.2
United States	313.9	33	0.5	1.9	79	20	13	3.1
Indonesia	241.0	127	1.3	2.3	43	27	6	−0.8
Brazil	194.3	23	1.0	1.9	84	24	7	−0.2
Pakistan	180.4	227	2.1	3.6	35	35	4	−1.4
Nigeria	170.1	184	2.6	5.6	51	44	3	−0.4
Bangladesh	152.9	1,062	1.6	2.3	25	31	5	−0.1
Russia	143.2	8	−0.1	1.6	74	15	13	1.2
Japan	127.6	338	−0.2	1.4	86	13	24	0.4

[a]Net Migration Rate from the United Nations, Population Division, *World Population Prospects: The 2010 Revision Population Database.*
Source: Population Reference Bureau, *World Population Data Sheet,* 2012.

In contrast, Europe has only 16 percent of its population under 15, and North America has 19 percent.

The other end of the age spectrum is also important, and it is measured by the percentage of a population over age 65. This number is useful for inferring the needs of a society in providing social services for its senior citizens and pensioners. In Japan, for example, about a quarter of the population is over the age of 65.

Population Pyramids The best graphical indicator of a population's age and gender structure is the **population pyramid**. This graph depicts the percentage of a population (or, in some cases, the raw number) that is male or female in different age classes, from young to old (Figure 1.27). If a country has higher numbers of young people than old, the graph has a broad base and a narrow tip, thus taking on a pyramidal shape that commonly forecasts rapid population growth. In contrast, slow-growth or no-growth populations are top-heavy, with a larger number of seniors than younger age classes.

Not only are population pyramids useful for comparing different population structures around the world at a given point in time, but also they can capture the structural changes of a population in time if it transitions from fast to slow growth. Population pyramids are also useful for displaying gender differences within a population, showing whether or not there is a disparity in the numbers of males and females. In the mid-20th century, for example, population pyramids for those countries that fought in World War II (such as the United States, Germany, France, and Japan) showed a distinct deficit of males, indicating those lost to warfare. Similar patterns are found today in those countries experiencing widespread conflict and civil unrest.

Cultural preferences for one sex or another, such as the preference for male infants in China and India, also show up in population pyramids. Because of their usefulness in showing different population structures, comparative population pyramids are found throughout the regional chapters of this book.

FIGURE 1.26 Total Fertility Rate Birthrates and death rates vary widely around the world. Fertility rates result from an array of variables, including state family-planning programs and the level of a woman's education. This family lives in the state of Rajasthan, India.

Life Expectancy Another demographic indicator that contains information about health and well-being in a society is *life expectancy*, which is the average length of a life expected at the birth of a typical male or female in a specific country. Because a large number of social factors—such as health services, nutrition, and sanitation—influence life expectancy, these data are often used as an indicator of the level of social development in a country. In this book, life expectancy is used as a social indicator; thus, these data are found in the economic and social development tables instead of in the population data tables.

Not surprisingly, because social conditions vary widely around the world, so do life expectancy figures. In general, though, life expectancy has been increasing over the decades, implying that the conditions supporting life and longevity are improving. To illustrate, in 1975 the average life expectancy figure for the world was 58 years.

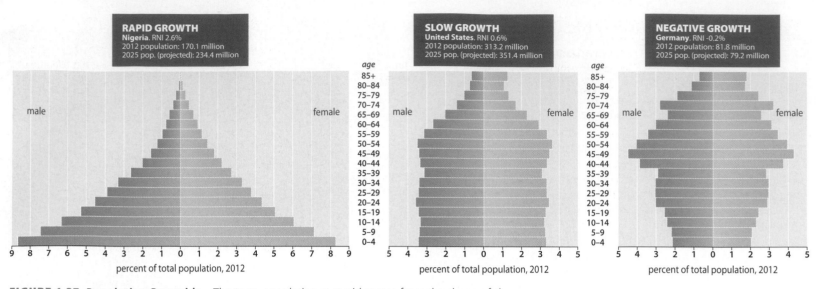

RAPID GROWTH
Nigeria. RNI 2.6%
2012 population: 170.1 million
2025 pop. (projected): 234.4 million

SLOW GROWTH
United States. RNI 0.6%
2012 population: 313.2 million
2025 pop. (projected): 351.4 million

NEGATIVE GROWTH
Germany. RNI -0.2%
2012 population: 81.8 million
2025 pop. (projected): 79.2 million

percent of total population, 2012

FIGURE 1.27 Population Pyramids The term *population pyramid* comes from the shape of the graph assumed by a rapidly growing country such as Nigeria, when data for age and sex are plotted as percentages of the total population. The broad base illustrates the high percentage of young people in the country's population, which indicates that rapid growth will probably continue for at least another generation. This pyramidal shape contrasts with the narrow bases of slow- and negative-growth countries, such as the United States and Germany, which have fewer people in the childbearing years.

whereas today it is 70. In Sub-Saharan Africa, however, life expectancy has changed very little over the last 30 years because of the HIV/AIDS epidemic. As a result, the life expectancy for the region is just about the same (55) as it was in 1975 (52).

The Demographic Transition The historical record suggests that population growth rates slow down similarly over time. More specifically, in Europe, North America, and Japan, population growth slowed as

countries became increasingly industrialized and urbanized. From these historical data, demographers generated the **demographic transition model**, a conceptualization that tracked the changes in birthrates and death rates over time. Originally, this model had four stages; however, a fifth stage is now commonly added to characterize the fact that many countries have slowed still further to a no-growth point (Figure 1.28).

In the original demographic transition model, Stage 1 is characterized by both high birthrates and high death rates, resulting in a very low

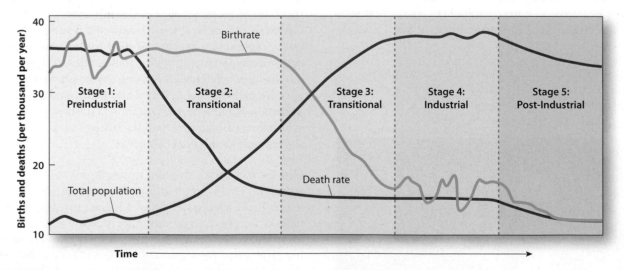

FIGURE 1.28 Demographic Transition As a country goes through industrialization, its population moves through the five stages in this diagram, referred to as the *demographic transition*. In Stage 1, population growth is low because high birthrates are offset by high death rates. Rapid growth takes place in Stage 2, as death rates decline. Stage 3 is characterized by a decline in birthrates. The transition was initially thought to end with low growth once again in Stage 4, resulting from a relative balance between low birthrates and low death rates. But with a large number of developed countries now showing no natural growth, demographers have recently added a fifth stage to the traditional demographic transition model, one that shows no or even negative natural growth.

rate of natural increase. Historically, this stage is associated with Europe's preindustrial period, a time that predated common public health measures such as sewage treatment, an understanding of disease transmission, and the most fundamental aspects of modern medicine. Not surprisingly, death rates were high and life expectancy was short. Tragically, these conditions are still found today in some parts of the world.

In Stage 2, death rates fall dramatically, while birthrates remain high, thus producing a rapid rise in the RNI. In both historical and contemporary times, this decrease in death rates is commonly associated with the development of public health measures and modern medicine. Additionally, one of the assumptions of the demographic transition model is that these health services become increasingly available only after some degree of economic development and urbanization takes place.

However, even as death rates fall and populations increase, it takes time for people to respond with lower birthrates, which happens in Stage 3. This, then, is the transitional stage in which people become aware of the advantages of smaller families in an urban and industrial setting, contrasted with the earlier need for large families in rural, agricultural settings or where children worked at industrial jobs (both legally and illegally).

Then, in Stage 4, a low RNI results from a combination of low birthrates and very low death rates. Until recently, this stage was assumed be the static end point of change of a developing, urbanizing population. However, as mentioned earlier, that does not seem to be the case. In many highly urbanized developed countries, such as those in Europe, the death rate now exceeds the birthrate. As a result, the RNI falls below a replacement level, as attested to by a negative number. This negative growth state argues for the addition of a fifth stage to the traditional demographic transition model. Of the world's largest countries, Russia and Japan fall into this category.

Remember, though, that the RNI is just that—the rate of natural increase. Thus, it does not capture a country's growth from in-migration. In the United States, for example, although the RNI is below the replacement level, the overall population continues to grow because of immigration from other countries, both legal and illegal. The same is true of many other developed countries, particularly those in Europe. This topic is discussed further in Chapter 8.

Global Migration

Never before in human history have so many people been on the move. Today more than 190 million people live outside the country of their birth and thus are officially designated as migrants by international agencies. Much of this international migration is directly linked to the new globalized economy because half of the migrants live either in the developed world or in developing countries with vibrant industrial, mining, or petroleum extraction economies. In the oil-rich countries of Kuwait and Saudi Arabia, for example, the labor force is composed primarily of foreign migrants (Figure 1.29). In total numbers, fully one-third of the world's migrants live in seven industrial countries: Japan, Germany, France, Canada, the United States, Italy, and the United Kingdom.

Moreover, most of these migrants have moved to cities; in fact, 20 percent of migrants live in just 20 world cities. Further, because industrial countries usually have very low birthrates, immigration accounts for a large proportion of their population growth. For example, about one-third of the annual growth in the United States is due to in-migration.

FIGURE 1.29 Net Migration Rate While this photo of autos and people waiting to cross from Mexico to the United States at the Tijuana border station exaggerates the case since they're mostly U.S. day-trippers returning home, nevertheless data on the net migration rate is an important demographic indicator because in- and out-migration is generally more important to the growth of developed countries than the natural birth rate.

However, not all migrants move for economic reasons. War, persecution, famine, and environmental destruction cause people to flee to safe havens elsewhere. Accurate data on refugees are often difficult to obtain for several reasons (such as individuals not legally crossing international boundaries or countries deliberately obscuring the number for political reasons), but UN officials estimate that some 35 million people should be considered refugees. More than half of these are in Africa and western Asia.

Push and Pull Forces Although its causes may be complicated, the migration process can be understood through three interactive concepts. First, *push* forces, such as civil strife, environmental degradation, or unemployment, drive people from their homelands. Second, *pull* forces, such as better economic opportunities or health services, attract migrants to certain locations, within or beyond their national boundaries. Connecting the two are the *informational networks* of families, friends, and sometimes labor contractors who provide information on the mechanics of migration, transportation details, and housing and job opportunities.

Net Migration Rates The amount of immigration (in-migration) and emigration (out-migration) is measured by the **net migration rate**, a statistic that indicates whether more people are entering or leaving a country. A positive figure means the population is growing because of in-migration, whereas a negative number means more people are leaving than arriving. As with other demographic indicators, the net migration rate is provided for the number of migrants per 1,000 of a base population. To illustrate, the net migration rate for the United States is 2 per 1,000 people, whereas Canada, which receives even more immigrants than the United States, has a net migration rate of 7. In contrast, Mexico, the source of many migrants to North America, has a net migration rate of −2.5.

Some of the highest net migration rates are found in countries that depend heavily on migrants for their labor force. These include Qatar, with a net migration rate of 19, and Kuwait, at 15. Countries with the highest negative migration rates are Tonga, −15.4;

Samoa, −13.5; Micronesia, −12.8, and several Caribbean islands with rates close to −10.

Settlement Geography

Population Density   The average number of people per area unit (square mile or square kilometer) is referred to as **population density**. This statistic conveys important information about settlement patterns and landscape in a specific area or country. In Table 1.1, you can see the striking difference between the high population density of India and the much lower figure for the United States. Flying over these two countries and looking down at the settlement patterns explains this contrast (Figure 1.30). Much of the United States is covered by farms covering hundreds of acres, with houses and barns several miles from their neighbors. In contrast, the landscape of India is made up of small villages, distanced from each other by only a mile

FIGURE 1.30 Contrasting Settlement Density: U.S. and India   In the U.S. the population density is 33 per square kilometer, contrasted with 383 per square kilometer in India. Besides the difference in the size of each country's population, another factor in the contrasting densities is the difference in settlement patterns. In the United States people commonly settle on dispersed farms on large acreage, as illustrated by the landscape in Iowa, whereas as in India there is dense settlement in both towns and rural landscapes.

or so. This results in a population density three times higher in India than in the United States. Japan has a similar settlement pattern to India, even though it is primarily an urban, industrial country. Bangladesh, one of the most densely settled countries in the world, must somehow squeeze its large and rapidly growing population into a limited amount of dry land built by the delta of two large rivers, the Ganges and the Brahmaputra.

Because population densities differ considerably between rural and urban areas, the gross national figure can be a bit misleading. Many of the world's largest cities, for example, have densities of more than 30,000 people per square mile (10,300 per square kilometer), with the central areas of Mumbai (Bombay) and Shanghai easily twice as dense because of the prevalence of high-rise apartment buildings. In contrast, most North American cities have densities of fewer than 10,000 people per square mile (3,800 per square kilometer), due largely to the cultural preference for single-family dwellings on individual urban lots.

An Urbanizing World   Cities are the focal points of the contemporary, globalizing world—the fast-paced centers of deep and widespread economic, political, and cultural change. Because of this vitality and the options cities offer to impoverished and uprooted rural peoples, they are also magnets for migration. The scale and rate of growth of some world cities are absolutely staggering. Estimates are that between natural growth and in-migration, Mumbai (Bombay) will add over 7 million people by 2020, which, assuming growth is constant throughout the period (perhaps a questionable assumption), would mean that the urban area would add over 10,000 new people each week. The same projections would have Lagos, Nigeria, which currently has the highest annual growth of any megacity, adding almost 15,000 per week.

Based upon data on the **urbanized population**, which is the percentage of a country's population living in cities, as mentioned, at least half the world's population now lives in cities. Further, demographers predict that the world will be 60 percent urbanized by 2025.

Tables in this book's regional chapters include data on the urbanization rate for each country. To illustrate, more than 80 percent of the populations of Europe, Japan, Australia, and the United States live in cities. Generally speaking, most countries with such high rates of urbanization are also highly industrialized because manufacturing tends to cluster around urban centers. In contrast, the urbanized rate for developing countries is usually less than 50 percent, with figures closer to 40 percent not uncommon. Urbanization figures also show where there is high potential for urban migration. If the urbanized population is relatively small, as in Zimbabwe (Africa), where only 29 percent of the population lives in cities, the probability of high rates of urban migration in the next decades is high.

REVIEW QUESTIONS

1. How is the rate of natural increase calculated? Given an example.
2. What is the total fertility rate?
3. Describe and explain the demographic transition model.
4. How is a population pyramid constructed, and what kind of information does it convey?

CULTURAL COHERENCE AND DIVERSITY: THE GEOGRAPHY OF CHANGE AND TRADITION

Social scientists like to say that culture binds together the world's diverse social fabric. If this is true, one glance at the daily news suggests this complex global tapestry could be unraveling because of widespread cultural tensions and conflict. As noted earlier, with the recent rise of global communication systems (satellite TV, films, videos, etc.), stereotypical Western culture is spreading at a rapid pace. Although some cultures accept these new cultural influences willingly, others resist and push back against these new forms of cultural imperialism through local protests, censorship, and even terrorism.

The geography of cultural cohesion and diversity, then, entails an examination of tradition and change; of the new cultural forms produced by interactions between cultures; of gender issues; and of global languages and religions (Figure 1.31).

Culture in a Globalizing World

Given the diversity of cultures around the world, coupled with the dynamic changes connected with globalization, traditional definitions of culture must be stretched somewhat to provide a viable conceptual framework. A very basic definition provides a starting point. **Culture** is learned, not innate, and is behavior held in common by a group of people, empowering them with what is commonly called a "way of life."

In addition, culture has both abstract and material dimensions: speech, religion, ideology, livelihood, and value systems, but also technology, housing, foods, and music. These varied expressions of culture are relevant to the study of world regional geography because they tell us much about the way people interact with their environment, with one another, and with the larger world. Not to be overlooked is that culture is dynamic and ever changing, not static. Thus, culture is a process, not a condition—an abstract, yet useful concept that is constantly adapting to new circumstances. As a result, there are always tensions between the conservative, traditional elements of a culture and the newer forces promoting change.

FIGURE 1.31 Ethnic Tensions Unfortunately, much contemporary cultural change is characterized by violence among different ethnic groups. In this photo, a commuter in the Indian state of Gujarat passes by a car burned during rioting between Muslims and Hindus.

When Cultures Collide Cultural change often takes place within the context of international tensions. Sometimes, one cultural system will replace another; at other times, resistance by one group to another's culture will stave off change. More commonly, however, a newer, hybrid form of culture results from an amalgamation of two cultural traditions. Historically, colonialism was the most important perpetuator of these cultural collisions; today, though, globalization in its varied forms is a major vehicle of cultural tensions and change.

The active promotion of one cultural system at the expense of another is called **cultural imperialism**. Although many expressions of cultural imperialism still exist today, the most severe examples occurred in the colonial period. In those years, European cultures spread worldwide, often overwhelming, eroding, and even replacing indigenous cultures. During this period, Spanish culture spread widely in South America, French culture diffused into parts of Africa and Southeast Asia, and British culture overwhelmed South and Southwest Asia. New languages were mandated, new educational systems were implanted, and new administrative institutions replaced the old. Foreign dress styles, diets, gestures, and organizations were added to existing cultural systems. Many vestiges of colonial culture are still evident today. In India, the makeover was so complete that pundits are fond of saying, with only slight exaggeration, that "the last true Englishman will be an Indian."

Today's cultural imperialism is seldom linked to an explicit colonizing force, but more often comes as a fellow traveler with economic globalization. Though many expressions of cultural imperialism carry a Western (even U.S.) tone—such as McDonald's, MTV, KFC, Marlboro cigarettes, and the widespread use of English as the dominant language of the Internet—these facets result more from a search for new consumer markets than from deliberate efforts to spread modern U.S. culture throughout the world.

The reaction against cultural imperialism is **cultural nationalism**. This is the process of protecting and defending a cultural system against diluting or offensive cultural expressions, while at the same time actively promoting national and local cultural values. Often cultural nationalism takes the form of explicit legislation or official censorship that simply outlaws unwanted cultural traits. Examples of cultural nationalism are common. France has long fought the Anglicization of its language by banning "Franglais" in official governmental French, thereby exorcising commonly used words such as *weekend*, *downtown*, *chat*, and *happy hour*. France has also sought to protect its national music and film industries by legislating that radio DJs play a certain percentage of French songs and artists each broadcast day (40 percent currently). Many Muslim countries limit Western cultural influences by restricting or censoring international TV, an element they consider the source of many undesirable cultural influences. Most Asian countries as well are increasingly protective of their cultural values, and many are demanding changes to tone down the sexual content of MTV and other international TV networks.

Cultural Hybrids As noted above, the most common product of cultural collision is the blending of forces to form a new, synergistic form of culture, a process called **cultural syncretism or hybridization** (Figure 1.33). To characterize India's culture as British, for example, is to grossly oversimplify and exaggerate England's colonial influence. Instead, Indians have adapted many British traits to their own circumstances, infusing them with their own meanings. India's use of English, for example, has produced a unique form of "Indlish" that often befuddles

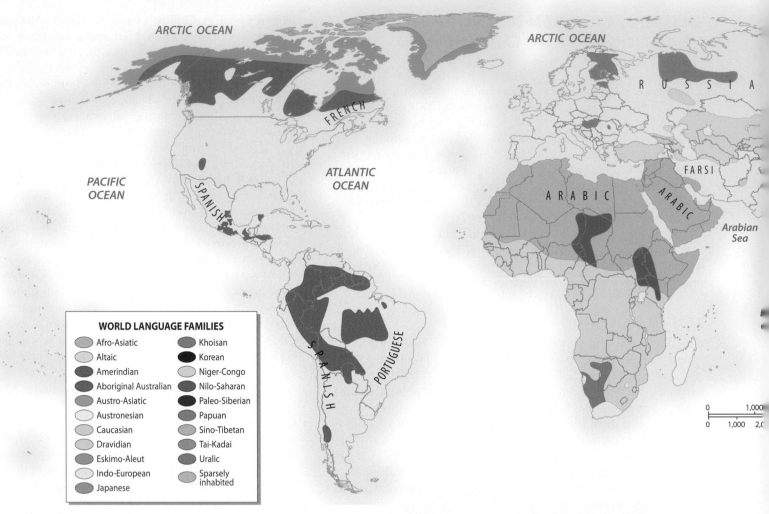

FIGURE 1.32 **World Language Families**   Most languages of the world belong to a handful of major language families. About 50 percent of the world's population speaks a language belonging to the Indo-European language family, which includes languages common to Europe, but also major languages in South Asia, such as Hindi. They are in the same family because of their linguistic similarities. The next largest family is the Sino-Tibetan family, which includes languages spoken in China, the world's most populous country.

visitors to South Asia. Nor should we forget that India has added many words to our English vocabulary—*khaki*, *pajamas*, *veranda*, and *bungalow*, among others. Clearly, both the Anglo and the Indian cultures have been changed by the British colonial presence in South Asia.

Other examples of cultural hybrids abound: Australian-rules football, hip-hop music, fusion cuisine, Tex-Mex fast food, and on and on.

Gender and Globalization

Gender is a sociocultural construct, linked to the values and traditions of specific cultural groups that differentiate the characteristics of the two biological sexes, male and female. Central to this concept are **gender roles**, which are the cultural guidelines that define appropriate behavior for each gender within a specific context. In a traditional tribal or ethnic group, for example, gender roles might rigidly define the difference between women's work (which often consists of domestic tasks) and men's work (which is done mostly outside the home). Similarly, gender roles might guide all other social behaviors within a group, such as child rearing, education, marriage, and even recreational activities (Figure 1.34).

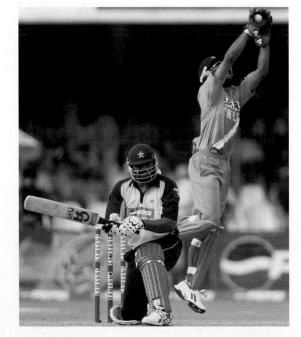

FIGURE 1.33 **Cultural Hybrid**   The hybridization of culture is clearly shown in this photo of South Asian (India vs Pakistan) cricket, a game that before the colonial period was a uniquely British sport played only in the United Kingdom. Today, though, it's played throughout the world in those countries with a British colonial history. Note, too, that that commercial globalization has also become part of cricket as the playing field and the players themselves display the logos of global culture.

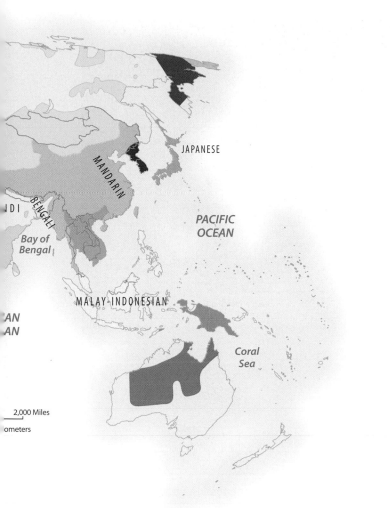

FIGURE 1.34 Women Protesting Muslim women played a major role in the 2011 Arab Spring protests against autocratic rulers, as shown here in this photo of Egyptian women during a Cairo demonstration.

cultural groups and societies that practice blatant discrimination against women. This topic is discussed later in the chapter within the section on measures of social development.

Language and Culture in Global Context

Language and culture are so intertwined that often language is the major characteristic that differentiates and defines cultural groups (Figure 1.32). Furthermore, because language is the primary means for communication, it folds together many other aspects of cultural identity, such as politics, religion, commerce, and customs. Language is fundamental to cultural cohesiveness and distinctiveness, for language not only brings people together, but also sets them apart from nonspeakers of that language. Therefore, language is an important component of national or ethnic identity, as well as a means for creating and maintaining boundaries for group and regional identity.

The explicit and often rigid gender roles of a traditional social unit contrast greatly with the less rigid, more implicit, and often flexible gender roles of a large, modern, urban industrial society (Figure 1.35). More to the point, globalization in its varied expressions is causing significant changes to traditional gender roles throughout the world. Global apparel firms, for example, often locate sewing factories in areas of the less developed world. They do this to gain economic advantage by paying low wages to female workers, who are often the first generation to work outside the home. However, because of traditional gender roles, men do not always pick up the domestic tasks formerly done by these new factory workers; instead, the women often put in what is commonly called a double day, returning to their homes from the factory to then put in hours of work doing cooking, child rearing, and housecleaning.

In contrast to providing new work opportunities for women in less developed countries, there is also a gender dimension to the economic effects of globalization in many developed countries. In the United States, for example, male workers have suffered more from unemployment than have females, as industrial and technology jobs have been outsourced to China and India. To compensate, many females who previously did not work outside their homes have now taken full- and part-time jobs to provide support for their families. Many males, meanwhile, have taken on new roles in domestic activities.

Globalization has also spread the notion of gender equality (and inequality) around the globe, calling into question and exposing those

FIGURE 1.35 Traditional Dress and Gender Roles Women's gender roles—much more so than for men—are reinforced by appropriate dress and traditional clothing, as shown here for these Muslim women in Zanzibar, Tanzania.

Because most languages have common historical (and even pre-historical) roots, linguists have grouped the thousands of languages found throughout the world into a handful of language families. This is simply a first-order grouping of languages into large units, based on common ancestral speech. For example, about half of the world's people speak languages of the Indo-European family, a large group that includes not only European languages such as English and Spanish, but also Hindi and Bengali, the dominant languages of South Asia.

Within language families, smaller units also give clues to the common history and geography of peoples and cultures. *Language branches and groups* (also called *subfamilies*) are closely related subsets within a language family, usually sharing similar sounds, words, and grammar. Well known are the similarities between German and English and between French and Spanish.

Additionally, individual languages often have very distinctive *dialects* associated with specific regions and places. Think of the distinctive differences, for example, among British, North American, and Australian English or the city-specific dialects that set apart New Yorkers from residents of Dallas, Berliners from inhabitants of Munich, Parisians from villagers of rural France, and so on.

When people from different cultural groups cannot communicate directly in their native languages, they often agree on a third language to serve as a common tongue, a **lingua franca**. Swahili has long served that purpose for speakers of the many tribal languages of eastern Africa, and French was historically the lingua franca of international politics and diplomacy. Today English is increasingly the common language of international communications, science, and air transportation (Figure 1.36).

The Geography of World Religions

Another important defining trait of cultural groups is religion (Figure 1.37). Indeed, in this era of a comprehensive global culture, religion is becoming increasingly important in defining cultural identity. Recent ethnic violence and unrest in far-flung places such as the Balkans, Iraq, and Indonesia illustrate the point.

Universalizing religions, such as Christianity, Islam, and Buddhism, attempt to appeal to all peoples, regardless of location or culture. These religions usually have a proselytizing or missionary program that actively seeks new converts throughout the world. In contrast are **ethnic religions**, which are identified closely with a specific ethnic, tribal, or national group. Judaism and Hinduism, for example, are usually regarded as ethnic religions because they normally do not actively seek new converts; instead, people are born into ethnic religions.

Christianity, because of its universalizing ethos, is the world's largest religion in both areal extent and number of adherents. Although fragmented into separate branches and churches, Christianity as a whole has 2.1 billion adherents, encompassing about one-third of the world's population. The largest numbers of Christians can be found in Europe, Africa, Latin America, and North America. Islam, which has spread from its origins on the Arabian Peninsula east to Indonesia and the Philippines, has about 1.3 billion members.

Although not as severely fragmented as Christianity, Islam should not be thought of as a homogeneous religion because it is also split into separate groups. The two major branches are **Shi'a Islam**, which constitutes about 11 percent of the total Islamic population and represents a majority in Iran and southern Iraq, and the more dominant **Sunni Islam**, which is found from the Arab-speaking lands of North Africa to Indonesia. Probably in response to Western

FIGURE 1.36 English as Global Language A bilingual traffic sign in Dubai, United Arab Emirates, is a reminder that English has become a universal form of communication for transportation, business, and science.

influences connected to globalization, both of these forms of Islam are currently experiencing fundamentalist revivals in which proponents are interested in maintaining purity of faith, separate from these Western influences.

Judaism, the parent religion of Christianity, is also closely related to Islam. Although tensions are often high between Jews and Muslims, these two religions, along with Christianity, actually share historical and theological roots in the Hebrew prophets and leaders. Judaism now numbers about 14 million adherents, having lost perhaps one-third of its total population due to the systematic extermination of Jews during World War II.

Hinduism, which is closely linked to India, has about 900 million adherents. Outsiders often regard Hinduism as polytheistic because Hindus worship many deities. Most Hindus argue, however, that all of their faith's gods are merely representations of different aspects of a single divine, cosmic unity. Historically, Hinduism is linked to the caste system, with its segregation of peoples based on ancestry and occupation. Today, however, because India's democratic government is committed to reducing the social distinctions among castes, the connections between religion and caste are now much less explicit than in the past.

Buddhism, which originated as a reform movement within Hinduism 2,500 years ago, is widespread in Asia, extending from Sri Lanka to Japan and from Mongolia to Vietnam (Figure 1.38). There are two major branches of Buddhism, *Theravada*, found throughout Southeast

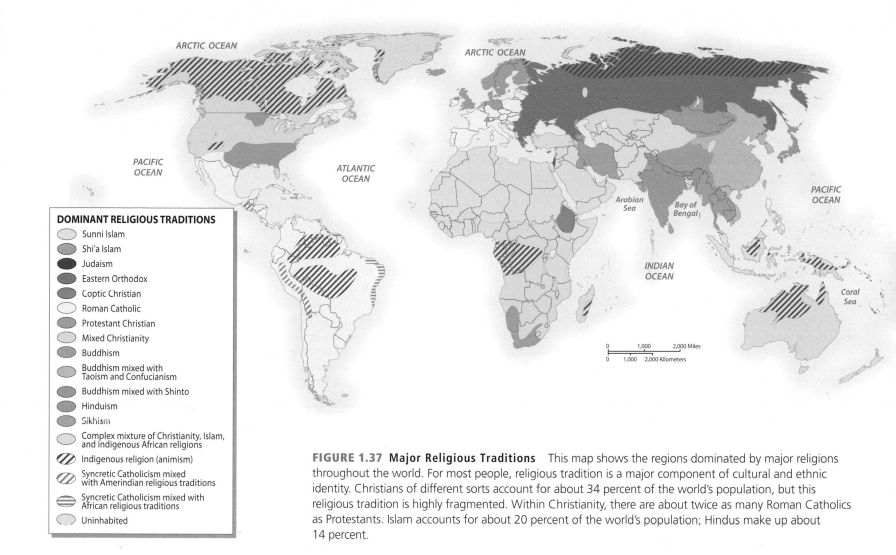

DOMINANT RELIGIOUS TRADITIONS

- Sunni Islam
- Shi'a Islam
- Judaism
- Eastern Orthodox
- Coptic Christian
- Roman Catholic
- Protestant Christian
- Mixed Christianity
- Buddhism
- Buddhism mixed with Taoism and Confucianism
- Buddhism mixed with Shinto
- Hinduism
- Sikhism
- Complex mixture of Christianity, Islam, and indigenous African religions
- Indigenous religion (animism)
- Syncretic Catholicism mixed with Amerindian religious traditions
- Syncretic Catholicism mixed with African religious traditions
- Uninhabited

FIGURE 1.37 Major Religious Traditions This map shows the regions dominated by major religions throughout the world. For most people, religious tradition is a major component of cultural and ethnic identity. Christians of different sorts account for about 34 percent of the world's population, but this religious tradition is highly fragmented. Within Christianity, there are about twice as many Roman Catholics as Protestants. Islam accounts for about 20 percent of the world's population; Hindus make up about 14 percent.

Asia and Sri Lanka, and *Mahanyana*, found in Tibet and East Asia. In its spread, Buddhism came to coexist with other faiths in certain areas, making it difficult to accurately estimate the number of its adherents. Estimates of the total Buddhist population range from 350 million to 900 million people.

Finally, in some parts of the world, religious practice has declined significantly, giving way to **secularism**, in which people consider themselves either nonreligious or outright atheistic. Though secularism is difficult to measure, social scientists estimate that about 1.1 billion people fit into this category worldwide. Perhaps the best example of secularism comes from the former communist lands of Russia and eastern Europe, where, historically, overt hostility occurred between government and church, dating from the time of the Russian Revolution of 1917. Since the demise of Soviet communism in the 1990s, however, many of these countries have experienced religious revivals.

Recently, secularism has also grown more pronounced in western Europe. Although historically and to some extent still culturally a Roman Catholic country, France quite possibly has more people attending Muslim mosques on Fridays than it has attending Christian churches on Sundays. Japan and the other countries of East Asia are also noted for their high degree of secularization.

FIGURE 1.38 Buddhist Landscapes An array of buildings—temples, monasteries, and shrines—produce a distinctive landscape throughout Asia, as is illustrated by this photo from Chiang Mai in northern Thailand.

GEOPOLITICAL FRAMEWORK: UNITY AND FRAGMENTATION

The term **geopolitics** is used to describe the close link between geography and politics. More specifically, geopolitics focus on the interactivity between political power and territory at all scales, from the local to the global. Unquestionably, one of the global characteristics of the last several decades has been the speed, scope, and character of political change in various regions of the world; thus, discussions of geopolitics are central to world regional geography.

With the demise of the Soviet Union in 1991 came opportunities for self-determination and independence in eastern Europe and Central Asia, resulting in fundamental changes to economic, political, and even cultural alignments. Religious freedom helped drive national identities in some new Central Asian republics, whereas eastern Europe was primarily concerned with new economic and political links to western Europe. Russia itself still wavers perilously between different geopolitical pathways. All of these topics are discussed further in Chapters 8, 9, and 10.

The Nation-State Revisited

A map of the world shows an array of 200 or so countries, ranging in size from microstates like Vatican City and Andorra to the huge geopolitical expanses of Russia, the United States, Canada, and China. All of these countries are regulated by governmental systems, ranging from democratic to autocratic. Commonly, these different forms of government share a concern with sovereignty, which can be defined geopolitically as the ability (or its lack) of a government to control activities within its borders.

The notion of sovereignty is closely linked to the concept of the **nation-state**. In this hyphenated term, *nation* describes a large group of people with shared sociocultural traits, such as language, religion, and shared identity. The word *state* refers to an internal political entity with clearly delimited boundaries, control over its internal space, and recognition by external political states. Historically, France and England are often used as the archetypal examples of nation-states. Contemporary countries such as Albania, Egypt, Bangladesh, Japan, and the two Koreas are more modern examples of countries where there is close overlap between nation and state. The related term *nationalism* is the sociopolitical expression of identity and allegiance to the shared values and goals of the nation-state.

Globalization, however, has weakened the vitality of the nation-state concept because today most of the world's countries are questionable fits with the traditional definition of nation-state. International migration, for example, has led to large populations of ethnic minorities within a country who do not necessarily share the national culture of the majority. In England, for example, large numbers of South Asians form their own communities, speak their own languages, have their own religions, and dress to their own standards. Similarly, France is home to a mosaic of peoples from its former colonial lands in Africa and Asia. In North America, Canada and the United States also have large immigrant populations who are legal citizens of the political state, yet who have changed the very nature of the national culture by their presence. The fact that states such as the United States, Canada, and Germany officially embrace cultural diversity with policies declaring themselves as multicultural states underscores these changes.

Also, within many nation-states are groups of people who seek autonomy from the central government by advocating the right to govern themselves. This is the case with the French-speaking people of Quebec Province in Canada, the Catalonians and Basques of Spain, and numerous tribes of Native Americans within the United States (Figure 1.39). A closer look at France, Italy, and the United Kingdom would also reveal separatist groups within the state who seek some sort of autonomy or even independence from the larger political unit.

Not to be overlooked is how giant multinational firms and regional political organizations have eclipsed the power of traditional political states. This is certainly the case for the 27 member states of the European Union, a topic discussed in Chapter 8.

Finally, some cultural groups lack political voice and representation due to the way political borders have been drawn. In Slovakia, for instance, a group that traces its cultural heritage to adjacent Hungary agitates for redrawing the political border between the two states. In Southwest Asia, the Kurdish people have long been considered a nation without a state because they are divided amongst the political borders of Turkey, Syria, Iraq, and Iran (Figure 1.40).

FIGURE 1.39 Ethnic Separatism A major aspect of contemporary geopolitics is the way ethnic groups are demanding recognition, autonomy, and often independence from larger political units. These Basque women in southwestern France are protesting the outlawing of a Basque youth group that French and Spanish authorities suspected of aiding Basque terrorists.

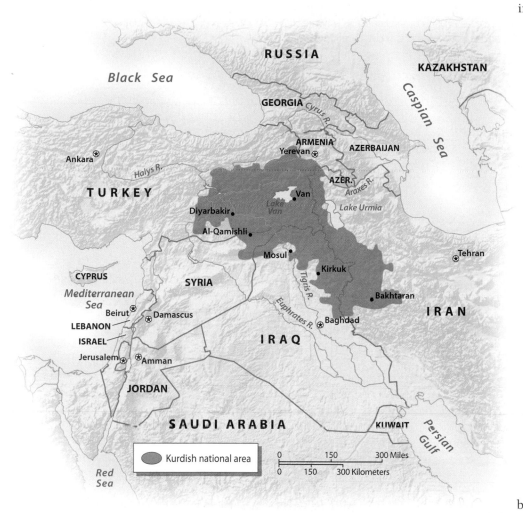

FIGURE 1.40 A Nation Without a State   Not all nations or large cultural groups control their own political territories. As this map shows, the Kurdish people of Southwest Asia occupy a large cultural territory that lies in four different political states—Turkey, Iraq, Syria, and Iran. As a result of this political fragmentation, the Kurds are considered a minority in each of these four countries.

Colonialism, Decolonialization, and Neocolonialism

One of the overarching themes in world geopolitics is the waxing and waning of European colonial power in South America, Asia, and Africa. **Colonialism** refers to the formal establishment of rule over a foreign population. A colony has no independent standing in the world community, but instead is seen only as an appendage of the colonial power. The historic Spanish presence and rule over parts of the United States, Mexico, and South America is an example. Generally speaking, the main period of colonialization by European countries was from 1500 through the mid-1900s, with the major players being England, Belgium, the Netherlands, Spain, and France (Figure 1.41).

Decolonialization refers to the process of a colony's gaining (or, more correctly, regaining) control over its own territory and establishing a separate, independent government. As was the case with the Revolutionary War in the United States, this process often involves violent struggle. Similar wars of independence became increasingly common in the mid-20th century, particularly in Southeast Asia and Africa. Consequently, most European colonial powers recognized the inevitable and began working toward peaceful disengagement from their colonies. In the late 1950s and early 1960s, for example, Britain and France granted independence to their former African colonies. This period of European colonialization was symbolically completed in 1997 when England turned over Hong Kong to China.

However, decades and even centuries of colonial rule are not easily erased. The influences of colonialism are still commonly found in the culture, government, educational systems, and economic life of the former colonies. Examples are the many contemporary manifestations of British culture in India and the continuing Spanish influences in South America.

In the 1960s, the term **neocolonialism** came into popular usage to characterize the many ways that newly independent states, particularly those in Africa, felt the continuing control of Western powers, especially in economic and political matters. To receive financial aid from the World Bank, for example, former colonial countries were often required to revise their internal economic structures to become better integrated with the emerging global system. This economic restructuring may seem warranted from a global perspective, but the dislocations caused at national and local scales led critics of globalization to characterize these external influences as no better than the formal control by historical colonial powers—hence the currency of the term *neocolonialism*.

Global Conflict

As mentioned earlier, challenges to a centralized political state or authority have long been part of global geopolitics as rebellious and separatist groups seek independence, autonomy, and territorial control. These actions are referred to as **insurgency**. Armed conflict has also been part of this process; the American and Mexican revolutions were both successful wars for independence fought against European colonial powers. **Terrorism**, which can be defined as violence directed at nonmilitary targets, has also been common, albeit to a far lesser degree than today.

Until the September 2001 terrorist attacks on the United States by Al Qaeda, terrorism was usually directed at specific local targets and committed by insurgents with focused goals. The Irish Republican Army (IRA) bombings in Great Britain and Basque terrorism in Spain are illustrations. The attacks on the World Trade Center and the Pentagon (as well as the thwarted attack on the Capitol), however, went well beyond conventional geopolitics as a small group of religious extremists attacked the symbols of Western culture, finance,

FIGURE 1.41 The Colonial World, 1914   This world map shows the extent of colonial power and territory just prior to World War I. At that time, most of Africa was under colonial control, as were Southwest Asia, South Asia, and Southeast Asia. Australia and Canada were very closely aligned with England. Also note that in Asia, Japan controlled colonial territory in Korea and northeastern China, which was known as Manchuria at that time.

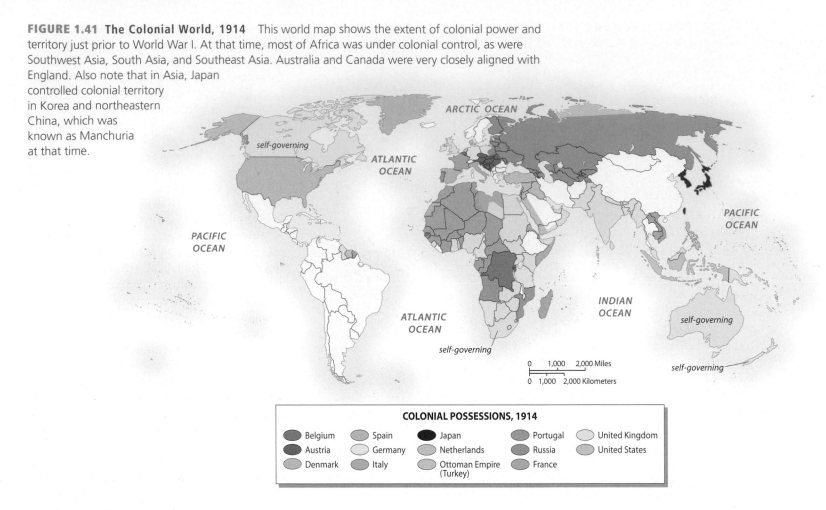

COLONIAL POSSESSIONS, 1914

Belgium Spain Japan Portugal United Kingdom
Austria Germany Netherlands Russia United States
Denmark Italy Ottoman Empire (Turkey) France

and power (Figure 1.42). Experts believe the goal of the Al Qaeda terrorists was less to disrupt world commerce and politics and more to make a statement about the strength of their own convictions and power. Regardless of motives, those acts of terrorism underscore the need to expand our conceptualization of the linkages between globalization and geopolitics.

More to the point, many experts argue that global terrorism is both a product of and a reaction to globalization. Unlike earlier geopolitical conflicts, the geography of global terrorism is not defined by a war between well-established political states. Instead, the Al Qaeda terrorists appear to belong to a web of small, well-organized cells located in many different countries. These cells are linked in a decentralized network that provides guidance, financing, and political autonomy and that has used the tools and means of globalization to its advantage. Members communicate instantaneously via mobile phones and the Internet. Transnational members travel between countries quickly and frequently. The network's activities are financed through a complicated array of holding companies and subsidiaries that traffic in a range of goods, including honey, diamonds, and opium. To recruit members and political support, the network feeds on the unrest and inequities (real and imagined) resulting from economic globalization. The network's terrorist acts then target symbols of those modern global values and activities it opposes. Even though the 9/11 attacks focused terror on the United States, the casualties and resultant damage were international, as citizens from more than 80 countries were killed in the World Trade Center tragedy.

Although Al Qaeda is the most visible and possibly the most threatening of contemporary global terrorist groups, the U.S. State Department names 45 groups on its list of foreign terrorist organizations. While most of these groups are clustered in North Africa and Southwest and Central Asia, this list also includes insurgent groups in all other regions of the world.

FIGURE 1.42 Global Terrorism   The September 11, 2001, attacks by Al Qaeda terrorists on the World Trade Center and the Pentagon and the thwarted plan to destroy the U.S. Capitol building resulted in more than 3,000 deaths. These attacks were not targeted at local places of dispute, but were aimed at global symbols of Western power.

FIGURE 1.43 The 2009 Global Recession Two indicators of global economic activity—the price of crude oil and the Baltic Dry Index (BDI), which is linked to world shipping activity—show the expansion of the global economy since 2000, then the sudden contraction in 2009. More recent data show the slow recovery from the 2009 collapse.

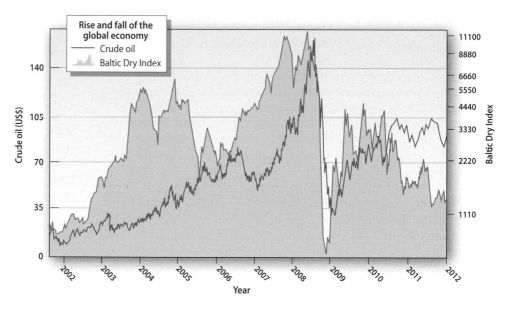

The military responses to global terrorism and insurgency involve several components, ranging from the neutralization of terrorist activities, known as counterterrorism, to **counterinsurgency**. The latter is a more complicated, multifaceted strategy that combines military warfare with social and political service activities, designed to win over the local population and deprive insurgents of a political base. Counterinsurgency activities include, first, clearing and then holding territory held by insurgents, followed by building schools, medical clinics, and a viable (and legal) economy. Often these nonmilitary activities are referred to as nation-building, since the goal is to replace the separatist insurgency with a viable social, economic, and political fabric more complementary to the larger geopolitical state. This is the strategy employed by the United States recently in Iraq and Afghanistan.

REVIEW QUESTIONS

1. Why is it common to use two different concepts—nation and state—to describe political entities?

2. Explain the differences between colonialism and neocolonialism.

3. Describe the differences between counterterrorism and counterinsurgency.

ECONOMIC AND SOCIAL DEVELOPMENT: THE GEOGRAPHY OF WEALTH AND POVERTY

The pace of global economic change and development has accelerated dramatically in the past several decades. It ascended rapidly at the start of the 21st century and then slowed precipitously in 2008 as the world fell into an economic recession. Currently, the world economy has returned unsteadily to a postrecession state (Figure 1.43). Given the far-ranging global ramifications of these recent economic fluctuations, the overarching question remains: Do the positive changes of economic globalization outweigh the negative? Answers vary considerably, as discussed earlier in this chapter. Consequently, returning to some basic assumptions about economic and social development is an appropriate starting point.

Economic development is commonly accepted as desirable because it generally brings increased prosperity to people, regions, and nations. Following conventional thinking, this economic development usually translates into social improvements such as better health care, improved educational systems, and more progressive labor practices. One of the most troubling expressions of global economic growth, however, has been the geographic unevenness of prosperity and social improvement. That is, while some regions and places in the world prosper, others languish and, in fact, apparently fall farther behind the more developed countries. As a result, the gap between rich and poor regions has actually increased over the past several decades in many areas. This economic and social unevenness has, unfortunately, become one of the signatures of globalization. Today, according to the World Bank, about half the people in the world live on less than $2 a day, the commonly accepted definition of poverty (Figure 1.44).

These inequities are problematic because of their inseparable interaction with political, environmental, and social issues. For example, political instability and civil strife within a nation are often driven by the economic disparity between a poor periphery and an affluent industrial core—between the haves and have-nots. Such instability throughout a country can strongly influence international economic interactions.

FIGURE 1.44 Living on Less than $2 a Day More than half the world's people live on less than $2 a day, which is the UN's definition of poverty. In South Asia, where these workers are making bricks, about three-quarters of the population is considered impoverished.

More and Less Developed Countries

Until the 20th century, economic development was centered in North America, Japan, and Europe, with most of the rest of the world gripped in poverty. This uneven distribution of economic power led scholars to devise a **core–periphery model** of the world. According to this scheme, the United States, Canada, western Europe, and Japan constituted the global economic core, centered in the northern hemisphere, whereas most of the areas in the southern hemisphere made up a less developed periphery. Although oversimplified, this core–periphery dichotomy does contain some truth. All the G8 countries—the exclusive club of the world's major industrial nations, made up of the United States, Canada, France, England, Germany, Italy, Japan, and Russia—are located in the northern hemisphere. (China—which is unquestionably an industrial power, located in the northern hemisphere—is currently excluded from the G8.) In addition, many critics postulate that the developed countries achieved their wealth primarily by exploiting the poorer countries of the southern periphery, historically through colonial relationships and today with various forms of neocolonialism and economic imperialism.

Following this core–periphery model, much has been made of "north–south tensions," a phrase implying that the rich and powerful countries of the northern hemisphere are still at odds with the poor and less powerful countries of the south. However, this model demands revision because over recent decades the global economy has grown much more complicated. A few former colonies of the south—most notably, Singapore—have become very wealthy. A few northern countries—notably, Russia—have experienced very uneven economic growth since 1989, with some parts of the country actually experiencing economic decline. Additionally, the well-developed southern hemisphere countries of Australia and New Zealand never fit into the north–south division. For these reasons, many global experts conclude that the designation *north–south* is now outdated and should be avoided. We agree.

The *third world* is another term often erroneously used as a synonym for the developing world. This phrase implies a low level of economic development, unstable political organizations, and a rudimentary social infrastructure. Historically, the term was part of the Cold War vocabulary used to describe countries that were independent and not part of either the capitalist first world or the communist second world, dominated by the Soviet Union and China. In short, in its original sense the term *third world* spoke to a political and economic orientation (capitalist vs. communist), not to the level of economic development. Today, because the Soviet Union no longer exists and China has changed its economic orientation considerably, the term *third world* has lost its original political meaning. Therefore, in this book we avoid that term. Instead, we prefer relational terms that capture a complex spectrum of economic and social development—*more developed country (MDC)* and *less developed country (LDC)*. This global pattern of MDCs and LDCs can be inferred from a map of gross national income (Figure 1.45), one of several indicators commonly used to assess development and economic wealth.

Indicators of Economic Development

The terms *development* and *growth* are often used interchangeably when referring to international economic activities. There is, however, value in keeping them separate. *Development* has both qualitative and quantitative dimensions. Common dictionary definitions use phrases such as "expanding or realizing potential" and "bringing gradually to a fuller or better state." When we talk about economic development, then, we usually imply structural changes, such as a shift from agricultural to manufacturing activity that also involves changes in the allocation of labor, capital, and technology. Along with these changes are assumed improvements in standard of living, education, and political organization. The structural changes experienced by Southeast Asian countries such as Thailand and Malaysia in the past several decades capture this process.

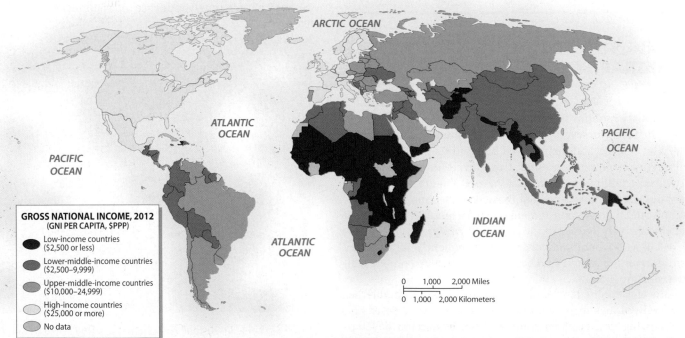

FIGURE 1.45 More and Less Developed Countries Based on GNI per capita, PPP adjusted, you can see the global pattern of more and less developed countries (MDCs and LDCs). Africa and the different parts of Asia stand out as the regions with the greatest number of developing countries.

TABLE 1.2   DEVELOPMENT INDICATORS

Country	GNI per Capita, PPP 2010	GDP Average Annual % Growth 2000–10	Human Development Index 2011[1]	Percent Population Living Below $2 a Day	Life Expectancy 2012[2]	Under Age 5 Mortality Rate 1990	Under Age 5 Mortality Rate 2010	Adult Literacy (% ages 15 and older)	Gender Inequality Index (2011)[3,1]
China	7,640	10.8	.687	29.8	75	48	18	94	0.209
India	3,400	8.0	.547	68.7	65	115	63	63	0.617
United States	47,310	1.8	.910	—	79	11	8	—	0.299
Indonesia	4,200	5.3	.617	46.1	72	85	35	92	0.505
Brazil	11,000	3.7	.718	10.8	74	59	19	90	0.449
Pakistan	2,790	5.1	.504	60.2	65	124	87	56	0.573
Nigeria	2,170	6.7	.459	84.5	51	213	143	61	—
Bangladesh	1,810	5.9	.500	76.5	69	143	48	56	0.550
Russia	19,240	5.4	.755	<2	69	27	12	100	0.338
Japan	34,610	0.9	.901	—	83	6	3	—	0.123

[1]United Nations, *Human Development Report*, 2011.

[2]Population Reference Bureau, *World Population Data Sheet*, 2012.

[3]Gender Equality Index—A composite measure reflecting inequality in achievements between women and men in three dimensions: reproductive health, empowerment and the labor market that ranges between 0 and 1. The higher the number, the greater the inequality.

Source: World Bank, *World Development Indicators*, 2012.

Growth, in contrast, is simply the increase in the size of a system. The agricultural or industrial output of a country may grow, as it has for India in the past decade, and this growth may—or may not—have positive implications for development. Many growing economies, in fact, have actually experienced increased poverty with economic expansion. When something grows, it gets bigger; when it develops, it improves. Critics of the world economy often say that we need less growth and more development.

In this book, each of the regional chapters includes a table of economic and development indicators (Table 1.2). However, a few introductory comments are necessary to explain these data.

Gross Domestic Product and Income

The traditional measure of the size of a country's economy is the value of all final goods and services produced within its borders, which is called the **gross domestic product (GDP)**. When combined with net income from outside its borders through trade and other forms of income, this constitutes a country's **gross national income (GNI)** (formerly referred to as gross national product [GNP]). Although the term is widely used, GNI is an incomplete and sometimes misleading economic indicator because it completely ignores nonmarket economic activity, such as bartering or household work, and also does not take into account ecological degradation or depletion of natural resources. For example, if a country were to clear-cut its forests—an activity that would probably limit future economic growth if forest resources were in short supply—this resource usage would actually increase the GNI for that particular year. Diverting educational funds to purchase military weapons might also increase a country's GNI in the short run, but the economy would likely suffer in the future because of its less-well-educated population. In other words, GNI is a snapshot of a country's economy at a specific moment in time, not a reliable indicator of continued vitality or social well-being.

Because GNI data vary widely among countries and are commonly expressed in mind-boggling numbers such as billions and trillions of dollars, a better comparison is obtained by dividing GNI by the country's population, thereby generating a **gross national income (GNI) per capita** figure. This way we can compare large and small economies in terms of how they may (or may not) be benefiting the population. For example, the annual GNI for the United States is over 15 trillion. Dividing that figure by the population of 313.9 million results in a GNI per capita of around $47,310. Japan, in contrast, has a GNI about half the size of that of the United States; however, because Japan has a much smaller population of 127.6 million, its unadjusted GNI per capita is about $48,150. Thus, we conclude that the two different economies are comparable (which is true).

An important qualification to these GNI per capita data is the concept of adjustment through **purchasing power parity (PPP)**, which takes into account the strength or weakness of local currencies (Figure 1.46). When not adjusted by PPP, GNI data are based on the market

FIGURE 1.46 PPP and Local Currencies   The concept of purchasing power parity (PPP) is used to generate a sense of the true cost of living for different countries. Thus, economic data can be adjusted for local conditions based upon a common market basket of goods. This outdoor produce market is in Chichicastenango, Guatemala.

FIGURE 1.47 Human Development Index This map depicts the most recent rankings assigned to four categories that make up parts of the Human Development Index (HDI). In the numerical tabulation, Norway, Australia, and New Zealand have the highest rankings, while several African countries are lowest in the scale.

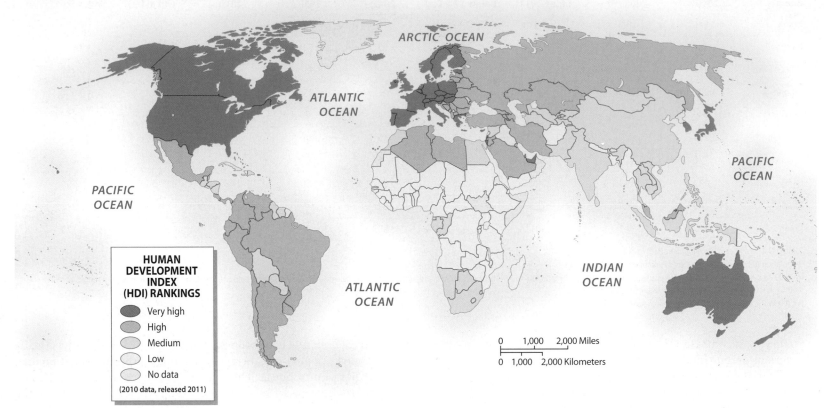

exchange rate for a country's national currency compared to the U.S. dollar. As a result, the GNI data might be inflated or undervalued, depending on the strength or weakness of that currency. If the Japanese yen were to fall overnight against the dollar as a result of currency speculation, Japan's GNI would correspondingly drop, despite the fact that Japan had experienced no real decline in economic output. Because of these possible distortions, the PPP adjustment provides a more accurate sense of the local cost of living. To illustrate, when Japan's GNI per capita is adjusted for PPP, which takes out the inflationary factor, the figure is $36,610, somewhat lower than the unadjusted GNI per capita figure.

Economic Growth Rates A country's rate of economic growth is measured by the average annual growth of its GDP over a five-year period, a statistic called *GDP average annual percent growth*. Looking at Table 1.2, you can see that the average growth rates for developing countries such as China, India, and Nigeria are considerably higher than those of the developed countries such as the United States and Japan. This difference in economic growth rates is expected, given that developing countries are just that—developing. We would expect a higher annual growth rate from a developing country than from a mature, developed economy like that of the United States.

Indicators of Social Development

Although economic growth is a major component of development, equally important are the conditions and quality of human life. As noted earlier, the standard assumption is that economic development will spill over into the social infrastructure, leading to improvements in public health, **gender equity**, and education. Unfortunately, even the briefest glance at the world reveals that poverty, disease, illiteracy, and gender inequity are still widespread, despite a growing global economy. Even in China, which has experienced unprecedented economic growth in recent decades, almost one-third of the population is still impoverished; in Pakistan, two-thirds live below the poverty line of $2 per day; and in some African countries, that figure approaches 90 percent.

Hints of social improvement appear, however, in many developing countries. For example, the percentage of those living in deep poverty, which the UN defines as living on less $1 per day, has fallen from 28 percent in 1990 to just under 20 percent today. Life expectancy, too, has increased in those countries, rising from 60 in 1990 to 65.

The Human Development Index For the past three decades, the UN has tracked social development in the world's countries through the **Human Development Index (HDI)**, which combines data on life expectancy, literacy, educational attainment, gender equity, and income (Figure 1.47). In a 2011 analysis, the 194 countries that provided data to the UN are ranked from high to low, with Norway achieving the highest score, Australia in second place, New Zealand in third, and the United States in fourth. At the lowest end of the HDI are a handful of small Pacific island nations—Vanuatu, Tuvalu, and Samoa—as well as the African country of Somalia.

Although the HDI can be criticized for using national data that overlook the diversity of development within a country, overall, the HDI conveys a reasonably accurate sense of a country's human and

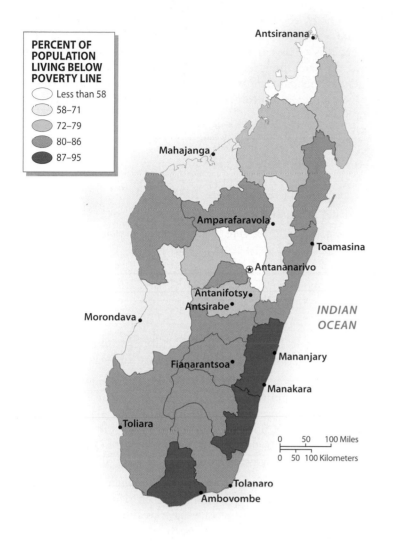

PERCENT OF POPULATION LIVING BELOW POVERTY LINE

- Less than 58
- 58–71
- 72–79
- 80–86
- 87–95

Antsiranana

Mahajanga

Amparafaravola

Toamasina

Antananarivo

Antanifotsy
Antsirabe

Morondava

INDIAN OCEAN

Mananjary

Fianarantsoa

Manakara

Toliara

0   50   100 Miles
0 50 100 Kilometers

Tolanaro
Ambovombe

FIGURE 1.48 The Landscape of Poverty   As is true of most other countries, the distribution of poverty within Madagascar is uneven, with clusters of abject poverty contrasting with regions that are less poor. This map shows that the highest rates of poverty are in the central highlands, where the country's population is concentrated and the density is highest. In contrast, the northern and eastern lowlands are less poor.

social development. Thus, we include HDI data in our social development tables for each regional chapter.

Poverty and Mortality   As noted earlier, the international definition of *poverty* is living on less than $2 per day, with the category of *deep poverty* defined as existing on less than $1 per day. While the cost of living varies greatly around the world, the UN has found that these definitions work well for measuring both poverty and its associated social conditions. While poverty data are usually presented at the country level, the UN and other agencies are also attempting to compile data at a local scale in order to better understand the uneven economic and social landscape within a country. The patterns of poverty in the African country of Madagascar provide an instructive example (Figure 1.48).

Another widely used indicator of social development are data on *under age 5 mortality*, which is the number of children in that age bracket who die per 1,000 of the general population. Aside from the tragedy of

FIGURE 1.49 Child Mortality   The mortality rate of children under the age of 5 is an important indicator of social conditions such as health services, food availability, and public sanitation. Here kindergarten children in Jakarta, Indonesia, receive an oral polio vaccine.

infant death, child mortality also reflects the wider conditions of a society, such as the availability of food, health services, and public sanitation. If those factors are lacking, children under age five suffer most; therefore, their death rate is taken as an indication of whether a country has the necessary social infrastructure to sustain life (Figure 1.49). In the social development tables throughout this book, child mortality data are given for two points in time, 1990 and 2010, to indicate whether the social structure has improved over the intervening years.

Adult Literacy   Reading and writing are crucial skills in this world, yet current data show that about 20 percent of the world's adult population (those 15 years and over) lack those skills. Of those who cannot read or write, two-thirds of them are women. While in most developed countries 99 percent of the people are literate, of the 183 countries where literacy data were gathered, 11 countries record literacy rates below 50 percent; with the exception of Haiti, all of these countries are in sub-Saharan Africa. The greatest disparity between male and female literacy is in South Asia, an artifact of long-standing cultural favoritism towards males.

Gender Inequity   Discrimination against women takes many forms from not allowing them to vote to discouraging school attendance (Figure 1.50). Given the importance of this topic, the United Nations calculates gender inequality among countries in order to measure the relative position of women to men in terms of employment, empowerment, and reproductive health (in terns of maternal mortality and adolescent fertility). The UN index ranges from 0 to 1, which expresses the highest level of gender inequality. Sweden, for example, has the lowest score for inequality with 0.049, while Yemen has one of the highest at 0.769. The UN gender inequality scores are found on Development Indicators table in the regional chapters.

Of interest is that in some cases country may register reasonably high on the Human Development Index (HDI), which is positive,

FIGURE 1.50 Women and Literacy Gender inequities in education lead to higher rates of illiteracy for women. However, when there is gender equity in education, female literacy has several positive outcomes in a society. For example, educated women have a higher participation rate in family planning, which usually results in lower birthrates.

yet also be given a relatively high Gender Inequality score (which is not so good) by the United Nations. Saudi Arabia, for example is a rich country that uses assets from its oil resources to provide many social benefits to its citizens, which explains its high HDI ranking, while at the same time its conservative Muslim culture produces a high gender inequality rating. Readers are advised to look carefully at the Development Indicators data for these kinds of contradictions and inconsistencies.

> ### REVIEW QUESTIONS
>
> 1. Explain the difference between GDP and GNI.
> 2. What is PPP and why is it useful?
> 3. How does the UN measure gender inequity? Explain why this is a useful metric for social development.

Summary

- Because globalization involves both positive and negative transformations, it is a controversial and contentious topic. Proponents argue that economic globalization benefits everyone as regional activities become more efficient in the face of global competition. Opponents counter this argument with evidence that only the rich are benefiting and that poorer countries are actually falling even farther behind.

- In most regions of the developing world, population and settlement issues revolve around four issues: rapid population growth, family planning (or its absence), migration to new centers of economic activity (both within and outside the region), and rapid urbanization.

- A major theme in cultural geography is the tension between the forces of cultural homogenization and the countercurrents of local cultural and ethnic identity. Throughout the world, small groups are setting themselves apart from larger national cultures with renewed interest in ethnic traits, languages, religion, territory, and shared histories.

- As the world becomes increasingly connected through international economic and political alliances, the power of the traditional nation-state lessens, with doors opening for separatist groups seeking autonomy and even independence from larger political entities. In addition, terrorist groups such as Al Qaeda wage their own battles against Western interests and ways of life.

- The theme of economic and social development is dominated by one issue—the increasing disparity between rich and poor, between countries and regions that already have wealth and are getting even richer through globalization and those that remain impoverished. Often, blatant inequities in social development, education, gender roles, health care, and working conditions accompany these disparities in wealth.

Key Terms

areal differentiation 11
areal integration 12
bubble economy 10
choropleth map 17
colonialism 33
core–periphery model 36
counterinsurgency 35
cultural imperialism 27
cultural landscape 13
cultural nationalism 27
cultural syncretism or hybridization 27
culture 27
decolonialization 33
demographic transition model 24
diversity 4
economic convergence 9
ethnic religion 30
formal region 12
functional region 12

gender 28
gender equity 38
gender roles 28
geographic information systems (GIS) 18
geopolitics 32
globalization 4
global positioning systems (GPS) 15
graphic or linear scale 16
gross domestic product (GDP) 37
gross national income (GNI) 37
gross national income (GNI) per capita 37
Human Development Index (HDI) 38
insurgency 33
lingua franca 30
map projections 15
map scale 15
meridians (lines of longitude) 15
nation-state 32
neocolonialism 33

net migration rate 25
parallels (lines of latitude) 15
population density 26
population pyramid 23
prime meridian 15
purchasing power parity (PPP) 37
rate of natural increase (RNI) 21
region 12
remote sensing 17
representative fraction 16
secularism 31
Shi'a Islam 30
Sunni Islam 30
terrorism 33
total fertility rate (TFR) 22
universalizing religion 30
urbanized population 26
vernacular region 13

Thinking Geographically

1. Select an economic, political, or cultural activity in your city or town, and discuss how it has been influenced by globalization.

2. Choose a specific country or region of the world and examine the benefits, and liabilities that globalization has posed for that country or region. Remember to look at different facets of globalization, such as the environment, cultural change and conflict, and the economic effects on different segments of the population.

3. Drawing on information in current newspapers and magazines, as well as TV and the Internet, apply the concepts of cultural imperialism, nationalism, and cultural syncretism to a region or place experiencing cultural tensions.

4. Select an African country with a colonial past. (a) Trace its pathway of decolonialization; (b) describe and analyze its contemporary relations with its former colonial overseer, being sensitive to the matter of whether a neocolonial relationship has been established.

5. Using the tables of social indicators in the regional chapters of this book, identify traits shared by countries that have a high percentage of female illiteracy. Based on your inquiry, what general conclusions do you reach?

MasteringGeography™

Looking for additional review and test prep materials? Visit the Study Area in MasteringGeography™ to enhance your geographic literacy, spatial reasoning skills, and understanding of this chapter's content by accessing a variety of resources, including **MapMaster** interactive maps, videos, RSS feeds, flashcards, web links, self-study quizzes, and an eText version of *Globalization and Diversity*.

Scan to visit the author's blog for chapter updates.

http://gad4blog.wordpress.com/

Authors' Blogs

Scan now to access the authors' blogs for up-to-date information on Concepts of World Geography.

Scan to visit the GeoCurrents blog.

http://geocurrents.info/

Globalization and Diversity

Our world's environment is unique in the solar system, thus a common starting point for studying world geography is learning more about Earth's physical environment—its geology, climate, hydrology, and bioregions.

GEOLOGY: A RESTLESS EARTH

Earth's surface is comprised of numerous tectonic plates that move about slowly, driven by convection currents deep within the mantle.

GLOBAL CLIMATES

The varied mosaic of global climates can be explained by a set of physical processes known as climate controls. Human activities, however, are changing Earth's climate in worrisome ways.

WATER: A SCARCE RESOURCE

All life, human and otherwise, needs water to exist; however, at a global scale freshwater is a scarce resource resulting in many areas of water stress.

BIOREGIONS

Earth is covered by a cloak of different kinds of vegetation, resulting in a variety of ecosystems within larger, more general bioregions.

➤ The picturesque landscape of Blue Lake in Washington state's North Cascade National Park blends together all the elements of Earth's physical geogaphy—geology, climate, hydrology, and vegetation.

THE CHANGING GLOBAL ENVIRONMENT 2

The immense physical diversity of the global environment, with its deep oceans, towering mountain ranges, dry deserts, wet tropics, and cold polar regions, makes Earth absolutely unique in our solar system. It is also unique in providing the necessary conditions for life. In turn, life forms of all different sorts—plant, animal, and human—have interacted in countless ways with each other and with the environment to produce the diversity of landscapes and habitats, human and otherwise, that we see today. Thus, a logical starting point for the study of world regional geography is knowing more about Earth's physical environment—its geology, climate, hydrology, and vegetation (Figure 2.1).

GEOLOGY: A RESTLESS EARTH

Geology shapes Earth's surface and gives character to its diverse landscapes with a physical fabric of mountains, hills, valleys, and plains. This geologic foundation is also central to a wide array of human activities, such as agriculture, mining, transportation, and settlement patterns. Not to be overlooked, however, is that the geologic environment also presents serious challenges in the form of earthquakes, landslides, and volcanic eruptions. These geologic complexities make understanding the physical processes shaping Earth's landscapes crucial to appreciating its geography.

Plate Tectonics

The common starting point is **plate tectonics**, a geophysical theory that Earth is comprised of large geologic platforms, or plates, that move slowly across its surface. Driving these tectonic plates is a heat exchange deep within Earth; Figure 2.2 illustrates this complicated process.

On top of these tectonic plates sit continents and ocean basins. Note that continents are not identical to the underlying tectonic plates. Instead, most continents and ocean basins commonly straddle several different tectonic plates. This is important because most earthquakes and volcanoes are found along the plate boundaries. Figures 2.3 and 2.9 show those relationships. In western North America, for example, coastal California lies atop two different tectonic plates—the Pacific and the North American plates—on a **colliding plate boundary**. Here two tectonic plates are converging, or being forced together by convection cells circulating in different directions deep within Earth's

FIGURE 2.1 The World's Physical Geography The interaction between humans and their physical environment—geology, climate, hydrology, and vegetation—is central to the understanding of world regional geography. Here, the Tukhela waterfall tumbles down the escarpment of the Drakensberg Mountain in South Africa.

mantle. The infamous San Andreas Fault is the actual plate boundary, which explains why the large cities of Los Angeles and San Francisco are vulnerable to destructive earthquakes (Figure 2.4).

Often in these collision zones, one tectonic plate sinks below another, creating a **subduction zone**. Such a region is characterized by deep trenches where the ocean floor has been pulled downward by regional tectonic warping. Subduction zones exist off the western coast of South America, off the northwest coast of North America, offshore of eastern Japan, and near the Philippines, where the Mariana Trench forms the world's deepest ocean depths at 35,000 feet (10,700 meters). These subduction zones are also the locations of Earth's most powerful earthquakes, as witnessed by the magnitude 9.0 earthquake and accompanying tsunami that devastated coastal Japan in 2011 and the magnitude 8.8 quake in Chile in 2010 (Figure 2.5).

LEARNING OBJECTIVES

After reading this chapter you should be able to:

- Explain tectonic plate theory by describing those aspects responsible for shaping Earth's surface.

- Identify those parts of the world where earthquakes and volcanoes are hazardous to human settlement and explain why the casualty rates from those hazards often differ from place to place.

- List the factors that control the world's weather and climate.

- Describe the major characteristics and global location of the world's major climate regions.

- Explain the greenhouse effect and how it relates to anthropogenic global warming.

- Describe the major issues underlying the international controversy over reducing global warming emissions.

- Identify the causes of global water stress.

- Describe the characteristics and distribution of the world's major bioregions.

- Explain the reasons behind deforestation in both tropical and higher-latitude forests.

FIGURE 2.2 Plate Tectonics   According to plate tectonics theory, large convection cells circulate molten rock in different directions within Earth's mantle. Near the crust, the slow movement of the cells drags tectonic plates away from the mid-oceanic ridges, resulting in the collision of plates in convergent plate boundaries. While the rate of movement differs between convection cells, in general it is only a few inches (or centimeters) per year.

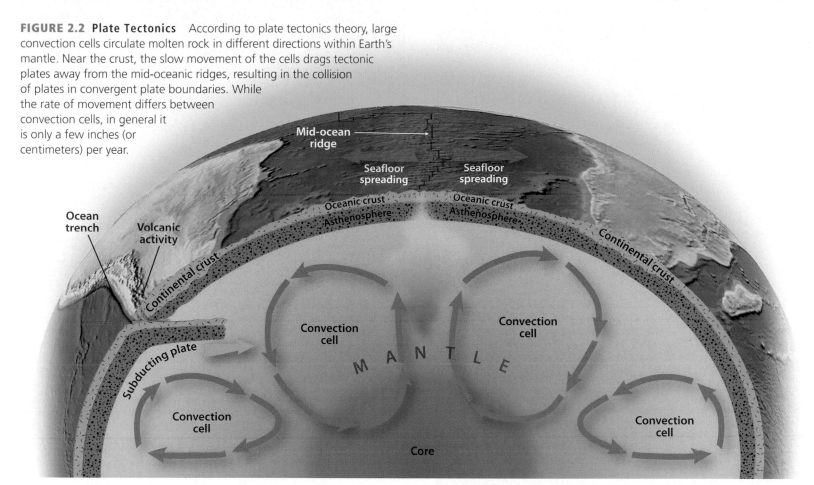

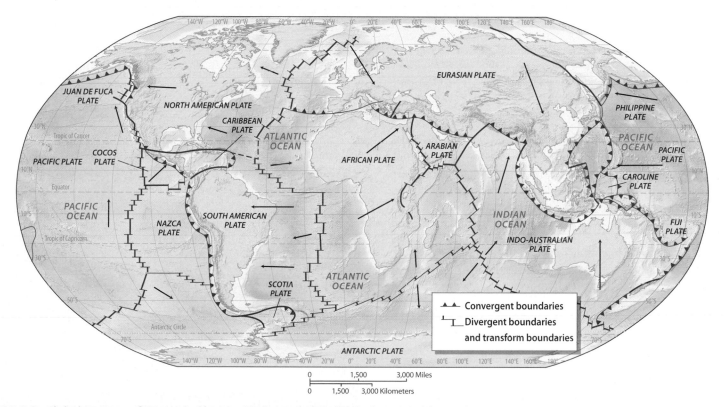

FIGURE 2.3 Global Pattern of Tectonic Plates   Similar to the pieces of a jigsaw puzzle, Earth's tectonic plates vary in size and shape. Where plates collide, earthquakes and volcanoes are common, as are folded and faulted mountain ranges. In other areas, where one plate dives below another, deep oceanic trenches result, as is the case off the coast of South America and in the western Pacific Ocean.

FIGURE 2.4 The San Andreas Fault Zone   This fault zone, which is actually 40 miles (64 km) wide, may be the transform boundary where the North American and Pacific plates slide past each other. More recently, however, geologists have determined that the small Sierran Microplate lies east of the San Andreas Fault, with the actual border of the North American Plate lying just east of the Sierra-Nevada Mountains. This photo is of the main fault trace of the San Andreas in the dry Carrizo Plain area halfway between San Francisco and Los Angeles in the California Coast Ranges.

In other parts of the world, tectonic plates move away from each other in opposite directions, forming **divergent plate boundaries**. As these plates diverge, magma flows from Earth's interior, creating mountain ranges and active volcanoes. In the North Atlantic, Iceland lies on the divergent plate boundary that bisects the Atlantic Ocean (Figure 2.6). In other places, however, divergent boundaries form deep depressions, or rift valleys, such as that occupied by the Red Sea between northern Africa and Saudi Arabia. To the west of the Red Sea, in Africa, a splinter of this plate boundary has created the African rift valley, part of which forms the Olduvai Gorge, where tectonic activity has preserved the earliest traces of human ancestors in volcanic soils.

Geologic evidence suggests that some 250 million years ago all the world's plates were tightly consolidated into a supercontinent centered on present-day Africa. Over time, this large area, called Pangaea, was broken up as convection cells moved the tectonic plates apart. A hint of this former continent can be seen in the jigsaw puzzle fit of South America with Africa and of North America with Europe.

Although tectonic plate theory explains many of the world's large mountain ranges, it does not account for all highlands. While both the Himalaya and the Alpine mountain ranges were created by the forces of

FIGURE 2.5 Japanese Earthquake and Tsunami   On March 11, 2011 at 2:46 pm local time a magnitude 9.0 undersea subduction zone earthquake struck the northeastern coast of Japan with its epicenter 43 miles (70 km) east of the Oshika Peninsula, generating a massive tsunami wave that reached heights of 133 feet (40.5 meters) and traveled 6 miles (10 km) inland, causing about 20,000 deaths. This photo shows the devastation in Miyagi Prefecture.

FIGURE 2.6 Iceland's Divergent Plate Boundary   Volcanic activity is common in Iceland because of the divergent tectonic border that bisects the island (and the North Atlantic ocean). Here, people watch the 2010 eruption of the Eyjafjallajokull eruption, which was most notable for the major disruption its volcanic ash caused to both trans-Atlantic air travel as well as air traffic within Europe.

FIGURE 2.7 The Ural Mountains of Russia Although this extensive mountain range in Russia lies far from a current tectonic boundary, the mountains were probably created from the collision of ancient plates when the global tectonic configuration was much different. Mt. Narodnaya, pictured here, the highest point in the Urals, in the Republic of Komi, Russia.

FIGURE 2.8 Refugees from Haiti Earthquake Years after the January 2010 earthquake that killed over 230,000 people, half a million people reportedly remain homeless, living in refugee camps composed of makeshift shelters. In this photo taken in the suburbs of Port au Prince, a small market stand offers food and other necessities.

colliding from tectonic plates, many mountain ranges are far removed from tectonic boundaries. In North America, the Rocky Mountains serve as an example, as do the Ural Mountains in the center of Russia (Figure 2.7). This reminds us that geologic forces over vast periods of time have shaped Earth's topography, and that not all landscapes can be explained by the current pattern of plate tectonics.

Geologic Hazards

Although floods and tropical storms commonly take a higher toll of human life each year, earthquakes and volcanoes nonetheless can have a major effect on human settlement and activities. An estimated 20,000 people died in March 2011 from the combination of an earthquake and a tsunami in coastal Japan, while the year before (January 2010) over 230,000 people were killed in a magnitude 7.0 earthquake in Haiti (Figure 2.8). The vastly different effects of those two quakes underscore the fact that vulnerability to geologic hazards differs considerably around the world, depending on local building standards, population density, housing traditions, and, not unimportant, the effectiveness of search, rescue, and relief organizations.

In addition to earthquakes, volcanic eruptions are found along most tectonic plate boundaries and can also cause major destruction (Figure 2.9). In 1985, for example, a volcanic eruption

FIGURE 2.9 Global Earthquakes and Volcanoes The distribution of earthquakes and volcanoes is closely associated with tectonic plate boundaries. The circum-Pacific zone of activity, from the western Americas (both North and South) to East Asia, is particularly active and is often referred to as the Pacific Rim of Fire. Another dangerous zone runs east-west, from northern South Asia to Mediterranean Europe. Between these two active earthquake zones, billions of the world's population are at risk from destructive earthquakes.

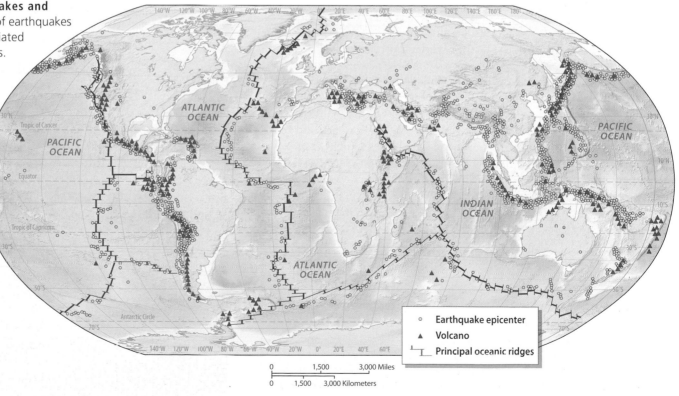

FIGURE 2.10 Pinatubo Volcano Although volcanic ash can be beneficial for certain kinds of agriculture by enriching the soil, in large amounts, as shown here in the Mt Pinatubo region of the Philippines, it becomes burdensome.

resulted in 23,000 deaths in Colombia, South America. In some cases, eruptions can be predicted days in advance, which usually provides enough time for evacuation. During the 1991 eruption of Mt. Pinatubo in the Philippines, 60,000 people were evacuated, although 800 did die in the disaster (Figure 2.10). Because volcanoes usually generate warning activity before they erupt, the loss of life from volcanoes is generally a fraction of that from earthquakes. More to the point, in the 20th century an estimated 75,000 people were killed by volcanic eruptions, while approximately 1.5 million died in earthquakes.

Unlike earthquakes, volcanoes provide some benefits to people. In Iceland, New Zealand, and Italy, geothermal activity produces energy to heat houses and power factories. In other parts of the world, such as the islands of Indonesia, volcanic ash has enriched soil fertility for agriculture. Additionally, local economies benefit from tourists attracted by scenic volcanic landscapes in such places as Hawaii, Japan, and the Pacific Northwest.

REVIEW QUESTIONS

1. What are the three kinds of tectonic plate boundaries?
2. What drives tectonic plate movement?
3. Where are most of the world's earthquakes and volcanoes? Why are they located where they are?

GLOBAL CLIMATES: A WORRISOME FORECAST

Untold numbers of human activities are closely linked to weather and climate, from farming that depend on moisture and warmth to city dwellers who suffer from heat waves and disruptive storms. In fact, much of the world's diversity results from the many different ways in which people adapt to local weather and climate. While some desert areas in California are covered with high-value irrigated agriculture, producing vegetables for the global marketplace year-round, most of the world's arid regions are barren and support very little farming. Also, severe weather events such as storms and droughts often ripple through interconnected world trade and food systems with major repercussions. When drought strikes Russia's grain belt or Africa's Sahel, for example, the socioeconomic consequences are felt throughout the world.

Aggravating these interconnections is the fact that the world's climates are changing due to global warming. Just what the future holds is not entirely clear, but even if the forecast has some uncertainty, there is little question that all forms of life—including humans—will have to adjust to vastly different climatic conditions by the middle of the 21st century (Figure 2.11).

Climate Controls

Even though the world's climates differ considerably from place to place and seasonally with highly varying temperature and precipitation regimes, most of these differences can be explained by a set of physical processes that influence weather and climate throughout the world. Let's examine each of these factors.

FIGURE 2.11 Polar Bears Threatened by Global Warming Ice floes are an important part of the polar bear's habitat. Because arctic ice is melting more rapidly now due to global warming, the 20,000 bears living in the wild face an uncertain future. Recently, polar bears have been reported drowning trying to swim long distances between ice sheets.

Solar Energy Both Earth's surface and the atmosphere immediately above it are heated by energy from the Sun, making solar energy the most important factor affecting world climates. Not only does this energy explain the differences in temperature between warm equatorial zones and the cold polar regions, but also it drives other important processes, including global pressure systems, winds, and ocean currents.

Incoming short-wave solar energy, or **insolation**, passes through the atmosphere and is absorbed by Earth's land and water surfaces. As these surfaces warm, they reradiate heat back into the lower atmosphere as infrared, long-wave energy that, in turn, is absorbed by water moisture and atmospheric gases such as carbon dioxide (CO_2) in the lower atmosphere. This absorbed energy produces the envelope of warmth that makes life possible on our planet. Because there is some similarity between this heating process and the way a garden greenhouse traps sunlight, warming the structure's interior, this natural process of atmospheric heating is called the **greenhouse effect** (Figure 2.12). Were it not for this process, Earth's climate would average about 60°F (33°C) colder, resulting in conditions much like Mars.

Latitude Because of the curvature of the globe, insolation strikes Earth at a true right angle only in the tropics, meaning that each unit of solar energy is more intense in the tropics than at higher latitudes both north and south of the equator. Thus, the Sun is more effective at heating surface ground and water at the equator than at higher latitudes (Figure 2.13). You can feel the difference in solar intensity when standing in direct sunlight near the equator in Singapore, contrasted to the welcome warmth provided by the summer sun in a high-latitude city like Tromso, Norway.

Not only does this difference in solar intensity result in warm tropical climates that contrast with cooler climates in the middle or high latitudes, but also this difference in solar intensity builds up heat in the equatorial regions. This heat must then be redistributed through other physical processes such as global pressure and wind systems, ocean currents, and even massive tropical typhoons and hurricanes (Figure 2.14).

Interactions Between Land and Water Because land and water differ in their abilities to absorb and reradiate insolation, the global arrangement of oceans and land areas is a major influence on world climates (Figure 2.15). Basically, land areas heat and cool faster than do bodies of water, which explains why the temperature extremes of hot summers and cold winters, such as experienced in the interior United States and Canada, are always found away from coastal areas. These differences are also found at smaller

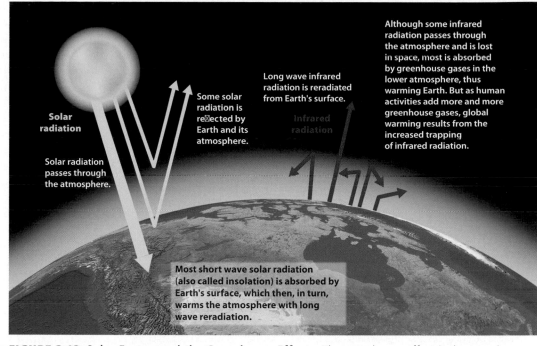

FIGURE 2.12 Solar Energy and the Greenhouse Effect The greenhouse effect is the trapping of solar radiation in the lower atmosphere, resulting in a warm envelope surrounding Earth.

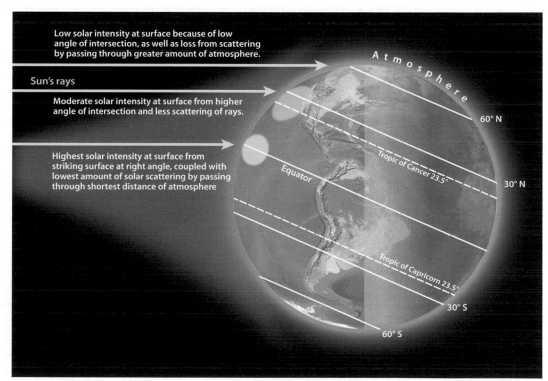

FIGURE 2.13 Latitude as a Climate Control Because of Earth's curvature, solar radiation is more effective at warming the surface in the tropics than in higher latitudes.

scales, within hundreds of miles of each other. For example, the average high July temperature in San Francisco, on the California coast, is 60°F (15.6°C), whereas in Sacramento, the inland state capital—only 80 miles away—the average high temperature in July is 92.4°F (33.5°C).

Global Pressure Systems   The uneven heating of Earth due to latitudinal differences and the arrangement of oceans and continents produces a regular pattern of high- and low-pressure cells. These cells drive the world's wind and storm systems. To illustrate, the interaction between high- and low-pressure systems in the North Pacific produces storms that are driven onto the North American continent in both winter and summer. Similar processes in the North Atlantic also produce winter and summer weather for Europe.

Farther south, in the subtropical zones, large oceanic cells of high pressure cause very different conditions. These high-pressure cells expand during the warm summer months because of the subsidence of warm air from the equatorial regions. As they enlarge, the cells produce the warm, rainless summers of Mediterranean climate areas in Europe and California. In the equatorial zone itself, summer weather spawns the strong tropical storms known as typhoons in Asia and hurricanes in North America and the Caribbean.

FIGURE 2.14 Hurricane Katrina, 2005   This satellite photo shows Hurricane Katrina just before it moved northward and onshore into the Mississippi delta area of Louisiana, devastating New Orleans and adjacent areas on August 29, 2005. As a result of flooding and wind damage to urban and coastal areas, Katrina remains the costliest natural disaster in U.S. history. Some 2000 people lost their lives in the storm. A week earlier Katrina had crossed Florida as a category 1 hurricane, but then increased to category 3 in the Gulf of Mexico before moving making its second landfall in Louisiana.

The term **continentality** describes inland climates with hot summers and cold, snowy winters, such as those found in interior North America and Europe. In contrast **Maritime climates** are those close to the ocean, with moderate temperatures in both summer and winter. Western Oregon and the British Isles are good examples of regions with maritime climates.

Global Wind Patterns   High- and low-pressure systems also produce global wind patterns at local, regional, and global scales. As a rule, winds flow away from high-pressure and toward low-pressure cells (just as water flows from high to low elevations). This explains the monsoon in India, for example, which arrives in June as moisture-laden air masses flow from high pressure over the Indian Ocean, across the South Asian subcontinent, and toward the low-pressure area of northern India and Tibet, where summer heat has created a massive low-pressure system. In the winter, wind flow is opposite: As high pressure builds over these inland areas with winter cooling, winds flow outward from cold Tibet and the snowy Himalayas toward low pressure over the warm Indian Ocean (see Chapter 12 for more detail). Similar seasonal pressure and wind regimes are found in many other parts of the world as well. The U.S. Southwest is one example, where the summer monsoon brings moisture to the arid lands of Nevada, Arizona, and New Mexico (Figure 2.16).

Topography   Weather and climate are affected by topography in two ways: Cooler temperatures are associated with higher elevations, and topography influences precipitation (rain and snow) patterns.

Because the lower atmosphere is heated by solar energy reradiated from Earth's surface, air temperatures are warmer close to the surface (and at sea level) and become cooler as you move up in elevation. As a general rule, the atmosphere cools by 3.5°F for every 1,000 feet gained in elevation

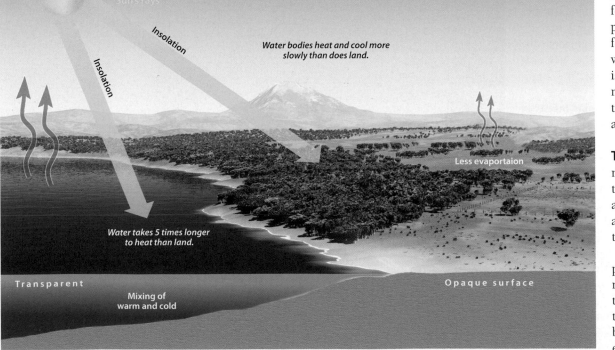

FIGURE 2.15 Differential Heating of Land and Water   Because land heats and cools faster than do bodies of water, temperatures are usually both warmer in the summer and colder in the winter inland than on the coasts.

Summer heating of interior creates low pressure that then draws in moist, unstable air from the Gulf of Mexico, resulting in thunderstorms (and rainfall) in southwestern U.S.

FIGURE 2.16 Summer Monsoon in the Southwest As the Southwest U.S. warms in June and July, this heating creates thermal lows that draw in moist air from the Gulfs of Mexico and California, resulting in cloudiness (as shown here), thunderstorms, and much-needed rainfall.

(in the metric system, the cooling rate is 1°C per 100 meters). This is called the **adiabatic lapse rate**, which is the rate of cooling with increasing altitude within the lower atmosphere. Thus, on a hot summer day in Phoenix, AZ, elevation 1,100 feet (335 meters), the temperature often reaches 100°F (37.7°C). Just 140 miles away in the mountains of northern Arizona at 7,100 feet (2,160 meters) near Flagstaff, the temperature is usually 21°F lower at a pleasant 79°F (26°C). The difference of 21°F equals 6 (thousands of feet difference in elevation) times 3.5, the adiabatic lapse rate. Cooler temperatures, then, are always found at higher elevations. This also explains why in mountainous areas the precipitation that falls as rain in the lowlands will probably fall as snow in the uplands.

Topography also wrings moisture out of storms by forcing moving air masses to cool as they are forced up and over mountain ranges

in what is known as the **orographic effect** (Figure 2.17). The cooler air cannot hold as much moisture, which falls as precipitation. This explains the common pattern of wet mountains and nearby dry lowlands. These dry areas are said to be in the **rain shadow** of the adjacent mountains. They are areas of reduced rainfall because downslope winds warm (the opposite of upslope winds, which cool), thus increasing an air mass's ability to retain moisture. Rain shadow areas are common in the mountainous areas of western North America, as they are in Andean South America, as well as in many parts of South and Central Asia.

World Climate Regions

Even though the world's weather and climate vary greatly from place to place, we can generalize about similarities in temperature, moisture, and seasonality that facilitate mapping global climate regions (Figure 2.18). Before going further, though, it is important to note the difference between these two terms. *Weather* is the short-term, day-to-day expression of atmospheric processes: Weather can be rainy, cloudy, sunny, hot, windy, calm, or stormy, all within a short time period. Thus, weather is measured at regular intervals each day (usually hourly). These data are then compiled over a 30-year period to generate statistical averages that describe the common meteorological conditions of a specific place. From these averages, we can place the weather of a specific locale in the worldwide scheme of climate regions. Simply stated, then, *weather* is the short term expression of atmospheric processes while *climate* is the long term averages from daily weather measurements. As pundits like to say, climate is what you expect while weather is what you get.

Knowing the climate type for a specific part of the world not only conveys a clear sense of average rainfall and temperatures, but also allows inferences about a wider array of human activities. If an area is categorized as desert, then we infer that rainfall is so limited that any agricultural activities require irrigation. In contrast, when an area is characterized as tropical monsoon, we know it has warm temperatures and adequate amounts of rainfall for certain kinds of farming.

We use a standard scheme of climate types throughout this book, and each regional chapter contains a map showing the different climate types of the area. In addition, these maps contain **climographs**, which are graphic representations of monthly average high and low temperatures, along with monthly precipitation (rain and snow) amounts for a specific location. Figure 2.18 shows climographs for Tokyo, Japan, and Cape Town, South Africa. Two lines for temperature data are presented on each climograph: The upper line plots average high temperatures for each month, while the lower line shows average low temperatures. Besides temperatures, climographs contain bar graphs depicting average monthly precipitation.

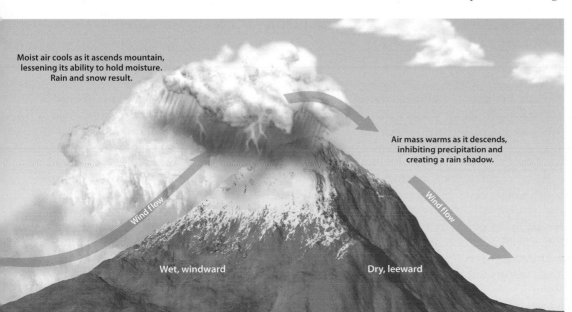

Moist air cools as it ascends mountain, lessening its ability to hold moisture. Rain and snow result.

Air mass warms as it descends, inhibiting precipitation and creating a rain shadow.

Wind flow

Wind flow

Wet, windward

Dry, leeward

FIGURE 2.17 The Orographic Effect Upland and mountainous areas are usually wetter than the adjacent lowland areas because of the orographic effect whereby rising air is cooled and loses its ability to hold moisture as it flows up and over mountains, resulting in rain and snowfall. In contrast, the leeward or downwind side of the mountains is usually drier because warming air increases its ability to hold moisture.

FIGURE 2.18 World Climate Regions   A standard scheme, called the *Köppen system,* after the Austrian geographer who devised it in the early 20th century, is used to describe the world's diverse climates. Combinations of upper and lower-case letters describe the general climate type along with precipitation and temperature characteristics. Specifically, the *A* climates are tropical, the *B* climates are dry, the *C* climates are generally moderate and are found in the middle latitudes, and the *D* climates are associated with continental and high-latitude locations.

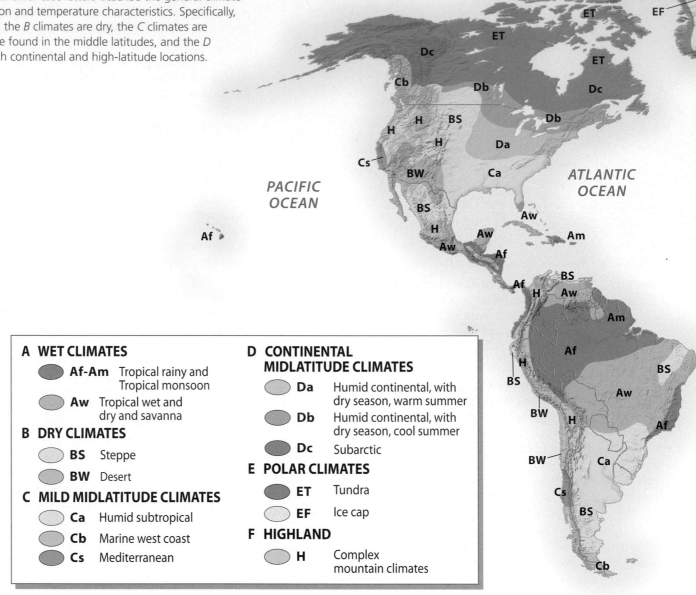

A WET CLIMATES

Af-Am   Tropical rainy and Tropical monsoon

Aw   Tropical wet and dry and savanna

B DRY CLIMATES

BS   Steppe

BW   Desert

C MILD MIDLATITUDE CLIMATES

Ca   Humid subtropical

Cb   Marine west coast

Cs   Mediterranean

D CONTINENTAL MIDLATITUDE CLIMATES

Da   Humid continental, with dry season, warm summer

Db   Humid continental, with dry season, cool summer

Dc   Subarctic

E POLAR CLIMATES

ET   Tundra

EF   Ice cap

F HIGHLAND

H   Complex mountain climates

Not only is the total amount of rain and snowfall important, but also the seasonality of precipitation provides valuable information about the link between moisture and agricultural growing seasons.

Global Climate Change

Human activities, primarily those connected with economic development and industrialization, are changing the world's climate in ways that have significant consequences for all living organisms, whether plants, animals, or humans. More specifically, **anthropogenic** (human-caused) pollution of the lower atmosphere is increasing the natural greenhouse effect, so that worldwide **global warming**—an increase in Earth's average atmospheric temperature—is taking place. This warming, in turn, could lead to climate changes with major consequences. Rainfall patterns, for example, could change, so that existing agricultural production in traditional breadbasket areas such as the U.S. Midwest and Canadian prairies is threatened; low-lying coastal settlement in places like Florida and Bangladesh could be flooded as sea levels rise from warming oceans and melting polar ice caps; an increase in searing heat waves could cause higher human death tolls in the world's cities; and water will become an increasingly scarce resource in many areas of the world. Although the world's nations recognize the seriousness of global warming, only minor progress

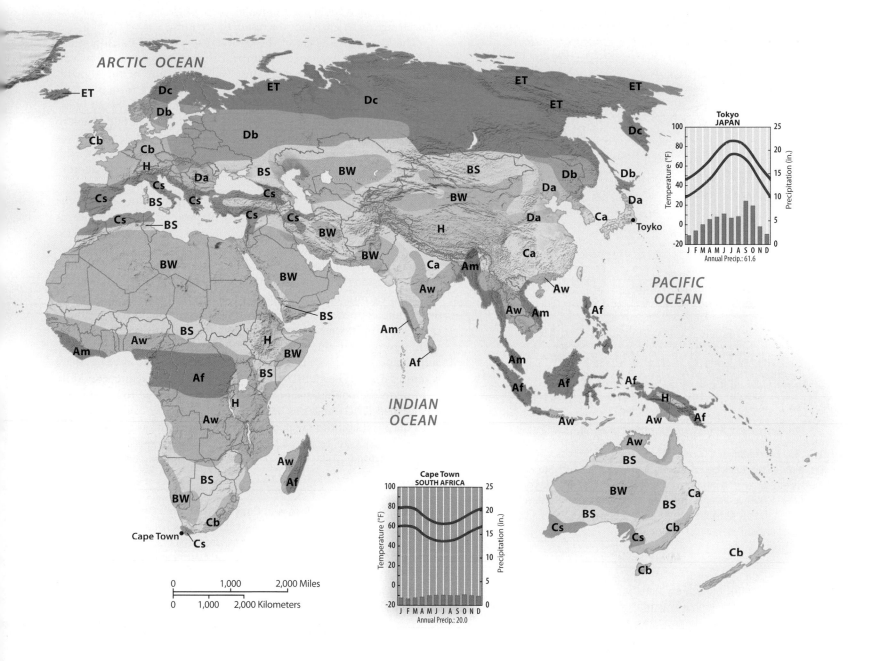

Tokyo
JAPAN
Annual Precip.: 61.6

Cape Town
SOUTH AFRICA
Annual Precip.: 20.0

has been made in the last decades in crafting international agreements that could mitigate the problem.

Causes of Global Warming   As noted, the natural greenhouse effect provides Earth with a warm atmospheric envelope that supports life, and this warmth comes from the trapping of incoming and outgoing solar radiation by an array of natural greenhouse gases (GHGs)—water vapor, carbon dioxide (CO_2), methane (CH_4), and ozone (O_3). While the composition of these natural greenhouse gases has varied somewhat over long periods of geologic time, they have been relatively stable since the last ice age ended 20,000 years ago (more detail on GHGs in Mastering Geography).

However, with the widespread consumption of coal and petroleum associated with global industrialization, a huge increase has occurred in atmospheric carbon dioxide and methane (by-products of burning oil and coal). This increase has greatly magnified the natural greenhouse effect and produced global warming (Figure 2.19). More specifically, in 1860 atmospheric CO_2 was 280 parts per million (ppm); today it is more than 390 ppm. More troubling is that, unless global emissions are reduced considerably in the near future, the CO_2 level will reach 450 ppm by 2020, a level at which climate scientists project that Earth's climate will be changed irrevocably. Although the complexity of the global climate system leaves some uncertainty about exactly how the world's climates may change, climate scientists using

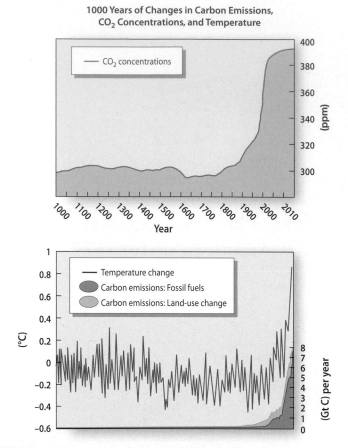

FIGURE 2.19   Increase in CO$_2$ and Temperature   These two graphs show the relationship between the rapid increase of CO$_2$ in the atmosphere and the associated rise in average annual temperature for the world. The graphs go back 1,000 years and show both CO$_2$ and temperature to have been relatively stable until the recent industrial period, when the burning of fossil fuels (coal and oil) began on a large scale.

high-powered computer models are reaching consensus on what can be expected from continued global warming. Basically, computer models predict that average global temperatures will increase almost 4°F (2°C) by 2020, a change in temperature that is the same magnitude as the cooling that caused ice-age glaciers to cover much of Europe and North America 30,000 years ago. If emissions continue at the current rate, this temperature increase is projected to double by 2100. As noted earlier, the consequences of future global warming are indeed sobering, with widespread coastal flooding from sea-level rise, worldwide droughts that impact food systems, widespread water shortages, devastating heat waves, more frequent wildfires, and catastrophic animal and plant extinctions (Figure 2.20).

The International Debate on Limiting Emissions   The international debate on limiting GHG emissions has passed through two phases in the last 30 years. It is now in a crucial

third stage, which could determine the extent of global warming and climate change experienced for the rest of the century.

The first phase began at the UN-sponsored 1992 Earth Summit in Rio de Janeiro, when 167 countries signed an agreement to voluntarily limit their GHG emissions. However, none of the signatories reached its emission reduction targets, so a more formal second phase began with a 1997 meeting in Kyoto, Japan. Here the 30 Western industrialized countries agreed to reduce their emissions back to 1990 levels by the year 2012. Unlike the Rio agreement, which was voluntary, the Kyoto Protocol had the force of international law, with penalties for those countries not reaching their emission reduction targets.

But not all has gone well with the Kyoto Treaty. Not only did the United States refuse to ratify the treaty because of political concerns about possible injury to the U.S. economy, but also it became increasingly clear that most industrial countries were not going to achieve their emission reduction targets. Further complications arose from increasing tensions between developing and developed countries over the matter of **carbon inequity**. This term refers to the position taken by developing countries such as China and India, which argue that, because Western industrial countries in North America and Europe have been burning large amounts of fossil fuels since the mid-19th century and because CO$_2$ stays in the atmosphere for hundreds of years, these countries caused the global warming problem and therefore should fix it. More to the point, developed countries should take the lead in reducing their GHG emissions before asking developing countries to take steps that could hinder their economic development. As a result of this concern, the Kyoto Treaty of 1997 did not include emission reduction targets for China, India, Brazil, or any other developing country.

FIGURE 2.20   Sea-Level Rise and Coastal Flooding   One consequence of global warming will be a rise in the world's sea level due to a combination of polar ice cap melting and the thermal expansion of warmer ocean water. Forecasts are that at the current rate of climate warming sea levels will rise about 4 feet by year 2100, causing considerable flooding of low lying coastal areas such as this region of the Netherlands near Amsterdam. Other densely-settled coastal areas throughout the world will also be vulnerable.

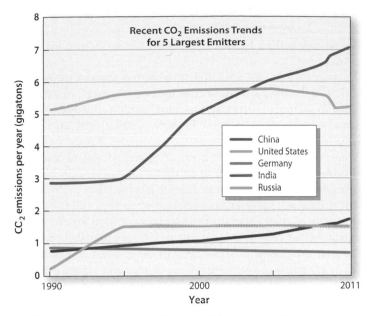

FIGURE 2.21 Largest CO$_2$ Emitters China's yearly CO$_2$ emissions continue to grow as hundreds of new coal-fire power plants come online. In contrast, emissions in the U.S. have stabilized recently; more power plants switch from coal to natural gas as the price of that cleaner source of energy has become increasingly competitive. The flat line of Russia's emissions after the collapse of the Soviet Union is somewhat of a mystery and may be a reporting problem. In contrast, Germany's flat line is a result of an increasing reliance on renewable energy, namely wind and solar. India's every-increasing emissions continues to be of concern.

However, by 2008, China's annual GHG emissions surpassed those of the United States, which until that year had historically been the world's largest emitter (Figure 2.21). Adding to the issue was fast-growing India's stated ambition to follow China's example of rapid industrialization. Brazil, too, was a problem because of rapidly increasing emissions from deforestation of the Amazon rain forest and from its expansive livestock economy.

With the Kyoto Protocol scheduled to expire in 2012, the third phase of the international global warming emission debate began in December 2011, with a meeting of 194 countries in Durban, South Africa. At that conference, a conceptual road map was agreed on that guides current discussions on limiting emissions. The key points are as follows:

- The Kyoto Protocol was extended beyond its 2012 expiration date to 2015, which is the target date for a completely new international agreement.

- Unlike Kyoto, this new agreement must include all countries, developed and developing, which clearly includes the large CO$_2$ emitters China and India, as well as those countries like Brazil whose emission levels are projected to grow significantly in the next decades. Not surprisingly, this component of the Durban Agreement is extraordinarily contentious because of China's and India's strong opposition to emission reduction targets. Whether these two giant countries will lessen their resistance over the next several years is crucial to the extent of future global warming (Figure 2.22).

- By 2015, new emission reduction goals will be set for all the world's countries, and these will take effect no later than 2020.

- A $100 billion Green Fund was created to help Less Developed Countries (LDCs) meet the challenges of global warming. This fund will finance a wide range of activities, from renewable energy development to flood control structures for areas threatened by sea level rise.

- A Reducing Emissions from Deforestation and Degradation (REDD) program was established to finance preservation of the world's forests by paying developing countries to reduce cutting their forests and instead treat them as carbon storehouses.

Will the Durban Agreement significantly reduce the global warming problem? Not really, since, as noted earlier, the current rate of global CO$_2$ emissions will most likely reach 450 ppm by the time the new Durban reduction targets take effect. Consequently, the amount of CO$_2$ in the atmosphere will reach the point where, even if all emissions were halted overnight, the world's climate processes would still be irrevocably changed.

FIGURE 2.22 China's Global Warming Emissions China surpassed the United States as the largest emitter of greenhouse gases in 2008. A major component of this global warming pollution is that China is the world's largest consumer of coal, which is used primarily to generate electricity from coal-fired power plants. Because of this growing need experts predict that China's emissions could double in the next 20 years, a worrisome prospect that will have profound effects on the world's climate.

WATER: A SCARCE WORLD RESOURCE

Water is central to human life and our supporting activities—agriculture, industry, transportation, even recreation—yet water is unevenly distributed around the world, being plentiful in some areas while distressingly scarce in others. These water problems are not simply the product of varied global climates that produce wet or dry conditions; they are also due to complex factors such as political control over river basins and storage facilities. In addition, different cultures and different economies have highly varied water usage and consumption patterns.

A world map shows that more than 70 percent of the surface area of Earth is covered by oceans. As a result, 97 percent of the total global water supply is saltwater, with only 3 percent freshwater. Of that minuscule amount of freshwater on Earth, almost 70 percent is locked up in polar ice caps and mountain glaciers. Furthermore, groundwater supplies account for almost 30 percent of the world's freshwater, which leaves less than 1 percent easily accessible from surface rivers and lakes.

Another way to conceptualize this limited amount of freshwater is to think of the total global water supply as 100 liters, or 26 gallons. Of that amount, only 3 liters (0.8 gallons) would be freshwater; and of that small supply, only a mere 0.003 liters, or about half a teaspoon, is available to people.

Water planners use the concept of **water stress** to map where water problems exist and also to predict where future problems will occur (Figure 2.23). Water stress data are generated by calculating the amount of freshwater available in relation to current and future population. Northern Africa stands out as a region of high water stress; hydrologists predict that three-quarters of Africa's population will experience water shortages by 2025. Other problem areas will be northern China, India, much of Southwest Asia, Mexico, the western U.S., and even parts of Russia. Although global warming may actually increase rainfall in some parts of the world, scientists forecast that climate change will probably aggravate global water problems. Three areas of concern occupy water planners: scarcity, sanitation, and access.

Water Scarcity

Currently, about half the world's population lives in areas where water shortages are common. As the population in these areas increases, these water problems will become more acute. Also, since 70 percent of the world's freshwater usage is for agriculture, food production will probably decline as water becomes increasingly scarce.

FIGURE 2.23 Global Water Stress This map shows those parts of the world forecast by the UN to have serious water shortages by 2025. In many of the areas—namely, in northern Africa and Southwest Asia—water stress is already evident. Global warming will aggravate the water situation as temperatures increase. Africa, in particular, will suffer greatly; projections are that fully 75 percent of its population will face water shortages by 2025.

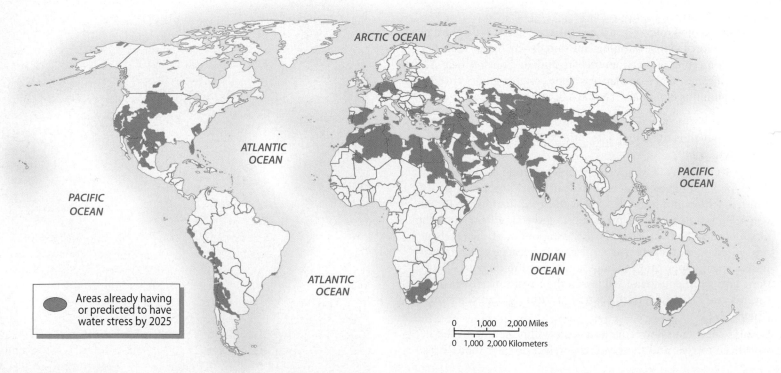

FIGURE 2.24 Women and Water In many parts of the world, women and young girls spend much of their day providing water for their homes and villages. For young girls, this task often interferes with their schooling. These women and girls are fetching water in the Punjab region of India.

Water Sanitation

Where clean water is not available, people use polluted water for their daily needs, resulting in a high rate of sickness and death. More specifically, the United Nations reports that over half of the world's hospital beds are occupied by people suffering from illnesses linked to contaminated water, and more people die each year from polluted water than are killed in all forms of violence including wars. This toll from polluted water is particularly high for infants and children who have not yet developed any sort of resistance or tolerance for contaminated water. In Haiti, for example, even before the 2010 earthquake, almost 20% of all deaths of children under five were directly tied to waterborne diseases. That toll is probably much higher now because of shortages of clean water in the earthquake refugee camps.

Water Access

By definition, when a resource is scarce, access is problematic, and these hardships take many forms. Women and children, for example, often bear the burden of providing water for family use, and this can mean walking long distances to pumps and wells and then waiting in long lines to draw water. The result is that their daily time budget for other activities, such as school or work, is severely curtailed (Figure 2.24). Given the amount of human labor involved in providing water for crops, it is not surprising that some studies have shown that in certain areas people expend as many calories of energy irrigating their crops as they gain from the food itself.

Ironically, some recent international efforts to increase people's access to clean water have gone astray and have actually aggravated access problems instead. Historically, domestic water supplies have been public resources, organized and regulated—either informally by common consent or more formally as public utilities—resulting in free or low-cost water. In recent decades, however, the World Bank and the International Monetary Fund have promoted the privatization of water systems

as a condition for providing loans and economic aid to developing countries. While the agency's goals have been laudable, the means have been controversial because, commonly, the international engineering firms that have upgraded rudimentary water systems by installing modern water treatment and delivery technology usually have increased the costs of water delivery to recoup their investment. Although the people may now have access to cleaner and more reliable water, in many cases the price is higher than they can afford, forcing them to either do without or go to other unreliable and polluted sources.

In Cochabamba, Bolivia, for example, the privatization of the water system resulted in a 35 percent increase in water costs. In response, the people rebelled and rioted, with demonstrations that became tragically violent. Eventually, the water system was returned to public control. Reportedly, however, today half the city's population is still without a reliable water source.

REVIEW QUESTIONS

1. How much water is there on Earth, and how available is it for human usage? Use the concept in your answer that Earth's water budget is just 100 liters.

2. What are the three major issues that cause water stress?

BIOREGIONS: THE GLOBALIZATION OF NATURE

One aspect of Earth's uniqueness is the rich diversity of plants and animals covering its continents. Geographers and biologists think of this cloak of vegetation as the "green glue" that binds together life, land, and atmosphere. Humans are very much a part of this interaction. Not only are we evolutionary products of a specific **bioregion**—that is, an assemblage of local plants and animals—but also our human prehistory includes the domestication of specific plants and animals that led to modern agriculture.

A brief overview of the most important bioregions is provided below. The discussion begins with the warm equatorial tropics and then moves poleward, describing the dry desert and grassland regions, followed by two different kinds of temperate forests (Figure 2.25). Note that these bioregions are closely correlated to the climate regions discussed previously, since the major traits of a climate region—temperature, precipitation, and seasonality—are the same factors influencing the distribution of the array of flora and fauna making up a particular bioregion.

Tropical Rain Forests

Tropical rain forests are found along the equator in the Af climate region, with its high average annual temperatures, abundant sunlight, and copious rainfall occurring throughout the year. Rainfall amounts of 70 to 150 inches (180 to 380 cm) are common.

This bioregion covers about 7 percent of the world's land area (roughly the size of the contiguous United States) in Central and South America, Sub-Saharan Africa, Southeast Asia, and Australia and on many tropical Pacific islands. More than half of the world's known

FIGURE 2.25 World Bioregions   Although global vegetation has been greatly modified by clearing land for agriculture and settlements and by cutting forests for lumber and paper pulp, there is still a recognizable pattern to the world's bioregions, ranging from tropical forests to arctic tundra. Each bioregion has its own array of ecosystems, containing plants, animals, and insects.

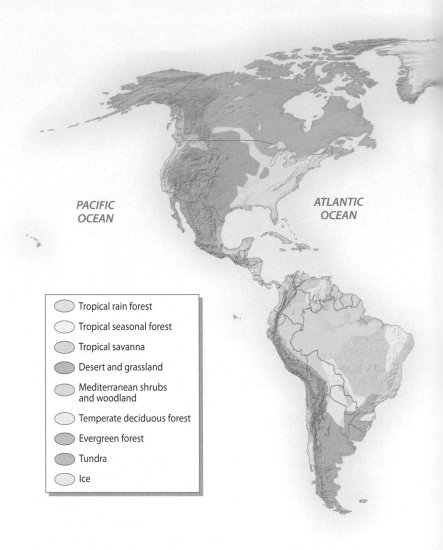

Tropical rain forest

Tropical seasonal forest

Tropical savanna

Desert and grassland

Mediterranean shrubs and woodland

Temperate deciduous forest

Evergreen forest

Tundra

Ice

plant and animal species live in the tropical rain forest bioregion, making it the most diverse of all bioregions.

The dense tropical forest vegetation is usually arrayed in three distinct levels that are adapted to decreasing amounts of sunlight, from the treetops to the darker forest floor (Figure 2.26). The tallest trees, around 200 feet (60 meters) high, receive open sunlight; the middle level, around 100 feet (30 meters) high, gets filtered sunlight; and the lowest level, the forest floor, is where plants survive with very little direct sunlight. Even though much organic material accumulates on the forest floor in the form of falling leaves, rain forest soils are not particularly well suited to intensive agriculture because most nutrients are absorbed by extensive tree root systems.

Tropical Seasonal Forests

Bracketing the true tropical rain forest, north and south of the equator, are the tropical seasonal forests, found in those climates that have a distinct dry season, typically four months or so. This is the Am or monsoonal climate region. Because of seasonal droughts, rainfall is only about half of that found in the true rain forest, resulting in a more open landscape with trees spaced farther apart than where rain falls throughout the year (Figure 2.27). Additionally, seasonal rain forest trees are **deciduous**, meaning they shed their leaves during the harsh dry season in order to slow or completely halt growth. With a pronounced dry season, this bioregion is also subject to regular fires,

FIGURE 2.26 Tropical Rain Forest   As fragile as they are diverse, tropical rain-forest environments feature a complex, multilayered canopy of vegetation. Plants on the forest floor are well adapted to receiving very little direct sunlight.

both natural and human-set, a process that thins the forest even further by killing young trees and expanding the grass cover. Humans often set these fires to make this bioregion more suitable for grazing animals and farming.

Tropical Savannas

The tropical savanna bioregions are found in the Aw climate region, which has a dry season lasting half the year or longer. In these equatorial areas, even fewer trees are found in a landscape dominated by grasslands (Figure 2.28). Annual rainfall amounts are commonly around 50 inches (125 cm), which is a third of what falls in the tropical rain forests. As is true for the tropical seasonal forests, humans have expanded the tropical savanna by using fire to make these bioregions more suitable for grazing and farming. This change has been so extensive that ecologists are hard-pressed to differentiate between a *climatic savanna*, one created by natural climatic conditions, and a *derived savanna*, produced by human manipulation.

Deforestation of Tropical Forests

Tropical forests are being devastated at an unprecedented rate, creating a crisis that tests our political, economic, and ethical systems.

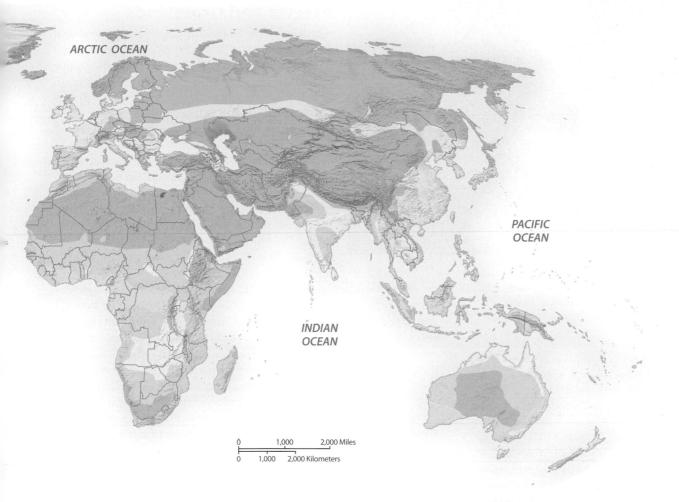

Although deforestation rates differ from region to region, each year an area about the size of Wisconsin or Pennsylvania is denuded. Geographically, almost half of this activity is in the Amazon Basin of South America; however, deforestation appears to be occurring faster in Southeast Asia, where reportedly logging is taking place at rates three times faster than in the Amazon (Figure 2.29).

Besides destroying the habitat of crucial and highly diverse plants and animals, the other major concern is that tropical deforestation releases huge amounts of CO_2 into the atmosphere. Current estimates suggest that fully 20 percent of all human-caused GHG emissions result from cutting and burning tropical forests.

Driving tropical deforestation is the recent globalization of commerce in international wood products. Currently, about one-half of all tropical forest timber is destined for China, where much of the wood is used for throwaway items such as chopsticks and newspapers. Another

FIGURE 2.27 Seasonal Tropical Forest Unlike the true tropical rain forest that receives rainfall throughout the year, the seasonal tropical forest has a distinct dry season during which many of the trees and bushes lose their leaves and hibernate until the rains return. This photo is from the Panna National Park in India.

FIGURE 2.28 The Tropical Savanna With a much longer dry season, there are fewer trees and more open grasslands in the tropical savanna compared to the seasonal rain forest. This photo was taken in Brazil.

factor contributing to the rapid destruction of tropical rain forests is the world's seemingly insatiable appetite for beef. Cattle species originally bred to survive in the hot weather of India are now raised on grassland pastures created worldwide by converting tropical forests and savannas into rangeland.

More recently, tropical forests have been cleared away to make room for palm oil plantations in response to a growing demand for this popular cooking oil. Also, tropical forest areas are often thought of as a settlement solution for rapidly growing populations of developing countries, places where national policies allow settlers to clear and homestead tropical forestlands as a way (real or imagined) of relieving pressure on overcrowded cities. (See "Working Toward Sustainability: The Complexities of Tropical Rain Forest Management.")

FIGURE 2.29 Tropical Forest Destruction There are many ill effects resulting from tropical rainforest destruction, including the loss of plants and animals, the loss of homeland for native peoples, and the release of large amounts of global-warming greenhouse gases. This photo is of a new cacao plantation on Borneo Island, East Malaysia.

Deserts and Grasslands

Poleward of the tropics lie large areas of arid and semiarid climates (BW and BS climate regions) where the world's extensive deserts and steppe grasslands are found. In fact, fully one-third of Earth's land area qualifies as either true desert, where annual rainfall is less than 5 inches (13 cm), or steppes, where the yearly rainfall is between 5 and 10 inches (13 and 25 cm).

Grassy plants appear in these semiarid areas, forming a lush cover during the wet season. In North America, the midsections of both Canada and the United States are covered by grasslands known as **prairie**, characterized by thick, long grasses. In other parts of the world, such as Central Asia, Russia, and Southwest Asia, shorter, less dense grasslands form **steppes**.

The boundary between desert and grassland has always varied naturally because of fluctuations in rainfall. During wet periods, grasslands might expand, only to contract once again during drier years. The transition zone between the two is a precarious environment for people, as the United States learned during the 1930s, when the semiarid grasslands of the western prairie lands turned into the notorious "Dust Bowl." At that time, thousands of farmers watched their fields become devastated by wind erosion and drought—a disaster that led to an exodus of people from these once-productive lands. Farming these semiarid lands may lead to **desertification**, which is the creation of truly arid desert lands in what were formerly grasslands (Figure 2.30). This has happened on a large scale throughout the world—in Africa, Australia, South Asia, and North America.

Mediterranean Shrubs and Woodlands

In the Mediterranean climate regions (Cs), where precipitation exceeds 10 inches (25 cm), a prolonged summer-season drought of three or four months produces a unique array of grasses, shrubs, and trees.

FIGURE 2.30 Desertification Climatic fluctuations and human misuse combine to produce desertification in certain localities, where marginal lands are overcropped or grazed heavily, resulting in expansion of nearby deserts, as in the Sahel region of northern Nigeria.

The Complexities of Tropical Rain Forest Management

There are at least a hundred good reasons to save the world's tropical rain forests—and probably a hundred more as to why it won't happen in a truly sustainable manner.

Lee Talbot, former director of the International Union for Conservation of Nature (IUCN), put it this way: "Sustainability is a smoke screen to cover destruction of irreplaceable forests for financial gain. In practical terms, no commercial logging of tropical forests has proven to be sustainable from the standpoint of the forest ecosystem, and any such logging must be recognized as mining, not sustaining the basic forest resource."

Why such harsh criticism? To begin with, any truly sustainable cutting of the tropical rain forest must be capable of being repeated later in time and provide the same level of resource value for a future generation as it does in the present. While this notion of sustained yield (as it's called in the timber business) makes sense for managing a forest that matures in 20 or 30 years, tropical rain forests take far longer to recover completely (Figure 2.1.1). True, some fast-growing tropical species could be harvested again in 30–50 years, but the most valuable trees take 100 or even 200 years to mature.

Further, even if future generations are willing to wait that long before cutting the forest again, this purely economic notion of sustainability completely overlooks the ecological losses of rain forest cutting. Once trees are cut, hundreds (perhaps even thousands) of tropical plant and animal species are deprived of their natural habitat. Along with the loss of specific species and ecological communities, regional ecological services that provide value to humans are also lost: soil fertility, watershed management, and even local microclimates. When rain forests are cut, soils erode more quickly, local climates become drier and warmer, and watersheds tend to dry up.

What is the solution to this vexatious issue? A first step, say environmental groups such as Friends of the Earth, is to be more candid and careful about how we apply the concept of sustainability. For example, if forest cutting is being managed on a sustained yield basis,

FIGURE 2.1.1 Tropical Rain Forest Plantations
Like any natural forest, a tropical rain forest is also a complex mixture of young and old trees of different species. Today, however, when tropical rain forests are cut they are often replaced with planted trees of the same age and the same species, like in this plantation in Indonesia. And while this may meet the commercial needs of the lumber industry it is questionable whether this process is truly sustainable.

focusing solely on the economic bottom line of a timber company, let's call that "fiscal sustainability." This is in contrast to bio- or eco-sustainability, which emphasize retaining the ecological services and integrity of tropical forests. A second step is for consumers of tropical forest products (and that can include everything from furniture to biofuel) to cut through the smoke screen by demanding more information when buying products that make general claims about their sustainability.

This is the Mediterranean shrub and woodland bioregion found in only five areas of the world—around the Mediterranean Sea; in parts of South Africa, Australia, and Chile; and along the U.S. West Coast (mainly in California, but also extending into southern Oregon). In these bioregions, the typical landscape is a combination of grasslands, low drought-resistant shrubs (called *chaparral* in Spain and California and *maquis* in southern France), and woodlands consisting of oak and pine (Figure 2.31). In the late 19th century, Australia's fast-growing eucalyptus trees were introduced into these Mediterranean bioregions and are now so widespread that they completely dominate many summer-dry landscapes throughout the world.

Temperate Deciduous Forests

This bioregion is associated with the widespread temperate C climate, where precipitation falls year-round in amounts of 30–60 inches (75–150 cm), summers are warm, and winters are cold, yet no month averages below freezing. In North America, temperate deciduous forests are the major habitat from the Gulf Coast to New England, as well as in parts of the Midwest; in Europe, they were the natural vegetation from the British isles eastward to Germany, as well as in the more temperate parts of Russia and Asia.

FIGURE 2.31 California chaparral In the chaparral plant and shrub communities common in the Mediterranean climate area, both plants and animals have evolved strategies to survive the long dry summer. Plants, for example, have leaf structures that reduce evapotranspiration during the hot daylight hours, and, as well, are adapted to frequent fire. This photo of a typical chaparral community was taken in the Topanga State Park near Los Angeles, California.

As mentioned earlier, a *deciduous* tree loses its leaves during the harsh season to slow metabolic activity. In the temperate deciduous forest, this occurs during the winter season, when sunlight is scarce and temperatures are too cold for the tree's nutrient circulation system. To make up for a period of winter hibernation, deciduous trees usually have large leaves to take full advantage of warm season photosynthesis; they then drop these leaves as winter approaches, often providing a colorful display as the leaves wither and die (Figure 2.32).

Common tree species in this temperate deciduous forest are maple, oak, elm, ash, and beech. The interior cellular structures of these broadleaf species make them difficult to mill for lumber; thus, they are grouped into the category of *hardwoods*. This category contrasts with the evergreen, needleleaf trees (discussed below), the *softwoods*, which are the preferred lumber species in most areas of the world.

Because the fallen tree leaves of the deciduous forest produce a rich soil, this bioregion has been cleared extensively for agriculture in North America and Europe, thus leaving only remnants in areas that were previously richly forested.

Evergreen Forests

In the colder winter D climate regions, where temperatures average below freezing for at least one month and precipitation is around 40 inches (100 cm), evergreen, needleleaf trees replace the deciduous broadleaf trees because they are better adapted to cold temperatures and low sunlight. In this bioregion, the dominant tree species are fir, pine, and spruce; however, several deciduous species such as the cold-loving aspen and alder are also found amongst the evergreens, thus providing splashes of autumnal color in the otherwise monotonous tracts of evergreen.

FIGURE 2.33 Evergreen Trees In contrast to broadleaf deciduous trees that lose their leaves and shut down during the harsh winter season, evergreen needle leaf trees are adapted to cold temperatures through their small needles (in contrast to the larger leaves of deciduous trees). In terms of evolution needle leaf trees are thought to be an older—even archaic—plant form contrasted to deciduous trees. This evergreen needle leaf forest is in the Austrian Alps.

In North America, evergreen, needleleaf forests are found north of the temperate deciduous forests, from New England to the upper Midwest and then west to the Pacific Coast (Figure 2.33). They are also found in all mountainous environments of western North America, from the Mexican border northward. In Canada and Alaska, these forests are called **boreal** forests, referring to their near arctic location. Evergreen forests dominate in continental and northern Europe, as well as in the mountains, and then form vast forest tracts eastward through cold-winter Asia, where the Russian word **taiga** is commonly applied.

In the needleleaf forests of western North America, the struggle between timber harvesting and environmental groups remains contentious. Timber interests argue that increased cutting is the only way to meet the high domestic demand for lumber and other wood products, whereas environmental groups are concerned about the protection of habitat for endangered species like the northern spotted owl. This conflict has led the government to close large tracts of evergreen forest to commercial logging (Figure 2.34). Further complicating the future of western forests are global market forces. Many Japanese and Chinese timber firms pay premium prices for logs cut from U.S. and Canadian forests, outbidding domestic firms for these scarce resources. Adding yet another element of controversy is that many of these trees are cut from public lands in both Canada and the United States. This aspect of globalization raises troublesome questions about whether North American public forests

FIGURE 2.32 New England Deciduous Forests As they prepare for a long, cold winter by dropping their leaves and hibernating during the harsh season, deciduous trees such as these in New Hampshire often put on a colorful display. The trees shut down their nutrient systems and starve their foliage.

hand, cutting this extensive Siberian forest would release massive amounts of CO_2 and methane into the atmosphere, increasing Russia's global warming emissions.

Tundra

The **tundra** bioregion has two versions: the expansive *arctic tundra* of the far northern hemisphere and, sharing many similar traits, the *alpine tundra* found at high elevations in mountainous areas worldwide. In both cases, the tundra landscape is primarily treeless because of a very short growing season lasting only a month or two. Moisture is also limited, with arctic tundra areas receiving the equivalent of only 6–10 inches (15–25 cm) of precipitation, which falls mostly as snow. As a result of these severe conditions, the tundra bioregions consist primarily of low shrubs, reindeer moss, sedge, and grasses (Figure 2.35). For a brief time in summer, a colorful flower display breaks the monotony of this otherwise bleak landscape.

Despite the low biotic productivity of the tundra bioregion, it stores vast amounts of methane (as do all boggy environments), which is being increasingly released into the atmosphere because of global warming. Because methane is 20 times more effective than CO_2 at trapping atmospheric heat, the thawing of the tundra is a very worrisome factor of the global warming picture.

REVIEW QUESTIONS

1. How and why do the three tropical bioregions differ?
2. What are the causes of tropical forest deforestation?
3. What is desertification?
4. Why are some trees deciduous?

FIGURE 2.34 Clear-Cut Forest Commercial logging in Washington's Olympic Peninsula has dramatically reshaped this landscape. Throughout the Pacific Northwest, environmental lobbies have successfully restricted logging to protect habitat for endangered species and recreation.

FIGURE 2.35 Tundra High in the mountains and closer to the poles where harsh winter seasons inhibit tree growth, the tundra dominates with grasses and shrubs adapted to very short summers and very long winters. This photo is taken north of the Arctic Circle in Norway.

should be cut to meet the lumber and wood-product needs of foreign countries (see "Exploring Global Connections: Can China Revive Pacific Northwest Lumber Towns?").

The Siberian *taiga* forest is a resource that may become an attractive source of income for Russia's hard-pressed economy. Some observers argue that, if lumber from the Siberian forests is put on the market for global trade, it will reduce logging pressure on North America's western forests, which could make it easier to enact and enforce comprehensive environmental protection in the United States and Canada. On the other

EXPLORING GLOBAL CONNECTIONS

Can China Revive Pacific Northwest Lumber Towns?

Despite the stagnating U.S. economy, things are looking a bit better in Longview, Washington, since China started buying Pacific Northwest forest products. In this logging town of 35,000 people, thousands of mill workers were laid off several years ago when the national housing market collapsed. But they are now back at work because of a complicated series of events that took place thousands of miles away from Longview. These interactions illustrate what geographers call the "local-to-global" aspects of today's globalized world.

Historically, the Pacific Northwest lumber economy has been tied to cycles of boom and bust, linked to the ups and downs of the U.S. housing construction market. More recently, however, international politics have played an increasingly important role as to whether mill towns in Oregon and Washington prosper or wilt. About a decade ago, for example, a timber war broke out between the United States and Canada over the fact that Canadian wood products were heavily subsidized by the Canadian government and, consequently, could be sold in U.S. markets at lower prices than domestic lumber. Although a temporary lull in those tensions has occurred, still today about a quarter of all lumber sold in the United States comes from Canada. Now Pacific Northwest mill towns are looking to

China for their economic salvation.

Until very recently, most of China's softwood lumber (lumber from coniferous trees) came from Russia's boreal forest, which is nearby and easily accessible. But this lumber was not always legal. Reportedly, half of this Russian lumber was cut and milled illegally by gangs of workers allegedly organized and controlled by the Russian mafia. Conservation groups in Europe were particularly outraged by this illegal logging for two reasons: first, because the environmental consequences are high due to habitat destruction and CO_2 release in the *taiga,* and second, because some of this illegal lumber found its way into European markets. This wood appeared not just as lumber, but also in finished wood furniture, which violated European Union guidelines on certifiable sustainable forestry. As a result, Europe put considerable pressure on both Russia and China to halt this illegal trafficking of wood products.

Simultaneously with this crackdown on illegal activities (which may or not be

FIGURE 2.2.1 Logs to Asia Sending logs and lumber from Washington State to Japan and China is a global connection that has helped Pacific Northwest logging towns survive the recent downturn in domestic markets due to the lack of new home constructions. These logs are being shipped to Japan from Port Olympia, Washington.

effective), Russia recently raised the taxes on legal exports of lumber, thus increasing the costs of lumber to China. Today, with its former supply of cheap softwood from Russia reduced and becoming more expensive, China has turned to the boreal forests of North America for its needs. Currently, only 1 percent of U.S. lumber is shipped to China (Canada exports about 10 times as much), but this new export market may become an important factor in reviving the Pacific Northwest's lumber economy (Figure 2.2.1). Whether this revival is just one more cycle in the Pacific Northwest's long history of boom and bust economics or becomes more sustainable remains to be seen.

Summary

- The arrangement of tectonic plates on Earth is responsible for diverse global landscapes, as well as earthquake and volcanic hazards that threaten the well-being of millions of people, particularly in the large cities of North America, Asia, and South America.

- Climate change and global warming, resulting from pollution of the atmosphere by greenhouse gases, is a by-product of industrialization, both past and present. A major facet of this problem is the burning of fossil fuels—petroleum and coal—which releases CO_2 into the atmosphere. Historically, the developed countries of Europe and North America were the major GHG producers; today, however, the developing economies of China and India have become major polluters as well, resulting in controversy and tensions that hinder international agreements on limiting global warming emissions.

- Water, a necessity for all life, is becoming an increasingly scarce resource in this world because of limited natural freshwater, the pollution of available supplies, and political and economic complexities that limit access to water.

- Plants and animals throughout the world face an extinction crisis because of habitat destruction from varied human activities. Tropical forests are a focus of these problems because they contain the most plant and animal species of any bioregion, yet they are threatened by numerous forces, including the world's demands for wood products, cattle ranching for beef production, and resettlement away from cities of fast-growing populations.

- Globalization is both a help and a hindrance to world environmental problems. As a positive force, globalization is central to sharing information and increasing public awareness of environmental issues. In addition, many people argue that globalization facilitates a new willingness of countries to work together under the umbrella of international agreements to resolve environmental problems. Such cooperation has led to international treaties on ocean pollution, global warming, and protection of wildlife species.

Key Terms

adiabatic lapse rate 51
anthropogenic 52
bioregion 57
boreal 62
carbon inequity 54
climograph 51
colliding plate boundary 44
continentality 50

deciduous 58
desertification 60
divergent plate boundary 46
global warming 52
greenhouse effect 49
insolation 49
maritime climate 50
orographic effect 51

plate tectonics 44
prairie 60
rain shadow 51
steppe 60
subduction zone 44
taiga 62
tundra 63
water stress 56

Thinking Geographically

1. Working in a small group, discuss what natural hazards your part of the country might experience. Then contact public agencies to see what sort of disaster preparation plans are in place to get local people ready for such an event and to help them after disaster strikes. Finally, evaluate those plans and put together a document with suggestions as to how they might be improved.

2. Write a short paper researching your local weather and climate. What climate region are you in? What are the major weather and climate problems faced in your area? How do people adjust and adapt to those problems? What are the major sources of global warming emissions in your area?

3. Working with a small group, put together a debate on the issue of carbon inequity—specifically, on whether developing countries should be held to the same emission reduction targets as developed countries.

4. Research a part of the world that has water stress problems, answering these questions: What is the climate of the area, and what is the natural availability of water? Which is used primarily, surface or groundwater? Then discuss in detail the causes of water stress. Is it because the supply is contaminated? Are there access problems? Is the water being monopolized by a powerful person, agency, or firm?

5. How has the vegetation in your area been changed by human activities in the past 100 years? Have these changes caused any plants or animals to become extinct or to be placed on the endangered species list? If so, what is being done to protect them or restore their habitat?

6. Study the globalization of wood products by acquainting yourself with the source areas for different items, such as building lumber, paper, furniture, or other items found in a local import store. More specifically, does the lumber used in construction in your area come from Canada or the United States? Which items in local stores come from tropical forests?

Mastering Geography™

Looking for additional review and test prep materials? Visit the Study Area in MasteringGeography™ to enhance your geographic literacy, spatial reasoning skills, and understanding of this chapter's content by accessing a variety of resources, including MapMaster interactive maps, videos, RSS feeds, flashcards, web links, self-study quizzes, and an eText version of *Globalization and Diversity*.

Scan to visit the author's blog for chapter updates.

http://gad4blog.
wordpress.com/
category/global-
environment/

Authors' Blogs

Scan now to access the authors' blogs for up-to-date information on The Changing Global Environment.

Scan to visit the GeoCurrents blog.

http://geocurrents.
info/category/
physical-geography

Globalization and Diversity

North America plays a pivotal role in economic globalization, both driving global change and participating in an increasingly interconnected international economy. The region's role as a destination for immigrants makes it one of the world's most diverse cultural settings, home to a relatively new amalgam of people from every corner of Earth.

ENVIRONMENTAL GEOGRAPHY

Stretching from Texas to the Yukon, the North American region is home to an enormously varied natural setting and to an environment that has been extensively modified by human settlement and economic development.

POPULATION AND SETTLEMENT

Settlement patterns in North American cities reflect the diverse needs of an affluent, highly mobile population. The region's sprawling suburbs are designed around automobile travel and mass consumption, while many traditional city centers struggle to redefine their role within the decentralized metropolis.

CULTURAL COHERENCE AND DIVERSITY

Cultural pluralism remains strong in North America. Currently, more than 46 million immigrants live in the region, more than double the total in 1990. The tremendous growth in Hispanic and Asian immigrants since 1970 has fundamentally reshaped the region's cultural geography.

GEOPOLITICAL FRAMEWORK

Cultural pluralism continues to shape political geographies in the region. Immigration policy remains hotly contested in the United States, and Canadians confront persistent regional and native peoples' rights issues.

ECONOMIC AND SOCIAL DEVELOPMENT

The geographic impacts of the recent economic crisis in North America were very uneven, but hit both the industrial Midwest and the Southwest particularly hard. Popular protests, exemplified by the Occupy Wall Street movement, highlighted growing income disparities between the region's rich and poor.

➤ **Wyoming's Black Thunder coal mine is one of the largest in the Powder River Basin and in all of North America.**

NORTH AMERICA 3

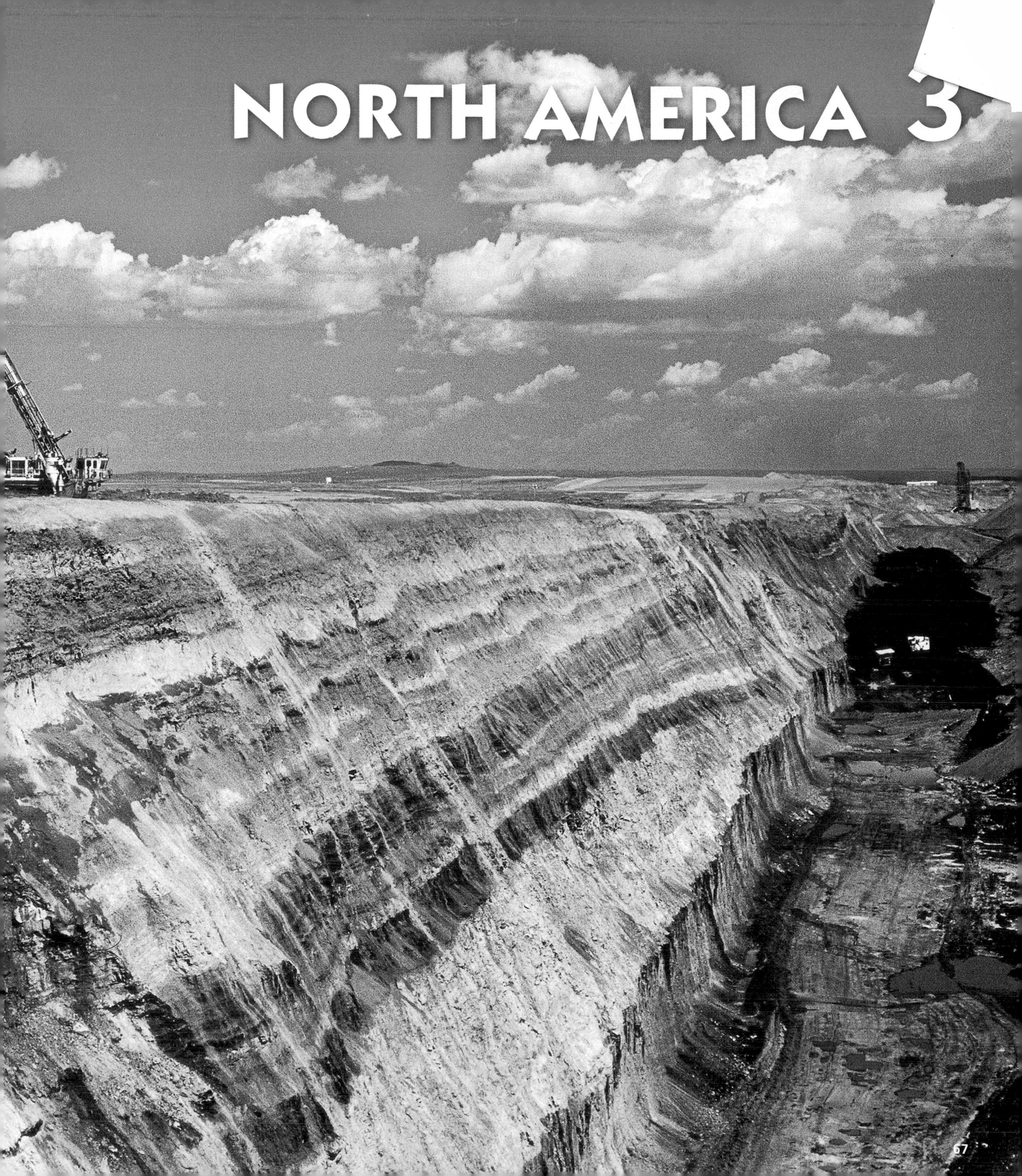

Wyoming's remote Powder River Basin seems an unlikely place to encounter globalization in North America. But as Earth's most productive coal mining area, the basin has certainly gained the attention of global economic interests. Global investors are gobbling up coal reserves. Coal consumption in both China and India continues to soar. Ambitious plans for large coal export terminals on the Pacific Coast are moving forward, focused on shipping Powder River coal west to Asia. Meanwhile, dozens of domestic coal trains leave the area daily, accounting for 40 percent of U.S. coal production. Small towns such as Gillette, Wyoming, try to manage the boom the best they can. Despite growing environmental objections (coal is a high-carbon fuel), the Powder River Basin's future seems firmly tied to growing global energy demands, mainly driven by the next generation of Asian consumers half a world away.

Globalization has fundamentally refashioned many other portions of North America. A walk down any busy street in Toronto, Tucson, or Toledo reveals how international products, foods, culture, and economic connections shape the everyday scene. Large foreign-born populations also provide direct links to every part of the world. Tourism brings in millions of additional foreign visitors and billions of dollars, which are spent everywhere from Las Vegas to Disney World. In more subtle ways, North Americans see globalization in their everyday lives. They eat ethnic foods, enjoy the sounds of salsa and Senegalese music, and surf the Internet from one continent to the next.

Globalization is a two-way street, and North American capital, culture, and power are ubiquitous. By any measure of multinational corporate investment and global trade, the region plays a role that far outweighs its population of 345 million residents. North American automobiles, consumer goods, information technology, and investment capital circle the globe. North American popular culture has also spread rapidly around the world.

North America includes the United States and Canada, a culturally diverse and resource-rich region that has seen tremendous human modification of its landscape and extraordinary economic development over the past two centuries (Figure 3.1). The result is that North America is one of the world's wealthiest regions, where two highly urbanized and mobile populations help drive the processes of globalization and have the highest rates of resource consumption on Earth.

Indeed, the region exemplifies a **postindustrial economy**, in which human geographies are shaped by modern technology, innovative information services, and a popular culture that dominates both North America and the world beyond.

Politically, North America is home to the United States, the last remaining global superpower. Such status brings the country onto center stage in times of global tensions, whether they are in the Middle East, South Asia, or West Africa. In addition, North America's largest metropolitan area, New York City (22 million people), is home to the United Nations and other global political and financial institutions. North of the United States, Canada is the other political unit within the region. Although slightly larger in area than the United States (3.83 million square miles [9.97 million square kilometers] versus 3.68 million square miles [9.36 million square kilometers]), Canada's population is only about 10 percent that of the United States.

The United States and Canada are commonly referred to as "North America," but that regional terminology can be confusing. Some geography textbooks call the region "Anglo America" because of its close and abiding connections with Britain and its Anglo-Saxon cultural traditions. The increasing cultural diversity of the region, however, has discouraged the use of this term in recent years. While more culturally neutral, the term North America also has problems. As a physical feature, the North American continent commonly includes Mexico, Central America, and often the Caribbean. Culturally, however, the United States–Mexico border seems a better dividing line, although the growing Hispanic presence in the southwestern United States, as well as ever-closer economic links across the border, makes even that regional division problematic. In addition, while Hawaii is a part of the United States (and thus included in Chapter 3), it is also considered a part of Oceania (and thus discussed in Chapter 14). Finally, Greenland (population 56,000), which often appears on the North American map, is actually an autonomous country within the Kingdom of Denmark and is mainly known for its valuable, but diminishing, ice cap. Our coverage of the "North American" region concentrates on Canada and the United States, two of the world's largest and most affluent nation-states.

Contemporary North America displays both the bounty and the price of economic development. On one hand, the region shares the benefits of modern agriculture, globally competitive industries,

LEARNING OBJECTIVES

After reading this chapter you should be able to:

- Identify key environmental issues facing North Americans in the 21st century and understand how these relate to the region's resource base and economic development.

- Explain the major ways in which people have modified the North American environment.

- Describe North America's major landform and climate regions.

- Summarize the three most important periods of European settlement in North America.

- Identify major migration flows in North American history.

- Explain the processes that shape contemporary urban and rural settlement patterns.

- List the five phases of immigration shaping North America and describe the recent importance of Hispanic and Asian immigration.

- Provide examples of how cultural globalization has shaped the region.

- Describe how the United States and Canada developed distinctive federal political systems and identify each nation's current political challenges.

- Indicate the role of key location factors in explaining why economic activities are located where they are in North America.

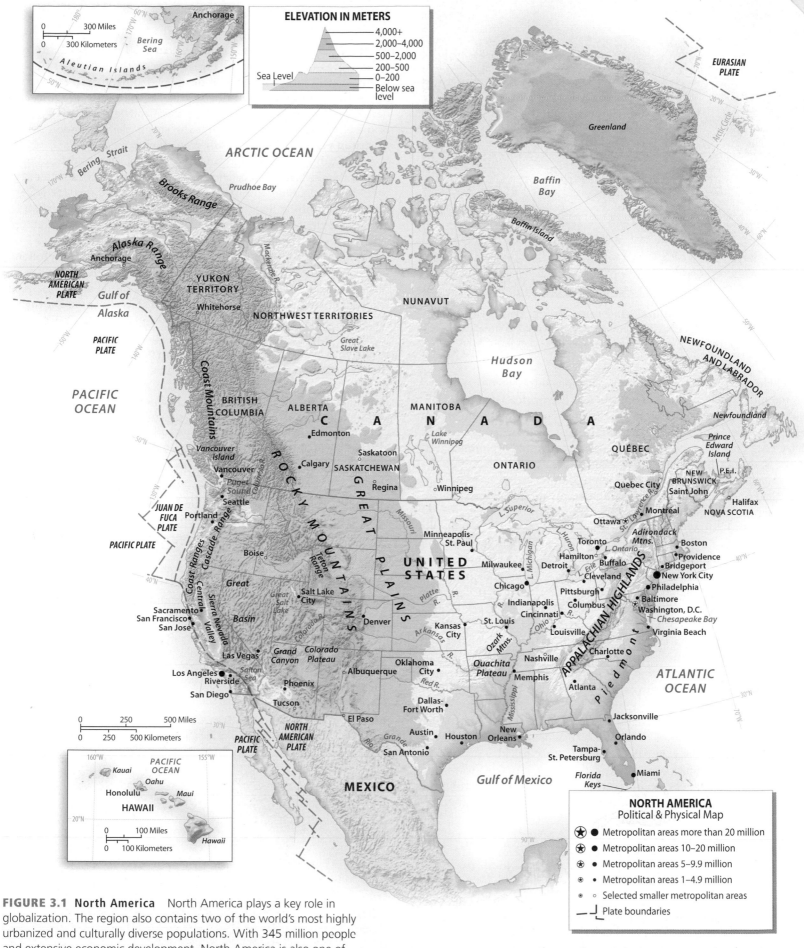

ELEVATION IN METERS

- 4,000+
- 2,000–4,000
- 500–2,000
- 200–500
- 0–200
- Sea Level
- Below sea level

FIGURE 3.1 North America North America plays a key role in globalization. The region also contains two of the world's most highly urbanized and culturally diverse populations. With 345 million people and extensive economic development, North America is also one of the largest consumers of natural resources on the planet.

NORTH AMERICA
Political & Physical Map

- ⊛ ● Metropolitan areas more than 20 million
- ⊛ ● Metropolitan areas 10–20 million
- ⊛ ● Metropolitan areas 5–9.9 million
- ⊛ ● Metropolitan areas 1–4.9 million
- ⊛ ○ Selected smaller metropolitan areas
- ⌐ Plate boundaries

FIGURE 3.2 Seattle's Cultural Landscape Practitioners of Falun Dafa, a Chinese meditation practice, begin their day in Seattle's culturally diverse International District, just southeast of downtown.

excellent transportation and communications infrastructure, and two of the most highly urbanized societies in the world. The cost of development, however, has been high: Native populations were all but eliminated by European settlers, forests have been logged, grasslands have been converted into farms, valuable soils have been eroded, numerous species have been made extinct, great rivers have been diverted, and natural resources have often been wasted.

Nevertheless, economic growth has vastly improved the standard of living for many North Americans, who enjoy high rates of consumption and varied amenities that are the envy of the less developed world. Wi-Fi, sushi, and shopping malls are within easy reach of most North Americans. Amid this material abundance, however, are continuing differences in income and quality of life. Poor rural and inner-city populations struggle to match the affluence of wealthier neighbors. Recent elections in 2012 saw the return of Barack Obama to a second term as President of the United States, but the Congress remained split between Democrats (who controlled the Senate) and Republicans (who controlled the House of Representatives), suggesting that no broad national consensus has emerged regarding key issues in the economy such as health care or taxation.

North America's cultural fabric also defines the region, including common processes of colonization, a heritage of Anglo dominance, and a shared set of civic beliefs in representative democracy and individual freedom. But the region's history has also juxtaposed Native Americans, Europeans, Africans, and Asians in fresh ways, and the results are two societies unlike any other (Figure 3.2). Adding to the mix is a popular culture that today exerts a powerful homogenizing influence on North American society.

ENVIRONMENTAL GEOGRAPHY: A THREATENED LAND OF PLENTY

North America's physical and human geographies are enormously diverse and intricately linked. Louisiana's lower Mississippi River valley exemplifies this close relationship between people and the land, having recently been the scene of some of the continent's most challenging environmental issues. Hurricane Katrina's tumultuous arrival in

August 2005 brought winds of more than 120 miles per hour. New Orleans was by far the largest urban area in the path of the storm. Unfortunately, city planners and developers downplayed the hazards of the city's low-lying, bowl-like setting. Aging levees were breached near Lake Pontchartrain and along several canals, putting 80 percent of the city under water. Years after the storm, the city's poor, African American neighborhoods are still struggling to recover.

In 2010, the Deepwater Horizon rig explosion and leaking oil well in the Gulf of Mexico delivered a different sort of disaster (Figure 3.3). The inability of British Petroleum (BP) and the U.S. government to control the flow of oil allowed tens of millions of gallons to escape into the Gulf, damaging sea and bird life, sensitive coastal ecosystems, and the region's resource and tourist economies.

The spring and summer of 2011 brought historically high floodwaters from the opposite direction as engineers tried to manage the Mississippi River's seasonal runoff. Much of the lower valley is already a humanized landscape of levees and ship channels. During the floods, huge portions of Louisiana were intentionally inundated to allow water through the system and to rebuild coastal mud deposits in precious wetland areas. Unfortunately, the freshwater devastated the region's commercial oyster beds. Louisiana's trials are a reminder of how the costs and impacts of both "natural" and "human" environmental disasters are inevitably intertwined with a region's broader cultural, social, and economic characteristics.

The Costs of Human Modification

Louisiana's story is also a reminder that North Americans have modified their physical setting in many ways. The processes of globalization and accelerated urban and economic growth have transformed North America's landforms, soils, vegetation, and climate. Indeed, problems such as acid rain, nuclear waste storage, groundwater depletion, and toxic chemical spills are manifestations of a way of life unimaginable only a century ago (Figure 3.4). Energy consumption in the region remains extremely high (the United States is still the source of almost 20 percent of Earth's greenhouse gas emissions), imposing a growing

FIGURE 3.3 Gulf Oil Spill North Americans witnessed the area's greatest environmental disaster in history in 2010 as millions of barrels of oil from a leaking well in the Gulf of Mexico damaged the region's sensitive ecosystems, fishing industry, and tourism economy. This view shows oil washing up on the beach at Gulf Shores, Alabama.

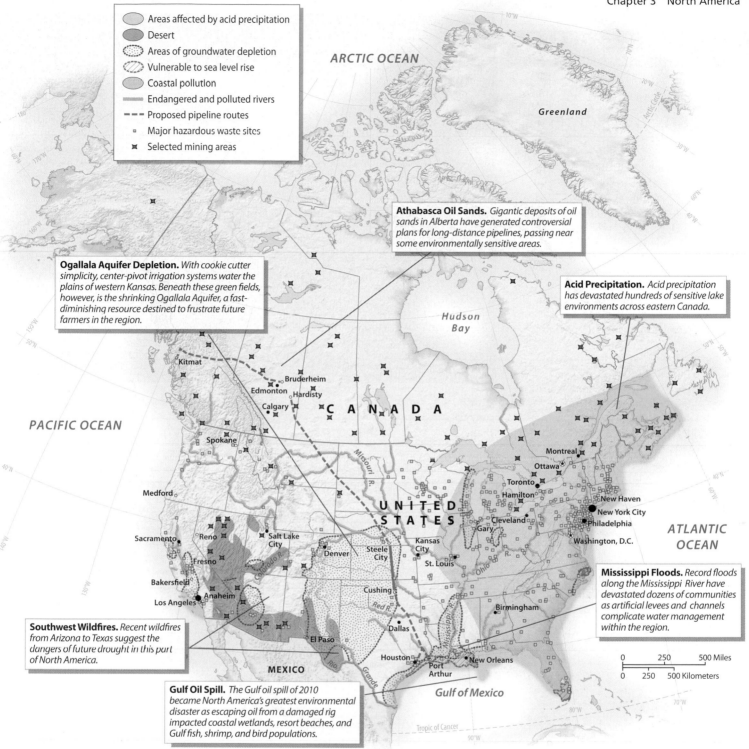

Athabasca Oil Sands. *Gigantic deposits of oil sands in Alberta have generated controversial plans for long-distance pipelines, passing near some environmentally sensitive areas.*

Ogallala Aquifer Depletion. *With cookie cutter simplicity, center-pivot irrigation systems water the plains of western Kansas. Beneath these green fields, however, is the shrinking Ogallala Aquifer, a fast-diminishing resource destined to frustrate future farmers in the region.*

Acid Precipitation. *Acid precipitation has devastated hundreds of sensitive lake environments across eastern Canada.*

Mississippi Floods. *Record floods along the Mississippi River have devastated dozens of communities as artificial levees and channels complicate water management within the region.*

Southwest Wildfires. *Recent wildfires from Arizona to Texas suggest the dangers of future drought in this part of North America.*

Gulf Oil Spill. *The Gulf oil spill of 2010 became North America's greatest environmental disaster as escaping oil from a damaged rig impacted coastal wetlands, resort beaches, and Gulf fish, shrimp, and bird populations.*

Legend:
- Areas affected by acid precipitation
- Desert
- Areas of groundwater depletion
- Vulnerable to sea level rise
- Coastal pollution
- Endangered and polluted rivers
- Proposed pipeline routes
- Major hazardous waste sites
- Selected mining areas

FIGURE 3.4 Environmental Issues in North America Many environmental issues threaten North America. Acid rain damage is widespread in regions downwind from industrial source areas. Elsewhere, air and water pollution presents health dangers and economic costs to residents of the region. Since 1970, however, both Americans and Canadians have become increasingly responsive to these environmental challenges.

list of environmental and economic costs both within North America and beyond. Controversial energy pipelines such as the Keystone XL project (Alberta to the Gulf of Mexico) and the Northern Gateway Pipeline (Alberta to the Pacific Coast), designed to tap into Canada's rich Athabasca oil sands, illustrate the ongoing tensions between increasing energy use and job creation and the potentially dangerous environmental consequences of moving fossil fuels across long distances (Figure 3.4). In addition, the growing use of hydraulic fracturing (or

fracking) drilling technologies (which inject a mix of water, sand, and chemicals underground in order to release natural gas) may lead to polluted groundwater and potentially hazardous environmental conditions for nearby residents.

Transforming Soils and Vegetation The arrival of Europeans in North America affected the region's flora and fauna as countless new species were introduced, including wheat, cattle, and horses. As

EXPLORING GLOBAL CONNECTIONS

North America's Neo-Asian Environment

North America has seen impressive recent gains in its Asian population, particularly in the West. However, an array of newly-arrived Asian plants and animals has had a much broader ecological impact across the larger regional landscape. Both good intentions and accidental introductions have transferred a growing array of exotic plants and animals to the North American continent. The environmental consequences of these new Asian plants and animals can be just as profound and enduring as the well-known Columbian Exchange between Europe and North America.

Take kudzu, one of the most famous examples of the global diffusion of an Asian plant (Figure 3.1.1). A native of Japan, kudzu (a member of the pea family) appeared in the United States in Philadelphia at the Japanese pavilion during the Centennial Exposition in 1876. Later, farmers in the South (it thrives in subtropical settings) planted the fast-growing vine to control soil erosion. Kudzu thrived, virtually taking over many settings. Today it is considered a major pest and invader, literally the star of an ecological horror movie that can envelop trees, barns, and utility poles in its grasp.

Flying carp are equally mobile migrants. Imported from East Asia by southern catfish farmers in the 1970s, carp removed algae from ponds in the lower Mississippi Valley. But

FIGURE 3.1.1 Kudzu Kudzu, a plant native to East Asia, has spread widely in the Southeast, often as an unwanted invasive plant.

many carp escaped during floods in the early 1990s. Twenty years later, carp have migrated northward and are edging toward the Great Lakes as they discover artificial ship canals that link the lakes with the Mississippi system. The Army Corps of Engineers has spent millions to slow an invasion that could decimate the Great Lakes fishing industry.

Or consider the carnage unfolding across the continent's population of northern white ash trees. Since 2002, the emerald ash borer, an insect pest native to China, has killed more than 60 million trees (many planted along streets and in yards) in a broad 15-state zone across the Midwest (Figure 3.1.2). The invasion ranks as the most destructive forest insect ever to invade the continent, and the death toll has already surpassed that of Dutch elm disease. The borer arrived accidentally, probably hidden in a wooden shipping pallet, but the continental consequences for the nation's 8 billion ash trees appear bleak.

the number of settlers increased, forest cover was removed from millions of acres. Grasslands were plowed under and replaced with grain and forage crops not native to the region. Widespread soil erosion was increased by unsustainable cropping and ranching practices, and many areas of the Great Plains and South suffered lasting damage. More recent globalization has also brought new varieties of plants and animals to North America, often radically transforming the region's environment (see "Exploring Global Connections: North America's Neo-Asian Environment").

Managing Water North Americans consume huge amounts of water. While conservation efforts and technology have slightly reduced per capita rates of water use over the past 25 years, city dwellers still use an average of more than 175 gallons daily. Metropolitan areas such as New York City struggle with outdated municipal water supply systems.

Beneath the Great Plains, the waters of the Ogallala Aquifer are being depleted. Center-pivot irrigation systems have steadily lowered water tables across much of the region by as much as 100 feet (30 meters) in the past 50 years. Farther west, California's complex system of water management is a reminder of that state's huge demands.

Water quality is also a major issue (Figure 3.4). North Americans are exposed to water pollution every day, and even environmental laws and guidelines, such as the U.S. Clean Water Act and Canada's Green Plan, do not eliminate the problem. Mining operations and industrial users—such as chemical, paper, and steel plants—generate toxic wastewater or metals that enter surface water and groundwater supplies. America's most toxic places include the petroleum-rich Texas and Louisiana Gulf coasts, older industrial centers of the Northeast and Midwest, and nuclear fuel and chemical warfare storage areas such as Hanford, Washington.

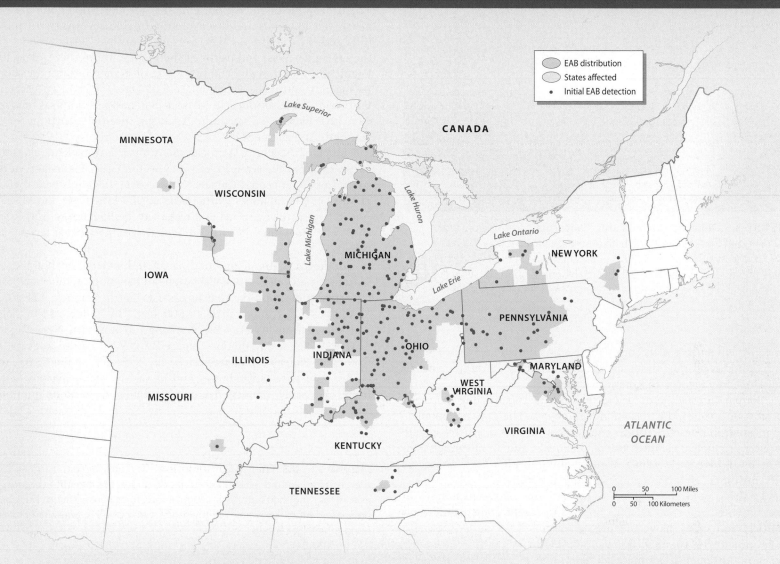

FIGURE 3.1.2 Spread of the Emerald Ash Borer, 2011 First discovered in 2002 near Detroit, the ash borer, native to China, has rapidly diffused through much of the Middle West.

Altering the Atmosphere North Americans modify the very air they breathe; in doing so, they change local and regional climates, as well as the composition of the atmosphere. For example, built-up metropolitan areas create an **urban heat island** effect, in which development associated with cities often produces nighttime temperatures some 9 to 14°F (5 to 8°C) warmer than those of nearby rural areas. At the local level, industries, utilities, and automobiles contribute carbon monoxide, sulfur, nitrogen oxides, hydrocarbons, and particulates to the urban atmosphere. While some of the region's worst offenders are U.S. cities such as Houston and Los Angeles, Canadian cities such as Toronto, Hamilton, and Edmonton experience significant air-quality problems.

On a broader scale, North America is plagued by **acid rain**—industrially produced sulfur dioxide and nitrogen oxides in the atmosphere that damage forests, poison lakes, and kill fish. Many acid rain producers are located in the Midwest and southern Ontario, where industrial plants, power-generating facilities, and motor vehicles contribute atmospheric pollution. Prevailing winds transport pollutants and deposit damaging acid rain and snow across the Ohio Valley, Northeast, and eastern Canada (Figure 3.4).

Air pollution is also going global, drifting into Canada and the United States on prevailing winds. One estimate suggests at least 30 percent of the region's ozone comes from beyond its borders. Both China and Mexico are major contributors of airborne pollutants.

Growing Environmental Awareness

Many environmental initiatives in the United States and Canada have addressed local and regional problems. For example, the improved water quality of the Great Lakes over the past 30 years is

FIGURE 3.5 Tehachapi Wind Farm, California Joshua trees and spinning turbines mingle in the high desert at the Tehachapi Wind Farm west of Mojave. The area's 5,000 turbines generate enough electricity to serve the residential needs of 350,000 Californians annually. Wind energy represents a growing percentage of North America's energy budget.

an achievement to which both nations contributed and that benefits both. Tougher air-quality standards have also reduced certain types of emissions in many North American cities.

Perhaps most important, North America is providing increased support for green industries and technologies. The growing popularity of **sustainable agriculture** exemplifies the trend, where organic farming principles, a limited use of chemicals, and an integrated plan of crop and livestock management combine to offer both producers and consumers environmentally friendly alternatives.

In the United States alone, Americans have invested billions in alternative energy sources since 2006, and recent court rulings and political initiatives have encouraged more environmentally friendly innovation (Figure 3.5). Although fossil fuels will continue to dominate U.S. energy consumption in the early 21st century, the growing technological and economic appeal of **renewable energy sources**—such as hydroelectric, solar, wind, and geothermal—are likely to fundamentally rework North America's economic geography in coming years. Policymakers, industrial innovators, and consumers are becoming increasingly attracted to their enduring availability and lower environmental impacts (Figure 3.6).

A Diverse Physical Setting

The North American landscape is dominated by vast interior lowlands bordered by more mountainous topography in the western portion of the region (see Figure 3.1). In the eastern United States, extensive coastal plains stretch from southern New York to Texas and include a sizable portion of the

lower Mississippi Valley. The Atlantic coastline is complex, made up of drowned river valleys, bays, swamps, and low barrier islands. The nearby Piedmont region, which is the transition zone between nearly flat lowlands and steep mountain slopes, consists of rolling hills and low mountains that are older and less easily eroded than the lowlands. West and north of the Piedmont are the Appalachian Highlands, an internally complex zone of higher and rougher country reaching altitudes from 3,000 to 6,000 feet (900 to 1,830 meters). Farther to the southwest, Missouri's Ozark Mountains and the Ouachita Plateau of northern Arkansas resemble portions of the southern Appalachians.

Much of the North American interior is a vast lowland, extending east–west from the Ohio River valley to the Great Plains and north–south from west central Canada to the coastal lowlands near the Gulf of Mexico. Glacial forces, particularly north of the Ohio and Missouri rivers, have actively carved and reshaped the landscapes of this lowland zone, including the environmentally complex Great Lakes Basin (Figure 3.7).

In the West, mountain building (including large earthquakes and volcanic eruptions), alpine glaciation, and erosion produce a regional topography quite unlike that of eastern North America. The Rocky Mountains reach more than 10,000 feet (3,050 meters) in height and stretch from Alaska's Brooks Range to northern New Mexico's Sangre de Cristo Mountains (Figure 3.8). West of the Rockies, the Colorado Plateau is characterized by colorful sedimentary rock eroded into spectacular buttes and mesas. Nevada's sparsely settled basin and range country features north–south-trending mountain

FIGURE 3.6 U.S. Energy Consumption The growing popularity of fossil fuels is evident in U.S. energy consumption during the late 19th century as coal, oil, and then natural gas supplanted wood consumption. Nuclear power and other renewable energy sources are poised to play a larger role during the 21st century.

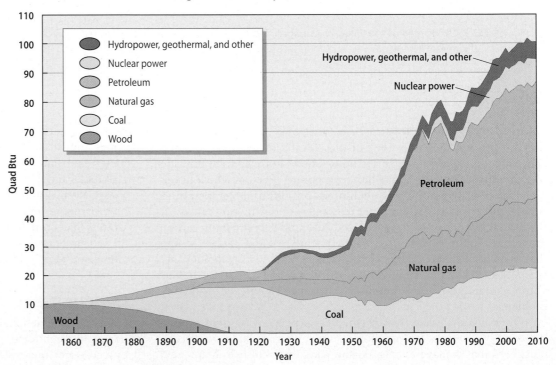

Patterns of Climate and Vegetation

North America's climates and vegetation are diverse, mainly due to the region's size, latitudinal range, and varied terrain (Figure 3.9). As the climographs for Dallas, Texas, and Columbus, Ohio, suggest, much of North America south of the Great Lakes and east of the Rocky Mountains is characterized by a long growing season, 30 to 60 inches (75 to 150 centimeters) of precipitation annually, and a deciduous broadleaf forest (later cut down and replaced by crops). From the Great Lakes north, the coniferous evergreen forest, or **boreal forest**, dominates the continental interior. Near Hudson Bay and across harsher northern tracts, trees give way to **tundra**, a mixture of low shrubs, grasses, and flowering herbs that grow briefly in the short growing seasons of the high latitudes. Drier continental climates found from west Texas to Alberta feature large seasonal ranges in temperature and unpredictable precipitation, averaging between 10 and 30 inches (25 and 75 centimeters) annually. The soils of much of this region are fertile and originally supported **prairie** vegetation, dominated by tall grasslands in the East and by short grasses and scrub vegetation in the West.

Western North American climates and vegetation are greatly complicated by the region's mountain ranges. The Rocky Mountains and the intermontane interior experience the typical seasonal variations of the middle latitudes, but patterns of climate and vegetation are greatly modified by the effects of topography. Many arid interior settings lie in the dry rain shadow of the Cascade Range and Sierra Nevada. Farther west, marine West Coast climates dominate north of San Francisco, while a dry-summer Mediterranean climate occurs across central and southern California. Contrast the temperature and precipitation patterns for Los Angeles and Vancouver in Figure 3.9.

FIGURE 3.7 Satellite Image of the Great Lakes North America's Great Lakes region features one of the most environmentally complex political boundaries in the world. Both Canada and the United States share responsibility (at a variety of local, state/provincial, and federal levels) for managing the ecological health of the five Great Lakes (W to E: Superior, Michigan, Huron, Erie, and Ontario).

ranges alternating with structural basins with no outlet to the sea. North America's western border is marked by the mountainous and rain-drenched coasts of southeast Alaska and British Columbia; the Coast Ranges of Washington, Oregon, and California; the lowlands of Puget Sound (Washington), Willamette Valley (Oregon), and Central Valley (California); and the complex uplifts of the Cascade Range and Sierra Nevada.

Climate Change and North America

Global warming has already profoundly reshaped North America. High-latitude and alpine environments are particularly vulnerable to global warming. Changes in arctic temperatures, sea ice, and sea levels have increased coastal erosion, affected migrating whale and polar

FIGURE 3.8 Rocky Mountains Montana's Glacier National Park reveals the characteristic signatures of alpine glaciation found in many portions of the Rocky Mountain region, in both the United States and Canada.

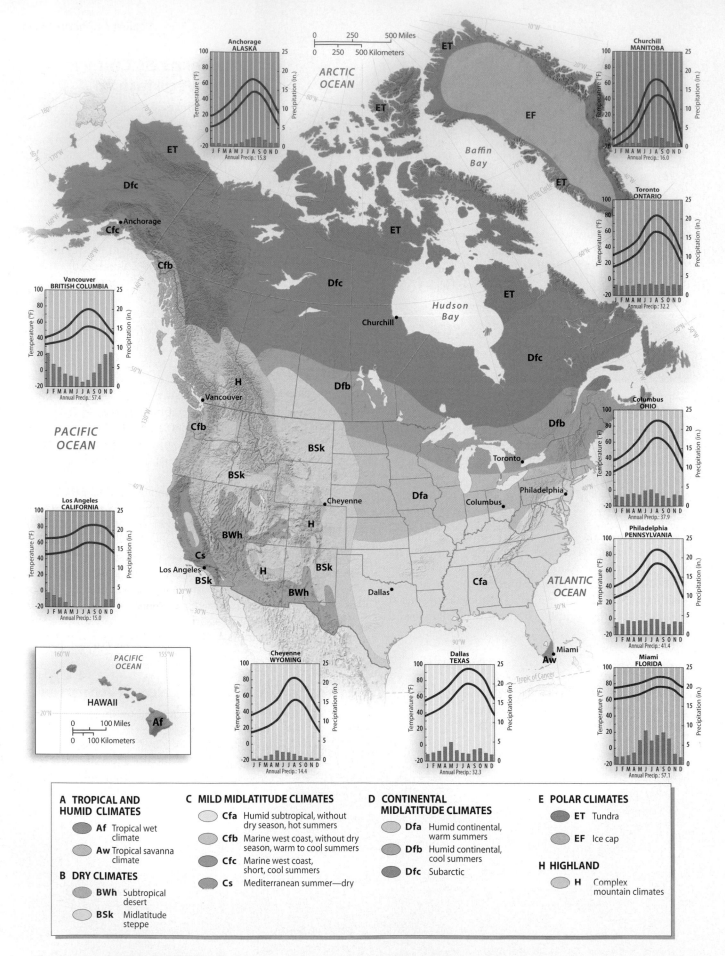

FIGURE 3.9 Climate Map of North America North American climates include everything from tropical savanna (Aw) to tundra (ET) environments. Most of the region's best farmland and densest settlements lie in the mild (C) or continental (D) midlatitude climate zones.

A TROPICAL AND HUMID CLIMATES

- **Af** Tropical wet climate
- **Aw** Tropical savanna climate

B DRY CLIMATES

- **BWh** Subtropical desert
- **BSk** Midlatitude steppe

C MILD MIDLATITUDE CLIMATES

- **Cfa** Humid subtropical, without dry season, hot summers
- **Cfb** Marine west coast, without dry season, warm to cool summers
- **Cfc** Marine west coast, short, cool summers
- **Cs** Mediterranean summer—dry

D CONTINENTAL MIDLATITUDE CLIMATES

- **Dfa** Humid continental, warm summers
- **Dfb** Humid continental, cool summers
- **Dfc** Subarctic

E POLAR CLIMATES

- **ET** Tundra
- **EF** Ice cap

H HIGHLAND

- **H** Complex mountain climates

bear populations, and stressed traditional ways of life for Eskimo and Inuit populations. In western North American mountains, expanding mountain pine beetle populations are rapidly infesting lodgepole and whitebark pine forests. A 2009 study (SCIENCE 323, 23 January, pp. 521–24) of 76 stands of old-growth forest throughout western North America confirmed that tree mortality rates rose dramatically in the past 15 years and cited higher temperatures as the primary culprit.

The cumulative long-term consequences of climate change for North Americans are enormous. Many coastal localities, especially low-lying zones in the arctic, along the East Coast, and in the Gulf of Mexico, are vulnerable to rising sea levels. Basic redistributions of plants, animals, and crops are already under way. For example, high-latitude distributions of tundra and permafrost environments may shift more than 300 miles (480 kilometers) poleward by 2050, dramatically impacting wildlife populations and human settlements in these regions. Many of North America's spectacular alpine glaciers are rapidly disappearing, often with important implications for downstream fisheries and populations (Figure 3.10). For northern U.S. and Canadian farmers, longer growing seasons may open new possibilities for agriculture, but weeds and plant pathogens will migrate north with the crops. The Great Lakes region may become wetter by the end of the century. On the other hand, recent droughts and wildfires in the Southwest and Texas may portend similar conditions in the years ahead (see Figure 3.4).

Domestic politics have shaped North American responses to global warming. The 1997 Kyoto Protocol, designed to reduce greenhouse gas emissions among industrialized nations, was never approved by the United States, and Canada withdrew from the agreement late in 2011. While environmentalists in both countries complain about the failure of their leaders to address the issue, conservative political interests argue that such agreements limit national sovereignty and future economic growth in North America, while not effectively addressing the issue in settings such as China and India.

FIGURE 3.10 Glaciers in Retreat   With prospects for further global warming, the outlook is bleak for many of North America's alpine glaciers. This pair of images records changes at Grinnell Glacier, in Montana's Glacier National Park, between 1940 and 2006.

REVIEW QUESTIONS

1. Describe North America's major landform regions, and suggest ways in which the region's physical setting has shaped patterns of human settlement.

2. Identify key ways in which humans have transformed the North American environment since 1600.

POPULATION AND SETTLEMENT: RESHAPING A CONTINENTAL LANDSCAPE

The present-day North American landscape is the product of four centuries of extraordinary human change. During that period, Europeans, Africans, and Asians arrived in the region, disrupted Native American peoples, and created new patterns of human settlement. Today more than 345 million people live in the region, and they are some of the world's most affluent and highly mobile populations (Table 3.1).

Table 3.1   POPULATION INDICATORS

Country	Population (millions) 2012	Population Density (per square kilometer)	Rate of Natural Increase (RNI)	Total Fertility Rate	Percent Urban	Percent <15	Percent >65	Net Migration (Rate per 1000) 2010–15[a]
Canada	34.9	3	0.4	1.7	80	16	14	5.6
United States	313.9	33	0.5	1.9	79	20	13	3.1

[a]Net Migration Rate from the United Nations, Population Division, *World Population Prospects: The 2010 Revision Population Database.*
Source: Population Reference Bureau, *World Population Data Sheet,* 2012.

Modern Spatial and Demographic Patterns

Major decennial censuses in the United States (April 2010) and Canada (May 2011) gathered an unprecedented amount of information on regional household size, age, ethnicity, and a variety of other social and economic characteristics. Statistics Canada (**www.statcan.gc.ca**) and the U.S. Census Bureau (**www.census.gov**) are two of the world's largest, most efficient data-gathering organizations.

These data reflect how large metropolitan areas (including both central cities and suburbs) dominate North America's population geography, producing uneven regional patterns of settlement (Figure 3.11). Canada's "Main Street" corridor contains most of that nation's urban population, led by Toronto (5.7 million) and Montreal (3.9 million). **Megalopolis**, the largest settlement cluster in the United States, includes Baltimore/Washington, DC (8.6 million), Philadelphia (6.5 million), New York City (22 million), and Boston (7.6 million) (Figure 3.12). Beyond these two core areas, other sprawling urban centers cluster around the southern Great Lakes (Chicago, 9.7 million), in various parts of the South (Dallas, 7.4 million), and along the Pacific Coast (Los Angeles, 17.9 million; Vancouver, 2.4 million).

North America's population has increased greatly since European colonization. Before 1900, high rates of natural increase produced large families and immigrants swelled settlement. In Canada,

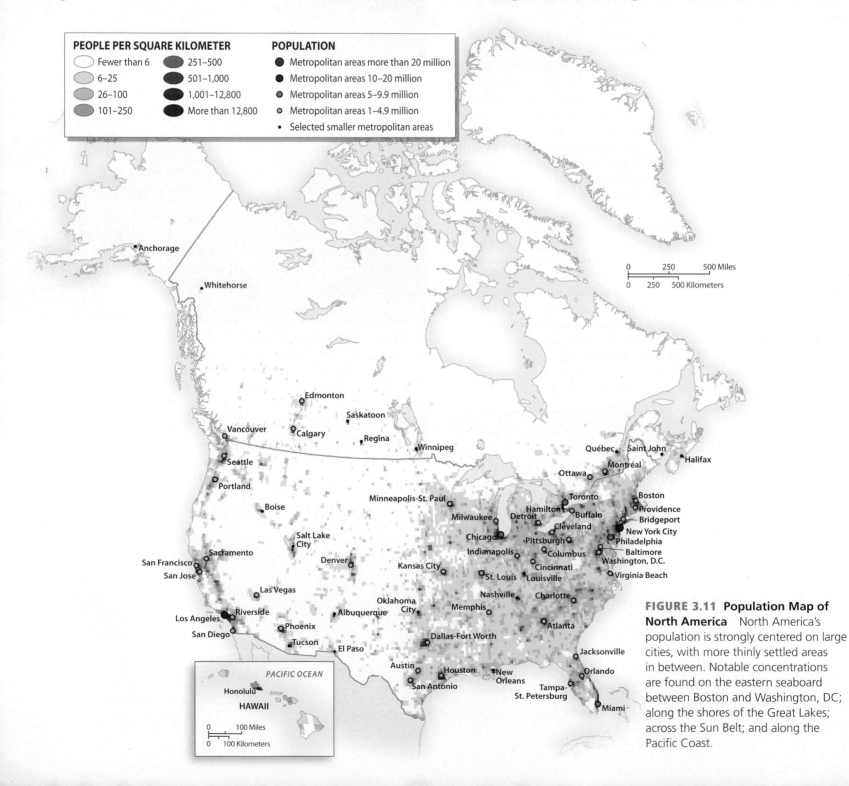

FIGURE 3.11 Population Map of North America North America's population is strongly centered on large cities, with more thinly settled areas in between. Notable concentrations are found on the eastern seaboard between Boston and Washington, DC; along the shores of the Great Lakes; across the Sun Belt; and along the Pacific Coast.

FIGURE 3.12 New York Skyline New York City's mid-town Manhattan remains one of North America's most dramatic urban skylines.

for at least 12,000 years by people as culturally diverse as the Europeans who came to conquer them. Native Americans were broadly distributed across the region and adapted to its many natural environments. Cultural geographers estimate Native American populations in 1500 C.E. at 3.2 million for the continental United States and another 1.2 million for Canada, Alaska, Hawaii, and Greenland. In many areas, European diseases, wars, and economic disruptions reduced these Native American populations by more than 90 percent as contacts increased.

The first stage of European settlement created a series of colonies, mostly in the coastal regions of eastern North America (Figure 3.14). Established between 1600 and 1750, these regionally distinct societies were anchored in the north by the French settlement of the St. Lawrence Valley and extended south along the Atlantic Coast, including several separate English colonies. Scattered developments along the Gulf Coast and in the Southwest also appeared before 1750.

The second stage in the Europeanization of North America occurred between 1750 and 1850, highlighted by settlement of better agricultural lands in the eastern half of the continent. Pioneers surged westward across the Appalachians following the American Revolution (1776) and a series of Indian conflicts. They found the Interior Lowlands region almost ideal for agricultural settlement. Southern Ontario, or Upper Canada, was also opened to development after 1791.

The third stage in North America's settlement accelerated after 1850 and continued until just after 1910. During this period, most of the region's remaining agricultural lands were settled by a mix of native-born and immigrant farmers. Farmers were challenged by drought, mountains, and short growing seasons. In the American West, settlers were attracted by opportunities in California, Oregon, Utah, and the Great Plains. In Canada, thousands occupied southern portions of Manitoba, Saskatchewan, and Alberta. Gold and silver discoveries led to development in areas such as Colorado, Montana, and British Columbia's Fraser Valley.

a population of fewer than 300,000 Native Americans and Europeans in the 1760s grew to an impressive 3.2 million a century later. For the United States, a late colonial (1770) total of around 2.5 million increased over 10-fold to more than 30 million by 1860. Both countries saw even higher rates of immigration in the late 19th and 20th centuries, although birthrates gradually fell after 1900. After World War II, birthrates rose once again in both countries, resulting in the "baby boom" generation born between 1946 and 1965. Today, however, rates of natural increase in North America are below 1 percent annually, and the overall population is growing older, particularly in states such as Iowa (Figure 3.13). Still, the region attracts immigrants. These growing numbers, along with higher birthrates among immigrant populations (exemplified by Texas), recently led experts to increase long-term population projections (Figure 3.13). Predictions by the U.S. Census Bureau that by 2050 the region's population will reach 464 million (423 million in the United States and 41 million in Canada) may prove conservative.

Occupying the Land

Europeans began occupying North America about 400 years ago. They were not settling an empty land; North America was populated

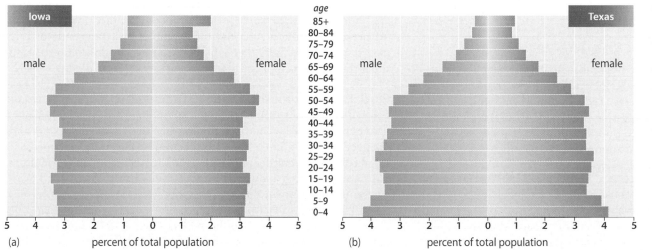

FIGURE 3.13 Population Pyramids (2010): Iowa (a) and Texas (b) Iowa's aging population stands in contrast to the larger proportion of young people in Texas, reflecting that state's higher birthrate and sizable influx of young immigrants.

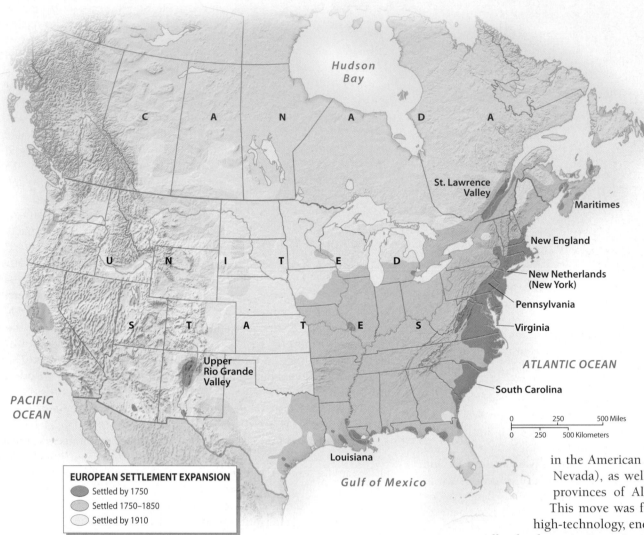

FIGURE 3.14 **European Settlement Expansion** Large portions of North America's East Coast and the St. Lawrence Valley were occupied by Europeans before 1750. The most remarkable surge of settlement, however, occurred during the next century as vast areas of land were opened to European dominance and Native American populations were exterminated or expelled from their former homelands.

EUROPEAN SETTLEMENT EXPANSION
- Settled by 1750
- Settled 1750–1850
- Settled by 1910

Incredibly, in a mere 160 years, much of the North American landscape was occupied as expanding populations sought new land to settle and as the global economy demanded resources to fuel its growth. It was one of the largest and most rapid transformations of the landscape in the history of the human population. This European-led advance forever reshaped North America in its own image. In the process, it also changed the larger globe in lasting ways by creating a "New World" destined to reshape the old.

North Americans on the Move

From the legendary days of Davy Crockett and Daniel Boone to the 20th-century sojourns of John Steinbeck and Jack Kerouac, North Americans have been on the move. Indeed, almost one in every five Americans moves annually, suggesting that residents of the region are quite willing to change residence in order to improve their income or their quality of life. Several trends dominate the picture.

Westward-Moving Populations The most persistent regional migration trend in North America has been the tendency for people to move west. By 1990, more than half of the population of the United States lived west of the Mississippi River, a dramatic shift from colonial times. Since 1990, some of the fastest-growing states have been

in the American West (including Arizona and Nevada), as well as in the western Canadian provinces of Alberta and British Columbia. This move was fueled by new job creation in high-technology, energy, and service industries, as well as by the region's scenic, recreational, and retirement attractions. With new population, demands for water in this arid region continue to grow. Also, the area's increasing political and economic power is redefining its place in national affairs. The 2010 census data paved the way for the West to gain power in the House of Representatives and play a larger role in national elections.

Recently, however, the economic slowdown hit the region hard (Figure 3.15). The contraction of construction and leisure-time industries in settings such as Las Vegas, Phoenix, and Tucson slowed growth in these metropolitan areas to their lowest rates in decades, and home values in some neighborhoods declined by more than 50 percent. Southern Arizona, for example, has thousands of foreclosed houses and more than a half million lots approved for development that remain empty.

Black Exodus from the South African Americans have generated distinctive patterns of interregional migration. Most blacks remained economically tied to the rural South after the Civil War. Conditions changed, however, in the early 20th century. Many African Americans migrated because of declining demands for labor in the agricultural South and growing industrial opportunities in the North and West. Migrants ended up in cities where jobs were located. Boston, New York, Philadelphia, Detroit, Chicago, Los Angeles, and Oakland became key destinations for southern blacks. Since 1970, however, more blacks have moved from North to South. Sun Belt jobs and federal

FIGURE 3.15 Unfinished Subdivision, near Tucson, Arizona Many homes in this attractive subdivision west of Tucson remain unsold or unfinished amid the housing crisis, which hit the Southwest particularly hard.

Settlement Geographies: The Decentralized Metropolis

Settlement landscapes of North American cities are characterized by **urban decentralization**, in which metropolitan areas sprawl in all directions and suburbs take on many of the characteristics of traditional downtowns. Although both Canadian and U.S. cities have experienced decentralization, the impact has been particularly profound in the United States, where inner-city problems, poor public transportation, widespread automobile ownership, and fewer regional-scale planning initiatives have encouraged middle-class urban residents to move beyond the central city.

Historical Evolution of the City in the United States Changing transportation technologies decisively shaped the evolution of the city in the United States (Figure 3.16). The pedestrian/horsecar city (pre-1888) was compact, essentially limiting urban growth to a 3- or 4-mile-diameter ring around downtown. The invention of the electric trolley in 1888 expanded the urbanized landscape farther into new "streetcar suburbs," often 5 or 10 miles from the city center. A star-shaped urban pattern resulted, with growth extending outward along and near the streetcar lines. The biggest technological revolution came after 1920, with the mass production of cars. The automobile city (1920–1945) continued the expansion of middle-class suburbs. Following World War II, growth in the outer city (1945 to the present) promoted more decentralized settlement along commuter routes as built-up areas appeared 40 to 60 miles from downtown.

Urban decentralization also reconfigured land-use patterns, producing metropolitan areas today that are strikingly different from

civil rights guarantees now attract many northern urban blacks to growing southern cities. The net result is still a major change from 1900: In that year, more than 90 percent of African Americans lived in the South, while today only about 55 percent of the nation's 42 million blacks reside within the region.

Rural-to-Urban Migration Another continuing trend in North American migration (and a growing global phenomenon) has taken people from the country to the city. Two centuries ago only 5 percent of North Americans lived in urban areas (cities of more than 2,500 people), whereas today about 80 percent of the North American population is urban. Shifting economic opportunities account for much of the transformation: As mechanization on the farm reduced the demand for labor, many young people left for new employment opportunities in the city.

Growth of the Sun Belt South Particularly since the 1970s, southern states from the Carolinas to Texas have grown much more rapidly than states in the Northeast and Midwest. During the 1990s, Georgia, Florida, Texas, and North Carolina each grew by more than 20 percent. The South's expanding economy, modest living costs, adoption of air conditioning, attractive recreational opportunities, and appeal to snow-weary retirees have all contributed to its growth. Dallas–Fort Worth (23%), Houston (26%), and Atlanta (24%) enjoyed some of the nation's fastest metropolitan growth rates between 2000 and 2010.

Nonmetropolitan Growth During the 1970s, some areas in North America beyond its large cities witnessed significant population gains, including rural settings that had previously lost population. Selectively, this pattern of **nonmetropolitan growth**—in which people leave large cities and move to smaller towns and rural areas—continues today. Some participants in the process are part of the growing retiree population in both Canada and the United States, but a substantial number are younger, so-called *lifestyle migrants*. Given our electronically connected world, they find or create employment in affordable smaller cities and rural settings rich in amenities and often removed from the perceived problems of urban America.

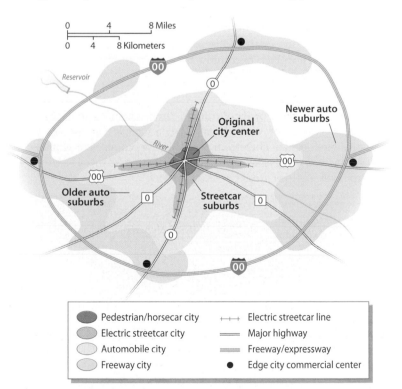

FIGURE 3.16 Growth of the American City Many U.S. cities became increasingly decentralized as they moved through the pedestrian/horsecar, electric streetcar, automobile, and freeway eras. Each era left a distinctive mark on metropolitan America, including the recent growth of edge cities on the urban periphery.

their counterparts of the early 20th century. In the city of 1920, urban land uses were generally organized in rings around a highly focused central business district (CBD) that contained much of the city's retailing and office functions. Residential districts beyond the CBD were added as the city expanded, with higher-income groups seeking more desirable locations on the outside edge of the urbanized area.

The modern pattern has shifted. Today's suburbs feature a mix of peripheral retailing (such as commercial strips, shopping malls, and big-box stores), industrial parks, office complexes, and entertainment facilities. These nodes of activity, called **edge cities**, have fewer functional connections with the central city than they have with other suburban centers. Tysons Corner, Virginia, located west of Washington, DC, is an excellent example of the edge-city landscape on the expanding periphery of a North American metropolis (Figure 3.17).

The Consequences of Sprawl

As suburbanization increased in the 1960s and 1970s, many inner cities, especially in the Northeast and Midwest, suffered absolute losses in population, increased levels of crime and social disruption, and a shrinking tax base. Poverty rates average almost three times those of nearby suburbs. Unemployment rates remain above the national average. Central cities in the United States also remain places of racial tension, the product of decades of discrimination, segregation, and poverty.

Even with these challenges, inner-city landscapes enjoy selective improvement. Referred to as **gentrification**, the process involves the displacement of lower-income residents of central-city neighborhoods by higher-income residents, the improvement of deteriorated inner-city landscapes, and the construction of shopping complexes, sports and entertainment attractions, and convention centers in selected downtown locations. Seattle's Pioneer Square, Toronto's Yorkville, and Baltimore's Harborplace exemplify how such public and private investments shape the city.

FIGURE 3.18 Pittsburgh's SouthSide Works Neighborhood New investments have transformed Pittsburgh's SouthSide Works Neighborhood into an upscale office, shopping, and entertainment district.

Many city planners and developers involved in such efforts advocate **new urbanism**, an urban design movement stressing higher-density, mixed-use, pedestrian-scaled neighborhoods where residents can walk to work, school, and entertainment. Pittsburgh's urban renaissance offers an affordable housing market, an older and highly skilled workforce, and mixed-use neighborhoods such as the SouthSide Works development, a 34-acre assortment of residences, offices, and stores on the site of an old steel plant (Figure 3.18). Urban rooftops have even been incorporated into this re-imagined, more sustainable version of the North American city (see "Working Toward Sustainability: Urban Rooftops Go Green").

The suburbs are also changing. Construction of corporate office centers, fashion malls, and industrial facilities has created true "suburban downtowns" in suburbs that are no longer simply bedroom communities for central-city workers. Indeed, such localities have been growing players in the continent's globalization process. For example, many of North America's key internationally connected corporate offices (IBM, Microsoft), industrial facilities (Boeing, Cisco, Oracle), and entertainment complexes (Disney World, Universal Studios, and the Las Vegas Strip) are located in such settings, intimately tied to global information, technology, capital, and migration flows.

Settlement Geographies: Rural North America

Rural North American cultural landscapes trace their origins to early European settlement. Over time, these immigrants from Europe showed a clear preference for a dispersed rural settlement pattern as they created new farms on the North American landscape. In portions of the United States settled after 1785, the federal government surveyed and sold much of the rural landscape. Surveys were organized around the simple, rectangular pattern of the federal government's township-and-range survey system (Canada's system is similar), which offered a convenient method of dividing and selling the public domain in 6-mile-square townships (Figure 3.19).

FIGURE 3.17 Tysons Corner, Virginia North America's edge-city landscape is nicely illustrated by Tysons Corner, Virginia. Far from a traditional metropolitan downtown, this sprawling complex of suburban offices and commercial activities reveals how and where many North Americans will live their lives in the 21st century.

Urban Rooftops Go Green

Take a stroll in one of New York City's newest parks or explore the top of Chicago's City Hall. You may find yourself surrounded by a rooftop verdure of grass, trees, and shrubs. The budding interest in rooftop parks is all part of a larger move on the part of urban planners, architects, and local neighborhoods to make their everyday environments more livable and energy efficient. While these pioneering patches of metropolitan greenery make up only a small percentage of the built

FIGURE 3.2.1 Rooftop Greenery, Chicago City Hall This garden is atop Chicago's City Hall. Chicago has some of the most extensive rooftop gardens of any North American city.

environment, they are a harbinger of more ambitious plans to transform thousands of rooftops, in an effort to make North American cities more sustainable and to add variety to the surrounding concrete jungle. These rooftop refuges reduce expensive heating bills, help control urban runoff, and lower the cumulative impact of the urban heat island effect, which makes cities much warmer than surrounding rural lands.

Chicago's City Hall roof was planted in 2000 and includes about half an acre of plants, shrubs, and small trees high above the bustling downtown landscape (Figure 3.2.1). Part of a larger plan to reduce the city's greenhouse gas emissions, the experimental venture has caught the eye of other architects and municipal officials.

In New York City, multiple efforts aim at making old settings into modern public places, true parks that are social as well as ecological assets. New York's High Line Park transformed the route of an old elevated railway track (closed in 1980) into an amazing 1½-mile-long open space high above the city's grit (Figure 3.2.2). At nearby Lincoln Center, the

FIGURE 3.2.2 High Line Park, New York City New York's High Line Park makes creative use of one of the city's abandoned elevated railway corridors.

Hypar Pavillion features an elevated slanted green roof atop a glass-enclosed restaurant. In 2011, this project won one of the city's key architectural design awards. Diners below graze on salads, while park visitors above enjoy different greenery.

In San Francisco, St. Mary's Square is already a major up-in-the-air park in the Financial District. Plans for the city's Transbay Transit Center include a 5.4-acre green roof complete with trees, walking paths, and an entertainment amphitheater for outside music performances. Potentially, these additions to the urban landscape portend a greening of North American cities that will unfold one rooftop at a time to ultimately produce environments as livable as they are productive.

FIGURE 3.19 Iowa Settlement Patterns The regular rectangular look of this Iowa town and the nearby rural setting is a common cultural landscape feature across North America. In the United States, the township-and-range survey system stamped such predictable patterns across vast portions of the North American interior.

Commercial farming and technological changes further transformed the settlement landscape. Railroads opened corridors of development, provided access to markets for commercial crops, and helped to establish towns. By 1900, several transcontinental lines spanned North America, radically transforming the farm economy and the pace of rural life. After 1920, however, even greater change accompanied the arrival of the automobile, farm mechanization, and better rural road networks. The need for farm labor declined with mechanization, and many smaller market centers became unnecessary as farmers equipped with automobiles and trucks could travel farther and faster to larger, more diverse towns.

Today many areas of rural North America face population declines as they adjust to changing economic conditions. Both U.S. and Canadian farm populations fell by more than two-thirds during the last half of the 20th century. Typically, a smaller number of farms (but larger in acreage) dot the modern rural scene, and many young people leave the land to obtain employment, often in more urban settings.

Elsewhere, rural settings show signs of growth. Some places begin to experience the effects of expanding edge cities. Other growing rural settings lie beyond direct metropolitan influence, but are seeing new populations that seek amenity-rich environments, removed from city pressures. These trends are shaping the settlement landscape from British Columbia's Vancouver Island to Michigan's Upper Peninsula.

CULTURAL COHERENCE AND DIVERSITY: SHIFTING PATTERNS OF PLURALISM

North America's cultural geography exerts global influence. At the same time, it is internally diverse. History and technology have produced a contemporary North American cultural force that is second to none in the world. Yet North America is also a collection of different peoples who retain part of their traditional cultural identities. In fact, North Americans celebrate their varied roots and acknowledge the region's multicultural character.

The Roots of a Cultural Identity

Powerful historical forces formed a common dominant culture within North America. Although both the United States (1776) and Canada (1867) became independent from Great Britain, the two countries remained closely tied to their Anglo roots. Key Anglo legal and social institutions solidified core values that North Americans shared with

Britain and, eventually, with one another. Traditional Anglo beliefs emphasized representative government, separation of church and state, liberal individualism, privacy, pragmatism, and social mobility. From those shared foundations, particularly within the United States, consumer culture blossomed after 1920, producing a shared set of experiences oriented around convenience, consumption, and the mass media.

But North America's cultural unity coexists with pluralism—the persistence and assertion of distinctive cultural identities. Closely related is the concept of **ethnicity**, in which people with a common background and history identify with one another, often as a minority group within a larger society. For Canada, the French colonization of Quebec and the enduring power of its native peoples complicate its modern cultural geography. Within the United States, a greater diversity of ethnic groups exists, and differences in cultural geography are found on both local and regional scales.

Peopling North America

North America is a region of immigrants. Decisively displacing Native Americans in most portions of the region, immigrant populations created a new cultural geography of ethnic groups, languages, and religions. Early migrants had considerable cultural influence, even though their numbers were small. Over time, immigrant groups and their changing destinations produced a varied cultural geography across North America. Also varying between groups was the pace and degree of **cultural assimilation**, the process in which immigrants were absorbed by the larger host society.

Migration to the United States   In the United States, variations in the number and source regions of migrants produced five distinctive chapters in the country's history (Figure 3.20). In Phase 1 (prior to

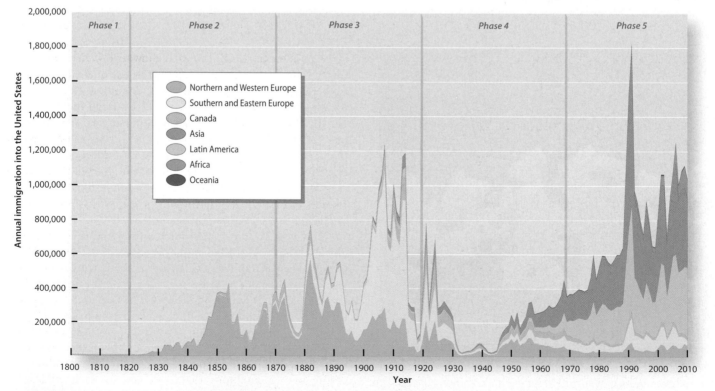

FIGURE 3.20 U.S. Immigration, by Year and Group   Annual immigration rates peaked around 1900, declined in the early 20th century, and then surged again, particularly beginning in 1970. The source areas of these migrants have also shifted. Note the decreased role Europeans currently play versus the growing importance of Asians and Latin Americans.

1820), English and African influences dominated. Slaves, mostly from West Africa, contributed additional cultural influences in the South. Northwest Europe served as the main source region of immigrants between 1820 and 1870 (Phase 2). The emphasis, however, shifted away from English migrants. Instead, Irish and Germans dominated the flow and provided more cultural variety.

As Figure 3.20 shows, immigration reached a much higher peak around 1900, when almost 1 million foreigners entered the United States *annually*. During Phase 3 (1870–1920), the majority of immigrants were southern and eastern Europeans. Political strife and poor economies in Europe existed during this period. News of available land and expanding industrialization in the United States offered an escape from such difficult conditions. By 1910, almost 14 percent of the nation was foreign-born. Very few of these immigrants, however, targeted the job-poor U.S. South, creating a cultural divergence that still exists.

Between 1920 and 1970 (Phase 4), more immigrants came to the United States from neighboring Canada and Latin America, but overall totals fell sharply, a function of more-restrictive federal immigration policies (the Quota Act of 1921 and the National Origins Act of 1924), the Great Depression, and the disruption caused by World War II. After 1970 (Phase 5), the number of immigrants increased, particularly from Latin America and Asia, and total numbers matched those of the early 20th century. Since 2008, however, the pace of undocumented immigration has slowed appreciably, mostly because of declining job opportunities in the United States, as well as an increased number of U.S. border patrol agents. Today approximately 11–12 million undocumented immigrants live in the United States.

The nation's Hispanic population continues to grow (Figure 3.21). An estimated 12 million Mexican-born residents (more than 10 percent of Mexico's population) now live in the United States. In the next 25 years, most of the projected increase in the U.S. Hispanic population will be fueled by births within the country, rather than new immigrants. Almost half of U.S. Hispanics live in California (27% of California's population is foreign-born) or Texas, but they are

FIGURE 3.21 Alley Murals, Mission District, San Francisco, California One of the Bay Area's largest Hispanic communities features dozens of street-side murals that help define the area's cultural identity. A portrait of César Chávez celebrates the importance of the farm labor-union movement within the state.

increasingly moving to other areas (Figure 3.22). States such as South Carolina, Alabama, Wisconsin, Georgia, Kansas, and Arkansas have witnessed dramatic increases in Hispanic populations. Many settlements across the Great Plains are also home to Hispanic immigrants, who bring new churches, taquerías, and school-aged children to once-dying communities.

In percentage terms, migrants from Asia constitute the fastest-growing immigrant group, and various Asian ethnicities, both native and foreign-born, account for 6 percent of the U.S. population. Chinese is the third most common spoken language in the United States (behind English and Spanish). California remains a key entry point for migrants and is home to one-third of the nation's Asian population, while Hawaii has the highest percentage

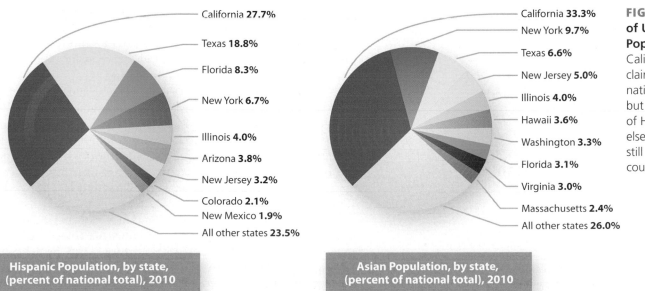

California **27.7%**
Texas **18.8%**
Florida **8.3%**
New York **6.7%**
Illinois **4.0%**
Arizona **3.8%**
New Jersey **3.2%**
Colorado **2.1%**
New Mexico **1.9%**
All other states **23.5%**

Hispanic Population, by state, (percent of national total), 2010

California **33.3%**
New York **9.7%**
Texas **6.6%**
New Jersey **5.0%**
Illinois **4.0%**
Hawaii **3.6%**
Washington **3.3%**
Florida **3.1%**
Virginia **3.0%**
Massachusetts **2.4%**
All other states **26.0%**

Asian Population, by state, (percent of national total), 2010

FIGURE 3.22 Distribution of U.S. Hispanic and Asian Populations, by State, 2010 California, Texas, and Florida claim more than half of the nation's Hispanic population, but a growing number of Hispanics are locating elsewhere. California alone is still home to one-third of the country's Asian population.

of Asian immigrants (Figure 3.22). The largest Asian groups in the United States include Chinese (3.8 million), Filipino (3.2 million), Asian Indian (2.8 million), Vietnamese (1.7 million), and Korean (1.6 million).

The future cultural geography of the United States will be dramatically redefined by these recent immigration patterns. By 2050, Asians may total almost 10 percent of the U.S. population, and almost one American in three will be Hispanic. Indeed, it is likely that the U.S. non-Hispanic white population will achieve minority status by that date (Figure 3.23).

The Canadian Pattern   French arrivals in the St. Lawrence Valley dominated the early European immigration to Canada. After 1765, many migrants came from Britain, Ireland, and the United States. Canada also experienced the same surge and reorientation in migration flows seen in the United States around 1900. Between 1900 and 1920, more than 3 million foreigners ventured to Canada, an immigration rate far higher than for the United States, given Canada's smaller population. Eastern Europeans, Italians, Ukrainians, and Russians dominated. Today about 60 percent of Canada's immigrants are Asians, and its 19 percent foreign-born population is among the highest in the developed world. In Toronto, the city's 44 percent foreign-born population reveals a slight bias toward European backgrounds, whereas Asian migrants (especially Chinese) dominate the West Coast metropolis of Vancouver (38 percent foreign-born).

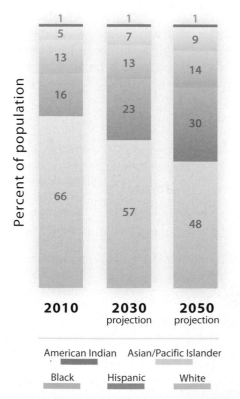

FIGURE 3.23 Projected U.S. Ethnic Composition, 2010 to 2050
By the middle of the 21st century, almost one in three Americans will be Hispanic, and non-Hispanic whites will achieve minority status amid an increasingly diverse U.S. population.

Culture and Place in North America

Cultural and ethnic identity is often strongly tied to place. North America's cultural diversity is expressed geographically in two ways. First, people with similar backgrounds congregate near one another and derive meaning from the territories they occupy together. Second, these distinctive cultures leave their mark on the everyday scene: The landscape is filled with the artifacts, habits, language, and values of different groups. Boston's Italian North End simply looks and smells different from nearby Chinatown.

Cultural Homelands   French-Canadian Quebec is an excellent example of a **cultural homeland**: It is a culturally distinctive settlement in a well-defined geographic area, and its ethnicity has survived over time, stamping the landscape with an enduring personality (Figure 3.24). More than 80 percent of Quebec's population speaks French, and language remains the "cultural glue" that holds the homeland together. Policies adopted after 1976 strengthened the French language within the province by requiring French instruction in schools and national bilingual programming by the Canadian Broadcasting Corporation (CBC). Many Quebeçois feel that the greatest cultural threat may come not from Anglo-Canadians, but rather from recent immigrants to the province. Southern Europeans and Asians in Montreal, for example, show little desire to learn French, preferring instead to put their children in English-speaking private schools.

Another well-defined cultural homeland is the Hispanic Borderlands (Figure 3.24). It is similar in size to French-Canadian Quebec, but significantly larger in total population and more diffuse in its cultural and political expression. Historical roots of the homeland are deep, extending back to the 17th century, when Spaniards opened the region to the European world. A rich legacy of Spanish place names, earth-toned Catholic churches, and traditional Hispanic settlements dots the rolling highlands of northern New Mexico and southern Colorado. From California to Texas, other historical sites and place names also reflect the Hispanic legacy.

Unlike Quebec, however, large 20th-century migrations from Latin America brought an entirely new wave of Hispanic settlement to the Southwest. About 50 million Hispanics live in the United States, with more than half in California, Texas, and Florida combined. Indeed, in 2015, Hispanics will likely outnumber non-Hispanic whites in California. Cities such as San Antonio, Phoenix, and Los Angeles are major areas for the Hispanic presence in the Southwest. New York City, Chicago, and Cuban South Florida serve as key points of Hispanic influence beyond the cultural homeland.

African Americans also retain a cultural homeland in the South, but it has become less important because of out-migration (Figure 3.24). Dozens of rural counties in the region still have large black majorities, and the South remains home to many black folk traditions, including music such as black spirituals and the blues, now popular far beyond their rural origins. Beyond the South, African Americans have created large, enduring communities in largely urban settings in the Northeast, Midwest, and West (Figure 3.25).

A second rural homeland in the South is Acadiana, a zone of persisting Cajun culture in southwestern Louisiana (Figure 3.24). This homeland was founded in the 18th century, when French settlers were expelled from eastern Canada (an area known as Acadia) and relocated to Louisiana. Nationally popularized today through food and music, Cajuns have a long-lasting attachment to the bayous and swamps of southern Louisiana.

FIGURE 3.24 Selected Cultural Regions of North America From northern Canada's Nunavut to the Southwest's Hispanic Borderlands, different North American cultural groups strongly identify with traditional local and regional homelands. The map displays a sampling of these regions across North America.

Native American populations are also strongly tied to their homelands. Over 5 million Indians, Inuits, and Aleuts live in North America, and they claim allegiance to more than 1,100 tribal bands. Particularly in the American West and the Canadian and Alaskan North, native peoples control sizable reservations, including the 16-million-acre (6.5-million-hectare) Navajo Reservation in the Southwest and self-governing Nunavut in the Canadian North (Figure 3.24). Although these homelands preserve ties to place, they are also settings for pervaive poverty (Figure 3.26). Within the United States, many Native American tribes have taken advantage of the special legal

FIGURE 3.25 African-American Church, Oakland, California
Oakland's First African Methodist Episcopal Church has been a Bay Area cultural institution for more than a century. It remains a focus of the community today.

FIGURE 3.26 Native American Poverty Navajo youngsters enjoy a game of basketball on the reservation. Poor housing, low incomes, and persistent unemployment plague many Native American settings across the rural West.

status of reservations and built gambling casinos and tourist facilities that bring in capital, but also challenge traditional ways of life.

A Mosaic of Ethnic Neighborhoods North America's cultural mosaic is characterized by smaller-scale ethnic signatures that shape both rural and urban landscapes. When much of the agricultural interior was settled, immigrants often established close-knit communities. Among others, German, Scandinavian, Slavic, Dutch, and Finnish neighborhoods took shape, held together by common origins, languages, and religions. Rural landscapes in Wisconsin, Minnesota, the Dakotas, and the Canadian prairies still display these cultural imprints. Folk architecture, distinctive settlement patterns, ethnic place names, and rural churches survive as signatures of cultural diversity across rural North America.

Ethnic neighborhoods are also a part of the urban landscape and reflect both global-scale and internal North American migration patterns. The ethnic geography of Los Angeles is an example of both economic and cultural forces at work (Figure 3.27). Because most of its economic expansion took place during the 20th century, the city's ethnic patterns reflect the movements of more recent migrants.

African-American communities on the city's south side (Compton and Inglewood) represent the legacy of black population movements out of the South. Hispanic (East Los Angeles) and Asian (Alhambra and Monterey Park) neighborhoods are a reminder that about 40 percent of the city's population is foreign-born.

The Globalization of American Culture

Simply put, North America's cultural geography is becoming more global at the same time that global cultures are becoming more North American (influenced particularly by the United States). But processes of cultural globalization are complex. We can no longer think of simple flows of foreign influences into North America or of U.S. cultural dominance invading every traditional corner of the globe. In the 21st century, the story of cultural globalization increasingly features mixed influences flowing in many directions at once, resulting in new hybrid cultural creations.

North Americans: Living Globally More than ever, North Americans in their everyday lives are exposed to people from beyond the region. With 46 million foreign-born migrants living across the region, diverse global influences are free to mingle in new ways. In 2010, the United States recorded more than 60 million international visitors (over half from Canada and Mexico). At U.S. colleges and universities, several hundred thousand international students add global flavor to the classroom.

Globalization presents cultural challenges for North Americans. In the United States, one key issue revolves around the English language, which some have described as the "social glue" holding the nation together. Since 1980, the continuing flow of non-English-speaking immigrants into the country has sharpened the debate over the role English should play in U.S. culture. Evidence indicates that North America's immigrants are learning English more rapidly than ever before, seeing it as a powerful way to gain entrance into the economic mainstream, both in the United States and in Canada. The growing popularity of **Spanglish**, a hybrid combination of English and Spanish spoken by Hispanic Americans, also illustrates the complexities of North American globalization. Spanglish includes interesting hybrids such as *chatear*, which means "to have an online conversation."

North Americans are going global in other ways. By 2010, the vast majority of Americans and Canadians had Internet access, opening the door for far-reaching journeys in cyberspace. Social media such as Facebook and Twitter have for many North Americans redefined the kinds of communities and networks that shape their daily lives. North Americans also travel more widely. Within North America, the popularity of ethnic restaurants has peppered the region with a bewildering variety of Cuban, Ethiopian, Basque, and Pakistani eateries. Americans consume more than 125 million cases of imported beer annually (Figure 3.28). In another example of globalization, Heineken, famed for its fine Dutch brew, now owns the rights to resell in the United States both the Tecate and the Dos Equis brands, two popular Mexican beers. In fashion, *Gucci*, *Brioni*, and *Prada* are household words for millions who keep their eyes on European styles. The beat of German techno bands, Gaelic instrumentals, and Latin rhythms is an increasingly seamless part of daily life. Indeed, from acupuncture and massage therapy to soccer

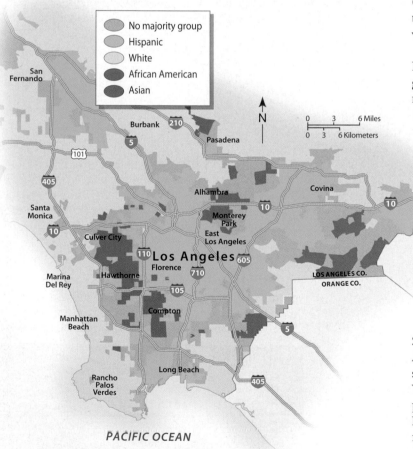

FIGURE 3.27 Ethnicity in Los Angeles Economic opportunities have attracted a wide variety of migrants to North American cities, producing an ethnic mixture of distinctive neighborhoods and communities. In Los Angeles, several cycles of economic expansion have attracted a diverse collection of residents from around the globe.

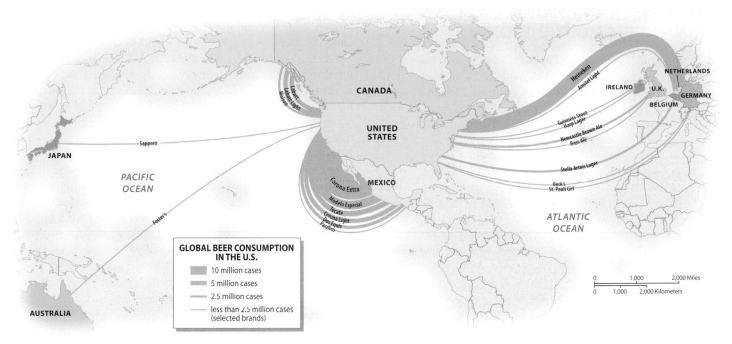

FIGURE 3.28 Annual Beer Imports to the United States, 2010   Whether they are aware of it or not, North Americans are increasingly eating and drinking globally. Rising beer imports, including many more expensive foreign brands, exemplify the pattern. The nation's beer drinkers know no bounds to their thirsts. Mexico dominates, along with varied European, Asian, and Australian producers.

and New Age religions, North Americans are tirelessly borrowing, adapting, and absorbing the larger world around them.

The Global Diffusion of U.S. Culture   In parallel fashion, U.S. culture has forever changed the lives of billions of people beyond the region. Although the economic and military power of the United States was notable by 1900, it was not until after World War II that the country's popular culture fundamentally reshaped global human geographies. The Marshall Plan and Peace Corps initiatives exemplified the growing presence of the United States on the world stage, even as European colonialism waned. Rapid improvements in global transportation and information technologies also brought the world more surely under the region's spell. Perhaps most critical was the marriage between growing global demands for consumer goods and the rise of the multinational corporation, which was superbly structured to meet and cultivate those needs. The results of these connections are not a simple Americanization of traditional cultures or a single, synthesized global culture shaped in the U.S. image. Still, millions of people, particularly the young, are attracted by the North American emphasis on individualism, consumption, youth, and mobility. The popularity of English-language teaching programs in places from China to Cuba is testimony to the cultural power of the United States.

The United States shapes contemporary culture around the world. In the built landscape, central-city skylines become indistinguishable from one another, suburban apartment blocks take on a global sameness, and one airport hotel looks the same as another eight time zones away. Global corporate advertising, distribution networks, and mass consumption bring Cokes and Big Macs to Moscow and Beijing, golf courses to Thai jungles, and Mickey and Minnie Mouse to Tokyo

and Paris. Western-style business suits have become the professional uniform of choice, while T-shirts and jeans offer standardized global comfort on days away from work.

But U.S. cultural control has not gone unchallenged, illustrating the varied consequences of globalization. Hollywood's dominance within the global film industry has declined dramatically as filmmakers build their own movie business in India, Latin America, China, and elsewhere. As worldwide use of the Internet has grown, the online dominance of English-speaking users has dramatically declined globally from more than 71 percent in 1998 to only 27 percent in 2010. Not surprisingly, given the rapid diffusion of the Internet and China's growing influence, Mandarin may soon surpass English as the leading global language of Internet users. Active resistance to U.S. cultural influence is also notable. For example, Canadian government agencies routinely chastise their radio, television, and film industries for letting in too much U.S. cultural influence. The French also criticize U.S. dominance in such media as the Internet. Elsewhere, Iran has banned satellite dishes and many U.S. films, although illegal copies of top box-office hits often find their way through national borders.

REVIEW QUESTIONS

1. What are the distinctive eras of immigration in U.S. history, and how do they compare with those of Canada?

2. Identify four enduring North American cultural regions, and describe their key characteristics.

GEOPOLITICAL FRAMEWORK: PATTERNS OF DOMINANCE AND DIVISION

North America is home to two of the world's largest states. The creation of these states, however, was neither simple nor preordained, but rather the result of historical processes that might have created quite a different North American map. Once established, these two states have coexisted in a close relationship of mutual economic and political interdependence.

Creating Political Space

The United States and Canada have very different political roots. The United States broke cleanly and violently from Great Britain. Canada, in contrast, was a country of convenience, born from a peaceful separation from Britain and then assembled as a collection of distinctive regional societies that only gradually acknowledged their common political destiny.

In the case of the future United States, Europe imposed its own political boundaries. The 13 English colonies, sensing their common destiny after 1750, united two decades later in the Revolutionary War. The Louisiana Purchase (1803) nearly doubled the national domain, and by the 1850s the remainder of the West had been added. The acquisition of Alaska (1867) and Hawaii (1898) rounded out what became the 50 states.

Canada was created under quite different circumstances. After the American Revolution, England's remaining territories in the region were controlled by administrators in British North America. In 1867, the British North America Act united the provinces of Ontario, Quebec, Nova Scotia, and New Brunswick in an independent Canadian Confederation. Soon, the Northwest Territories (1870), Manitoba (1870), British Columbia (1871), and Prince Edward Island (1873) joined this confederation, and the continental dimensions of the country took shape. Later infilling added Alberta, Saskatchewan, and Newfoundland. The creation of Nunavut Territory (1999) represents the latest change in Canada's political geography.

Continental Neighbors

Geopolitical relationships between Canada and the United States have always been close: Their common 5,525-mile (8,900-kilometer) boundary requires both nations to pay attention to one another. During the 20th century, the two countries lived largely in harmony. In 1909, the Boundary Waters Treaty created the International Joint Commission, an early step in the common regulation of cross-boundary issues involving water resources, transportation, and environmental quality. The St. Lawrence Seaway (1959) opened the Great Lakes region to better global trade connections. The two nations also joined in cleaning up Great Lakes pollution and in reducing acid rain.

Close political ties also have strengthened trade. The United States receives 73 percent of Canada's exports and supplies 63 percent of its imports. Conversely, Canada accounts for roughly 20 percent of U.S. exports and 15 percent of its imports. A bilateral Free Trade Agreement, signed in 1989, paved the way five years later for the larger **North American Free Trade Agreement (NAFTA)**, which extended the alliance to Mexico. Paralleling the success of the European Union (EU), NAFTA has forged the world's largest trading bloc, including more than 450 million consumers and a huge free-trade zone that stretches from beyond the Arctic Circle to Latin America.

Political conflicts still divide the two countries (Figure 3.29). Environmental issues produce cross-border tensions, especially when environmental degradation in one nation affects the other. For example, Montana's North Flathead River flows out of British Columbia, where Canadian logging and mining operations periodically threaten fisheries and recreational lands south of the border. Agricultural and natural resource competition also causes occasional controversy. The appearance of mad cow disease in Canadian livestock curtailed exports to the United States and elsewhere. Tensions also rose when Canadian wheat and potato growers were accused of dumping their products into U.S. markets, thus depressing prices and profits for U.S. farmers. Similar issues have arisen in the logging industry. In the far north, the two countries disagree on the maritime boundary between Yukon Territory and Alaska. In addition, the United States disputes assertions that a potential Northwest Passage opening across a more ice-free Arctic Ocean would essentially be within Canada's territorial waters.

More generally, tighter U.S. regulations since 2009 have made it more difficult to cross the border in either direction. Reflecting security concerns in the United States, the regulations demand that persons crossing the border present a passport or other approved form of identification, just as they would on the Mexican border. Canadians cite the rules as potentially harmful to tourism between the two countries. Late in 2011, the Beyond the Border Agreement reached between the two nations was designed to ease these tensions and facilitate the sharing of information.

The Legacy of Federalism

The United States and Canada are **federal states** in that both nations allocate considerable political power to units of government beneath the national level. Other nations, such as France, have traditionally been **unitary states**, in which power is centralized at the national level. Federalism leaves many political decisions to local and regional governments and often allows distinctive cultural and political groups to be recognized within a country. The U.S. Constitution (1787) limited centralized authority, giving all unspecified powers to the states or the people. In contrast, the Canadian Constitution (1867), which created a federal state under a parliamentary system, reserved most powers to central authorities. Ironically, the evolution of the United States produced an increasingly powerful central government, while Canada's geopolitical balance of power shifted toward more provincial autonomy and a relatively weak national government.

Quebec's Challenge The political status of Quebec remains a major issue in Canada (see Figure 3.29). Economic disparities between the Anglo and French populations have reinforced cultural differences between the two groups, with the French Canadians often suffering when compared with their wealthier neighbors in Ontario. Beginning in the 1960s, a separatist political party in Quebec (the Parti Quebecois) increasingly voiced French-Canadian concerns. When the party won provincial elections in 1976, it declared French the official language of Quebec. Formal provincial votes over the question of remaining within Canada were held in 1980 and 1995.

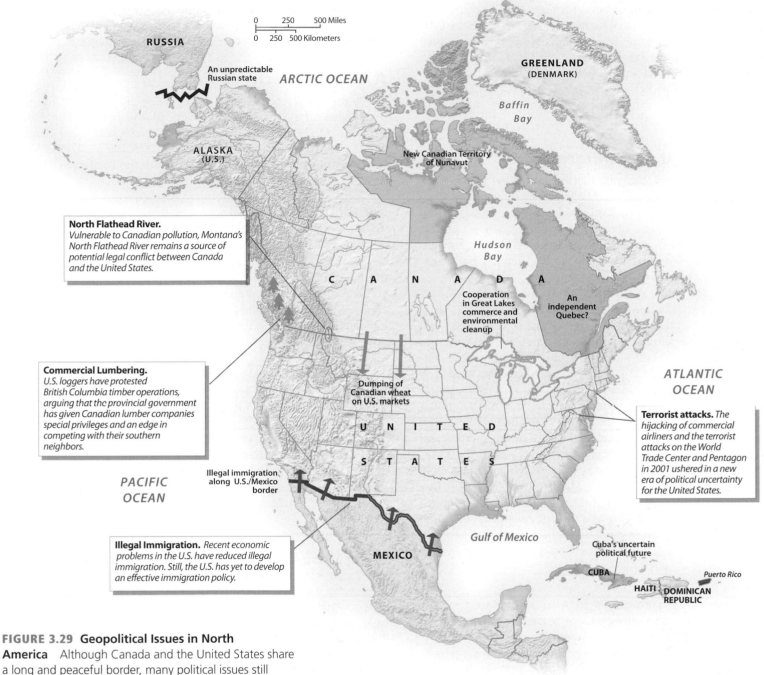

FIGURE 3.29 Geopolitical Issues in North America Although Canada and the United States share a long and peaceful border, many political issues still divide the two countries. In addition, internal political conflicts, particularly in bicultural Canada, cause tensions.

Both measures failed. Since then, support for separation has ebbed in favor of a more modest strategy of increased "autonomy" within Canada.

Native Peoples and National Politics Another challenge to federal political power has come from North American Indian and Inuit populations, in both Canada and the United States. Within the United States, Native Americans asserted their political power in the 1960s, marking a decisive turn away from policies of assimilation. Since passage of the Indian Self-Determination and Education Assistance Act of

1975, the trend has been toward increased Native American control of their economic and political destiny. The Indian Gaming Regulatory Act (1988) offered potential economic independence for many tribes. By 2011, Indian gaming operations (primarily gambling casinos) nationally netted tribes about $25 billion annually. In Alaska and the interior West, tribes are also developing natural resources and buying up additional acreage.

In Canada, ambitious challenges by native peoples have yielded dramatic results. Canada established the Native Claims Office in 1975. Agreements with native peoples in Quebec, Yukon, and

British Columbia turned over millions of acres of land to aboriginal control and increased native participation in managing remaining public lands. By far the most ambitious agreement created Nunavut out of the eastern portion of the Northwest Territories in 1999 (Figure 3.29). Nunavut is home to 30,000 people (85 percent Inuit) and is the largest territorial/provincial unit in Canada. Its creation represents a new level of native self-government in North America. Agreements between the Parliament and British Columbia tribes (the Nisga'a) have made similar moves toward more native self-government (see Figure 3.24).

The Politics of U.S. Immigration

Immigration policies are hotly contested in the United States. Four key issues remain at the center of the debate. First, there are disagreements concerning how many legal immigrants should be allowed into the country. Some suggest that sharply reduced numbers protect American jobs and allow for gradual assimilation of existing foreigners, but others argue that looser restrictions could actually boost economic growth and business expansion.

A second major issue, particularly along the U.S. border with Mexico, is tightening daily flows of undocumented immigrants. Many argue the country's southern border is a national security issue. Recent federal legislation mandates an increased number of border patrol agents, and more than 20,000 officers presently monitor the boundary. More than 700 miles (1,125 kilometers) of fencing also have been built or improved (Figure 3.30).

Third, U.S. relations with Mexico have soured with the growth of drug-related violence near the border between the two countries. Mexico remains the leading source of methamphetamine, heroin, and marijuana for the United States and is a key transit nation for northward-bound cocaine originating in South America. In addition, more than 30,000 deaths, mostly in northern Mexico, were tied to the illegal drug business between 2007 and 2011. Violence has spilled north into places such as El Paso and Phoenix, causing American officials to worry that the Mexican government has lost effective political control of its northern border.

Finally, there is no political consensus on a policy to deal with existing undocumented workers. Some policymakers advocate strict penalties for undocumented immigrants, while others propose loosening requirements for citizenship or some form of amnesty to enable immigrants to more easily enter the American mainstream.

A Global Reach

The geopolitical reach of the United States, in particular, has taken its influence far beyond the borders of the region. The Monroe Doctrine (1824) asserted that U.S. interests extended throughout the western hemisphere. Between 1890 and 1920, the United States became increasingly involved in both the Pacific and the Latin American regions.

World War II and its aftermath forever redefined the U.S. role in world affairs. The United States emerged from the conflict as the world's dominant political power. It also developed multinational political and military agreements, such as the North Atlantic Treaty Organization (NATO) and the Organization of American States (OAS). Conflicts in Korea (1950–1953) and Vietnam (1961–1975) pitted U.S. political interests against communist attempts to extend control beyond the Soviet Union and China. The Cuban missile crisis (1962) reminded Americans that traditional political boundaries provide little defense in a world uneasily brought closer together by technologies of mass destruction.

Even as the Cold War gradually faded during the late 1980s, the global political reach of the United States expanded. Direct involvement in conflicts within Central America, the Middle East, Serbia, and Kosovo exemplified the country's global agenda. Recent controversial wars in Iraq (2003–2011) and Afghanistan (from 2001) offer continuing instances of America's global political presence (Figure 3.31). Defense expenditures of more than $680 billion in 2011 (nearly as much as the rest of the world combined) suggest the country will continue to play a highly visible role in global affairs.

FIGURE 3.31 U.S. War in Afghanistan For more than a decade, U.S. troops in Afghanistan have struggled to increase stability in the region. Projection of American power into Asia is seen as part of a larger global war on terror, sparked by the September 11, 2001, attacks on the United States.

FIGURE 3.30 International Border North America's southwestern landscape is boldly bifurcated by an increasingly hardened international border that separates the United States and Mexico.

REVIEW QUESTIONS

1. How do the political origins of the United States and Canada differ?
2. What are four of the key issues surrounding U.S. immigration policy?

ECONOMIC AND SOCIAL DEVELOPMENT: GEOGRAPHIES OF ABUNDANCE AND AFFLUENCE

North America possesses the world's most powerful economy and its wealthiest population. Its 345 million people consume huge quantities of global resources, but also produce some of the world's most sought-after manufactured goods and services. The region's human capital—the skills and diversity of its population—has enabled North Americans to achieve high levels of economic development (Table 3.2). Even so, residents of the region suffered in the global recession of the late 2000s as household incomes fell and unemployment levels rose. Manufacturing states such as Michigan saw already high jobless rates rise even higher. Some of the worst-hit states, such as California and Arizona, included areas that had witnessed some of the highest growth—particularly in home prices—earlier in the decade.

An Abundant Resource Base

North America is blessed with numerous natural resources that provide diverse raw materials for development. Indeed, the direct extraction of natural resources still makes up 3 percent of the U.S. economy and more than 6 percent of the Canadian economy. Some of these North American resources are then exported to global markets, while other raw materials are imported to the region.

Opportunities for Agriculture North Americans have created one of the most efficient food-producing systems in the world, and agriculture remains a dominant land use across much of the region (Figure 3.32). Farmers practice highly commercialized, mechanized, and specialized agriculture. The system emphasizes the importance of efficient transportation, global markets, and large capital investments in farm machinery. Today agriculture employs only a small percentage of the labor force in both the United States (1 percent)

and Canada (2 percent). Changes in farm ownership have sharply reduced the number of operating units, while average farm sizes have steadily risen.

The geography of North American farming represents the combined impacts of (1) diverse environments, (2) varied continental and global markets for food, and (3) historical patterns of settlement and agricultural evolution. In the Northeast, dairy operations and truck farms take advantage of proximity to major cities in Megalopolis and southern Canada. Corn and soybeans dominate the Midwest and western Ontario, where mixed-farming traditions combine growth of feed grains with production of livestock. To the south, only remnants of the old Cotton Belt remain, largely replaced by subtropical specialty crops; poultry, catfish, and livestock production; and commercial logging. Farther west, extensive, highly mechanized commercial grain-growing operations stretch from Kansas to Saskatchewan and Alberta. Depending on surface and groundwater resources, irrigated agriculture across western North America also offers opportunities for farming. Indeed, California's agricultural output, nourished by large agribusiness operations in the irrigated Central Valley, accounts for more than 10 percent of the nation's farm economy.

Energy and Industrial Raw Materials North Americans produce and consume huge quantities of other natural resources. The region consumes 40 percent more oil than all of the European Union. While regional production of fossil fuels is on the rise (particularly with large increases in Canada), the United States, in particular, must still import more than 10 million barrels of oil a day (about half of the total consumption). Within the region, areas of oil and gas production are the Gulf Coast, the Central Interior, Alaska's North Slope, and Central Canada (especially Alberta's oil sands). The most abundant fossil fuel in the United States is coal (27 percent of the world's total), but its relative importance in the overall energy economy declined in the 20th century as industrial technologies changed and environmental concerns grew. In addition, wind, solar, nuclear, and biofuel energy sources appear likely to make up larger portions of the region's future energy budget as the region gradually shifts away from its dependence on fossil fuels.

Creating a Continental Economy

The timing of European settlement in North America was critical in its rapid economic transformation. The region's abundant resources came under the control of Europeans possessing new technologies that reshaped the landscape and reorganized its economy. By the 19th

Table 3.2 DEVELOPMENT INDICATORS

Country	GNI per capita, PPP 2010	GDP Average Annual % Growth 2000–10	Human Development Index (2011)[1]	Percent Population Living Below $2 a Day	Life Expectancy (2012)[2]	Under Age 5 Mortality Rate (1990)	Under Age 5 Mortality Rate (2010)	Adult Literacy (% ages 15 and older)	Gender Inequality Index (2011)[3]
Canada	38,370	2.0	0.908	—	81	8	6	—	0.140
United States	47,310	1.8	0.910	—	79	11	8	—	0.299

[1]United Nations, *Human Development Report, 2011.*

[2]Population Reference Bureau, *World Population Data Sheet, 2012.*

[3]Gender Inequality Index—A composite measure reflecting inequality in achievements between women and men in three dimensions: reproductive health, empowerment and the labor market that ranges between 0 and 1. The higher the number, the greater the inequality.

Source: World Bank, *World Development Indicators, 2012.*

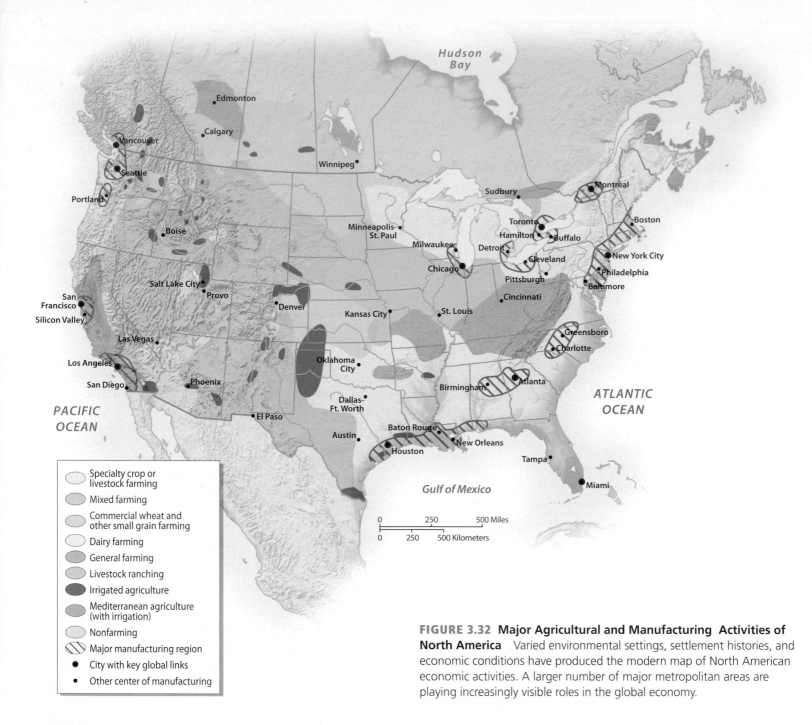

FIGURE 3.32 Major Agricultural and Manufacturing Activities of North America Varied environmental settings, settlement histories, and economic conditions have produced the modern map of North American economic activities. A larger number of major metropolitan areas are playing increasingly visible roles in the global economy.

Legend:
- Specialty crop or livestock farming
- Mixed farming
- Commercial wheat and other small grain farming
- Dairy farming
- General farming
- Livestock ranching
- Irrigated agriculture
- Mediterranean agriculture (with irrigation)
- Nonfarming
- Major manufacturing region
- City with key global links
- Other center of manufacturing

century, North Americans actively contributed to those technological changes. In addition, new natural resources were developed in the interior, and new immigrant populations arrived in large numbers. In the 20th century, although natural resources remained important, industrial innovations and more jobs in the service sector added to the economic base and extended the country's global reach.

Connectivity and Economic Growth   Dramatic improvements in North America's transportation and communication systems laid the foundation for urbanization, industrialization, and the commercialization of agriculture. Indeed, the region's economic success was a function of its **connectivity**, or how well its different locations became linked with one another through improved transportation and communications networks. Those links greatly facilitated the interaction

between locations and dramatically reduced the cost of moving people, products, and information.

Technological breakthroughs revolutionized North America's economic geography between 1830 and 1920. By 1860, more than 30,000 miles (48,000 kilometers) of railroad track had been laid in the United States, and the network grew to more than 250,000 miles (400,000 kilometers) by 1910. Farmers in the Midwest and Plains found ready markets for their products in cities hundreds of miles away. Industrialists collected raw materials from faraway places, processed them, and shipped manufactured goods to their final destinations. The telegraph brought similar changes to information: Long-distance messages flowed across eastern North America by the late 1840s, and 20 years later, undersea cables linked the region to Europe, another milestone in the process of globalization.

North America's transportation and communications systems were modernized further after 1920. Automobiles, mechanized farm equipment, paved highways, commercial air links, national radio broadcasts, and dependable transcontinental telephone service reduced the cost of distance across the region. Perhaps most important, the region has taken the lead in the global information age, integrating computer, satellite, telecommunications, and Internet technologies in a web of connections that facilitates the flow of knowledge both within the region and beyond.

The Sectoral Transformation   Changes in employment structure signaled North America's economic modernization just as surely as its increasingly interconnected society. The **sectoral transformation** refers to the evolution of a nation's labor force from one dependent on the *primary* sector (natural resource extraction) to one with more employment in the *secondary* (manufacturing or industrial), *tertiary* (services), and *quaternary* (information processing) sectors. For example, with agricultural mechanization, lower demands for primary-sector workers are replaced by new opportunities in the growing industrial sector. In the 20th century, new services (trade, retailing) and information-based activities (education, data processing, research) created other employment opportunities. Today the tertiary and quaternary sectors employ more than 70 percent of the labor force in both Canada and the United States.

Regional Economic Patterns   The locations of North America's industries show important regional patterns. **Location factors** are the varied influences that explain *why* an economic activity is located where it is. Many influences, both within and beyond the region, shape patterns of economic activity. Patterns of industrial location illustrate the concept (see Figure 3.32). The historical manufacturing core includes Megalopolis (Boston, New York, Philadelphia, Baltimore, and Washington, DC), southern Ontario (Toronto and Hamilton), and the industrial Midwest. The region's proximity to *natural resources* (farmland, coal, and iron ore); its increasing *connectivity* (canals and railroad networks, highways, air traffic hubs, and telecommunications centers); its ready supply of *productive labor*; and a growing national, then global, *market demand* for its industrial goods encouraged continued *capital investment* within the core. Traditionally, the core dominated in steel, automobiles, machine tools, and agricultural equipment and played a key role in financial and insurance services.

In the last half of the 20th century, industrial- and service-sector growth shifted to the South and West. Cities of the South's Piedmont manufacturing belt (Greensboro to Birmingham) grew after 1960, partly because lower labor costs and Sun Belt amenities attracted new investment. The North Carolina "research triangle" area, encompassing Raleigh, Durham, and Chapel Hill, has emerged as the nation's third largest biotech cluster, behind California and Massachusetts. The Gulf Coast industrial region is strongly tied to nearby fossil fuels that provide raw materials for the energy-refining and petrochemical industries (Figure 3.33).

The varied West Coast industrial region stretches from Vancouver, British Columbia, to San Diego, California (and beyond into northern Mexico), and it demonstrates the increasing importance of Pacific Basin trade. Large western aerospace operations also suggest the role of *government spending* as a location factor. Silicon Valley is now one of North America's leading regions of manufacturing exports. Its proximity to Stanford, Berkeley, and other universities

FIGURE 3.33 Gulf Coast Petroleum Refining   Petroleum-related manufacturing has transformed many Gulf Coast settings. Much of Houston's 20th-century growth was fueled by the dramatic expansion of oil-related industries. The port of Houston remains a major center of North America's refining and petrochemical operations.

demonstrates the importance of *access to innovation and research* for many fast-changing high-technology industries. Silicon Valley's location also shows the advantages of *agglomeration economies,* in which companies with similar, often integrated manufacturing operations locate near one another. Smaller places such as Provo, Utah, and Austin, Texas, specialize in high-technology industries and demonstrate the growing role of *lifestyle amenities* in shaping industrial location decisions, both for entrepreneurs and for skilled workers who are attracted by such opportunities. We will see in later chapters that these same factors affect the growth of modern industries in other nations, such as China, Japan, and India.

Persistent Social Issues

Profound economic and social problems shape the human geography of North America. Even with its continental wealth, great differences persist between rich and poor. In addition, both nations face issues related to gender inequity and challenges related to aging populations.

Wealth and Poverty   The global economic downturn of the late 2000s rippled through both the United States and Canada. Especially in the United States, unemployment levels soared between 2008 and 2011 and hovered between 8.5 and 10 percent. Young and poor African Americans and Hispanics fared the worst. Many people lost health-care coverage. Real estate values and home ownership rates fell as many people lost their homes to foreclosure. In 2011, 1 in 20 homes in Nevada was in the foreclosure process. Poverty rates rose for the first time in years. States such as California, Nevada, and Arizona, once home to real estate and construction booms, witnessed some of the most dramatic economic declines. The Occupy Wall Street movement recently highlighted the growing gap between the wealthiest Americans and a middle-class population increasingly strained by elevated unemployment, stagnant wages, and rising health-care and education costs (Figure 3.34).

FIGURE 3.34 Occupy Wall Street Protests A series of public protests in many U.S. cities in 2011 highlighted growing income disparities in the region.

The regional landscape displays contrasting scenes of wealth and poverty. Elite northeastern suburbs, gated California neighborhoods, upscale shopping malls, and posh alpine ski resorts are all expressions of private and exclusive landscape settings that characterize wealthier North American communities (Figure 3.35). In contrast, substandard housing, abandoned property, aging infrastructure, and unemployed workers are reminders of the gap between rich and poor. The problems of the rural poor remain major regional social issues in the Canadian Maritimes, Appalachia, the Deep South, the Southwest, and agricultural California. Most poor people in the United States, however, live in central-city locations, and links between ethnicity and poverty are strong in these communities. Specifically, in the United States, black household incomes remain only 67 percent of the national norm, while Hispanic incomes average 80 percent of the national average.

Gender Equity Since World War II, both the United States and Canada have seen great improvements in the role that women play in society. However, the **gender gap** is yet to be closed when it comes

FIGURE 3.35 Gated America An automatic gate protects the entrance of a community in Apollo Beach, near Tampa, Florida.

to differences in salary, working conditions, and political power. Women widely participate in the workforce and are as educated as men, but they still earn only about 78 cents for every dollar that men earn. Measured by corporate power, women still play modest roles. A 2011 survey reveals that women comprise only 16 percent of corporate board membership of large publicly held companies in the United States (versus 35 percent in Norway). Women also head the vast majority of poorer single-parent families in the United States, and almost 40 percent of all births in the country are to unwed mothers. Canadian women, particularly single mothers who work full-time, are also greatly disadvantaged, averaging only about 65 to 70 percent of the salaries of Canadian men. In addition, political power remains largely in male hands, even though women make up the majority of the electorate. Although Canadian women have voted since 1918 and U.S. women since 1920, females in the early 21st century remain minorities in the Canadian Parliament and the U.S. Congress.

Health Care and Aging Although U.S. residents spend more than 15 percent of gross domestic product on health care (Canadians spend slightly less), more than 40 million Americans are without health-care insurance. A recent report on aging in the United States predicted that 20 percent of the nation's population will be older than 65 by 2050 and that the most elderly (age 85+) are the fastest-growing part of the population. The geographic consequences of aging are already abundantly clear. Whole sections of the United States—from Florida to southern Arizona—have become increasingly oriented around retirement (Figure 3.36). Communities cater to seniors with special assisted-living arrangements, health-care facilities, and recreational opportunities.

but in periods of international instability, globalization means that the region is more vulnerable to economic downturns.

Creating the Modern Global Economy   The United States, with Canada's firm support, played a formative role in creating much of the new global economy and in shaping its key institutions. In 1944, allied nations met at Bretton Woods, New Hampshire, to discuss economic affairs. Under U.S. leadership, the group set up the International Monetary Fund (IMF) and the World Bank and gave these global organizations the responsibility for defending the world's monetary system. The United States was also the driving force for the creation (in 1948) of the General Agreement on Tariffs and Trade (GATT). Renamed the **World Trade Organization (WTO)** in 1995, its 157 member states are dedicated to reducing global trade barriers. In addition, the United States and Canada participate in the **Group of Eight (G8)**, a collection of economically powerful countries (the United States, Canada, Japan, Germany, Great Britain, France, Italy, and Russia) that regularly meets for discussions on key global economic and political issues.

Attracting Skilled Immigrants   North America's role in the global economy attracts thousands of skilled workers from other countries, adding to the region's supply of human capital. Statistics gathered by the U.S. Department of Homeland Security point to the unique contributions of highly skilled immigrants. H-1B visas are granted to special "temporary skilled workers" to encourage computer programmers, doctors, and other professionals to work in the United States. More than 400,000 such visas were issued in 2010, enabling these individuals to work within the United States. The geography of the top 20 contributing countries offers another snapshot of the economic impact of globalization and is a powerful example of another way in which the United States benefits in the process (Figure 3.37).

FIGURE 3.36 Tomorrow's Baby Boom Landscape?   Hundreds of golf resorts and retirement communities have been built across North America's Sun Belt since 1980 and now cater increasingly to the Baby Boom generation.

North America and the Global Economy

Together with Europe, Japan, and China, North America plays a key role in the global economy. The region is home to a growing number of truly "global cities" that serve as key connecting points and decision-making centers in the world economy (see Figure 3.32). When the economy is thriving, the region benefits from global economic growth,

FIGURE 3.37 Origins of Temporary Skilled Workers (H-1B Visas) in the United States, 2010 (Top 20 Contributing Countries)   India's contribution to the skilled human capital of the United States is particularly notable. Both Mexico and Canada contribute many skilled workers, as do other nations of East Asia and Western Europe.

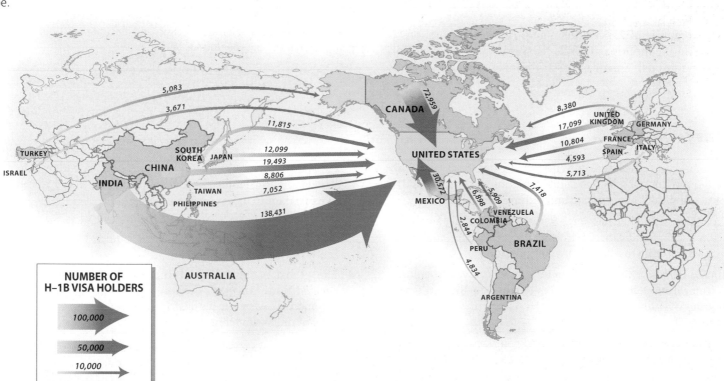

FIGURE 3.38 Container Shipping, Port of Seattle Major North American ports such as Seattle are key links in facilitating global trade between the region and the rest of the world. Standard-sized container-shipping modules can easily be stored, stacked, and moved in such settings.

While ties to Europe are predictable, linkages with India (particularly through the computer, software, and electronics industries) are more dramatic. The pattern offers a powerful reminder that the economic evolution of both the United States and Canada remains intimately connected to skilled immigrant populations.

Doing Business Globally Patterns of capital investment and corporate power place North America at the center of global trade and money flows (Figure 3.38). The region attracts inflows of foreign capital, both as investments in North American companies and as foreign direct investment (FDI) by international companies. Since 1980, aging U.S. baby boomers also have poured billions of pension fund and investment dollars into Japanese, European, and "emerging" stock markets; U.S. investments in foreign countries are also made directly by multinational corporations based in the United States.

But the geography of 21st-century multinational corporations is changing, illustrating three recent shifts in broader patterns of globalization. First, traditional American-based multinational corporations are adopting a new, more globally integrated model. For example, IBM now has more than 50,000 employees in India. Second, multinational corporations based elsewhere in the world are buying companies and assets once controlled by North American or European capital. In 2011, investors from China, the United Arab Emirates, and Singapore were the leading acquirers of developed-world firms. Third, many of these same global multinational companies are making sizable investments of their own in other portions of the less developed world, from Africa to Southeast Asia, bypassing North American control altogether. Today more than one-third of foreign direct investment in emerging market nations comes from other emerging market nations. Simply put, the late-20th-century top-down model of multinational corporate control and investment, traditionally based in North America, Europe, and Japan, is being replaced by a more globally distributed model of corporate control. This new model has many origins, many destinations, and new patterns of labor, capital, production, and consumption.

North Americans experience direct consequences from these shifts in global capitalism. A growing reaction has arisen in the United States, for example, to corporate **outsourcing**, a business practice that transfers portions of a company's production and service activities to lower-cost settings, often located overseas. In addition, millions of jobs in manufacturing, textiles, semiconductors, and electronics have effectively migrated to places such as China, India, and Mexico, as those localities offer low-cost, less regulated settings for production, for both local and foreign firms. The results are complex: North American consumers benefit from buying cheap imports, but they may find that their own jobs are threatened in the corporate restructurings that make such bargains possible.

REVIEW QUESTIONS

1. What is sectoral transformation, and how does it help explain economic change in North America?

2. Cite five types of location factors, and illustrate each with examples from your local economy.

Summary

- North America's future remains bright: It possesses a highly educated population and considerable natural resources, and it remains a seat of global corporate and political power.

- North America's affluence has come with a stiff price tag, and today the region faces significant environmental challenges, including soil erosion, acid rain, and air and water pollution.

- In a remarkably short time period, a unique mix of varied cultural groups from around the world has contributed to the settlement of a huge and resource-rich continent that is now the world's most urbanized region.

- North Americans produced two closely intertwined societies that still face distinctive national political and cultural issues. In Canada, the nation's identity remains problematic as the country struggles with its multicultural character and the costs and benefits of its proximity to its continental neighbor.

- For the United States, social problems linked to ethnic diversity, immigration issues, health-care costs, and enduring poverty remain central concerns.

- The global economic downturn of the late 2000s profoundly affected North America's economic geography, particularly in many regions that were hit hardest by the housing crisis and by rising rates of unemployment.

Key Terms

acid rain 73
boreal forest 75
connectivity 94
cultural assimilation 84
cultural homeland 86
edge city 82
ethnicity 84
federal state 90
gender gap 96
gentrification 82

Group of Eight (G8) 97
location factor 95
Megalopolis 78
new urbanism 82
nonmetropolitan growth 81
North American Free Trade Agreement
  (NAFTA) 90
outsourcing 98
postindustrial economy 68
prairie 75

renewable energy source 74
sectoral transformation 95
Spanglish 88
sustainable agriculture 74
tundra 75
unitary state 90
urban decentralization 81
urban heat island 73
World Trade Organization (WTO) 97

Thinking Geographically

1. Explain how "natural hazards" can be "culturally" defined. In other words, what role do humans play in shaping the distribution of hazards?

2. Map the ethnic background and migration history of your own family. Write an essay exploring how these patterns parallel or depart from larger North American trends.

3. Divide the class into two groups and develop opposing arguments to debate the following proposition: "While the environmental price for North American development has been steep, the economic rewards have been well worth the cost." Suggest why it may or may not be worth the price, and debate your responses.

4. Who will the United States' leading trade partner be in 2050? Explain the reasons for your choice.

5. In groups of three students, plan a field trip through your local community, and identify key features of the North American urban landscape described in the text. Report back to the class, and share your results and examples.

MasteringGeography™

Looking for additional review and test prep materials? Visit the Study Area in MasteringGeography™ to enhance your geographic literacy, spatial reasoning skills, and understanding of this chapter's content by accessing a variety of resources, including MapMaster🌐 interactive maps, videos, RSS feeds, flashcards, web links, self-study quizzes, and an eText version of *Globalization and Diversity*.

Scan to visit the author's blog for chapter updates.

http://gad4blog.
wordpress.com/
category/north-
america/

Authors' Blogs

Scan now to access the authors' blogs for up-to-date information on North America.

Scan to visit the GeoCurrents blog.

http://geocurrents.
info/category/place/
north-america

Globalization and Diversity

Neoliberal policies that emphasize free trade, foreign investment, and expansion of the private sector have profoundly changed the way Latin American economies and societies function. Foreign investment and international trade have intensified and diversified, deepening connections with regions other than North America, especially East Asia and Europe.

ENVIRONMENTAL GEOGRAPHY

Tropical forests in Latin America, especially in the Amazon Basin, are one of the planet's greatest reserves of biological diversity. How this diversity is managed is a critical question, especially with increasing pressure to extract mineral wealth or convert forests into farms or pastures.

POPULATION AND SETTLEMENT

Latin America is the most urbanized region of the developing world, with 77 percent of the population living in cities. Four megacities (10 million or more) are found here. Yet it is also a region with high rates of emigration, especially to North America.

CULTURAL COHERENCE AND DIVERSITY

Amerindian activism is on the rise in Latin America. Indigenous peoples from Central America to the Andes are finding their political voice and demanding cultural and territorial recognition.

GEOPOLITICAL FRAMEWORK

As many Latin American governments mark 200 years of independence from Spain, most are fully democratic. Recent elections in the region have seen liberal democrats and populists gain power, promising to reduce income inequality through government programs. Women are also becoming political actors in Latin America, holding nearly one-quarter of all seats in national parliaments.

ECONOMIC AND SOCIAL DEVELOPMENT

Economic growth, increased trade, and fewer people in extreme poverty are all positive trends for the region, but serious income inequality persists. Government programs such as Brazil's Bolsa Familia have sought to address both social and economic development for poor families and reduce income inequality.

➤ **A monumental celebration in Mexico City of the 200th anniversary of the Mexican battle for independence from Spain held in the city's main square (the zocalo) on September 16, 2010. Most states in Latin America have been independent countries for 200 years.**

LATIN AMERICA 4

In September 2010, Mexico City had a massive street party, marking Mexico's 200th anniversary of its battle for independence from Spain. In the large central plaza under a massive Mexican flag—a place where Aztec kings once stood upon stone pyramids—tens of thousands celebrated with a laser light show that bounced off of the walls of the colonial-era buildings. That same year two Mexicans achieved prominence: Carlos Slim (a telecom mogul) was named the richest man in the world and Ximena Navarrete became Miss Universe. Mexico City itself, a metropolitan area of 20 million, is the political, cultural, and economic core of the country, and as Mexico prospers, the city's prominence rises as an international business center. Not surprisingly, it is a city of extremes where the elite inhabit the glimmering towers of modern neighborhoods such as Santa Fe to the west, while the poorest residents build informal settlements to the east such as Ciudad Neza with over 1.5 million people. Like Mexico, most of the nations of Latin America are marking their bicentennials as independent and democratic countries, with a growing middle class and urban majorities. Yet extremes in income inequality and rising levels of violence in some states have led many observers to question how neoliberal policies and economic globalization might be modified to address pressing social and economic needs.

Beginning with Mexico and extending to the tip of South America, Latin America's regional unity stems largely from its shared colonial history, rather than from different levels of development seen today. More than 500 years ago, the Iberian countries of Spain and Portugal began their conquest of the Americas. Iberia's mark is still visible throughout Latin America: Officially, two-thirds of the population speaks Spanish, and the rest speaks Portuguese. Iberian architecture and town design add homogeneity to the colonial landscape. The vast majority of the population is Catholic. These European traits blended with those of various Amerindian peoples. The Indian presence remains especially strong in Bolivia, Peru, Ecuador, Guatemala, and southern Mexico, where large and diverse Amerindian populations maintain their native languages, dress, and traditions. After the initial colonial conquest, other cultural groups from Africa and Asia were added to this mix of native and Iberian peoples, making it one of the world's most racially mixed regions.

The concept of Latin America as a distinct region has been popularly accepted for nearly a century. The boundaries of this region are straightforward, beginning at the Rio Grande (called the Rio Bravo in Mexico) and ending at Tierra del Fuego (Figure 4.1). French geographers are credited with coining the term *Latin America* in the 19th century to distinguish the Spanish- and Portuguese-speaking republics of the Americas plus Haiti from the English-speaking territories. There is nothing particularly "Latin" about the area, other than the predominance of romance languages. The term stuck because it was vague enough to include different colonial histories, while also offering a clear cultural boundary from Anglo-America, the region referred to as North America in this book.

This chapter discusses Latin America, consisting of the Spanish- and Portuguese-speaking countries of Central and South America, including Mexico. This division emphasizes the important Amerindian and Iberian influences affecting mainland Latin America and separates it from the unique colonial and demographic history of the Caribbean and the Guianas, discussed in Chapter 5.

Through colonialism, immigration, and trade, the forces of globalization have been embedded in the Latin American landscape. The early Spanish Empire concentrated on extracting precious metals, sending galleons laden with silver and gold back home across the Atlantic. The Portuguese became important producers of dyewoods, sugar products, gold, and later coffee. In the late 19th and early 20th centuries, exports to North America and Europe fueled the region's economy. Most countries specialized in one or two products: bananas and coffee, meat and wool, wheat and corn, petroleum and copper. Such a primary export tradition, according to Latin American economists, led to an unhealthy economic dependence. They argued in the 1960s that Latin American economies were too specialized and faced unequal terms of trade that inhibited overall development.

Since then, the countries of the region have industrialized and diversified their production, but they continue to be major producers of primary goods for North America, Europe, and East Asia. Today neoliberal policies that encourage foreign investment, export production, and privatization have been adopted by many states. The results are mixed, with some states experiencing impressive economic growth, but increased disparity between rich and poor.

LEARNING OBJECTIVES

After reading this chapter you should be able to:

- Explain the relationships among elevation, climate, and agricultural production, especially in tropical highland areas.

- Identify the major environmental issues of Latin America and how countries are addressing them.

- Summarize the demographic issues impacting this region, such as rural-to-urban migration, urbanization, smaller families, and emigration.

- Describe the cultural mixing of European and Amerindian groups in this region and indicate where Amerindian cultures thrive today.

- Explain the global reach of Latino culture through immigration, sport, music, and television.

- Describe the Iberian colonization of the region and how it affected the formation of today's modern states.

- Identify the major trade blocs in Latin America and how they are influencing development.

- Summarize the significance of primary exports from Latin America, especially agricultural commodities, minerals, wood products, and fossil fuels.

- Describe the neoliberal economic reforms that have been applied to Latin America and how they have influenced the region's development.

- Identify major Amerindian groups today and their efforts towards territorial and political recognition.

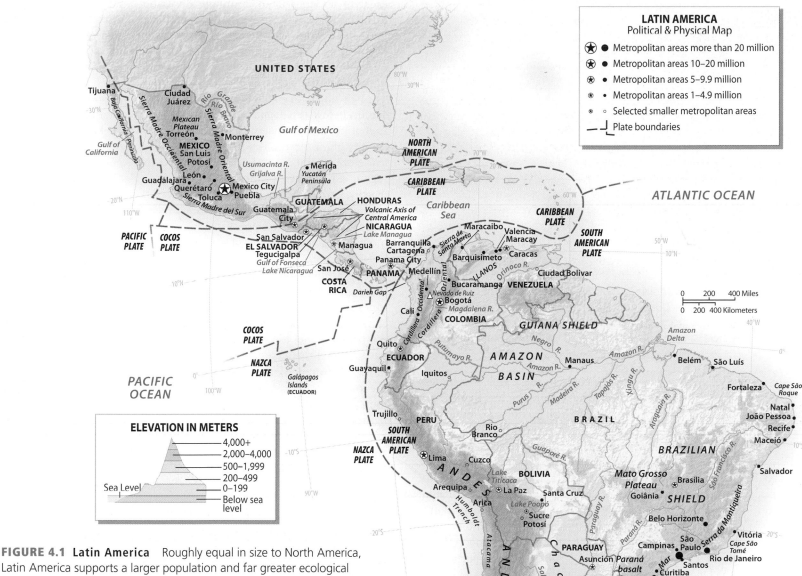

FIGURE 4.1 Latin America Roughly equal in size to North America, Latin America supports a larger population and far greater ecological diversity. The 17 countries included in this region share a history of Iberian colonization. Three-quarters of the region's 555 million people live in cities, making it the most urbanized region of the developing world. It is noted for its production of primary exports and manufactured goods, although rates of economic development vary greatly among states.

Intraregional trade within Latin America, stimulated by **Mercosur** (the Southern Cone Common Market), as well as the impact of the North American Free Trade Agreement (NAFTA; see Chapter 3) and the **Central American Free Trade Association (CAFTA)**, is an indicator of heightened economic integration in the western hemisphere.

Roughly equal in area to North America, Latin America has a much larger and faster-growing population of 555 million people. Its most populated state, Brazil, has nearly 200 million people, making it the fifth largest country in the world. The next largest state, Mexico, has a population of 116 million. Collectively, many Latin American states fall into the middle-income category and support a significant middle class. Yet poverty is a major concern for this region.

FIGURE 4.2 Tropical Wilderness The biological diversity of Latin America—home to the world's largest rain forest—is increasingly seen as a genetic and an economic asset. This forested mesa (called a *tepui*) in southern Venezuela is representative of the vast wilderness areas that many conservationists seek to protect.

It is estimated that 12 percent of the people in the region live on less than $2 per day. In the 1990s, the percentage living on less than $2 per day was nearly twice as large, which shows that efforts to reduce extreme poverty are working.

Unlike other developing regions today, Latin America is decidedly urban. Prior to World War II, most people lived in rural settings and worked as farmers. Today, three-quarters of Latin Americans are city dwellers. Even more startling is the number of **megacities**. São Paulo, Mexico City, Buenos Aires, and Rio de Janeiro all have more than 10 million residents. In addition, more than 40 cities have at least 1 million residents. The residents of these cities aspire for global recognition. A dozen Brazilian cities will host the World Cup in 2014, and Rio de Janeiro will proudly host the Olympics in 2016.

Despite the region's growing industrial capacity, extractive industries will continue to prevail, in part because of the area's impressive resource base. Latin America is home to Earth's largest rain forest, the greatest river by volume, and massive reserves of natural gas, oil, and copper. With its extensive territory, its tropical location, and its relatively low population density (Latin America has half the population of India in nearly seven times the area), the region is also recognized as one of the world's great reserves of biological diversity (Figure 4.2). How this diversity will be managed in the face of global demand for natural resources is an increasingly important question for the countries of this region.

ENVIRONMENTAL GEOGRAPHY: NEOTROPICAL DIVERSITY AND DEGRADATION

Much of Latin America is characterized by its tropicality. Travel posters of the region showcase lush forests and brightly colored parrots. The diversity and uniqueness of the **neotropics** (tropical ecosystems of the western hemisphere) have long been attractive to naturalists eager to understand their unique flora and fauna. It is no accident that Charles Darwin's theory of evolution was inspired by his two-year journey in tropical America. Even today, scientists throughout the region work to understand complex ecosystems, discover and protect new species, and interpret the impact of human settlement, especially in neotropical forests.

Not all of the region is tropical. Important population centers extend below the Tropic of Capricorn—most notably, Buenos Aires, Argentina, and Santiago, Chile. Much of northern Mexico, including the city of Monterrey, is north of the Tropic of Cancer. Yet Latin America's tropical climate and vegetation affect most popular images of the region. Given the territory's large size and relatively low population density, Latin America has not experienced the same levels of environmental degradation witnessed in East Asia and Europe. The region's biggest environmental concerns are related to deforestation, loss of biodiversity, and livability of urban areas (Figure 4.3).

Huge areas of Latin America remain relatively untouched, supporting an incredible diversity of plant and animal life. Throughout the region, national parks offer some protection to unique communities of plants and animals. A growing environmental movement in countries such as Costa Rica and Brazil has yielded both popular and political support for conservation efforts. In short, Latin Americans have entered the 21st century with a real opportunity to avoid many of the environmental mistakes seen in other regions of the world. At the same time, global market forces are driving governments to exploit minerals, fossil fuels, forests, shorelines, and soils. The region's biggest natural resource management challenge is to balance the economic benefits of extraction with the principles of sustainable development. Another major challenge is to improve the environmental quality of Latin American cities.

The Destruction of Tropical Rain Forests

Perhaps the environmental issue most commonly associated with Latin America is deforestation. The Amazon Basin and portions of the eastern lowlands of Central America and Mexico still maintain unique and impressive stands of tropical forest. Other woodland areas, such as the Atlantic coastal forests of Brazil and the Pacific forests of Central America, have nearly disappeared as a result of agriculture, settlement, and ranching. The coniferous forests of northern Mexico are also falling, in part because of a bonanza for commercial logging stimulated by NAFTA. In Chile, the ecologically unique evergreen rain forest (the Valdivian forest) in the midlatitudes is being cleared for wood chip exports to Asia.

The loss of tropical rain forests is most critical in terms of biological diversity. Tropical rain forests cover only 6 percent of Earth's landmass, but at least 50 percent of the world's species are found in this biome. Moreover, the Amazon contains the largest undisturbed stretches of rain forest in the world. Unlike Southeast Asian forests, where hardwood extraction drives deforestation, Latin American forests are usually seen as an agricultural frontier. State governments divide areas in an attempt to give land to the landless and reward political elites. Thus, forests are cut and burned, with settlers and politicians carving them up to create permanent settlements, slash-and-burn plots, or large cattle ranches. In addition, some tropical forest cutting has been motivated by the search for gold (Brazil, Venezuela, and Costa Rica) and the production of coca leaf for cocaine (Peru, Bolivia, and Colombia).

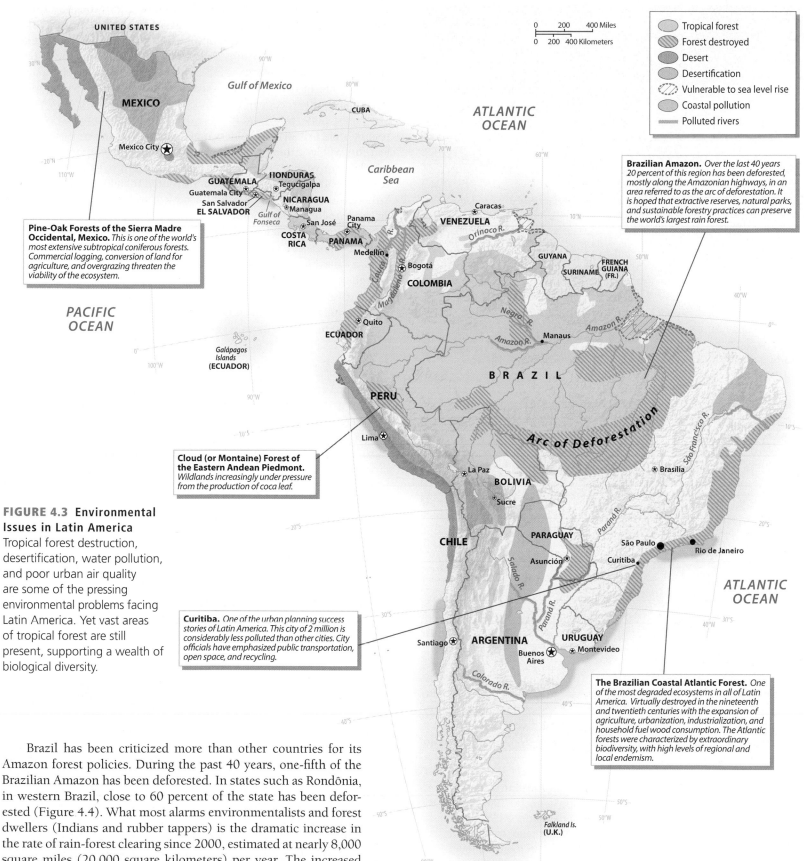

FIGURE 4.3 Environmental Issues in Latin America Tropical forest destruction, desertification, water pollution, and poor urban air quality are some of the pressing environmental problems facing Latin America. Yet vast areas of tropical forest are still present, supporting a wealth of biological diversity.

Legend:
- Tropical forest
- Forest destroyed
- Desert
- Desertification
- Vulnerable to sea level rise
- Coastal pollution
- Polluted rivers

Pine-Oak Forests of the Sierra Madre Occidental, Mexico. *This is one of the world's most extensive subtropical coniferous forests. Commercial logging, conversion of land for agriculture, and overgrazing threaten the viability of the ecosystem.*

Brazilian Amazon. *Over the last 40 years 20 percent of this region has been deforested, mostly along the Amazonian highways, in an area referred to as the arc of deforestation. It is hoped that extractive reserves, natural parks, and sustainable forestry practices can preserve the world's largest rain forest.*

Cloud (or Montaine) Forest of the Eastern Andean Piedmont. *Wildlands increasingly under pressure from the production of coca leaf.*

Curitiba. *One of the urban planning success stories of Latin America. This city of 2 million is considerably less polluted than other cities. City officials have emphasized public transportation, open space, and recycling.*

The Brazilian Coastal Atlantic Forest. *One of the most degraded ecosystems in all of Latin America. Virtually destroyed in the nineteenth and twentieth centuries with the expansion of agriculture, urbanization, industrialization, and household fuel wood consumption. The Atlantic forests were characterized by extraordinary biodiversity, with high levels of regional and local endemism.*

Brazil has been criticized more than other countries for its Amazon forest policies. During the past 40 years, one-fifth of the Brazilian Amazon has been deforested. In states such as Rondônia, in western Brazil, close to 60 percent of the state has been deforested (Figure 4.4). What most alarms environmentalists and forest dwellers (Indians and rubber tappers) is the dramatic increase in the rate of rain-forest clearing since 2000, estimated at nearly 8,000 square miles (20,000 square kilometers) per year. The increased rates of deforestation in the Brazilian Amazon are due to the expansion of industrial mining and logging, the growth of corporate farms, the development of new road networks, the incidence of human-ignited wildfires, and continued population growth.

(a) July 30, 2000

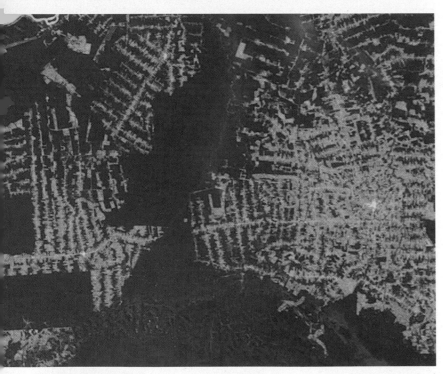

(b) August 2, 2010

FIGURE 4.4 Tropical Forest Settlement in the Amazon   These satellite images of Rondônia, Brazil, illustrate the dramatic change in forest cover in just 10 years near the settlement of Buritis and road BR-364. Intact forest is dark green, whereas cleared areas are light green (crops) or tan (bare ground). Typically, the first clearings appear off of roads, forming a fishbone pattern. Over time, as more forest is cleared and settlements grow, the fishbone pattern collapses into a mosaic of pasture, farmland, and forest fragments.

Under the Advance Brazil program started in 2000, some $40 billion has gone to new highways, railroads, gas lines, hydroelectric projects, power lines, and river canalization projects that will reach into remote areas of the basin. In an effort to slow deforestation rates, the Brazilian government has created new conservation areas, many of them alongside the "arc of deforestation"—a swath of agricultural development along the southern edge of the Amazon Basin (see Figure 4.3). Yet Brazil's Forest Code, as revised in 2012, has reduced the amount of "forest reserve" that private landholders must maintain. Many conservationists fear this will lead to more forest clearing and fragmentation.

The conversion of tropical forest into pasture, called **grassification**, is another practice that has contributed to deforestation. Particularly in southern Mexico, Central America, and the Brazilian Amazon, an assortment of development policies from the 1960s through the 1980s encouraged deforestation to make room for cattle in areas of frontier settlement (Figure 4.5). Although many natural grasslands such as the Llanos (Venezuela and Colombia), the Chaco (Bolivia, Paraguay, and Argentina), and the Pampas (Argentina) are suitable for grazing, the rush to convert forest into pasture made ranching so environmentally destructive. Ironically, cattle ranching in remote tropical frontiers is seldom economically self-sustaining.

Urban Environmental Challenges

For most Latin Americans, air pollution, water availability and quality, and garbage removal are the pressing environmental problems of everyday life. Consequently, many environmental activists from the region focus their efforts on making urban environments cleaner by introducing "green" legislation and calling people to action. In this most urbanized region of the developing world, city dwellers do have better access to water, sewers, and electricity than their counterparts in Asia and Africa. Moreover, the density of urban settlement seems to encourage the widespread use of mass transportation; both public and private bus and van routes make getting around cities fairly easy.

FIGURE 4.5 Converting Forest into Pasture   Cattle graze in northern Guatemala's Petén region. Clearing of this tropical forest lowland began in the 1960s and continues today. Ranching is a status-conferring occupation in Latin America with serious ecological costs. The beef produced from this region is for domestic and export markets.

However, the usual environmental problems that come from dense urban settings ultimately require expensive remedies, such as new power plants and modernized sewer and water lines. The money for such projects is never enough, due to currency devaluation, inflation, and foreign debt. Because many urban dwellers tend to reside in unplanned squatter settlements, servicing these communities with utilities after they are built is difficult and costly.

Air Pollution Air pollution is a concern for most major cities, and this is especially true for Santiago and Mexico City. A combination of geographical factors (basin settings) and meteorological factors (winter inversion layers), along with dense human settlement and automobile dependence, has led to these two cities having some of the highest recorded concentrations of particulate matter and ozone, major contributors to air pollution. Air pollution is not just an aesthetic issue—the health costs of breathing such contaminated air are real, as elevated death rates due to heart disease, asthma, influenza, and pneumonia suggest. The burden of air pollution is not evenly distributed among city residents, as the elderly, the very young, and the poor are more likely to suffer the negative health effects of contaminated air. Fortunately, both cities have taken steps to address this vexing problem.

Santiago, a prosperous city of nearly 7 million in Chile's Central Valley, has an elevation of 1,700 feet (520 meters). Although not as high as Mexico City, this basin setting regularly produces thermal inversions, when warm air traps a layer of cold air near the surface. This trapped surface layer becomes filled with engine exhaust, industrial pollution, garbage, and even fecal matter (Figure 4.6). The inversion layers happen year round, but can be especially bad in the winter months from May through August, often forcing schools to suspend all sports when smog emergencies are called. Santiago officials began addressing this problem in the late 1980s by restricting vehicular traffic. On a given weekday, 20 percent of all buses, taxis, and cars are restricted from driving, based on license plate numbers. During smog emergencies, up to 40 percent of vehicles can be restricted from driving. In addition, numerous private buses were replaced by a large fleet of clean-running public buses. These buses, combined with the city's subway system, can move over 2 million riders a day. By 2000, air quality had noticeably improved, and public support for restrictive measures and public transport has solidified.

The smog of Mexico City has been so bad that most modern-day visitors have no idea that mountains surround them. Air quality has been a major issue for Mexico City since the 1960s, driven in part by the city's unusually high rate of growth. (Between 1950 and 1980, the city's annual rate of growth was 4.8 percent.) It is difficult to imagine a better setting for creating air pollution. The city sits in a bowl 7,400 feet (2,200 meters) above sea level, and thermal inversions regularly form. Steps were finally taken in the late 1980s to reduce emissions from factories and cars. Unleaded gas is now widely available for the 4 million cars in the metropolitan area, and cars manufactured for the Mexican market must have catalytic converters. Also, some of the worst polluting factories in the Valley of Mexico have closed. In the last few years, the mayor of Mexico City has expanded a low-emissions bus system, eliminating thousands of tons of carbon monoxide. In 2007, the decision was made to close the elegant Paseo de Reforma to traffic on Sunday mornings and open it to bike riders. This change was so popular that now bike lanes have been introduced to some downtown areas in an effort to encourage bike ridership. For longer commutes, a suburban train system is being built that will

FIGURE 4.6 Air Pollution in Santiago Smog blankets Santiago, with the Andes in the background. During the winter months (May through August), thermal inversion layers form that trap pollutants near ground level, causing a spike in pollution-related health problems. By reducing vehicular traffic and greatly expanding public transportation, the city has experienced improved air quality.

complement the existing subway system. The payoff is real: Mexico City no longer ranks among the top polluted cities in the world, and it appears to have cut most of its pollutants by at least half.

Water Mexico City's other significant environmental problem is water. When Vicente Fox was president of Mexico, he declared water (both scarcity and quality) a national security issue, not just for the capital, but for the entire country. Ironically, it was the abundance of water that made this site attractive for settlement initially. Large shallow lakes once filled the valley, but over the centuries, most were drained to expand agricultural land. As surface water became scarce, wells were dug to tap the basin's massive freshwater aquifer. Today approximately 70 percent of the water used in the metropolitan area is drawn from the valley's aquifer. There is troubling evidence that the aquifer is being overdrawn and at risk of contamination, especially in areas where unlined drainage canals can leak pollutants into the surrounding soil, which then leach into the aquifer. To reduce reliance on the aquifer, the city now pumps water nearly a mile uphill from more than 100 miles (160 kilometers) away.

Andean Cities such as Bogotá, Quito, and La Paz are increasingly experiencing water scarcity and rationing. Some of this is due to increased demands on aging water systems due to population growth. However, changes in precipitation patterns due to El Niño years or global climate change make these large urban centers especially vulnerable. La Paz, Bolivia, for example, gets much of its water from glacial runoff. A major Bolivian glacier, Chacaltaya, has lost 80 percent of its area in the past 20 years. Thus, as average temperatures increase in the highlands and glaciers recede, there is widespread concern about future drinking-water supplies in this metropolitan area of nearly 2 million people.

Working Toward Sustainable Cities Other Latin American cities offer important examples of urban sustainability. Curitiba is the celebrated "green city" of Brazil because of some relatively simple, yet progressive, planning decisions. More than 2 million people live

FIGURE 4.7 Bogotá's TransMilenio A modern fleet of some 1,500 articulated red buses glides through Bogotá, carrying over 1.5 million riders per day. Made efficient by building dedicated bus lanes and creating quick-loading platforms, this bus system is used by the rich and poor in this Andean city. Commuting times have been cut in half, and air quality has improved. The one complaint is that the system is too crowded because so many use it.

in this industrial and commercial center, yet it is significantly less polluted than similar-sized cities. Because the city's location was vulnerable to flooding, city planners built drainage canals and set aside the remaining natural drainage areas as parks in the 1960s, well before explosive growth would have made such a policy difficult. This action added green space and reduced the negative impacts of flooding. Also, public transportation became a top priority. An extensive bus system that featured rapid loading and unloading of passengers made Curitiba a model for transportation in the developing world. Finally, a low-tech, but effective recycling program has greatly reduced solid waste.

In the past decade, the city of Bogotá, Colombia, has been transformed through a rapid-transit bus system called TransMilenio. Opened in 2000 and still expanding, this system links most areas of the city with modern articulated buses that use dedicated bus lanes and rapid-loading platforms that mimic subways (Figure 4.7). Since the buses have their own lanes, taking the bus is faster than driving on the city's congested roads, so most bus riders have shorter commutes. In addition, air quality has improved because fewer cars are on the roads. Like many European cities, Bogotá's urban planners integrated bicycle use into the city's transportation system. In the past decade, bike lanes have been built throughout the city, and bike stations are found at suburban TransMilenio bus stops. Bogotá has a high elevation of 8,600 feet (2,600 meters), but it is relatively flat, so the bicycle is a practical and clean form of urban transport, making Bogotá one of Latin America's more bike-friendly cities.

Cities such as Curitiba and Bogotá demonstrate that sustainable planning efforts not only improve the environment, but also can address social needs as well. To appreciate the diversity of environments Latin Americans must cope with when making development decisions, an understanding of the region's physical geography is important.

Western Mountains and Eastern Lowlands

Latin America is a region of diverse landforms, including high mountains, extensive upland plateaus, and vast river basins. The movement of tectonic plates explains much of the region's basic topography, including the formation of its geologically young western mountain ranges, such as the Andes and the Volcanic Axis of Central America (see Figure 4.1). This area is also geologically active, especially with regard to earthquakes that threaten people and damage property. In February 2010, for example, a massive 8.8 magnitude earthquake struck off the coast near the Chilean city of Concepción, killing some 400 people and unleashing tsunami warnings across the Pacific. Because the epicenter was not near a major population center and Chilean buildings are engineered to withstand earthquakes, the death toll was remarkably light, given the intensity of the event. In contrast, the Atlantic side of South America is characterized by humid lowlands interspersed with large upland plateaus called **shields**. The Brazilian Shield is the largest, followed by the Patagonian and Guiana shields. Across these lowlands meander some of the great rivers of the world, including the Amazon, Plata, and Orinoco.

Historically, the most important areas of settlement in tropical Latin America were not along the region's major rivers, but across its shields, plateaus, and fertile mountain valleys. In these places, the combination of arable land, mild climate, and sufficient rainfall produced the region's most productive agricultural areas and its densest settlement. The Mexican Plateau, for example, is a massive upland area ringed by the Sierra Madre mountains. The southern end of the plateau is where the Valley of Mexico is located. Similarly, the elevated and well-watered basins of Brazil's southern mountains provide an ideal setting for agriculture. These especially fertile areas are able to support high population densities, so it is not surprising that the region's two largest cities, Mexico City and São Paulo, emerged in these settings. The Latin American highlands also lend a special character to the region. Lush tropical valleys nestled below snow-covered mountains hint at the diversity of ecosystems found near one another. The most dramatic of these highland areas, the Andes, runs like a spine down the length of the South American continent.

The Andes Beginning in northwestern Venezuela and ending at Tierra del Fuego, the Andes are relatively young mountains that extend nearly 5,000 miles (8,000 kilometers). They are an ecologically and geologically diverse mountain chain, with some 30 peaks higher than 20,000 feet (6,000 meters). Created by a collision of oceanic and continental plates, the Andes are a series of folded and faulted sedimentary rocks with intrusions of crystalline and volcanic rock. Many rich veins of precious metals and minerals are found in these mountains. In fact, the initial economic wealth of many Andean countries came from mining silver, gold, tin, copper, and iron.

Given the length of the Andes, the mountain chain is typically divided into northern, central, and southern components. In Colombia, the northern Andes split into three distinct mountain ranges before merging near the border with Ecuador. High-altitude plateaus and snow-covered peaks distinguish the central Andes of Ecuador, Peru, and Bolivia. The Andes reach their greatest width here. Of special interest is the treeless high plain of Peru and Bolivia, called the **Altiplano**. The floor of this elevated plateau ranges from 11,800 feet (3,600 meters) to 13,000 feet (4,000 meters) in altitude, and it has limited usefulness for grazing. Two high-altitude lakes—Titicaca on the Peruvian and Bolivian border and the smaller Poopó in Bolivia—are located in the

FIGURE 4.8 Bolivian Altiplano The Altiplano is an elevated plateau straddling the Bolivian and Peruvian Andes. One of its striking features is beautiful Lake Titicaca, at an elevation of 12,500 feet (3,800 meters). Amerindians inhabit this stark and windswept land, which is one of the most scenic areas of the Andes.

Altiplano, as are many mining sites (Figure 4.8). The highest peaks are found in the southern Andes, shared by Chile and Argentina, including the highest peak in the western hemisphere, Aconcagua, at almost 23,000 feet (7,000 meters).

The Uplands of Mexico and Central America The Mexican Plateau and the Volcanic Axis of Central America are the most important Latin American uplands in terms of settlement. Most major cities of Mexico and Central America are found here. The Mexican Plateau is a large, tilted block that has its highest elevations, about 8,000 feet (2,500 meters), in the south, around Mexico City, and its lowest, just 4,000 feet (1,200 meters), at Ciudad Juárez. The southern end of the plateau, the Mesa Central, contains several flat-bottomed basins interspersed with volcanic peaks. It also contains Mexico's megalopolis—a concentration of the largest population centers such as Mexico City and Puebla.

Along the Pacific Coast of Central America lies a chain of volcanoes that stretches from Guatemala to Costa Rica, called the Volcanic Axis of Central America. It is a handsome landscape of rolling green hills, elevated basins with sparkling lakes, and volcanic peaks. More than 40 volcanoes are found here, many of them still active, which have produced a rich volcanic soil that yields a wide variety of domestic and export crops. Most of Central America's population is also concentrated in this zone, in the capital cities or the surrounding rural villages. The bulk of the agricultural land is tied up in large holdings that produce beef, cotton, and coffee for export. However, in terms of numbers, most of the farms are small subsistence properties that produce corn, beans, squash, and assorted fruits.

The Eastern Shields South America has three major shields—large upland areas of exposed crystalline rock that are similar to upland plateaus found in Africa and Australia. (The Guiana Shield will be discussed in Chapter 5.) The Brazilian and Patagonian shields vary in elevation between 600 and 5,000 feet (200 and 1,500 meters). The Brazilian Shield is larger and more important in terms of natural resources and settlement. Far from a uniform land surface, the Brazilian Shield covers much of Brazil from the Amazon Basin in the north to the Plata Basin in the south. In the southeast corner of the plateau is the city of São Paulo, the largest urban conglomeration in South America. The other major population centers are on the coastal edge of the plateau, where large protected bays made the sites of Rio de Janeiro and Salvador attractive to Portuguese colonists. Finally, the Paraná basalt plateau, located on the southern end of the Brazilian Shield, is famous for its fertile red soils (*terra roxa*), which yield coffee, oranges, and soybeans. So fertile is this area that the economic rise of São Paulo is attributed to the expansion of commercial agriculture, especially coffee, into this area (Figure 4.9).

The Patagonian Shield lies in the southern tip of South America. Beginning south of Bahia Blanca and extending to Tierra del Fuego, the region to this day is sparsely settled and hauntingly beautiful. It is treeless, covered by scrubby steppe vegetation, and home to wildlife such as the guanaco (Figure 4.10). Sheep were introduced to Patagonia in the late 19th century, spurring a wool boom. More recently, offshore oil production has renewed the economic importance of Patagonia.

River Basins Three great river basins drain the Atlantic lowlands of South America: the Amazon, Plata, and Orinoco. The Amazon drains an area of roughly 2.3 million square miles (5.9 million square kilometers), making it the largest river system in the world by volume and area and the second largest by length. Home to the world's largest rain forest, annual rainfall is more than 60 inches (150 centimeters) everywhere in the basin and close to 100 inches (250 centimeters) in the basin's largest city, Belem. The mighty Amazon drains eight countries, but two-thirds of the watershed is within Brazil. Active settlement of the Brazilian portion of the Amazon since the 1960s has boosted the population. Roughly 15 million people live in the Brazilian Amazon, and the population is increasing at 4 percent a year. The development of the basin—most notably through towns, roads, dams, farms, and mines—is forever changing what was viewed as a vast tropical wilderness just a half century ago.

FIGURE 4.9 Brazilian Oranges The states of São Paulo and Paraná have some of the finest soils in Brazil. Most estate-grown oranges in Brazil are processed into frozen concentrate and exported. In addition to oranges, coffee and soybeans are widely cultivated.

FIGURE 4.10 Patagonian Wildlife Guanacos, native to South America, thrive on the steppe vegetation found throughout Patagonia. The numbers of guanacos fell dramatically due to hunting and competition with introduced livestock but thrive today in protected areas such as Torres del Paine in Chile.

The region's second largest watershed, the Plata Basin, begins in the tropics and discharges into the Atlantic in the midlatitudes near Buenos Aires. Several major rivers make up this system: the Paraná, the Paraguay, and the Uruguay. Unlike the Amazon, much of the Plata Basin is now economically productive through large-scale mechanized agriculture, especially soybean production. The basin contains several major dams, including the region's largest hydroelectric plant, the Itaipú on the Paraná, which generates electricity for all of Paraguay and much of southern Brazil (Figure 4.11). As agricultural output in the watershed grows, sections of the Paraná River have been canalized and dredged to enhance the river's capacity for barge and boat traffic.

The third largest basin by area is the Orinoco in northern South America. The Orinoco River meanders through much of southern Venezuela and part of eastern Colombia, giving character to the sparsely settled tropical grasslands called the *Llanos*. Although it is only one-seventh the size of the Amazon watershed, its discharge

FIGURE 4.11 Itaipú Dam The largest dam in Latin America, the Itaipú blocks the flow of the Paraná River on the border between Paraguay and Brazil. The power station at Itaipú generates all of Paraguay's electricity needs and much of southern Brazil's power.

roughly equals that of the Mississippi River. Since the colonial era, these grasslands have supported large cattle ranches. Although cattle are still important, the Llanos are also a dynamic area of petroleum production from both Colombia and Venezuela.

Climate and Climate Change in Latin America

In tropical Latin America, average monthly temperatures in settings such as Managua (Nicaragua), Quito (Ecuador), and Manaus (Brazil) show little variation (see the climographs in Figure 4.12). Precipitation patterns, however, are variable and create distinct wet and dry seasons. In Managua, for example, January is typically a dry month, and October is a wet one. The tropical lowlands of Latin America, especially east of the Andes, are usually classified as tropical humid climates that support forest or savanna, depending on the amount of rainfall. The region's desert climates are found along the Pacific coasts of Peru and Chile, Patagonia, northern Mexico, and the Bahia of Brazil. Thus, a city such as Lima, Peru, which is clearly in the tropics, averages only 1.5 inches (4 centimeters) of rainfall a year due to the extreme aridity of the Peruvian coast.

Midlatitude climates, with hot summers and cold winters, prevail in Argentina, Uruguay, and parts of Paraguay and Chile (see the climographs in Figure 4.12 for Buenos Aires and Punta Arenas). Recall that the midlatitude temperature shifts in the southern hemisphere are the opposite of those in the northern hemisphere (cold Julys and warm Januarys). In the mountain ranges, complex climate patterns result from changes in elevation. To appreciate how humans adapt to tropical mountain ecosystems, the concept of **altitudinal zonation**, which is the relationship between cooler temperatures at higher elevations and changes in vegetation, is important.

Altitudinal Zonation First described in the scientific literature by Alexander von Humboldt in the early 1800s, altitudinal zonation has practical applications that are intimately understood by all the region's native inhabitants. Humboldt systematically recorded declines in temperature as he ascended to higher elevations, a phenomenon known as the **environmental lapse rate**. According to Humboldt's description of the environmental lapse rate, temperature declines approximately 3.5°F for every 1,000 feet in higher elevation, or 6.5°C for every 1,000 meters. Humboldt also noted changes in vegetation by elevation, demonstrating that plant communities common to the midlatitudes could thrive in the tropics at higher elevations. These different altitudinal zones are commonly referred to as the *tierra caliente* (hot land), from sea level to 3,000 feet (900 meters); the *tierra templada* (temperate land), at 3,000 to 6,000 feet (900 to 1,800 meters); the *tierra fría* (cold land), at 6,000 to 12,000 feet (1,800 to 3,600 meters); and the *tierra helada* (frozen land), above 12,000 feet (3,600 meters). Exploitation of these zones allows agriculturists, especially in the uplands, access to a great diversity of domesticated and wild plants (Figure 4.13).

The concept of altitudinal zonation is most relevant for the Andes, the highlands of Central America, and the Mexican Plateau. For example, traditional Andean farmers might use the high pastures of the Altiplano for grazing llamas and alpacas, the tierra fría for potato and quinoa production, and the lower temperate zone for corn production. All the great pre-contact civilizations, especially the Incas and the Aztecs, systematically extracted resources from these zones, thus ensuring a diverse and abundant resource base. Yet these complex ecosystems are extremely fragile and have become important areas of research for the effects of climate change in the tropics.

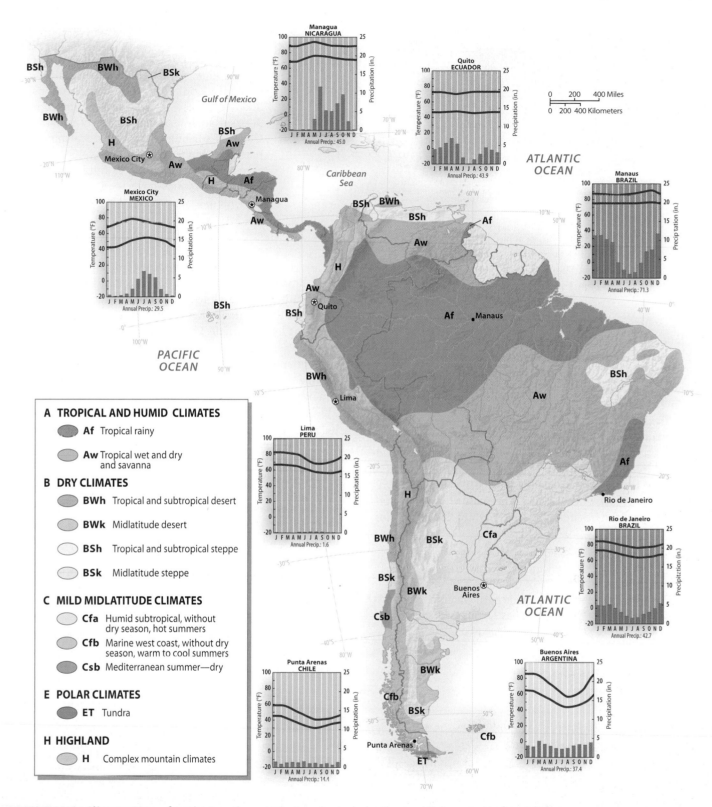

FIGURE 4.12 Climate Map of Latin America Latin America includes the world's largest rain forest (Af) and driest desert (BWh), as well as nearly every other climate classification. Latitude, elevation, and rainfall play important roles in determining the region's climates. Note the contrast in rainfall patterns between humid Quito and arid Lima.

El Niño One of the most studied weather phenomena in Latin America, **El Niño** (referring to the Christ child), occurs when a warm Pacific current arrives along the normally cold coastal waters of Ecuador and Peru in December, around Christmastime. This change in ocean temperature, which happens every few years, produces torrential rains, signaling the arrival of an El Niño year. The 2009–2010 El Niño was especially bad for Latin America; scores of people were killed by floods or storms attributed to El Niño–related disturbances.

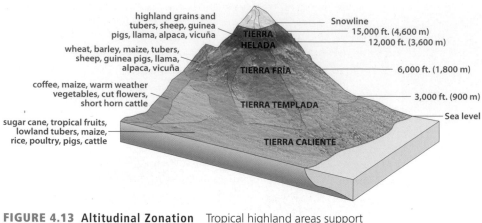

FIGURE 4.13 **Altitudinal Zonation** Tropical highland areas support a complex array of ecosystems. In the tierra fría zone for example, midlatitude crops such as wheat and barley can be grown. The diagram depicts the range of crops and animals found at different elevations in the Andes.

Devastating floods occurred in Peru and Brazil. In Peru, heavy rains and flooding damaged the railroad leading to the ancient Incan site of Machu Picchu, temporarily limiting access to this popular tourist destination until the railroad could be rebuilt.

Other than flooding, the less-talked-about result of El Niño is drought. While the Pacific Coast of South and North America experienced record rainfall in the 1997–1998 El Niño, Colombia, Venezuela, northern Brazil, Central America, and Mexico battled drought. In addition to crop and livestock losses, estimated to be in the billions of dollars, hundreds of brush and forest fires left their mark on the region's landscape. Scientists are not yet sure whether or how global climate change will impact the frequency and strength of El Niño cycles.

FIGURE 4.14 **Glacial Retreat** Research in areas such as the Peruvian Andes over the past four decades demonstrates a dramatic decline in the size of Andean glaciers. This is an indicator of the pace of climate change in the tropics. Highland glaciers are also sources of water for mountain villages and cities; their demise has serious repercussions for the communities that depend on them.

Impacts of Climate Change Global climate change has both immediate and long-term implications for Latin America. Of greatest immediate concern is how climate change will influence agricultural productivity, water availability, changes in the composition and productivity of ecosystems, and incidence of vector-borne diseases such as malaria and, especially, dengue fever. Changes attributable to global warming are already apparent in higher-elevation regions, making these concerns more pressing. For example, coffee growers in the Colombian Andes have seen a decline in productivity over the past five years, which they attribute to higher temperatures and longer dry spells. The long-term effects of global climate change on lowland tropical forest systems is less clear; for example, some areas may experience more rainfall, others less.

Climate change research indicates that highland areas are particularly vulnerable to global warming. Tropical mountain systems are projected to experience temperature increases of 2 to 6°F (1 to 3°C), as well as lower rainfall. This will raise the altitudinal limits of various ecosystems, impacting the range of crops and arable land available to farmers and pastoralists. Research over the past 50 years has documented the dramatic retreat of Andean glaciers; some no longer exist, and others will cease to exist in the next 10 to 15 years (Figure 4.14). This is a visible indicator of global warming, and it also has pressing human repercussions, since many mountain communities, as well as large cities, rely upon water from glacial runoff.

Another immediate concern brought on by warmer temperatures is the sudden rise in dengue fever, a mosquito-borne virus. Dengue fever was once considered relatively uncommon in highland Latin America, but the number of cases has risen sharply in the past decade. Tens of thousands now suffer from its fever, headache, nausea, joint pain, and, in rare cases, external and internal bleeding that can result in death. In Latin America, the sudden rise in cases of dengue fever suggests that warmer highland temperatures have placed millions more at risk.

POPULATION AND SETTLEMENT: THE DOMINANCE OF CITIES

Latin America did not have great river basin civilizations like those in Asia. In fact, the great rivers of the region are surprisingly underused as areas of settlement or corridors for transportation. The major population clusters of Central America and Mexico are in the interior plateaus and valleys, and the interior lowlands of South America are relatively empty. Historically, the highlands supported most of the region's population during the pre-Hispanic and colonial eras. In the 20th century, population growth and migration to the Atlantic lowlands of Argentina and Brazil, along with continued growth of Andean coastal cities such as Guayaquil, Barranquilla, and Maracaibo, have reduced the demographic importance of the highlands. Major highland cities such as Mexico City, Guatemala City, Bogotá, and La Paz still dominate their national economies, but the majority of large cities are on or near the coasts (Figure 4.15).

Like the rest of the developing world, Latin America experienced dramatic population growth in the 1960s and 1970s. In 1950, its population totaled 150 million people, which equaled the population of the United States at that time. By 1995, the population had tripled to 450 million; in comparison, the United States only reached 300 million people in 2006. Latin America outpaced the United States because its infant mortality rate declined and its life expectancy soared, while its birthrate remained higher than that of the United States. In 1950, Brazilian life expectancy was only 43 years; by the 1980s, it was 63, and by 2010 it was 73. In fact, most countries in the region experienced a 15- to 20-year improvement in life expectancy between 1950 and 1980, which pushed up growth rates. Four countries account for 70 percent of the region's population: Brazil with 194 million, Mexico with 116 million, Colombia with 47 million, and Argentina with 41 million (Table 4.1).

The Latin American City

A quick glance at the population map of Latin America shows a concentration of people in cities (see Figure 4.15). One of the most significant demographic shifts has been the movement out of rural areas to cities, which began in earnest in the 1950s. In 1950, just one-quarter of the region's population was urban; the rest lived in small villages and the countryside. Today the pattern is reversed, with three-quarters of the population living in cities. In the most urbanized countries, such as Argentina, Chile, Uruguay, and Venezuela, nearly 90 percent of the population lives in cities. This preference for urban life is attributed to cultural as well as economic factors. Under Iberian rule, people residing in cities had higher social status and greater economic opportunity.

TABLE 4.1   POPULATION INDICATORS

Country	Population (millions) 2012	Population Density (per square kilometer)	Rate of Natural Increase (RNI)	Total Fertility Rate	Percent Urban	Percent <15	Percent >65	Net Migration (Rate per 1000) 2010–15[a]
Argentina	40.8	15	1.1	2.4	91	25	10	−0.5
Bolivia	10.8	10	1.9	3.3	66	36	5	−3.0
Brazil	194.3	23	1.0	1.9	84	24	7	−0.2
Chile	17.4	23	1.0	1.9	87	23	9	0.3
Colombia	47.4	42	1.3	2.1	76	29	6	−0.5
Costa Rica	4.5	88	1.1	1.8	62	24	7	2.7
Ecuador	14.9	52	1.6	2.5	66	30	6	−1.6
El Salvador	6.3	298	1.4	2.3	63	32	7	−7.3
Guatemala	15.0	138	2.4	2.6	50	41	4	−1.0
Honduras	8.4	75	2.2	3.2	50	38	4	−1.3
Mexico	116.1	59	1.5	2.3	77	29	6	−2.3
Nicaragua	6.0	46	1.9	2.6	57	35	5	−4.0
Panama	3.6	48	1.5	2.4	65	29	7	0.6
Paraguay	6.6	16	1.9	3	59	34	5	−1.2
Peru	30.1	23	1.5	2.6	74	30	6	−2.8
Uruguay	30.4	19	0.4	2.0	94	23	14	−1.8
Venezuela	29.7	33	1.5	2.5	88	29	6	0.3

[a]Net Migration Rate from the United Nations, Population Division, *World Population Prospects: The 2010 Revision Population Database.*
Source: Population Reference Bureau, *World Population Data Sheet, 2012.*

FIGURE 4.15 Population Map of Latin America The concentration of population in urban and coastal settlements is evident in this map. Population density in central and southern Mexico, as well as in Central America, is quite high. In South America, the majority of people live on or near the coasts, leaving the interior of the continent lightly populated.

Initially, only Europeans were allowed to live in the colonial cities, but this exclusivity was not strictly enforced. Over the centuries, colonial cities became the hubs for transportation and communication, making them the primary centers for economic and social activities.

Latin American cities are noted for high levels of **urban primacy**, a condition in which a country has a primate city three to four times larger than any other city in the country. Examples of primate cities are Lima, Caracas, Guatemala City, Panama City, Santiago, Buenos Aires, and Mexico City. Primacy is often viewed as a problem because too many national resources are concentrated into one urban center. In three cases, the growth of urbanized regions has led to the emergence of a megalopolis: the Mexico City–Puebla–Toluca–Cuernavaca area on Mexico's Mesa Central; the Niterói–Rio de Janeiro–Santos–São Paulo–Campinas axis in southern Brazil; and the Rosario–Buenos Aires–Montevideo–San Nicolás corridor in Argentina and Uruguay's lower Rio Plata Basin (see Figure 4.15).

Urban Form Latin American cities have a distinct urban form that reflects both their colonial origins and their present-day growth (Figure 4.16). Usually, a clear central business district exists in the old colonial core. Radiating out from the central business district is older middle- and lower-class housing found in the zones of maturity and *in situ* accretion (an area of mixed levels of housing and services). In this model, residential quality declines moving from the center to the periphery. The exception is the elite spine, a newer commercial and business strip that extends from the colonial core to newer parts of the city. Along the spine are superior services, roads, and transportation. The city's best residential zones, as well as shopping malls, are usually on either side of the spine. Close to the elite residential sector, a limited area of middle-class housing is typically found. Most major urban centers also have a *periférico* (a ring road or beltway highway) that encircles the city. Industry is located in isolated areas of the inner city and in larger industrial parks outside the ring road.

Straddling the *periférico* is a zone of peripheral **squatter settlements**, where many of the urban poor live in self-built housing on land that does not belong to them. Services and infrastructure are extremely limited: Roads are unpaved, water is often trucked in, and sewer systems are nonexistent.

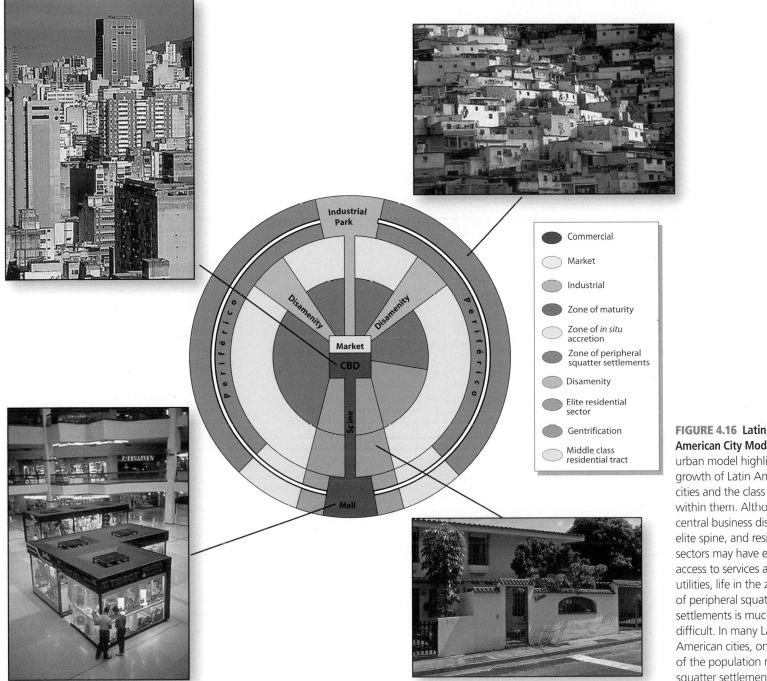

FIGURE 4.16 Latin American City Model   This urban model highlights the growth of Latin American cities and the class divisions within them. Although the central business district, elite spine, and residential sectors may have excellent access to services and utilities, life in the zone of peripheral squatter settlements is much more difficult. In many Latin American cities, one-third of the population resides in squatter settlements.

The dense ring of squatter settlements that encircles Latin American cities reflects the speed and intensity with which these zones were created. In some cities, more than one-third of the population lives in these self-built homes of marginal or poor quality. These kinds of dwellings are found throughout the developing world, yet the practice of building homes on the "urban frontier" has a longer history in Latin America than in most Asian and African cities. The combination of a rapid inflow of migrants (at times reaching 1,000 people per day), the inability of governments to meet pressing housing needs, and the eventual official recognition of many of these neighborhoods with land titles and utilities meant that this housing strategy was seldom discouraged. Each successful settlement on the urban edge encouraged more.

Patterns of Rural Settlement

Throughout Latin America, a distinct rural lifestyle exists, especially among peasant subsistence farmers. Although the majority of people live in cities, some 125 million people do not. In Brazil alone, at least 35 million people live in rural areas. Interestingly, the absolute number of people living in rural areas today is roughly equal to the number in the 1960s. Yet rural life has definitely changed. The links between rural and urban areas are much improved, making rural areas less isolated. In addition to village-based subsistence production, highly mechanized, capital-intensive farming occurs in most rural areas. Much like the region's cities, the rural landscape is divided by extremes of poverty and wealth. A source of social and economic tension in the countryside is the uneven distribution of arable land.

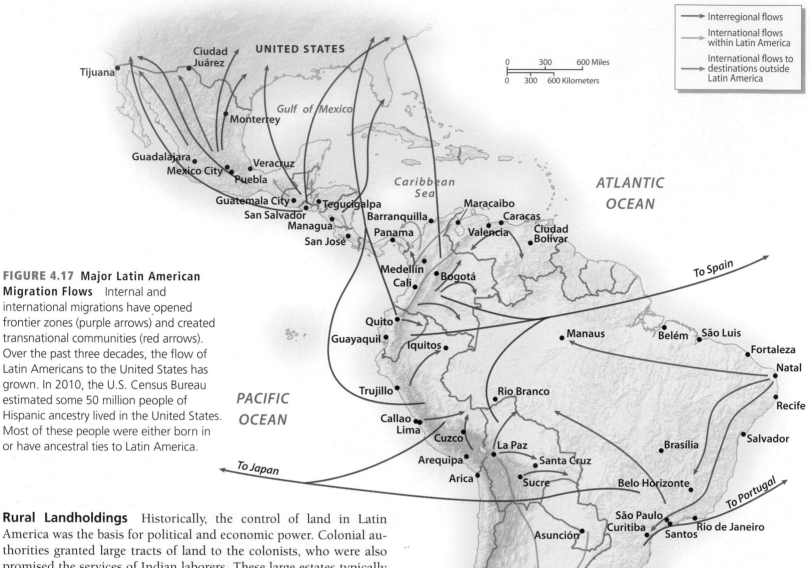

FIGURE 4.17 Major Latin American Migration Flows Internal and international migrations have opened frontier zones (purple arrows) and created transnational communities (red arrows). Over the past three decades, the flow of Latin Americans to the United States has grown. In 2010, the U.S. Census Bureau estimated some 50 million people of Hispanic ancestry lived in the United States. Most of these people were either born in or have ancestral ties to Latin America.

Rural Landholdings Historically, the control of land in Latin America was the basis for political and economic power. Colonial authorities granted large tracts of land to the colonists, who were also promised the services of Indian laborers. These large estates typically took up the best lands along the valley bottoms and coastal plains. The owners were often absentee landlords, spending most of their time in the city and relying on a mixture of hired and slave labor to run their rural operations. Passed down from one generation to the next, many estates can trace their ownership back a century or more. The establishment of large blocks of estate land meant that peasants were denied access to territory of their own, so they were forced to work for the estates. This long-observed practice of maintaining large estates is called **latifundia**.

Although the pattern of estate ownership is well documented, peasants have always farmed small plots for their subsistence. This practice of **minifundia** can lead to permanent or shifting cultivation. Small farmers typically plant a mixture of crops for subsistence, as well as for trade. Peasant farmers in Colombia or Costa Rica, for example, grow corn, fruits, and various vegetables alongside coffee bushes that produce beans for export. Strains on the minifundia system occur when rural populations grow and land becomes scarce, forcing farmers to divide their properties into smaller and less-productive parcels or seek out new parcels on steep slopes.

Much of the turmoil in 20th-century Latin America surrounded the issue of land ownership, with peasants demanding its redistribution through the process of **agrarian reform**. Governments have addressed these concerns in different ways. The Mexican Revolution in 1910 led to a system of communally held lands called *ejidos*. In the 1950s, Bolivia crafted agrarian reform policies that led to the government appropriating estate lands and redistributing them to small farmers. As part of the Sandinista revolution in Nicaragua in 1979, lands were taken from the political elite and converted into collective farms. In 2000, President Hugo Chavez ushered in a new era of agrarian reform in Venezuela, and Bolivian President Evo Morales introduced an agrarian reform program in 2006 to give land title to indigenous communities in the eastern lowlands. Each of these programs has met with resistance and proved to be politically and economically difficult.

Eventually, the path chosen by most governments has been to make frontier lands available to land-hungry peasants.

Agricultural Frontiers   The creation of agricultural frontiers served several purposes: providing peasants with land, tapping unused resources, and filling in blank spots on the map with settlers. Although the dominant demographic trend has been a rural-to-urban movement, an important rural-to-rural flow has changed undeveloped areas into agricultural communities (Figure 4.17).

The opening of the Brazilian Amazon for settlement was the most ambitious frontier colonization scheme in the region. In the 1960s, Brazil began its frontier expansion by constructing several new Amazonian highways, a new capital (Brasília), and state-sponsored mining operations. The Brazilian military directed the opening of the Amazon to provide an outlet for landless peasants and to extract the region's many resources. However, the generals' plans did not deliver as intended. Throughout the basin, nutrient-poor tropical soils could not support permanent agricultural colonies. Government-promised land titles, agricultural subsidies, and credits were slow to reach small farmers. Instead, too much money went to subsidizing large cattle ranches through tax breaks and improvement deals in which "improved" land meant cleared land. Today five times more people live in the Amazon than in the 1960s; thus, continued human modification of this region is inevitable.

Population Growth and Movement

The high growth rates in Latin America throughout the 20th century have been attributed to natural increases, as well as to immigration in the early part of the century. The 1960s and 1970s were decades of tremendous growth, resulting from high fertility rates and increasing life expectancy. In the 1960s, for example, a typical Latin American woman had six or seven children. By the 1980s, family sizes were half as big. Several factors explain this: more urban families, which tend to be smaller than rural ones; increased participation of women in the workforce; higher education levels of women; state support of family planning; and better access to birth control. The exceptions to this trend are the poor and more rural countries, such as Guatemala and Bolivia, where the average woman has four or five children. Cultural factors may also be at work, as Amerindian peoples in the region tend to have more children than others.

Even with family sizes shrinking and nearing replacement rates in Uruguay and Chile, built-in potential for continued growth exists because of the relative demographic youth of these countries. The average percentage of the population below age 15 is 28 percent. In North America, that same group is 19 percent of the population, and in Europe it is just 16 percent. This means that a proportionally larger segment of the population has yet to enter into its childbearing years.

The population pyramids of two countries, Uruguay and Guatemala, contrast the profile of a country with a stable population size

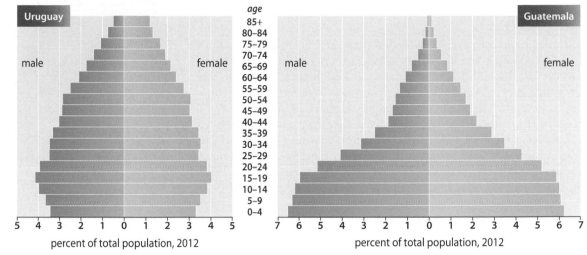

FIGURE 4.18   Population Structure of Uruguay and Guatemala These two pyramids contrast the population structure of the more developed and demographically stable Uruguay with that of the youthful and rapidly growing Guatemala. The average Uruguayan woman has 2 children, whereas Guatemalan women have a total fertility rate (TFR) of 3.6 children. Due to differences in natural increase and TFR, Uruguay is projected to have about the same population size in 2050, but Guatemala is projected to be twice as large, with 27.4 million people.

versus a demographically growing state (Figure 4.18). Uruguay is a small, but prosperous country with a high human development index ranking and relatively little poverty. Uruguayan women average two children, which is slightly below replacement level. Life expectancy is also high, but most population projections have the country growing very slowly between now and 2050. In contrast, Guatemala has a wide-based population pyramid and is considerably poorer. Total fertility rates have declined in Guatemala, but are still considered high at 3.6. Due to the youthfulness of the population and the increase in life expectancy, Guatemala's population is expected to double between 2010 and 2050.

In addition to natural increase, waves of immigrants into Latin America and migrant streams within Latin America have influenced population size and patterns of settlement. Beginning in the late 19th century, new immigrants from Europe and Asia added to the region's size and ethnic complexity. Important population shifts within countries have also occurred in recent decades, as witnessed by the growth of Mexican border towns and new settlements in eastern Bolivian lowlands. In an increasingly globalized economy, even more Latin Americans live and work outside the region, especially in the United States and Europe.

European Migration   After gaining their independence from Spain and Portugal, Latin America's new leaders sought to develop their territories through immigration. Firmly believing that "to govern is to populate," many countries set up immigration offices in Europe to attract hard-working peasants to till the soils and "whiten" the *mestizo* (people of mixed European and Indian ancestry) population. The Southern Cone countries of Argentina, Chile, Uruguay, Paraguay, and sou Brazil were the most successful in attracting European immi from the 1870s until the depression of the 1930s. During this some 8 million Europeans arrived (more than came during th colonial period), with Italians, Portuguese, Spaniards, and Germans being the most numerous.

Asian Migration

Less well known than the European immigrants to Latin America are the Asian immigrants, who also arrived during the late 19th and 20th centuries. Although considerably fewer, over time they established an important presence in the large cities of Brazil, Peru, Argentina, and Paraguay. Beginning in the mid-19th century, the Chinese and Japanese who settled in Latin America were contracted to work on the coffee estates in southern Brazil and the sugar estates and coastal mines of Peru. Over time, Asian immigrants became prominent members of society; for example, Alberto Fujimori, a son of Japanese immigrants, was president of Peru from 1990 to 2000.

Between 1908 and 1978, one-quarter of a million Japanese immigrated to Brazil; today the country is home to 1.3 million people of Japanese descent (Figure 4.19). As a group, the Japanese have been closely associated with the expansion of soybean and orange production. Increasingly, second- and third-generation Japanese have taken professional and commercial jobs in Brazilian cities; many of them have married outside their ethnic group and are losing their fluency in Japanese. South America's economic turmoil in the 1990s encouraged many ethnic Japanese to emigrate to Japan in search of better wages. Nearly one-quarter of a million ethnic Japanese left South America in the 1990s (mostly from Brazil and Peru) and now work in Japan.

Latino Migration and Hemispheric Change

Migration within Latin America and between Latin America and North America has had a significant impact on both sending and receiving communities. Within Latin America, international migration is shaped by shifting economic and political realities. For example, Venezuela's oil wealth attracted Colombian immigrants, who tended to work as domestics or agricultural laborers. Argentina has long been a destination for

FIGURE 4.20 Salvadorans in the Suburbs Day laborers wait for employment in a Maryland suburb outside Washington, DC. War and economic hardship drove many Salvadorans from their country in the 1980s and 1990s. Many Salvadorans send remittances back to their country of birth, representing 17 percent of El Salvador's gross national income. Today Salvadorans are the largest immigrant group in the Washington, DC, metropolitan area.

Bolivian and Paraguayan laborers. And farmers in the United States have depended on Mexican laborers for more than a century.

Political turmoil has also sparked waves of international migrants. The bloody civil wars in El Salvador and Guatemala, for example, sent waves of refugees into neighboring countries, such as Mexico and the United States. Violence and low-intensity conflict have internally displaced over 2.5 million Colombians in the past two decades, and official statistics suggest another 3 million Colombians live abroad. With democratization on the rise in the region, many of today's immigrants are classified as economic migrants, not political asylum seekers.

Presently, Mexico is the country of origin for the most legal immigrants to the United States. In the 2010 census, there were 50 million Hispanics in the United States, two-thirds of whom claimed Mexican ancestry, and approximately 12 million were born in Mexico. Mexican labor migration to the United States dates back to the late 1800s, when relatively unskilled labor was recruited to work in agriculture, mining, and railroads. Mexican immigrants are most concentrated in California and Texas, but increasingly they are found throughout the country. Although Mexicans continue to have the greatest presence among Latinos in the United States, the number of immigrants from El Salvador, Guatemala, Nicaragua, Colombia, Ecuador, and Brazil has steadily grown. Most of the Hispanic population has ancestral ties with peoples from Latin America and the Caribbean (see Chapter 5 on Caribbean migration).

Today Latin America is seen as a region of emigration, rather than one of immigration. Both skilled and unskilled workers from Latin America are an important source of labor in North America, Europe, and Japan (Figure 4.20). Many of these immigrants send monthly **remittances** (monies sent back home) to sustain family members. In 2008, it was estimated that immigrants sent more than $70 billion to Latin America. By 2011, that figure had declined to $61 billion, which shows the lingering impact of the economic recession on both remittance and migrant flows. Much of this money comes from workers in the United States, but Latino immigrants in Spain, Japan, Canada, and Italy also sent money back to the region.

FIGURE 4.19 Japanese-Brazilians Brazilian youth of Japanese ancestry perform in Curitiba, Brazil, to mark the 100th anniversary of Japanese immigration to Brazil in 2008. In 1908, the first Japanese immigrants arrived as agricultural workers, choosing Brazil as a destination after countries such as the United States and Canada had banned Japanese immigration. Today there are over 1.3 million ethnic Japanese in Brazil, especially in the states of São Paulo and Paraná, and they have distinguished themselves as large-scale farmers and urban professionals.

REVIEW QUESTIONS

1. What are the historical and economic explanations for urban dominance and urban primacy in Latin America?

2. How have policies such as agrarian reform and frontier colonization impacted the patterns of settlement and primary resource extraction in the region?

3. Demographically, Latin American has grown much faster than North America. What factors contribute to faster growth, and is it likely to continue?

CULTURAL COHERENCE AND DIVERSITY: REPOPULATING A CONTINENT

The Iberian colonial experience (1492 to the 1800s) imposed a political and cultural coherence on Latin America that makes it recognizable today as a world region. Yet this was not a simple transplanting of Iberia across the Atlantic. Instead, a process unfolded in which European and Indian traditions blended as indigenous groups were added into either the Spanish or the Portuguese empire. In some areas, such as southern Mexico, Guatemala, Bolivia, Ecuador, and Peru, Indian cultures have shown remarkable resilience, as evidenced by the survival of Amerindian languages. However, the prevailing pattern was one of forced assimilation in which European religion, languages, and political organization were imposed on surviving Amerindian societies. Later, other cultures—especially more than 10 million African slaves—added to the cultural mix of Latin America, the Caribbean, and North America. The legacy of the African slave trade will be examined in greater detail in Chapters 5 and 6. For Latin America, perhaps the single most important factor in the dominance of European culture was the demographic collapse of native populations.

The Decline of Native Populations

It is difficult to grasp the enormity of cultural change and human loss due to this encounter between the Americas and Europe. Throughout the region, archaeological sites are reminders of the complexity of Amerindian civilizations prior to European contact. Dozens of stone temples found throughout Mexico and Central America, where the Mayan and Aztec civilizations flourished, attest to the ability of these societies to thrive in the area's tropical forests and upland plateaus. The Mayan city of Tikal flourished in the lowland forests of Guatemala, supporting tens of thousands, before its mysterious collapse centuries before the arrival of Europeans (Figure 4.21). In the Andes, the complexity of Amerindian civilizations can be seen in political centers such as Cuzco and Machu Picchu. The Spanish, too, were impressed by the sophistication and wealth they saw around them, especially in Tenochtitlán, where Mexico City sits today. Tenochtitlán was the political and ceremonial center of the Aztecs, supporting a complex metropolitan area with some 300,000 residents. The largest city in Spain at the time was considerably smaller.

The Demographic Toll
The most telling figures of the impact of European expansion in Latin America are demographic. It is widely

FIGURE 4.21 Tikal, Guatemala This ancient Mayan city, located in the lowland forests of the Petén, was part of a complex network of cities located in the Yucatan and northern Guatemala. At its height, Tikal supported over 100,000 people before its collapse in the late 10th century. Today it is a major tourist destination.

believed that the pre-contact Americas had 54 million inhabitants; by comparison, western Europe in 1500 had approximately 42 million people. Of the 54 million, about 47 million were in what is now Latin America, and the rest were in North America and the Caribbean. By 1650, after a century and a half of colonization, the indigenous population was one-tenth its pre-contact size. The human tragedy of this population loss is hard to comprehend. The relentless elimination of 90 percent of the native population was largely caused by epidemics of influenza and smallpox, but warfare, forced labor, and starvation due to a collapse of food-production systems also contributed to the rapid population decline.

Indian Survival Presently, Mexico, Guatemala, Ecuador, Peru, and Bolivia have the largest indigenous populations. Not surprisingly, these areas had the densest native populations at the time of European contact. Indigenous survival also occurs in isolated settings where the workings of national and global economies are slow to break through, such as in eastern Panama, the Miskito Coast of Honduras, or the roadless sections of western Amazonia.

In many cases, Indian survival comes down to one key resource—land. Indigenous peoples who are able to maintain a territorial home, either formally through land title or informally through long-term occupancy, are more likely to preserve a distinct ethnic identity. Because of this close association between identity and territory, native peoples are increasingly insisting on recognized spaces within their countries.

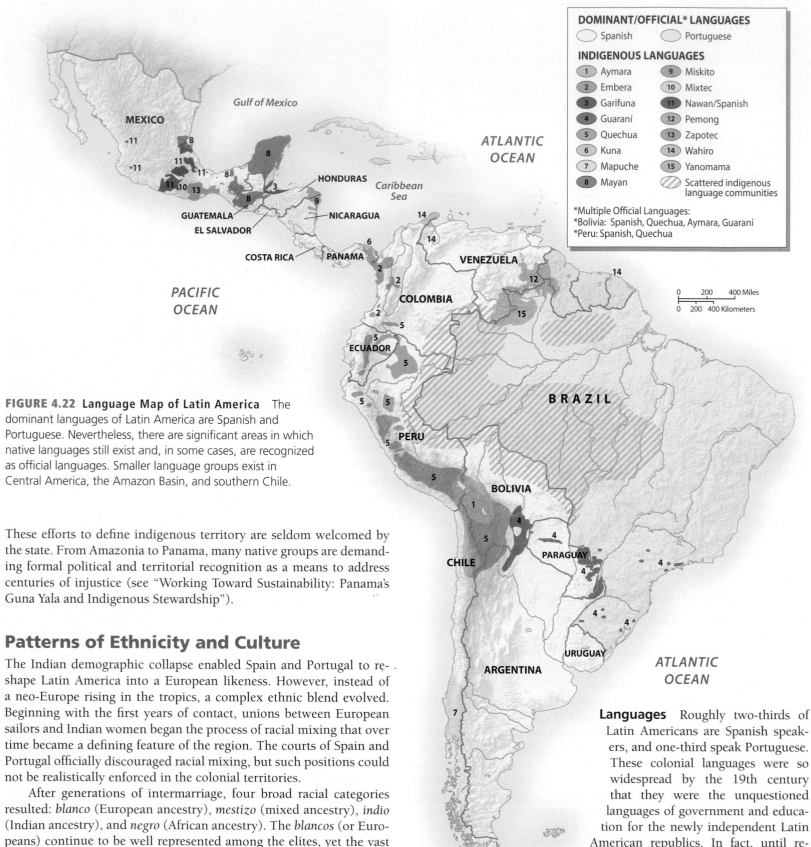

FIGURE 4.22 Language Map of Latin America The dominant languages of Latin America are Spanish and Portuguese. Nevertheless, there are significant areas in which native languages still exist and, in some cases, are recognized as official languages. Smaller language groups exist in Central America, the Amazon Basin, and southern Chile.

These efforts to define indigenous territory are seldom welcomed by the state. From Amazonia to Panama, many native groups are demanding formal political and territorial recognition as a means to address centuries of injustice (see "Working Toward Sustainability: Panama's Guna Yala and Indigenous Stewardship").

Patterns of Ethnicity and Culture

The Indian demographic collapse enabled Spain and Portugal to re-shape Latin America into a European likeness. However, instead of a neo-Europe rising in the tropics, a complex ethnic blend evolved. Beginning with the first years of contact, unions between European sailors and Indian women began the process of racial mixing that over time became a defining feature of the region. The courts of Spain and Portugal officially discouraged racial mixing, but such positions could not be realistically enforced in the colonial territories.

After generations of intermarriage, four broad racial categories resulted: *blanco* (European ancestry), *mestizo* (mixed ancestry), *indio* (Indian ancestry), and *negro* (African ancestry). The *blancos* (or Europeans) continue to be well represented among the elites, yet the vast majority of people are of mixed racial ancestry. *Dia de la Raza*, the region's observance of Columbus Day, recognizes the emergence of a new *mestizo* race as the legacy of European conquest. Throughout Latin America, more than other regions of the world, miscegenation (or racial mixing) is the norm, and it makes the process of mapping racial or ethnic groups especially difficult.

Languages Roughly two-thirds of Latin Americans are Spanish speakers, and one-third speak Portuguese. These colonial languages were so widespread by the 19th century that they were the unquestioned languages of government and education for the newly independent Latin American republics. In fact, until recently, many countries actively discouraged, and even repressed, Indian tongues. It took a constitutional amendment in Bolivia in the 1990s to legalize native-language instruction in primary schools and to recognize the country's multiethnic heritage (more than half the population is Indian, and Quechua, Aymara, and Guaraní are widely spoken) (Figure 4.22).

WORKING TOWARD SUSTAINABILITY

Panama's Guna Yala and Indigenous Stewardship

People associate Panama with its canal, which turned this small nation of 3.5 million people into a global center for trade and banking. In 2014, Panama will mark the 100th anniversary of the opening of the Panama Canal with a new set of locks that will accommodate larger ships and more cargo. Since the U.S. handover in 2000, this vital waterway has been fully under the control of Panamanians, who are determined to make it more profitable and modern.

Yet Panama is also a nation where nearly one-tenth of the population is Amerindian, and indigenous politics have, at times, taken their own revolutionary turns. One group, the Guna (formally known as the Kuna) (Figure 4.1.1), live in the eastern half of the country and are well known for their political activism and insistence upon autonomy. Their struggle for territorial recognition and indigenous resource management of a lowland tropical forest area is instructive.

The Guna live along the Caribbean coast of eastern Panama, occupying some three dozen of the nearly 400 San Blas Islands and using the nearby mainland for subsistence farming. This remote and nearly roadless area has fantastic beaches, beautiful islands, and intact forests, something that is increasingly rare in Central America. The Guna economy is based on subsistence farming, fishing, artisan craft production (the women sew colorful *molas*, which are embroidered fabrics used in women's blouses), and small-scale indigenous-run tourism.

In the 1920s, a Guna revolution established Guna Yala (formerly Kuna Yala) (Figure 4.1.2), an indigenous-controlled territory that is governed by an elected Guna chief (*cacique*). By the 1950s, the Guna system of government, the *comarca*, was put into place, allowing for self-government and direct communication with Panamanian authorities. By the 1970s, the Guna were anxious about colonizers illegally entering their territory for cattle ranching. They engaged national authorities and international nongovernmental organizations for support and created an innovative Guna wildlands project to better enforce their boundaries and limit resource extraction from the area.

FIGURE 4.1.2 A Guna woman with her four children in front on her home on one of the San Blas islands in Guna Yala.

Today there are many unique aspects of this region. Although much of the Central American forests has been cleared for cattle ranching, there are no cattle in Guna Yala. The forests are used by the Guna for subsistence slash-and-burn farming and wood collection, which places decidedly less stress on them. Also, the Guna maintain control of their coast, not allowing foreign companies or owners to build tourist facilities, although they have been approached many times. The tourist facilities that do exist are modest and are owned and operated by the Guna, but they generally are on smaller islands and not within the Guna villages themselves. Guna villages are densely settled, with villagers traveling by boat to fish or to tend crops on the mainland. Most villages do not have electricity, or if they do, it's for a few hours a day through gasoline-powered generators.

Guna Yala is home to some 40,000 Guna who have used political means and ethnic solidarity to maintain control of their land. Yet they are not untouched by global change. The Guna are deeply concerned about sea-level rising (induced through climate change) undermining their way of life and fear that in the next decade or two they will have to abandon their island homes for the mainland.

FIGURE 4.1.1 Guna Yala Territory The Guna (formerly known as the Kuna) have maintained their territory in Eastern Panama since the 1920s.

Because Spanish and Portuguese dominate, there is a tendency to overlook the influence of indigenous languages in the region. Mapping the use of native languages, however, reveals important areas of Indian resistance and survival. In the Central Andes of Peru, Bolivia, and southern Ecuador, more than 10 million people still speak Quechua and Aymara, along with Spanish. In Paraguay and lowland Bolivia, there are 4 million Guaraní speakers, and in southern Mexico and Guatemala at least 6 to 8 million speak Mayan languages. Small groups of native-language speakers are found scattered throughout the sparsely settled interior of South America and the more isolated forests of Central America. However, many of these languages have fewer than 10,000 speakers.

Blended Religions   Like language, the Roman Catholic faith appears to have been imposed upon the region without challenge. Most countries report 90 percent or more of their population as Catholic. Every major city has dozens of churches, and even the smallest village maintains a graceful church on its central square (Figure 4.23). In some countries, such as El Salvador and Uruguay, a sizable portion of the population attends Protestant churches, but the Catholic core of this region is still intact.

Exactly what native peoples absorbed of the Christian faith is complex. Throughout Latin America, **syncretic religions**—blends of different belief systems—enabled animist practices to be included in Christian worship. These blends took hold and endured, in part because Christian saints were easy replacements for pre-Christian gods and because the Catholic Church tolerated local variations in worship as long as the process of conversion was under way. The Mayan practice of paying tribute to spirits of the underworld seems to be replicated today in Mexico and Guatemala through the practice of building small cave shrines to favorite Catholic saints and leaving offerings of fresh flowers and fruits. One of the most celebrated religious symbols in Mexico is the Virgin of Guadalupe—a dark-skinned virgin seen by an Indian shepherd boy—who became the patron saint of Mexico.

Syncretic religious practices also evolved and endured among African slaves. By far the greatest concentration of slaves was in the Caribbean, where slaves were used to replace the indigenous population after it was wiped out by disease (see Chapter 5). Within Latin

FIGURE 4.23   The Catholic Church   Churches, such as the Dolores Church in Tegucigalpa, are important religious and social centers. The majority of people in Latin America define themselves as Catholic. Many churches built in the colonial era are valued as architectural treasures and are beautifully preserved.

America, the Portuguese colony of Brazil received the most Africans. The blend of Catholicism with African traditions is most obvious in the celebration of carnival, Brazil's most popular festival and one of the major components of Brazilian national identity. The three days of carnival known as the Reign of Momo combine pre-Lenten celebrations with African musical traditions represented by the rhythmic samba bands. Today the festival—which is most associated with Rio de Janeiro—draws thousands of participants from all over the world.

The Global Reach of Latino Culture

Latin American culture, vivid and diverse as it is, is widely recognized throughout the world. Whether it is the sultry pulse of the tango or the fanaticism with which Latinos embrace soccer as an art form, aspects of Latin American culture have been absorbed into a globalizing world culture. A dramatic example of the reach of Latino culture can be seen every Saturday on the Univision television network, based in Miami. There, a charismatic Chilean, Don Francisco, has hosted *Sabado Gigante* since 1986. This 4-hour-long Spanish variety show is viewed in 28 countries and draws a weekly audience of 120 million viewers. In the arts, Latin American writers such as Jorge Luis Borges, Gabriel García Marquez, and Isabel Allende have obtained worldwide recognition. In terms of popular culture, musical artists such as Colombia's Shakira and Brazil's hip-hop samba singer Max de Castro have international audiences. Latino movie stars such as Salma Hayek and Jennifer Lopez are icons of world pop culture. Through music, literature, art, and even *telenovelas* (soap operas), Latino culture is being transmitted to an eager worldwide audience.

Soccer   Perhaps the quintessential global sport, soccer has a fanatical following throughout much of the world. Yet it is Latin America, and especially South America, where *futbol* is considered a cultural necessity. Still largely a male game, although girls are beginning to play, young boys and men are constantly seen on fields, beaches, and blacktops playing soccer, especially in late afternoons and on weekends. The great soccer stadiums of Buenos Aires (Bombonera) and Rio de Janeiro (Maracaña) are regarded as shrines to the game. Many individuals use the victories and losses of their national soccer teams as important chronological markers of their lives.

Pelé, the Brazilian soccer star of the 1960s and 1970s, introduced the free-flowing acrobatic style that became known as "the beautiful game." Today Latino soccer stars such as Argentine Lionel Messi and Brazilian Kaká play for corporate clubs in Europe and earn millions (Figure 4.24). Latin Americans also fill up the few slots allotted to foreign players on the U.S. Major League Soccer teams. Yet the dream of many Latino soccer players is to be on the national team and bring home the World Cup. Visit any Latin American country when their team is playing a World Cup qualifying match and the streets are eerily quiet. Of the 19 World Cups awarded between 1930 and 2010, South American teams have won 9. In 2014, Brazil will host the World Cup, with the final match to be held in the recently renovated Maracaña stadium. Somehow, as if by magic, it is believed that a World Cup victory will make things better, improve the economy, and even reduce crime. In short, soccer is regarded with near religious significance. As Latin Americans emigrate (both as players and as laborers), they bring their enthusiasm for the sport with them.

Telenovelas   Popular nightly soap operas are a mainstay of Latin American television. These tightly plotted series are filled with intrigue and double-dealing. Unlike their counterparts in the United

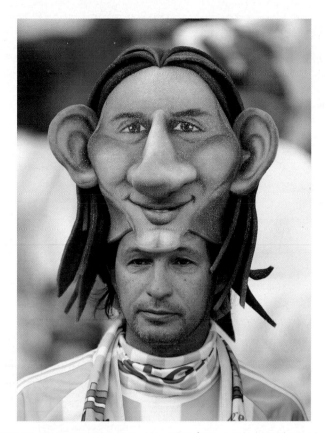

FIGURE 4.24 Messi Mania An Argentine fan wears a Lionel Messi mask on his head and the national jersey of Argentina at the World Cup Quarter Final match in Cape Town, South Africa in 2010.

States, they end, usually after 100 episodes. Once standard fare for the working class, many telenovelas take hold and absorb an entire nation. During particularly popular episodes, the streets are noticeably calm as millions of people tune in to catch up on the lives of their favorite heroines. Brazil, Venezuela, and Mexico each produce scores of telenovelas, but the Mexican ones are international megahits.

Televisa, a Mexican production agency, has aggressively marketed its inventory of soap operas to an eager global public. Mexican telenovelas are avidly watched in countries as diverse as Croatia, Russia, China, South Korea, Iran, the United States, and France, as well as throughout Latin America. Predictably scripted as Mexican Cinderella stories, these sagas of poor underclass women (often domestics) falling in love with members of the elite, battling jealous rivals, and ultimately emerging triumphant seem to resonate with fans around the world. In addition to their broad appeal, telenovelas are big business, perhaps Mexico's largest cultural export. Hollywood and Mumbai grind out movies, but much of Mexico's entertainment industry is geared toward producing this popular art form.

REVIEW QUESTIONS

1. What factors contributed to racial mixing in Latin America, and where are the areas of strongest Amerindian survival?

2. What are the cultural legacies of Iberia in Latin America and how are they expressed?

GEOPOLITICAL FRAMEWORK: REDRAWING THE MAP

Latin America's colonial history, more than its present condition, unifies this region geopolitically. For the first 300 years after the arrival of Columbus, Latin America was a territorial prize sought by various European countries, but was effectively settled by Spain and Portugal. By the 19th century, the independent states of Latin America had formed, but they continued to experience foreign influence and, at times, overt political pressure, especially from the United States. At various times, a more neutral hemispheric vision of American relations and cooperation has held sway, represented by the formation of the **Organization of American States (OAS)**. The present OAS was officially formed in 1948, but its origins date to 1889. Yet there is no doubt that U.S. policies toward trade, economic assistance, political development, and at times military intervention are often seen as undermining the independence of these states. Today the geopolitical influence of the United States in the region is declining, especially in South America. Brazil's largest trading partner is now China. In many South American countries, trade with the European Union, China, and Japan is as important as, if not more important than, trade with the United States.

Within Latin America, cycles of intraregional cooperation and antagonism have occurred. Neighboring countries have fought over territory, closed borders, imposed high tariffs, and cut off diplomatic relations. The last 20 years have witnessed a revival in the trade bloc concept, with the formation of Mercosur (the Southern Cone Common Market) in South America, NAFTA in North America, and CAFTA, the Central American Free Trade Association. In 2008, Brazil proposed the formation of **UNASUR**, the Union of South American Nations, which includes all the states of South America except French Guiana. Modeled more like the European Union than a free trade association, UNASUR marks a significant change in the region's political geography but its long-term impact remains to be seen.

Iberian Conquest and Territorial Division

When Christopher Columbus claimed the Americas for Spain, the Spanish became the first active colonial agents in the western hemisphere. In contrast, the Portuguese presence in the Americas was the result of the **Treaty of Tordesillas** in 1493–1494. By that time, Portuguese navigators had charted much of the coast of Africa in an attempt to find a water route to the Spice Islands (Moluccas) in Southeast Asia. With the help of Christopher Columbus, Spain sought a western route to the Far East. When Columbus discovered the Americas, Spain and Portugal asked the pope to settle how these new territories should be divided. Without consulting other European powers, the pope divided the Atlantic world in half: The eastern half, containing the African continent, was awarded to Portugal; the western half, with most of the Americas, was given to Spain. The line of division established by the treaty actually cut through the eastern part of South America, placing it under Portuguese rule. The treaty was never recognized by the French, English, or Dutch, who also claimed territory in the Americas, but it did provide the legal justification for the creation of Portuguese Brazil. This state would later become the largest and most populous in Latin America (Figure 4.25).

Six years after the treaty was signed, Portuguese navigator Alvares Cabral accidentally reached the coast of Brazil on a voyage to southern

Latin America in 1830

States with date of independence

MEXICO
1821

Mexico City • Veracruz

UNITED PROVINCES OF
CENTRAL AMERICA
1823–1839

Caracas

Bogotá

GRAN COLOMBIA
1819–1830

Quito •

Manaus •

PERU
1821

Lima •

La Paz •

BOLIVIA
1825

PARAGUAY
1811
Asunción •

BRAZIL
1822

• Natal

• Salvador

Rio de Janeiro •
São Paulo •

CHILE
1817
Santiago •

URUGUAY
1828

Buenos Aires •
• Montevideo

UNITED
PROVINCES
OF LA PLATA
1816

0   500   1,000 Miles
0   500   1,000 Kilometers

Latin America in 1650

Viceroyalty of New Spain
Viceroyalty of Peru
Brazil

SANTO
DOMINGO
1511
New Spain

MEXICO
1529

GUATEMALA
1544

NUEVA
GALICIA
1549

PANAMA
1538 & 1567

SANTA FE
1549

New Granada

*Unexplored Spanish
Territory*

QUITO
1563

Treaty of
Tordesillas,
1494

LIMA
1542

*Unexplored Spanish
Territory*

CHARCAS
1559

La Plata

CHILE
1565 &
1609

*Unexplored
Spanish
Territory*

0   500   1,000 Miles
0   500   1,000 Kilometers

FIGURE 4.25 Shifting Political Boundaries
The evolution of political boundaries in Latin America began with the 1494 Treaty of Tordesillas, which gave much of the Americas to Spain and a slice of South America (Brazil) to Portugal. The larger Spanish territory was gradually divided into viceroyalties and audiencias, which formed the basis for many modern national boundaries. The 1830 borders of these newly independent states were far from fixed. Bolivia would lose its access to the coast, Peru would gain much of Ecuador's Amazon, and Mexico would be stripped of its northern territory by the United States.

Africa. The Portuguese soon realized that this territory was on their side of the Tordesillas line. Initially, they were unimpressed by what Brazil had to offer; there were no spices or major native settlements. Over time, they came to appreciate the utility of the coast as a provisioning site, as well as a source for brazilwood, used to produce a valuable dye. Portuguese interest in the territory intensified in the late 16th century, with the development of sugar estates and the expansion of the slave trade, and in the 17th century, with the discovery of gold in the Brazilian interior.

Spain, in contrast, aggressively pursued the conquest and settlement of its new American territories from the very start. After

discovering little gold in the Caribbean, by the mid-16th century, Spain's energy was directed toward developing the silver resources of Central Mexico and the Central Andes (most notably Potosí in Bolivia). Gradually, the economy diversified to include some agricultural exports, such as cacao (for chocolate) and sugar, as well as a variety of livestock. In terms of foodstuffs, the colonies were virtually self-sufficient. Manufacturing, however, was forbidden in the Spanish-American colonies in order to keep them dependent on Spain.

Revolution and Independence   Not until the rise of revolutionary movements between 1810 and 1826 was Spanish authority on the mainland challenged. Ultimately, European elites born in the Americas gained control, displacing leaders loyal to the crown. In Brazil, the evolution from Portuguese colony to independent republic was a slower and less violent process that spanned eight decades (1808–1889). In the 19th century, Brazil was declared a separate kingdom from Portugal, with its own king, and later became a republic.

The territorial division of Spanish and Portuguese America into administrative units provided the legal basis for the modern states of Latin America (see Figure 4.25). The Spanish colonies were first divided into two viceroyalties (the administrative units of New Spain and Peru), and within these were various subdivisions that later became the basis for the modern states. (In the 18th century, the Viceroyalty of Peru, which included all of Spanish South America, was divided to form three viceroyalties: La Plata, Peru, and New Granada.) Unlike Brazil, which evolved from a colony into a single republic, the former Spanish colonies experienced fragmentation in the 19th century.

Today the former Spanish mainland colonies include 16 states (plus 3 Caribbean islands), with a total population of over 370 million. If the Spanish colonial territory had remained a unified political unit, it would now have the third largest population in the world, following China and India.

Persistent Border Conflicts

As the colonial administrative units turned into states, it became clear that the territories were not clearly demarcated, especially the borders that stretched into the sparsely populated interior of South America. This would later become a source of conflict, as new states struggled to define their territorial limits. Numerous border wars erupted in the 19th and 20th centuries, and the map of Latin America had to be redrawn many times. Some of the most notable conflicts were the War of the Pacific (1879–1882), in which Chile expanded to the north and Bolivia lost its access to the Pacific; warfare between Mexico and the United States in the 1840s, which resulted in the present border under the Treaty of Hidalgo (1848); and the War of the Triple Alliance (1864–1870), the bloodiest war of the postcolonial period, which occurred when Argentina, Brazil, and Uruguay allied themselves to defeat Paraguay in its claim to control the upper Paraná River Basin. In the 1980s, Argentina lost a war with Great Britain over control of the Falklands, or Malvinas, Islands in the South Atlantic. And as recently as 1998, Peru and Ecuador fought over a disputed boundary in the Amazon Basin.

The Trend Toward Democracy

Most of the 17 countries in Latin America have or will soon celebrate their bicentennials. Compared with most of the rest of the developing world, Latin Americans have been independent for a long time. Yet political stability is not a characteristic of the region. Among the countries in the region, some 250 constitutions have been written since independence, and military takeovers have been alarmingly frequent. Since the 1980s, however, the trend has been toward democratically elected governments, the opening of markets, and broader public participation in the political process. Where dictators once outnumbered elected leaders, by the 1990s each country in the region had a democratically elected president. (Cuba, the one exception, will be discussed in Chapter 5.)

Democracy may not be enough for the millions frustrated by the slow pace of political and economic reform. In survey after survey, Latin Americans reveal their dissatisfaction with politicians and governments. Many of the democratic leaders have also been free-market reformers who are quick to eliminate state-backed social safety nets, such as food subsidies, government jobs, and pensions. Many of the poor and members of the middle class have grown doubtful about whether this brand of neoliberal democracy can make their lives better. Even in prosperous Chile, widespread student protests in 2012 demanding a fairer society and free university education have left politicians scrambling. Under such conditions, it is not surprising that left-leaning politicians have won presidential elections in Brazil, Bolivia, Nicaragua, Ecuador, Peru, and Venezuela. These leaders have not rolled back neoliberal trade reforms, but many have tried to improve social services and attempted to reduce income inequalities.

Regional Organizations

At the same time that democratically elected leaders struggle to address the pressing needs of their countries, political developments at the supranational and subnational levels pose new challenges to their authority. The most discussed **supranational organizations** (governing bodies that include several states) are the trade blocs, the newest one being UNASUR—The Union of South American Nations. **Subnational organizations** (groups that represent areas or people within the state) form along ethnic or ideological lines or can support organized crime. Subnational organizations can have positive or destabilizing impacts. Examples include native groups that seek territorial recognition (such as the Guna in Panama), insurgent groups (such as the FARC [Revolutionary Armed Forces of Colombia] or the Zapatistas in Mexico) that have challenged the authority of the state, and more recently drug cartels such as the *Zetas* and *Sinaloa*, who have terrorized Mexican society with extreme violence since 2006.

Trade Blocs   Beginning in the 1960s, countries formed regional trade alliances in an effort to promote internal markets and reduce trade barriers. The Latin American Integration Association (formerly LAFTA), the Central American Common Market (CACM), and the Andean Group have existed for decades, but their ability to influence economic trade and growth has been limited at best. In the 1990s, Mercosur and NAFTA emerged as supranational structures that could influence development (Figure 4.26). For Latin America, the lessons from Mercosur in particular led Brazil to propose UNASUR in 2008, uniting virtually all of South America.

NAFTA took effect in 1994 as a free-trade area that would gradually eliminate tariffs and ease the movement of goods among the member countries (Mexico, the United States, and Canada). NAFTA has increased intraregional trade, but has provoked considerable controversy about costs to the environment and to employment (see Chapter 3). NAFTA did prove, however, that a free-trade area combining industrialized and developing states was possible. In 2004, the United States, five Central American countries—Guatemala, El Salvador, Nicaragua, Honduras, and Costa Rica—and the Dominican Republic signed CAFTA. CAFTA, like NAFTA, aims to increase trade and reduce tariffs among member countries. The treaty became fully ratified in 2009 but much debate surrounds the question of whether such a treaty would lead to more economic development in Central America.

Mercosur was formed in 1991 with Brazil and Argentina—the two largest economies in South America—and the smaller states of Uruguay and Paraguay as members. Since its formation, trade among these countries has grown tremendously, so much so that eventually Chile, Bolivia, Peru, Ecuador, and Colombia joined the group as associate members and Venezuela received ratification as a full member in 2012.

FIGURE 4.26 Geopolitics and Trade Blocs in Latin America Of the four economic trade blocs shown, Mercosur and NAFTA are the most dynamic. In fact, several members of the Andean Group have joined Mercosur. Meanwhile, Central American states signed an agreement in 2004 to form CAFTA.

This success is significant in two ways: It reflects the growth of these economies and also the willingness to put aside old rivalries (especially long-standing antagonisms between Argentina and Brazil) for the economic benefits of cooperation.

In 2008, Brazil initiated the formation of the 12-member Union of South American Nations (UNASUR). Some see this as an assertion of Brazil's greater political and economic clout in the region, since it is Latin America's largest country and the sixth largest economy in the world. UNASUR includes all countries in South America, minus French Guiana, which is a territory of France. It has formally organized with a permanent secretariat, and it has responded to political crises in Bolivia (2008), Ecuador (2010), and Paraguay (2012). Significantly, unlike NAFTA or CAFTA, UNASUR was a Brazilian-led effort, not one led by the United States. Brazil's lead in this initiative underscores its larger geopolitical ambitions to influence South American development and to secure a permanent seat on the United Nations Security Council. Brazil's ambitions have not always been well received by its neighbors, but the strengthening of UNASUR suggests a shift in the region's geopolitical alignment.

Insurgencies and Drug Cartels Guerilla groups such as the FARC in Colombia have controlled large territories of their countries through the support of those loyal to the cause, along with theft, kidnapping, and violence. The FARC, along with the ELN (National Liberation Army), gained wealth and weapons through the drug trade. The level of violence in Colombia has escalated further with the rise of paramilitary groups—armed private groups that terrorize those sympathetic to insurgency. The paramilitary groups have been blamed for hundreds of politically motivated murders each year. As many as 2.5 million Colombians have been internally displaced by violence since the late 1980s, most fleeing rural areas for towns and cities. Fortunately, the situation has improved considerably in the last decade.

Under President Uribe, the police presence has increased throughout the state, and negotiations with insurgencies have stopped, ultimately reducing their power. Still, Colombia remains the world's largest producer of cocaine.

Drug cartels and gangs in states as diverse as Mexico, Guatemala, El Salvador, Honduras, and Brazil have been blamed for increases in violence and lawlessness. The spike in violence and corruption in Mexico has been especially destabilizing. Profiting from the illegal production and/or shipment of cocaine, marijuana, methamphetamine, and heroin, the cartels generate billions of dollars. Beginning in 2006, the Mexican government brought in the army to quell the violence, kidnapping, and intimidation brought on by cartel groups, especially in the border region, but now extending throughout Mexico and into Central America (Figure 4.27). The Mexican government reported some 50,000 cartel-related murders from 2006 to 2012, and finding ways to stem the violence (rather than the flow of drugs) is one of the biggest issues in the 2012 Mexican election. Tragically, the most violent states in Latin America are in Central America. Honduras,

El Salvador, and Guatemala have some of the highest homicide rates in the world outside of war zones. These three impoverished countries have long been in the transshipment zone of cocaine produced in Colombia, Peru, and Bolivia. More recently, however, Mexico's Sinaloa, Gulf, and Zetas cartels have been active in the isthmus, paying locals in drugs, creating drug-processing centers, and, in the process, driving up the murder rate.

REVIEW QUESTIONS

1. How did the colonization of Latin America by Iberia lead to the formation of the modern states of Latin America?

2. How are trade blocs reshaping the region's geopolitics?

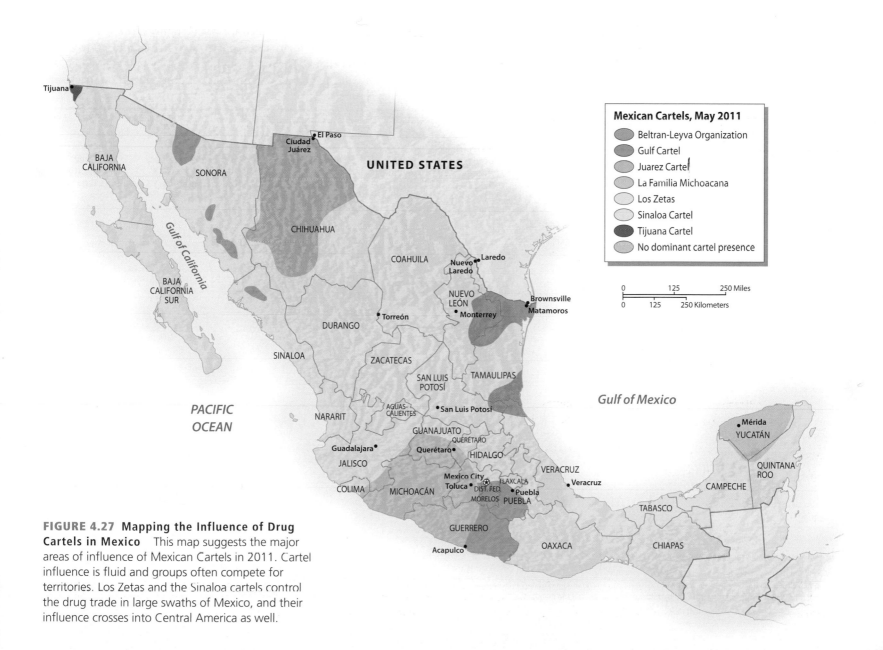

FIGURE 4.27 Mapping the Influence of Drug Cartels in Mexico This map suggests the major areas of influence of Mexican Cartels in 2011. Cartel influence is fluid and groups often compete for territories. Los Zetas and the Sinaloa cartels control the drug trade in large swaths of Mexico, and their influence crosses into Central America as well.

Mexican Cartels, May 2011
- Beltran-Leyva Organization
- Gulf Cartel
- Juarez Cartel
- La Familia Michoacana
- Los Zetas
- Sinaloa Cartel
- Tijuana Cartel
- No dominant cartel presence

ECONOMIC AND SOCIAL DEVELOPMENT: FOCUSING ON NEOLIBERALISM

Most Latin American economies fit into the broad middle-income category set by the World Bank. Clearly part of the developing world, Latin American people are much better off than those in Sub-Saharan Africa, South Asia, and most of China. Still, the economic contrasts are sharp, both between states and within them. Generally, the Southern Cone states (including southern Brazil and excluding Paraguay) and Mexico are the richest. The poorest countries in terms of per capita purchasing power parity (PPP) are Nicaragua, Bolivia, and Honduras. Although per capita incomes in Latin America are well below levels of developed countries, the region has witnessed steady improvements in various social indicators, such as life expectancy, child mortality, and literacy. Also, some small states, such as Costa Rica, do very well in the human development index (Table 4.2).

The economic engines of Latin America are the two largest countries, Brazil and Mexico. According to the International Monetary Fund, Brazil's economy in 2011 was the 6th largest in the world, and Mexico's was the 14th largest, based on gross domestic product. The region has also seen a reduction in extreme poverty, as the percentage of people living on less than $2 per day dropped from 22 percent in 1999 to 12 percent in 2008, in part because of conditional cash transfer programs such as Brazil's Bolsa Familia.

Development Strategies

The path toward economic development in Latin America has been a volatile one. In the 1960s, Brazil, Mexico, and Argentina all seemed poised to enter the ranks of the developed world. Multilateral agencies such as the World Bank and the Inter-American Development Bank loaned money for big development projects: continental highways, dams, mechanized agriculture, and power plants. All sectors of the economy were radically transformed. Agricultural production increased with the application of "green revolution" technology and mechanization. State-run industries reduced the need for imported goods, and the service sector ballooned as a result of new government and private-sector jobs. In the end, most countries in Latin America made the transition from predominantly rural and agrarian economies, dependent on one or two commodities, to more economically diversified and urbanized countries with mixed levels of industrialization.

The modernization dreams of Latin American countries were trampled in the 1980s, when debt, currency devaluation, hyperinflation, and falling commodity prices undermined the aspirations of the region. By the 1990s, most Latin American governments had radically changed their economic development strategies. State-run national industries and tariffs were jettisoned for policy reforms that emphasized privatization, direct foreign investment, and free trade, collectively labeled **neoliberalism**. Through tough fiscal policy, increased trade, privatization, and reduced government spending, most countries saw

TABLE 4.2   DEVELOPMENT INDICATORS

Country	GNI per capita, PPP 2010	GDP Average Annual %Growth 2000–10	Human Development Index (2011)[1]	Percent Population Living Below $2 a Day	Life Expectancy (2012)[2]	Under Age 5 Mortality Rate (1990)	Under Age 5 Mortality Rate (2010)	Adult Literacy (% ages 15 and older)	Gender Inequality Index (2011)[3,1]
Argentina	15,570	5.6	.797	<2	76	27	14	98	0.372
Bolivia	4,640	4.1	.668	24.9	67	121	54	91	0.473
Brazil	11,000	3.7	.718	10.8	74	59	19	90	0.449
Chile	14,640	4.0	.805	2.7	79	19	9	99	0.374
Colombia	9,060	4.5	.710	15.8	73	37	22	93	0.482
Costa Rica	11,270	4.9	.744	6.0	79	17	10	96	0.361
Ecuador	7,880	4.8	.720	10.6	75	52	20	84	0.469
El Salvador	6,550	2.2	.674	16.9	72	62	16	84	0.487
Guatemala	4,650	3.6	.574	26.3	71	78	32	74	0.542
Honduras	3,770	4.6	.625	29.8	73	58	24	84	0.511
Mexico	14,340	2.1	.770	5.2	77	49	17	93	0.448
Nicaragua	2,790	3.6	.589	31.7	74	68	27	78	0.506
Panama	12,770	6.8	.768	13.8	76	33	20	94	0.492
Paraguay	5,080	3.8	.665	13.2	72	50	25	95	0.476
Peru	8,930	6.1	.725	12.7	74	78	19	90	0.415
Uruguay	13,620	3.6	.783	<2	76	23	11	98	0.352
Venezuela	12,150	4.7	.735	12.9	74	33	18	95	0.447

[1]United Nations, *Human Development Report, 2011.*

[2]Population Reference Bureau, *World Population Data Sheet, 2012.*

[3]Gender Inequality Index—A composite measure reflecting inequality in achievements between women and men in three dimensions: reproductive health, empowerment, and the labor market, that ranges between 0 and 1. The higher the number, the greater the inequality.

Source: World Bank, *World Development Indicators, 2012.*

their economies grow and poverty decline. In aggregate, the economies of the region averaged an annual growth rate of 4 percent from 2000 to 2010. However, sporadic economic downturns have made these neoliberal policies highly unpopular with the masses, at times causing major political and economic turmoil. In particular, the value of increased trade and direct foreign investment has been criticized as benefiting a minority of the people in the region, while not adequately addressing the region's long-standing problem of gross income disparities.

Maquiladoras and Foreign Investment The growth in foreign investment and the presence of foreign-owned factories are examples of neoliberalism. The Mexican assembly plants, called **maquiladoras**, that line the border with the United States are characteristic of manufacturing systems in an increasingly globalized economy. These plants began to be constructed in the 1960s as part of a border industrialization program. Today more than 3,000 maquiladoras exist, employing over 1 million people, who assemble automobiles, consumer electronics, and apparel. Between 1994 and 2000, 3 out of every 10 new jobs in Mexico were in the maquiladoras along the border, which account for nearly half of Mexico's exports. Since 2003, China has become a favorite destination for the labor-intensive assembly work that Mexico has specialized in for the past four decades. As Mexican wages have gone up, some companies have relocated factories to East Asia. Northern Mexico is still an attractive location, but competition from China and even Central America may erode Mexico's various locational and structural advantages. Mexico's competitive advantage with its location along the U.S. border and its membership in NAFTA is still significant.

Considerable controversy on both sides of the border surrounds this form of industrialization. Organized labor in the United States complains that well-paying manufacturing jobs are being lost to low-cost competitors, whereas environmentalists point out serious industrial pollution resulting from lax government regulation of the maquiladoras. Mexicans worry that these plants are poorly integrated with the rest of the economy and that many factories choose to hire young, unmarried women because they are viewed as docile laborers (Figure 4.28). In Ciudad Juarez, where for many years women outnumbered men in factory jobs, this led to many women outearning men. And, shockingly, beginning in the 1990s, a surge in female homicides occurred in this border city that went unprosecuted and generated popular protests by women's groups. With NAFTA, foreign-owned manufacturing plants are no longer restricted to the border zone and are increasingly being built near the population centers of Monterrey, Puebla, and Veracruz. Aguascalientes, in central Mexico, has emerged as the country's auto city, although most of the cars produced there are by foreign companies and are destined for export.

Other Latin American states are attracting foreign companies through tax incentives and low labor costs. Assembly plants in Honduras, Guatemala, and El Salvador are drawing foreign investors, especially in the apparel industry. A recent report from El Salvador claims that not one of its apparel factories has a union. Making goods for major American labels, many Salvadoran garment workers complain that they do not make a living wage, work 80-hour weeks, and will lose their jobs if they become pregnant. The situation in Costa Rica, which has been a major computer chip manufacturer for Intel since 1998, is quite different. With a well-educated population, low crime rate, and stable political scene, Costa Rica is now attracting other high-tech firms. Hopeful officials claim that Costa Rica is transitioning from a

FIGURE 4.28 Mexican Maquiladora Workers Many maquiladoras rely upon women for tedious and demanding assembly work. These women manufacture car radios at Delphi Delco Electronics in Matamoros, Mexico. Delphi, which makes parts for General Motors cars, has about 11,000 Mexican workers in seven factories near Matamoros. Matamoros is across the border from Brownsville, Texas, near the Gulf of Mexico.

banana republic (bananas and coffee were the country's long-standing exports) to a high-tech manufacturing center. As a result, the Costa Rican economy averaged 5 percent annual growth from 2000 to 2010.

The Informal Sector Even in prosperous San José, Costa Rica, a short drive to the urban periphery shows large neighborhoods of self-built housing filled with street traders and family-run workshops. Such activities make up the **informal sector**, which is the provision of goods and services without the benefit of government regulation, registration, or taxation. Most people in the informal economy are self-employed and receive no wages or benefits except the profits they clear. The most common informal activities are housing construction (in many cities, one-third of all residents live in self-built housing), manufacturing in small workshops, street vending, transportation services (messenger services, bicycle delivery, and collective taxis), garbage picking, street performing, and even paid line-waiting (Figure 4.29).

No one is sure how big this economy is, in part because it is difficult to separate formal activities from informal ones. Visit Lima, Belém, Guatemala City, or Guayaquil, and it is easy to get the impression that the informal economy *is* the economy. From self-help housing that dominates the landscape to hundreds of street vendors that crowd the sidewalks, it is impossible to avoid. The informal sector has some advantages—hours are flexible, children can work with their parents, and there are no bosses. As important as this sector may be, however, widespread dependence on it signals Latin America's poverty, not its wealth. It reflects the inability of the formal economies of the region, especially in industry, to provide enough jobs for the many people seeking employment.

Primary Export Dependency

Historically, Latin America's abundant natural resources were its wealth. In the colonial period, silver, gold, and sugar generated great wealth for the colonists. With independence in the 19th century, a series of export booms introduced commodities such as bananas, coffee, cacao, grains, tin, rubber, copper, wool, and petroleum to an expanding world market. One of the legacies of this export-led development was a tendency to specialize in one or two major commodities, a pattern that continued into the 1950s. During that decade, 90 percent of Costa Rica's export earnings came from bananas and coffee, 70 percent of Nicaragua's came from coffee and cotton, 85 percent of Chile's came from copper, and half of Uruguay's came from wood. Even Brazil generated 60 percent of its export earnings from coffee in 1955. However, by 2000, coffee accounted for less than 5 percent of the country's exports, even though Brazil remained the world leader in coffee production.

Agricultural Production   Since the 1960s, the trend in Latin America has been to diversify and to mechanize agriculture. Nowhere is this more evident than in the Plata Basin, which includes southern Brazil, Uruguay, northern Argentina, Paraguay, and eastern Bolivia. Soybeans, used for oil and animal feed, transformed these lowlands in the 1980s and early 1990s. Brazil is now the second largest producer of soy in the world (following the United States) and is already the world's largest exporter of soy. Argentina is the third largest, and production is still increasing. Between the late 1990s and 2010, soy production tripled in Argentina. The speed with which the Plata and Amazon basins are being converted into soy fields alarms many; it is eliminating forest and savanna, negatively impacting biodiversity and the natural storage of greenhouse gases. However, with soy priced high, the rush to plant continues (Figure 4.30).

Similar large-scale agricultural frontiers exist along the piedmont zone of the Venezuelan Llanos (mostly grains) and the Pacific slope of Central America (cotton and some tropical fruits). In northern Mexico, water supplied from dams along the Sierra Madre Occidental has turned the valleys in Sinaloa into intensive agricultural centers of fruits and vegetables for consumers in the United States. The relatively mild winters in northern Mexico allow growers to produce strawberries and tomatoes during the winter months.

In each of these cases, the agricultural sector is capital-intensive. By using machinery, hybrid crops, chemical fertilizers, and pesticides, many corporate farms are extremely productive and profitable. What these operations fail to do is employ many rural people, which is especially problematic in countries where one-third or more of the population depends on agriculture for its livelihood. Interestingly, a few traditional Amerindian foods, such as quinoa, are gaining consumers thanks to a growing appetite for organic and healthy foods. Recently, Peru and Bolivia have experienced a boom in quinoa production and exports, with much of the crop being grown in small and medium-sized highland farms. Bolivian President Evo Morales even declared 2012 the "year of quinoa" in an effort to promote traditional "Indian" foods.

Mining and Forestry   The mining of silver, zinc, copper, iron ore, bauxite, and gold is an economic mainstay for many countries in the region. Moreover, many commodity prices reached record levels in the last decade, boosting foreign exchange earnings. Chile is the world leader in copper production, far outproducing the next two largest producers, Peru and the United States. Peru, however, led in global silver production in 2009, and regional rivals Chile, Bolivia, and Mexico were each top-10 producers. Peru was also Latin America's top gold producer.

Lithium metal is gaining world interest. This soft, silver-white metal is used in making lightweight batteries, like those in cell phones and laptops. It is also a key metal for electric car batteries. Today the largest producer of lithium is Chile, but the world's largest reserves are in Bolivia under the Solar de Uyuni in the Altiplano. These reserves are so immense that Bolivia has been dubbed the Saudi Arabia of lithium. But it remains to be seen how and under what terms this critical resource will be extracted from this remote region.

Logging is another important, and controversial, resource-based activity. Several countries rely on plantation forests of introduced species of pine, teak, and eucalyptus to supply domestic fuel wood, pulp, and board lumber. These plantation forests grow single species and fall far short of the complex ecosystems occurring in natural forests. Still, growing trees for paper or fuel reduces the pressure on other forested areas. Leaders in plantation forestry are Brazil, Venezuela, Chile, and Argentina. Considered one of Latin America's economic stars, Chile exports timber and wood chips to add to its mineral export earnings. Thousands of hectares of exotics (especially pine) have been planted, systematically harvested, and cut into boards or chipped for wood pulp. Japanese capital is heavily invested in this sector of the Chilean economy. The recent expansion of the wood chip business, however, led to a dramatic increase in the logging of native forests.

FIGURE 4.29 Peruvian Street Vendors   Street vendors selling produce in Huancayo, Peru. Street vending plays a critical role in the distribution of goods and the generation of income. Some aspects of street vending (such as access to space) are regulated by local governments and by the vendors themselves.

FIGURE 4.30 Soy Production in Brazil Fartura Farm in the state of Mato Grosso, Brazil, embodies the large-scale industrial agriculture that has transformed much of South America into one of the world's largest producers and exporters of soy products.

The Energy Sector The oil-rich countries of Mexico, Venezuela, Ecuador, and now Brazil are able to meet most of their own fuel needs and also earn vital revenues from oil exports. In 2010, Mexico was the 7th largest producer of oil in the world, Brazil was 9th, and Venezuela was 11th (Figure 4.31). Of the three, Venezuela is the most dependent on energy revenues, earning up to 90 percent of its foreign exchange from petroleum and natural gas products. The largest new oil discovery in recent years has been off the coast of Brazil. In the past decade, oil production in Brazil has doubled, moving Brazil into the ranks of oil exporters and drawing increased foreign investment. Although Latin American oil producers do not receive as much attention as those in the Middle East, they are significant. Venezuela was one of the five founding members of the Organization of the Petroleum Exporting Countries (OPEC), and Ecuador joined the group in 1973.

Natural gas production is also on the rise in this region. Venezuela and Bolivia have the largest proven reserves of natural gas in Latin America, but Mexico and Argentina are by far the largest producers. In recent years, Argentina's natural gas production has been boosted by new finds in Patagonia. In 2012, Argentina made headlines when President Christina Fernandez de Kirchner seized majority control of YPF, the major producer of oil and natural gas in Argentina, from a Spanish-owned company, claiming frustration over unnecessary declines in output. Argentina is especially dependent upon natural gas for its urban markets and exports; the production declines produced shortages and price increases in the domestic liquefied natural gas market that required action.

In the area of biofuels, Brazil offers a story of sweet success. In the 1970s, when oil prices skyrocketed, then oil-poor Brazil decided to convert its abundant sugarcane harvest into ethanol. Over the years, even when oil prices plummeted, Brazil continued to invest in ethanol, building mills and a distribution system that delivered ethanol to gas stations. One of Brazil's major technological successes was inventing flex-fuel cars that run on any combination of ethanol and gasoline. At the time, Brazil was motivated by its limited oil reserves, but today, with interest in biofuels growing as a way to reduce CO_2 emissions, Brazil's support of its ethanol program looks visionary.

Latin America in the Global Economy

To conceptualize Latin America's place in the world economy, **dependency theory** was advanced in the 1960s by scholars from the region. The premise of the theory is that expansion of European capitalism created the region's underdevelopment. For the developed "cores" of the world to prosper, the "peripheries" became dependent and impoverished. Dependent economies, such as those in Latin America, were export-oriented and vulnerable to fluctuations in the global market. Even when they experienced economic growth, it was secondary to the economic demands of the core (North America and Europe).

Economists who accepted this interpretation of Latin America's history were convinced that economic development could occur only through self-sufficiency, growth of internal markets, agrarian reform, and greater income equality. In short, they argued for strong state intervention and an uncoupling from the core economies of Western Europe and North America. Policies such as import substitution industrialization (developing a country's industrial sector by making imported products extremely expensive and domestic manufactured goods cheaper) and nationalization of key industries were partially influenced by this view. Dependency theory has its detractors. In its simplest form, it becomes a means to blame forces

FIGURE 4.31 Oil and Gas Production A gas compression plant located on Lake Maracaibo in western Venezuela. Below the surface of the shallow lake are some of Venezuela's oldest and largest reserves of oil and natural gas.

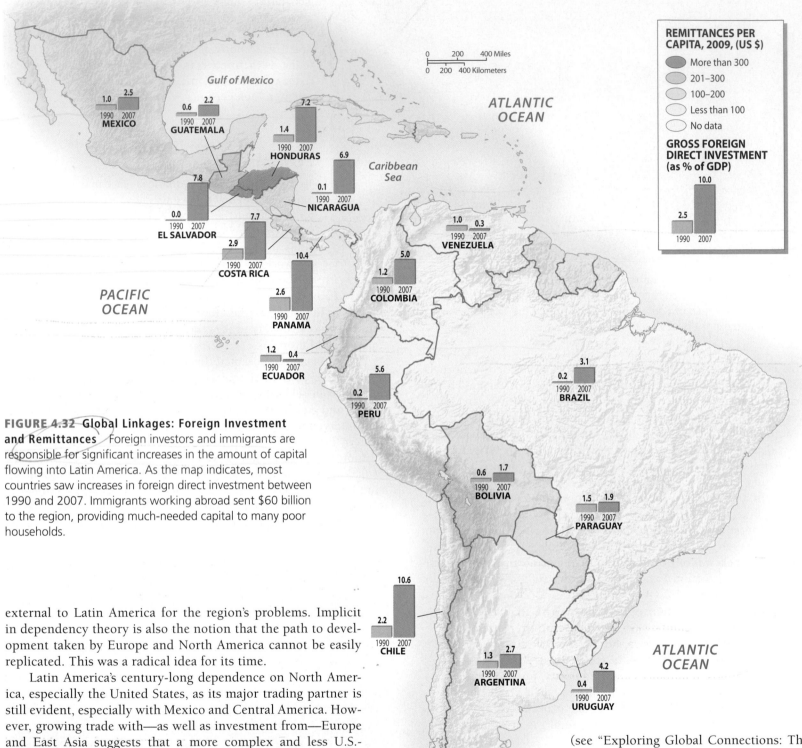

FIGURE 4.32 Global Linkages: Foreign Investment and Remittances Foreign investors and immigrants are responsible for significant increases in the amount of capital flowing into Latin America. As the map indicates, most countries saw increases in foreign direct investment between 1990 and 2007. Immigrants working abroad sent $60 billion to the region, providing much-needed capital to many poor households.

external to Latin America for the region's problems. Implicit in dependency theory is also the notion that the path to development taken by Europe and North America cannot be easily replicated. This was a radical idea for its time.

Latin America's century-long dependence on North America, especially the United States, as its major trading partner is still evident, especially with Mexico and Central America. However, growing trade with—as well as investment from—Europe and East Asia suggests that a more complex and less U.S.-dependent pattern of trade is emerging. Latin America is linked to the world economy in ways other than trade. Figure 4.32 shows the changes in foreign direct investment (FDI) as a percentage of gross domestic product (GDP) from 1990 to 2007. For nearly every country in the region, the value of foreign investment in terms of percentage of GDP went up, Ecuador and Venezuela being the exceptions. In 1995, Brazil's FDI was less than $5 billion, and Mexico's was $9.5 billion. By 2010, FDI in Brazil had soared to $48 billion, and Mexico's was $20 billion. Much of this foreign investment was from Europe and Asia

(see "Exploring Global Connections: The Growth of European and Asian Investment in Latin America.")

Remittances are another important financial flow that reflects the integration of Latin American migrants into labor markets around the world, especially in North America. Scholars debate whether this flow of capital can actually lead to sustained development or whether it is simply a survival strategy of last

EXPLORING GLOBAL CONNECTIONS

The Growth of European and Asian Investment in Latin America

Throughout the 20th century, the United States was Latin America's largest trading partner, and for many countries this is still true. But trade and investment from Europe and Asia are diversifying Latin America's trade relationships and lessening its dependence upon U.S. markets. These trends also suggest an erosion of Washington's influence in the region. This is especially true in South America, where trade with the European Union and China is surpassing trade with the United States.

Collectively, the European Union is Latin America's second largest trading partner after the United States. Between 1990 and 2006, trade between Latin America and the European Union doubled. In 2008, the trade volume between Latin America and the European Union amounted to €178 billion, representing nearly 15 percent of Latin American's trade. Latin American countries still tend to export agricultural and energy products to Europe. They import European machinery, transport equipment, and chemicals.

The European Union is now the leading investor in the region. The presence of European businesses in Latin America has been an important source of new foreign direct investment. Perhaps not surprisingly, given its colonial ties to the region, Spain has been the single largest European investor, especially in the 1990s. Under neoliberal reforms, as state-owned industries in Latin America were sold, Spain was poised to invest in telecommunications, banking, public utilities, oil, and natural gas. Over the last few years, the European Union has established free-trade agreements with Chile and Mexico.

Coming up fast is China, whose investments in Latin America soared in the past decade. In terms of individual states, China is now Latin America's second largest trading partner after the United States. Chinese investors are focused on energy and mineral resources. During the economic crisis of 2009, the Chinese brought loans to the region, brokering deals with Venezuela, Ecuador, Argentina, and Brazil. These monies gave Venezuela access to hard currency; in exchange, the volume of Venezuelan oil shipments to China nearly tripled. In 2012, China became Brazil's single largest trading partner. China and Brazil,

FIGURE 4.2.1 Brazilian President Dilma Rousseff (L) shakes hands with Chinese Premier Wen Jiabao at a 2012 meeting in Rio de Janeiro, Brazil.

the so-called BRIC nations along with Russia and India, have recognized their global strategic partnership. Not only is Brazil exporting grains, minerals, and energy resources, but also it has signed an agreement that will allow Brazilian airplane maker Embraer to manufacture and sell its regional jets in China (see Figure 4.2.1).

---

resort. The economic impact of remittances shown on a per capita basis is real (see Figure 4.32). Mexico is the regional leader, receiving over $21 billion in remittance income in 2009 (which is equal to over $200 per capita). But for smaller countries such as El Salvador and Honduras, remittances contribute far more to the domestic economy. El Salvador, a country of 7 million people, received over $3.5 billion in remittances in 2009, which is nearly $500 per capita. For many Latinos, remittances are the surest way to poverty alleviation, although they depend upon an international migration system that is constantly changing and includes both legal and illegal channels of movement.

Neoliberalism as Globalization By the 1990s, governments and the World Bank had become champions of neoliberalism as a sure path to economic development. Neoliberal policies accentuate the forces of globalization by turning away from policies that emphasize state intervention and self-sufficiency. Most Latin American political leaders embrace neoliberalism and the benefits that come with it, such as increased trade and more favorable terms for debt repayment. However, there are signs of discontent with neoliberalism throughout the region. Recent protests in Peru and Chile reflect the popular anger against trade policies that seem to benefit

only the elite. The case of Chile is especially interesting because that nation's leaders were early and outspoken defenders of neoliberalism and it is also one of the more prosperous nations in the region.

Dollarization As financial crises spread throughout Latin America in the late 1990s, governments began to consider the economic benefits of **dollarization**, a process by which a country adopts—in whole or in part—the U.S. dollar as its official currency. In a totally dollarized economy, the U.S. dollar becomes the only medium of exchange, and the country's national currency ceases to exist. This was the radical step taken by Ecuador in 2000 to address the dual problems of currency devaluation and hyperinflation rates of more than 1,000 percent annually. El Salvador adopted dollarization in 2001 as a means to reduce the cost of borrowing money. Dollarization is not a new idea; in 1904, Panama dollarized its economy, a year after it gained independence from Colombia. Until 2000, however, Panama was the only fully dollarized state in Latin America.

A more common strategy in Latin America is limited dollarization, in which U.S. dollars circulate and are used alongside the country's national currency. Limited dollarization exists in many countries around the world, but is widespread in Latin America. Because the economies of Latin America are prone to currency devaluation and hyperinflation, limited dollarization is a type of insurance. Many banks in Latin America, for example, allow customers to maintain accounts in dollars to avoid the problem of capital flight should a local currency be devalued.

Social Development

Over the past three decades, Latin America has experienced marked improvements in life expectancy, child survival, and educational equity. One telling indicator is the steady decline between 1990 and 2010 in mortality rates for children below age five (Table 4.2). This indicator is important because an increase in the number of children younger than five years surviving suggests that basic nutritional and health-care needs are being met. We can also conclude that resources are being used to sustain women and their children. Despite economic downturns, the region's social networks have been able to lessen the negative effects on children.

A combination of government policies and grassroots and nongovernmental organizations (NGOs) plays a fundamental role in contributing to social well-being. In the past few years, conditional cash transfer programs, such as **Bolsa Familia**, have reduced extreme poverty. Poor Brazilian families who qualify for Bolsa Familia receive a monthly check from the state, but are required to keep their children in school and take them to clinics for health checkups. Such programs have both the immediate impact of giving poor families cash and the long-term impact of improving the educational attainment and health care of their children. Mexico has adopted a similar program. For states with far fewer resources than Brazil or Mexico, international humanitarian organizations, church organizations, and community activists provide many services that state and local governments cannot. Catholic Relief Services and Caritas, for example, work with rural poor throughout the region to improve their water supplies, health care, and education. Other groups lobby local governments to build schools and recognize squatters' claims. Grassroots organizations also develop cooperatives that market everything from sweaters to cheeses.

Other important indicators for social development are life expectancy, gender educational equity, and access to improved water sources. In aggregate, 84 percent of the people in the region have access to an adequate amount of water from an improved source, slightly more girls receive secondary education than boys (Figure 4.33), and life expectancy (for both men and women) is 74 years. Masked by these combined data are extreme variations between rural and urban areas, between regions, and along racial and ethnic lines.

Race and Inequality   There is much to admire about race relations in Latin America. The complex racial and ethnic mix that was created in Latin America fostered tolerance for diversity. That said, Indians and blacks are more likely to be counted among the poor of the region. More than ever before, racial discrimination is a major political issue in Brazil. Reports of organized killings of street children, most of them Afro-Brazilian,

make headlines. For decades, Brazil championed its vision of a color-blind racial democracy. True, residential segregation by race is rare in Brazil, and interracial marriage is common, but certain patterns of social and economic inequality seem best explained by racial discrimination.

Assessing racial inequalities in Brazil is problematic. The Brazilian census asks few racial questions, and all are based on self-classification. In the 2000 census, less than 11 percent of the population called itself black. Some Brazilian sociologists, however, claim that more than half the population is of African ancestry, making Brazil the second largest "African state" after Nigeria. Racial classification is always highly subjective and relative, but some patterns support the existence of racism. Evidence from northeastern Brazil, where Afro-Brazilians are the majority, shows death rates approaching those of some of the world's poorest countries. Throughout Brazil, blacks suffer higher rates of homelessness, landlessness, illiteracy, and unemployment than others. To address this problem, various affirmative action measures have been implemented (along with the Bolsa Familia program). From federal ministries to public universities, various

FIGURE 4.33 School Children in Panama   Uniformed public school children walk to school in Panama City. Latin American states have seen steady improvements in youth literacy, with 97 percent of youth (between the ages of 15 and 24) being literate. Per capita expenditures on education and access to postsecondary education still lag behind levels in Europe and North America.

FIGURE 4.34 Development Issues: Women's Participation in National Politics Women are active players in Latin American politics. On average, 23 percent of seats in national parliaments are held by women. (In contrast, the figure for the United States is 19 percent.) In Brazil, only 9 percent of the seats are held by women, but in Mexico 28 percent are held by women, and in Argentina the figure is 41 percent. Moreover, six countries have had or have women presidents. The most recently elected female president is Dilma Rousseff of Brazil.

PERCENT OF SEATS HELD BY WOMEN IN NATIONAL PARLIAMENTS, 2009

Fewer than 10
10–20
21–30
31–40
More than 40

● Countries that have or have had women presidents

quota systems are being tried to improve the condition of Afro-Brazilians.

In areas of Latin America where Indian cultures are strong, indicators of low socioeconomic position are also present. In most countries, areas where native languages are widely spoken regularly correspond with areas of persistent poverty. In Mexico, the Indian south lags behind the booming north and Mexico City. Prejudice is embedded in the language. To call someone an *indio* (Indian) is an insult in Mexico. In Bolivia, women who dress in the Indian style of full, pleated skirts and bowler hats are called *cholas*, a descriptive term referring to the rural *mestizo* population that suggests a backwardness and even cowardice. No one of high social standing, regardless of skin color, would ever be called a *chola* or *cholo*.

It is difficult to separate status divisions based on class from those based on race. From the days of conquest, being European meant an immediate elevation in status over the Indian, African, and *mestizo* populations. Race does not necessarily determine one's economic standing, but it certainly influences it. In the case of Amerindian people, they are politically asserting themselves. For example, Evo Morales was inaugurated as president of Bolivia in 2006, making him the first Indian leader in that country.

The Status of Women Many contradictions exist with regard to the status of women in Latin America. Many Latina women work outside the home. In most countries, the formal figures are between 30 and 40 percent of the workforce, not far off from many European countries, but lower than in the United States. Legally speaking, women can vote, own property, and sign for loans, although they are less likely to do so than men, reflecting the patriarchal (male-dominated) tendencies in the society. Even though Latin America is predominantly Catholic, divorce is legal and family planning is promoted. In most countries, however, abortion remains illegal.

Overall, access to education in Latin America is good, compared to other developing regions, and thus illiteracy rates tend to be low. The rates of adult illiteracy are slightly higher for women than for men, but usually by only a few percentage points. Throughout higher education in Latin America, male and female students are equally represented today. Consequently, women are regularly employed in the fields of education, medicine, and law.

The biggest changes for women are the trends toward smaller families, urban living, and educational parity with men. These factors have greatly improved the participation of women in the labor force. In the countryside, however, serious inequalities remain. Rural women are less likely to be educated and tend to have larger families. In addition, they are often left to care for their families alone, as husbands leave in search of seasonal employment. In most cases, the conditions facing rural women have been slow to improve.

Women are increasingly playing an active role in politics. In 1990, Nicaragua elected the first woman president in Latin America, Violeta Chamorro, the owner of an opposition newspaper. Nine years later, Panamanians voted Mireya Moscoso into power. In 2005, South America had its first woman president when Dr. Michelle Bachelet, a pediatrician and single mother, took the oath of office in Chile. Two years later, Cristina Fernández de Kirchner was elected president of Argentina, following the presidency of her husband Néstor Kirchner. In 2010, voters in Costa Rica and Brazil elected female heads of state. As shown in Figure 4.34, many Latin American states have larger percentages of women in seats of national parliaments than does the United States.

Across the region, women and indigenous groups are active organizers and participants in cooperatives, small businesses, and unions and are elected to national office. In a relatively short period, they have won a formal place in the economy and a political voice. Moreover, evidence suggests that this trend will continue.

REVIEW QUESTIONS

1. How has the export of primary products (food, fiber, and energy) shaped the economies of Latin America?

2. What explains some of the positive indicators of social development in Latin America?

Summary

- Latin America and the Caribbean were the first world regions to be fully colonized by Europe. In the process, perhaps 90 percent of the native population died from disease, cruelty, and forced resettlement. The slow demographic recovery of native peoples and the continual arrival of Europeans and Africans resulted in an unprecedented level of racial and cultural mixing.

- Unlike in other developing areas, most people in Latin America live in cities. This trend started early and reflects a cultural bias toward urban living with roots in the colonial past. The cities are large and combine aspects of the formal industrial economy with an informal one.

- Compared to Europe and Asia, Latin America is still rich in natural resources and relatively lightly populated. However, as populations continue to grow and trade in natural resources increases, there is growing concern for the state of the environment. Of particular concern is the relentless cutting of tropical rain forests.

- Uneven development and economic frustration have led many Latin Americans (both highly skilled and low skilled) to consider emigration as an economic option. Today Latin America is a region of emigration, with emigrants going to work in North America, Europe, and Japan. Collectively, they send billions of dollars in remittances back to Latin America each year.

- Most Latin American governments have embraced neoliberal policies in an attempt to foster economic development. As a result, exports have surged, along with direct foreign investment in the region. Although the region experienced growth from 2000 to 2010, there is still stubbornly high income inequality.

- Latin American governments were early adopters of neoliberal economic policies. Although some states prospered, others faltered, sparking popular protests against the negative effects of globalization. It does seem, however, that new political actors are emerging—from indigenous groups to women—who are challenging old ways of doing things.

Key Terms

agrarian reform 116
Altiplano 108
altitudinal zonation 110
Bolsa Familia 134
Central American Free Trade Association
(CAFTA) 103

dependency theory 131
dollarization 133
El Niño 111
environmental lapse rate 110
grassification 106
informal sector 129

latifundia 116
maquiladora 129
megacity 104
Mercosur 103
mestizo 117
minifundia 116

Thinking Geographically

1. Discuss the processes driving tropical deforestation in Latin America and how they compare with those causing deforestation in other areas of the world.

2. How is neoliberalism influencing the way Latin America interacts with the rest of the world? What are the social and environmental costs of neoliberalism? How is this development approach related to globalization?

3. Given the dominance of cities in this region, describe the particular urban environmental problems facing cities in the developing world. How might Latin America's megacities use their size and density to reduce the environmental problems associated with urbanization?

4. Discuss the social, environmental, and economic consequences behind the modernization of agriculture. How does Latin America's experience with modern agricultural systems compare with North America's?

5. Latin America is a region of emigration. Discuss the impact mass emigration has had on this region. Why is it higher here than in other regions of the developing world? Which countries are producing the most immigrants and where are they going?

MasteringGeography™

Looking for additional review and test prep materials? Visit the Study Area in MasteringGeography™ to enhance your geographic literacy, spatial reasoning skills, and understanding of this chapter's content by accessing a variety of resources, including **MapMaster** interactive maps, videos, RSS feeds, flashcards, web links, self-study quizzes, and an eText version of *Globalization and Diversity*.

Scan to visit the author's blog for chapter updates.

http://gad4blog.
wordpress.com/
category/
latin-america/

Authors' Blogs

Scan now to access the authors' blogs for up-to-date information on Latin America.

Scan to visit the GeoCurrents blog.

http://geocurrents.
info/category/place/
latin-america

Globalization and Diversity

Named for some of its former inhabitants—the Carib Indians—the modern Caribbean is a culturally complex and economically peripheral world region. The region was settled by various colonial powers, and its current residents are mostly a mix of African, European, and South Asian peoples. Greatly impacted by the global economy, but with relatively little influence on it, the livelihoods of Caribbean peoples are filled with uncertainty.

ENVIRONMENTAL GEOGRAPHY
Climate change threatens the Caribbean, with the potential for stronger and more frequent hurricanes, loss of territory due to rising sea levels, and destruction of coral reefs. In 2010, a devastating earthquake flattened the capital of Haiti, initiating the region's worst natural disaster in decades.

POPULATION AND SETTLEMENT
Having experienced its demographic transition, the region now has slow population growth. In addition, large numbers of Caribbean peoples have emigrated from the region, leaving in search of economic opportunity and sending back billions of dollars.

CULTURAL COHERENCE AND DIVERSITY
Creolization—the blending of African, European, and Amerindian elements—has resulted in many unique Caribbean expressions of culture, such as rara, reggae, and steel drum bands.

GEOPOLITICAL FRAMEWORK
The first area of the Americas to be extensively explored and colonized by Europeans, the region has seen many rival European claims and, since the early 20th century, has experienced strong U.S. influence. Many Caribbean territories have been independent states for 50 years or less.

ECONOMIC AND SOCIAL DEVELOPMENT
Environmental, locational, and economic factors make tourism a vital component of this region's economy, particularly in Puerto Rico, Cuba, the Dominican Republic, Jamaica, and the Bahamas. Offshore manufacturing and banking are also significant in the region's modern economic development.

➤ **Visitors gather at Bob Marley's birthplace in St. Ann Parish, Jamaica. Bob Marley was one of the Caribbean's first pop superstars when he toured with his band the Wailers in the 1970s. Images of Marley, as well as his reggae music, continue to have global appeal.**

138

THE CARIBBEAN 5

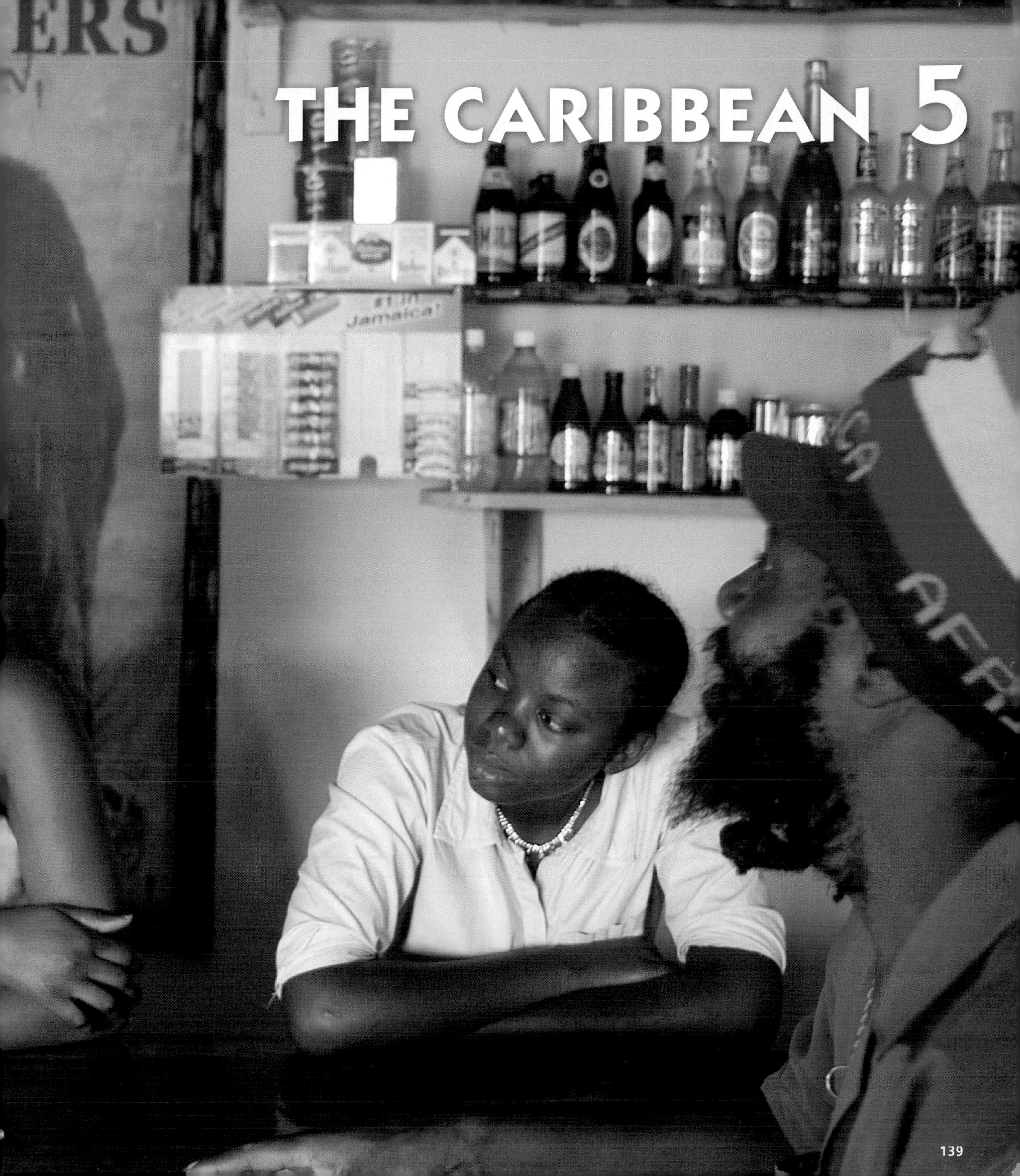

Say the word "Caribbean" and images of white sand, palm trees, turquoise water, and dancing to reggae music come to mind. Jamaica was home to Bob Marley, one of the developing world's first pop superstars, who introduced the musical sounds of his island to the world by the 1970s. Marley's image is still found in the streets of Kingston, the lively, but poor and sometimes violent city that inspired him. Kingston was a minor city in the Caribbean, built to serve the many sugar plantations that made the island profitable for the British. Most of its residents today are of African ancestry, linked to the legacy of slavery that permeates the entire region. Jamaica is emblematic of the Caribbean region—it was an island originally settled by Arawaks (an Amerindian group), then conquered by the Spanish, and ultimately colonized by the British. Like other territories in the region, it received its independence in the 1960s. Jamaica is economically poor and too small to influence global politics, but it is connected to the world through emigration, tourism, and cultural products such as its music. Understanding the creation of Jamaica, and other territories like it in the Caribbean, offers a critical window to the long-term processes influencing globalization in the region today.

The Caribbean was the first region of the Americas to be extensively explored and colonized by Europeans. Yet its modern regional identity is unclear, often merged with Latin America, but also viewed as apart from it. Today the region is home to 44 million people, scattered across 26 countries and dependent territories. They range from the small British dependency of Turks and Caicos, with 12,000 people, to the island of Hispaniola, with nearly 20 million. In addition to the Caribbean Islands, Belize of Central America and the three Guianas—Guyana, Suriname, and French Guiana—of South America are included as part of the Caribbean. For historical and cultural reasons, the peoples of these mainland states identify with the island nations and are thus included in this chapter (Figure 5.1).

Historically, the Caribbean was a battleground between rival European powers competing for territorial control of these tropical lands. In the early 1900s, the United States took over as the dominant geopolitical force in the region, maintaining what some have called a neocolonial presence. As in many developing areas, external control of the Caribbean by foreign governments and companies produced highly dependent and inequitable economies. In the Caribbean, extreme inequalities were accentuated by the area's reliance upon slave labor, plantation agriculture, and the prosperity that sugar promised in the 18th century. Over time, other products began to have economic importance, as did the international tourist industry. Increasingly, regional governments have sought to diversify their economies, expanding into banking services and manufacturing to reduce the region's dependence on agriculture and tourism.

The basis for treating the Caribbean as a distinct area lies within its particular cultural and economic history. Culturally, the region can be distinguished from the largely Iberian-influenced mainland of Latin America because of its more diverse European colonial history and its strong African imprint due to the region's historical reliance upon slavery. In terms of economic production, the dominance of export-oriented plantation agriculture explains many of the social, economic, and environmental patterns in the region.

Still a developing area, most states in the region have achieved life expectancies in the 70s, low child mortality, and high rates of literacy. Millions of tourists visiting the Caribbean view it is an international playground for sun, sand, and fun. However, there is another Caribbean that is far poorer and economically more dependent than the one portrayed on travel posters. Haiti, by many measures the poorest country in the Western Hemisphere, has 10 million people. The Dominican Republic also has 10 million, and Cuba has 11 million. Each of these major Caribbean nations suffers from serious economic problems and widespread poverty.

The majority of Caribbean people are poor, living in the shadow of North America's vast wealth. The concept of **isolated proximity** has been used to explain the region's unique position in the world. The *isolation* of the Caribbean sustains the area's cultural diversity (Figure 5.2) and also explains its limited economic opportunities. Caribbean writers note that this isolation fosters a strong sense of place and a tendency to focus inward. Yet the relative *proximity* of the Caribbean to North America (and, to a lesser extent, Europe) ensures its international connections and economic dependence. For example, each year Dominican workers in the United States send nearly $3 billion to family and friends in the Dominican Republic, who rely on this money for their sustenance. Through the years, the Caribbean has evolved as a distinct, but economically marginal world region. This position expresses itself today as workers flee the region in search of employment, while foreign companies are attracted to the Caribbean for its cheap labor. The economic well-being of most Caribbean countries is uncertain. Despite such uncertainty, the people of the Caribbean maintain an enduring cultural richness and attachment to place that may explain a growing countercurrent of immigrants back to the region.

LEARNING OBJECTIVES

After reading this chapter you should be able to:

- Differentiate island and rimland environments and the environmental issues that affect these areas.

- Summarize the demographic shifts in the Caribbean as population growth slows, settlement in cities intensifies, emigration abroad continues, and a return migration begins.

- Explain the reasons why European colonists so aggressively sought control of the Caribbean and why independence in the region came about more gradually than in neighboring Latin America.

- Identify the demographic and cultural implications of the massive transfer of African peoples to the Caribbean, as well as other smaller labor flows from South Asia.

- Describe how the Caribbean is linked to the global economy through offshore banking, emigration, and tourism.

- Explain the difference between social and economic development and suggest reasons why the Caribbean does better in social indicators of development, compared to economic indicators.

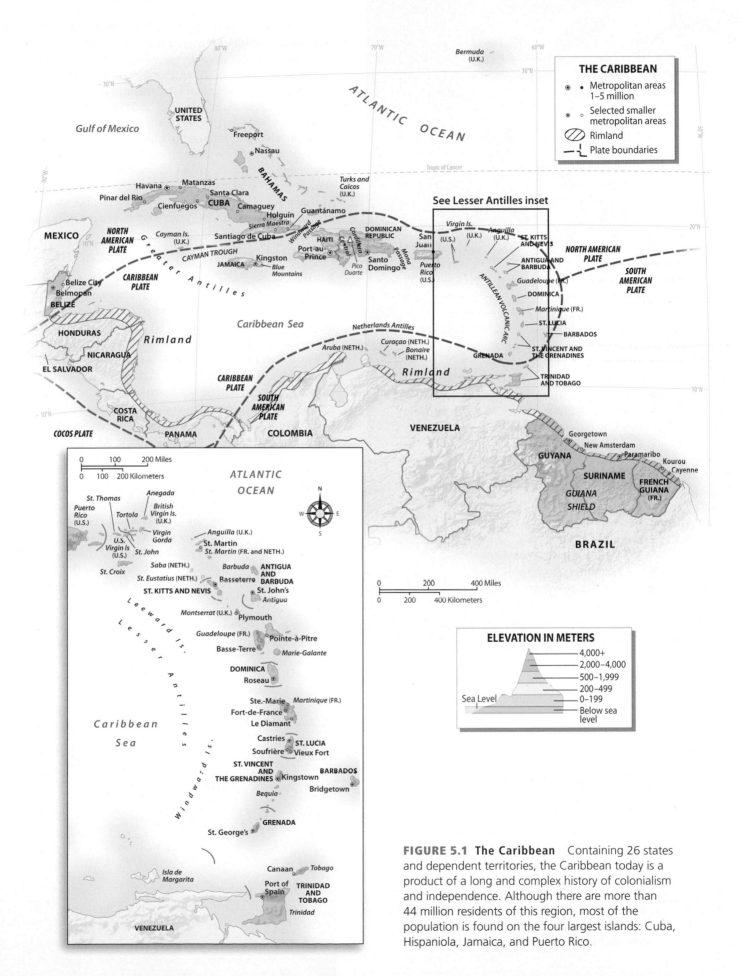

FIGURE 5.1 The Caribbean Containing 26 states and dependent territories, the Caribbean today is a product of a long and complex history of colonialism and independence. Although there are more than 44 million residents of this region, most of the population is found on the four largest islands: Cuba, Hispaniola, Jamaica, and Puerto Rico.

FIGURE 5.2 **Carnival Dancers** Carnival participants take a break from the festivities in Port of Spain, Trinidad. Trinidad and Tobago celebrate carnival before the beginning of Lent with elaborate parades and costuming. Carnival events and associated music have become important symbols of Caribbean identity.

ENVIRONMENTAL GEOGRAPHY: PARADISE UNDONE

Tucked between the Tropic of Cancer and the equator, with year-round temperatures averaging in the high 70s, the hundreds of islands and picturesque waters of the Caribbean have often inspired comparisons to paradise. Columbus began the tradition by describing the islands of the New World as the most marvelous, beautiful, and fertile lands he had ever known, filled with flocks of parrots, exotic plants, and friendly natives. Writers today are still lured by the sea, sands, and swaying palms of the Caribbean.

Ecologically speaking, it is difficult to picture a landscape that has been more completely altered than that of the Caribbean. For nearly five centuries, the destruction of forests and the unrelenting cultivation of soils resulted in the extinction of many endemic (native) Caribbean plants and animals, including various shrubs and trees, songbirds, large mammals, and monkeys. This severe depletion of biological resources helps explain some of the present economic and social instability of the region. Most of the environmental problems in the region are associated with agricultural practices, soil erosion, excessive reliance on wood and charcoal for fuel, and the threat of global climate change. However, because many countries rely upon tourism as a vital source of income, the region has also experienced a growth in protected areas, both on the land and in maritime areas.

Island and Rimland Landscapes

It is the Caribbean Sea itself—the body of water enclosed between the *Antillean* islands (the arc of islands that begins with Cuba and ends with Trinidad) and the mainland of Central and South America—that links the states of the region. Historically, the sea connected people through its trade routes and sustained them with its marine resources of fish, green turtle, manatee, lobster, and crab. The Caribbean Sea is noted for its clarity and biological diversity, but it has never supported large commercial fishing because the quantities of any one species are not large. The surface temperature of the sea ranges from 73° to 84°F (23° to 29°C), over which forms a warm tropical marine air mass that influences daily weather patterns. This warm water and tropical setting continue to be key resources for the region, as millions of tourists visit the Caribbean each year (Figure 5.3).

The arc of islands that stretches across the Caribbean Sea is its most distinguishing physical feature. The Antillean islands are divided into two groups: the Greater and Lesser Antilles. The majority of the region's population lives on these islands. The **rimland** (the Caribbean coastal zone of the mainland) includes Belize and the Guianas, as well as the Caribbean coast of Central and South America. In contrast to the islands, the rimland has low population densities (refer to Figure 5.1).

FIGURE 5.3 **Caribbean Sea** Noted for its calm turquoise waters, steady breezes, and treacherous shallows, the Caribbean Sea has both sheltered and challenged sailors for centuries. This aerial photograph shows the southern Caribbean islands of Los Roques, off the Venezuelan coast.

Most of the islands, with the exception of Cuba, are on the Caribbean tectonic plate, wedged between the South American and North American plates. Generally, this is not one of the most tectonically active zones, although earthquakes and volcanic eruptions do happen. In January 2010, however, a 7.0 earthquake leveled Port-au-Prince in one of the most tragic natural disasters to strike the Caribbean.

The Enriquillo Fault, near the densely settled and extremely poor capital city of Haiti, had been inactive for more than a century. When it violently shifted in 2010, the epicenter of the resulting earthquake was just a few miles away from the city (Figure 5.4). The disaster affected nearly 3 million people, as homes were rendered unsafe and water and electricity supplies were disrupted. Shockingly, some 200,000 people died as a result of the earthquake, and 1 million people were made homeless. Many of the government agencies that would have assisted in the relief response were also destroyed. The tragedy of the Haitian earthquake was compounded by the state's poverty and corruption, as most buildings were not built to standards that could withstand an earthquake of this magnitude. The international community, as well as the large diaspora of Haitians living abroad, immediately offered financial aid and assistance. Yet two years after the earthquake, relatively little rebuilding had occurred, half a million people remained in tent cities, and insecurity and unemployment undermined recovery efforts.

Greater Antilles The four large islands of Cuba, Jamaica, Hispaniola (shared by Haiti and the Dominican Republic), and Puerto Rico make up the **Greater Antilles**. These islands contain the bulk of the region's population, arable lands, and large mountain ranges. Many people are knowledgeable about the Caribbean coasts, but are surprised to learn that Pico Duarte in the Cordillera Central of the Dominican Republic is more than 10,000 feet (3,000 meters) tall, Jamaica's Blue Mountains top 7,000 feet (2,100 meters), and Cuba's Sierra Maestra is more than 6,000 feet (1,800 meters). Historically, the mountains of the Greater Antilles were of little economic interest because plantation owners preferred the coastal plains and valleys. However, the mountains were an important refuge for runaway slaves and subsistence farmers and thus are important in the cultural history of the region.

The best farmlands are found in the Greater Antilles, especially in the central and western valleys of Cuba, where a limestone base contributes to the formation of a fertile red clay soil (locally called *mantanzas*), and in Jamaica, where a gray or black soil type called

FIGURE 5.4 Earthquake in Haiti The earthquake that struck Port-au-Prince on January 12, 2010, was one of the worst natural disasters in the region's history, killing over 200,000 people. In this satellite image, Port-au-Prince (in grey and white) rests just north of the Enriquillo Fault line and the nearby epicenter of the 7.0-magnitude quake.

rendzinas is found. Surprisingly, given the region's agricultural history, many of the other soils are nutrient-poor, heavily leached, and acidic.

Lesser Antilles The **Lesser Antilles** form a double arc of small islands stretching from the Virgin Islands to Trinidad. Smaller in size and population than the Greater Antilles, they were important early footholds for rival European colonial powers. The islands from St. Kitts to Grenada form the inner arc of the Lesser Antilles. These mountainous islands, with peaks ranging from 4,000 to 5,000 feet (1,200 to 1,500 meters), have volcanic origins. Erosion of the island peaks and accumulation of ash from volcanic eruptions have created small pockets of arable soils, although the steepness of the terrain places limits on agricultural development.

Just east of this volcanic arc are the low-lying islands of Barbados, Antigua, Barbuda, and the eastern half of Guadeloupe. Covered in limestone that overlays volcanic rock, these lands were much more inviting for agriculture. In particular, such soils were ideal for growing sugarcane, causing these islands to become important early settings for the plantation economy that diffused throughout the region.

Rimland States Unlike the rest of the Caribbean, the rimland states of Belize and the Guianas still contain significant amounts of forest cover. The rimland refers to the coastal zone of the mainland, beginning with Belize and extending along the coast of Central America to northern South America. As on the islands, agriculture in these states is closely tied to local geology and soils. Much of low-lying Belize is limestone. Sugarcane is the dominant crop in the drier north, whereas citrus is produced in the wetter central portion of the state. The Guianas are characterized by the rolling hills of the Guiana Shield, whose crystalline rock is responsible for the area's overall poor soil quality. Thus, most agriculture in the Guianas occurs on the narrow coastal plain, with sugar and rice as the primary crops. French Guiana, which is an overseas territory of France, relies mostly on French subsidies, but exports shrimp and timber. It is also home to the European Space Center at Kourou (Figure 5.5).

FIGURE 5.5 Kourou, French Guiana The European Space Agency regularly launches rockets from its center in Kourou, French Guiana. This French territory, near the equator and on the coast, makes an ideal launch site. Here the *Ariane 5* rocket is being moved to the launch pad.

Caribbean Climate and Climate Change

Much of the Antillean islands and rimland receive more than 80 inches (200 centimeters) of rainfall annually, which is enough to support tropical forests. Average temperatures are typically highs of 80 degrees and lows of 70. Distinctly dry areas exist as well, such as the rain-shadow basin in western Hispaniola (Figure 5.6).

As in many other tropical lowlands, seasonality in the Caribbean is defined by changes in rainfall more than temperature. Although rain falls throughout the year, the rainy season is from July to October. This is also when unstable atmospheric conditions sometimes cause the formation of hurricanes. During the slightly cooler months of December through March, rainfall declines (see the climographs in Figure 5.6 for Havana, Port-au-Prince, and Bridgetown). This time of year corresponds with the peak tourist season.

The Guianas have a different rainfall cycle. On average, these territories receive more rain than the Antillean islands. In Cayenne, French Guiana, an average of 126 inches (320 centimeters) falls each year (see the climograph in Figure 5.6). Unlike the Antilles, the Guianas experience a brief dry period in late summer (September to October). Also, January tends to be a wet period for the mainland, while it is a dry time for the islands. Climatically, the Guianas also are different from the rest of the region because they are not affected by hurricanes.

Hurricanes Each year several **hurricanes** pound the Caribbean, as well as Central and North America, with heavy rains and fierce winds. Beginning in July, westward-moving low-pressure disturbances form off the coast of West Africa, picking up moisture and speed as they move across the Atlantic. Usually no more than 100 miles across,

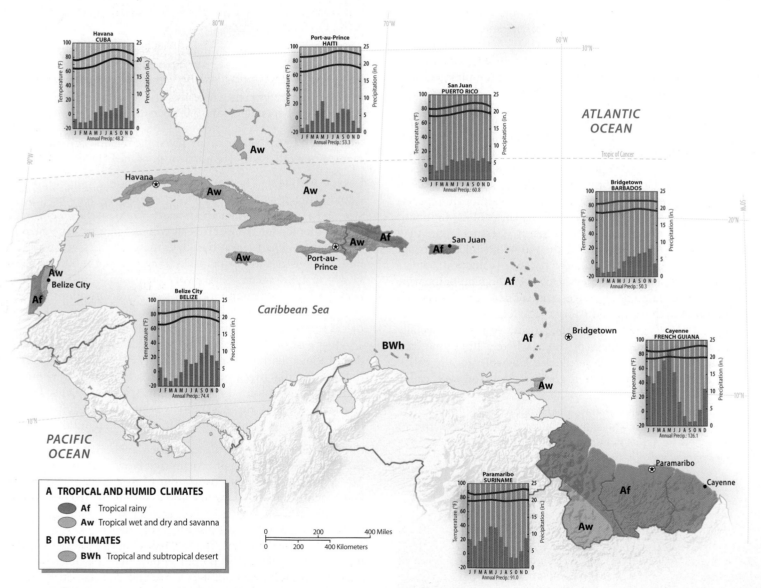

FIGURE 5.6 Climate Map of the Caribbean Most of the region is classified as having either a tropical wet (Af) or a tropical savanna (Aw) climate. Temperature varies little across the region, with highs slightly above 80°F (27°C) and lows around 70°F (21°C). Important differences in total rainfall and the timing of the dry season distinguish different places. In the Guianas, for example, the dry season is September through October, whereas the drier months for the islands are December through March.

FIGURE 5.7 Hurricane Dennis Strikes Cuba   A Cuban man looks at what remains of his house in the village of Casilda after Hurricane Dennis pounded the island with drenching rains and devastating winds. Each season hurricanes roll through the Caribbean. In many climate change scenarios, the intensity of hurricanes in this region are expected to increase.

these disturbances achieve hurricane status when wind speeds reach 74 miles per hour. Hurricanes may take several paths through the region, but they typically enter through the Lesser Antilles. They then curve north or northwest and collide with the Greater Antilles, Central America, Mexico, or southern North America before moving to the northeast and dissipating in the Atlantic Ocean. The hurricane zone lies just north of the equator on both the Pacific and Atlantic sides of the Americas. Typically, a half dozen to a dozen hurricanes form each season and move through the region, causing limited damage.

There are, of course, exceptions, and most longtime residents of the Caribbean have felt the full force of at least one major hurricane in their lifetimes. The destruction caused by these storms is not just from the high winds, but also from the heavy downpours, which can cause severe flooding and deadly coastal storm surges. Modern tracking equipment has improved hurricane forecasting and reduced the number of fatalities, primarily by allowing early evacuation of areas in a hurricane's path. Improved forecasting saves lives, but it cannot reduce the damage to crops, forests, or infrastructure. In 2005, Hurricane Dennis struck Cuba twice, flattening houses, downing power lines, and dropping more than 40 inches of rain in some areas before moving north (Figure 5.7). Cuba's primary citrus growing areas were hit hard, drastically reducing harvests that year. In 2010, Antigua in the Lesser Antilles was hit hard by Hurricane Earl, which destroyed homes and caused serious flooding.

Climate Change   Of all the issues facing the Caribbean, one of the most difficult to address is climate change. The Caribbean has not been a major contributor of greenhouse gases, but this maritime region is extremely vulnerable to the negative impacts of climate change, including sea-level rise, increased intensity of storms, variable rainfall leading to both floods and droughts, and loss of biodiversity (both in forests and in coral reefs). The scientific consensus is that climate change could cause a sea-level rise of 3 to 10 feet (1 to 3 meters) in this century. In terms of land loss due to inundation, the low-lying Bahamas would be the most impacted country in the region—it would

lose nearly 30 percent of its land with a 10-foot (3-meter) sea-level rise. In terms of people affected by inundation, Suriname, French Guiana, Guyana, Belize, and The Bahamas would be the most severely impacted: Just a 3-foot (1-meter) sea-level rise would be devastating because most of the population lives near the coast. With a sea-level rise of 10 feet (3 meters), 30 percent of Suriname's population and 25 percent of Guyana's would be displaced (Figure 5.8).

In addition to land loss and population displacement due to sea-level rise, other concerns include changes in rainfall patterns, leading to declines in agricultural yields and freshwater supplies, and increases in storm intensity—especially hurricanes—that cause destruction of infrastructure and other problems. All these changes would negatively affect tourism and thus the gross domestic income of countries in the region. Some of the worst-case scenarios are catastrophic.

In terms of biodiversity, continued warming of ocean temperatures will further negatively impact the Caribbean's coral reefs, which are the most biologically diverse ecosystems of the marine world. These reefs, particularly those of the rimland, are already threatened by water pollution and subsistence fishing practices. Recently, evidence is mounting of coral bleaching and die-off due to higher sea temperatures. Coral reefs are diverse and productive ecosystems that function as nurseries for many marine species. Healthy reefs also serve as barriers to protect populated coastal zones, as well as mangroves and wetlands. As the reefs become more ecologically vulnerable, so, too, do the human populations that depend on the many benefits that the reefs provide.

Throughout the Caribbean, protecting the environment and preparing for the effects of climate change are increasingly being recognized not as a luxury, but as a question of economic livelihood. In fact, the Caribbean Community and Common Market (CARICOM) has monitored the threat of climate change for over a decade. To address the issue of greenhouse gases regionally, Guyana entered into an innovative agreement with Norway in 2009. Norway provided Guyana with an initial payment of $30 million into its Reducing Emissions from Deforestation and Forest Degradation (REDD) development fund. If this initial investment succeeds in reducing emissions and tackling poverty, Norway has agreed to invest up to $250 million in this project.

Environmental Issues

Climate change is both a medium- and a long-term concern for the Caribbean region. However, other environmental issues, such as soil erosion and deforestation, have preoccupied the region due to its long-standing dependence upon agriculture. Also, as the Caribbean has become more urbanized and more reliant upon tourism, governments have realized that protection of local ecosystems is not just good for the environment, but also good for the overall economy of the region (Figure 5.8).

Agriculture's Legacy of Deforestation   Prior to the arrival of Europeans, much of the Caribbean was covered in tropical forests. The great clearing of these forests began on European-owned plantations on the smaller islands of the eastern Caribbean in the 17th century and spread westward. The island forests were removed not only to make room for sugarcane, but also to provide the fuel necessary to turn the cane juice into sugar, as well as to provide lumber for housing, fences, and ships. Primarily, however, tropical forests were removed because

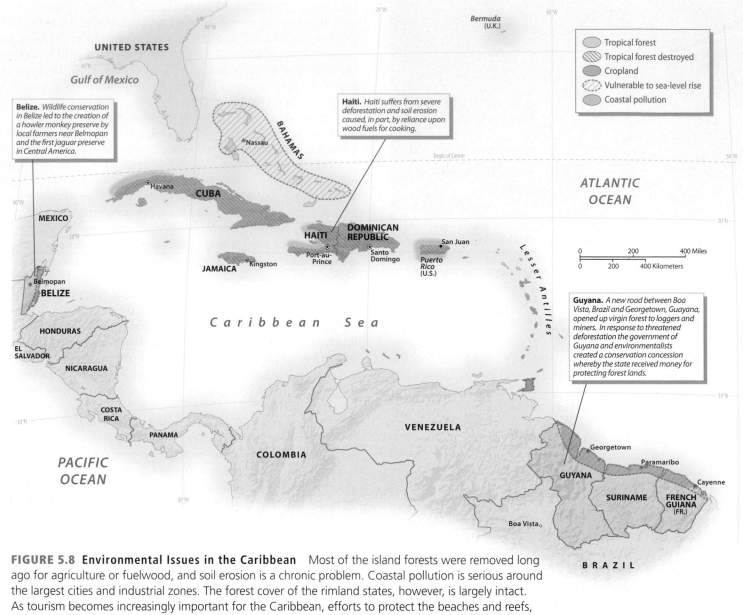

FIGURE 5.8 Environmental Issues in the Caribbean Most of the island forests were removed long ago for agriculture or fuelwood, and soil erosion is a chronic problem. Coastal pollution is serious around the largest cities and industrial zones. The forest cover of the rimland states, however, is largely intact. As tourism becomes increasingly important for the Caribbean, efforts to protect the beaches and reefs, along with the fauna and flora, are growing. Areas vulnerable to sea-level rises are noted on this map.

they were seen as unproductive; the European colonists valued cleared land. The newly exposed tropical soils easily eroded and ceased to be productive after several harvests, a situation that led to two distinct land-use strategies. On the larger islands of Cuba and Hispaniola, as well as on the mainland, new lands were constantly cleared and older ones abandoned or fallowed in an effort to keep up sugar production. On the smaller islands, where land was limited, such as Barbados and Antigua, labor-intensive efforts to conserve soil and maintain fertility were employed. In either case, the island forests were replaced by a landscape devoted to crops for world markets.

Haiti's problems with deforestation were accentuated throughout the 20th century as a destructive cycle of environmental and economic impoverishment was established. Half of Haiti's people are peasants who work small hillside plots and seasonally labor on larger estates. As the population grew, people sought more land. They cleared the remaining hillsides, subdivided their plots into smaller units, and abandoned the practice of fallowing land in an

effort to eke out an annual subsistence. When the heavy tropical rains came, the exposed and easily eroded mountain soils were washed away (Figure 5.9). As sediments collected in downstream irrigation ditches and behind dams, agriculture suffered, electricity production declined, and water supplies were degraded. Deforestation was further aggravated by reliance upon charcoal (made from trees) by many poor Haitians for household needs. It is estimated that less than 3 percent of Haiti remains forested.

While Haiti has lost most of its forest cover, on Jamaica and Cuba nearly one-third of the land is still forested. In the case of the Dominican Republic, about 40 percent of the country is forested, and more than half of Puerto Rico is forested. For these more forested islands, the decline in agriculture has caused some fields to be abandoned and forests to recover. In the case of Puerto Rico, which is a territory of the United States, its national parks—such as El Yunque on the eastern side of the island—contribute greatly to the biological diversity protected by the U.S. National Forest System (Figure 5.10).

FIGURE 5.9 Deforestation in Haiti Denuded slopes on the outskirts of Port-au-Prince are indicative of how much Haitians rely on their island for fuelwood and food. Poverty and land scarcity force many Haitians to farm the hillsides. The search for wood fuels has led to widespread deforestation. With so many trees removed, soils are more vulnerable to erosion, and agricultural yields tend to decline.

Conservation Efforts In general, biological diversity and stability are less threatened in the rimland states than in the rest of the Caribbean. Thus, current conservation efforts could produce important results. Even though much of Belize was selectively logged for mahogany in the 19th and 20th centuries, healthy forest cover still supports a diversity of mammals, birds, reptiles, and plants. Public awareness of the negative consequences of deforestation is also greater now. Many protected areas have been established in Belize. In the mid-1980s, villagers in Bermudian Landing, Belize, established a community-run sanctuary for black howler monkeys (locally referred to as baboons). The villagers banded together to maintain habitat for the monkeys and commit to land management practices that accommodate this gregarious species. The success of the project has resulted in tourists visiting the villages to see these indigenous primates up close (Figure 5.11).

Slowly, the territorial waters surrounding the Caribbean nations have gained protection, although more could be done. Here again, Belize has been a leader in creating over a dozen marine reserves and national parks along its barrier reef and outer atolls. The country has also created a substantial coastal wildlife sanctuary to protect mangroves. The Caribbean island of Bonaire, which attracts large numbers of scuba divers, maintains the Bonaire Marine Park, recognized as one of the most effectively managed marine reserves in the region.

REVIEW QUESTIONS

1. What environmental issues currently impact the Caribbean and why?
2. Describe the locational, environmental, and climate factors that together help make the Caribbean a major international tourist destination.

FIGURE 5.10 El Yunque National Forest Puerto Rico's largest remaining rainforest is in El Yunque National Forest. Visitors enjoy the park's hiking trails and its rich biological diversity.

FIGURE 5.11 Protecting Habitat and Wildlife Tourists visit the Community "Baboon" Sanctuary in Bermudian Landing, Belize. The sanctuary is a community-run project to preserve the habitat and increase the number of black howler monkeys (locally referred to as baboons). The sanctuary, established in 1985, attracts domestic and foreign visitors.

POPULATION AND SETTLEMENT: DENSELY SETTLED ISLANDS AND RIMLAND FRONTIERS

In the Caribbean, the population density is generally quite high and, as in neighboring Latin America, increasingly urban. Eighty-five percent of the region's population is concentrated on the four islands of the Greater Antilles (Figure 5.12). Add to this Trinidad's 1.3 million and Guyana's 800,000, and most of the population of the Caribbean is accounted for by six countries and one U.S. territory (Puerto Rico).

In terms of total population, few people inhabit the Lesser Antilles; nevertheless, some of these island states are densely settled. The small island of Barbados is an extreme example. With only 166 square miles (430 square kilometers) of territory, it has nearly 1,700 people per square mile (650 per square kilometer). Bermuda, which is one-third the size of the District of Columbia, has more than 3,000 people per square mile (1,200 per square kilometer). Population densities on St. Vincent, Martinique, and Grenada, while not as high, are still more than 700 people per square mile (270 people per square kilometer). Because arable land is scarce on some of these islands, access to land is a basic resource problem for many island states. The growth in the region's population, coupled with the scarcity of land, has forced many people into cities or abroad.

In contrast to the islands, the mainland territories of Belize and the Guianas are lightly populated: Guyana averages 10 people per

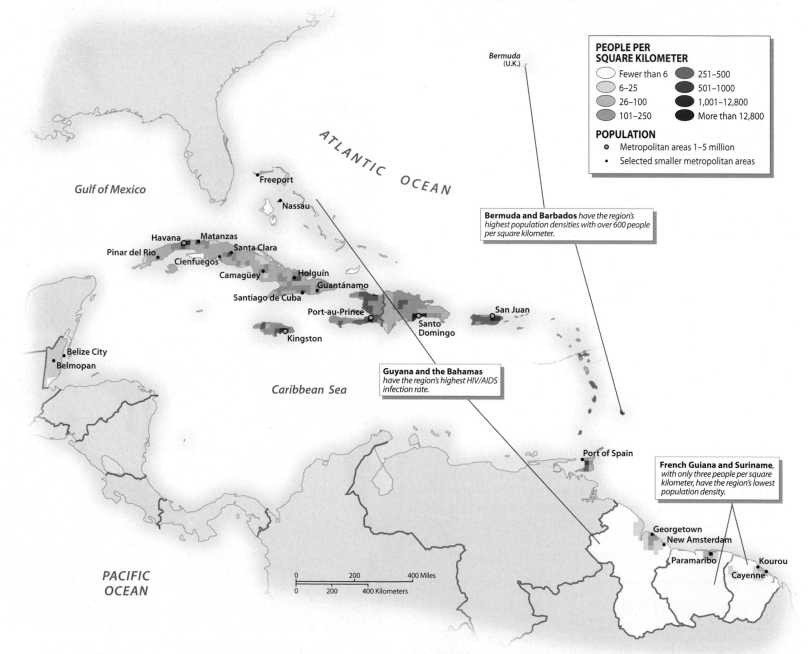

FIGURE 5.12 Population of the Caribbean The major population centers are on the islands of the Greater Antilles. The pattern here, as in the rest of Latin America, is a tendency toward greater urbanism. The largest city of the region is Santo Domingo, followed by Havana and San Juan. In comparison, the rimland states are very lightly settled.

square mile (4 per square kilometer), Suriname only 8 per square mile (3 per square kilometer), and Belize 36 per square mile (14 per square kilometer). These areas are sparsely settled in part because the relatively poor quality and accessibility of arable land made them less attractive to colonial enterprises.

Demographic Trends

During the years of slave-based sugar production, mortality rates were extremely high due to disease, inhumane treatment, and malnutrition.

Consequently, the only way population levels could be maintained was by continuing to import African slaves. With the end of slavery in the mid- to late 19th century and the gradual improvement of health and sanitary conditions on the islands, populations began to increase naturally. In the 1950s and 1960s, many states achieved peak growth rates of 3.0 or higher, causing population totals and densities to soar. Over the past 30 years, however, growth rates have come down or stabilized. As noted earlier, the current population of the Caribbean is 44 million. The population is now growing at an annual rate of 1.1 percent, and the projected population in 2025 is 47 million (see Table 5.1).

TABLE 5.1   POPULATION INDICATORS

Country	Population (millions) 2012	Population Density (per square kilometer)	Rate of Natural Increase (RNI)	Total Fertility Rate	Percent Urban	Percent <15	Percent >65	Net Migration (Rate per 1000) 2010–15[a]
Anguilla	0.01*	169*		1.8*	100*	24*	8*	13*
Antigua and Barbuda	0.1	197	0.8	1.7	30	28	7	2.3*
Bahamas	0.4	26	0.9	1.9	84	25	6	1.8
Barbados	0.3	644	0.5	1.7	45	19	10	0.0
Belize	0.3	14	2.1	2.8	44	34	4	−0.6
Bermuda	0.07*	1171*		2*	100*	18*	15*	2.0*
Cayman Islands	0.05*	199*		1.9*	100*	19*	10*	15.4*
Cuba	11.2	101	0.3	1.7	75	17	13	−2.5
Dominica	0.1	95	0.5	1.8	67	23	10	−5.4*
Dominican Republic	10.1	208	1.6	2.6	66	31	6	−2.7
French Guiana	0.2	3	2.3	3.4	81	35	4	4.9
Grenada	0.1	334	1.3	2.2	40	28	7	−9.5
Guadeloupe	0.4	236	0.6	2.1	98	22	14	−1.5
Guyana	0.8	4	1.5	2.5	29	33	1	−9.5
Haiti	10.3	370	1.8	3.4	47	36	4	−4.2
Jamaica	2.7	247	1.0	2.1	52	27	9	−7.2
Martinique	0.4	359	0.6	2.1	89	20	15	−1.0
Montserrat	0.005*	51*		1.3*	14*	27*	7*	0 *
Netherlands Antilles	0.2	444	0.5	2.2	—	20	12	1.7
Puerto Rico	3.7	416	0.3	1.6	99	20	15	−5.4
St. Kitts and Nevis	0.1	207	0.7	1.8	32	23	8	
St. Lucia	0.2	314	0.7	1.6	28	24	9	−1.1
St Vincent and the Grenadines	0.1	278	1.2	2.3	40	31	7	−9.1
Suriname	0.5	3	1.2	2.3	67	29	6	−1.9
Trinidad and Tobago	1.3	256	0.6	1.8	13	25	7	−2.9
Turks and Caicos	0.05*	49*		1.7*	93*	23*	4*	17.3*

[a]Net Migration Rate from the United Nations, Population Division, *World Population Prospects: The 2010 Revision Population Database.*

*Additional data from the *CIA World Factbook, 2012.*

Source: Population Reference Bureau, *World Population Data Sheet, 2012.*

Fertility Decline and Longer Lives

The most significant demographic trends in the Caribbean are the decline in fertility and the increase in life expectancy. Cuba and Puerto Rico have the region's lowest rates of natural increase (0.4). In socialist Cuba, due to the education of women, combined with the availability of birth control and abortion, the average woman has 1.7 children (compared to 2.1 in the United States). Yet in capitalist Puerto Rico, low rates of natural increase have also been achieved, along with a total fertility rate of 1.6. In general, educational improvements, urbanization, and a preference for smaller families have contributed to slower growth rates. Even states with relatively high total fertility rates, such as Haiti, have seen a decline in family size. Haiti's total fertility rate fell from 6.0 in 1980 to 3.4 in 2012.

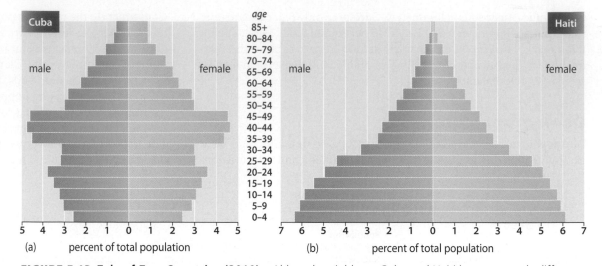

FIGURE 5.13 Tale of Two Countries (2010) Although neighbors, Cuba and Haiti have extremely different population profiles. Cuba's population is stable and older, with a notable decline in family size. Haiti's population is much younger and growing which is reflected in its broad based pyramid.

Figure 5.13 provides a stark contrast in the population profiles of Cuba and Haiti. Although both are poor Caribbean countries, Haiti has the more classic, broad-based pyramid of a developing country, where more than one-third of the population is under the age of 15. Also, there are very few old people, due to the relatively low life expectancy (62 years) in Haiti. In contrast, Cuba's population pyramid is more diamond-shaped, bulging in the 35- to 49-year-old age cohort and tapering down after that. Here the impact of the Cuban revolution and socialism is evident. Family size came down sharply after education improved and modern contraception became readily available. With better health care, Cuba's population also lives longer, having nearly the same life expectancy as those in the United States (78 years). Cuba has 12 percent of its population over 65 and just 17 percent under 15; thus, it has an extremely low rate of natural increase, similar to many developed countries in the world.

The Rise of HIV/AIDS The rate of HIV/AIDS infection in the Caribbean has come down in the past few years, but it is still twice that of North America, making the disease an important regional issue. Although nowhere near the infection rates in Sub-Saharan Africa (see Chapter 6), 1 percent of the Caribbean population between the ages of 15 and 49 had HIV/AIDS in 2009. In Haiti, one of the earliest locations where AIDS was detected, 1.9 percent of the population between the ages of 15 and 49 is infected with the virus. The infection rate is 1.7 percent in Jamaica, 2.3 percent in Belize, and 3.1 percent in The Bahamas. AIDS is already the largest single cause of death among young men in the English-speaking Caribbean.

Officials in The Bahamas have introduced drug treatments to help prevent mother-to-child transmission. Also, nearly every country has launched educational campaigns to bring infection rates down. Cuba, which witnessed a surge in both tourism and prostitution in the 1990s, has a very low infection rate of only 0.1 percent among its 15- to 49-year-old population. Education programs and an effective screening and reporting system for the disease have kept Cuba's infection rate down.

Emigration Driven by the region's limited economic opportunities, a pattern of emigration to other Caribbean islands, North America, and Europe began in the 1950s. Over more than 50 years, a **Caribbean diaspora**—the economic flight of Caribbean peoples across the globe—has become a way of life for much of the region (Figure 5.14). Barbadians generally choose to move to England. In contrast, one out of every three Surinamese has moved to the Netherlands, with most residing in Amsterdam. As for Puerto Ricans, only slightly more live on the island than reside on the U.S. mainland. In the 1980s, roughly 10 percent of Jamaica's population legally emigrated to North America (some 200,000 to the United States and 35,000 to Canada). Cubans have made the city of Miami their destination of choice since the 1960s; today they are a large percentage of that city's population.

As a region, the Caribbean has one of the highest negative rates of net migration in the world, at −3.0. This means that for every 1,000 people, three leave the Caribbean each year for other world regions. Individual countries have much higher rates, such as Guyana and Grenada at −9.5 per 1,000 and Jamaica at −7.2 per 1,000 (see Table 5.1). The economic implications of this labor-related migration are significant and will be discussed later.

The Rural-Urban Continuum

Initially, plantation agriculture and subsistence farming shaped Caribbean settlement patterns. Low-lying arable lands were dedicated to export agriculture and controlled by the colonial elite. Only small amounts of land were set aside for subsistence production. Over time, villages of freed or runaway slaves were established, especially in remote island interiors. However, the vast majority of people lived on estates as owners, managers, or slaves. Cities were formed to serve the administrative and social needs of the colonizers, but most were small, containing a small fraction of a colony's population. The colonists who linked the Caribbean to the world economy saw no need to develop major urban centers.

Plantation America Anthropologist Charles Wagley coined the term **plantation America** to designate a cultural region that extends from midway up the coast of Brazil through the Guianas and the Caribbean into the southeastern United States. Ruled by a European elite

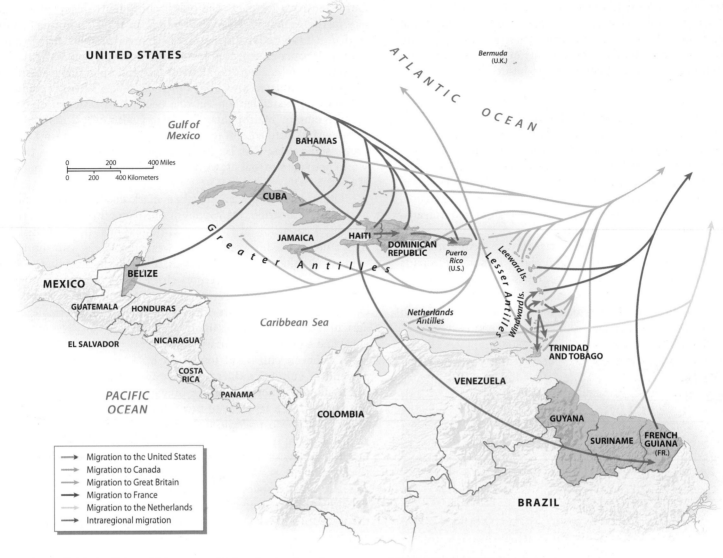

FIGURE 5.14 **Caribbean Diaspora**   Emigration has long been a way of life for Caribbean peoples. With relatively high education levels, but limited professional opportunities, the region commonly loses residents to North America, the United Kingdom, France, and the Netherlands. Intraregional migrations also occur between Haiti and the Dominican Republic and between the Dominican Republic and Puerto Rico (see orange arrows).

dependent on an African labor force, this society was primarily coastal and produced agricultural exports. It relied on **mono-crop production** (a single commodity, such as sugar) under a plantation system that concentrated land in the hands of elite families. Such a system created rigid class lines, as well as forming a multiracial society in which people with lighter skin were privileged. The term *plantation America* is not meant to describe a race-based division of the Americas, but rather a production system that brought about specific ecological, social, and economic relations (Figure 5.15).

Even today, the structure of Caribbean communities reflects this plantation legacy. Many of the region's subsistence farmers are descendants of former slaves who continue to farm their small plots

FIGURE 5.15 **Sugar Plantation**   A historical illustration (1823) of slaves harvesting sugarcane on a plantation in Antigua. Sugar production was profitable but physically demanding work. Several million Africans were enslaved and forcibly relocated to the region.

and work part-time as wage laborers on estates. The social and economic patterns generated by slavery still mark the landscape. Rural communities tend to be loosely organized, labor is temporary, and small farms are scattered on available pockets of land. Because men tend to leave home for seasonal labor, female-headed households are common.

Caribbean Cities Since the 1960s, the mechanization of agriculture, offshore industrialization, and population growth have caused a surge in rural-to-urban migration. Cities have grown accordingly, and today 66 percent of the region is classified as urban. Of the large countries, Puerto Rico is the most urban (99 percent) and Haiti the least (47 percent). Caribbean cities are not large by world standards, as only five have 1 million or more residents: Santo Domingo, Havana, Port-au-Prince, San Juan, and Kingston.

Like their counterparts in Latin America, the Spanish Caribbean cities were laid out on a grid with a central plaza. Vulnerable to raids by rival European powers and pirates, these cities were usually walled

FIGURE 5.16 Old San Juan Tourists roam the cobbled streets of Old San Juan. An UNESCO World Heritage site, many of the 18th and 19th century structures of this port city have been handsomely restored.

and extensively fortified. The oldest continually occupied European city in the Americas is Santo Domingo in the Dominican Republic, settled in 1496; today the metropolitan area has more than 2 million people. Havana emerged as the most important colonial city in the region, serving as a port for all incoming and outgoing Spanish ships. Strategically situated on Cuba's north coast at a narrow opening to a natural deep-water harbor, Havana became an essential city for the Spanish empire. Metropolitan San Juan is also over 2 million people. It was a vital fortified port for Spain and is a major port for Caribbean cruise ships today. Old San Juan has been restored and is a UNESCO world heritage site that draws in tourists from Puerto Rico and abroad (Figure 5.16).

Other colonial powers left their mark on the region's cities. For example, Paramaribo, the capital of Suriname, has been described as a tropical extension of Holland, without tulips. In the British colonies, a preference for wooden whitewashed cottages with shutters was evident. Yet the British and French colonial cities tended to be unplanned afterthoughts; these port cities were built to serve the rural estates, not the other way around. Most of them have grown dramatically over the past 40 years. These cities are no longer small ports for agricultural exports; increasingly, their focus is on welcoming cruise ships and sun-seeking tourists.

Caribbean cities and towns do have their charms and reflect a blend of cultural influences. Throughout the region, houses are often simple structures (made of wood, brick, or stucco), raised off the ground a few feet to avoid flooding, and painted in pastels (Figure 5.17). Most people still get around by foot, bicycle, or public transportation; neighborhoods are filled with small shops and services that are within easy walking distance. Streets are narrow, and the pace of life is markedly slower than in North America and Europe. Even when space is tight in town, most settlements are close to the sea and its cooling breezes. An afternoon or evening stroll along the waterfront is a common activity. (See "Working Toward Sustainability: Urban Agriculture in Havana.")

FIGURE 5.17 Belize City Cottage Residents of Belize City build their wooden cottages on stilts as protection against flooding. Shuttered cottages with metal roofs are typical throughout the English Caribbean.

Urban Agriculture in Havana

FIGURE 5.1.1 **Urban Agriculture** A worker harvests lettuce in Havana.

Many cities around the world have seen a renewed interest in urban gardening as a way to build community unity, reduce food insecurity, improve nutrition, create income opportunities, and enhance urban environments by converting brown spaces into green ones. Cuba is a global leader in urban agriculture, and these farming efforts are especially evident in metropolitan Havana. Scattered throughout this city of 2 million are thousands of small and large plots where urban residents are producing vegetables on raised beds; harvesting fruit trees; raising rabbits, chickens, and goats for meat, eggs, and milk (Figure 5.1.1). Admittedly, this experiment in urban agriculture was driven by necessity due to food scarcity. Also, the Cuban context is unique—a socialist planned economy with fixed prices and limited exposure to market forces—yet some of the successes of Havana farmers are transferable to other cities in the world.

In 1989, the Cuban government officially recognized the potential for urban gardens as a means to address the pressing food shortages provoked by drastic cuts in food and energy supports from the Soviet Union. Within a couple of years after the collapse of the USSR, average per capita daily caloric intake in Cuba plummeted by about 1,000 calories (from 2,800 to 1,800). Cubans needed a creative, fast, and durable solution to their food problem. The Ministry of Agriculture responded by creating the first coordinated urban agriculture program, which included providing access to land, especially small urban lots; extension services for training and research; supply stores; and sales outlets. In addition to government actions, nongovernmental organizations from Germany, Canada, and the United States were consulted for best urban agricultural practices and innovative organic techniques. From the start, intensive organic farming techniques and the use of biological agents for pest control were emphasized—in part, due to the expense of imported fertilizers and pesticides.

By the early 1990s, it was also clear that the entire agricultural system in Cuba required radical reform. Large state-run farms that typically grew sugarcane or citrus were partitioned into Basic Units of Cooperative Production. The producers on these smaller farms have use rights to the land for an indefinite period and the freedom to choose the crops they grow and to sell their products at market prices. In cities, however, residents interested in growing food on empty lots or other open spaces (say, public parks) can receive use of the land free of charge as long as they keep it in production. An Urban Agriculture Department was formed to change city laws so that gardeners would have legal priority for all unused space and also to set up consulting centers in each of the administrative districts of the city. These consulting centers are a key innovation of the program, as they provide tools, seeds, compost, and advice for a population that before these reforms was unaccustomed to farming.

These efforts have transformed Havana, improving access to a quantity and variety of fresh foods, creating new forms of self-employment, and forging new green spaces. Empty plots are now filled with raised beds growing eggplants, tomatoes, or strawberries that are tended by local residents. Residents have better access to fresh eggs, milk, and meats than before the initiative began. Even portions of Havana's *Parque Metropolitano* have been converted into food-producing plots. The food shortages are less severe today in Havana because of urban and rural agricultural practices, as well as other market-based reforms that have stimulated economic growth and small-scale entrepreneurship. Interestingly, a new form of tourism has emerged as international visitors book special tours to view the agricultural innovations found in Havana. Cuba was clearly forced to reinvent itself when it lost its major benefactor, the Soviet Union. Certainly, the experiment in Havana was driven by desperate need, and state control of land, food prices, and support systems facilitated the surge in urban farming. Also, it is possible that if Cuba becomes more of a market-driven economy, then new pressures will arise to convert urban farms into other uses. Yet today, many city governments beyond Cuba are looking at urban farming as a way to make cities greener and the citizens within them healthier. In this respect, Havana's urban gardeners have much to share with a rapidly urbanizing world.

REVIEW QUESTIONS

1. What are the major demographic trends for this region, and what factors explain these patterns?

2. How did the long-term reliance on a plantation economy influence patterns of settlement in the Caribbean?

CULTURAL COHERENCE AND DIVERSITY: A NEO-AFRICA IN THE AMERICAS

Linguistic, religious, and ethnic differences abound in the Caribbean. The presence of several former European colonies, millions of descendants of ethnically distinct Africans and indentured workers from India, and isolated Amerindian communities on the mainland challenges any notion of cultural coherence.

Common historical and cultural processes hold this region together. In particular, this section focuses on three cultural influences shared throughout the Caribbean: the European colonial presence, African influences, and the mix of European and African cultures referred to as *creolization*.

The Cultural Impact of Colonialism

The arrival of Columbus in 1492 triggered a devastating chain of events that depopulated the Caribbean region within 50 years. A combination of Spanish brutality, enslavement, warfare, and disease reduced the densely settled islands, which supported up to 3 million Caribs and Arawaks, into an uninhabited territory ready for the colonizer's

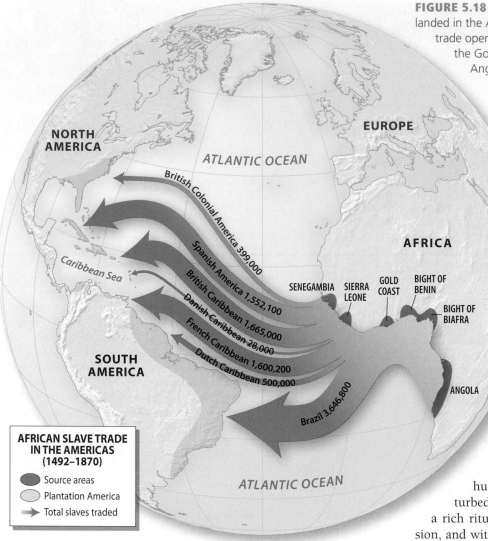

FIGURE 5.18 Transatlantic Slave Trade   At least 10 million Africans landed in the Americas during the four centuries in which the Atlantic slave trade operated. Most of the slaves came from West Africa, especially the Gold Coast (now Ghana) and the Bight of Biafra (now Nigeria). Angola, in southern Africa, was also an important source area.

The African source areas extended from Senegal to Angola, and slave purchasers intentionally mixed tribal groups in order to weaken ethnic identities. Consequently, intact transfer of religion and languages into the Caribbean rarely occurred; instead, languages, customs, and beliefs were blended.

Maroon Societies   Communities of runaway slaves—termed **maroons** in English, *palenques* in Spanish, and *quilombos* in Portuguese—offer interesting examples of African cultural diffusion across the Atlantic. Hidden settlements of escaped slaves existed wherever slavery was practiced. While many of these settlements were short-lived, others have endured and allowed for the survival of African traditions, especially farming practices, house designs, community organization, and language.

The maroons of Suriname still show clear links to West Africa. Whereas other maroon societies gradually blended into their local populations, to this day the Suriname maroons maintain a distinct identity. Six tribes formed, ranging in size from a few hundred to 20,000 (Figure 5.19). Living relatively undisturbed for 200 years, these rain-forest inhabitants fashioned a rich ritual life for themselves, involving prophets, spirit possession, and witch doctors. Recently, pressures to modernize and extract resources have placed Suriname's maroons in direct conflict with the state and private business.

FIGURE 5.19 Maroon Village in Suriname   A maroon woman carries a pail of water to her home in the village of Stonuku, Suriname. Maroon communities existed throughout the Caribbean as slaves ran away from plantations and formed villages in remote locations. The maroon communities in Suriname still retain many African traditions.

hand. By the mid-16th century, as rival European states competed for Caribbean territory, the lands they fought for were virtually uninhabited. In many ways, this simplified their task, as they did not have to recognize native land claims or work amid Amerindian societies. Instead, the colonizers reorganized the Caribbean territories to serve a plantation-based production system. The critical missing element was labor. Once slave labor from Africa, and later contract labor from Asia, was secured, the small Caribbean colonies became surprisingly profitable.

Creating a Neo-Africa   The introduction of African slaves to the Americas began in the 16th century and continued into the 19th century. This forced migration of Africans to the Americas was only part of a much more complex **African diaspora**—the forced removal of Africans from their native areas. The slave trade also crossed the Sahara to include North Africa and linked East Africa with a slave trade in the Middle East (see Chapter 6). The best-documented slave route was the transatlantic one—at least 10 million Africans landed in the Americas, and it is estimated that another 2 million died en route. More than half of these slaves were sent to the Caribbean (Figure 5.18).

This influx of slaves, combined with the elimination of nearly all the native inhabitants, remade the Caribbean as the area with the greatest concentration of relocated African people in the Americas.

African Religions Linked to maroon societies, but more widely diffused, was the transfer of African religious and magical systems to the Caribbean. These patterns, another reflection of neo-Africa in the Americas, are most closely associated with northeastern Brazil and the Caribbean. Millions of Brazilians practice the African-based religions of Umbanda, Macuba, and Candomblé, along with Catholicism. Likewise, Afro-religious traditions in the Caribbean have evolved into unique forms that have clear ties to West Africa. The most widely practiced are Voodoo (also Vodoun) in Haiti, Santería in Cuba, and Obeah in Jamaica. Each of these religions has its own priesthood and unique pattern of worship. Their impact is considerable; the father-and-son dictators of Haiti, the Duvaliers, were known to hire Voodoo priests to scare off government opposition.

Indentured Labor from Asia By the mid-19th century, most colonial governments in the Caribbean had begun to free their slaves. Fearful of labor shortages, they sought **indentured labor** (workers contracted to labor on estates for a set period of time, often several years) from South, Southeast, and East Asia.

The legacy of these indentured arrangements is clearest in Suriname, Guyana, and Trinidad and Tobago. In Suriname, a former Dutch colony, more than one-third of the population is of South Asian descent, and 16 percent is Javanese (from Indonesia). Guyana and Trinidad were British colonies, and most of their contract labor came from India. Today half of Guyana's population and 40 percent of Trinidad and Tobago's claim South Asian ancestry. Hindu temples are found in the cities and villages, and many families speak Hindi at home. The current prime minister of Trinidad and Tobago, Kamla Persad-Bissessar, is of Indian ancestry (Figure 5.20).

Most of the former English colonies have Chinese populations of not more than 2 percent. Once these East Asian immigrants fulfilled their agricultural contracts, they often became merchants and small-business owners, positions they still hold in Caribbean society.

Creolization and Caribbean Identity

Creolization refers to the blending of African, European, and some Amerindian cultural elements into the unique cultural systems found in the Caribbean. The Creole identities that have formed over time are complex; they illustrate the dynamic cultural and national identities of the region. Today Caribbean writers (V. S. Naipaul, Derek Walcott, and Jamaica Kinkaid), musicians (Bob Marley, Ricky Martin, and Juan Luís Guerra), and athletes (Dominican baseball player David Ortiz and Jamaican sprinter Usain Bolt) are internationally regarded. Collectively, these artists and athletes represent their individual islands and Caribbean culture as a whole.

Language The dominant languages in the region are European: Spanish (25 million speakers), French (11 million), English (7 million), and Dutch (about half a million) (Figure 5.21). Yet these figures tell only part of the story. In Cuba, the Dominican Republic, and Puerto Rico, Spanish is the official language, and it is universally spoken. As for the other countries, local variants of the official language exist, especially in spoken form, that can be difficult for a nonnative speaker to understand. In some cases, completely new languages have emerged. In the islands of Aruba, Bonaire, and Curaçao, Papiamento (a trading language that blends Dutch, Spanish, Portuguese, English, and African languages) is the lingua franca, with use of Dutch declining. Similarly,

FIGURE 5.20 South Asian Influences Trinidad and Tabogo's Prime Minister, Kamla Persad-Bissessar attends a celebration to mark the 165th anniversary of Indian Arrival Day in Trinidad. The Prime Minister is of Indian descent.

French Creole, or *patois*, in Haiti has been given official status as a distinct language. In practice, French is used in higher education, government, and the courts, but *patois* (with clear African influences) is the language of the street, the home, and oral tradition.

With the independence of Caribbean states from European colonial powers in the 1960s, Creole languages became politically and culturally charged with national meaning. While most formal education is taught using standard language forms, the richness of vernacular expression and its ability to instill a sense of identity are appreciated. Locals rely on their ability to switch from standard to vernacular forms of speech. Thus, a Jamaican can converse with a tourist in standard English and then switch to a Creole variant when a friend walks by, effectively excluding the outsider from the conversation. This ability to switch is evident in many cultures, but it is widely used in the Caribbean.

Music The rhythmic beats of the Caribbean might be the region's best-known product. This small area is the home of reggae, calypso, merengue, rumba, zouk, and scores of other musical forms. The roots of modern Caribbean music reflect a combination of African rhythms with European forms of melody and verse. These diverse influences, coupled with a long period of relative isolation, sparked distinct local sounds. As movement among the Caribbean population increased, especially during the 20th century, musical traditions were blended, but characteristic sounds remained.

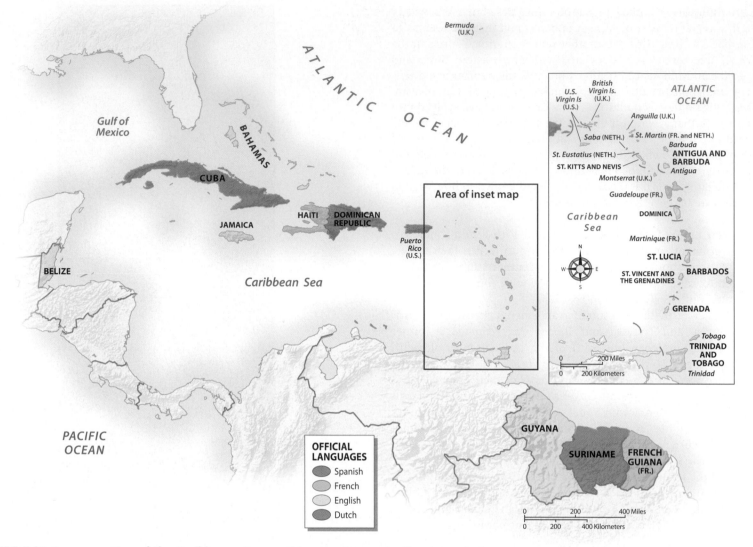

FIGURE 5.21 Language Map of the Caribbean   Because this region has no significant Amerindian population (except on the mainland), the dominant languages are European: Spanish, French, English, and Dutch. However, many of these languages have been creolized, making it difficult for outsiders to understand them.

The famed steel pan drums of Trinidad were created from oil drums discarded from a U.S. military base there in the 1940s. The bottoms of the pans are pounded with a sledgehammer to create a concave surface that produces different tones. During Carnival (a pre-Lenten celebration), racks of steel pans are pushed through the streets by dancers while drummers play. So skilled are these musicians that they even perform classical music, and government agencies encourage troubled teens to learn steel pan (see "Exploring Global Connections: Caribbean Carnival Around the World").

The distinct sound and the ingenious rhythms have made Caribbean music popular. When Jamaican Bob Marley and the Wailers crashed the international music scene with their soulful reggae sound, it was the lyrics about poverty, injustice, and freedom that resonated with the world. More than good dancing music, Caribbean music can be closely tied to Afro-Caribbean religions and is a popular form of political protest. In Haiti, *rara* music mixes percussion instruments, saxophones, and bamboo trumpets, weaving in funk and reggae bass lines. The songs are always performed in French Creole and typically celebrate Haiti's African ancestry and the use of Voodoo. The lyrics deal with difficult issues, such as political oppression and poverty (Figure 5.22).

FIGURE 5.22 Haiti's Rara Music   Performed in procession, rara music is sung in *patois*. Considered the music of the poor, it is used to express risky social commentary. This rara band performs at a folk festival in Washington, DC.

Caribbean Carnival Around the World

The origins of modern carnival are in Christian Lenten beliefs and African musical traditions. The term is Latin in origin, a reference to giving up eating meat (*carnivale* means to put meat away) in observance of Lent (the 40-day period prior to Easter). In the context of the Caribbean and Brazil, former slaves imbued carnival with special meaning because it became their opportunity to break the monotony of their daily lives, as well as their chance to mock the white elite while appearing to observe Christian traditions. In the Americas, carnival was always associated with African rhythms, dancing, and costuming. Afro-Trinidadians, for example, might dress as European royalty and mawkishly put on whiteface while dancing to rhythmic music. In the 19th century, local governments made several attempts to ban carnival, but the music and the celebration were hard to resist. Over time,

carnival celebrations became part of the national identity of Brazil and Trinidad and, to a lesser extent, of the Dominican Republic and Cuba. In each of these cases, the musical styles and methods of celebration were quite different. Brazilians preferred samba, Trinidadians liked steel pan, and Dominicans moved to merengue (Figure 5.2.1).

Today carnival is celebrated on nearly every island of the Caribbean as a national street party that can go on for weeks and attract thousands of tourists. Many official carnivals are no longer tied to Lent. In Barbados, carnival runs from late May to early August; in the Cayman Islands, it takes place in May; and in Cuba, it is in July. Stripped of its religious significance, carnival is now a celebration of life in the Caribbean in which people express themselves through music, costumes, dance, beauty contests, and parades. Over time, the celebrations have become more and

more alike, with popular musical styles and elaborate revealing costumes.

Carnival celebrations have followed the Caribbean diaspora to new settings in North America and Europe. More than a dozen U.S. cities—such as Atlanta, Baltimore, and Boston—have carnivals that last a day or two in the summer. One of the biggest carnivals in North America is in Toronto, where a large Caribbean immigrant population maintains the tradition every July. In London, Birmingham, and Leicester, Caribbean carnivals are celebrated annually. The same is true for the Netherlands, where a large number of Surinamese have introduced carnival to Amsterdam and Rotterdam. Carnival has even caught on in places where there are relatively few Caribbean migrants. Helsinki, Finland, now has a summer carnival with samba bands, steel drums, and women parading around in sequined outfits (Figure 5.2.2). As a cultural measure of globalization, the diffusion of carnival shows that a great party easily translates into many cultural settings.

FIGURE 5.2.2 Finnish Samba Band   A sign of globalization, Finns celebrate carnival in Helsinki with their own samba band and carnival queen.

FIGURE 5.2.1 Carnival Drummer   A steel pan drummer performs while his drum cart is pushed through the streets during carnival in Port of Spain, Trinidad.

From Baseball to Béisbol   Latin Americans are known for their love of soccer, but baseball is the dominant sport for much of the Caribbean. A by-product of early U.S. influence in the region, baseball is the sport of choice in Cuba, Puerto Rico, and the Dominican Republic. Even in socialist Cuba, baseball is embraced with a fervor that would humble many U.S. fans. Yet it is the Dominican Republic that sends more players to the major leagues than any other country outside of the United States. In 2009, Dominican-born players made up 17 percent of the major and minor league rosters.

The Dominican Republic became a talent pipeline for major league baseball due to a complex mix of boyhood dreams, economic

inequality, and greed. This small country has produced many baseball legends, and over the decades franchises have invested millions in training camps there. In the past two decades, however, Dominican pride in its baseball prowess has been tinged by the realities of a merciless feeder system that depends on impoverished kids, performance-enhancing drugs, fake documents, and scouts who skim a percentage of the signing bonuses. Yet the reality is that more and more young boys, who can sign contracts at age 16, see their future in baseball, rather than in schooling. Even a modest signing bonus of $10,000 to $20,000 can build a nice home for a boy's family (Figure 5.23).

FIGURE 5.23 Caribbean Baseball A young Cuban batter takes aim during a pickup game in rural Cuba. Several Caribbean states have adopted baseball as their national sport—most notably, the Dominican Republic and Cuba.

San Pedro de Macoris, not far from Santo Domingo, epitomizes this field of dreams. This humble sugarcane town has produced many baseball legends. It is a place of cane fields, kids on bicycles with bats and gloves, sugarcane factories, dusty baseball diamonds, and large homes of former players, such as George Bell, Pedro Guerrero, and Sammy Sosa. These houses are silent testaments to what is possible through baseball. In an effort to clean up baseball's image, Major League Baseball has officials in the Dominican Republic investigating drug use and fraudulent papers. However, as long as there are families pushing their teenage boys and a talent pool that delivers, this transnational system is self-perpetuating.

REVIEW QUESTIONS

1. What kinds of African influences exist in the Caribbean, and how do they express themselves?
2. What is meant by creolization, and how does it explain different cultural influences and patterns found in the Caribbean?

GEOPOLITICAL FRAMEWORK: COLONIALISM, NEOCOLONIALISM, AND INDEPENDENCE

Caribbean colonial history is a patchwork of competing European powers, fighting over profitable tropical territories. By the 17th century, the Caribbean had become an important proving ground for European ambitions. Spain's grip on the region was slipping, and rival European nations felt confident that they could gain territory by gradually pushing Spain out. Many territories, especially islands in the Lesser Antilles, changed European rulers several times.

Europeans viewed the Caribbean as a strategically located and profitable region in which to produce sugar, rum, and spices. Geopolitically, rival European powers also felt that their presence in the Caribbean limited Spanish authority there. However, Europe's geopolitical dominance in the Caribbean began to diminish by the mid-19th century, just as the U.S. presence increased. Inspired by the **Monroe Doctrine**, which claimed that the United States would not tolerate European military involvement in the western hemisphere, the U.S. government made it clear that it considered the Caribbean to be within its sphere of influence. This view was highlighted during the Spanish-American War in 1898. Even though several English, Dutch, and French colonies remained after this date, the United States indirectly (and sometimes directly) asserted its control over the region, bringing in a period of **neocolonialism**. In an increasingly global age, however, even neocolonial interests can be short-lived or sporadic. The Caribbean has not attracted the level of private foreign investment seen in other regions. As the Caribbean's strategic importance in a post–Cold War era fades, new geopolitics are shaping the region. Taiwan began wooing small Caribbean islands in the 1990s with strategic investments, in the hopes of winning United Nations votes for its cause. Not surprisingly, China has invested still more money in the region, in part to convince nations that supported Taiwan, such as Dominica and Grenada, to switch their support to China.

Life in the "American Backyard"

To this day, the United States exerts considerable influence in the Caribbean, which was commonly referred to as the "American backyard" in the early 20th century. The initial foreign policy objectives were to free the region from European authority and encourage democratic governance. Yet time and again, American political and economic ambitions undermined those goals. President Theodore Roosevelt made his priorities clear with imperialistic policies that extended the influence of the United States beyond its borders. Policies and projects such as the construction of the Panama Canal and the maintenance of open sea-lanes benefited the United States, but did not necessarily support social, economic, or political gains for the Caribbean people. The United States later offered benign-sounding development packages, such as the Good Neighbor Policy (1930s), the Alliance for Progress (1960s), and the Caribbean Basin Initiative (1980s). The Caribbean view of these initiatives has been wary at best. Rather than feeling liberated, many residents believe that one kind of political dependence was being traded for another—colonialism for neocolonialism.

In the early 1900s, the role of the United States in the Caribbean was overtly military and political. The Spanish-American War (1898) secured Cuba's freedom from Spain and also resulted in Spain's giving up the Philippines, Puerto Rico, and Guam to the United States; the latter two are still U.S. territories. The U.S. government also purchased the Danish Virgin Islands in 1917, renaming them the U.S. Virgin Islands and developing the harbor of St. Thomas. French, English, and Dutch colonies were tolerated as long as these allies recognized the supremacy of the United States in the region. Outwardly against colonialism, the United States had become much like an imperial force.

One of the requirements of an empire is the ability to impose one's will, by force if necessary. When a Caribbean state refused to abide by U.S. trade rules, U.S. Navy vessels would block its ports. Marines landed, and U.S.-backed governments were installed throughout the Caribbean Basin. These were not short-term engagements: U.S. troops occupied the Dominican Republic from 1916 to 1924, Haiti from 1913 to 1934, and Cuba from 1906 to 1909 and 1917 to 1922 (Figure 5.24). Even today, the United States maintains several important military bases in the region, including Guantánamo in eastern Cuba.

Many critics of U.S. policy in the Caribbean complain that business interests overwhelm democratic principles when foreign policy is determined. The U.S. banana companies that settled the coastal plain of the Caribbean rimland operated as if they were independent states. Sugar and rum manufacturers from the United States bought the best lands in Cuba, Haiti, and Puerto Rico. Meanwhile, truly democratic institutions remained weak, and there was little improvement in social development. True, exports increased, railroads were built, and port facilities were improved, but levels of income, education, and health remained dreadfully low throughout the first half of the 20th century.

The Commonwealth of Puerto Rico Puerto Rico is both within the Caribbean and apart from it because of its status as a commonwealth of the United States. Throughout the 20th century, various Puerto Rican independence movements sought to separate the island from the United States. Even today, residents of the island are divided about their island's political future. At the same time, Puerto Rico depends on U.S. investment and welfare programs; U.S. food stamps are a major source of income for many Puerto Rican families. Commonwealth status also

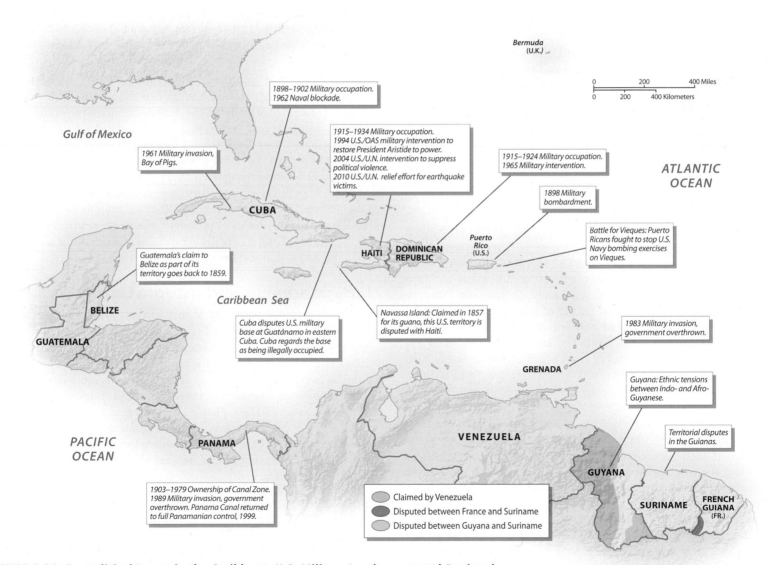

FIGURE 5.24 Geopolitical Issues in the Caribbean: U.S. Military Involvement and Regional Disputes The Caribbean was regarded as the geopolitical backyard of the United States, and U.S. military occupation was a common occurrence in the first half of the 20th century. Border and ethnic conflicts also exist—most notably, in the Guianas.

means that Puerto Ricans can freely move between the island and the U.S. mainland, a right they actively assert. In other ways, Puerto Ricans symbolically display their independence; for example, they support their own "national" sports teams and send a Miss Puerto Rico to international beauty pageants.

Puerto Rico led the Caribbean in the transition from an agrarian economy to an industrial one, beginning in the 1950s. For some U.S. officials, Puerto Rico became the model for the rest of the region. Puerto Rican President Muñoz Marín advocated an industrialization program called "Operation Bootstrap." Drawn by tax incentives and cheap labor, hundreds of U.S. textile and clothing firms relocated to Puerto Rico. Over the next two decades, 140,000 industrial jobs were added, resulting in a marked increase in per capita gross national product. In the 1970s, when Puerto Rico faced stiff competition from Asian clothing manufacturers, the government encouraged petrochemical and pharmaceutical plants to relocate to the island. By the 1990s, Puerto Rico was one of the most industrialized places in the region, with a significantly higher per capita income than its neighbors. Yet it still showed many signs of underdevelopment, including extensive out-migration, low rates of education, and widespread poverty.

Cuba and Regional Politics   The most profound challenge to U.S. authority in the region came from Cuba and its superpower ally, the former Soviet Union. In the 1950s, a revolutionary effort began in Cuba, led by Fidel Castro against the pro-American Batista government. Cuba's economic productivity had soared, but its people were still poor, uneducated, and increasingly angry. The contrast between the lives of average sugarcane workers and the foreign elite was sharp. Castro tapped a deep vein of Cuban resentment against six decades of American neocolonialism. In 1959, Castro took power.

After Castro's government nationalized American industries and took ownership of all foreign-owned properties, the United States responded by refusing to buy Cuban sugar and ultimately ending diplomatic relations with the state. Various U.S. embargoes (laws forbidding trade with a particular country) against Cuba have existed for five decades. When Cuba established strong diplomatic relations with the Soviet Union in 1960, at the height of the Cold War, the island state became a geopolitical enemy of the United States. With the Soviet Union financially and militarily backing Castro, a direct U.S. invasion of Cuba was too risky. The fall of 1962 produced one of the most dangerous episodes of the Cold War, when Soviet missiles were discovered on Cuban soil. Ultimately, the Soviet Union removed its weapons; in return, the United States promised not to invade Cuba.

Even with the end of the Cold War, when Cuba lost its financial support from the Soviet Union, it managed to reinvent itself by growing its tourism sector and courting foreign investment, especially from Spain. More recently, Castro and Venezuelan President Hugo Chavez became close political allies, signing an important trade agreement in 2004 to exchange medical personnel (from Cuba) for petroleum (from Venezuela). Still today, the United States maintains its tough trade sanctions against Cuba and forbids U.S. tourists to visit the island.

Cuba is widely recognized as a poor country with good health care, so it intends to expand its medical vision beyond its borders. Today 30,000 Cuban doctors and health-care professionals work overseas in some 40 countries among the rural and urban poor. Thousands of Cuban doctors work in Venezuela, El Salvador, Nicaragua,

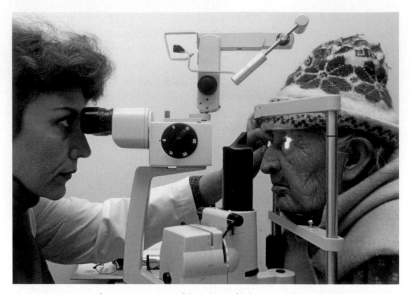

FIGURE 5.25 Cuban Doctor Working in Bolivia   A Cuban doctor examines the eyes of a Bolivian woman in a Bolivian public hospital as part of the Operation Miracle outreach program. Many Bolivian doctors, however, protested this form of Cuban assistance, claiming that it hurt their business. Meanwhile, Bolivia's urban poor eagerly took advantage of this free medical service.

Bolivia, and Haiti (Figure 5.25). Cuba runs the Latin American School of Medicine (ELAM), which has trained thousands of doctors. The majority of students are from the Americas, including students from the U.S. mainland and Puerto Rico. After completing six years of demanding training, in Spanish and English, in rural and urban areas, the doctors are sent back out to the world. In an ironic twist, the U.S. government recently recognized medical degrees from Cuba.

A new political era for Cuba appears imminent. In 2008, Fidel Castro, then 82 and in poor health, left office, and his younger brother, Raúl, assumed the duties of president. Although many consider Raúl Castro more willing to encourage private enterprise in Cuba, as of 2012, no major political or economic changes had occurred.

Independence and Integration

Given the repressive colonial history of the Caribbean, it is no wonder that the struggle for political independence began more than 200 years ago. Haiti was the second colony in the Americas to gain independence, in 1804, after the United States in 1776. However, the political independence of many states in the region has not guaranteed economic independence. Many Caribbean states struggle to meet the basic needs of their people. Today some Caribbean territories maintain their colonial status as an economic asset. For example, the French territories of Martinique, Guadeloupe, and French Guiana are overseas departments of France; residents have full French citizenship and social welfare benefits.

Independence Movements   Haiti's revolutionary war began in 1791 and ended in 1804. During this conflict, the island's population was cut in half by casualties and emigration; ultimately, the former slaves became the rulers. Independence, however, did not allow this crown jewel of the French Caribbean to prosper. Slowed by economic and political problems, it was ignored by the European powers and never fully accepted by the Latin American countries on the mainland once they gained their political independence in the 1820s.

Several revolutionary periods followed in the 19th century. In the Greater Antilles, the Dominican Republic finally gained independence in 1844, after taking control of the territory from Spain and Haiti. Cuba and Puerto Rico were freed from Spanish colonialism in 1898, but their independence was weakened by greater U.S. involvement. The British colonies also faced revolts, especially in the 1930s; it was not until the 1960s that independent states emerged from the English Caribbean. First, the larger colonies of Jamaica, Trinidad and Tobago, Guyana, and Barbados gained independence. Other British colonies followed throughout the 1970s and early 1980s. Suriname, the only Dutch colony on the rimland, became a self-governing territory in 1954, but remained part of the Netherlands until 1975, when it declared itself an independent republic.

Limited Regional Integration Perhaps the most difficult task facing the Caribbean is to increase economic integration. Scattered islands, a divided rimland, different languages, and limited economic resources hinder the formation of a meaningful regional trade bloc. Economic cooperation is more common between groups of islands with a shared colonial background than between, for example, former French and English colonies.

During the 1960s, the Caribbean began to experiment with regional trade associations as a means to improve its economic competitiveness. The goal of regional cooperation was to improve employment rates, increase intraregional trade, and ultimately reduce economic dependence. The countries of the English Caribbean took the lead in this development strategy. In 1963, Guyana proposed an economic integration plan with Barbados and Antigua. In 1972, the integration process intensified, with the formation of the **Caribbean Community and Common Market (CARICOM)**. Representing the former English colonies, CARICOM proposed an ambitious regional industrialization plan and the creation of the Caribbean Development Bank to assist the poorer states. As important as this trade group is as an economic symbol of regional identity, it has produced limited improvements in intraregional trade. It has 13 full member states—all of the English Caribbean and French-speaking Haiti. Other dependencies, such as Anguilla, Turks and Caicos, Bermuda, and the British Virgin Islands, are associate members. Still, the predominance of English-speaking territories in CARICOM illustrates the deep linguistic divides in the Caribbean.

The dream of regional integration as a way to produce a more stable and self-sufficient Caribbean has never been realized. One scholar of the region argues that a limiting factor is a "small-islandist ideology." For example, islanders tend to keep their backs to the sea, oblivious to the needs of neighbors. At times, such isolationism results in suspicion, distrust, and even hostility toward nearby states. Yet economic necessity dictates engagement with partners outside the region. This peculiar status of isolated proximity unfolds in the Caribbean, expressing itself in uneven social and economic development trends.

REVIEW QUESTIONS

1. Which countries have had colonial or neocolonial influences in the Caribbean, and why have they engaged with the region?
2. What are the obstacles to Caribbean political or economic integration?

ECONOMIC AND SOCIAL DEVELOPMENT: FROM CANE FIELDS TO CRUISE SHIPS

Collectively, the population of the Caribbean, although poor by U.S. standards, is economically better off than most of Sub-Saharan Africa, South Asia, and even China. Despite periods of economic stagnation in the Caribbean, social gains in education, health, and life expectancy are significant (Table 5.2). Historically, the Caribbean's links to the world economy were through tropical agricultural exports, but several specialized industries—such as tourism, offshore banking, and assembly plants—have challenged the dominance of agriculture. These industries grew because of the region's proximity to North America and Europe, the availability of cheap labor, and the presence of policies that created a nearly tax-free environment for foreign-owned companies. Unfortunately, growth in these industries does not employ all the region's displaced rural laborers, so the lure of jobs in North America and Europe is still strong.

From Fields to Factories and Resorts

Agriculture used to dominate the economic life of the Caribbean. However, decades of unstable crop prices and a decline in special trade agreements with former colonial states have produced more hardship than prosperity. Ecologically, the soils are overworked, and there are no frontier areas to expand production except for areas of the rimland. Moreover, agricultural prices have not kept pace with rising production costs, so wages and profits remain low. With the exception of a few mineral-rich territories, such as Trinidad, Guyana, Suriname, and Jamaica, most countries have tried to diversify their economies, relying less on their soils and more on manufacturing and services.

Comparing export figures over time demonstrates the shift away from mono-crop dependence. In 1955, Haiti earned more than 70 percent of its foreign exchange through the export of coffee; by 1990, coffee accounted for only 11 percent of its export earnings. Similarly, in 1955, the Dominican Republic earned close to 60 percent of its foreign exchange through sugar, but 35 years later sugar accounted for less than 20 percent of the country's foreign exchange, and pig iron exports nearly equaled exports of sugar.

Sugar The economic history of the Caribbean cannot be separated from the production of sugarcane. Even relatively small territories such as Antigua and Barbados yielded fabulous profits because there was no limit to the demand for sugar in the 18th century. Once considered a luxury crop, it became a popular necessity for European and North American laborers by the 1750s. It sweetened tea and coffee and made jams a popular spread for stale bread. In short, it made the meager and bland diets of ordinary people tolerable, and it also boosted caloric intake. Distilled into rum, sugar produced a popular intoxicant. Though it is difficult to imagine today, consumption of a pint of rum a day was not uncommon in the 1800s.

Sugarcane is still grown throughout the region for domestic consumption and export. Its economic importance has declined, however, mostly due to increased competition from corn and sugar beets grown in the midlatitudes. The Caribbean and Brazil are the world's major sugar exporters. Until 1990, Cuba alone accounted for more than 60 percent of the value of world sugar exports. The country earned 80 percent of its foreign exchange through sugar production. However, Cuba's dominance in sugar exports had more to do with its subsidized and

TABLE 5.2 DEVELOPMENT INDICATORS

Country	GNI per capita, PPP 2010	GDP Average Annual % Growth 2000–10	Human Development Index (2011)[1]	Percent Population Living Below $2 a Day	Life Expectancy (2012)[2]	Under Age 5 Mortality Rate (1990)	Under Age 5 Mortality Rate (2010)	Adult Literacy (% ages 15 and older)	Gender Inequality Index (2011)[3,1]
Anguilla	12,200*	—		—	81*	—		—	—
Antigua and Barbuda	20,400	—	.764	—	75	26	8	99	—
Bahamas	24,800	—	.771	—	75	22	16	—	0.332
Barbados	19,000	—	.793	—	74	18	20	—	0.364
Belize	6,200	—	.699	22.0	76	44	17	—	0.493
Bermuda	69,900*	—		—	81*	—		—	—
Cayman Islands	43,800*	—		—	81*	—		99	—
Cuba	9,900*	6.7	.776	—	78	13	6	100	0.337
Dominica	11,940	—	.724	—	76	17	12	—	—
Dominican Republic	9,030	5.6	.689	9.9	73	62	27	88	0.480
French Guiana	—	—		—	79	—		—	—
Grenada	9,930	—	.748	—	76	21	11	—	—
Guadeloupe	—	—		—	80	—		—	—
Guyana	3,450	—	.633	18.0	70	66	30	—	0.511
Haiti	1,180	0.6	.454	77.5	62	151	165	49	0.599
Jamaica	7,310	1.2	.727	5.4	73	38	24	86	0.450
Martinique	—	—		—	81	—		—	—
Montserrat	8,500*	—		—	73*	—		—	—
Netherlands Antilles	15,000*	—		—	77			—	—
Puerto Rico	16,300*	0.0		—	79	—		90	—
St. Kitts and Nevis	15,970	—	.735	—	74	28	8	—	—
St. Lucia	10,520	—	.723	40.6	73	23	16	—	—
St Vincent and the Grenadines	10,870	—	.717	—	72	27	21	—	—
Suriname	7,680	—	.680	27.2	71	52	31	95	—
Trinidad and Tobago	24,050	6.5	.760	13.5	71	37	27	99	0.331
Turks and Caicos	11,500*	—		—	79*	—		98*	—

*Additional data from the *CIA World Factbook, 2012.*

[1]United Nations, *Human Development Report, 2011.*

[2]Population Reference Bureau, *World Population Data Sheet, 2012.*

[3]Gender Inequality Index—A composite measure reflecting inequality in achievements between women and men in three dimensions: reproductive health, empowerment, and the labor market, that ranges between 0 and 1. The higher the number, the greater the inequality.

Source: World Bank, *World Development Indicators, 2012.*

guaranteed markets in Eastern Europe and the Soviet Union than with exceptional productivity. Since the breakup of the Soviet Union in 1991, the value and volume of the Cuban sugar harvest have plummeted.

The Banana Wars   The major banana exporters are in Latin America, not the Caribbean. In fact, the success of banana plantations is mixed in this region, as banana plants are especially vulnerable to hurricanes. Still, several small states in the Lesser Antilles—most notably Dominica, St. Vincent, and St. Lucia—have become economically dependent on bananas, making as much as 60 percent of their export earnings from the yellow fruit. Bananas have not made people rich, but their production for export has led to greater economic and

FIGURE 5.26 Caribbean Bananas A laborer in St. Lucia harvests bananas on a small farm. Farmers in the eastern Caribbean have long depended on preferential markets in Europe to sell their bananas. A recent trade decision by the World Trade Organization court opens up Caribbean banana farmers to new competition, jeopardizing their economic viability.

social development. In the eastern Caribbean, where most bananas are grown on small farms of 5 acres, the landowners are the laborers and thus earn two to four times more than banana plantation workers in Ecuador and Central America. Moreover, for these small states, banana exports are a link with the global economy. However, with legal pressure on the European Union (EU) to stop the guaranteed prices paid to banana growers from the former colonies, the economic viability of banana production in the Caribbean is in doubt (Figure 5.26).

The case of the eastern Caribbean shows what happens to the losers in globalization. In 1996, the United States, Ecuador (the world's leading banana exporter), Mexico, Guatemala, and Honduras challenged the EU's banana trade agreement in the World Trade Organization (WTO) court. The agreement was denounced as unfair, and the EU was told to eliminate it by 1998. To make matters worse, consumer tastes had changed, with buyers preferring the uniformly large and unblemished yellow bananas typical of the Latin American plantations, rather than those grown in places like St. Lucia. To survive in this newly competitive environment, the banana growers of the eastern Caribbean will have to produce a more standardized fruit and increase their yield per acre. Although island governments are trying to aid in this transition, local farmers have experimented with new crops, such as okra, tomatoes, avocados, and even marijuana. Reportedly, a few well-tended marijuana plants will earn 30 times more per pound than bananas. Just what will happen to family-run banana farms is hard to tell, but many Caribbean growers fear that the days of the banana economy are numbered. The banana story also reflects another major shift for the region: the decline in the economic importance of agriculture, as well as the number of people employed by it.

Assembly-Plant Industrialization One regional strategy to deal with the unemployed agricultural workers was to invite foreign investors to set up assembly plants and thus create jobs. This was first tried successfully in Puerto Rico in the 1950s and was copied throughout the region. During Puerto Rico's Operation Bootstrap, island leaders encouraged U.S. investment by offering cheap labor, local tax breaks, and, most importantly, federal tax exemptions (something only Puerto Rico can do because of its special status as a commonwealth of the United States). Initially, the program was a tremendous success, and by 1970 nearly 40 percent of the island's gross domestic product (GDP) came from manufacturing. Today 25 percent of the male labor force and 11 percent of the female labor force in Puerto Rico are employed in industry, and this sector accounts for nearly half of the island's GDP. However, competition from other states with even lower wages and the 1996 decision of the U.S. Congress to phase out many of the tax exemptions may threaten Puerto Rico's ability to maintain its specialized industrial base.

Through the creation of **free-trade zones (FTZs)**—duty-free and tax-exempt industrial parks for foreign corporations—the Caribbean has become an increasingly attractive location for assembling goods for North American consumers. The Dominican Republic took advantage of tax incentives and guaranteed access to the U.S. market offered through the Caribbean Basin Initiative. The Dominican Republic now has 50 FTZs. The majority of them are clustered around the outskirts of Santo Domingo and Santiago, the country's two largest cities (Figure 5.27). Firms from the United States and Canada are the most frequent investors in these zones, followed by Dominican, South Korean, and Taiwanese firms. Traditional manufacturing on the island was tied to sugar refining, whereas production in the FTZs focuses on garments and textiles. These manufacturing centers now account for three-quarters of the country's exports.

FIGURE 5.27 Free-Trade Zones in the Dominican Republic A sign of globalization is the increase in duty-free and tax-exempt industrial parks in the Caribbean. The Dominican Republic, which is also a member of the Central American Free Trade Association, has 50 FTZs with foreign investors from the United States, Canada, South Korea, and Taiwan.

The growth in manufacturing depends on national and international policies that support export-led development through foreign investment. Certainly, new jobs are being created, and national economies are diversifying in the process, but critics believe that foreign investors gain more than the host countries. Because most goods are assembled from imported materials, there is little development of national suppliers. Although wages are often higher than local averages, they are still low compared to those in the developed world—sometimes just a few dollars a day.

Offshore Banking and Online Gambling The rise of offshore banking in the Caribbean is most closely associated with The Bahamas, which began this industry in the 1920s. **Offshore banking** centers appeal to foreign banks and corporations by offering specialized services that are confidential and tax-exempt. Places that provide offshore banking make money through registration fees, not taxes. The Bahamas was so successful in developing this sector that by 1976 the country was the third largest banking center in the world. Its dominance began to decline because competitors from the Caribbean, Hong Kong, and Singapore appeared and because there was no longer a tax advantage in arranging large international loans offshore. Concerns about corruption and laundering of drug money also hurt the islands' financial status in the 1980s, and major reforms were introduced to reduce the presence of funds gained from illegal activities. By 1998, the ranking of The Bahamas in terms of global financial centers had dropped to 15th. Still, offshore banking remains an important part of the Bahamian economy. In the 1990s, the Cayman Islands emerged as the region's leader in financial services. With a population of 45,000, this crown colony of Britain has some 50,000 registered companies and a per capita purchasing power parity of $45,000. In 2007, the Caymans were the fifth largest banking center in the world after New York, London, Hong Kong, and Tokyo.

Each of the offshore banking centers in the Caribbean tries to develop special financial services to attract clients, such as banking, functional operations, insurance, and trusts. Bermuda, for example, is a global leader in the reinsurance business, which makes money from underwriting part of the risk of other insurance companies (Figure 5.28). The Caribbean is an attractive location for such services because of its closeness to the United States (home of many of the registered firms), client demand for these services in different countries, and the steady improvement in telecommunications that makes this industry possible. The resource-poor islands of the region see providing financial services as a way to bring foreign capital to state treasuries. Envious of the economic success of The Bahamas, Bermuda, and the Cayman Islands, countries such as Antigua, Aruba, Barbados, and Belize have attempted to establish connections with international banking, but with less success.

Online gambling is the newest industry for the microstates of the Caribbean. Antigua and St. Kitts were the leaders of the region, beginning legal online gambling services in 1999. Other states soon followed; as of 2003, Dominica, Grenada, Belize, and the Cayman Islands had gambling domain sites. Although the online gambling business is illegal in the United States, nothing can stop Americans from betting in cyberspace. In only a decade, online gambling services had taken root throughout the Caribbean. The countries in the region have also offered casino-based resort gambling for decades.

In 2007, the WTO deemed restrictions imposed on overseas Internet gambling sites by the United States to be illegal. The tiny nation

FIGURE 5.28 Financial Services in Bermuda Front Street in Hamilton, Bermuda, is a reflection of the territory's ties to the United Kingdom and its prosperity. Tourism and financial services in the reinsurance business explain Bermuda's wealth.

of Antigua is currently seeking $3 billion in compensation from the United States for lost revenue due to illegal restrictions placed on Antigua's business. Meanwhile, sensing a lucrative business opportunity, lobby efforts are under way to legalize Internet gambling in the United States, which would have repercussions for the Internet gambling business in the Caribbean.

Tourism Environmental, locational, and economic factors converge to support tourism in the Caribbean. The earliest visitors to this tropical sea admired its clear and sparkling turquoise waters. By the 19th century, wealthy North Americans were fleeing winter to enjoy the healing warmth of the Caribbean during the dry season. Developers later realized that the simultaneous occurrence of the Caribbean's dry season and the northern hemisphere's winter was ideal for beach resorts. By the 20th century, tourism was well established, with both destination resorts and cruise lines. By the 1950s, the leader in tourism was Cuba, and The Bahamas was a distant second. Castro's rise to power, however, eliminated this sector of the island's economy for nearly three decades and opened the door for other islands to develop tourism economies.

Six countries or territories hosted two-thirds of the 21 million international tourists who came to the Caribbean in 2007: Puerto Rico,

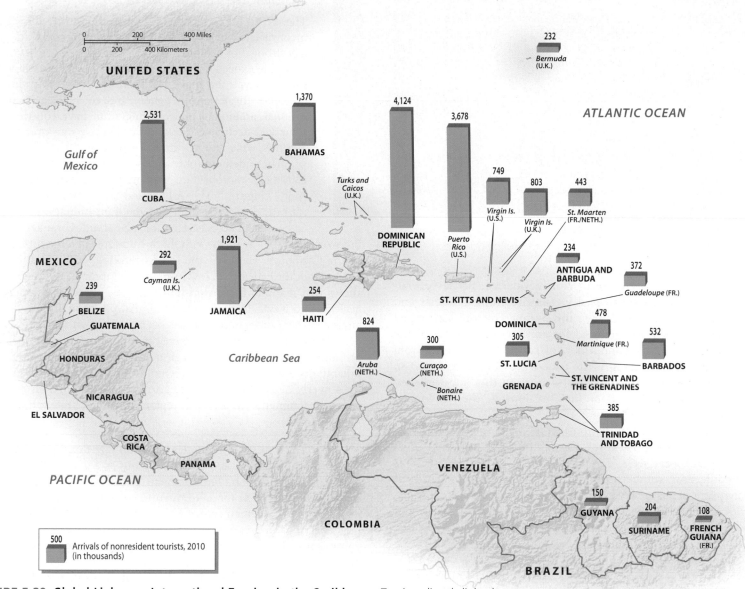

FIGURE 5.29 Global Linkages: International Tourism in the Caribbean Tourism directly links the Caribbean to the global economy. Each year more than 20 million tourists come to the islands, mostly from North America, Latin America, and Europe. The most popular destinations are the Dominican Republic, Puerto Rico, Cuba, The Bahamas, and Jamaica.

the Dominican Republic, Cuba, the Bahamas, Jamaica, and the British Virgin Islands (Figure 5.29). Puerto Rico saw its tourist sector begin to grow with commonwealth status in 1952. San Juan is now the largest home port for cruise lines and the second largest cruise-ship port in the world in terms of total visitors. The Dominican Republic is the region's largest tourist destination, receiving 4 million visitors in 2007, many of them Dominican nationals who live overseas. Since 1980, tourist receipts have increased 20-fold, making tourism a leading foreign exchange earner.

The Bahamas attributes most of its economic development and high per capita income to tourism. With more than 1.5 million hotel guests in 2007 and 1.7 million cruise-ship passengers, The Bahamas is another major hub for tourism in the region. Some 30 percent of the Bahamian population is employed in tourism, and tourism represents nearly half the country's GDP (Figure 5.30).

After years of neglect, Cuba is reviving tourism in an attempt to earn badly needed foreign currency. Tourism contributed very little to the national economy in the early 1980s. By 2007, more than 2 million tourists (mostly Canadians and Europeans) yielded gross receipts approaching $2 billion. Conspicuous in their absence are travelers from the United States, forbidden to travel to Cuba as tourists because of the U.S.-imposed sanctions. With U.S. investors out of the picture, Spanish and other European developers are busy building up Cuba's tourist capacity, anticipating that someday the U.S. ban will be lifted.

As important as tourism is for the larger islands, it is often the principal source of income for smaller ones. The Virgin Islands, Barbados, Turks and Caicos, and, recently, Belize all greatly depend on international tourists. To show how quickly this sector can grow, consider this example: When Belize began promoting tourism in the early 1980s, it had just 30,000 arrivals per year. An

FIGURE 5.30 Caribbean Cruising Tourists leave a cruise ship in Nassau Harbour, the Bahamas. Tourism is a major industry for the Caribbean.

English-speaking country close to North America, Belize specialized in ecotourism that showcased its interior tropical forests and coastal barrier reef. By the 1990s, the number of tourists topped 300,000, and tourism was credited for employing one-fifth of the workforce. Belize City became a port of call for day visitors from cruise ships in 2000, making it the fastest-growing tourist port in the Caribbean. Yet the influx of day visitors to this impoverished coastal town of 60,000 has done little to improve the city's infrastructure or high unemployment.

For more than four decades, tourism has been the foundation of the Caribbean economy. However, this regional industry has grown more slowly in recent years, compared to other tourist destinations in the Middle East, southern Europe, and even Central America. It seems that Americans are favoring domestic destinations, such as Hawaii, Florida, and Las Vegas, or are going to more "exotic" settings, such as Costa Rica. European tourists also seem to be either sticking closer to home or venturing to new locations, such as Dubai on the Persian Gulf or Goa in India. Increasingly, foreign tourists are opting to experience the Caribbean from the decks of cruise ships rather than land-based resorts. This trend undermines the local benefits of tourism, directing capital to large cruise lines, rather than island economies.

Tourism-led development has detractors for other reasons. For example, it is subject to the overall health of the world economy and current political affairs. Thus, if North America experiences a recession or international tourism declines due to heightened fears of terrorism, the flow of tourist dollars to the Caribbean dries up. Where tourism is on the rise, local resentment may build as residents confront the differences between their own lives and those of the tourists. There is also a serious problem of **capital leakage**, which is the huge gap between gross income and the total tourist dollars that remain in the Caribbean. Because many guests stay in hotel chains or cruise ships with corporate headquarters outside the region, leakage of profits is expected. On the plus side, tourism tends to promote stronger environmental laws and regulations. Countries quickly learn that their physical environment is the foundation for success. Also, although tourism does have its costs (higher energy and water consumption, as well as demand for more imports), it is environmentally less destructive than traditional export agriculture and at present more profitable.

Social Development

The record for economic growth in the Caribbean region is inconsistent, but measures of social development are generally strong. For example, most Caribbean peoples have an average life expectancy of more than 70 years (see Table 5.2). Literacy levels are high, and there is nearly parity in terms of school enrollment by gender. Indeed, high levels of educational attainment and out-migration have contributed to a marked decline in the natural increase rate over the past 30 years, which hovers around 1 percent.

These demographic and social indicators explain why Caribbean nations fare well in the Human Development Index (Figure 5.31). All ranked states (territories are not ranked) are in the high and medium human development categories. The island nation of Barbados, ranked 47th in the world in 2011, is in the very high human development category. In the Caribbean, Haiti has the lowest ranking, 158th in the world, placing it in the category of low human development. Figure 5.31 also shows that many of these well-ranked states, especially Jamaica and St. Kitts and Nevis, have significant annual per capita flows of remittances entering the economy. It has been argued that remittances have become extremely important in boosting the overall level of social and economic development in the region. Despite real social gains, many inhabitants are chronically underemployed, poorly housed, and perhaps overly dependent upon foreign remittances. For rich and poor alike, the temptation to leave the region in search of better opportunities remains.

Status of Women The matriarchal (female-dominated) basis of Caribbean households is often singled out as a distinguishing characteristic of the region. The rural custom of men leaving home for seasonal employment tends to nurture strong and self-sufficient female networks. Women typically run the local street markets. With men absent for long stretches, women tend to make household and community decisions. Although giving women local power, this position does not always imply higher status. In rural areas, female status is often undermined by the relative exclusion of women from the cash economy—men earn wages, while women provide subsistence.

As Caribbean society urbanizes, more women are being employed in assembly plants (the garment industry, in particular, prefers to hire women), in data-entry firms, and in tourism. With new employment opportunities, female participation in the labor force has surged; in countries such as Barbados, Haiti, Jamaica, Puerto Rico, and Trinidad and Tobago, more than 40 percent of the workforce is female. Increasingly, women are the principal earners of cash, and they are more likely to complete secondary education than men. There are also signs of greater political involvement by women. In recent years Jamaica, Dominica, and Guyana have all had female prime ministers.

Education Many Caribbean states have excelled in educating their citizens. Literacy is the norm, and the expectation is for most people to receive at least a high school diploma. In many respects, Cuba's educational accomplishments are the most impressive given the size of the country and its high illiteracy rates in the 1960s. Today nearly all adults are literate. Hispaniola is the obvious contrast to Cuba's success. Although the Dominican Republic has made strides in improving adult

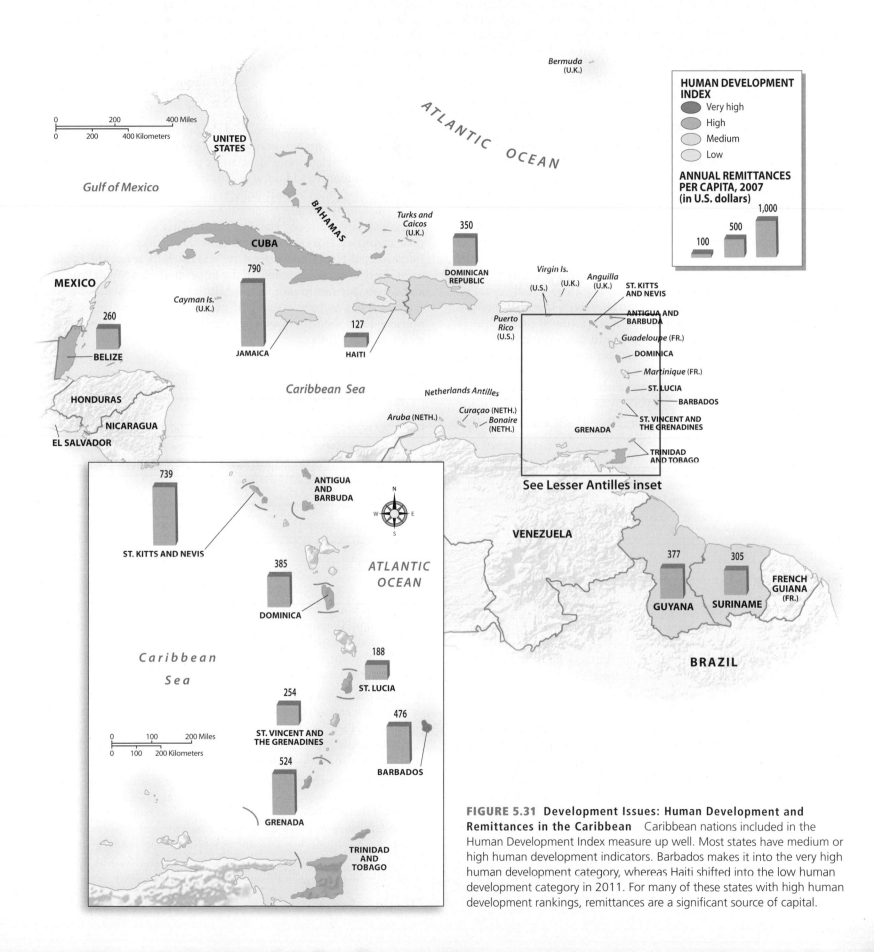

FIGURE 5.31 Development Issues: Human Development and Remittances in the Caribbean   Caribbean nations included in the Human Development Index measure up well. Most states have medium or high human development indicators. Barbados makes it into the very high human development category, whereas Haiti shifted into the low human development category in 2011. For many of these states with high human development rankings, remittances are a significant source of capital.

literacy (88 percent of adults are literate), half of all Haitian adults are illiterate. Political stability and economic growth have helped the Dominican Republic better its social conditions over the past decade. In fact, many Haitians have crossed the border into the Dominican Republic because conditions, although far from ideal, are much better there than in their homeland.

Education is expensive for these nations, but it is considered essential for development. Ironically, many states express frustration about training professionals for the benefit of developed countries in a phenomenon called **brain drain**. In the early 1980s, the prime minister of Jamaica complained that 60 percent of his country's newly trained workers left for the United States, Canada, and Britain, representing a subsidy to these economies far greater than the foreign aid Jamaica received from them. A 2007 World Bank study of skilled migrants revealed that 40 percent of Caribbean migrants living abroad were college-educated. For countries such as Guyana, Grenada, Jamaica, St. Vincent and the Grenadines, and Haiti, over 80 percent of the college-educated population will emigrate. No other region in the world has proportionately this many educated people leaving. Brain drain occurs throughout the developing world, especially between former colonies and the mother countries. Given the small population of many Caribbean territories, each professional person lost to emigration can negatively impact local health care, education, and enterprise. However, although the outflow of professionals continues to be high, many countries are experiencing a return migration of Caribbean peoples from North America and Europe as a **brain gain**. Brain gain refers to the potential of returnees to contribute to the social and economic development of a home country with the experiences they have gained abroad.

Labor-Related Migration Given the region's high educational rates and limited employment opportunities, Caribbean countries have seen their people emigrate for decades. After World War II, better transportation and political developments in the Caribbean produced a surge of migrants to North America. This trend began with Puerto Ricans going to New York in the early 1950s and intensified in the 1960s, with the arrival of nearly half a million Cubans. Since then, large numbers of Dominicans, Haitians, Jamaicans, Trinidadians, and Guyanese have also migrated to North America, typically settling in Miami, New York, Los Angeles, and Toronto. There are also substantial numbers of Caribbean migrants in the United Kingdom, France, and the Netherlands.

Crucial in this exchange of labor is the flow of cash **remittances** (monies sent back home). Immigrants are expected to send something back, especially when immediate family members are left behind. Collectively, remittances add up; it is estimated that nearly $3.5 billion is sent annually to the Dominican Republic by immigrants in the United States, making remittance income the country's second leading source of income. Jamaicans and Haitians remit nearly $2 billion annually to their countries. Governments and individuals alike depend on these transnational family networks. Families carefully select the member most likely to succeed abroad, in the hopes that money will flow back and a base for future immigrants will be established. Labor-related migration has become a standard practice for tens of thousands of households in the region, as well as a clear expression of the globalization of labor flows.

REVIEW QUESTIONS

1. As the Caribbean has shifted out of agricultural dependence, what other economic sectors have emerged in the region?

2. What explains the relatively high levels of social development in the Caribbean given the region's relative poverty?

Summary

- The Caribbean is more integrated into the global economy than most other developing nations, albeit as a dependent economic periphery. Although small in area, the region offers some of the clearest examples of the long-term efforts of globalization, from plantation agriculture to offshore banking.

- This tropical region has exploited its environment to produce export commodities such as sugar and bananas. The region's warm waters and mild climate attract millions of tourists. However, serious problems with deforestation and soil erosion have degraded urban and rural environments throughout the Caribbean.

- Population growth in the Caribbean has slowed over the past two decades. The average woman now has two or three children. Social measures of development such as life expectancy and literacy are very good in the region, yet countries struggle with supplying adequate employment. Emigration is a way of life in the region, as many people leave for employment opportunities abroad.

- The Caribbean was forged through European colonialism and the labor of millions of Africans. The blending of African and European elements, referred to as creolization, has resulted in many unique cultural expressions, especially in the religions, music, and languages of the region.

- Today the region contains 26 independent countries and territories. With the end of the Cold War, many microstates in the region fear that their lack of strategic significance may result in neglect by the United States and Europe, which may limit their ability to participate in world trade.

- In terms of development, the Caribbean has gradually shifted from being an exporter of primary agricultural resources (especially sugar) to a service and manufacturing economy. Employment opportunities in assembly plants, tourism, and offshore banking are steadily replacing jobs in agriculture. The region's strides in social development, especially in education, health, and the status of women, distinguish it from other developing areas.

Key Terms

African diaspora 154
brain drain 168
brain gain 168
capital leakage 166
Caribbean Community and Common Market
  (CARICOM) 161
Caribbean diaspora 150
creolization 155

free-trade zone
  (FTZ) 163
Greater Antilles 143
hurricane 144
indentured labor 155
isolated proximity 140
Lesser Antilles 143
maroon 154

mono-crop production 151
Monroe Doctrine 158
neocolonialism 158
offshore banking 164
plantation America 150
remittances 168
rimland 142

Thinking Geographically

1. Contrast the historical African diaspora to the contemporary Caribbean one. What patterns are formed by these two distinct population movements? What social and economic forces are behind them? Are they comparable?

2. Will agricultural exports continue to be important for Caribbean economies in the 21st century? What are the environmental consequences of agricultural production in this tropical region?

3. What advantages might Caribbean free-trade zones retain over other competitors in Southeast and East Asia? What disadvantages might they face?

4. Are remittances a sign of the Caribbean's isolation or integration with the global economy? How might remittances be used to foster development?

5. Why have U.S. actions in the region been considered neocolonial? Are there other regions in the world where the United States exerts neocolonial tendencies?

MasteringGeography™

Looking for additional review and test prep materials? Visit the Study Area in MasteringGeography™ to enhance your geographic literacy, spatial reasoning skills, and understanding of this chapter's content by accessing a variety of resources, including MapMaster interactive maps, videos, RSS feeds, flashcards, web links, self-study quizzes, and an eText version of *Globalization and Diversity*.

Scan to visit the author's blog for chapter updates.

http://gad4blog.
wordpress.com/
category/
the-caribbean/

Authors' Blogs

Scan now to access the authors' blogs for up-to-date information on the Caribbean.

Scan to visit the GeoCurrents blog.

http://geocurrents.
info/category/place/
caribbean

Globalization and Diversity

Sub-Saharan Africans have yet to experience many of the benefits of globalization. However, since 2000, foreign assistance, private investment, and exports have all climbed. Mobile phone technology has transformed communication in the region, and new cell phone applications facilitate sending remittances, obtaining micro-finance, and accessing information.

ENVIRONMENTAL GEOGRAPHY

Wood is a main source of energy for this region. The Green Belt Movement, led by the late Kenyan Wangari Maathai, resulted in the planting of millions of trees by rural women throughout the region. In areas such as the Sahel, policy changes that provided ownership or incentives for the protection of trees have resulted in an increase in tree cover.

POPULATION AND SETTLEMENT

As a region, Sub-Saharan Africa is demographically young and growing. With over 900 million people, its rate of natural increase is 2.5, making it the fastest-growing world region in terms of population. It is also the region hit hardest by HIV/AIDS, which has lowered overall life expectancies in many countries.

CULTURAL COHERENCE AND DIVERSITY

Religious life is important in this region, with large and growing numbers of Muslims and Christians. With a few notable exceptions, religious diversity and tolerance have been distinctive features of this region. However, religious conflict, especially in the Sahel region, has been on the rise.

GEOPOLITICAL FRAMEWORK

Most countries gained their independence in the 1960s. Since then, many ethnic conflicts have taken place, as governments have struggled for national unity within the boundaries drawn by European colonialists. The newest country in the region is South Sudan, which gained its independence from Sudan in 2011.

ECONOMIC AND SOCIAL DEVELOPMENT

The Millennium Development Goals established by the United Nations to reduce extreme poverty by 2015 will not be met by most states of the region, but progress is being made. International assistance has increased, and with the growing global demand for natural resources, companies from Asia, Europe, and North America are investing in the extraction of the region's metals and fossil fuels.

➤ A flock of sacred ibis fly above downtown Nairobi in the late afternoon. Kenya's bustling capital city is also a regional hub for all of East Africa.

SUB-SAHARAN AFRICA 6

Nairobi is the political and economic center of Kenya, as well as East Africa's hub for transportation, finance, and communication. This tropical metropolis of 3.5 million people spreads across a high plateau at an elevation of 5,450 feet (1,660 meters), so that the daytime high temperatures are pleasantly in the 70s. The city has grown rapidly, and like many African urban centers, it has sprawling slums, downtown high rises, and even gated communities. Nairobi also has a massive wildlife refuge within its boundaries, as well as other parklands, that are signature features.

Kenya is representative of many Sub-Saharan states—poor by global standards, but experiencing major gains. The largest economy in East Africa, Kenya has averaged over 4 percent annual growth over the past decade. Although there are still many poor people, there have also been significant improvements in life expectancy, literacy, and income over the past 15 years. The country has a strong agricultural sector based on tea, coffee, and floriculture and a growing manufacturing sector. Due to its tropical setting, Indian Ocean resorts, northern deserts, and world-class wildlife, tourism has long been an important component of the economy. Two of the country's latest developments include its first superhighway, built with Chinese capital and engineers, and a massive geothermal project that will provide much of the country's future energy needs. Kenya is a country increasingly integrated into the regional and global economy and experiencing economic and social benefits from this greater inclusion.

Africa south of the Sahara is poorer, more rural, and its population is much younger when compared to Latin America and the Caribbean. Nearly 900 million people reside in this region, which includes 48 states and 1 territory (Reunion, off the coast of Madagascar). Demographically, this is the world's fastest-growing region (with a 2.5 percent rate of natural increase); in most countries, nearly half the population (43 percent) is younger than 15 years of age. Income levels are extremely low: 69 percent of the population lives on less than $2 per day. Life expectancy is only 55 years. Such statistics and the all-too-frequent negative headlines about violence, disease, and poverty might lead to despair. Yet this is also a region of resilience, where many Africans are optimistic about the future. Local and international nongovernmental organizations, diaspora-led groups, and various government agencies are improving the quality of life in many parts of the region. In the process, many countries have reduced infant mortality, expanded basic education, and increased food production in the past two decades. One of the most transformational changes has been the rapid diffusion of cell phones, along with innovative applications that improve communication and information sharing.

Sub-Saharan Africa—that portion of the African continent lying south of the Sahara Desert—is a commonly accepted world region (Figure 6.1). The unity of this region has to do with similar livelihood systems and a shared colonial experience. No common religion, language, philosophy, or political system ever united the area. Instead, loose cultural bonds developed from a variety of lifestyles and idea systems that evolved here. The impact of outsiders also helped to determine the region's identity. Slave traders from Europe, North Africa, and Southwest Asia treated Africans as property; up until the mid-1800s, millions of Africans were taken from the region and sold into slavery. In the late 1800s, the entire African continent was divided by European colonial powers, imposing political boundaries that remain to this day. In the postcolonial period, which began in the 1960s, most Sub-Saharan African countries faced many of the same economic and political challenges.

When setting this particular regional division, the major question is how to treat North Africa. Some scholars argue for treating the African continent as one world region. Regional organizations such as the African Union are modern expressions of this continental unity. However, North Africa is generally considered more closely linked, both culturally and physically, to Southwest Asia. Arabic is the dominant language, and Islam is the leading religion of North Africa. Consequently, North Africans feel more closely connected to the Arab hearth in Southwest Asia than to the Sub-Saharan world.

In this chapter, we focus on the states south of the Sahara. We preserve political boundaries when delimiting the region so that the Mediterranean states of North Africa, along with Western Sahara and Sudan, are discussed with Southwest Asia (Chapter 7). Before the independence of South Sudan from Sudan in 2011, Sudan was Africa's largest state by area. In the more populous and powerful north, Muslim leaders have crafted an Islamic state that is culturally and politically oriented toward North Africa and Southwest Asia. The new country of South Sudan, however, has more in common with the Christian and animist groups in Sub-Saharan Africa. South Sudan, along with the Sahelian states of Mauritania, Mali, Niger, and Chad, form the northern boundary of the countries discussed in this chapter.

The region is culturally complex, with dozens of languages spoken in some states. Consequently, most Africans understand and speak several languages. Ethnic identities do not follow the political

LEARNING OBJECTIVES

After reading this chapter you should be able to:

- List the characteristics that make Sub-Saharan Africa a distinct world region.

- Summarize the major ecosystems in the region and how humans have adapted to living in them.

- Describe the factors that have made wildlife conservation and tourism important aspects of the region's economy.

- Explain the region's rapid demographic growth and describe the differential impact of HIV/AIDS upon the region.

- Summarize various cultural influences of African peoples within the region and globally.

- Describe the relationship between ethnicity and conflict in this region and the strategies for maintaining peace.

- Assess the roots of African poverty and explain why many of the fastest-growing economies in the world today are in Sub-Saharan Africa.

- List the major resources of the region, especially metals and fossil fuels, and describe how they are impacting the region's development.

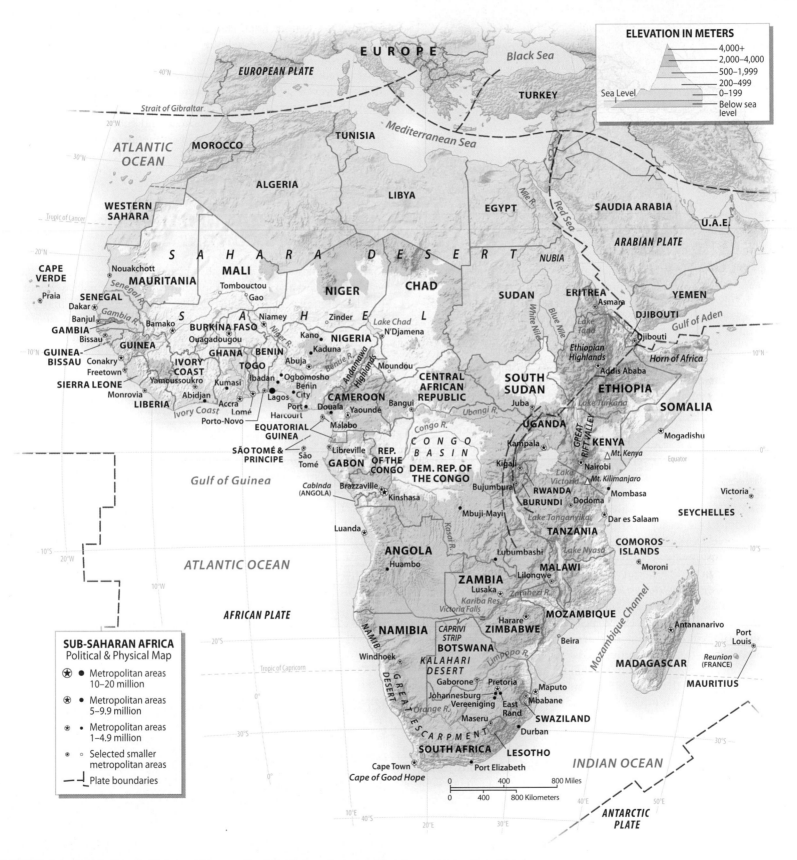

FIGURE 6.1 Sub-Saharan Africa   Africa south of the Sahara includes 48 states and 1 territory. This large region of rain forests, tropical savannas, and deserts is home to nearly 900 million people. Much of the region consists of broad plateaus ranging in elevation from 1,600 to 6,500 feet (500 to 2,000 meters). Although the population is growing rapidly, the overall population density of Sub-Saharan Africa is low. Considered one of the least developed regions of the world, it remains an area rich in natural resources.

FIGURE 6.2 Women Farmers in Senegal Senegalese women grow manioc, a drought-resistant staple introduced from the Americas. Throughout Sub-Saharan Africa, subsistence agriculture is still widely practiced, and women are often responsible for tending these crops.

divisions of Africa, sometimes resulting in deadly ethnic conflict, such as in Rwanda in the mid-1990s and in Nigeria today. Nevertheless, throughout the region peaceful coexistence between distinct ethnic groups is the norm. The cultural significance of European colonizers cannot be ignored: European languages, religions, educational systems, and political ideas were adopted and modified. Yet the daily rhythms of life are far removed from the industrial world. Most Africans still engage in subsistence and cash-crop agriculture (Figure 6.2). The influence of African peoples outside the region is great, especially considering that human origins are traceable to this part of the world. In historic times, the legacy of the slave trade resulted in the transfer of African peoples, religious systems, and musical traditions throughout the western hemisphere. Even today, African-based religious systems are widely practiced in the Caribbean and Latin America.

The African economy, however, is marginal compared with the rest of the world. According to the World Bank, Sub-Saharan Africa's economic output in 2010 amounted to just 2 percent of global output, even though the region contains 12 percent of the world's population. Moreover, the gross domestic product (GDP) of just one country, South Africa, accounts for more than one-third of the region's total GDP.

Many scholars feel that Sub-Saharan Africa has benefited little from its integration (both forced and voluntary) into the global economy. Slavery, colonialism, and export-oriented mining and agriculture served the needs of consumers outside the region, while failing to improve domestic food supplies, infrastructure, and standards of living.

Foreign assistance in the postcolonial years initially improved agricultural and industrial output, but also led to mounting foreign debt and corruption, which over time have undercut the region's economic gains. Ironically, many of the scholars and politicians that worry about the lack of benefit from economic integration also worry that negative global attitudes about the region have produced a pattern of neglect. Private capital investment in Sub-Saharan Africa lags far behind that in other developing regions, but it is growing, especially in oil-rich states such as Angola, Chad, and Equatorial Guinea. The idea of debt forgiveness for Africa's poorest states is steadily being applied as a strategy to improve access to basic services and education.

The past few years have witnessed a surge in philanthropic outreach to the region. Rock star Bono of U2 and the Bill and Melinda Gates Foundation have led the One Campaign, directing millions of dollars to support health care, disease prevention, education, and poverty reduction in Sub-Saharan Africa and other developing areas. Many observers believe that through a combination of internal reforms, better governance, foreign assistance, and foreign investment in infrastructure and technology, social and economic gains are possible for Sub-Saharan Africa. Such investments in social development are paying off, as a dramatic decline in child mortality was achieved between 1990 and 2010.

ENVIRONMENTAL GEOGRAPHY: THE PLATEAU CONTINENT

The largest landmass straddling the equator, Sub-Saharan Africa is vast in scale and remarkably beautiful. Sometimes called the plateau continent, Africa is dominated by extensive areas of geologic uplift that formed huge elevated plateaus. The highest areas are found on the eastern edge of the continent, where the Great Rift Valley forms a complex upland area of lakes, volcanoes, and deep valleys. In contrast, lowlands prevail in West Africa (see Figure 6.1). Despite this region's immense biodiversity, vast water resources, and wealth of precious minerals, it also has relatively poor soils, widespread disease, and vulnerability to drought.

Plateaus and Basins

A series of plateaus and elevated basins dominates the African interior and explains much of the region's unique physical geography (refer to Figure 6.1). Generally, elevations increase toward the south and east of the continent. Most of southern and eastern Africa lies well above 2,000 feet (600 meters), and sizable areas sit above 5,000 feet (1,500 meters). These regions are typically referred to as High Africa; Low Africa includes West Africa and much of Central Africa. Steep escarpments form where plateaus abruptly end, as illustrated by the majestic Victoria Falls on the Zambezi River (Figure 6.3). Much of

FIGURE 6.3 Victoria Falls The Zambezi River descends over Victoria Falls in the southern part of the region. A fault zone in the African plateau explains the existence of the 360-foot (110-meter) drop. The Zambezi has never been important for transportation, but it is a vital source of hydroelectricity for Zimbabwe, Zambia, and Mozambique.

FIGURE 6.4 Mount Kilimanjaro Rising from the tropical plains and capped in snow, Africa's highest peak is a popular destination for tourists who hope to reach its lofty heights.

FIGURE 6.5 Congo River Africa's largest river by volume, the mighty Congo River flows through the Ituri rain forest in the Democratic Republic of the Congo.

southern Africa is rimmed by a landform called the **Great Escarpment** (a high cliff separating two comparatively level areas), which begins in southwestern Angola and ends in northeastern South Africa. Such landforms proved to be a barrier for European colonial settlement in the interior of the continent and, in part, explain the prolonged colonization period.

Though Sub-Saharan Africa is an elevated landmass, it has few significant mountain ranges. The one extensive area of mountainous topography is in Ethiopia, which lies in the northern portion of the Rift Valley zone. Receiving heavy rains in the wet season, the Ethiopian Plateau forms the headwaters of several important rivers—most notably the Blue Nile, which joins the White Nile at Khartoum, Sudan. A discontinuous series of volcanic mountains, some of them quite tall, is associated with the southern half of the Rift Valley, such as Tanzania's Mount Kilimanjaro—the region's highest peak at 19,000 feet (5,900 meters) (Figure 6.4). However, the dominant features of the high plateaus are often their deep valleys, rather than dramatic mountain ranges.

Watersheds Africa south of the Sahara does not have the broad, alluvial lowlands that influence patterns of settlement throughout other regions. The four major river systems are the Congo, Niger, Nile, and Zambezi (the Nile will be discussed in Chapter 7). Smaller rivers—such as the Orange in South Africa; the Senegal, which divides Mauritania and Senegal; and the Limpopo in Mozambique—are locally important, but drain much smaller areas. Ironically, most people think of Africa south of the Sahara as suffering from water scarcity and tend to discount the size and importance of the watersheds (or catchment areas) drained by these river systems.

The Congo River (or Zaire) is the largest watershed in the region, in terms of both the drainage area and the volume of river flow produced. It is second only to South America's Amazon River in terms of annual flow. The Congo flows across a relatively flat basin that lies more than 1,000 feet (300 meters) above sea level, meandering through Africa's largest tropical forest, the Ituri (Figure 6.5). Entry from the Atlantic into the Congo Basin is prevented by a series of rapids and falls, making the Congo River only partially navigable. Despite these limitations, the Congo River has been the major corridor for travel within the Republic of the Congo and the

Democratic Republic of the Congo (formerly Zaire); the capitals of both countries, Brazzaville and Kinshasa, rest on opposite sides of the river.

The Niger River is the critical source of water for all of West Africa, but especially for the arid countries of Mali and Niger. Beginning in the humid Guinea highlands, the Niger flows to the northeast and then spreads out to form a huge inland delta in Mali before making a great bend southward at the margins of the Sahara near Gao. On the banks of the Niger River are the capitals of Mali (Bamako) and Niger (Niamey), as well as the historic city of Tombouctou (Timbuktu). After flowing through the Sahel, the Niger River returns to the humid lowlands of Nigeria, where the Kainji Reservoir temporarily blocks its flow to produce electricity for Africa's most heavily populated state. At the river's end lies the Niger Delta, which is also the center of Nigeria's oil industry. This fertile delta region, home to ethnic groups such as the Igbo and Ogoni, is extremely poor. For decades, conflict in the region has centered on who benefits from the region's oil and the serious environmental degradation that has resulted from oil and gas extraction.

The considerably smaller Zambezi River begins in Angola and flows east, spilling over an escarpment at Victoria Falls before finally reaching Mozambique and the Indian Ocean. More than other rivers in the region, the Zambezi is a major supplier of commercial energy. Sub-Saharan Africa's two largest hydroelectric installations are located on this river.

Soils With a few major exceptions, Sub-Saharan Africa's soils are relatively infertile. Generally speaking, fertile soils are young soils, those deposited in recent geologic time by rivers, volcanoes, glaciers, or windstorms. In older soils—especially those located in moist tropical environments—natural processes tend to wash out most plant nutrients over time. Over most of Sub-Saharan Africa, the agents of soil renewal have largely been absent.

Portions of Sub-Saharan Africa are, however, noted for their natural soil fertility, and, not surprisingly, these areas support denser settlement. Some of the most fertile soils are in the Rift Valley, made productive by the volcanic activity associated with the area. The population densities of rural Rwanda and Burundi, for example, are partially explained by the highly productive soils. The same can be said for highland Ethiopia, which supports the region's second largest population. The Lake Victoria lowlands and central highlands of Kenya are also noted for their sizable populations and productive agricultural bases.

The drier grassland and semidesert areas feature a soil type called *alfisols*. High in aluminum and iron content, these red soils have greater fertility than comparable soils found in wetter zones. This helps to explain the tendency of farmers to plant in drier areas, such as the Sahel, even though they risk exposure to drought. Many agronomists suggest that with irrigation, the southern African countries of Zambia and Zimbabwe could greatly increase commercial grain production on these soils.

Climate and Vegetation

Most of Sub-Saharan Africa lies in the tropical latitudes, and it is the largest tropical landmass on the planet. Only the far south of the continent extends into the subtropical and temperate belts. Much of the region averages high temperatures from 70 to 80°F (22 to 28°C) year-round (Figure 6.6). Rainfall, more than temperature, determines the different vegetation belts that characterize the region. Addis Ababa, Ethiopia, and Walvis Bay, Namibia, have similar average temperatures (see the climographs in Figure 6.6), but Addis Ababa is in the moist highlands and receives nearly 50 inches (127 centimeters) of rainfall annually, whereas Walvis Bay rests on the Namibian Desert and receives less than 1 inch (2.5 centimeters).

The three main biomes of the region are tropical forests, savannas, and deserts.

Tropical Forests The center of Sub-Saharan Africa falls in the tropical wet climate zone. The world's second-largest expanse of humid equatorial rain forest, the Ituri, lies in the Congo Basin, extending from the Atlantic Coast of Gabon two-thirds of the way across the continent, including the northern portions of the Republic of the Congo and the Democratic Republic of the Congo (Zaire). The conditions here are constantly warm to hot, and precipitation falls year-round (see the climograph for Kisangani in Figure 6.6).

Commercial logging and agricultural clearing have degraded the western and southern fringes of the Ituri, but much of this vast forest is still intact. The Ituri has, so far, fared much better than Southeast Asia and Latin America in terms of tropical deforestation. Major

national parks such as Okapi and Virunga have been created in the Democratic Republic of the Congo. Virunga National Park—one of the oldest in Africa—is on the eastern fringe of the Ituri forest and is home to the endangered mountain gorillas. Poor infrastructure and political chaos in the Democratic Republic over the past 18 years have made large-scale logging impossible, but they have also made conservation difficult. Due to regional conflict, parks such as Virunga have been repeatedly taken over by rebel groups, and park rangers have been killed; poaching has become a means of survival for people in the region. In the future, it seems likely that Central Africa's rain forest, and the wildlife within it, could suffer the same kind of degradation experienced in other tropical forests.

Savannas Wrapped around the Central African rain-forest belt in a great arc lie Africa's vast tropical wet and dry savannas. Savannas are dominated by a mixture of trees and tall grasses in the wetter zones immediately adjacent to the forest belt and by shorter grasses with fewer trees in the drier zones. North of the equatorial belt, rain generally falls only from May to October. The farther north you travel, the less the total rainfall and the longer the dry season. Climatic conditions south of the equator are similar, only reversed, with the wet season occurring between October and May and precipitation generally decreasing toward the south (see the climograph for Lusaka in Figure 6.6). A larger area of wet savanna exists south of the equator, with extensive woodlands located in southern portions of the Democratic Republic of the Congo, Zambia, Zimbabwe, and eastern Angola. These savannas are also a critical habitat for the region's large fauna (Figure 6.7).

Deserts Major deserts exist in the southern and northern boundaries of the region. The Sahara, the world's largest desert and one of its driest, crosses the landmass from the Atlantic coast of Mauritania all the way to the Red Sea coast of Sudan. A narrow belt of desert extends to the south and east of the Sahara, wrapping around the **Horn of Africa** (the northeastern corner that includes Somalia, Ethiopia, Djibouti, and Eritrea) and pushing as far as eastern and northern Kenya. An even drier zone is found in southwestern Africa. In the striking red dunes of the Namib Desert of coastal Namibia, rainfall is a rare event, although temperatures are usually mild (Figure 6.8). Inland from the Namib lies the Kalahari Desert. Because it receives slightly more than 10 inches (25 centimeters) of rain a year, most of the Kalahari is not dry enough to be classified as a true desert. Its rainy season, however, is brief, and most of the precipitation is immediately absorbed by the underlying sands. Surface water is thus scarce, giving the Kalahari a desert-like appearance for most of the year.

Africa's Environmental Issues

The prevailing perception of Africa south of the Sahara is one of environmental scarcity and degradation, no doubt reinforced by televised images of drought-ravaged regions and starving children. Single explanations such as rapid population growth or colonial exploitation cannot fully capture the complexity of Africa's environmental issues or the ways that people have adapted to living in marginal ecosystems with debilitating tropical diseases. Because most of Sub-Saharan Africa's population is rural, earning its livelihood directly from the land, sudden environmental changes can have devastating effects on household income and food consumption.

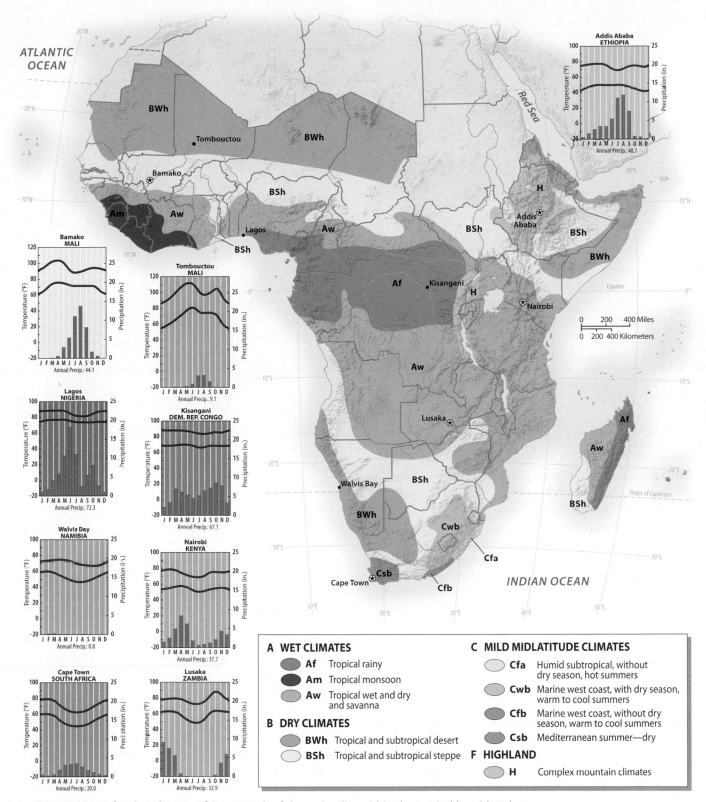

FIGURE 6.6 Climate Map of Sub-Saharan Africa Much of the region lies within the tropical humid and tropical dry climatic zones; thus, the seasonal temperature changes are not great. Precipitation, however, varies significantly from month to month. Compare the distinct rainy seasons in Lusaka and Lagos: Lagos is wettest in June, and Lusaka receives most of its rain in January. Although there are important tropical forests in West and Central Africa (coinciding with the tropical wet and monsoon climate zones), vegetation in much of the region is tropical savanna.

FIGURE 6.7 **African Savannas**   Buffalo gather at a watering hole in Zambezi National Park in Zimbabwe. The savannas of southern Africa are a noted habitat for the region's large mammals, such as buffalo, elephant, zebra, and lion.

FIGURE 6.8 **Namib Desert**   A hiker explores the coastal dunes of the Namib Desert which can reach 300 feet in height. Namibia has the region's lowest population density; much of the country is desert.

As Figure 6.9 illustrates, deforestation and **desertification**—the expansion of desert-like conditions as a result of human-induced degradation—are commonplace. Sub-Saharan Africa is also vulnerable to drought, most notably in the Horn of Africa, parts of southern Africa, and the Sahel. Many scientists fear that drought will come more often and be prolonged under global climate change scenarios. Yet the region is also home to some of the most impressive wildlife conservation in the world, which is a source of pride and revenue for many African nations.

The Sahel and Desertification   The **Sahel** is a zone of ecological transition between the Sahara to the north and the wetter savannas and forest in the south (see Figure 6.9). In the 1970s, the Sahel became a popular symbol for the dangers of unchecked population growth and human-induced environmental degradation when a relatively wet period came to an abrupt end. Six years of drought (1968–1974) were followed by a second prolonged drought during the mid-1980s, ravaging the land. During these droughts, rivers in the area diminished, and desert-like conditions began to move south. Unfortunately, tens of millions of people lived in this area, and farmers and pastoralists whose livelihoods had come to depend on the more abundant precipitation of the relatively wet period were temporarily forced out.

Life in the Sahel depends on a delicate balance of limited rain, drought-resistant plants, and a pattern of animal **transhumance** (the movement of animals between wet-season and dry-season pasture). What appears to be desert wasteland in April or May is transformed into a lush garden of millet, sorghum, and peanuts after the drenching rains of June. Relatively free of the tropical diseases found in the wetter zones to the south, Sahelian soils are quite fertile, which helps to explain why people continue to live there despite the unreliable rainfall patterns (Figure 6.10).

Considerable disagreement continues over the basic causes of desertification and drought in the Sahel. Has it been due to too many humans degrading their environment, or unsound settlement schemes encouraged by European colonizers, or a failure to understand global atmospheric cycles? Certainly, human activity in the region greatly increased in the mid-20th century, making the case for human-induced desertification more likely. But parts of the Sahel were important areas of settlement long before European colonization.

The main practices cited in the desertification of the Sahel are the expansion of agriculture and overgrazing, leading to the loss of natural vegetation and declines in soil fertility. For example, French colonial authorities forced villagers to grow peanuts as an export crop, a

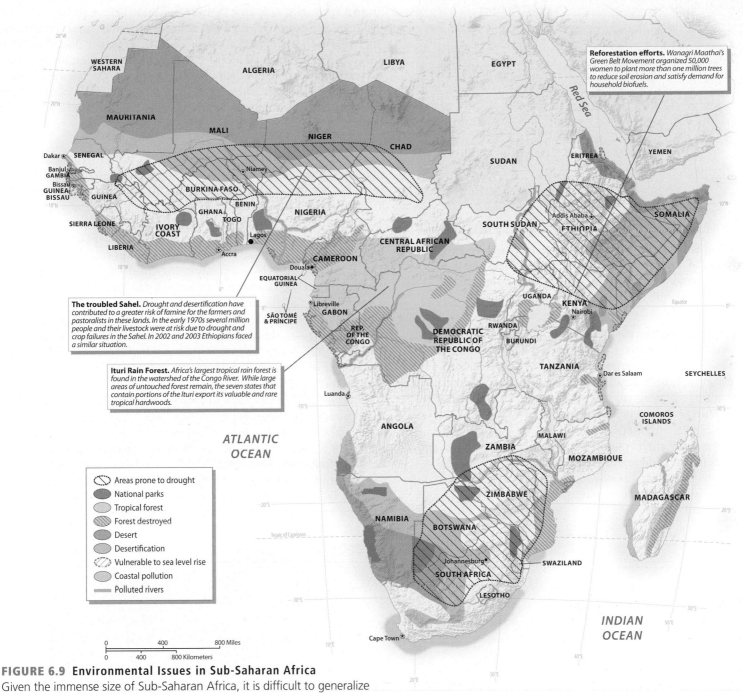

FIGURE 6.9 Environmental Issues in Sub-Saharan Africa
Given the immense size of Sub-Saharan Africa, it is difficult to generalize
about environmental problems. Dependence on trees for fuel places strain on forests and wooded savannas
throughout the region. In semiarid regions, such as the Sahel, population pressures and land-use practices
seem to have led to desertification. Yet Sub-Saharan Africa also supports the most impressive array of wildlife,
especially large mammals, on Earth.

policy continued by the newly independent states of the region. However, peanuts tend to deplete several key soil nutrients, which means
that peanut fields are often abandoned after a few years as cultivators
move on to fresh sites. Harvesting this crop also turns up the soil at
the onset of the dry season, leading to accelerated wind erosion of
valuable topsoil.

Overgrazing by livestock, another traditional product of the region, has also been implicated in Sahelian desertification. Development agencies hoping to increase livestock production introduced

deep wells into areas that previously had been unused by herders
through most of the year. The new supplies of water, in turn, allowed
year-round grazing in places that, over time, could not withstand it.
Large barren circles around each new well began to appear even on
satellite images.

Importantly, some Sahelian areas are experiencing some veg-
etative recovery thanks to simple actions taken by farmers, a change
in government policy, and better rainfall. In the Sahelian portion of
Niger, local agronomists have recently documented an unanticipated

FIGURE 6.10 The Sahel in Bloom A woman prepares millet grains grown near the city of Maradi, Niger. The soils of the Sahel are fertile, and peasant farmers can produce a surplus with adequate rain. However, in times of drought, crop failures can lead to famine in this region.

increase in tree cover over the past 35 years (Figure 6.11). More interesting still, increases in tree cover have occurred in some of the most densely populated rural areas. After a drought in 1984, farmers began to actively protect saplings instead of clearing them from their fields, including the nitrogen-fixing *goa* tree, which had disappeared from many villages. During the rainy season, the goa tree loses its leaves, so it does not compete with crops for water or sun. The leaves themselves fertilize the soil. Sahelian farmers also use branches, pods, and leaves from trees for fuel and for animal fodder.

Until the 1990s, all trees were considered property of the state of Niger, thus giving farmers little incentive to protect them. Since then, the government has recognized the value of allowing individuals to own trees. Not only can farmers sell branches, pods, or fruit, but they can also conserve them to ensure sustainable rural livelihoods. The villages that protect their trees are much greener than those that do not. In the village of Moussa Bara, where regeneration has been successful, not one child died of malnutrition in the famine of 2005. The Sahel is still poor and prone to drought, but as the case of Niger shows, relatively simple conservation practices can have a positive impact.

Deforestation Although Sub-Saharan Africa still contains extensive forests, much of the region is either grasslands or agricultural lands that were once forest. Lush forests that existed in places such as highland Ethiopia were long ago reduced to a few remnant patches. Throughout history, local populations have relied on such woodlands for their daily needs. Tropical savannas, which cover large portions of the region to the north and south of the tropical rain forest zone, are dotted with woodlands. For many people of the region, deforestation of the savanna woodlands is of greater local concern than the commercial logging of the rain forest. This is because of the importance of **biofuels**—wood or charcoal used for household energy needs, especially cooking—as the leading source of energy for many rural settlements. Loss of woody vegetation has resulted in extensive hardship, especially for women and children who must spend many hours a day looking for wood.

In some countries, village women have organized into community-based nongovernmental organizations (NGOs) to plant trees and create greenbelts to meet ongoing fuel needs. One of the most successful efforts is that in Kenya, led by Wangari Maathai. Maathai's Green Belt Movement has more than 50,000 members, mostly women, organized into 2,000 local community groups. Since the group's beginning in 1977, millions of trees have been successfully planted. In those areas, village women now spend less time collecting fuel,

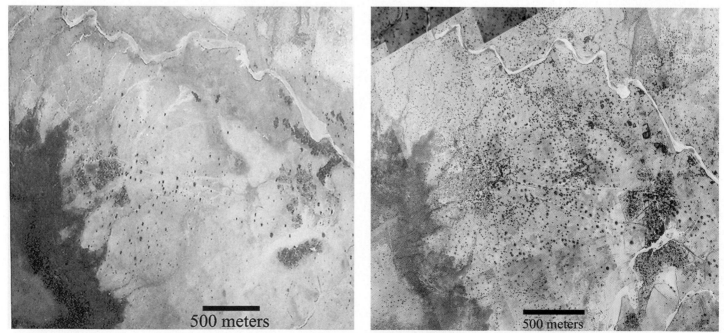

(a)

(b)

FIGURE 6.11 More Trees in Southern Niger These two satellite images demonstrate an increase in tree cover in Niger. In 1975 (a), there were relatively few trees in this part of the Sahel. By 2003 (b), tree cover had increased substantially.

and local environments have improved. Kenya's success has drawn interest from other African countries, spurring a Pan-African Green Belt Movement largely organized through NGOs interested in biofuel generation, protection of the environment, and the empowerment of women. In 2004, Professor Maathai was awarded a Nobel Peace Prize for her contribution to sustainable development, democracy, and peace. She died in 2011, but the Green Belt Movement remains a powerful force in the region.

Destruction of tropical rain forests for logging is most evident in the fringes of Central Africa's Ituri (refer to Figure 6.9). Given the vastness of this forest and the relatively small number of people living there, however, it is less threatened than other forest areas. Two smaller rain forests—one along the Atlantic coast from Sierra Leone to western Ghana and the other along the eastern coast of the island of Madagascar—have nearly disappeared. These rain forests have been severely degraded by commercial logging and agricultural clearing. Madagascar's eastern rain forests, as well as its western dry forests, have suffered serious degradation in the past three decades (Figure 6.12). Deforestation in Madagascar is especially worrisome because the island forms a unique environment with a large number of native species—most notably, the charismatic lemurs.

Wildlife Conservation   Sub-Saharan Africa is famous for its wildlife. No other region of the world has such abundance and diversity of large mammals. The survival of wildlife here reflects, to some extent, the historically low human population density and the fact that sleeping sickness and other diseases have kept people and their livestock out of many areas. In addition, many African peoples have developed various ways of successfully coexisting with wildlife, and about 12 percent of the region is included in nationally protected lands.

However, as is true elsewhere in the world, wildlife is declining in much of Sub-Saharan Africa. The most noted wildlife reserves are in East Africa (Kenya and Tanzania) and southern Africa (South Africa, Zimbabwe, Namibia, and Botswana). These reserves are vital

FIGURE 6.13 South African Wildlife   A white rhinoceros grazes the savannas of Kruger National Park in South Africa, one of the region's oldest wildlife sanctuaries. A protected and endangered species, white rhinos are found in South Africa, Zimbabwe, Botswana, Angola, and Kenya. The total wild rhino populations are estimated at 10,000 to 15,000.

for wildlife protection and are major tourist attractions. Wildlife reserves in southern Africa now seem to be the most secure. In fact, elephant populations are considered to be too large for the land to sustain in countries such as Zimbabwe. Yet throughout the region, population pressure, political instability, and poverty make the maintenance of large wildlife reserves difficult. Poaching is a major problem, particularly for rhinoceroses and elephants; the price of a single horn or tusk in distant markets represents several years' wages for most Africans (Figure 6.13). Ivory is sought in East Asia, especially in Japan. In China, powdered rhino horn is used as a traditional medicine, whereas in Yemen rhino horn is prized for dagger handles. Some wildlife experts contend that herds could be reduced to prevent overgrazing and that the ivory and rhino horn should be legally sold in the international market in order to generate revenue for further conservation. Not surprisingly, many environmentalists disagree strongly.

In 1989, a worldwide ban on ivory trade was imposed as part of the Convention on International Trade in Endangered Species (CITES). Although several African states, such as Kenya, lobbied hard for the ban, others, such as Zimbabwe, Namibia, and Botswana, complained that their herds were growing and the sale of ivory helped to pay for conservation efforts. Conservationists feared that lifting the ban would bring on a new wave of poaching and illegal trade. However, in the late 1990s, the ban was lifted so that some southern African states could sell down their inventories of elephant ivory confiscated from poachers, and limited sales continued. The last legal auction of elephant ivory was in 2008; officials have been reluctant to hold more auctions as there has been a corresponding spike in poaching. The ivory controversy shows how differences in animal distribution in the region, global markets, and international conservation policies are impacting the long-term survival of some 400,000 elephants in Sub-Saharan Africa's protected areas.

Climate Change and Vulnerability in Sub-Saharan Africa

Global climate change poses extreme risks for Sub-Saharan Africa due to the region's poverty, recurrent droughts, and overdependence on

FIGURE 6.12 Deforestation in Madagascar   A man fills a bag with charcoal made from burning tropical forest. Rates of deforestation in Madagascar are high due to a reliance on slash-and-burn agriculture and biofuels. Loss of forest cover has led to serious problems with soil erosion and species extinction for this island nation. About 3 percent of the national territory is in protected areas.

rain-fed agriculture. Sub-Saharan Africa is the lowest emitter of greenhouse gases in the world, but it is likely to experience greater-than-average human vulnerability to global warming because of the region's limited resources to both respond and adapt to environmental change. The areas most vulnerable are arid and semiarid regions such as the Sahel and the Horn of Africa, some grassland areas, and the coastal lowlands of West Africa and Angola.

Climate change models suggest that parts of highland East Africa and equatorial Central Africa may receive more rainfall in the future. Thus, some lands that are currently marginal for farming might become more productive. These effects are likely to be offset, however, by the decline in agricultural productivity in the Sahel, as well as in the grasslands of southern Africa, especially in Zambia and Zimbabwe. Drier grassland areas could deplete wildlife populations, which are a major factor behind the growing tourist economy. As in Latin America, higher temperatures in the tropics may result in the expansion of vector-borne diseases such as malaria and dengue fever into the highlands, where they have until now been relatively rare. Given the relatively high elevations in the region, the negative consequences of a rising sea level would be mostly felt on the West African coast (Senegal, Gambia, Sierra Leone, Nigeria, Cameroon, and Gabon).

Even without the threat of climate change, famine stalks many areas of Africa. The Famine Early Warning Systems (FEWS) network is in place to monitor food insecurity throughout the developing world, but especially in Sub-Saharan Africa. By tracking rainfall, vegetation cover, food production, food prices, and regional conflict, the network maps food security along a continuum from food secure to famine. In 2005, crops in Niger failed due to drought, and an estimated 2.5 million people were vulnerable to famine; however, the early warning systems were taken seriously, and relief efforts resulted in relatively low mortality rates. Figure 6.14 shows the status of food security in eastern Africa during a relatively wet year (in 2010), when food insecurity conditions were expected to remain stable or improve slightly and there were no areas under famine status. Still, even in a good year, the FEWS report estimated that 19 million people faced food insecurity in eastern Africa, especially in Somalia and eastern Ethiopia.

REVIEW QUESTIONS

1. What environmental factors help explain the patterns of human settlement in the region?

2. Summarize the factors that make Sub-Saharan Africans especially vulnerable to climate change.

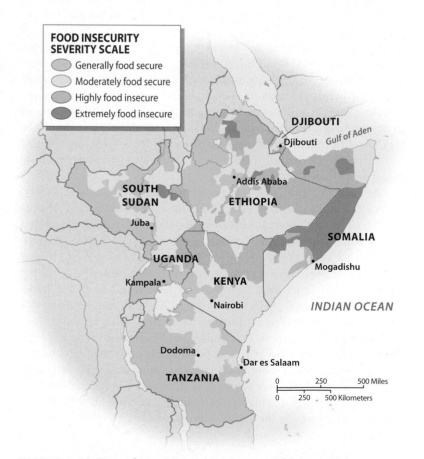

FIGURE 6.14 Map of Food Insecurity in East Africa Anticipating areas of food insecurity based upon the timing and amount of rainfall and changes in vegetation/crop cover is the mission of FEWS network. FEWS has existed since the late 1980s to map areas of potential famine, especially in Sub-Saharan Africa.

POPULATION AND SETTLEMENT: YOUNG AND RESTLESS

Sub-Saharan Africa's population is growing quickly. The global population is projected to increase by 36 percent by 2050, but the projected population growth for Africa south of the Sahara is 130 percent for the same time period. This means that Nigeria is forecast to be the fourth largest country in the world, right after the United States, with over 400 million people in 2050. The region also has a very young population, with 43 percent of the people younger than age 15, compared to just 16 percent for more developed countries. Only 3 percent of the region's population is older than age 65, whereas in Europe 16 percent is older than 65. Families tend to be large, with an average woman having five children (Table 6.1). However, high child and maternal mortality rates also exist, although child mortality rates have declined substantially in the past two decades. The most troubling indicator for the region is its low life expectancy, which dropped to 50 years in 2008 (in part due to the AIDS epidemic), but is currently estimated at 55 years. Life expectancy in other developing nations is much better: India's is 65 years and China's is 75 years. The growth of cities is also a major trend. In 1980, an estimated 23 percent of the population lived in cities; now the figure is 37 percent.

Behind these demographic facts lie complex differences in settlement patterns, livelihoods, belief systems, and access to health care. Although the region is experiencing rapid population growth, Sub-Saharan Africa is not densely populated. The entire region holds nearly 900 million people—roughly half the population that is crowded into the much smaller land area of South Asia. In fact, the overall population density of the region (38 people per square kilometer) is similar to that of the United States (33 people per square kilometer). Just six states—Nigeria, Ethiopia, the Democratic Republic of the Congo, South Africa,

TABLE 6.1 POPULATION INDICATORS

Country	Population (millions) 2012	Population Density (per square kilometer)	Rate of Natural Increase (RNI)	Total Fertility Rate	Percent Urban	Percent <15	Percent >65	Net Migration (Rate per 1000) 2010–15[a]
Angola	20.9	17	3.2	6.3	59	48	2	0.6
Benin	9.4	83	2.9	5.4	44	44	3	−0.2
Botswana	1.9	3	1.2	2.8	24	34	4	1.5
Burkina Faso	17.5	64	3.1	6.0	24	45	2	−1.4
Burundi	10.6	379	3.2	6.4	10	46	2	−0.5
Cameroon	20.9	44	2.7	5.1	49	43	4	−0.1
Cape Verde	0.5	126	2.0	2.5	62	32	6	−4.9
Central African Republic	4.6	7	1.9	4.6	38	40	4	0.4
Chad	11.8	9	2.8	6.0	28	46	3	−2.0
Comoros	0.8	346	2.9	4.9	28	43	3	−2.6
Congo	4.2	12	2.8	5.1	63	41	4	−2.1
Dem. Rep. of Congo	69.1	29	2.8	6.3	34	46	3	−0.2
Djibouti	0.9	40	1.9	3.8	76	36	3	0.0
Equatorial Guinea	0.7	26	2.2	5.2	40	39	3	5.3
Eritrea	5.6	47	2.8	4.5	22	42	2	1.9
Ethiopia	87.0	79	2.4	4.8	17	41	3	−0.2
Gabon	1.6	6	1.8	3.3	73	36	4	0.6
Gambia	1.8	162	2.9	4.9	59	44	2	−1.5
Ghana	25.5	107	2.4	4.2	59	39	4	−0.2
Guinea	11.5	47	2.6	5.2	44	43	3	−0.2
Guinea-Bissau	1.6	45	2.2	5.1	28	41	3	−0.6
Ivory Coast	20.6	64	2.3	4.6	50	41	4	0.1
Kenya	43.0	74	2.7	4.4	32	42	3	−0.2
Lesotho	2.2	73	1.2	3.2	23	37	4	−1.8
Liberia	4.2	38	2.9	5.4	47	43	3	−0.9
Madagascar	21.9	37	2.9	4.7	31	43	3	−0.0
Malawi	15.9	134	2.8	5.7	15	46	3	−0.2
Mali	16.0	13	3.2	6.3	33	47	2	−1.8
Mauritania	3.6	4	2.4	4.5	42	40	3	−1.1
Mauritius	1.3	633	0.4	1.4	42	22	7	0.0
Mozambique	23.7	30	2.8	5.9	31	45	3	−0.2
Namibia	2.4	3	1.8	3.3	39	46	4	−0.3
Niger	16.3	13	3.5	3.3	39	36	4	−0.3
Nigeria	170.1	184	2.6	5.6	51	44	3	−0.4
Reunion	0.9	341	1.2	2.5	94	25	8	0.0
Rwanda	10.8	411	2.2	4.6	17	42	2	0.0
São Tomé and Principe	0.2	190	2.8	4.9	63	43	3	−2.9
Senegal	13.1	67	2.9	5.0	42	44	2	−1.5
Seychelles	0.1	204	1.0	2.3	56	20	7	
Sierra Leone	6.1	85	2.3	5.0	40	43	2	−0.7
Somalia	10.1	16	2.8	6.4	34	45	2	−3.0
South Africa	51.1	42	0.9	2.4	62	31	5	−1.2
South Sudan	9.4	15	2.8	5.4	17	44	2	—
Swaziland	1.2	70	1.6	3.5	22	38	3	−1.0
Tanzania	47.7	50	3.0	5.4	26	45	3	—
Togo	6.0	106	2.8	4.7	37	41	3	−0.3
Uganda	35.6	148	3.3	6.2	15	48	3	−0.8
Zambia	13.7	18	3.0	6.3	39	46	3	−1.2
Zimbabwe	12.6	32	1.9	4.1	29	43	4	4.5

[a]Net Migration Rate from the United Nations, Population Division, *World Population Prospects: The 2010 Revision Population Database.*

Source: Population Reference Bureau, *World Population Data Sheet, 2012.*

Tanzania, and Kenya—account for half of the region's population. Some states have very high population densities (such as Rwanda and Mauritius), whereas others are sparsely settled. Namibia and Botswana have just 3 people per square kilometer. However, population density is not correlated with overall development. Mauritius is a densely settled island nation that is well governed and relatively prosperous, and the same could be said for the arid and sparsely settled state of Botswana. Population density, however, does provide an indicator of the relative population pressures of states in the region, so it is included in Table 6.1.

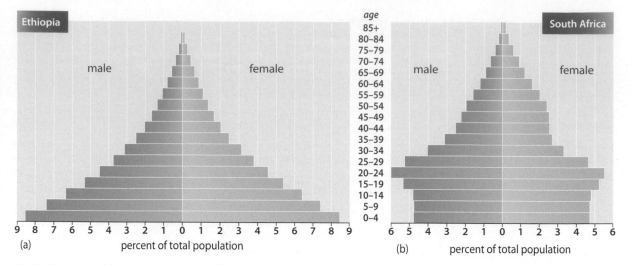

FIGURE 6.15 Ethiopian and South African Population Pyramids These two pyramids show the contrasting demographic profiles of the more rural and rapidly growing Ethiopia and the more urbanized and slower-growing South Africa. A lost generation of South African women in their 30s is also evident, due to the disproportionate impact of HIV/AIDS on women in this country.

Population Trends and Demographic Debates

African demography is a dynamic and controversial field of study. Some believe that Sub-Saharan Africa could support many more people than it presently does, especially if large investments were made in agricultural development. Pessimists, however, argue that the region is a demographic time bomb and that, unless fertility rapidly declines, Sub-Saharan Africa will face massive famines in the near future. The majority of African states officially support lowering rates of natural increase and are promoting modern contraception practices, both to reduce family size and to protect people from sexually transmitted diseases.

The demographic profile of the region is changing. One positive change is the decline in child mortality due to greater access to primary health care. Gone are the days when 1 in 5 children did not live past his or her fifth birthday. Today the child mortality figure is closer to 1 in 10—still high by world standards, but a considerable improvement. Also, life expectancy figures bottomed out in the 2000s due to the devastating impact of HIV/AIDS and are now on the rise. Finally, like populations in other world regions, Africans are moving to cities, which in most cases tends to result in declines in fertility. The family size in South Africa, one of the more urbanized large countries, is half the regional average.

Figure 6.15 compares the population pyramids of Ethiopia and South Africa. Ethiopia has the classic broad-based pyramid of a demographically growing and youthful country. In Ethiopia, most women have five children, and the rate of natural increase is 2.4 percent. There is a near even number of men and women, and only 3 percent of the people are over the age of 65 (life expectancy is 59). In contrast, the South African population pyramid tapers down, reflecting the country's smaller family size (the average woman has two or three children). One unusual aspect in this graphic is the smaller number of women in their 30s and early 40s when compared to men. This is due to the disproportionate impact of AIDS on women in Africa (which will be discussed later). South Africa has more people over the age of 65 (5 percent), but its life expectancy is only 54 years, less than that of Ethiopia. This, too, is the legacy of the AIDS epidemic, which hit

southern Africa with deadly force in the 1990s, but, thankfully, infection rates are now on the decline. Even with the region's low life expectancy, the population is still growing, due to large family size and improvements in child survival rates.

Family Size A continued preference for large families is the basis for the region's demographic growth. In the 1960s, many areas in the developing world had total fertility rates (TFRs) of 5.0 or higher. Today Sub-Saharan Africa, at 5.1, is the only region with such a high TFR. A combination of cultural practices, rural lifestyles, child mortality, and economic realities encourages large families (Figure 6.16). Yet average family sizes are coming down; as recently as 1996, the regional TFR was 6.0.

Throughout the region, large families guarantee a family's lineage and status. Even now, most women marry young, typically when they are teenagers, which increases their opportunity to have children. Demographers often point to the limited formal education available to

FIGURE 6.16 Large Families Large families are still common in Sub-Saharan Africa. The average total fertility rate for the region is 5.1 children per woman.

women as another factor contributing to high fertility. Religious affiliation has little bearing on the region's fertility rates; Muslim, Christian, and animist communities all have similarly high birthrates.

The everyday realities of rural life make large families an asset. Children are an important source of labor; from tending crops and livestock to gathering fuelwood, they add more to the household economy than they take. Also, for the poorest places in the developing world, such as Sub-Saharan Africa, children are seen as social security: When parents' health falters, they expect their grown children to care for them.

Beginning in the 1980s, a shift in national policies occurred. For the first time, government officials argued that smaller families and slower population growth were needed for social and economic development. Other factors are bringing down the growth rate. As African states slowly become more urban, there is a corresponding decline in family size—a pattern seen throughout the world. Tragically, declines in natural increase and life expectancy were also occurring as a result of AIDS.

The Impact of AIDS on Africa

Now it its fourth decade, HIV/AIDS has been one of the deadliest epidemics in modern human history, yet it is beginning to be tamed. At the 2012 International AIDS Conference, the consensus was that HIV infection is still a dangerous, but now a treatable disease. This is especially welcome news for Sub-Saharan Africa, where two-thirds of the 34 million people living with HIV/AIDS are found.

The virus is thought to have originated in the forests of the Congo, possibly crossing over from chimpanzees to humans sometime in the late 1950s. Yet it was not until the 1980s that the impact of the disease became felt. In Sub-Saharan Africa, as in much of the developing world, the virus is transmitted by unprotected heterosexual activity, or during the birth process between mother and child, or through breast feeding. A long pattern of seasonal male labor migration helped to spread the disease. So, too, did lack of education, inadequate testing early on in the epidemic, and the disempowerment of women. Consequently, women bear a disproportionate burden of the HIV/AIDS epidemic. They account for approximately 60 percent of the HIV infections, and they are usually the caregivers for those who are infected. Until the late 1990s, many African governments were unwilling to acknowledge publicly the severity of the situation or to discuss frankly the measures necessary for prevention.

Southern Africa is ground zero for the AIDS epidemic (Figure 6.17). The nine countries with the highest HIV prevalence are all located there.

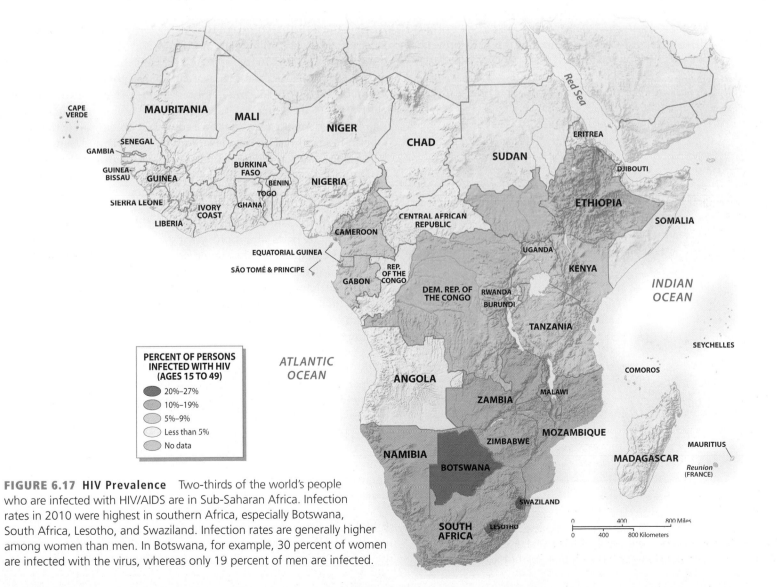

FIGURE 6.17 HIV Prevalence   Two-thirds of the world's people who are infected with HIV/AIDS are in Sub-Saharan Africa. Infection rates in 2010 were highest in southern Africa, especially Botswana, South Africa, Lesotho, and Swaziland. Infection rates are generally higher among women than men. In Botswana, for example, 30 percent of women are infected with the virus, whereas only 19 percent of men are infected.

In South Africa, the most populous state in southern Africa, 5.7 million people (nearly one in five people age 15–49) are infected with HIV/AIDS. The rate of infection in neighboring Botswana is 24 percent (down from a staggering 36 percent in 2001) for the same age group. Infection rates in other African states are lower, but still high by world standards. In Kenya, it is estimated that 6 percent of the 15–49 age group has HIV or AIDS. By comparison, only 0.6 percent of the same age group in North America is infected.

The social and economic implications of this epidemic have been profound. Life expectancy rates tumbled—in a few places even dropping to the early 40s. AIDS typically hits the portion of the population that is most economically productive. Time lost to care for sick family members and the outlay of workers' compensation benefits have reduced economic productivity and overwhelmed public services in hard-hit areas. The disease makes no class distinctions: Countries are losing both peasant farmers and educated professionals (doctors, engineers, and teachers). Many countries struggle to care for millions of children orphaned by AIDS.

After three devastating decades, there are finally hopeful signs. Prevention measures are widely taught, and treatment with a mix of available drugs means that HIV infection is now manageable and not a death sentence. Due to international financial support and national outreach efforts, Sub-Saharan Africa now has many more health facilities that offer HIV testing and counseling (Figure 6.18). Prevention services in prenatal settings in 2011 allowed nearly half of all pregnant HIV-positive women to get antiretroviral drugs to prevent

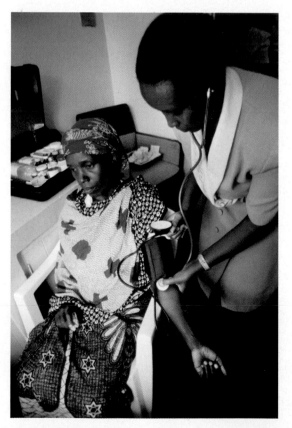

FIGURE 6.18 HIV/AIDS Clinic in Tanzania   A woman with AIDS receives a checkup in a Tanzanian clinic. Programs such as these offer new hope from the ravages of this disease. In 2011, it was estimated that 6 million people in the region were taking antiretroviral drugs, which allow them to manage the disease and greatly prolong their lives.

transmission of the virus to their babies. Changes in sexual practices, driven by educational campaigns, more condom use, and higher rates of male circumcision have prevented hundreds of thousands of new cases.

Patterns of Settlement and Land Use

Because of the dominance of rural settlements in Sub-Saharan Africa, people are widely scattered throughout the region (Figure 6.19). Population concentrations are highest in West Africa, highland East Africa, and the eastern half of South Africa. The first two areas have some of the region's best soils, and native systems of permanent agriculture developed there. In South Africa, the more densely settled east is a result of an urbanized economy based on mining, as well as the forced concentration of black South Africans into former homelands in the east.

As more Africans move to cities, patterns of settlement are becoming more concentrated. Towns that were once small administrative centers for colonial elites grew into major cities. The region even has its own megacity, Lagos, which recently topped 10 million residents. Throughout the continent, African cities are growing faster than rural areas. However, before we examine the Sub-Saharan urban scene, we need a more detailed discussion of rural areas.

Agricultural Subsistence   The staple crops over most of Sub-Saharan Africa are millet, sorghum, and corn (maize), as well as a variety of tubers and root crops such as yams. Irrigated rice is widely grown in West Africa and Madagascar. Geographer Judith Carney, in her book *Black Rice*, documents how African slaves introduced rice cultivation to the Americas. Corn, in contrast, was introduced to Africa from the Americas by the slave trade and quickly grew to become a basic food. In the higher elevations of Ethiopia and South Africa, wheat and barley are grown. Intermixed with subsistence foods is a variety of export crops—coffee, tea, rubber, bananas, cacao, cotton, and peanuts—that are grown in distinct ecological zones and often in some of the best soils.

In areas that support annual crop yields, population densities are greater. In parts of humid West Africa, for example, the yam became the dominant subsistence crop. The Igbos' mastery of yam farming allowed them to produce more food and live in denser permanent settlements in the southeastern corner of present-day Nigeria. Much of traditional Igbo culture is tied to the demanding tasks of clearing the fields, tending the delicate plants, and celebrating the harvest.

Over much of the continent, African agriculture remains relatively unproductive, and population densities tend to be low. Amid the poorer tropical soils, farming usually entails shifting cultivation (or **swidden**). This process involves burning the natural vegetation to release fertilizing ash and then planting crops such as maize, beans, sweet potatoes, bananas, papayas, manioc, yams, melons, and squash. Each plot is temporarily abandoned once its source of nutrients has been exhausted. Swidden cultivation is often a very finely tuned adaptation to local environmental conditions, but it is unable to support high population densities.

Export Agriculture   Agricultural exports, whether from large estates or small producers, are critical to the economies of many states. If African countries are to import the modern goods and energy

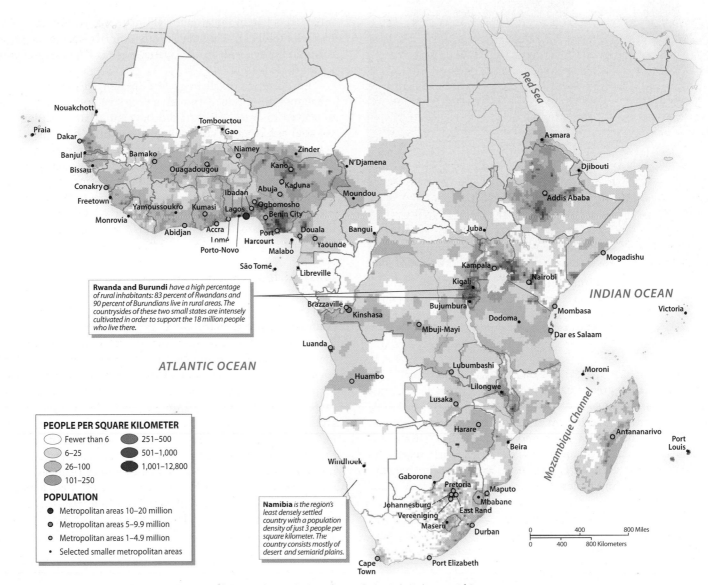

PEOPLE PER SQUARE KILOMETER

- Fewer than 6
- 6–25
- 26–100
- 101–250
- 251–500
- 501–1,000
- 1,001–12,800

POPULATION

- Metropolitan areas 10–20 million
- Metropolitan areas 5–9.9 million
- Metropolitan areas 1–4.9 million
- Selected smaller metropolitan areas

Rwanda and Burundi *have a high percentage of rural inhabitants: 83 percent of Rwandans and 90 percent of Burundians live in rural areas. The countrysides of these two small states are intensely cultivated in order to support the 18 million people who live there.*

Namibia *is the region's least densely settled country with a population density of just 3 people per square kilometer. The country consists mostly of desert and semiarid plains.*

FIGURE 6.19 Population Map of Sub-Saharan Africa   The majority of people in Sub-Saharan Africa live in rural areas. Some of these rural zones, however, are densely settled, such as West Africa and the East African highlands. Major urban centers, especially in South Africa and Nigeria, support millions. There is only one megacity in the region with more than 10 million residents (Lagos), but more than two dozen cities have more than 1 million residents.

resources they require, they must sell their own products on the world market. Because the region has few competitive industries, the bulk of its exports are primary products derived from farming, mining, and forestry.

Several African countries rely heavily on one or two export crops. Coffee, for example, is vital for Ethiopia, Kenya, Rwanda, Burundi, and Tanzania. Peanuts have historically been the primary source of income in the Sahel, whereas cotton is tremendously important for the Central African Republic and South Sudan. Ghana and the Ivory Coast have long been the world's main suppliers of cacao (the source of chocolate), Liberia produces plantation rubber, and many farmers in Nigeria specialize in palm oil. The export of such products can bring good money when commodity prices are high, but when prices collapse, as they periodically do, economic devastation may follow.

Nontraditional agricultural exports that depend upon significant capital inputs and refrigerated air transport have emerged in the last two to three decades. One industry is floriculture for the plant and cut flower industry. Here the highland tropical climate of Kenya, Ethiopia, and South Africa is advantageous. After Colombia in South America, Kenya is the largest exporter of cut flowers in the world, and most of its exports go to Europe (Figure 6.20). Similarly, the European market for fresh vegetables and fruits in the winter is being met by some African producers in West and East Africa.

Pastoralism   Animal husbandry (the care of livestock) is extremely important in Sub-Saharan Africa, particularly in semiarid zones. Camels and goats are the principal animals in the Sahel and the Horn of Africa, but farther south, cattle are most common (Figure 6.21). Many African peoples have traditionally specialized in cattle raising and are often tied in mutually beneficial relationships with neighboring farmers. Such **pastoralists** typically graze their stock on the stubble of harvested fields during the dry season and then move them to drier uncultivated areas during the wet season,

FIGURE 6.20 Kenyan Floriculture Workers Women cut roses for export at a greenhouse in Naivasha, Kenya. Floriculture is a vital contributor to Kenya's agricultural exports and has grown dramatically in the past two decades. One-third of the cut flowers sold in Europe, especially roses, come from Kenya.

when the pastures turn green. Farmers thus have their fields fertilized by the manure of the pastoralists' stock, while the pastoralists find good dry-season grazing. At the same time, the nomads can trade their animal products for grain and other goods of the sedentary world. Several pastoral peoples of East Africa, particularly the Masai of the Tanzanian-Kenyan borderlands, are noted for their extreme reliance on cattle and general (but never complete) independence from agriculture.

Large expanses of Sub-Saharan Africa have been off-limits to cattle because of infestations of **tsetse flies**, which spread sleeping sickness to cattle, humans, and some wildlife. In environments containing brush or woodland (which is necessary for the survival of tsetse flies), wild animals that harbor the disease, but are immune to it, could be present in large numbers, but cattle simply could not be raised. At present, tsetse fly eradication programs are reducing the threat, and cattle raising is spreading into areas where it was previously forbidden. This process is beneficial for African peoples, but it may endanger the continued survival of many wild animals. When people and their stock move into new areas in large numbers, wildlife almost inevitably declines.

Urban Life

Although Sub-Saharan Africa is considered the least urbanized region in the developing world, most Sub-Saharan cities are growing at twice the national growth rates. If present trends continue, half of the region's population may well be living in cities by 2030. One of the consequences of this surge in city living is urban sprawl. Rural-to-urban migration, industrialization, and refugee flows are forcing the cities of the region to absorb more people and use more resources. As in Latin America, the tendency is toward urban primacy—the condition in which one major city is dominant and at least three times larger than the next largest city. Luanda, the capital of Angola, was built for half a million people, but now has at least 4 million residents (many of them displaced during the country's long years of war). Most of the city's

inhabitants live in the growing slums that surround the colonial core. In such rapidly growing places, city officials struggle to build enough roads and provide electricity, water, trash collection, and employment for all these people.

European colonialism greatly influenced urban form and development in the region. Although a very small percentage of the population lived in cities, Africans did have an urban tradition prior to the colonial era. Ancient cities, such as Axum in Ethiopia, thrived 2,000 years ago. Similarly, in the Sahel, major trans-Saharan trade centers, such as Tombouctou (Timbuktu) and Gao, have existed for more than a millennium. In East Africa, an urban trading culture emerged that was rooted in Islam and the Swahili language. West Africa, however, had the most developed precolonial urban network, reflecting both native and Islamic traditions. It also supports some of the region's largest cities today.

West African Urban Traditions The West African coastline is dotted with cities, from Dakar, Senegal, in the far north to Lagos, Nigeria, in the east. Half of Nigerians live in cities, and in 2010 the country had seven metropolitan areas with populations of more than 1 million. Historically, the Yoruba cities in southwestern Nigeria have been the best documented. Developed in the 12th century, cities such as Ibadan were walled and gated, with a palace encircled by large rectangular courtyards at the city center. An important center of trade for an extensive surrounding area, Ibadan was also a religious and political center. Lagos was also a Yoruba settlement. Founded on a coastal island on the Bight of Benin, most of the modern city has spread onto the nearby mainland. Its coastal setting and natural harbor made this relatively small native city attractive to colonial powers. When the British took control in the mid-19th century, the city's size and importance grew. When Nigeria gained its independence from the United Kingdom in 1960, Lagos was a city of 1 million. Today it has well over 10 million residents, making it the largest city in the region.

FIGURE 6.21 Pastoralists in the Horn of Africa Two Ethiopian girls lead a herd of camels. Camels are well suited for the arid conditions of eastern Ethiopia and Somalia. Pastoralists rely upon their camels for milk, transport, and trade.

FIGURE 6.22 Global Accra This satellite image shows the elite neighborhoods, international businesses, superior roads, and airport access that make up the "Global" CBD of Accra, the capital city of Ghana. The Global CBD was fostered through international investment and is several miles to the east of the traditional CDB, which is considerably more congested, pedestrian-oriented, and serves the local population.

Most West African cities are hybrids, combining Islamic, European, and national elements such as mosques, Victorian architecture, and streets named after independence leaders. Accra is the capital city of Ghana and home to nearly 2 million people. Originally settled by the Ga people in the 16th century, it became a British colonial administrative center by the late 1800s. Here, again, the modern city is being transformed through neoliberal policies introduced in the 1980s that attracted international corporations. A growth in foreign investment in financial and producer services led to the creation of a "Global Central Business District (CBD)" on the east side of the city, away from the "National CBD" (Figure 6.22). Here foreign companies clustered in areas with secure land title, new modern roads, parking, and access to the airport. Upper-income gated communities have also formed near the Global CBD. Accra, like other cities in the region, is rapidly changing due to an influx of foreign capital. The result is highly segregated urban spaces that reflect a global phase in urban development in which world market forces, rather than colonial or national ones, are driving the change.

Urban Industrial South Africa The major cities of southern Africa, unlike those of West Africa, are colonial in origin. Most of these cities, such as Lusaka, Zambia, and Harare, Zimbabwe, grew as administrative

or mining centers. South Africa is one of the most urbanized states in the entire region, and it is certainly the most industrialized. The foundations of South Africa's urban economy rest largely on its incredibly rich mineral resources (diamonds, gold, chromium, platinum, tin, uranium, coal, iron ore, and manganese). Eight of its metropolitan areas have more than 1 million people; the largest of them are Johannesburg, Durban, and Cape Town.

The form of South African cities continues to be imprinted by the legacy of **apartheid** (an official policy of racial segregation that shaped social relations in South Africa for nearly 50 years). Even though apartheid was abolished in 1994, it is still evident in the landscape. Under apartheid rules, South African cities were divided into residential areas according to racial categories: white, **coloured** (a South African term describing people of mixed African and European ancestry), Indian (South Asian), and African (black). Whites occupied the largest and most desirable portions of the city. Blacks were crowded into the least desired areas, in *townships* in places such as Soweto outside of Johannesburg or Gugulethu outside of Cape Town. Today blacks, coloureds, and Indians are legally allowed to live anywhere they want. Yet the economic differences between racial groups, as well as long-standing prejudices, make residential integration difficult.

Johannesburg is the African metropolitan area that most consciously aspires to global city status, a dream underscored by its hosting of soccer's World Cup in 2010. However, as apartheid ended, Johannesburg became infamous for its high crime rate. Many white-owned businesses fled the CBD for the northern suburb of Sandton (Figure 6.23). By the late 1990s, Sandton was the new financial and business hub for the city of Johannesburg. It epitomizes the modern urban face of South Africa: While it is racially mixed, affluent whites are overrepresented. When the province of Gauteng invested $3 billion in a new high-speed commuter rail called Gautrain, it was no accident that the first functioning line linked Sandton with Tambo International Airport.

FIGURE 6.23 Sandton, Johannesburg Just north of Johannesburg's central business district, Sandton has emerged as the financial and business center of the new South Africa. Many businesses relocated here, fleeing high crime rates in the old CBD. With world-class shopping, hotels, a convention center, and even Nelson Mandela Square, Sandton has become the part of Johannesburg that most tourists and businesspeople experience.

REVIEW QUESTIONS

1. Explain the factors that contribute to high population growth rates in this region.
2. Where has the impact of HIV/AIDS been most pronounced, and what are governments and aid organizations doing to fight this disease?
3. What are the major rural livelihoods in this region?

CULTURAL COHERENCE AND DIVERSITY: UNITY THROUGH ADVERSITY

No world region is culturally homogeneous, but most have been partially unified in the past through widespread systems of belief and communication. The lack of traditional cultural and political coherence across Sub-Saharan Africa is not surprising, considering the region's huge size. Sub-Saharan Africa is more than four times larger than Europe or South Asia. Had foreign imperialism not impinged on the region, it is quite possible that West Africa and southern Africa would have developed into their own distinct world regions.

An African identity south of the Sahara was created through a common history of slavery and colonialism, as well as struggles for independence and development. More telling, the people of the region often define themselves as African, especially to the outside world. No one will argue the fact that Sub-Saharan Africa is poor. And yet the cultural expressions of its people—its music, dance, and art—are joyous. Africans share a resilience and optimism that visitors to the region often comment on. The cultural diversity of the region is obvious, yet there is unity among the people, drawn from surviving many adversities.

Language Patterns

In most Sub-Saharan countries, as in other former colonies, multiple languages are used that reflect tribal, colonial, and national affiliations. Indigenous languages, many from the Bantu subfamily, are often localized to relatively small rural areas. More widely spoken African trade languages, such as Swahili and Hausa, serve as a lingua franca over broader areas. Overlaying native languages are Indo-European (French and English) and Afro-Asiatic (Arabic and Somali) languages. Figure 6.24 illustrates the complex pattern of language families and major languages found in Africa today. A comparison of the larger map with the inset that shows current "official" languages indicates that most African countries are multilingual, which can be a source of tension within states. In Nigeria, for example, the official language is English, yet there are millions of Hausa, Yoruba, Igbo (or Ibo), Ful (or Fulani), and Efik speakers, as well as speakers of dozens of other languages.

African Language Groups Three of the six language groups mapped in Figure 6.24 (Niger-Congo, Nilo-Saharan, and Khoisan) are unique to the region, while the other three (Afro-Asiatic, Austronesian, and Indo-European) are more closely associated with other parts of the world. Afro-Asiatic languages, especially Arabic, dominate

North Africa and are understood in Islamic areas of Sub-Saharan Africa as well. Amharic in Ethiopia and Somali in Somalia are also Afro-Asiatic languages. The Austronesian language family is limited to the island of Madagascar, which many believe was first settled by seafarers from Indonesia some 1,500 years ago. Indo-European languages, especially French, English, Portuguese, and Afrikaans, are a legacy of colonialism and are widely used today.

The Niger-Congo language group is by far the most important one in the region. This linguistic group, which originated in West Africa, includes Mandingo (one of the many Mande languages), Yoruba, Ful (or Fulani), and Igbo (or Ibo), among others. Around 3,000 years ago, the Bantu, a people of the Niger-Congo language group, began to expand out of western Africa into the equatorial zone. In one of the most far-ranging migrations in human history, they introduced agriculture into large areas of central and southern Africa. One language in the Bantu subfamily, Swahili, eventually became the most widely spoken Sub-Saharan language. Swahili originated as a trade language on the East African coast, where merchants from Arabia established several colonies around 1100 CE. A hybrid society grew up in a narrow coastal band of modern Kenya and Tanzania, speaking a language of Bantu structure enriched with many Arabic words. Although Swahili became the primary language only in this narrow coastal belt, it spread far into the interior as the language of trade. After independence was achieved, both Kenya and Tanzania adopted Swahili as an official language, along with English. Swahili, with some 70 million speakers, is a lingua franca for East Africa. It has generated a fairly extensive literature and is often studied in other regions of the world.

Language and Identity Ethnic identity and linguistic affiliation have historically been highly unstable over much of Sub-Saharan Africa. The tendency was for new language groups to form when people threatened by war fled to less-settled areas, where they often mixed with peoples from other places. In such situations, new languages arise quickly, and divisions between groups are blurred. Nevertheless, distinct **tribes** formed that consisted of a group of families or clans with a common kinship, language, and definable territory. European colonial administrators were eager to establish a fixed social order to better control native peoples. In this process, a flawed cultural map of Sub-Saharan Africa evolved. Some tribes were artificially divided, meaningless names were applied, and cultural areas were often misinterpreted.

Social boundaries between different ethnic and linguistic groups have become more stable in recent years, and some individual languages have become particularly important for communication on a national scale. Wolof in Senegal; Mandingo in Mali; Mossi in Burkina Faso; Yoruba, Hausa, and Igbo in Nigeria; Kikuyu in Central Kenya; and Zulu, Xhosa, and Sotho in South Africa are all nationally significant languages spoken by millions (Figure 6.25). None, however, has the status of being the official language of any country. With the end of apartheid in South Africa, the country officially recognized 11 languages, although English is still the lingua franca of business and government. Indeed, a single language has a clear majority status in only a handful of countries. The more linguistically homogeneous states include Somalia (where virtually everyone speaks Somali) and the very small states of Rwanda, Burundi, Swaziland, and Lesotho.

European Languages In the colonial period, European countries used their own languages for administrative purposes in their African

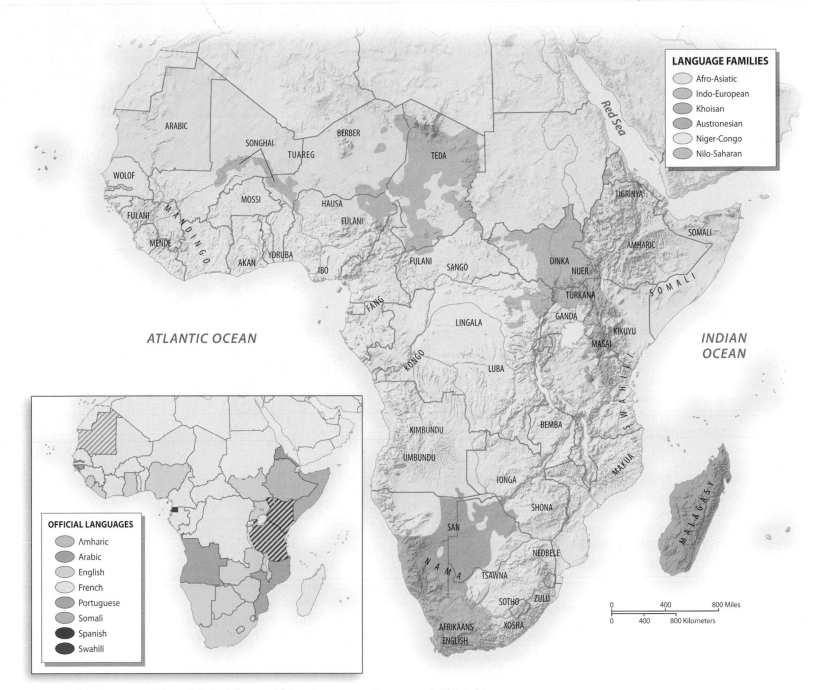

FIGURE 6.24 Language Map of Sub-Saharan Africa: Language Groups and Official Languages
Mapping language is a complex task for Sub-Saharan Africa. Some languages have millions of speakers, such as Swahili, and other languages are spoken by a few hundred people living in isolated areas. Six language families are represented in the region. Among these families are scores of individual languages (see the labels on the map). Because most modern states have many native languages, the colonial language often became the "official" language. English and French are the most common official languages in the region (see inset).

empires. Education in the colonial period also stressed literacy in the language of the imperial power. In the postindependence period, most Sub-Saharan African countries have continued to use the languages of their former colonizers for government and higher education. Few of these new states had a clear majority language that they could employ, and picking any minority tongue would have met with opposition from other peoples. The one exception is Ethiopia, which maintained its independence during the colonial era. The official language is Amharic, although other indigenous languages are also spoken.

Two vast blocks of European language exist in Africa today: Francophone Africa, including the former colonies of France and Belgium, where French serves as the main language of administration; and Anglophone Africa, where the use of English prevails (see the inset to Figure 6.24). Early Dutch settlement in South Africa resulted in the use of Afrikaans (a Dutch-based language) by several million South Africans. In Mauritania and Eritrea, Arabic serves as a main language. Interestingly, when South Sudan gained its independence from Sudan in 2011, it changed its official language from Arabic to English.

FIGURE 6.25 Multilingual South Africa A South African sign warns pedestrians not to cross a highway in three languages: English, Afrikaans, and Zulu. Throughout Sub-Saharan Africa, many people are multilingual because of the diversity of languages that exist in each state.

Religion

Native African religions are generally classified as animist. This is a somewhat misleading catchall term used to classify all local faiths that do not fit into one of the handful of "world religions." Most animist religions are centered on the worship of nature and ancestral spirits, but within the animist tradition there is great internal diversity. Classifying a religion as animist says more about what it is not than about what it actually is.

Both Christianity and Islam entered Sub-Saharan Africa early in their histories, but they advanced slowly for many centuries. Since the beginning of the 20th century, both religions have spread rapidly—more rapidly, in fact, than in any other part of the world. However, tens of millions of Africans still follow animist beliefs, and many others combine animist practices and ideas with their observances of Christianity and Islam.

The Introduction and Spread of Christianity Christianity came first to northeastern Africa. Kingdoms in both Ethiopia and central Sudan were converted by 300 CE—the earliest conversions outside the Roman Empire. The peoples of northern and central Ethiopia adopted the Coptic form of Christianity and have thus historically looked to Egypt's Christian minority for their religious leadership (Figure 6.26). At present, roughly half of the population in both Ethiopia and Eritrea is Coptic Christian; most of the rest is Muslim, but there are still some animist communities, especially in Ethiopia's western lowlands.

European settlers and missionaries introduced Christianity to other parts of Sub-Saharan Africa beginning in the 1600s. The Dutch, who began to colonize South Africa at that time, brought their Calvinist Protestant faith. Later European immigrants to South Africa brought Anglicanism and other Protestant beliefs, as well as Catholicism. Most black South Africans eventually converted to one or another form of Christianity as well. In fact, churches in South Africa were instrumental in the long fight against white racial supremacy. Religious leaders such as Bishop Desmond Tutu were outspoken critics of the injustices of apartheid and worked to bring down the system.

Elsewhere in Africa, Christianity came with European missionaries, most of whom arrived after the mid-1800s. As was true in the rest

FIGURE 6.26 Eritrean Christians at Prayer Coptic Christians gather in front of St. Mary's Church in Asmara, Eritrea, on Good Friday, a holy day for Christians. Half of the populations in Eritrea and Ethiopia belong to the Coptic Church, which has ties to the Christian minority in Egypt.

of the world, missionaries had little success where Islam had preceded them, but they eventually made numerous conversions in animist areas. As a general rule, Protestant Christianity prevails in areas of former British colonization, while Catholicism is more important in former French, Belgian, and Portuguese territories. In the postcolonial era, African Christianity has spread out, at times taking on a life of its own, independent from foreign missionary efforts. Still active in the region are various Pentecostal, Evangelical, and Mormon missionary groups, mostly from the United States. It is difficult to map the distribution of Christianity in Africa, however, because it has spread irregularly across the non-Islamic portion of the region.

The Introduction and Spread of Islam Islam began to advance into Sub-Saharan Africa 1,000 years ago (Figure 6.27). Berber traders from North Africa and the Sahara introduced the religion to the Sahel, and by 1050 the Kingdom of Tokolor in modern Senegal emerged as

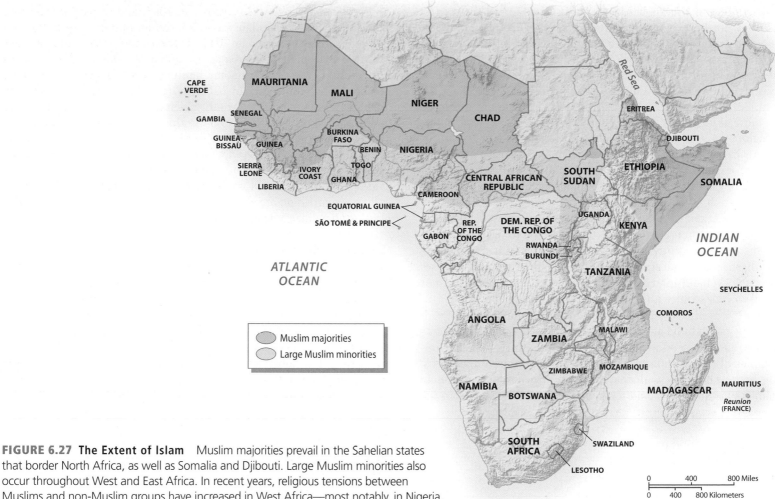

FIGURE 6.27 The Extent of Islam Muslim majorities prevail in the Sahelian states that border North Africa, as well as Somalia and Djibouti. Large Muslim minorities also occur throughout West and East Africa. In recent years, religious tensions between Muslims and non-Muslim groups have increased in West Africa—most notably, in Nigeria.

the first Sub-Saharan Muslim state. Somewhat later, the ruling class of the powerful Mande-speaking trading empires of Ghana and Mali converted as well. In the 14th century, the emperor of Mali astounded the Muslim world when he and his royal court made the pilgrimage to Mecca, bringing with them so much gold that they set off a brief period of high inflation throughout Southwest Asia.

Mande-speaking traders, whose networks spanned the area from the Sahel to the Gulf of Guinea, gradually introduced the religion to other areas of West Africa. There, however, many peoples remained committed to animism, and Islam made slow and unsteady progress. Today orthodox Islam prevails through most of the Sahel. Farther south, Muslims are mixed with Christians and animists, but their numbers continue to grow, and their practices tend to be orthodox as well (Figure 6.28).

Interaction Between Religious Traditions The southward spread of Islam from the Sahel, coupled with the northward spread of Christianity from the port cities, has generated a complex religious frontier across much of West Africa. In Nigeria, the Hausa are firmly Muslim, while the southeastern Igbo are largely Christian. The Yoruba of the southwest are divided between Christian and Muslim. In the more remote parts of Nigeria, animist traditions remain strong. Despite this religious diversity,

FIGURE 6.28 West African Muslims Villagers leave the Larabanga mosque after Friday prayers. The Muslim village of Larabanga in Northern Ghana is home to one of the oldest mosques in West Africa and was recently declared a UN World Heritage Site.

religious conflict in Nigeria has been relatively rare until recently. In 2000, seven northern Nigerian states imposed Muslim sharia laws, which has triggered intermittent violence ever since, especially in the northern city of Kaduna. In 2010, near the central city of Jos, hundreds of Christian and Muslim villagers were slaughtered in violent clashes and retaliatory killings. Many victims were women and children. Still, most of West Africa's regional conflicts continue to be framed more along ethnic terms than religious ones.

Religious conflict historically has been far more acute in northeastern Africa, where Muslims and Christians have struggled against each other for centuries. Such a clash eventually led to the creation of the region's newest state when South Sudan separated from the country of Sudan in 2011. Islam was introduced to Sudan in the 1300s by an invasion of Arabic-speaking pastoralists who destroyed the indigenous Coptic Christian kingdoms of the area. Within a few hundred years, northern and central Sudan had become completely Islamic. The southern equatorial province of Sudan, where tropical diseases and extensive wetlands prevented Arab advances, remained animist or converted to Christianity under British colonial rule.

In the 1970s, the Arabic-speaking Muslims of northern and central Sudan began to build an Islamic state. Experiencing both religious discrimination and economic exploitation, the Christian and animist peoples of the south launched a massive rebellion. In the 1980s, fighting became intense, with the government generally controlling the main towns and roads and the rebels maintaining power in the countryside. A peace was brokered in 2003, and as part of the peace agreement southern Sudan was promised an opportunity to vote on secession from the north in 2011. The vote took place, and the new nation was formed with Juba as its capital. Yet this landlocked territory is still not at peace, as some fighting with Sudanese troops along the border has continued. In 2012, South Sudan's access to Sudan's oil pipeline was temporarily cut off, which meant the new nation could not export its oil. Clearly, conflict in this area is likely to continue.

Sub-Saharan Africa is a land of religious vitality. Both Christianity and Islam are spreading rapidly, and devotional activities are part of the daily flow of life in cities and rural areas. Animism continues to hold widespread appeal as well, so that new and syncretic (blended) forms of religious expression are also emerging. With such a diversity of faiths, it is fortunate that religion is not typically the cause of overt conflict.

Globalization and African Culture   The slave trade that linked Africa to the Americas and Europe set in motion paths of cultural diffusion that transferred African peoples and cultures across the Atlantic. Tragically, slavery damaged the demographic and political strength of African societies, especially in West Africa, from where most slaves were taken. An estimated 12 million Africans were shipped to the Americas as slaves from the 1500s until 1870 (Figure 6.29). Slavery impacted all of Sub-Saharan Africa, sending Africans not just to the Americas, but also to Europe, North Africa, and Southwest Asia. The vast majority, however, worked on plantations in the Americas.

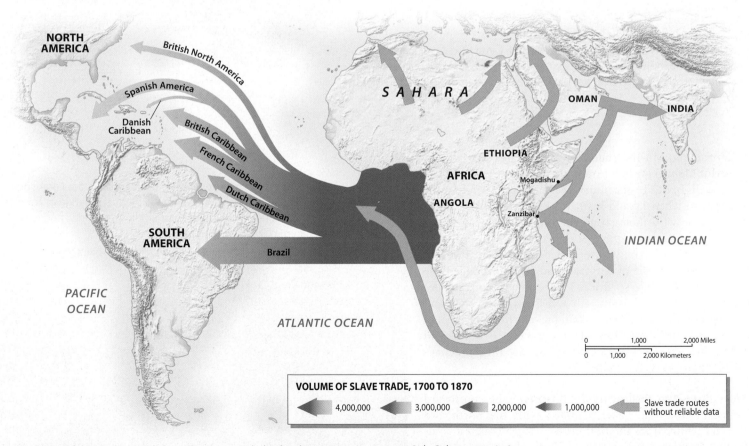

FIGURE 6.29 African Slave Trade   The slave trade had a devastating impact on Sub-Saharan societies. From ship logs, it is estimated that 12 million Africans were shipped to the Americas to work as slaves on sugar, cotton, and rice plantations; the majority went to Brazil and the Caribbean. Yet other slave routes existed, although the data on them are less reliable. Africans from south of the Sahara were used as slaves in North Africa. Others were traded across the Indian Ocean into Southwest Asia and South Asia.

EXPLORING GLOBAL CONNECTIONS

The New African Americans

Since the 1990s, tens of thousands of African immigrants have settled in the United States, making Africa one of the fastest-growing source regions for new immigrants. Like other immigrants, they come for economic opportunity, for family reunification, and as resettled refugees. The United States has about 1.5 million African-born immigrants, the majority of which (75 percent) have settled here since 1990. Nigeria and Ethiopia are the biggest sending countries, which is not terribly surprising, since they are also the biggest countries in terms of population.

The latest arrivals from Africa are a mix of highly educated college graduates and refugees seeking asylum. Over 40 percent of African immigrants have college degrees, a figure that is much higher than for the U.S. population as a whole. In particular, many women come as health care professionals, especially nurses. Although Africans are found throughout the country, one-third of them are concentrated in three metropolitan areas: New York, Washington, DC, and Atlanta. For example, metropolitan Washington, DC, hosts the largest concentration of Ethiopians in the United States. They have established scores of restaurants and businesses in the Shaw neighborhood, a historic African-American section of the District of Columbia that is now often referred to as "Little Ethiopia" (Figure 6.1.1). Similarly, in New York and Atlanta, African foods, music, and cultural events have been incorporated into the urban cultural landscape.

Refugees are also an important part of the immigrant flow. Through resettlement programs, concentrations of Africans have emerged in Minneapolis–St. Paul, Boston, and Columbus. In Minneapolis–St. Paul, where many Somali refugees were resettled, Africans account for 20 percent of all immigrants in the twin cities. In general, refugees are a more needy population, arriving with less social capital and fewer economic resources. Unlike economic migrants, refugees are assisted in their settlement and job placement as part of the federally funded Refugee Act of 1980. The major African refugee populations in the United States include Ethiopians, Eritreans, Somalis, Liberians, and Sudanese.

People of African ancestry have lived in the United States for nearly as long as people of European ancestry and account for 12 percent of the total U.S. population. The percentage of African Americans in the U.S. population was significantly higher in 1800. The percentage declined in part due to the abolition of slavery. Yet in the late 19th century, restrictive measures were also put into place that made it virtually impossible for Africans to emigrate to the United States until 1965, when the Immigration and Nationality Act opened up immigration to peoples from Africa and Asia.

In the United States, African immigrants often find the term *African American* confusing. As newcomers from Africa, some have proposed the term *Neo African Americans* because their cultural and social history is so different from that of native-born African

FIGURE 6.1.1 Ethiopian Businessman in Washington, DC An Ethiopian entrepreneur stands in front of his shop in the Shaw neighborhood of Washington, DC. Metropolitan Washington attracts many African immigrants and is home to the largest concentration of Ethiopians in the United States.

Americans. African immigrants have also maintained important ties with their countries of origin through remittances and transnational businesses. In fact, several African governments actively court this modern diaspora in the hope that some will return and invest in their home countries.

Out of this tragic displacement of people came a blending of African cultures with Amerindian and European cultures. African rhythms are at the core of various American musical styles, from rumba to jazz, the blues, and rock and roll. Brazil, the largest country in Latin America, is claimed to be the second largest "African state" (after Nigeria) because of its huge Afro-Brazilian population. Thus, the forced migration of Africans as slaves had a huge cultural influence on many areas of the world.

So, too, have contemporary movements of Africans influenced the cultures of many world regions. Perhaps one of the most celebrated persons of African ancestry today is President Barack Obama, the son of a Kenyan man and a woman from Kansas. Obama's heritage and upbringing embody the forces of globalization. In Kenya, he is hailed as part of the modern African diaspora—young professionals (and their offspring) who leave the continent and make their mark somewhere else. As an indicator of our global age, Kenyan President Mwai Kibaki declared a Kenyan national holiday after Obama won the

U.S. presidential election in November 2008 (see "Exploring Global Connections: The New African Americans").

Popular culture in Africa, like everywhere else in the world, is a dynamic mixture of global and local influences. Kwaito, the latest musical craze in South Africa, sounds a lot like rap from the United States. A closer listen, however, reveals an incorporation of local rhythms, lyrics in Zulu and Xhosa, and themes about life in postapartheid townships. Lagos, Nigeria, referred to as Nollywood, has emerged as Africa's film capital. More than 500 films a year are made there, most shot in a few days and with budgets of $10,000. These pulp movies are often criticized for being exploitative imitations of Hollywood B-movies. Yet they touch on themes that resonate with Africans, such as religion, ethnicity, political corruption, and witchcraft. They can be seen all over English-speaking Africa.

Music in West Africa Nigeria is the musical center of West Africa, with a well-developed and cosmopolitan recording industry. Modern

FIGURE 6.30 Festival in the Desert The Tuareg band Igbayen plays at the Mali Festival in the Desert. This annual winter festival in the oasis town of Essakane, Mali, draws thousands of Malian musicians, Tuareg nomads and western tourists.

Nigerian styles such as juju, highlife, and Afro-beat are influenced by jazz, rock, reggae, and gospel, and they are driven by an easily recognized African sound.

Farther up the Niger River lies the country of Mali. Bamako, the capital, is a music center that has produced scores of recording artists. Many Malian musicians descend from a caste of musical storytellers performing on either the traditional *kora* (a cross between a harp and a lute) or the guitar. The musical style is strikingly similar to that of blues from the Mississippi Delta, so much so that Ali Farka Touré, who was one of Africa's most renowned musicians, is still referred to as the Bluesman of Africa. Each January, not far from Timbuktu, music fans from West Africa and Europe gather for the Festival in the Desert. In this remote Saharan locale, a celebration of Malian music and Touareg nomadic culture draws together tourists from Europe, African musicians, and nomads (Figure 6.30). Even with the recent conflict in Mali (which we will discuss later), the festival was held in January 2012. However, since then, fighting intensified, and the military took control of the government in Bamako in March 2012, so the future of this cultural event—as well as the country overall—is uncertain.

Contemporary African music can be both commercially and politically important. Nigerian singer Fela Kuti (1938–1997) was an influential musician and a voice of political conscience for Nigerians struggling for true democracy. From an elite family and educated in England, Kuti borrowed from jazz, traditional, and popular music to produce the Afro-beat sound of the 1970s. The music was irresistible, but his searing lyrics also attracted attention. Acutely critical of the military government, he sang of police harassment, the inequities of the international economic order, and even Lagos's infamous traffic. Singing in English and Yoruba, his message was transmitted to a larger audience, but he also became a target of state harassment. Kuti died in Lagos in 1997 from complications related to AIDS; yet his music and politics later became the subject of the award-winning Broadway musical *Fela!*

Pride in East African Runners Ethiopia and Kenya have produced many of the world's greatest distance runners. Abebe Bikila won

Ethiopia's and Africa's first Olympic gold medal, running barefoot at the Rome games in 1960. Since then, nearly every Olympic games has yielded medals for Ethiopia and Kenya. At the London Olympics in 2012, Kenyan runners won 11 medals and Ethiopians 7. These states, along with South Africa, were the top medal winners for Sub-Saharan Africa in London.

Running is a national pastime in Kenya and Ethiopia, where elevation—Addis Ababa sits at 7,300 feet (2,200 meters) and Nairobi at 5,300 feet (1,600 meters)—increases oxygen-carrying capacity. Past medalists Haile Gebrselassie and Derartu Tulu are national celebrities in Ethiopia who are idealized by the country's youth. Tulu, the first black African woman to win a gold medal in distance running, is a forceful voice for women's rights in a country where women are discouraged from putting on running shorts. In the 2012 Olympics, Ethiopian women Tiki Gelana and Tirunesh Dibaba won gold in the women's marathon and 10,000-meter race, respectively (Figure 6.31).

FIGURE 6.31 East African Distance Runners Runners from Ethiopia and Kenya lead the pack in the 10,000-meter race during the 2012 London Olympics. Ethiopian Tirunesh Dibaba (third from the front) took gold, and Kenyans Sally Jepkosgei Kipyego and Vivian Jepkemoi Cheruiyot took silver and bronze.

REVIEW QUESTIONS

1. What are the dominant religions of Sub-Saharan Africa, and how have they diffused throughout the region?
2. What are the ways in which African peoples have influenced world regions beyond Africa?

GEOPOLITICAL FRAMEWORK: LEGACIES OF COLONIALISM AND CONFLICT

The duration of human settlement in Sub-Saharan Africa is unmatched by any other region. Evidence shows that humankind originated there, and many diverse ethnic groups have formed in the region over the past few thousand years. Although conflicts among these groups have occurred, cooperation and coexistence among different peoples have also continued over centuries.

Some 2,000 years ago the Kingdom at Axum arose in northern Ethiopia and Eritrea, strongly influenced by political models derived from Egypt and Arabia. The first wholly indigenous African states were founded in the Sahel around 700 CE. Over the next several centuries, a variety of other states emerged in West Africa. By the 1600s, the states located near the Gulf of Guinea took advantage of the opportunities presented by the slave trade—namely, the selling of slaves to Europeans (Figure 6.32).

Thus, prior to European colonization, Sub-Saharan Africa presented a complex mosaic of kingdoms, states, and tribal societies. With the arrival of Europeans, patterns of social organization and ethnic relations were changed forever. As Europeans rushed to carve up the continent to serve their imperial ambitions, they set up various administrations that heightened ethnic tensions and promoted hostility. Many of the region's modern conflicts can trace their roots back to the colonial era, especially the drawing of political boundaries.

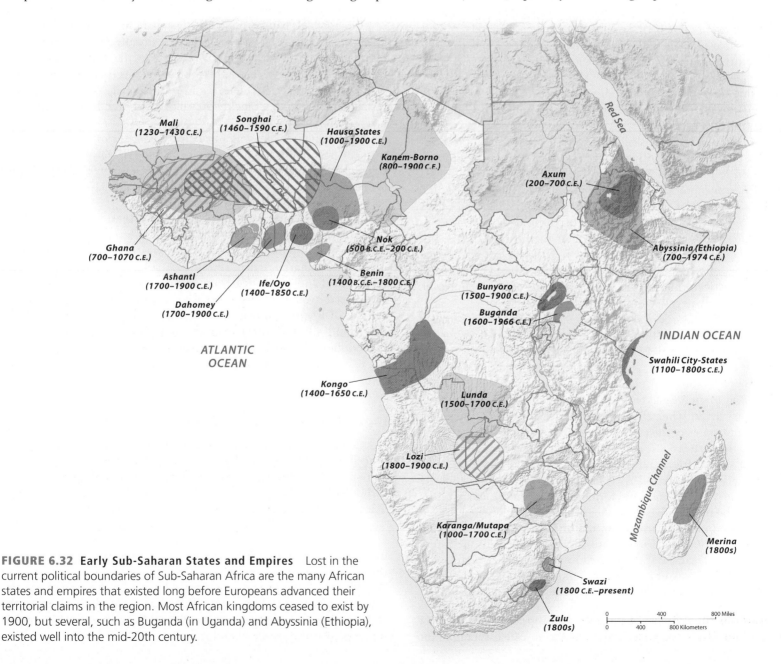

FIGURE 6.32 Early Sub-Saharan States and Empires   Lost in the current political boundaries of Sub-Saharan Africa are the many African states and empires that existed long before Europeans advanced their territorial claims in the region. Most African kingdoms ceased to exist by 1900, but several, such as Buganda (in Uganda) and Abyssinia (Ethiopia), existed well into the mid-20th century.

European Colonization

Unlike the relatively rapid colonization of the Americas, Europeans needed centuries to gain effective control of Sub-Saharan Africa. Portuguese traders arrived along the coast of West Africa in the 1400s, and by the 1500s they were established in East Africa as well. Initially, the Portuguese made large profits, converted a few local rulers to Christianity, established several defensive trading posts, and gained control over the Swahili trading cities of the east. They stretched themselves too thin, however, and failed in many of their colonizing activities. Only along the coasts of modern Angola and Mozambique, where a sizable population of mixed African and Portuguese peoples emerged, did Portugal maintain power. Along the Swahili, or eastern coast, the Portuguese were eventually expelled by Arabs from Oman, who then established their own trade-based empire in the area.

The Disease Factor One of the main reasons for the Portuguese failure was the disease environment of Africa. With no resistance to malaria and other tropical diseases, roughly half of all Europeans who remained on the African mainland died within a year. Protected both by their armies and by the diseases of their native lands, African states were able to maintain an upper hand over European traders and adventurers well into the 1800s. Unlike in the Americas, where European conquest was made easy by the introduction of Old World diseases that devastated native populations (see Chapters 4 and 5), in Sub-Saharan Africa, endemic disease limited European settlement until the mid-19th century.

The hazards of malaria, sleeping sickness, and other tropical diseases were compensated for by the lure of profit, and other European traders soon followed the Portuguese. By the 1600s, Dutch, British, and French firms dominated the profitable export of slaves, gold, and ivory from the Gulf of Guinea. The Dutch also established a settler colony in South Africa, safely outside the tropical disease zone, to supply their ships bound for Indonesia. For the next 200 years, European traders came and went, occasionally building fortified coastal posts, but they almost never ventured inland and seldom had any real influence on African rulers. By exporting millions of slaves, however, they had a strongly negative impact on African society.

In the 1850s, European doctors discovered that a daily dose of quinine would offer protection against malaria, radically changing the balance of power in Africa. Explorers immediately began to penetrate the interior of the continent, and merchants and expeditionary forces began to move inland from the coast. The first imperial claims soon followed. The French quickly grabbed power along the easily navigated Senegal River, while the British established protectorates over the indigenous states of the Gold Coast (modern-day Ghana).

Also in the early 1800s, two small territories were established in West Africa so that freed and runaway slaves would have a place to return to in Africa (Figure 6.33). The American Colonization Society set up a territory in 1822 to settle former African-American slaves; by 1847, it was the independent free state of Liberia. Sierra Leone served a similar function for ex-slaves from the British Caribbean, but it remained a protectorate of Britain until the 1960s. Despite the intentions behind the creation of these territories, they were colonies. Liberia, in particular, was imposed on existing indigenous groups who viewed their new "African" leaders with contempt.

The Scramble for Africa In the 1880s, European colonization of the region quickly accelerated, leading to the so-called scramble for Africa. By this time, due to the invention of the machine gun, no African state could long resist European force.

As the colonization of Africa intensified, tensions among the colonizing forces of Britain, France, Belgium, Germany, Italy, Portugal, and Spain mounted. Rather than risk war, 13 countries convened in Berlin, at the invitation of German Chancellor Bismarck in 1884, at a gathering known as the **Berlin Conference**. During the conference, which no African leaders attended, rules were established about what determined "effective control" of a territory, and Sub-Saharan Africa was carved into pieces that were traded like properties in a game of Monopoly (refer to Figure 6.33).

Although European weapons in the 1880s were far superior to anything found in Africa, several indigenous states organized effective resistance campaigns. For example, in South Africa, Zulu warriors resisted British invasion into their lands in what has been termed the Anglo-Zulu Wars (1879–1896). Eventually, European forces prevailed everywhere except Ethiopia. The Italians had conquered the Red Sea coast and the far northern highlands (modern Eritrea) by 1890, and they quickly set their sights on the large Ethiopian kingdom, called Abyssinia, which had been vigorously expanding for several decades. In 1896, however, Abyssinia defeated the invading Italian army, earning the respect of the European powers. In the 1930s, fascist Italy launched a major invasion of the country, by this time renamed Ethiopia, to redeem its earlier defeat, and with the help of poison gas and aerial bombardment, it quickly prevailed. However, by 1942, Ethiopia had regained its freedom.

Although Germany was a principal instigator of the scramble for Africa, it lost its own colonies after suffering defeat in World War I. Britain and France then partitioned most of Germany's African empire between themselves. Figure 6.33 shows the colonial status of the region in 1913, prior to Germany's territorial loss.

While the Europeans were cementing their rule over Africa, South Africa was inching toward political independence, at least for its white population. South Africa was one of the oldest colonies in Sub-Saharan Africa, and in 1910 it became the first to obtain its political independence from Europe. However, because of the formalized system of discrimination and racism, it was hardly a symbol of liberty. Ironically, as the Afrikaners tightened their political and social control over the nonwhite population through their policy of apartheid (or "separateness"), introduced in 1948, the rest of the continent was preparing for political independence from Europe.

Decolonization and Independence

Decolonization of Sub-Saharan Africa happened rather quickly and peacefully, beginning in 1957. Independence movements, however, had sprung up throughout the continent, some dating back to the early 1900s. Workers' unions and independent newspapers became voices for African discontent and the hope for freedom.

By the late 1950s, political demands from within Sub-Saharan Africa and changing attitudes within Europe made it clear to the leaders in Britain, France, and Belgium that they could no longer maintain their African empires. (Italy had already lost its colonies during World War II, and Britain gained Somalia and Eritrea.) Once started, the decolonization process moved rapidly. By the mid-1960s, virtually the entire region had achieved independence. In most cases, the transition was relatively peaceful and smooth.

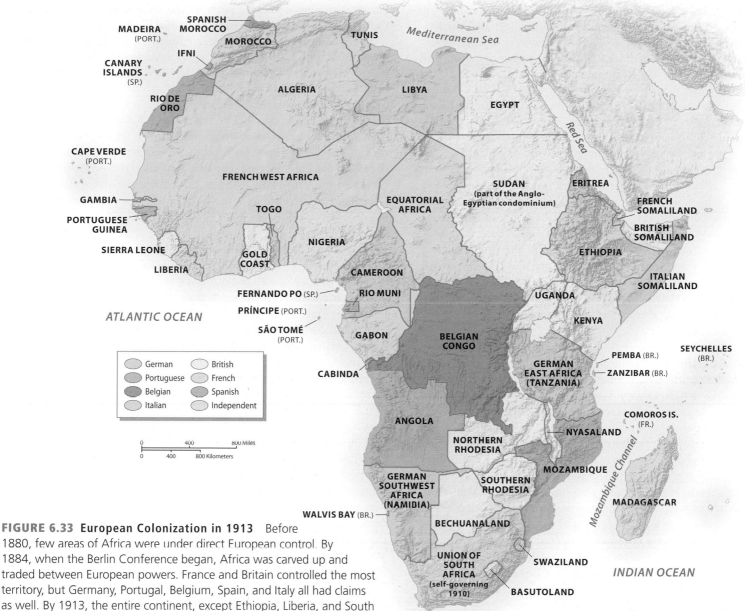

FIGURE 6.33 European Colonization in 1913   Before 1880, few areas of Africa were under direct European control. By 1884, when the Berlin Conference began, Africa was carved up and traded between European powers. France and Britain controlled the most territory, but Germany, Portugal, Belgium, Spain, and Italy all had claims as well. By 1913, the entire continent, except Ethiopia, Liberia, and South Africa, was under European colonial control.

Dynamic African leaders put their mark on the region during the early decades after independence. Men such as Kenya's Jomo Kenyatta, the Ivory Coast's Felix Houphouët-Boigny, and Ghana's Kwame Nkrumah became powerful father figures who molded their new nations. President Nkrumah's vision for Africa was the most expansive. After he helped to secure independence for Ghana in 1957, his ultimate aspiration was political unity of Africa. Although his dream was never realized, it set the stage for the founding of the Organization of African Unity (OAU) in 1963, which was renamed the **African Union (AU)** in 2002. The AU is a continent-wide organization whose main role has been to mediate disputes between neighbors. Certainly in the 1970s and 1980s, the AU was a constant voice of opposition to South Africa's minority rule, and it intervened in some of the most violent independence movements in southern Africa.

Southern Africa's Independence Battles   Independence did not come easily to southern Africa. In Southern Rhodesia (modern-day Zimbabwe), the problem was the presence of some 250,000 white residents, most of whom owned large farms. Unwilling to see power pass to the country's black majority, then some 6 million strong, these settlers declared themselves the rulers of an independent, white-supremacist state in 1965. The black population continued to resist, however, and in 1978 the Rhodesian government was forced to give up power. The renamed country of Zimbabwe was henceforth ruled by the black majority, although the remaining whites still form an economically privileged community. Since the mid-1990s, disputes over government land reform (splitting up the large commercial farms mostly owned by whites and giving the land to black farmers) and President Robert Mugabe's strongman politics have resulted in serious racial and political tensions, as well as the collapse of the country's economy.

In the former Portuguese colonies, independence came violently. Unlike the other imperial powers, Portugal refused to hand over its colonies in the 1960s. As a result, the people of Angola and Mozambique turned to armed resistance. The most powerful rebel movements adopted a socialist orientation and received support from the Soviet Union and Cuba. A new Portuguese government came to power in 1974, however, and it withdrew suddenly from its African colonies. At this point, Marxist regimes quickly came to power in both Angola and Mozambique. The United States, and especially South Africa, responded to this perceived threat by supplying arms to rebel groups that opposed the new governments. Fighting dragged on for three decades in Angola and Mozambique. The countryside in both states became so heavily laden with land mines that it was hard to use. With the end of the Cold War, however, outsiders lost their interest in continuing these conflicts, and sustained efforts began to negotiate a peace settlement. Mozambique has been at peace since the mid-1990s. After several failed attempts at peace in Angola, the Angolan army signed a peace treaty with rebels in 2002 that ended a 27-year conflict in which more than 300,000 Angolans died and 3 million more were displaced. With peace, Angola's impressive oil reserves have come online, and the economy has averaged annual growth rates of 13 percent over the past decade.

The End of Apartheid   While fighting continued in the former Portuguese zone, South Africa underwent a remarkable transformation. Through the 1980s, its government had remained firmly committed to white supremacy. Under apartheid, only whites enjoyed real political freedom, whereas blacks were denied even citizenship in their own country—technically, they were citizens of homelands.

The first major change came in 1990, when South Africa withdrew from Namibia, which it had controlled as a protectorate (a dependent political unit under the authority of another state) since the end of World War I. South Africa now stood alone as the single white-dominated state in Africa. A few years later, the leaders of the Afrikaner-dominated political party decided they could no longer resist the pressure for change. In 1994, free elections were held in which Nelson Mandela, a black leader who had been imprisoned for 27 years by the old regime, emerged as the new president. Black and white leaders pledged to put the past behind them and work together to build a new, multiracial South Africa. Since then, orderly elections have been held, and South Africans elected Thabo Mbeki for two terms (1999–2009) and Jacob Zuma in 2009.

Unfortunately, the legacy of apartheid is not so easily erased. Residential segregation is officially illegal, but neighborhoods are still sharply divided along racial lines. Under the multiracial political system, a black middle class emerged, but most blacks remain extremely poor (and most whites remain prosperous). Violent crime has increased, and rural migrants and immigrants have poured into South African cities, producing a xenophobic anti-immigrant backlash. Because the political change was not matched by significant economic transformation, the hopes of many people have been frustrated.

Enduring Political Conflict

Although most Sub-Saharan countries made a relatively peaceful transition to independence, virtually all of them immediately faced a difficult set of institutional and political problems. In several cases, the old authorities had done almost nothing to prepare their colonies for independence. Lacking an institutional framework for independent government, countries such as the Democratic Republic of the Congo faced a chaotic situation from the beginning. Only a handful of Congolese had received higher education, let alone been trained for administrative posts. The indigenous African political framework had been essentially destroyed by colonization, and in most cases very little had been built in its place.

Even more problematic in the long run was the political geography of the newly independent states. Civil servants could always be trained and administrative systems built, but little could be done to rework the region's basic political map. The problem was the fact that the European colonial powers had essentially ignored indigenous cultural and political boundaries, both in dividing Africa among themselves and in creating administrative subdivisions within their own imperial territories.

The Tyranny of the Map   All over Africa, different ethnic groups found themselves forced into the same state with peoples of distinct linguistic and religious backgrounds, many of whom had recently been their enemies. At the same time, several larger ethnic groups of the region found their territories split between two or more countries. The Hausa people of West Africa, for example, were divided between Niger (formerly French) and Nigeria (formerly British), each of which they had to share with several former ethnic rivals.

Given the imposed political boundaries, it is no wonder that many African countries struggled to generate a common sense of national identity or establish stable political institutions. **Tribalism**, or loyalty to the ethnic group rather than to the state, has emerged as the bane of African political life. Especially in rural areas, tribal identities are usually more important than national ones. Because nearly all of Africa's countries inherited an inappropriate set of colonial borders, some observers thought they would have been better off drawing a new political map based on indigenous identities. However, such a strategy was impossible, as all the leaders of the newly independent states realized. Any new territorial divisions would have created winners and losers and thus would have resulted in even more conflict. Moreover, because ethnicity in Sub-Saharan Africa was traditionally fluid and because many ethnic groups were intermixed, it would have been difficult to generate a clear-cut system of division. Finally, most African ethnic groups were considered too small to form viable countries. With such complications in mind, the new African leaders, meeting in 1963 to form the Organization of African Unity, agreed that colonial boundaries should remain. The violation of this principle, they argued, would lead to pointless wars between states and endless civil struggles within them.

Despite the determination of Africa's leaders to build their new nations within existing political boundaries, challenges to the states began soon after independence. Figure 6.34 maps the ethnic and political conflicts that have disabled parts of Africa since 1995. The human cost of this turmoil is several million refugees and internally displaced persons. **Refugees** are people who flee their state because of a well-founded fear of persecution based on race, ethnicity, religion, or political orientation. According to the United Nations, nearly 3 million Sub-Saharan Africans were considered refugees at the end of 2011, with Somalia and the Democratic Republic of the Congo accounting for half of the total number of refugees. Added to this figure are another 4.5 million **internally displaced persons (IDPs)**. IDPs have fled from conflict, but still reside in their country of origin. The

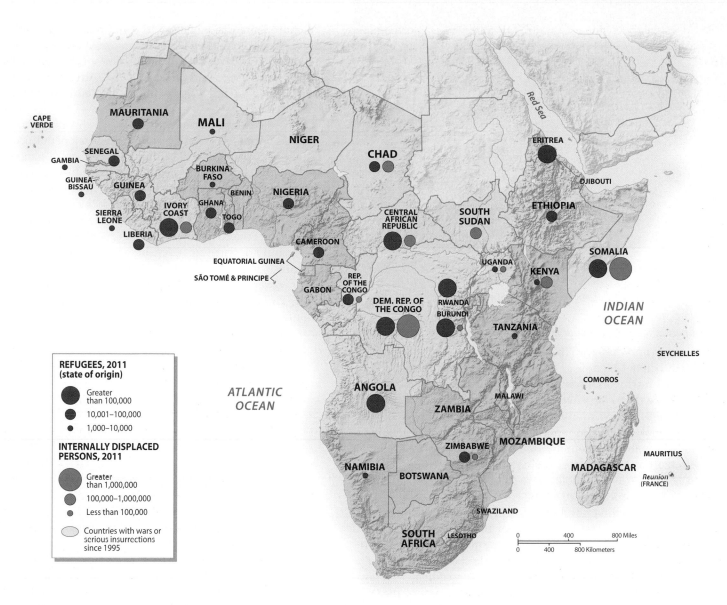

FIGURE 6.34 Geopolitical Issues in Sub-Saharan Africa   Many Sub-Saharan countries have experienced wars or serious insurrections since 1995. These same states are also likely to produce refugees (pink circles) and internally displaced persons (purple circles). As of 2011, 3 million Africans were refugees and 4.5 million were internally displaced.

Democratic Republic of the Congo has the largest number of IDPs (1.7 million), followed by Somalia (1.4 million). These populations are not technically considered refugees, making it difficult for humanitarian NGOs and the United Nations to assist them.

Ethnic Conflicts   As Figure 6.34 suggests, more than half the states in the region have experienced wars or serious insurrections since 1995. Fortunately, in the past few years, peace has returned in Sierra Leone, Liberia, and Angola, states that produced large numbers of refugees in the 1990s. In the Ivory Coast, where conflict began as violence spilled over from Liberia in 2002, a peace deal was brokered in 2007 between the New Forces rebel group in the north and the government-controlled south.

The deadliest ethnic and political conflict in the region has been in the Democratic Republic of the Congo. It is estimated that between 1998 and 2010, 5.4 million people died there, although many of the

deaths were from war-induced starvation and disease, rather than bullets or machetes. In addition, half a million refugees are living outside the country and four times as many IDPs within it (Figure 6.35). In 1996 and 1997, a loose alliance of armed groups from Rwanda (led by Tutsis) and Uganda joined forces with other militias in the Congo and marched their way across the country, installing Laurent Kabila as president. Under Kabila's rocky and ruthless leadership, which ended in assassination in 2001, rebel groups again invaded from Uganda and Rwanda and soon controlled the northern and eastern portions of the country, while the Kinshasa-based government loosely controlled the western and southern portions.

With Kabila's death, his son Joseph took power and signed a peace accord with the rebels in 2002. In 2003, rebel leaders were made part of a transitional government, and an unsteady peace was in place, with help from the UN, the AU, and Western donors. Remarkably, when elections were held in 2006, Joseph Kabila was elected president,

FIGURE 6.35 Internally Displaced Persons A camp of IDPs in the Nord Kivu District of the Democratic Republic of the Congo. Nearly two decades of conflict has resulted in 1.7 million people displaced form their homes.

and he was reelected in 2011. Yet Sub-Saharan Africa's largest state in terms of territory has only limited experience with democracy, its civil service barely functions, corruption is rampant, and there are few roads and little working infrastructure for the nearly 70 million people who live there. Moreover, armed groups scattered throughout the country continue to commit serious crimes, including mass rapes, torture, and murders. Due to years of conflict, the formal economy is small, and the informal economy dominates. Consequently, foreign aid accounted for 28 percent of the country's gross national income (GNI) in 2010. The level of violence is certainly lower now, but this is still one of the region's troubled spots.

A low-intensity ethnic conflict in northern Mali heated up in 2012 as a Touareg-based National Movement for the Liberation of Azawad (MNLA) proclaimed independence from the Bamako-based government in the south, largely controlled by Mande-speaking groups. The MNLA was formed in 2011, partly by armed Touareg fighters returning from Libya after the fall of Gaddafi's regime as part of the Arab Spring (see Chapter 7). As fighting intensified, a military coup occurred in the capital of Bamako, and President Toure was removed from office in March 2012. Leaders of the coup were dissatisfied with the state's inability to fight Touareg rebels emboldened by the arrival of weapons from Libya. The AU has denounced the actions and suspended Mali from the union, but the United Nations estimates that some 200,000 people have already been uprooted by the violence. The idea of Touareg independence, or autonomy, has existed for decades. Although this latest conflict seems to be driven by political events in North Africa and ethnic rivalries, others point to ecological pressures brought on by drought and/or climate change that undermine traditional patterns of resource sharing between farmers and pastoralists. There is no definitive connection between resource scarcity and ethnic clashes, but several of the region's current conflicts—from Nigeria to Mali, South Sudan to Somalia—are found in semiarid zones becoming more vulnerable to drought and famine.

Secessionist Movements Problematic African political boundaries have occasionally led to attempts by territories to secede and form new states. The Shaba (or Katanga) Province tried to leave what was then the state of Zaire soon after independence. The rebellion was crushed a couple of years after it started, with the help of France and

Belgium. Zaire was unwilling to give up its copper-rich territory, and the former colonialists had economic interests in the territory worth defending. Similarly, the Igbo in oil-rich southeastern Nigeria declared an independent state of Biafra in 1967. After a short, but brutal war, during which Biafra was essentially starved into submission, Nigeria was reunited.

In 1991, the government of Somalia disintegrated, and the territory has been in civil war ever since. The lack of political control facilitated the rise of piracy; Somali pirates raid vessels and extort ransom from ships from the Gulf of Aden to the Indian Ocean. For over two decades, the territory has been ruled by clan-based warlords and their militias, who have informally divided the country into clan-based units. **Clans** are social units that are branches of a tribe or an ethnic group larger than a family. Early in the conflict, the northern portion of the country declared its independence as a new country—Somaliland (Figure 6.36). Somaliland has a constitution, a functioning parliament, government ministries, a police force, a judiciary, and a president. The territory produces its own currency and passports. Yet no country has recognized this territory, in part because no government

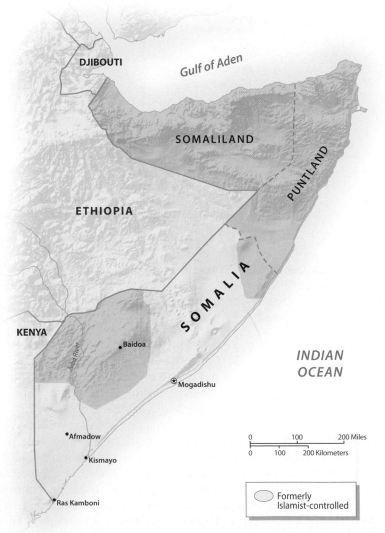

FIGURE 6.36 Somalia Divided A state in name only, Somalia has had no functioning central government since 1991. Two areas in the north—Somaliland and Puntland—seek political autonomy or independence. Meanwhile, UN Peacekeeping Forces and the Kenya military have fought in the south to rid the area of Islamic insurgents.

exists in Somalia to negotiate the secession. In 1998, neighboring Puntland also declared itself an autonomous state. Although it does not seek outright independence like Somaliland, Puntland is creating its own administration. Meanwhile, Islamic insurgents with their well-armed militias control the south around Mogadishu. In the past two years, UN Peacekeeping Forces and the Kenyan military have been fighting the insurgents, trying to reclaim the cities. The need for stability has been exacerbated by three years of drought, creating a humanitarian emergency. In 2011, the UN estimated that there were nearly 2 million IDPs in the territory of Somali, which is equal to 20 percent of the population.

Only two territories in the region have successfully seceded. In 1993, Eritrea gained independence from Ethiopia after two decades of civil conflict. This territorial secession is striking because Ethiopia gave up its access to the Red Sea, making it landlocked. Yet the creation of Eritrea still did not bring about peace. After years of fighting, the transition to Eritrean independence began remarkably well. Unfortunately, border disputes between the two countries erupted in 1998, resulting in the deaths of some 100,000 troops. In 2000, a peace accord was reached, and the fighting stopped. The second example is South Sudan, which gained its independence from Sudan in 2011 after some three decades of violent conflict between the largely Arab and Muslim north and the Christian and animist south. It is too early to measure South Sudan's success as an independent state, but its formation, along with Eritrea, may provide a model for territorial solutions in war-torn areas in the region. Still, major transformations of Africa's political map should not be expected.

REVIEW QUESTIONS

1. What are the processes behind Sub-Saharan Africa's political map, and why have there been relatively few boundary changes since the 1960s?

2. What are the present major conflicts in this region, and where are they occurring?

ECONOMIC AND SOCIAL DEVELOPMENT: THE STRUGGLE TO REBUILD

By almost any measure, Sub-Saharan Africa is the poorest world region. According to World Bank estimates, 69 percent of the population lives on less than $2 per day, although in 1993 the figure was 78 percent. Due to poverty and low life expectancy, nearly all the states in the region are ranked at the bottom of the Human Development Index. Although some demographically small or resource-rich states, such as Botswana, Equatorial Guinea, Mauritius, the Seychelles, and South Africa, have much higher per capita gross national income, adjusted for purchasing power parity (GNI-PPP), the regional average was about $2,100 in 2010 (Table 6.2). By way of comparison, the figure for South Asia, the next poorest region, was $3,100.

The decade of the 1990s was especially difficult for Sub-Saharan Africa, with most states registering low or negative growth rates. Not only was the AIDS crisis raging, but also an economic and debt crisis

of the 1980s and 1990s prompted the introduction of **structural adjustment programs**. Promoted by the International Monetary Fund (IMF) and the World Bank, structural adjustment programs typically reduce government spending, cut food subsidies, and encourage private-sector initiatives. Yet these same policies have caused immediate hardships for the poor, especially women and children, and have led to social protest, most notably in cities. The idea of debt forgiveness for Africa's poorest states has gradually gained acceptance as a strategy for reducing human suffering by redirecting monies that would have gone to debt repayment to the provision of services and the building of infrastructure. American economist Jeffrey Sachs argues that in order for the region to get out of the poverty trap, it will need substantial sums of new foreign aid and investment.

On the positive side, since 2000 there have been signs of economic growth. For most states, the average annual growth rates from 2000 to 2010 were positive. Several countries have seen average annual growth rates of 5 percent or more; Angola's annual growth rate averaged a stunning 13 percent for the decade (see Table 6.2). Although this is good news overall, such figures can be deceptive. Some are due to soaring oil prices (notably so for Angola, Chad, and Nigeria), whereas others are due to countries beginning from a very low base after years of conflict (Mozambique, Sierra Leone, and Rwanda). Still, for the first time in many years, Sub-Saharan African economies are growing at a faster rate than their populations. This is very good news.

Roots of African Poverty

In the past, observers often attributed Africa's poverty to its colonial history, poorly conceived development policies, and/or corrupt governance. For those who favored environmental explanations for poverty, the region's infertile soils, erratic patterns of rainfall, lack of navigable rivers, and virulent tropical diseases were all pointed to as reasons for underdevelopment. Most contemporary scholars, however, argue that such environmental handicaps are not prevalent everywhere throughout the region and that even where they do exist, they can be—and often have been—overcome by human labor and creativity. The favored explanations for African poverty now look much more to historical and institutional factors than to environmental circumstances.

Numerous scholars have singled out the slave trade for its debilitating effect on Sub-Saharan African economic life. Large areas of the region were depopulated, and many people were forced to flee into poor, inaccessible refuges. Colonization was another blow to Africa's economy. European powers invested little in infrastructure, education, and public health and were instead interested mainly in developing mineral and agricultural resources for their own benefit. Several plantation and mining zones did achieve some prosperity under colonial regimes, but strong national economies failed to develop. In almost all cases, the basic transport and communications systems were designed to link administration centers and zones of extraction directly to the colonial powers, rather than to their own surrounding areas. As a result, after achieving independence, Sub-Saharan African countries faced economic and infrastructural challenges that were as daunting as their political problems. Even today, only 18 percent of the region's roads are paved, which limits trade and transport. South Africa is the only African state with a fully developed modern road network. Recently, Kenya inaugurated its first superhighway, an

TABLE 6.2 DEVELOPMENT INDICATORS

Country	GNI per capita, PPP 2010	GDP Average Annual % Growth 2000–10	Human Development Index (2011)[1]	Percent Population Living Below $2 a Day	Life Expectancy (2012)[2]	Under Age 5 Mortality Rate (1990)	Under Age 5 Mortality Rate (2010)	Adult Literacy (% ages 15 and older)	Gender Inequality Index (2011)[3,1]
Angola	5,410	12.9	.486	70.2	54	243	161	70	—
Benin	1,590	4.0	.427	75.3	56	178	115	42	0.634
Botswana	13,700	4.1	.633	49.4	51	59	48	84	0.507
Burkina Faso	1,250	5.5	.331	72.6	55	205	176	29	0.596
Burundi	400	3.2	.316	93.5	58	183	142	67	0.478
Cameroon	2,270	3.2	.482	30.4	51	137	136	71	0.639
Cape Verde	3,820	—	.568	40.9	73	—	—	85	—
Central African Republic	790	1.0	.343	80.1	48	165	159	55	0.669
Chad	1,220	9.0	.328	83.3	49	207	173	34	0.735
Comoros	1,090	—	.433	65.0	61	—	—	74	—
Congo	3,220	4.3	.533	74.4	48	116	93	—	0.628
Dem. Rep. of Congo	320	5.3	.286	95.2	48	181	170	67	0.710
Djibouti	2,460	—	.430	41.2	—	—	—	—	—
Equatorial Guinea	23,760	—	.537	—	51	—	—	93	—
Eritrea	540	0.2	.349	—	61	141	61	67	—
Ethiopia	1,040	8.8	.363	77.6	59	184	106	30	—
Gabon	13,180	2.2	.674	19.6	62	93	74	88	0.509
Gambia	1,300	3.7	.420	55.9	58	165	98	46	0.610
Ghana	1,660	5.9	.541	51.8	64	122	74	67	0.598
Guinea	1,020	2.9	.344	69.6	54	229	130	39	—
Guinea-Bissau	1,180	1.5	.353	78.0	48	210	150	52	—
Ivory Coast	1,810	1.1	.400	46.3	55	151	123	55	0.655
Kenya	1,680	4.3	.509	67.2	62	99	85	87	0.627
Lesotho	1,960	3.5	.450	62.3	48	89	85	90	0.532
Liberia	340	0.9	.329	94.9	56	227	103	59	0.671
Madagascar	960	3.4	.480	92.6	66	159	62	64	—
Malawi	850	5.2	.400	90.5	53	222	92	74	0.594
Mali	1,030	5.2	.359	78.7	51	255	178	26	0.712
Mauritania	1,910	4.4	.453	47.7	58	124	111	57	0.605
Mauritius	13,960	3.9	.728	—	73	24	15	88	0.353
Mozambique	930	7.8	.322	81.8	52	219	135	55	0.602
Namibia	6,420	5.0	.625	51.1	62	73	40	89	0.466
Niger	720	4.2	.459	75.2	58	311	143	29	0.724
Nigeria	2,170	6.7	.459	84.5	51	213	143	61	—
Reunion	—	—	—	—	78	—	—	—	—
Rwanda	1,150	7.6	.429	82.4	54	163	91	71	0.453
São Tomé and Principe	1,930	—	.509	54.2	63	—	—	89	—
Senegal	1,910	4.2	.459	60.4	58	139	75	50	0.566
Seychelles	21,090	—	.773	<2	73	—	—	92	—
Sierra Leone	830	8.8	.336	76.1	47	276	174	41	0.662
Somalia	—	—	—	—	50	180	180	—	—
South Africa	10,360	3.9	.619	31.3	54	60	57	89	0.490
South Sudan	—	—	—	—	52	—	—	—	—
Swaziland	5,430	2.4	.522	60.4	48	96	78	87	0.546
Tanzania	1,430	7.1	.466	87.9	57	155	92	73	0.590
Togo	890	2.7	.435	69.3	62	147	103	57	0.602
Uganda	1,250	7.7	.446	64.7	53	175	99	73	0.577
Zambia	1,380	5.6	.430	82.6	48	183	111	71	0.627
Zimbabwe	—	−6.0	.376	—	48	78	80	92	0.583

[1]United Nations, *Human Development Report, 2011.*
[2]Population Reference Bureau, *World Population Data Sheet, 2012.*
[3]Gender Inequality Index—A composite measure reflecting inequality in achievements between women and men in three dimensions: reproductive health, empowerment, and the labor market that ranges between 0 and 1. The higher the number, the greater the inequality.
Source: World Bank, *World Development Indicators, 2012.*

FIGURE 6.37 Kenyan Highway   The country's first superhighway was inaugurated in 2012. Built from Nairobi to the northern town of Thika, the road will ease congestion on this heavily traveled route.

eight-lane 42-kilometer road from Nairobi to Thika (Figure 6.37). Not only has the highway transformed many of the towns along its route, but also it is an expression of growing Chinese investment in this region, since the Chinese firm Wu Yi Co. did much of the engineering and construction work.

Failed Development Policies   The first decade or so of independence was a time of relative prosperity and optimism for many African countries. Most of them relied heavily on the export of mineral and agricultural products, and through the 1970s commodity prices generally remained high. Some foreign capital was attracted to the region, and in many cases the European economic presence actually increased after decolonization.

In the 1980s, as most commodity prices began to decline, foreign debt began to weigh down many Sub-Saharan countries. By the end of the 1980s, most of the region was in serious economic decline. By the early 1990s, the region's foreign debt was around $200 billion. Although low compared to that of other developing regions (such as Latin America), as a percentage of its economic output, Sub-Saharan Africa's debt was the highest in the world.

Many economists argue that Sub-Saharan African governments enacted counterproductive economic policies and thus brought some of their misery on themselves. Eager to build their own economies and reduce their dependency on the former colonial powers, most African countries followed a course of economic nationalism. More specifically, they set about building steel mills and other forms of heavy industry that were simply not competitive (Figure 6.38). Local currencies were often maintained at artificially elevated levels, which benefited the elite who consumed imported products, but undercut exports.

The largest blunders made by Sub-Saharan leaders were in agricultural and food policies in the postindependence period. The main objective was to maintain a cheap supply of staple foods in the urban areas. Yet the majority of Africans are farmers, who could not make

money from their crops because prices were artificially low. Thus, they opted to grow food mainly for subsistence, rather than selling at a loss to national marketing boards. The end result was the failure to meet staple food needs at a time when the population was growing explosively. Many of these price subsidies have since been removed as part of structural adjustment policies.

Reflecting a neoliberal turn toward agricultural production, several Sub-Saharan states—namely, Uganda, Tanzania, Somalia, and Mozambique—have experienced a 21st-century land grab driven by food-insecure governments. Dutch geographer Annelies Zoomers describes food-insecure governments, such as China and the Gulf States, as those that seek to outsource their domestic food production by buying or leasing vast areas of farmland abroad for their own offshore food production. Ironically, the places that these countries invest in tend to be poor countries with their own food insecurity issues.

Corruption   Although prevalent throughout most of the world, corruption seems to have been particularly widespread in several African countries. Part of this is driven by a lack of transparent and representative governance and by a civil service class that lacks both resources and professionalism. According to a poll by an international business magazine, Nigeria ranks as the most corrupt country. (Skeptical observers, however, point out that several Asian nations with highly successful economies, such as China, are also noted for having high levels of corruption, so that corruption alone may not be the problem.)

With millions of dollars in loans and aid pouring into the region, officials at various levels have been tempted to take something for themselves. Some African states, such as the Democratic Republic of the Congo, were dubbed *kleptocracies*. A **kleptocracy** is a state in which corruption is so institutionalized that most politicians and

FIGURE 6.38 Industrialization   Heavy industry, such as this iron foundry in Zambia, failed to deliver Sub-Saharan Africa from poverty. In the worst cases, these industrial enterprises were unable to produce competitive products for world and domestic markets.

government bureaucrats siphon off a huge percentage of the country's wealth. President Mobutu, who ruled the Democratic Republic of the Congo (then Zaire), was a legendary kleptocrat. While his country was saddled with an enormous foreign debt, he reportedly skimmed several billion dollars from government funds and deposited them in Belgian banks during his presidency from 1965 to 1997.

Links to the World Economy

Sub-Saharan Africa's trade connection with the world is limited, accounting for 2 percent of global trade. The level of overall trade is low both within the region and outside it. Traditionally, most exports went to the European Union (EU), especially the former colonial powers of England and France. The United States is the second most common destination. That pattern is changing fast; China is now the single largest trading partner with the region, although collective trade between the EU nations and Sub-Saharan Africa is greater. Throughout the decade of the 2000s, China's trade with Sub-Saharan Africa grew on average 30 percent per year. During the same decade, India's and Brazil's levels of trade with Sub-Saharan Africa annually grew more than 20 percent. As of 2010, India's level of trade with Sub-Saharan Africa equaled that of the United States.

By most measures of connectivity, Sub-Saharan Africa has lagged behind other developing regions, but this is changing quickly due to cellular and digital technology. Admittedly, fixed telephone lines are scarce; the regional average is 1 line per 100 people. Cell phone usage, however, has soared. In 2007, the World Bank estimated 23 cell phone subscriptions per 100 people in Sub-Saharan Africa; by 2010, the figure had doubled to 45 per 100 people. Multinational providers are now competing for mobile-phone customers. Development specialists and entrepreneurs are exploring many new uses for cell phones and smartphone, with applications to secure not only micro-finance, but also educational tools and updates about health issues or weather patterns (Figure 6.39). The Internet is also facilitating the sharing of

alternative and positive narratives about Sub-Saharan Africa's future. Two new websites—See Africa Differently (based in the United Kingdom) and Africa Good News (based in South Africa)—focus on the new ways in which Sub-Saharan peoples are developing their communities and engaging with the world.

Aid Versus Investment   In many ways, Sub-Saharan Africa is linked to the global economy more through the flow of financial aid and loans than through the flow of goods. As Figure 6.40 shows, for several states this aid accounted for more than 20 percent of the GNI, and in Burundi it was 40 percent. Most of the aid came from a handful of developed countries (see the insert in Figure 6.40). The United States provided the most aid in 2005, followed by EU institutions and then individual states, such as France, the United Kingdom, and Germany. In contrast, net flows of private capital are extremely low. Countries such as South Africa, Equatorial Guinea, Gabon, and Angola are the exceptions because of their oil or mineral wealth.

Economic aid is extremely important for many African states, and foreign direct investment in the region substantially increased from only $4.5 billion in 1995 to $25 billion in 2010. Yet the overall level of foreign investment remains low when compared to that of other developing regions. In recent years, the largest recipients of foreign investment were the region's major mineral and oil producers such as South Africa, Ghana, the Democratic Republic of the Congo, Zambia, Angola, and Nigeria. China has been the leading investor in the region at a time when the United States and EU are more focused on offering aid or fighting terrorism. China wants to secure the oil and ore it needs for its massive industrial economy. In exchange, it offers Sub-Saharan nations money for roads, railways, housing, and schools, with relatively few strings attached. Some African leaders see China as a new kind of global partner, one that wants straight commercial relations without an ideological or political agenda. Angola, a country where China has invested heavily, is now one of China's top suppliers of oil. Yet not all investments have paid off. A massive real estate development outside of Luanda, Angola, called Nova Cidade de Kilamba, was largely unoccupied in 2012 because the cost of purchasing an apartment was well beyond the means of the average Angolan (Figure 6.41).

Debt Relief   Another development strategy making a difference in the region is debt relief. The World Bank and the IMF proposed in 1996 to reduce debt levels for heavily indebted poor countries, many of which are in Sub-Saharan Africa. Most Sub-Saharan states are indebted to official creditors such as the World Bank, not to commercial banks (as is the case in Latin America and Southeast Asia). Under this World Bank/IMF program, substantial debt reduction will be given to Sub-Saharan countries that are determined to have "unsustainable" debt burdens. Mauritania, for example, was spending six times more money on repaying its debts than it did on health care.

States qualify for different levels of debt relief, depending on their poverty-reduction strategies. Uganda was the first state to qualify for the program in 2000, using money it saved on debt repayment to expand primary schooling. Ghana, which qualified for debt relief in 2004, received a $3.5 billion relief package. Other countries that have benefited from debt reduction are Tanzania, Mozambique, Ethiopia, Mauritania, Mali, Niger, Nigeria, Senegal, Burkina Faso, and Benin. More African countries may also be able to redirect debt payments toward building infrastructure and improving basic health and educational services.

FIGURE 6.39 Mobile Phones for Africa   A woman on her cell phone in downtown Monrovia, Liberia. Cell phone subscriptions in Sub-Saharan Africa doubled between 2007 and 2010, greatly improving communication.

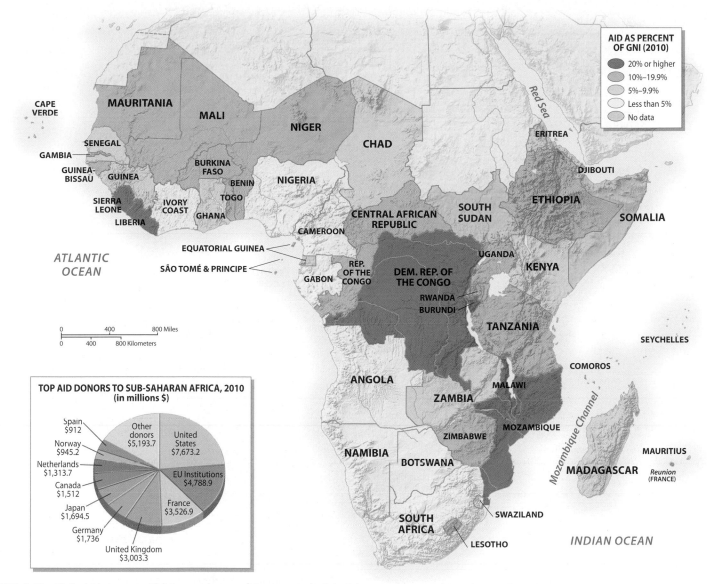

AID AS PERCENT OF GNI (2010)
- 20% or higher
- 10%–19.9%
- 5%–9.9%
- Less than 5%
- No data

TOP AID DONORS TO SUB-SAHARAN AFRICA, 2010
(in millions $)

- Spain $912
- Norway $945.2
- Netherlands $1,313.7
- Canada $1,512
- Japan $1,694.5
- Germany $1,736
- United Kingdom $3,003.3
- France $3,526.9
- EU Institutions $4,788.9
- United States $7,673.2
- Other donors $5,193.7

FIGURE 6.40 Global Linkages: Aid Dependency   Many states in Sub-Saharan Africa are dependent on foreign aid as their primary link to the global economy. This figure maps aid as a percentage of GNI, which ranges from less than 1 percent to 177 percent in Liberia.

FIGURE 6.41 Chinese Investment in Angola   Children play on the basketball courts at the Kilambi Kiaxi housing development, a massive Chinese-built project in the suburbs of Luanda, Angola. Most of these apartments remain vacant because they are far too expensive for Angolan workers to purchase.

Economic Differentiation Within Africa

As in most other regions, considerable differences in levels of economic and social development persist in Sub-Saharan Africa. In many respects, the small island nations of Mauritius and the Seychelles have little in common with the mainland. With high per capita GNI, life expectancies averaging in the low 70s, and economies built on tourism, they could more easily fit into the Caribbean were it not for their Indian Ocean location. In contrast, two-thirds of the population in mainland Sub-Saharan Africa subsists on less than $2 per day. Also, only a few states, such as Angola, Botswana, South Africa, Swaziland, Gabon, Namibia, and Equatorial Guinea, have a per capita GNI-PPP of over $5,000 (see Table 6.2).

Given the scale of the African continent, it is not surprising that groups of states have formed trade blocs to facilitate intraregional exchange and development. The two most active regional organizations are the Southern African Development Community (SADC) and the Economic Community of West African States (ECOWAS). Both were founded in the 1970s, but became more important in the 1990s (Figure 6.42).

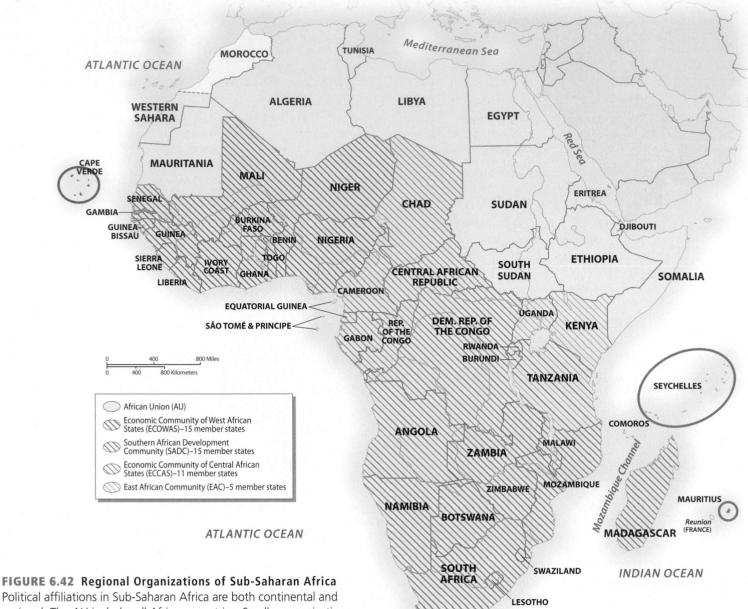

FIGURE 6.42 Regional Organizations of Sub-Saharan Africa
Political affiliations in Sub-Saharan Africa are both continental and regional. The AU includes all African countries. Smaller organizations, such as SADC, ECCAS, and ECOWAS, represent regional affiliations. Of these, SADC shows the most economic promise.

SADC and ECOWAS are anchored by the region's two largest economies: South Africa and Nigeria. Other regional trade blocs include the Economic Community of Central African States (ECCAS) and the smaller, but more effective East African Community (EAC).

South Africa

South Africa is the unchallenged economic powerhouse of Sub-Saharan Africa. The per capita GNI-PPP of Nigeria, the next largest economy, is just one-fifth that of South Africa. Only South Africa has a well-developed and well-balanced industrial economy. It also boasts a healthy agricultural sector, and, more importantly, it is one of the world's mining superpowers. South Africa remains unchallenged in gold production and is a leader in many other minerals and precious gems, including diamonds. In the summer of 2010, South Africa hosted the World Cup, the first African country to do so, symbolizing its arrival as a developed and modern nation.

South Africa is undeniably a wealthy country by African standards, and while its white minority is prosperous by any standard, it is also a country beset by severe and widespread poverty. In the townships lying on the outskirts of the major cities and in the rural districts of the former homelands, employment opportunities remain limited and living standards marginal. Despite the end of apartheid, South Africa continues to suffer from one of the most unequal distributions of income in the world.

Oil and Mineral Producers

Another group of relatively well-off Sub-Saharan countries benefits from large oil and mineral reserves and small populations. The prime example is Gabon, a country of noted oil wealth that is inhabited by 2 million people. Equatorial Guinea, a former Spanish colony, began producing significant amounts of oil in 1998. Today it has the distinction of having the region's highest per capita GNI at $23,760. Yet after over a decade of oil production, these new-found revenues have not been invested in the country's citizens; rather, as is often the case, they seem to have fallen into the pockets of a few elites.

Farther south, Namibia and Botswana also have the advantage of small populations and abundant mineral resources, especially diamonds. Both countries have also enjoyed stable governments over the past few years and have experienced solid economic growth. Angola now rivals Nigeria as one of Sub-Saharan Africa's major oil exporters. Its GNI per capita has risen to over $5,000, but 70 percent of the population still lives on less than $2 per day.

The Leaders of ECOWAS The most populous country in Africa, Nigeria is the core member of ECOWAS. Nigeria has the largest oil reserves in Sub-Saharan Africa, and it is a member of the Organization of Petroleum Exporting Countries (OPEC). Yet despite its natural resources, its per capita GNI-PPP is a low $2,170. It has been argued that oil money has helped to make Nigeria notoriously inefficient and corrupt. A small minority of its population has grown fantastically wealthy, more by manipulating the system than by engaging in productive activities. Eighty-five percent of Nigerians, however, remain trapped in poverty, earning less than $2 per day. Oil money also led to the explosive growth of the former capital of Lagos, which by the 1980s had become one of the most expensive—and least livable— cities in the region. As a result, the Nigerian government chose to build a new capital city in Abuja, located near the country's center, a move that has proved tremendously expensive. In 1991, Abuja became the national capital, and it now has 1.5 million inhabitants. The southwestern corner of the country, in the Niger Delta where most of the oil is produced, has seen few of the benefits of oil production, but bears the burden of serious environmental contamination, poverty, and related social unrest.

The second and third most populous states in ECOWAS, the Ivory Coast and Ghana, are also important West African commercial centers. These states rely upon a mix of agricultural and mineral exports. In the mid-1990s, the Ivorian economy began to take off. Boosters within the country called it an emerging "African elephant" (comparing it to the successful "economic tigers" of eastern Asia). However, a destructive civil war began in 2002, in which rebel forces controlled the northern half of the country and over half a million Ivorians were displaced. A peace agreement was signed in 2007, but the economic growth of the 1990s has yet to return. Ghana, a former British colony and an ECOWAS member, also began to see economic recovery in the 1990s. In 2001, it negotiated with the IMF and the World Bank for debt relief to reduce its nearly $6 billion foreign debt. Between 2000 and 2010, Ghana maintained an average annual growth rate of nearly 6 percent. In 2011, Ghana also became an emerging oil producer for Africa, with offshore wells being pumped near the city of Takoradi.

Life for the Region's Poorest Rose Shanzi, a mother of five, lives in the town of Maramba, Zambia, not far from the city of Livingstone, Zimbabwe, the site of Victoria Falls. Rose sells tomatoes in the local market, and if she sells enough, she and her children eat that day. Her daily goal is to earn 75 cents, which will enable her to purchase the vegetables and ground corn (called *mealie-meal*) needed to make the evening supper. She sits at her vegetable stand among the neatly piled mounds of tomatoes for 12 hours a day. Rose earns between $12 and $18 a month from her stand. Half of her earnings go to food. She pays

$2 a month for school fees, $2 a month for water, $1.50 for property tax, and 50 cents for government health insurance. Water, education, and health care were free before structural adjustment policies forced payment for these basic services. Rose used to be able to buy meat occasionally with her earnings, but not any longer because of her new expenses. There are luckless days when she returns home with no food. "You don't know what suffering is until you have watched your babies go hungry," says Rose. "I have suffered many times."

Livingstone and Maramba used to have three dozen clothing manufacturers that employed hundreds of people, including Rose's husband. The deluge of used clothing from the West killed this industry during the 1990s and plunged families like Rose's into stomach-tightening poverty. Then Rose's husband died. With no other job prospects, she joined the ranks of the informal sector.

What begins as charity in the West—giving old clothing away— blossoms into thousands of small businesses in villages and cities across Africa. It's called *mivumba* in Uganda and *salaula* in Zambia; throughout Africa, you can go to any market and find the vendors of used clothing. In a region where most people earn less than $2 per day, secondhand clothing is the norm. This is how a shirt sewn in a Honduras sweatshop and worn by a New Jersey teenager gets repackaged and sold by a Zambian street vendor. Securing sustainable livelihoods for Africa's poor majority is an ongoing challenge. Yet some of the poorest countries of the region, such as Rwanda, have experimented with innovative community development strategies in an effort to come back from the devastating civil war and genocide in 1994 (see "Working Toward Sustainability: The Rebirth of Rwanda").

Measuring Social Development

By global standards, measures of social development in Africa are extremely low. Yet there are some positive trends, especially with regard to child survival and literacy, that are cause for hope (see Table 6.2). Many governments in the region have reached out to the modern African diaspora (economic migrants and refugees who now live in Europe and North America). In other parts of the world, African immigrant organizations have worked to improve schools and health care, and former emigrants have returned to invest in businesses and real estate. The economic impact of remittances in this region is small, but growing.

Child Mortality and Life Expectancy Reductions in child mortality are often a surrogate measure for improved social development because, if most children make it to their fifth birthday, it usually indicates adequate primary care and nutrition. In a country where the child mortality rate is 200 per thousand, it means that one out of five children dies before his or her fifth birthday. As Table 6.2 shows, most of the states in the region saw modest to significant improvements in child survival between 1990 and 2010. Eritrea, Liberia, Madagascar, Malawi, and Rwanda actually experienced dramatic gains in child survival rates. However, countries with prolonged conflict (e.g., Somalia and the Democratic Republic of the Congo) have seen little improvement. In 1990, the regional child mortality rate was 175 per thousand; in 2010, it was down to 121. Thus, high child mortality is still a major concern for the region, but steady reductions in child mortality are significant for African families.

Life expectancy for Sub-Saharan Africa is only 55 years. Countries hit hard by HIV/AIDS or conflict have seen life expectancies

The Rebirth of Rwanda

In the densely settled and mostly rural country of Rwanda, most farms are small, but the mountainous volcanic soil is ideal for growing coffee. In an effort to rebuild the country after the ethnic genocide in 1994 left nearly 800,000 dead, the government targeted improving the quality of the country's coffee through the formation of cooperatives that would bring ethnic groups together and improve farmers' incomes. The cooperatives have taken off, and farmers, many of them war widows, have seen their incomes improve.

For decades, the country's coffee production languished, and farmers earned very little for their low-quality beans. Yet across Rwanda's hillsides were older varieties of coffee plants with high value in today's premium coffee market (Figure 6.2.1). Farmers just needed a better way to prepare the beans and market them. Through community-run cooperatives, the quality of their washed and sorted beans improved. So, too, did their ability to bargain with major buyers, such as Starbucks and Green Mountain,

leaving out the middleman. For many small farmers in Rwanda, their premium coffee is now a source of pride. Producing a better-quality product and marketing directly to international buyers allowed many Rwandan cooperative members to double their income.

Realizing that long-term sustainability means more than growing coffee, the U.S. Agency for International Development initiated the SPREAD program (Sustaining Partnerships to Enhance Rural Enterprise and Agribusiness Development). In the coffee zone, SPREAD works through the coffee cooperatives to improve soil fertility, pruning of the plants, and processing of the beans. It also provides local extension workers to disseminate information on maternal and child health, nutrition, gender issues, family planning, and how coffee income is spent within the household. By using community agents to nurture an integrated vision of environmental and social sustainability in rural Rwanda, a more sustainable rural economy is unfolding.

FIGURE 6.2.1 Rwandan Coffee Farmers Farmers sort coffee beans prior to processing. Efforts to improve the quality and marketing of Rwandan coffee through local cooperatives have lead to higher incomes for farmers.

tumble into the 40s. Despite these statistics, there are indications that access to basic health care is improving, and, eventually, so will life expectancies. Keep in mind that high infant- and child-mortality figures depress overall life expectancy figures; average life expectancies for people who make it to adulthood are much better.

The causes of low life expectancy are generally related to extreme poverty, environmental hazards (such as drought), and various environmental and infectious diseases (malaria, schistosomiasis, cholera, AIDS, and measles). Often these factors work in combination. Malaria, for example, kills a half million African children each year. The death rate is also affected by poverty, undernourished children being the most vulnerable to the effects of high fevers. Tragically, diseases that are preventable, such as measles, occur when people have no access to or cannot afford vaccines. National and international health agencies, along with NGOs such as the Gates Foundation, are working to improve access to vaccines, bed nets (to prevent malaria), and primary health care. These efforts are slowly making a difference.

Meeting Educational Needs Basic education is another obstacle confronting the region. The goal of universal access to primary education is a daunting one for a region in which 43 percent of the population is less than 15 years old. The UN estimates that 75 percent of African children are enrolled in primary school, but only 23 percent of the relevant population is in secondary school (high school or its equivalent). The region is home to one-sixth of the world's children under 15, but half of the world's uneducated children. Girls are still less likely than boys to attend school. In West African countries such

as Chad, Niger, and the Ivory Coast, girls are decidedly underrepresented.

A renewed focus on education since 2000 has been attributed to the **Millennium Development Goals**, a global UN effort to reduce extreme poverty by focusing on basic education, health care, and access to clean water. Although the region will not meet the goals set for 2015, more resources have been directed to schools from governments and from nonprofit organizations. More schools are being built across the region, and more children are attending them (Figure 6.43).

FIGURE 6.43 Educating African Youth Botswana students attend classes at a Lutheran theological seminary in Francistown, Botswana. Religious organizations as well as public institutions are critical in providing educational services.

Women and Development

Development gains cannot be made in Africa unless the economic contributions of African women are recognized. Officially, women are the invisible contributors to local and national economies. In agriculture, women account for 75 percent of the labor that produces more than half the food consumed in the region. Tending subsistence plots, taking in extra laundry, and selling surplus produce in local markets all contribute to household income. Yet because many of these activities are considered informal economic activities, they are not counted. For many of Africa's poorest people, however, the informal sector is the economy, and within this sector women dominate.

Status of Women   The social position of women is difficult to measure for Sub-Saharan Africa. Female traders in West Africa, for example, have considerable political and economic power. By such measures as female labor force participation, many Sub-Saharan African countries show relative gender equality. Also, women in most Sub-Saharan societies do not suffer the kinds of traditional social restrictions encountered in much of South Asia, Southwest Asia, and North Africa; in Sub-Saharan Africa, women work outside the home, conduct business, and own property. In 2006, Ellen Johnson-Sirleaf was sworn in as Liberia's president, making her Africa's first elected female leader. In 2012, she was joined by Joyce Banda, the recently elected president of Malawi. In fact, throughout the region, women occupied 18 percent of all seats in national parliaments in 2009. In the county of Rwanda, over half the parliamentary seats are filled by women (Figure 6.44).

By other measures, however, such as the prevalence of polygamy, the practice of the "bride-price," and the tendency for males to inherit

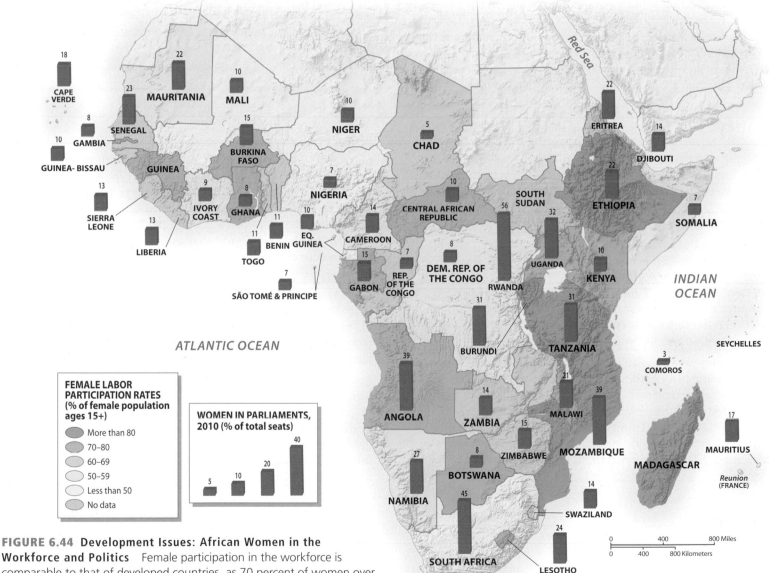

FIGURE 6.44 Development Issues: African Women in the Workforce and Politics   Female participation in the workforce is comparable to that of developed countries, as 70 percent of women over the age of 15 are in the labor force. Another significant change for the region is the increase of women holding seats in national parliaments. The regional average is 18 percent, but in South Africa 45 percent of parliamentary seats are held by women, and in Rwanda 56 percent of the seats are held by women.

property over females, African women do suffer discrimination. Perhaps the most controversial issue regarding women's status is the practice of female circumcision, or genital mutilation. In Ethiopia, Somalia, and Eritrea, as well as parts of West Africa, the majority of girls are subjected to this practice, which is extremely painful and can have serious health consequences. Yet because the practice is considered traditional, most African states are unwilling to ban it.

Regardless of their social position, most African women still live in remote villages where educational and wage-earning opportunities remain limited and caring for large families is time-consuming and demanding labor. As educational levels increase and urban society expands—and as reduced infant mortality provides greater security—we can expect fertility in the region to gradually decrease. Governments can speed up the process by providing birth control information and cheap contraceptives—and by investing more money in health and educational efforts aimed at women. As the economic importance of women receives greater attention from national and international organizations, more programs are being directed exclusively toward them.

Building from Within   Major shifts in the way development agencies view women and women view themselves have the potential to transform the region. All across the continent, support groups and networks have formed, raising women's awareness, offering women micro-credit loans for small businesses, and organizing their economic power. From farm-labor groups to women's market associations, investment in the organization of women has paid off. In Kenya, for example, hundreds of women's groups organize tree plantings to prevent soil erosion and ensure future fuel supplies.

Whether inspired by feminism, African socialism, or the free market, community organizations have made a difference in meeting basic needs in sustainable ways. No doubt the majority of the groups fall short of all of their objectives. But for many people, especially women, the message of creating local networks to solve community problems is an empowering one.

REVIEW QUESTIONS

1. What are the environmental, historical, structural, and institutional reasons offered to explain poverty in the region?

2. What technological and investment changes are impacting the region's social development?

3. How is the status of women characterized for this region?

Summary

- The largest landmass straddling the equator, Africa is called the plateau continent because it is dominated by extensive uplifted plains. Key environmental issues facing this tropical region are desertification, deforestation, and drought. At the same time, the region supports a tremendous diversity of wildlife, especially large mammals.

- With nearly 900 million people, Sub-Saharan Africa is the fastest-growing region in terms of population. Yet it is also the poorest region, with two-thirds of its people living on less than $2 a day. In addition, it has the lowest average life expectancy, at 55 years. HIV/AIDS has hit this region especially hard. Fortunately, efforts at disease prevention and access to treatment are saving lives.

- Culturally, Sub-Saharan Africa is an extremely diverse region, where multiethnic and multireligious societies are the norm. With a few exceptions, religious diversity and tolerance has been a distinctive feature of the region. Most states have been independent for 50 years, and in that time pluralistic, but distinct national identities have been forged. Many African cultural expressions, such as music, dance, and religion, have been influential beyond the region.

- Since 1995, numerous bloody ethnic and political conflicts have occurred in the region. Fortunately, peace now exists in many conflict-ridden areas, such as Angola, Sierra Leone, and Liberia. However, ongoing ethnic and territorial disputes in Somalia, the Democratic Republic of the Congo, and South Sudan and more recently in Mali have produced millions of internally displaced persons and refugees.

- In terms of contemporary economic globalization, Sub-Saharan Africa's connections to the global economy are weak. With 12 percent of the world's population, the region accounts for only about 2 percent of the world's economic activity. Most of the region's economic ties come through international aid and loans, rather than trade. However, foreign direct investment, especially from China, is on the rise.

- Poverty is the region's most pressing issue. Since 2000, Sub-Saharan economies have grown, led in part by higher commodity prices, greater investment, debt forgiveness, and the end of some of the longest-running conflicts in the region. Social indicators of development are also improving, due to greater attention from the international community and better access to health care and drugs to fight HIV/AIDS.

Key Terms

African Union (AU) 199
apartheid 189
Berlin Conference 198
biofuel 180
clan 202
coloured 189
desertification 178

Great Escarpment 175
Horn of Africa 176
internally displaced person (IDP) 200
kleptocracy 205
Millennium Development Goals 210
pastoralist 187
refugee 200

Sahel 178
structural adjustment program 203
swidden 186
transhumance 178
tribalism 200
tribe 190
tsetse fly 188

Thinking Geographically

1. What factors might explain why European conquest and settlement occurred much earlier in tropical America than in tropical Africa?

2. More than any other region, Sub-Saharan Africa is noted for its wildlife, especially large mammals. What environmental and historical processes explain the existence of so much fauna? Why are there relatively fewer large mammals in other world regions?

3. Compare and contrast the role of tribalism in Sub-Saharan Africa with the role of nationalism in Europe.

4. Discuss the contrasting development models put forward by the United States and Europe with that of China. How will Chinese influence in the region alter the course of development for Sub-Saharan Africa?

5. Is desertification a natural or human-induced process? Where does it occur? How might global climate change impact desertification trends?

MasteringGeography™

Looking for additional review and test prep materials? Visit the Study Area in MasteringGeography™ to enhance your geographic literacy, spatial reasoning skills, and understanding of this chapter's content by accessing a variety of resources, including **MapMaster** interactive maps, videos, RSS feeds, flashcards, web links, self-study quizzes, and an eText version of *Globalization and Diversity*.

Scan to visit the author's blog for chapter updates.

http://gad4blog.
wordpress.com/
category/
sub-saharan-africa/

Scan to visit the GeoCurrents blog.

http://geocurrents.
info/category/place/
subsaharan-africa

Authors' Blogs

Scan now to access the authors' blogs for up-to-date information on Sub-Saharan Africa.

Globalization and Diversity

Globalization has played out across Southwest Asia and North Africa in complex ways. Energy-rich nations have reaped great economic rewards in satisfying global demands for oil and natural gas, but other cultural and political elements of the outside world have been opposed by many residents of the region.

ENVIRONMENTAL GEOGRAPHY

Water shortages are likely to increase across this arid region in the 21st century as growing populations, rapid urbanization, and increasing demands for agricultural land consume limited supplies.

POPULATION AND SETTLEMENT

Rapid population growth in North African cities such as Algiers and Cairo is far outpacing the ability of these urban places to supply adequate housing and services.

CULTURAL COHERENCE AND DIVERSITY

The heart of the Islamic world, this region finds itself at the center of the global rise of Islamist movements that often come into conflict with Western values and traditions.

GEOPOLITICAL FRAMEWORK

Ongoing political instability is a fact of life. The Arab Spring uprisings jolted geopolitical realities in Tunisia, Egypt, Libya, Yemen, and Bahrain. Internal instability also produced extensive bloodshed in Syria. Prospects for peace between Israel and the Palestinians remain murky, and Iran's growing political role is seen by many as a threat both within and beyond the region.

ECONOMIC AND SOCIAL DEVELOPMENT

Unstable world oil prices and unpredictable geopolitical conditions have discouraged investment and tourism in many countries. The pace of social change, especially for women, has quickened, producing diverse regional responses.

➤ Cairo's Tahrir Square fills with demonstrators at a mass rally in November, 2011.

SOUTHWEST ASIA AND NORTH AFRICA

How does a *place* become a *symbol*? Cairo's Tahrir Square, oriented around a busy traffic circle and nearby public buildings and hotels, has long served as a downtown focal point in Egypt's largest metropolitan area. The modern version of the square emerged more than a century ago when portions of the city were remodeled to create a "Paris on the Nile," complete with great open spaces, grand boulevards, and key government buildings. Known as Tahrir (or Liberation) Square since the 1952 overthrow of the Egyptian monarchy, the location became a natural gathering point for protesters early in 2011, as the actions of the region's larger **Arab Spring** movement (a series of public protests, strikes, and rebellions, often facilitated by social media, that have called for fundamental government and economic reforms) swept across the country.

The square became the rallying point for hundreds of thousands of Egyptians, many of them young and frustrated with the nation's autocratic rule and sluggish economy. The square also became the focus of violence as tensions mounted and the government's hold on power slipped away. Finally, on February 11, 2011, long-time President Hosni Mubarak stepped down and transferred power to the army. Celebrations erupted on the square even as Egypt's political future remained uncertain. Regardless of how the nation's fortunes evolve, Tahrir Square has earned an enduring spot in Egypt's human geography. Its place identity will forever be wed to the hundreds of thousands of ordinary people who stopped traffic there for days and pressed their case for enduring political change.

Tahrir Square and Egypt itself sit almost at the geographical center of a much larger and more complex world region known as Southwest Asia and North Africa. Climate, culture, and oil resources help define the region, but this portion of the world also features amazing geographical diversity. Located at the historic meeting ground of Europe, Asia, and Africa, the region includes thousands of square miles of parched deserts, rugged plateaus, and oasis-like river valleys. It extends 4,000 miles (6,400 kilometers) between Morocco's Atlantic coastline and Iran's boundary with Pakistan. More than two dozen nations are included within its borders, with the largest populations located in Egypt, Turkey, and Iran (Figure 7.1).

"Southwest Asia and North Africa" is both an awkward term and a complex region. Often the same area is simply called the "Middle East," but some experts exclude the western parts of North Africa, as well as Turkey and Iran, from such a region. In addition, the "Middle East" suggests a European point of view—Lebanon is in the "middle of the east" only from the perspective of the western Europeans who colonized the region and still shape the names we give the world today. Instead, "Southwest Asia and North Africa" offers a straightforward way to describe the general limits of the region.

There are also problems with simply defining the geographical limits of the region. A small piece of northwest Turkey actually sits west of the Bosporus Strait, generally considered to be the dividing line between Europe and Asia (see Figure 7.1). To the northeast, the largely Islamic peoples of Central Asia share many religious and linguistic ties with Turkey and Iran, but those groups are treated in a separate chapter (Chapter 10). African borders also remain problematic. The conventional division of "North Africa" from "Sub-Saharan Africa" cuts through the middle of modern Mauritania, Mali, Niger, and Chad. These transitional countries are discussed in Chapter 6. Sudan's recent split also suggests the complexity. The new nation of South Sudan (created in 2011) is also treated in Chapter 6, while Sudan (an "Islamic republic") retains many ties to the Muslim world and remains in this chapter.

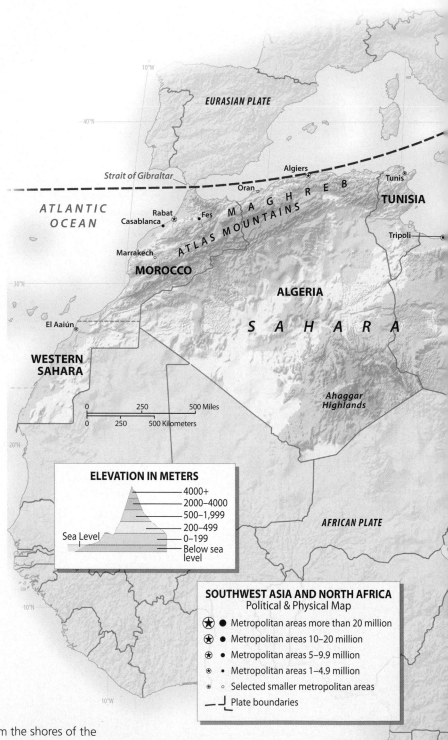

ELEVATION IN METERS

- 4000+
- 2000–4000
- 500–1,999
- 200–499
- 0–199
- Below sea level

Sea Level

SOUTHWEST ASIA AND NORTH AFRICA
Political & Physical Map

- ⊛ ● Metropolitan areas more than 20 million
- ⊛ ● Metropolitan areas 10–20 million
- ⊛ ● Metropolitan areas 5–9.9 million
- ⊛ • Metropolitan areas 1–4.9 million
- ⊛ ○ Selected smaller metropolitan areas
- ⌐⌐ Plate boundaries

FIGURE 7.1 Southwest Asia and North Africa This vast region extends from the shores of the Atlantic Ocean to the Caspian Sea. Within its boundaries, major cultural differences and globally important petroleum reserves have contributed to recent political tensions.

Diverse languages, religions, and ethnic identities have molded land and life within the region for centuries, strongly wedding people and place in ways that have had profound social and political implications (Figure 7.2). One traditional zone of conflict surrounds Israel, where Jewish, Christian, and Islamic peoples have yet to resolve long-standing cultural tensions and political differences. Iraq and Iran have also been settings for recent instability. Since 2010, Arab Spring movements across the region have overthrown some governments and pressured others to accelerate political and economic reforms, but it remains unclear in several countries what the long-term consequences of these dramatic changes will be.

No world region better exemplifies the theme of globalization than Southwest Asia and North Africa. The region is a key global **culture hearth**, producing many new cultural ideas that subsequently diffused widely. As an early center for agriculture, civilizations, and major world religions, the region has been a key human crossroads for thousands of years. Important long-distance trade routes have connected North Africa with the Mediterranean and Sub-Saharan Africa. Southwest Asia has also had historic ties to Europe, the Indian subcontinent, and Central Asia. As a result, new ideas within the region have often spread well beyond its bounds.

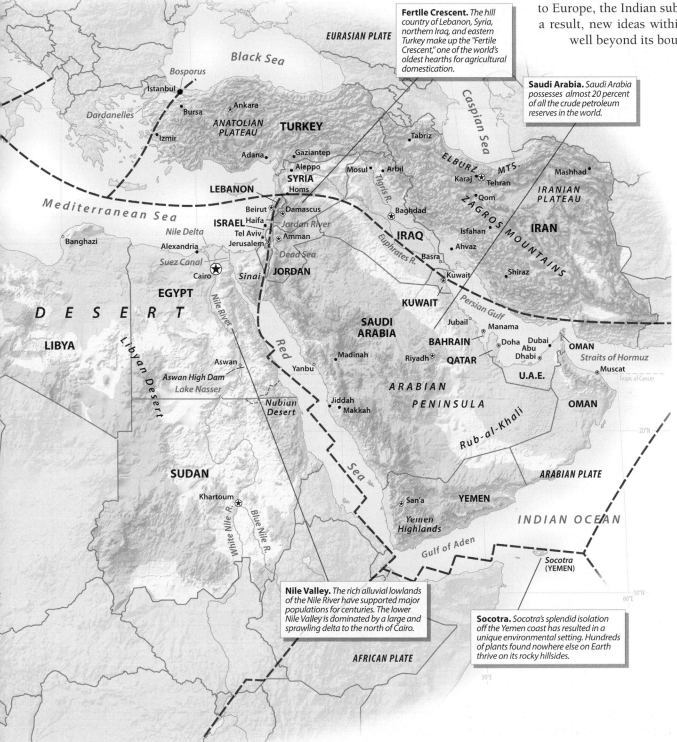

Fertile Crescent. *The hill country of Lebanon, Syria, northern Iraq, and eastern Turkey make up the "Fertile Crescent," one of the world's oldest hearths for agricultural domestication.*

Saudi Arabia. *Saudi Arabia possesses almost 20 percent of all the crude petroleum reserves in the world.*

Nile Valley. *The rich alluvial lowlands of the Nile River have supported major populations for centuries. The lower Nile Valley is dominated by a large and sprawling delta to the north of Cairo.*

Socotra. *Socotra's splendid isolation off the Yemen coast has resulted in a unique environmental setting. Hundreds of plants found nowhere else on Earth thrive on its rocky hillsides.*

FIGURE 7.2 Hebron, West Bank Home to a large Palestinian community, the West Bank city of Hebron is an important center of trade and industry within the region. It also contains key sites of religious significance for Muslim, Jewish, and Christian populations.

Particularly within the past century, processes of globalization and the region's strategic importance have made it increasingly open to outside influences. The 20th-century development of the petroleum industry, largely initiated by U.S. and European investment, had enormous consequences for economic development in oil-rich countries of the region. Many key members of **OPEC (Organization of the Petroleum Exporting Countries)** are found within the region, and these countries greatly influence global prices and production levels for petroleum.

Islamic fundamentalism in the region advocates a return to more traditional practices within the Islamic religion. Fundamentalists in any religion advocate a conservative adherence to enduring beliefs within their creed, and they strongly resist change. A related political movement within Islam, known as **Islamism**, challenges the encroachment of global popular culture and blames colonial, imperial, and Western elements for many of the region's political, economic, and social problems. Islamists resent the role they claim the West has played in creating poverty in their world, and many Islamists advocate merging civil and religious authority and rejecting many characteristics of modern, Western-style consumer culture.

ENVIRONMENTAL GEOGRAPHY: LIFE IN A FRAGILE WORLD

In the popular imagination, much of Southwest Asia and North Africa is a land of shifting sand dunes, searing heat, and scattered oases. Although examples of those stereotypes certainly exist, the actual physical setting is much more complex. One theme is dominant, however: A lengthy legacy of human settlement has left its mark on a fragile environment, and the entire region faces increasingly difficult ecological problems in the decades ahead.

Legacies of a Vulnerable Landscape

The environmental history of Southwest Asia and North Africa reflects both the short-sighted and the resourceful practices of its human occupants. Littered with examples of environmental problems, the region reveals the hazards of lengthy human settlement in a marginal land (Figure 7.3). The island of Socotra illustrates the region's fragile and vulnerable environment and suggests how processes of globalization threaten the area's ecological health. Socotra's stony slopes rise out of the shimmering waters of the Indian Ocean southeast of Yemen. Separated for millions of years from the Arabian Peninsula, Socotra's environment evolved in isolation. The island is home to hundreds of plants that are found nowhere else on Earth. Exotic dragon's blood trees dot many of the island's dry and rocky hillsides, and dozens of other species have only recently been catalogued by botanists (Figure 7.4).

LEARNING OBJECTIVES

After reading this chapter you should be able to:

- Describe how the region's fragile, often arid setting shapes the region's contemporary environmental challenges.

- Explain how latitude and topography produce the region's distinctive patterns of climate.

- Describe four distinctive ways in which people have learned to adapt their agricultural practices to the region's arid environment.

- Summarize the major forces shaping recent migration patterns within the region.

- List the major characteristics and patterns of diffusion of Islam.

- Identify the key modern religions and language families that dominate the region.

- Describe the local impacts of the Arab Spring rebellions in different regional settings.

- Identify the role of cultural variables in understanding key regional conflicts in Israel, Syria, and Iraq.

- Summarize the geography of oil and gas reserves in the region.

- Describe traditional roles for Islamic women and provide examples of recent changes.

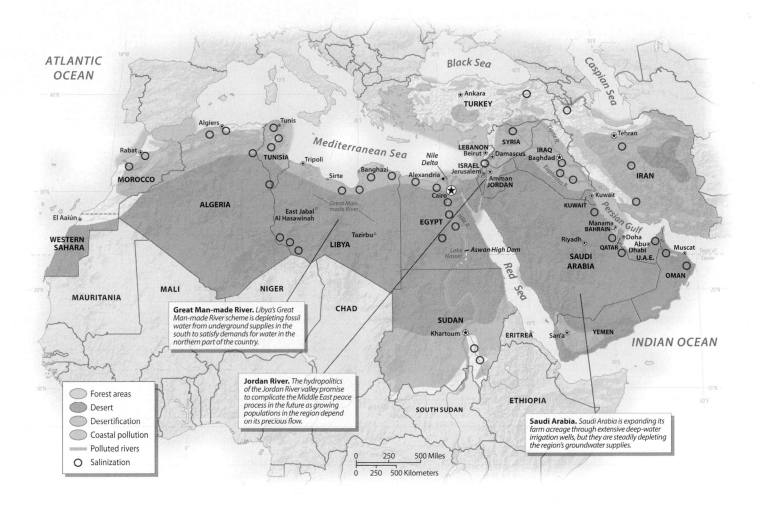

FIGURE 7.3 Environmental Issues in Southwest Asia and North Africa   Growing populations, pressures for economic development, and widespread aridity combine to create environmental hazards across the region. A long history of human activities has contributed to deforestation, irrigation-induced salinization, and expanding desertification. Saudi Arabia's deep-water wells, Egypt's Aswan High Dam, and Libya's Great Man-made River are all recent technological attempts to expand settlement, but they may carry a high long-term environmental price tag.

Socotra's unique environmental heritage now hangs in the balance. In 2008, the island was recognized as a United Nations World Natural Heritage Site, and the European Union has officially supported the idea of preserving the area's unique biogeography. At the same time, global pharmaceutical companies have harvested some of the island's unique plant species, and other plants and animals have been illegally taken by visitors. In addition, the Yemeni government has invited international petroleum companies to explore the island's offshore oil and gas potential, and it has considered a grand scheme for developing luxury tourist hotels that would forever change the island's character. Even well-meaning ecotourists on the island have damaged coral reefs and their growing numbers have led to increased firewood gathering and road construction. Socotra, often cited as the "Galápagos of the Indian Ocean," is an interesting case study for the battle of environmental preservation versus economic development. A recently developed mixed-use zoning plan and a newly paved airstrip suggest development is inevitable.

Deforestation and Overgrazing   Deforestation is an ancient problem in Southwest Asia and North Africa. Although much of the

FIGURE 7.4 Socotra's Dragon's Blood Tree   Unique to Socotra, the rare dragon's blood tree reflects the island's environmental isolation. Evolving as a part of the island's ecosystem, the tree survives in the region's dry tropical climate.

region is too dry for trees, the more humid and elevated lands that border the Mediterranean once supported heavy forests. Included in these woodlands are the cedars of Lebanon, cut down in ancient times and now reduced to a few scattered groves that survive in a largely unforested landscape.

Human activities have combined with natural conditions to reduce most of the region's forests to grass and scrub. Mediterranean forests often grow slowly, are highly vulnerable to fire, and usually fare poorly if subjected to heavy grazing. Browsing by sheep and goats in particular has often been blamed for much of the region's forest loss. Deforestation has resulted in a long, slow deterioration of the region's water supplies and in accelerated soil erosion. Several governments have launched reforestation drives and forest preservation efforts. Israel and Syria have expanded their forested lands since the 1980s, and in nearby Lebanon the Shouf Biosphere Reserve (created in 2005) protects rare cedar trees and endangered mammals, such as the wolf and the Lebanese jungle cat.

Salinization   Salinization, or the buildup of toxic salts in the soil, is another ancient environmental issue in an arid region where irrigation has been practiced for centuries (see Figure 7.3). Hundreds of thousands of acres of once-fertile farmland within the region have been destroyed or degraded by salinization. The problem has been particularly severe in Iraq, where centuries of canal irrigation along the Tigris and Euphrates rivers have seriously degraded land quality. Similar conditions affect central Iran, Egypt, and other irrigated portions of North Africa.

Managing Water   The people of the region have been modifying drainage systems and water flows for thousands of years. Traditional systems emphasized directing and conserving surface and groundwater resources at a local scale, but in the past half century the scope of environmental change has been greatly magnified. One remarkable example is Egypt's Aswan High Dam, completed in 1970 on the Nile River south of Cairo (see Figure 7.3). Increased storage capacity in the upstream reservoir made more water available for agriculture and generates clean electricity. But irrigation has increased salinization because water is not rapidly flushed from fields. In addition, the dam has led to more use of costly fertilizers, the infilling of Lake Nasser behind the dam with accumulating sediments, and the collapse of the Mediterranean fishing industry near the Nile Delta, an area previously nourished by the river's silt.

Elsewhere, other water-harvesting strategies have proven useful. **Fossil water**, or water supplies stored underground during earlier and wetter climatic periods, has been utilized. Libya's "Great Manmade River" scheme taps underground water in the southern part of the country, transports it 600 miles (965 kilometers) to northern coastal zones, and uses the precious resource to expand agricultural production in one of North Africa's driest countries (see Figure 7.3). Similarly, Saudi Arabia has invested huge sums to develop deep-water wells, allowing it to greatly expand its food output. Unfortunately, underground supplies are being depleted more rapidly than they are recharged, limiting their long-term sustainability.

A growing number of countries, including Saudi Arabia, Kuwait, and Israel, are making large investments in seawater desalination plants (the region accounts for half the world's total), but this technology consumes large amounts of energy and increases seawater salinity (Figure 7.5).

FIGURE 7.5 Desalination Plant, Israel   This facility, located south of Tel Aviv, produces about 13 percent of Israel's domestic freshwater and is one of the largest desalination plants in the world.

Most dramatically, **hydropolitics**, or the interplay of water resource issues and politics, has raised tensions between countries that share drainage basins. For example, Turkey's growing development of the upper Tigris and Euphrates rivers (the Southeast Anatolia Project) has periodically raised protests from Iraq and Syria. Hydropolitics have also played into negotiations among Israel, the Palestinians, and other neighboring states, particularly in the valuable Jordan River drainage, which runs through the center of the area's most hotly disputed lands (Figure 7.6). Israelis fear Palestinian and Syrian pollution, and nearby Jordanians argue for more water from Syria.

Aside from the use of water as a resource, it is also a vital transportation link in the area. The region's physical geography has produced enduring **choke points**, where narrow waterways are vulnerable to military blockade or disruption. For example, the Strait of Gibraltar (entrance to the Mediterranean), Turkey's Bosporus and Dardanelles, and the Suez Canal have all been key historical choke points within the region (see Figure 7.1). Iran's periodic threat to close the Straits of Hormuz (eastern end of the Persian Gulf) to world oil shipments suggests the strategic role water continues to play in the region (Figure 7.7).

Regional Landforms

A quick tour of Southwest Asia and North Africa reveals diverse environmental settings (see Figure 7.1). In North Africa, the **Maghreb** (meaning "western island") includes the nations of Morocco, Algeria,

FIGURE 7.7 Straits of Hormuz This satellite view shows the entrance to the Persian Gulf at the Straits of Hormuz, one of the region's most important choke points.

Southwest Asia is more mountainous than North Africa. In the **Levant**, or eastern Mediterranean region, mountains rise within 20 miles (30 kilometers) of the sea, and the highlands of Lebanon reach heights of more than 10,000 feet (3,000 meters). Farther south, the Arabian Peninsula forms a massive tilted plateau, with western highlands higher than 5,000 feet (1,500 meters) gradually sloping eastward to extensive lowlands in the Persian Gulf area. North and east of the Arabian Peninsula lie the two great upland areas of Southwest Asia: the Iranian and Anatolian plateaus (*Anatolia* refers to the large peninsula of Turkey, sometimes called Asia Minor; see Figure 7.1). Both of these plateaus, averaging between 3,000 and 5,000 feet (1,000 and 1,500 meters) in elevation, are geologically active and prone to earthquakes. One dramatic quake in western Turkey (1999) measured 7.8 on the Richter scale, killed more than 17,000 people, and left 350,000 residents homeless. Another quake near the Iranian city of Bam (2003) claimed more than 30,000 lives.

FIGURE 7.8 Atlas Mountains The rugged Atlas Mountains dominate a broad area of interior Morocco. Steep slopes and narrow canyons can make travel slow and hazardous across the region.

FIGURE 7.6 Hydropolitics in the Jordan River Basin Many water-related issues complicate the geopolitical setting in the Middle East. The Jordan River system has been a particular focus of conflict.

and Tunisia and is dominated near the Mediterranean coastline by the Atlas Mountains. The rugged flanks of the Atlas rise like a series of islands above the narrow coastal plains to the north and the vast stretches of the lower Saharan deserts to the south (Figure 7.8). South and east of the Atlas Mountains, interior North Africa varies between rocky plateaus and extensive desert lowlands. In northeast Africa, the Nile River shapes regional drainage patterns as it flows north through Sudan and Egypt (Figure 7.9).

FIGURE 7.9 Nile Valley This satellite image of the Nile Valley dramatically reveals the impact of water on the North African desert. Cairo lies at the southern end of the delta, where it begins to widen toward the Mediterranean Sea.

Smaller lowlands characterize other portions of Southwest Asia. Narrow coastal strips are common in the Levant, along both the southern (Mediterranean) and the northern (Black Sea) Turkish coastlines, as well as north of the Iranian Elburz Mountains near the Caspian Sea. Iraq contains the most extensive alluvial lowlands in Southwest Asia, dominated by the Tigris and Euphrates rivers, which flow southeast to empty into the Persian Gulf. Although much smaller, the Jordan River Valley is a notable lowland that straddles the strategic borderlands of Israel, Jordan, and Syria and drains southward to the Dead Sea (Figure 7.10).

Patterns of Climate

Although the region of Southwest Asia and North Africa is often termed the "dry world," a closer look reveals more complex patterns (Figure 7.11). Both latitude and altitude come into play. Aridity dominates large portions of the region (see the climographs for Cairo, Riyadh, Baghdad, and Tehran in Figure 7.11). A nearly continuous belt of desert land stretches eastward across interior North Africa, through the Arabian Peninsula, and into central and eastern Iran (Figure 7.12).

FIGURE 7.10 Jordan Valley This view of the Jordan Valley shows a fertile mix of irrigated vineyards and date palm plantations.

Throughout this zone, plant and animal life adapts to extreme conditions. Deep or extensive root systems allow desert plants to benefit from the limited moisture they receive. Similarly, animals adjust by efficiently storing water, hunting at night, or migrating seasonally to avoid the worst of the dry cycle.

Elsewhere, altitude and latitude produce a surprising amount of climatic variety. The Atlas Mountains and nearby lowlands of northern Morocco, Algeria, and Tunisia experience a Mediterranean climate, in which dry summers alternate with cooler, wet winters (see the climographs for Rabat and Algiers). In these areas, the landscape resembles that found in nearby southern Spain or Italy (Figure 7.13). A second zone of Mediterranean climate extends along the Levant coastline, into the nearby mountains, and northward across sizable portions of northern Syria, Turkey, and northwestern Iran (see the climographs for Jerusalem and Istanbul).

Climate Change in Southwest Asia and North Africa

Projected changes in global climate will aggravate several environmental issues within Southwest Asia and North Africa. Temperature changes will likely impact the region more than changes in precipitation. Warmer average temperatures produce higher overall evaporation rates and lower overall soil moisture, stressing crops, grasslands, and other vegetation. Semiarid lands are particularly vulnerable, especially dry-land cropping systems that cannot depend on irrigation. Even in irrigated zones, higher temperatures can reduce yields for wheat and maize. Warmer temperatures will likely reduce net runoff into the region's already stressed streams and rivers, potentially reducing hydroelectric potential and water availability for the region's increasingly urban population. Extreme weather events are also more likely, such as record-setting summer temperatures.

Sea-level changes will pose special threats to the Nile Delta. This portion of northern Egypt is a vast, low-lying landscape of settlements, farms, and marshland. Studies that simulate rising sea

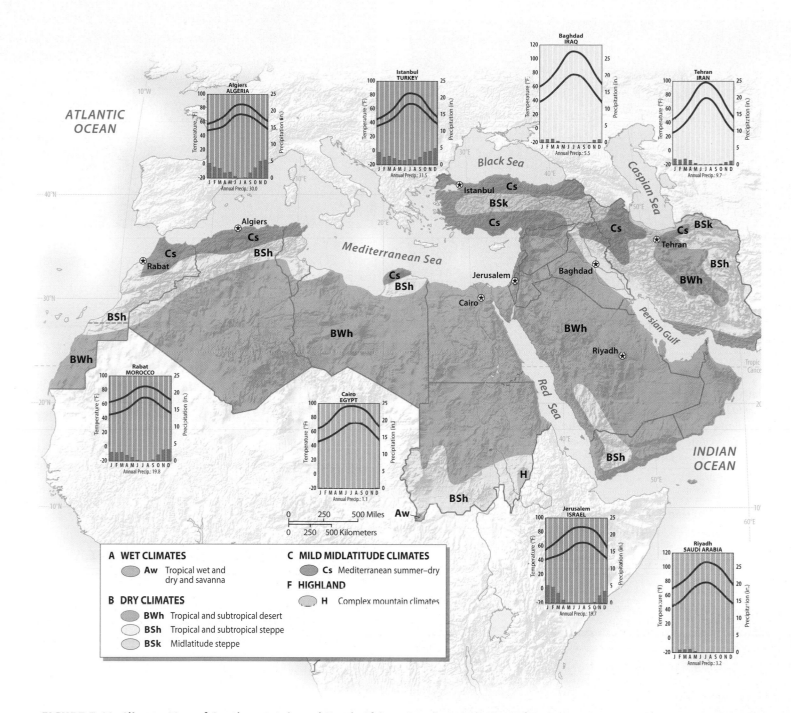

FIGURE 7.11 Climate Map of Southwest Asia and North Africa   Dry climates dominate from western Morocco to eastern Iran. Within these zones, subtropical high-pressure systems offer only limited opportunities for precipitation. Elsewhere, mild midlatitude climates with wet winters are found near the Mediterranean Basin and Black Sea. To the south, tropical savanna climates provide summer moisture to southern Sudan.

levels reveal that much of the region will be lost to inundation, erosion, or salinization. Farmland losses of more than 250,000 acres (100,000 hectares) are quite possible with even modest sea-level changes. For example, estimates are that a 1-meter (3.3-foot) sea-level rise could affect 15 percent of Egypt's habitable land and displace 8 million Egyptians in coastal and delta settings. In the Egyptian city of Alexandria, $30 billion in losses have been projected because sea-level changes will devastate the city's huge

resort industry, as well as nearby residential and commercial areas (Figure 7.14).

Some experts have also suggested broader political and economic costs associated with regional climate change. For example, given the political instability of the Middle East, even relatively small changes in water supplies, particularly where they might involve several nations, could significantly add to the potential for conflict within the region. Projected economic implications of climate change may depend on

FIGURE 7.12 **Arid Iran** Only sparse vegetation dots this arid scene from central Iran, a landscape characterized by isolated mountain ranges and dry interior plateaus.

FIGURE 7.14 **Alexandria, Egypt** This beachside view along northern Egypt's low-lying coastline at Alexandria could change significantly if global sea levels were to rise.

the relative importance of agriculture within different nations. In addition, wealthier nations such as Israel and Saudi Arabia may have more available resources to plan, adjust, and adapt to climate shifts and extreme events than the poorer, less developed countries such as Yemen, Syria, and Sudan.

FIGURE 7.13 **Algerian Orange Harvest** The Mediterranean moisture in northern Algeria produces an agricultural landscape similar to that of southern Spain or Italy. Winter rains create a scene that contrasts sharply with deserts found elsewhere in the region.

REVIEW QUESTIONS

1. Describe the climatic changes you might experience as you travel from the eastern Mediterranean coast to the highlands of Yemen. What are some of the key climatic variables that explain these variations?

2. Discuss five important human modifications of the Southwest Asian and North African environment, and assess whether these changes have benefited the region.

POPULATION AND SETTLEMENT: PATTERNS IN AN ARID LAND

The human geography of Southwest Asia and North Africa demonstrates the intimate tie between water and life in this part of the world. The pattern is complex: Large areas of the population map remain almost devoid of permanent settlement, whereas lands with available moisture suffer increasingly from problems of crowding and overpopulation (Figure 7.15).

The Geography of Population

Today about 500 million people live in Southwest Asia and North Africa (Table 7.1). The distribution of that population is strikingly varied (see Figure 7.15). In North Africa, the moist slopes of the Atlas Mountains and nearby better-watered coastal districts support dense populations, a stark contrast to thinly occupied lands southeast of the mountains. Egypt's zones of almost empty desert land differ dramatically from crowded, irrigated locations, such as those along the Nile River. In Southwest Asia, many residents live in well-watered coastal zones, moist highland settings, and desert localities where water is available from nearby rivers or subsurface aquifers. High population densities are found in better-watered portions of the eastern Mediterranean (Israel, Lebanon, and Syria), Turkey, and Iran. While overall population densities in such countries appear modest, the

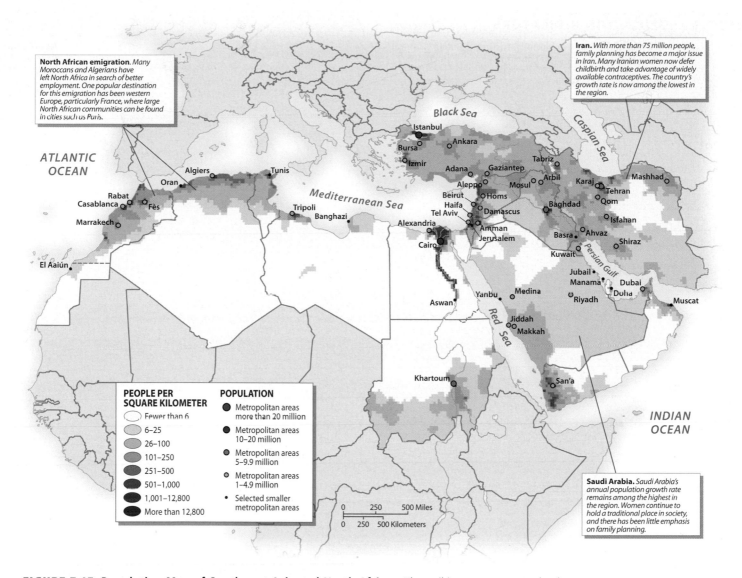

FIGURE 7.15 Population Map of Southwest Asia and North Africa The striking contrasts are clearly evident between large, sparsely occupied desert zones and much more densely settled regions where water is available. The Nile Valley and the Maghreb region contain most of North Africa's people, while Southwest Asian populations cluster in the highlands and along the better-watered shores of the Mediterranean.

physiological density, which is the number of people per unit area of arable land, is quite high by global standards. Although less than two-thirds of the overall population of the region is urban, many nations are dominated by huge cities (for example, Cairo in Egypt and Tehran in Iran) that produce the same problems of urban crowding found elsewhere in the developing world (Figure 7.16).

Water and Life: Rural Settlement Patterns

Water and life are closely linked across rural settlement landscapes of Southwest Asia and North Africa (Figure 7.17). Indeed, Southwest Asia is one of the world's earliest hearths of **domestication**, where plants and animals were purposefully selected and bred for their desirable characteristics. Beginning around 10,000 years ago, increased experimentation with wild varieties of wheat and barley led to agricultural settlements that later included domesticated animals, such

as cattle, sheep, and goats. Much of the early agricultural activity focused on the **Fertile Crescent**, an ecologically diverse zone that stretches from the Levant inland through the fertile hill country of northern Syria into Iraq. Between 5,000 and 6,000 years ago, better knowledge of irrigation techniques and increasingly powerful political states encouraged the spread of agriculture into nearby lowlands, such as the Tigris and Euphrates valleys (Mesopotamia) and North Africa's Nile Valley.

Pastoral Nomadism In the drier portions of the region, **pastoral nomadism** is a traditional form of subsistence agriculture where people move livestock seasonally. The settlement landscape of pastoral nomads reflects their need for mobility and flexibility as they move camels, sheep, and goats from place to place. Near highland zones such as the Atlas Mountains and the Anatolian Plateau, nomads practice **transhumance**—seasonally moving livestock to cooler, greener

Table 7.1 POPULATION INDICATORS

Country	Population (millions) 2012	Population Density (per square kilometer)	Rate of Natural Increase (RNI)	Total Fertility Rate	Percent Urban	Percent <15	Percent >65	Net Migration (Rate per 1000) 2010–15[a]
Algeria	37.4	16	2.0	2.9	72	28	5	−0.8
Bahrain	1.3	1,925	1.2	1.9	100	20	2	6.2
Egypt	82.3	82	2.0	2.9	43	32	4	−0.5
Gaza and West Bank	2.6*	448*		3*	72*	36*	4*	0*
Iran	78.9	48	1.3	1.9	69	24	5	−0.4
Iraq	33.7	77	2.9	4.6	67	43	3	1.8
Israel	7.9	357	1.6	3.0	92	28	10	1.6
Jordan	6.3	71	3.0	3.8	83	37	3	−0.9
Kuwait	2.9	162	1.5	2.3	98	27	3	9.7
Lebanon	4.3	414	1.6	1.9	87	25	7	−0.6
Libya	6.5	4	1.9	2.6	78	31	4	−9.9
Morocco	32.6	73	1.4	2.3	58	28	6	−3.0
Oman	3.1	10	2.0	2.9	73	32	3	6.0
Qatar	1.9	171	1.0	2.1	100	14	1	18.8
Saudi Arabia	28.7	13	1.8	2.8	81	30	3	3.6
Sudan	33.5	18	2.4	4.2	41	41	3	0.6
Syria	22.5	122	2.0	3.0	54	36	4	−1.6
Tunisia	10.8	66	1.3	2.1	66	24	7	−0.4
Turkey	74.9	96	1.2	2.0	77	26	7	−0.1
United Arab Emirates	8.1	97	1.2	1.8	83	17	0	10.9
Western Sahara	0.6	2	1.7	2.6	82	29	2	16.5
Yemen	25.6	48	3.1	5.2	29	44	3	−1.0

[a]Net Migration Rate from the United Nations, Population Division, *World Population Prospects: The 2010 Revision Population Database.*
*Additional data from the *CIA World Factbook, 2012*
Source: Population Reference Bureau, *World Population Data Sheet, 2012.*

high-country pastures in the summer and returning them to valley and lowland settings for fall and winter grazing. Elsewhere, seasonal movements often involve huge areas of desert that support small groups of a few dozen families. Fewer than 10 million pastoral nomads remain in the region today.

FIGURE 7.16 Tehran, Iran With a population of more than 8 million people, Iran's capital city of Tehran sprawls in all directions and is facing many of the same urban challenges as cities in North America.

Oasis Life Permanent oases exist where high groundwater levels or modern deep-water wells provide reliable water (Figures 7.17 and 7.18). Tightly clustered, often walled villages sit next to small, intensely utilized fields where underground water is applied to tree and cereal crops. In newer oasis settlements, concrete blocks and prefabricated housing add a modern look. Traditional oasis settlements contain families that work their own irrigated plots or, more commonly, work for absentee landowners (see "Working Toward Sustainability: Preserving Land and Life in Jordan's Azraq Basin"). Although oasis settlements are usually small, eastern Saudi Arabia's Hofuf oasis covers more than 30,000 acres (12,000 hectares). Although some crops are raised for local consumption, expanding world demand for products such as figs and dates increasingly includes even these remote locations in the global economy, as products end up on the tables of hungry Europeans or North Americans.

Exotic Rivers For centuries, the densest rural settlement of Southwest Asia and North Africa has been tied to its great irrigated river valleys and their seasonal floods of water and fertile nutrients. In such settings, **exotic rivers** transport much-needed water from more humid regions to drier regions that suffer from long-term moisture deficits (Figure 7.19). The Nile and the combined Tigris and Euphrates rivers are the largest regional examples of such activity, and both systems have large, densely settled deltas. Similar settlements are

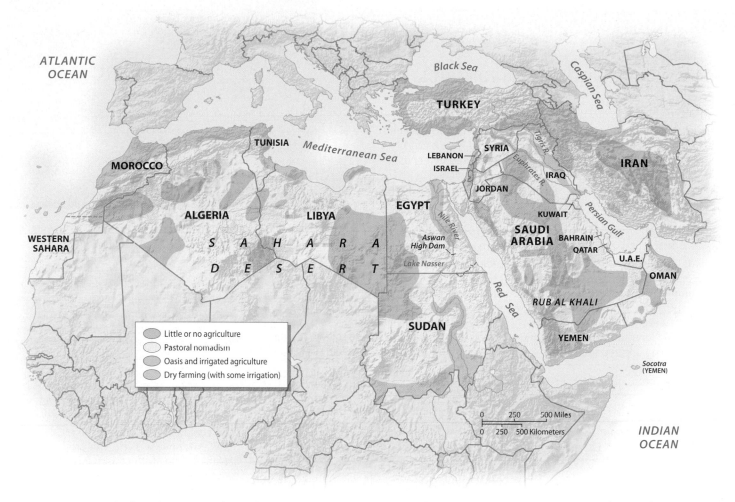

FIGURE 7.17 Agricultural Regions of Southwest Asia and North Africa Important agricultural zones include oases and irrigated farms where water is available. Elsewhere, dry farming supplemented with irrigation is practiced in midlatitude settings.

found along the Jordan River in Israel and Jordan, in the foothills of the Atlas Mountains, and on the peripheries of the Anatolian and Iranian plateaus. These settings, although capable of supporting sizable populations, are also vulnerable to overuse, particularly if irrigation produces salinization. Rural life is also changing in such settings. New dam- and canal-building schemes in Egypt, Israel, Syria, Turkey, and elsewhere are increasing the storage capacity of river systems, allowing for more year-round agriculture.

The Challenge of Dryland Agriculture Mediterranean climates in the region permit dryland agriculture that depends largely on seasonal moisture to support farming. These zones include the better-watered valleys and coastal lowlands of the northern Maghreb, lands along the shore of the eastern Mediterranean, and favored uplands across the Anatolian and Iranian plateaus. A mix of tree crops, grains, and livestock is raised in these settings. More mechanization, crop specialization, and fertilizer use are also transforming such agricultural settings, following a pattern set earlier in nearby areas of southern Europe.

FIGURE 7.18 Oasis Agriculture The green fields and trees of Morocco's Tinghir oasis contrast dramatically with the surrounding desert landscape.

WORKING TOWARD SUSTAINABILITY

Preserving Land and Life in Jordan's Azraq Basin

The Azraq Basin contains some of Jordan's largest supplies of groundwater (Figure 7.1.1). Associated surface springs and wetlands traditionally support both sedentary farmers (who depend on the basin for irrigation water for their desert grain and tree crops) and nomadic Bedouin populations (who use seasonal pastures to graze their camels, sheep, and goats). In addition, large flocks of migrating birds (including species from Scandinavia, Africa, and Siberia) use lakes in the area as stopovers on their multicontinental journeys (Figure 7.1.2).

Since 1980, however, unwelcome changes have threatened the long-term viability of the basin. Rapidly growing urban populations in nearby Amman and Zarqa have placed increased demands on regional aquifers. In rural areas, many farmers have shifted to more commercialized and intensive forms of irrigated agriculture. Large numbers of new wells have been dug, many of them illegal. Periodic regional droughts—potentially related to global warming—have also been on the rise, limiting surface recharge. The results have been dramatic and painful. Beginning in the 1990s, many springs and shallow wells have

FIGURE 7.1.2 Azraq Wetlands, Jordan This small freshwater resource is an invaluable part of the Azraq Wetlands Nature Reserve in Jordan.

seen lower flows or have dried up. The salinity of the remaining water resources has risen. Wetlands have shrunk greatly in size, and many migratory birds have disappeared.

In a coordinated response to the crisis, the Jordanian government established the Azraq Oasis Restoration Project, designed to study the problem and come up with sustainable solutions to reverse falling groundwater levels, preserve traditional farming and grazing lifestyles, and restore the ecological integrity of the critical wetlands environment. The project has also teamed up with the International Union for Conservation of Nature, the Arab Women Organization (focused on improving the lives of ordinary Jordanian women), and numerous stakeholders in the area. Project researchers have emphasized a *participatory approach*, in which they spend large amounts of time in local workshops and seminars and do extensive fieldwork partnering with residents. Although only about 5–10 percent of the surface wetlands have been temporarily restored (through increased groundwater pumping), project leaders hope that their efforts will slow the drilling of new wells, encourage farmers to rethink the mix of crops they grow (and adopt more efficient drip irrigation systems), and offer more examples of urban water consumption and recycling.

The fate of the Azraq Basin remains in doubt. Lower population growth and widespread local participation in water conservation efforts are crucial elements in restoring the basin and its groundwater to sustainable levels. Ultimately, the project's chances for long-term success may lie in a creative blending of tradition and innovation that combines local and global knowledge and in the process redefines water as a sustainable resource in a part of the world where it increasingly seems in short supply.

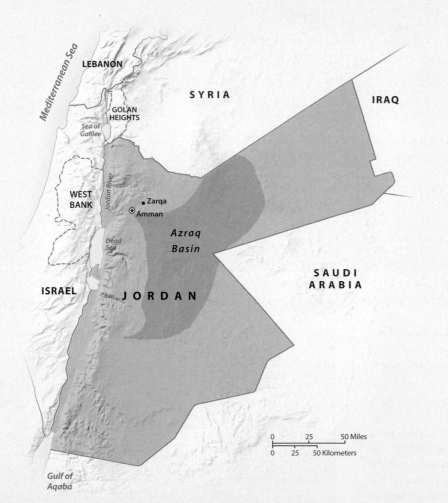

FIGURE 7.1.1 Azraq Basin, Jordan The map shows the size and centrality of Jordan's Azraq Basin. Note the proximity to Amman and Zarqa.

FIGURE 7.19 Nile River Valley These farmers are pulling clover for animal feed from these irrigated fields along the Nile River north of Cairo, Egypt.

Many-Layered Landscapes: The Urban Imprint

Cities have played a key role in the human geography of Southwest Asia and North Africa. Indeed, some of the world's oldest urban places are located in the region. Today continuing political, religious, and economic ties link the cities with the surrounding countryside.

A Long Urban Legacy Cities in the region have traditionally played important roles as centers of political and religious authority, as well as focal points of local and long-distance trade. Urbanization in Mesopotamia (modern Iraq) began by 3500 BCE, and cities such as Eridu and Ur reached populations of 25,000 to 35,000 residents. Similar centers appeared in Egypt by 3000 BCE, with Memphis and Thebes assuming major importance in the middle Nile Valley. By 2000 BCE, however, a different kind of city emerged along the eastern Mediterranean and along important overland trade routes. Centers such as Beirut, Tyre, and Sidon, all in modern Lebanon, as well as Damascus in nearby Syria, exemplified the growing role of trade in shaping the urban landscape. Expanding port facilities, warehouse districts, and commercial neighborhoods suggested how commerce shaped these urban settlements, and many early Middle Eastern trading towns survive to the present.

Islam also left an enduring mark on cities because urban centers traditionally served as places of Islamic religious power and education. Both Baghdad and Cairo were seats of religious authority. Urban settlements from North Africa to Turkey felt its influence. Indeed, the Moors carried Islam to Spain, where it shaped the architecture and culture of centers such as Córdoba and Málaga.

The traditional Islamic city features a walled core, or **medina**, dominated by the central mosque and its associated religious, educational, and administrative functions (Figure 7.20). A nearby bazaar, or *suq*, serves as a marketplace where products from city and countryside are traded (Figure 7.21). Housing districts feature a maze of narrow, twisting streets that maximize shade and emphasize the privacy of residents, particularly women. Houses have small windows, frequently are situated on dead-end streets, and typically open inward to private courtyards.

FIGURE 7.20 The Islamic Landscape Iranian mullahs (religious clerics or mosque leaders) in discussion beneath the minarets of the Moussavi Mosque in Qom. Islam has left a widespread mark on the region's cultural landscapes.

FIGURE 7.21 The Old City of Fes, Morocco These narrow streets in the old city of Fes are lined with shops and commercial stalls. The winding labyrinth of alleyways and cul-de-sacs forms an almost impenetrable maze to visitors, and it exemplifies the complex urban landscapes of the traditional Islamic city.

More recently, European colonialism has shaped selected cities. Particularly in North Africa, colonial builders added many architectural features from Great Britain and France. Victorian building blocks, French mansard roofs, suburban housing districts, and wide, European-style commercial boulevards remain landscape features in cities such as Algiers and Cairo.

Signatures of Globalization   Since 1950, cities in Southwest Asia and North Africa have become gateways to the global economy. As the region has been opened to new investment, industrialization, and tourism, the urban landscape has reflected the fundamental changes taking place. Expanded airports, commercial and financial districts, industrial parks, and luxury tourist facilities all reveal the influence of the global economy. Many cities, such as Algiers and Istanbul, have more than doubled in population in recent years. Crowded Cairo now has more than 15 million residents. The increasing demand for homes has produced ugly, cramped high-rise apartment houses, while elsewhere extensive squatter settlements provide little in the way of quality housing or public services.

Undoubtedly, the oil-rich states of the Persian Gulf have displayed the greatest changes in the urban landscape. Before the 20th century, urban traditions were relatively weak in the area, and even as late as 1950 only 18 percent of Saudi Arabia's population lived in cities. All that has changed, however, and today Saudi Arabia is more urban than many industrialized nations, including the United States. Particularly since 1970, cities such as Dubai (United Arab Emirates), Doha (Qatar), and Kuwait City (Kuwait) have mushroomed in size, featuring a mix of low-rent apartment houses and labor camps, which stand in dramatic contrast with central-city skylines offering futuristic architecture (Figure 7.22). Dubai's skyline is now home to the soaring Burj Dubai (completed 2010), a needle-like 160-story building that ascends a half mile into the dusty desert sky.

A Region on the Move

Although nomads have crisscrossed the region for ages, entirely new patterns of migration reflect the global economy and recent political events. The rural-to-urban shift seen widely in the less developed world is reworking population patterns across Southwest Asia and North Africa. The Saudi Arabian example is echoed in many other countries. Cities from Casablanca to Tehran are experiencing phenomenal growth rates, spurred by in-migration from rural areas.

Foreign workers have also migrated to areas within the region that have large labor demands. In particular, Persian Gulf nations support immigrant workforces that often comprise large proportions of the overall population. The influx has major economic, social, and demographic implications in nations such as Saudi Arabia (27% of the total population is foreign workers), Kuwait (68%), Qatar (87%), and the United Arab Emirates (81%) (Figures 7.22 and 7.23). Source regions for this Gulf-nation labor force are varied, but the largest communities are from South Asia and other Muslim countries within and beyond the region (Figure 7.24). In Dubai (United Arab Emirates), Pakistani cab drivers, Filipino nannies, and Indian shop clerks typify a foreign workforce that currently makes up a large majority of the city's 1.4 million inhabitants.

Other residents migrate to jobs elsewhere in the world. Because of its strong economy and close location, Europe is a powerful draw. More than 2 million Turkish guest workers live in Germany. Both Algeria and Morocco also have seen large out-migrations to western Europe, particularly France. Many Lebanese, often skilled workers and business owners, have also emigrated.

Political forces also encourage migration. Thousands of wealthier residents, for example, left Lebanon and Iran during the political turmoil of the 1980s and are today living in cities such as Toronto, Los Angeles, and Paris. More recent political instability has provoked other refugee movements. More than 35,000 Iraqis have resettled in the United States since 2003. Elsewhere, thousands of displaced Afghans remain in eastern Iran. In the Middle East, more than 500,000 Russian Jewish immigrants have migrated to Israel since 1991 and have reshaped the cultural geography of that country. Syria's recent civil war also has produced growing refugee populations in neighboring countries, especially Turkey. To the south, huge numbers of people in western Sudan's unsettled Darfur region have been forced to move to dozens of refugee camps in nearby Chad.

(a)

(b)

FIGURE 7.22 Contrasts in Dubai, United Arab Emirates   The soaring tower (a) of downtown Dubai's Burj Khalifa, the world's tallest structure, contrasts dramatically with a scene in a nearby labor camp in Shariah City (b), where South Asian workers enjoy a card game during their time off from work. A large majority of the country's population is foreign-born, and many immigrants like these work in the construction sector of the economy, building structures such as the Burj Khalifa.

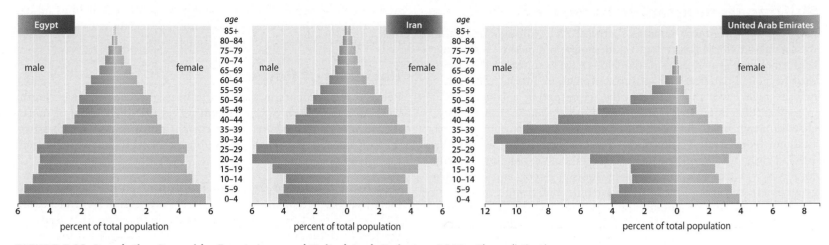

FIGURE 7.23 Population Pyramids: Egypt, Iran, and United Arab Emirates, 2012 Three distinctive demographic snapshots highlight regional diversity: Egypt's above-average growth rates differ sharply from those of Iran, where a focused campaign on family planning has reduced recent family sizes. Male immigrant laborers play a special role in skewing the pattern within the United Arab Emirates.

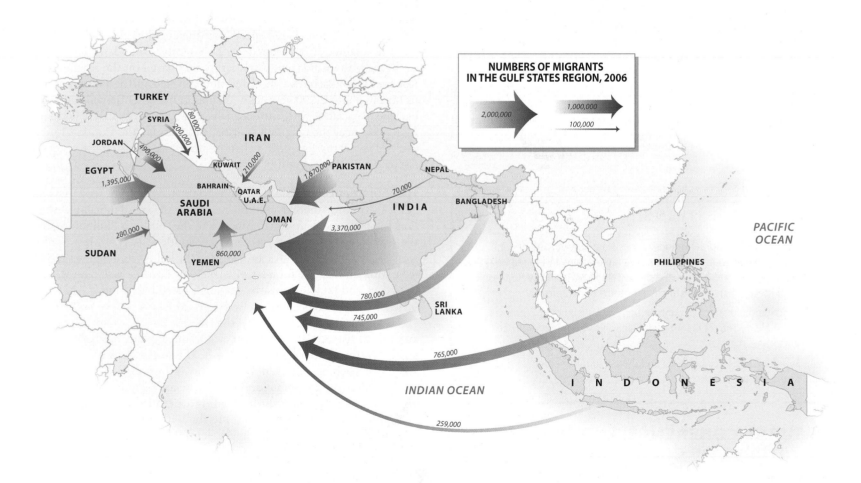

FIGURE 7.24 Estimated Migrant Populations in the Gulf States, 2006 The map shows the sources of immigrant populations working in the Persian Gulf states region in the early 21st century. Note the importance of nearby Arab countries, other Muslim nations, and South Asia.

Shifting Demographic Patterns

High population growth remains a critical issue throughout Southwest Asia and North Africa, but the demographic picture is shifting. Uniformly high growth rates in the 1960s have been replaced by more varied regional patterns. For example, women in Tunisia and Turkey now average fewer than three births, representing a large decline in total fertility rates (see Table 7.1). Various factors explain these changes. More urban, consumer-oriented populations opt for fewer children. Many Arab women now delay marriage into their middle 20s and early 30s. Family planning initiatives are expanding in many countries. For example, programs in Tunisia, Egypt, and Iran have greatly increased access to contraceptive pills, IUDs, and condoms.

Intriguingly, fundamentalist Iran has witnessed the fastest decline in fertility in the past two decades (Figure 7.23). Fertility has fallen by more than two-thirds since the mid-1970s (from an average of 6.6 births to 1.9 births per woman). While Iran's family-planning program was initially dismantled after the fundamentalist revolution in 1979 (it was seen as a Western idea), recent leaders have recognized the wisdom in containing the country's large population. Iran has one of the world's most successful family-planning programs, according to the Population Reference Bureau.

Still, areas such as the West Bank, Gaza, and Yemen are growing much faster than the world average. Poverty and traditional ways of rural life contribute to large rates of population increase, and even in more urban Saudi Arabia, growth rates remain near 2 percent. The increases result from high birthrates combined with low death rates. In Egypt, even though birthrates may decline, the labor market will need to absorb more than 500,000 new workers annually over the next 10 to 15 years just to keep up with the country's large youthful population (Figure 7.23). As that population ages in the mid-21st century, it will be increasingly urban and demand jobs, housing, and social services. Growing populations will also increase demands on the region's already limited water resources.

REVIEW QUESTIONS

1. Discuss how pastoral nomadism, oasis agriculture, and dryland wheat farming represent distinctive adaptations to the regional environments of Southwest Asia and North Africa. How do these rural lifestyles create distinctive patterns of settlement?

2. Describe the distinctive contributions of (a) Islam, (b) European colonialism, and (c) recent globalization to the region's urban landscape.

3. Summarize the key patterns and drivers of migration in and out of the region.

CULTURAL COHERENCE AND DIVERSITY: SIGNATURES OF COMPLEXITY

Although Southwest Asia and North Africa clearly define the heart of the Islamic and Arab worlds, cultural diversity also characterizes the region. Muslims practice their religion in varied ways, often disagreeing strongly on religious views. Elsewhere, other religions complicate cultural geography. Linguistically, Arabic languages are key, but non-Arab peoples, including Persians, Kurds, and Turks, also dominate portions of the region. These cultural geographies can help us understand the region's political tensions and appreciate why many of its residents resist processes of globalization.

Patterns of Religion

Religion is an important part of the lives of most people in Southwest Asia and North Africa. Whether it is the quiet ritual of morning prayers or discussions about current political and social issues, religion remains part of the daily routine of regional residents from Casablanca to Tehran.

Hearth of the Judeo-Christian Tradition   Both Jews and Christians trace their religious roots to an eastern Mediterranean hearth. The roots of Judaism lie deep in the past: Abraham, an early leader in the Jewish tradition who lived some 4,000 years ago, led his people from Mesopotamia to Canaan (modern-day Israel), near the shores of the Mediterranean. Jewish history, recounted in the Old Testament of the Holy Bible, focused on a belief in one God (or **monotheism**), a strong code of ethical conduct, and a powerful ethnic identity that continues to the present. During the Roman Empire, many Jews left the eastern Mediterranean to escape Roman persecution. The resulting forced migration, or diaspora, took Jews to the far corners of Europe and North Africa. Only in the past century have many of the world's far-flung Jewish people returned to the religion's place of origin, a process that gathered speed after the creation of Israel in 1948.

Christianity also emerged in the vicinity of modern-day Israel. An outgrowth of Judaism, Christianity was based on the teachings of Jesus and his disciples, who lived and traveled in the eastern Mediterranean about 2,000 years ago. Although many Christian traditions became associated with European history, forms of early Christianity remain near the religion's hearth. For example, the Coptic Church evolved in nearby Egypt. In Lebanon, Maronite Christians also retain a separate cultural identity.

The Emergence of Islam   Islam originated in Southwest Asia in 622 CE, forming another cultural hearth of global significance. Muslims can be found from North America to the southern Philippines, but the Islamic world remains centered on Southwest Asia. Most Southwest Asian and North African peoples still follow its religious teachings. Muhammad, the founder of Islam, was born in Makkah (Mecca) in 570 CE and taught in nearby Medinah (Medina) (Figure 7.25). His beliefs parallel Judeo-Christian traditions. Muslims believe both Moses and Jesus were prophets and that the Hebrew Bible (or Old Testament) and the Christian New Testament, while incomplete, are basically accurate. Ultimately, however, Muslims hold that the **Quran** (or Koran), a book of teachings received by Muhammad from Allah (God), represents God's highest religious and moral revelations to humanity.

Islam offers a blueprint for leading an ethical and religious life. Islam literally means "submission to the will of God," and the practice of the religion rests on five essential activities: (1) repeating the basic creed ("There is no God but God, and Muhammad is his prophet"); (2) praying facing Makkah five times daily; (3) giving charitable contributions; (4) fasting between sunup and sundown during the month of Ramadan; and (5) making at least one religious pilgrimage, or **Hajj**, to Muhammad's birthplace of Makkah (Figure 7.26). Islamic fundamentalists also argue for a **theocratic state**, such as modern-day Iran, in which religious leaders (ayatollahs) shape government policy.

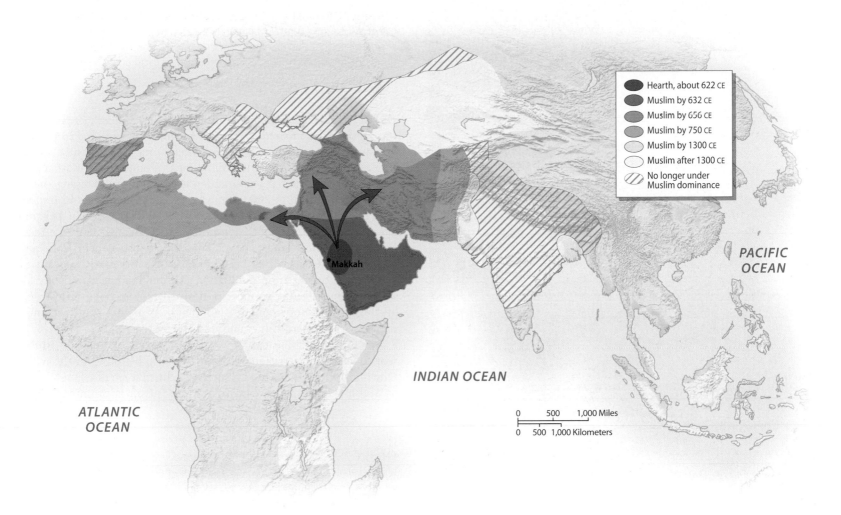

FIGURE 7.25 Diffusion of Islam The rapid expansion of Islam that followed its birth is shown here. From Spain to Southeast Asia, Islam's legacy remains strongest nearest its Southwest Asian hearth. In some settings, its influence has faded or has come into conflict with other religions, such as Christianity, Judaism, and Hinduism.

A major religious division split Islam early on and endures today. The breakup occurred almost immediately after the death of Muhammad in 632 CE. Questions surrounded who would inherit religious power. One group, now called **Shiites**, favored passing on power within Muhammad's family, specifically to Ali, his son-in-law. Most Muslims, later known as **Sunnis**, advocated passing down power through established clergy. This group was largely victorious. Ali was killed, and his Shiite supporters went underground. Ever since, Sunni Islam has formed the mainstream branch of the religion, to which Shiite Islam has presented a recurring and sometimes powerful challenge.

Islam quickly spread from the western Arabian Peninsula, following caravan routes and Arab military campaigns as it converted thousands to its beliefs (see Figure 7.25). By the time of Muhammad's death in 632 CE, peoples of the Arabian Peninsula were united under its banner. Shortly thereafter, the Persian Empire fell to Muslim forces, and the Eastern Roman (or Byzantine) Empire lost most of its territory to Islamic influences. By 750 CE, Arab armies swept across North Africa, conquered most of Spain and Portugal, and established footholds in Central and South Asia. By the 13th century, most people in the region were Muslims, and older religions such as Christianity and Judaism became minority faiths.

Between 1200 and 1500, Islamic influences expanded in some areas and contracted elsewhere. The Iberian Peninsula (Spain and Portugal) returned to Christianity in 1492, although Moorish (Islamic) cultural and architectural features remain today. At the same time, Muslims expanded southward and eastward into Africa. In addition, Muslim Turks largely replaced Christian Greek influences in Southwest Asia after 1100. One group of Turks moved into the Anatolian Plateau and conquered the Byzantine Empire in 1453. These Turks soon created the huge **Ottoman Empire** (named after one of its leaders, Osman), which included southeastern Europe (including modern-day Albania, Bosnia, and Kosovo) and most of Southwest Asia and North Africa. It provided a focus of Muslim political power until the empire's disintegration in the late 19th and early 20th centuries.

Modern Religious Diversity Today Muslims form the majority population in all the countries of Southwest Asia and North Africa except Israel, where Judaism dominates (Figure 7.27). Still, divisions

FIGURE 7.26 Makkah Thousands of faithful Muslims gather at the Grand Mosque in central Makkah (Mecca), part of the pilgrimage to this sacred place that draws several million visitors annually. A collection of hotels and portions of the city's commercial district can be seen in the distance.

within Islam create regional cultural differences. Most (73 percent) of the region is dominated by Sunni Muslims, but Shiites (23 percent) remain important elements in the contemporary cultural mix. In Iraq, for example, southern Shiite populations (around Najaf, Karbala, and Basra) asserted their cultural and political power following the fall of Saddam Hussein in 2003. Shiites also claim majorities in Iran and Bahrain. In addition, they form substantial minorities in Lebanon, Saudi Arabia, Yemen, and Egypt. Strongly associated with the recent flowering of the Islamist movement, Shiites also benefit from rapid growth rates because their brand of Islam is particularly appealing to the region's poorer rural population. Although some Sunnis are also Islamists, many reject its more radical cultural and political precepts and argue for a more modern Islam that incorporates some accommodation with Western values and traditions.

The Sunni–Shiite split is the great divide within the Muslim world, but other variations of Islam can also be found in the region. One division separates the mystically inclined form of Islam—known as *Sufism*—from mainstream traditions. Sufism is prominent in the

FIGURE 7.27 Modern Religions Islam continues to be the dominant religion across the region. Most Muslims are tied to the Sunni branch, whereas Shiites are found in places such as Iran and southern Iraq. In some locales, however, Christianity and Judaism remain important.

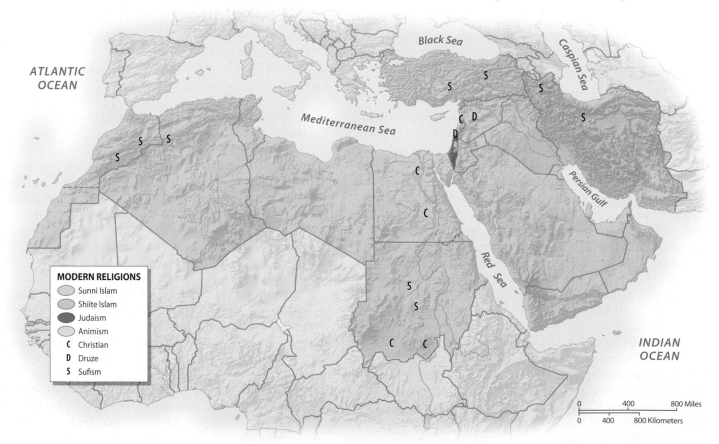

MODERN RELIGIONS
- Sunni Islam
- Shiite Islam
- Judaism
- Animism
- **C** Christian
- **D** Druze
- **S** Sufism

FIGURE 7.28 Old Jerusalem Jerusalem's historic center reflects its varied religious legacy. Sacred sites for Jews, Christians, and Muslims are all located within the Old City. The Western Wall, a remnant of the ancient Jewish temple, stands at the base of the Dome of the Rock and Islam's al-Aqsa Mosque.

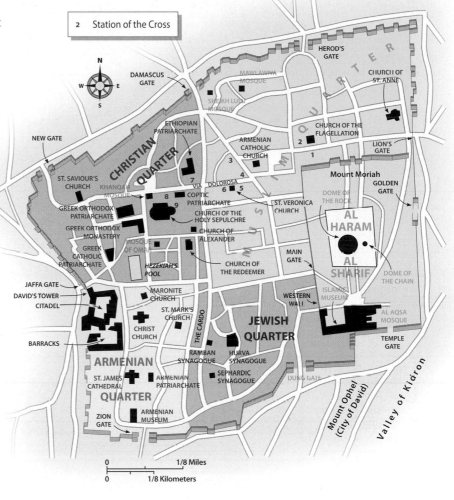

peripheries of the Islamic world, including the Atlas Mountains and across northwestern Iran and portions of Turkey. The Druze of Lebanon practice yet another variant of Islam.

Southwest Asia is also home to many non-Islamic communities. Israel has a Muslim minority (16 percent) that is dominated by that nation's Jewish population (77 percent). Even within Israel's Jewish community, differences divide Jewish fundamentalists from more reform-minded Jews. In neighboring Lebanon, a slight Christian (Maronite and Orthodox) majority was in evidence as recently as 1950. Christian out-migration and higher Islamic birthrates, however, have created a nation that today is about 60 percent Muslim.

Jerusalem (the Israeli capital) holds special religious significance for several groups and also stands at the core of the region's political problems (Figure 7.28). Indeed, the sacred space of this ancient Middle Eastern city remains deeply scarred and divided as different groups argue for more control of contested neighborhoods and nearby suburbs. Jews particularly honor the city's old Western Wall (the site of a Roman-era temple); Christians honor the Church of the Holy Sepulchre (the purported burial site of Jesus); and Muslims hold sacred religious sites in the city's eastern quarter (including the place from which the prophet Muhammad supposedly ascended to heaven).

Geographies of Language

Although the region is often referred to as the "Arab World," linguistic complexity creates many important cultural divisions across Southwest Asia and North Africa (Figure 7.29).

Semites and Berbers Afro-Asiatic languages dominate the region. Within that family, Arabic-speaking Semitic peoples are found from the Persian Gulf to the Atlantic and southward into Sudan. The Arabic language has religious significance for Muslims because it was the sacred language in which God delivered his message to Muhammad. Although most of the world's Muslims do not speak Arabic, the faithful often memorize prayers in the language, and many Arabic words have entered the other important languages of the Islamic world.

Hebrew, another Semitic language, was reintroduced into the region with the creation of Israel. Hebrew originated in the Levant and was used by the ancient Israelites 3,000 years ago. Today its modern version survives as the sacred tongue of the Jewish people and is the official language of Israel, although the country's non-Jewish population largely speaks Arabic.

Older Afro-Asiatic languages survive in more remote areas of North Africa. Collectively known as Berber, these languages are related to each other, but are not mutually intelligible. Most Berber languages have never been written, and none has generated a significant literature. Indeed, a Berber-language version of the Quran was not completed until 1999.

Persians and Kurds Although Arabic spread readily through portions of Southwest Asia, much of the Iranian Plateau and nearby mountains is dominated by older Indo-European languages. Here the principal tongue remains Persian, although, since the 10th century, the language has been enriched with Arabic words and written in the Arabic script.

Kurdish speakers of northern Iraq, northwest Iran, and eastern Turkey add further complexity to the regional pattern of languages. Kurdish, also an Indo-European language, is spoken by 10 to 15 million people in the region. The Kurds have a strong sense of shared cultural identity. Indeed, "Kurdistan" has sometimes been called the world's largest nation without its own political state. In the 2003 war in Iraq, the Kurds emerged as a cohesive group in the northern part of the country, and they have now gained more political autonomy. Kurds in eastern Turkey (about 20 percent of that nation's population), while enjoying more individual rights in that increasingly

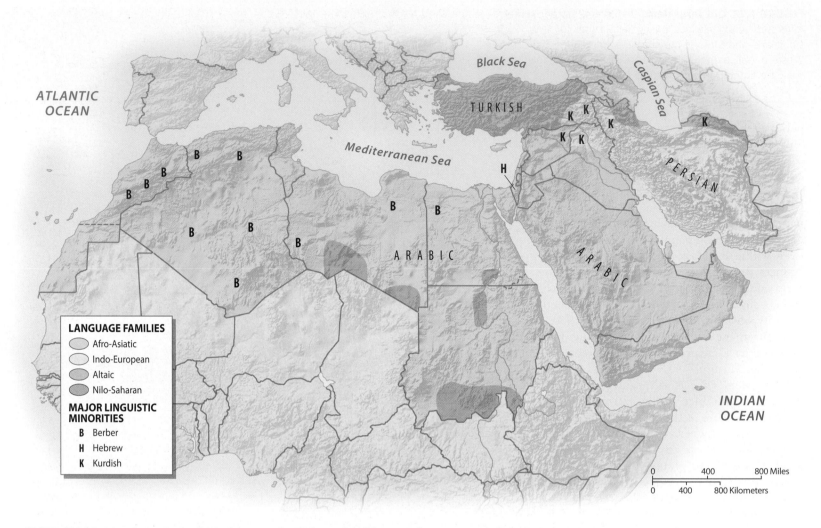

FIGURE 7.29 Language Map of Southwest Asia and North Africa   Arabic, a Semitic Afro-Asiatic language, dominates the region's cultural geography. Turkish, Persian, and Kurdish, however, remain important exceptions, and such differences within the region have often had long-lasting political consequences. Israel's more recent re-introduction of Hebrew further complicates the region's linguistic geography.

democratic country, have been stifled in their hopes for regional political autonomy. Violent street protests in 2012, building on similar efforts in Arab nations, called for increased Kurdish political control of their Turkish homelands.

The Turkish Imprint   Turkish languages provide variety across much of modern Turkey and in portions of far northern Iran. The Turkish languages are a part of the larger Altaic language family that originated in Central Asia. Turkey remains the largest nation in Southwest Asia dominated by this language family. Tens of millions of people in other countries of Southwest Asia and Central Asia speak related Altaic languages, such as Azeri, Uzbek, and Uyghur.

Regional Cultures in Global Context

Southwest Asian and North African peoples increasingly find themselves intertwined with cultural complexities and conflicts external to the region. Islam links the region with a Muslim population that now

lives all over the world, from North American cities to western China. Islam is a truly global religion. The religion's tradition of pilgrimage ensures that Makkah will continue as a city of ongoing global significance. People of the region also struggle to retain their traditional cultural values and still enjoy the benefits of global economic growth. Islamic fundamentalism is in many ways a reaction to the threat posed by external cultural influences, particularly those of western Europe and the United States.

Technology also contributes to cultural and political change. Particularly among the young, millions within and beyond the region found themselves linked by the Internet, cell phones, and various forms of social media during the Arab Spring uprisings of 2011 and 2012 (see "Exploring Global Connections: Social Media and Political Change: Lessons from the Arab Spring"). Widespread television viewing has transformed the region, with viewers offered everything from Islamist religious programming to American-style reality TV. Even in conservative Iran, where satellite dishes are officially banned, millions of people have access to them, beaming in multicultural programming from around the globe.

EXPLORING GLOBAL CONNECTIONS

Social Media and Political Change: Lessons from the Arab Spring

As civil unrest spread with lightning speed through much of the Middle East in 2011, and as several regional conflicts continued thereafter, discussion focused on the role played by social media in spreading the conflicts and in popularizing the underlying reform movements (Figure 7.2.1). So was the Arab Spring a child of the Facebook generation? Well, yes and no. Ultimately, the chain of events producing dramatic protests in settings such as Tunisia, Egypt, Libya, Bahrain, Yemen, and Syria was probably not about the tweets, viral videos, and cell-phone connections that were sensationalized in the global media. Rather, it was more about how complex *local* geographies of people (from diverse ethnic, class, and tribal backgrounds) found themselves dealing with an equally complex assortment of country-specific political issues.

Still, technology and the Arab Spring were inextricably intertwined, particularly in three ways. First, cell phones, blogs, email, and tweets *facilitated the flow of information* that helped protesters plan events and coordinate strategies with allies. Second, local videos from smart phones and pinhole cameras *documented government abuse* and often provided (in settings such as Syria) the only proof of widespread state-supported violence. Third, the global diffusion of this information promoted the *internationalization of political*

discourse and made it easier to spread the word about local conflicts and to identify common threads among different protest movements. A related point is that about 60 percent of the region's population is under 30—precisely the group most inclined to use these technologies and often the people most frustrated by unresponsive governments that refuse to change their ways.

But is this technology inherently more democratic or liberating? Critics argue that such notions are merely naïve cyber-utopianism. Consider what has happened in settings such as Iran, Syria, and China. In these countries, repressive governments have used the Internet and other forms of electronic communication (such as blogs and tweets) to compile dossiers on critics of the government, spy on online chatter, and even identify and arrest suspected enemies. In Iran, thousands were arrested after the 2009 uprisings as secret police made use of information from the Internet and social media. Similarly, Syria's repressive Assad regime made widespread use of Facebook to follow its opponents during its civil war. The

FIGURE 7.2.1 Communicating from the Front Lines, Cairo, Egypt, 2011   A young Egyptian woman talks on a mobile phone in Cairo's Tahrir Square during demonstrations in February, 2011.

Chinese government now employs thousands of workers to add pro-government online chatter to various Internet outlets.

So what lessons can we learn from the interplay of the Arab Spring and the new world of electronic and social media? No doubt, in that moment in time, tweets and cell phones mattered greatly. But equally clear is the fact that protesters have no monopoly on their knowledge and use of these communications technologies. Bloggers beware: Repressive governments can be fast learners. Finally, as events continue to unfold in the region, it appears that at least for now, local rivalries and shifting political alliances in particular locales may trump any potential for more pan-regional democratic and reform-minded discourse.

Hybrid forms of popular culture also reflect globalization. Recently, for example, both within the region and beyond, Arab hip-hop music has offered a form of artistic and political expression that represents the fusion of cultural traditions. Tune in MTV Arabia and you can hear Desert Heat (United Arab Emirates), Malikah (Lebanon), or MWR (Palestine) (Figure 7.30). These artists represent a generation of rappers that spits out lyrics challenging cultural and political stereotypes within the region. The global vibe is hard to miss: The African-American roots of rap remain a fixture in the strong beats and bops that set the tone amid the Arabic ouds (lutes), clangs of Asian pop, and soulful lyrics that might emerge in Arabic, English, or French. In other settings, millions of regional viewers tune in weekly to (and share tweets with) leading Islamic televangelists such as Amr Khaled (Egyptian) or Ahmad al-Shugairi (Saudi). Building on a Christian practice pioneered a half century earlier in the United States by Billy Graham, these broadcast preachers have become regional celebrities and iconic cultural figures.

FIGURE 7.30 Desert Heat   Rappers Salim and Abdullah Dahman from the United Arab Emirates created the group Desert Heat, and they celebrate many aspects of Arab culture in their music.

GEOPOLITICAL FRAMEWORK: NEVER-ENDING TENSIONS

Geopolitical tensions remain very high in Southwest Asia and North Africa (Figure 7.31). In the recent Arab Spring rebellions, governments fell in Tunisia (the regional movement began here in late 2010), Egypt, Libya, and Yemen; widespread protests shook once-stable states such as Bahrain; and a more protracted civil war decimated Syria. Many other countries witnessed shorter, more intermittent demonstrations against state authority. To varying degrees, these uprisings focused broadly on (1) charges of widespread government corruption; (2) limited opportunities for democracy and free elections; (3) rapidly rising food prices; and (4) the enduring reality of widespread poverty and high unemployment, especially for people under 30. When you add to this mix the complex local realities of traditional tribal rivalries, ethnic identities, and religious factionalism, it is no wonder that the future paths of particular countries appear as diverse as they are uncertain.

In addition to these recent conflicts, ongoing issues include the future of Israeli-Palestinian relations, Iran's saber rattling and nuclear ambitions, and Iraq's tentative emergence as an independent state. Some of these tensions relate to age-old patterns of cultural geography. European colonialism also contributes because modern boundaries were formed by colonial powers. Geographies of wealth and poverty also enter the geopolitical mix: Some residents profit from petroleum resources and industrial expansion, while others struggle to feed their families. Recently, Islamist elements in Iran, Egypt, Algeria, Tunisia, Turkey, Yemen, and Saudi Arabia have also changed the political atmosphere. The result is a political climate charged with tension, a region in which the sounds of bomb blasts and gunfire remain all-too-common characteristics of everyday life.

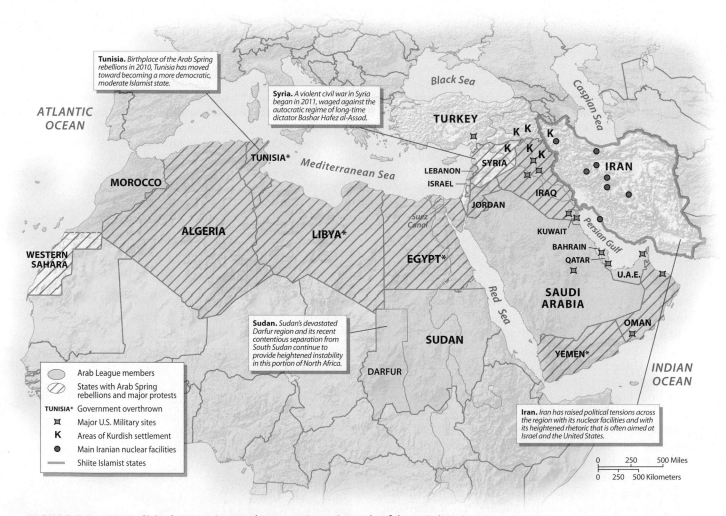

FIGURE 7.31 Geopolitical Issues in Southwest Asia and North Africa   Political tensions continue across much of the region. The Arab Spring rebellions shaped recent political changes in several regional settings. The Israeli-Palestinian conflict also remains pivotal.

The Colonial Legacy

European colonialism arrived relatively late in Southwest Asia and North Africa, but the era left an important imprint on the region's modern political geography. Between 1550 and 1850, the region was dominated by the Ottoman Empire, which expanded from its Turkish hearth to engulf much of North Africa, as well as nearby areas of the Levant, the western Arabian Peninsula, and modern-day Iraq. After 1850, Ottoman influences waned, and they were replaced with largely European colonial dominance after the dissolution of their empire in World War I (1918).

Both France and Great Britain were major colonial players within the region. French interests in North Africa included Tunisia and Morocco. French Algeria attracted large numbers of European immigrants. Following the defeat of the Ottoman Empire after World War I, France added more colonial territories in the Levant (Syria and Lebanon). The British loosely incorporated places such as Kuwait, Bahrain, Qatar, the United Arab Emirates, and Aden (in southern Yemen) into their empire to help control sea trade between Asia and Europe. Nearby Egypt also caught Britain's attention. Once the European-engineered **Suez Canal** linked the Mediterranean and Red seas in 1869, European banks and trading companies gained more influence over the Egyptian economy. In Southwest Asia, British and Arab forces joined to force out the Turks during World War I. The Saud family convinced the British that a country should be established on the Arabian Peninsula, and Saudi Arabia became fully independent in 1932. Britain divided its other territories into three entities: Palestine (now Israel) along the Mediterranean coast; Transjordan to the east of the Jordan River (now Jordan); and a third zone that became Iraq.

Persia and Turkey were never directly occupied by European powers. In Persia, the British and Russians agreed to establish two spheres of economic influence in the region (the British in the south, the Russians in the north), while respecting Persian independence. In 1935, Persia's modernizing ruler, Reza Shah, changed the country's name to Iran. In Turkey, European powers attempted to divide up the Ottoman Empire following World War I. The successful Turkish resistance to European control was based on new leadership provided by Kemal Ataturk. Ataturk decided to imitate the European countries and establish a modern, culturally unified, secular state.

European colonial powers began withdrawing from Southwest Asian and North African colonies before World War II. By the 1950s, most countries in the region were independent. In North Africa, Britain withdrew troops from Sudan and Egypt in 1956. Libya (1951), Tunisia (1956), and Morocco (1956) achieved independence peacefully during the same era, but the French colony of Algeria became a major problem. Several million French citizens resided there, and France had no intention of simply withdrawing. A bloody war for independence began in 1954, and France finally agreed to an independent Algeria in 1962.

Southwest Asia also lost its colonial status between 1930 and 1960. Iraq became independent from Britain in 1932, but its later instability resulted in part from its imposed borders, which never recognized its cultural diversity. Similarly, the French division of the Levant into Syria and Lebanon (1946) greatly angered local Arab populations and set the stage for future political instability. As a favor to the Lebanese Maronite Christian majority, France carved out a separate Lebanese state from largely Arab Syria, even guaranteeing the Maronites constitutional control of the government. The action created a culturally divided Lebanon as well as a Syrian state that has repeatedly asserted its influence over its Lebanese neighbors.

Modern Geopolitical Issues

The geopolitical instability in Southwest Asia and North Africa continues today. A quick regional transect from the shores of the Atlantic to the borders of Central Asia suggests how these forces are playing out in different settings early in the 21st century.

Across North Africa Varied North African settings have recently witnessed dramatic political changes (see Figure 7.31). In Tunisia, birthplace of the Arab Spring, a moderate Islamist government was elected to replace deposed dictator Zine el-Abidine Ben Ali. In nearby Libya, while many cheered the end of Colonel Muammar al-Qaddafi's rule in 2011, transitional government leaders and political parties were slow to establish a clear agenda and national consensus. Next door, following its 2011 overthrow of Hosni Mubarak, Egypt has moved ahead with the creation of more democratic institutions, including holding parliamentary and presidential elections in 2012. Complicating the story, however, is the interplay between the country's political and religious institutions. Many Christian Copts feel threatened if conservative Islamists gain too much power, including representatives of the increasingly political Islamist group called the Muslim Brotherhood. Even among Islamists, the popularity of the Salafists (a puritanical, fundamentalist Sunni group) is worrisome to moderates.

Elsewhere in North Africa, Islamist political movements have also reshaped the geopolitical landscape in several states. Most notably, recent violence has rocked both Algeria and Morocco. Protesters have pressed for some of the same democratic and economic reforms they see unfolding in nearby countries.

Sudan faces daunting political issues. A Sunni Islamist state since a military coup in 1989, Sudan imposed Islamic law across the country, antagonizing both moderate Sunni Muslims and the large non-Muslim (mostly Christian and animist) population in the south. Civil war between the north and south produced more than 2 million casualties (mostly in the south) between 1988 and 2004. A tentative peace agreement was signed in 2005, which opened the way for a successful vote on independence in South Sudan. Even though the two nations officially split in 2011, tensions remain, especially focused on an oil-rich and contested southern border zone between the two countries.

In addition, Sudan's western Darfur region remains in shambles (see Figure 7.31). Ethnicity, race, and control of territory seem to be at the center of the struggle in the largely Muslim region, as a well-armed Arab-led militia group (with many ties to the central government in Khartoum) has attacked hundreds of black-populated villages, killing more than 300,000 people (through violence, starvation, and disease) and driving 2.5 million more from their homes.

The Arab-Israeli Impasse The 1948 creation of the Jewish state of Israel produced another enduring zone of political tensions in the eastern Mediterranean. Jewish migration to Palestine increased after the defeat of the Ottoman Empire. In 1917, Britain issued the Balfour Declaration, a pledge to encourage the "establishment of Palestine as a home for the Jewish people." After World War II, the United Nations divided the region into two states: one mostly Jewish, the other mostly Muslim (Figure 7.32). Arab Palestinians rejected the partition, and war erupted. Jewish forces proved victorious, and by 1949, Israel had actually grown in size. The remainder of Palestine, including the West Bank and the Gaza Strip, passed to Jordan and Egypt, respectively. Hundreds of thousands of Palestinian refugees fled Israel to neighboring countries, where many remained in makeshift camps. Under these conditions, Palestinians nurtured the idea of creating their own state.

Israel's relations with neighboring countries remained poor. Supporters of Arab unity and Muslim solidarity sympathized with the Palestinians, while antipathy toward Israel grew. Israel fought additional wars in 1956, 1967, and 1973. In territorial terms, the Six-Day War of 1967 was the most important conflict (see Figure 7.32). In this struggle against Egypt, Syria, and Jordan, Israel occupied substantial new territories in the Sinai Peninsula, Gaza Strip, West Bank, and Golan Heights. Israel annexed the eastern part of the formerly divided city of Jerusalem, arousing particular bitterness among Palestinians because Jerusalem is a sacred city in the Muslim tradition (it contains the Dome of the Rock and the holy al-Aqsa Mosque) (see Figure 7.28). Jerusalem is sacred in Judaism as well (it contains the Temple Mount and the holy Western Wall of the ancient Jewish temple). A peace treaty with Egypt resulted in the return of the Sinai Peninsula in 1982, but tensions focused on other occupied territories remaining under Israeli control. To strengthen its claims, Israel built additional Jewish settlements in the West Bank and Golan Heights, further angering Palestinian residents.

Palestinians and Israelis negotiated a tentative settlement in the 1990s. Agreements called for a quasi-independent Palestinian state in which the ruling **Palestinian Authority (PA)** would have provisional control of Gaza and portions of the West Bank (Figure 7.33). However, a new cycle of heightened violence erupted late in 2000. Palestinian attacks against Jews increased, and Israelis continued building new settlements in occupied lands (especially in the West Bank)

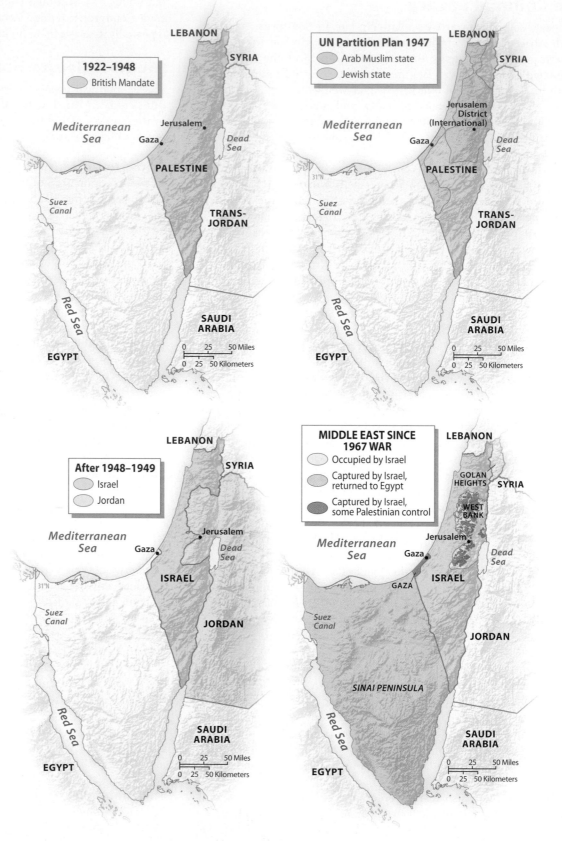

FIGURE 7.32 Evolution of Israel   Modern Israel's complex evolution began with an earlier British colonial presence and a United Nations partition plan in the late 1940s. Thereafter, multiple wars with nearby Arab states produced Israeli territorial victories in settings such as Gaza, the West Bank, and the Golan Heights. Each of these regions continues to be important in the country's recent relations with nearby states and with resident Palestinian populations.

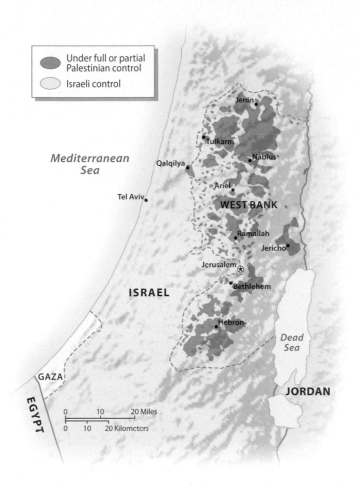

FIGURE 7.33 West Bank Portions of the West Bank were returned to Palestinian control in the 1990s, but Israel has partially reasserted its authority in some of these regions since 2000, citing the increased violence in the region. New Israeli settlements also are scattered throughout the West Bank in areas still under their nominal control.

FIGURE 7.34 Jewish Settlement, West Bank The new houses and well-planned neighborhoods of the West Bank Jewish settlement of Eli (foreground) contrast with the older Palestinian settlement in the distance. Many Palestinians resent the construction of these controversial Israeli settlements in the West Bank region.

(Figure 7.34). In 2010, with about 500,000 Israeli settlers already living in these West Bank settlements, the Israeli government pledged to continue supporting new construction, including developments near Jerusalem. They also enlarged their de facto control of many West Bank areas, especially near the Jordanian border.

Palestinian leaders harshly criticized continuing construction of these new Jewish settlements. Even more daunting has been construction of the Israeli security barrier, a partially completed 26-foot-high (8-meter-high) wall designed to separate Israelis from Palestinians across much of the West Bank region (Figure 7.35). Israeli supporters of the barrier (to be more than 400 miles [640 kilometers] long when completed) see it as a way to protect their citizens from suicide bombings and more terrorist attacks. Palestinians see it as a land grab, an "apartheid wall" designed to socially and economically isolate many settlements along the Israeli border. Nearby, Israel withdrew troops and settlers from the Gaza Strip in 2005, but political instability thereafter resulted in repeated Israeli incursions and reoccupations of the small war-torn region.

Political fragmentation of the Palestinians added further uncertainty. In 2006, control of the Palestinian government was split between the Fatah and Hamas political parties. Hamas has long been seen by many Israelis as an extremist and violent political party, whereas Fatah has shown more willingness to work peacefully with Israel. The split between these Palestinian factions became violent, with Hamas gaining effective control of the PA within Gaza, and with Fatah maintaining its greatest influence across the West Bank. Further complicating this political geography is the emergent Islamist movement among some Palestinians. Often, Islamists challenge the secular orientation of both Hamas and Fatah. These growing complexities within the Palestinian movement make an eventual peace treaty with Israel even more difficult to achieve.

One thing is certain: Geographical issues will remain at the center of the conflict. Israelis continue their search for secure borders to guarantee their political integrity. Most Palestinians still call for a "two-state solution," in which their autonomy is guaranteed, but a growing minority of Palestinians, frustrated with the stalemate, suggest considering a "one-state solution," in which Israel would be compelled to recognize the Palestinians as equals. Many Israelis, however, are wary of this option, seeing it as diluting a Jewish state they worked hard to create. Ultimately, the sacred geography of Jerusalem, a mere 220 acres (90 hectares) of land within the Old City, figures

241

SECURITY BARRIER

— Completed or planned in May 2013

○ Israeli settlements

FIGURE 7.35 Israeli Security Barrier The map (a) shows the completed and planned portions of the Israeli security barrier, as well as many of the Israeli settlements located in the West Bank region. The photo (b) shows a segment of the Israeli security barrier passing near Bethlehem, south of Jerusalem.

FIGURE 7.36 Violence in Homs, Syria Smoke rises from the Baba Amr neighborhood in Homs in June, 2012. The Syrian city became a focal point of rebel protests against the Assad government.

importantly in the conflict. Imaginative compromises in defining that political space will need to recognize its special value to both Jewish and Palestinian residents.

Other Regional Conflicts Elsewhere in the region, political instability in Syria erupted into civil war in 2011 and 2012. Rebel protests (mostly by Sunni Muslims) against the autocratic regime of President Bashar Hafez al-Assad (a member of the minority Alawite sect) reached a fever pitch, focused on the city of Homs, where government soldiers killed thousands of civilians in a series of violent confrontations (Figure 7.36). Rebels demand less corruption and a more democratic, religiously tolerant state. The larger Arab community also reacted against Assad, suspending Syria from the **Arab League** (a regional political and economic organization focused on

Arab unity and development; see Figure 7.31) and urging an international solution to the crisis. Thousands of Syrian refugees fled the country, burdening nations such as Turkey.

Iraq is another multinational state born during the colonial era that has yet to escape the consequences of its geopolitical origins. When the country was carved out of the British Empire in 1932, it contained the cultural seeds of its later troubles. Iraq remains culturally complex today (Figure 7.37). Most of the country's Shiites live in the lower Tigris and Euphrates valleys near and south of Baghdad. Indeed, the region near Basra contains some of the world's holiest Shiite shrines. In northern Iraq, the Kurds have their own ethnic identity and political aspirations. Many Kurds want complete independence from Baghdad, and they have managed to establish a federal region that already enjoys some autonomy from the central Iraqi government. A third major subregion is dominated by the Sunnis and includes part of the Baghdad area, as well as a triangle of territory to the north and west that includes strongholds such as Falujah and Tikrit. After enjoying a great deal of power under Saddam Hussein, many Sunnis feel they may get the short end of any future power-sharing agreement within the country.

Iraq was a major source of instability in the region before President Saddam Hussein was removed in 2003 by an American-led invasion of the country. When Iraqi leaders assumed control of their new state on June 28, 2004, growing religious and ethnic violence between different Iraqi factions threw portions of the nation into civil war. Rival Sunni and Shiite groups forced many Iraqis from their communities. Before largely leaving Iraq in 2011, American troops successfully worked with Iraqi officials to reduce the level of violence. Parliamentary elections, however, confirm the ongoing political uncertainties within Iraq. Both Shiite and Sunni candidates prove about equally popular, largely splitting the vote along regional lines (Shiites winning in the south and narrowly in Baghdad; Sunnis winning in central and western provinces). Complicating matters further, the Kurdish Alliance remains dominant in the north. An enduring geopolitical solution for a stable Iraqi state has yet to take shape, but it will need to recognize these cultural differences, as well as allow for an ongoing revival of the country's oil industry. By far, the country's dominant oil fields are located in Shiite- and Kurdish-controlled portions of Iraq, an uneasy fact of resource geography that has long troubled the nation's sizable Sunni population.

In the Persian Gulf states region, tiny Bahrain, while ruled by a Sunni monarchy, is a mostly Shiite country. Violent protests also erupted there in 2011, aimed at encouraging democratic reforms,

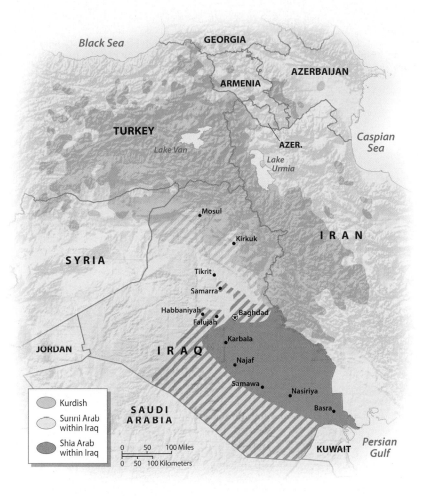

FIGURE 7.37 Multicultural Iraq Iraq's complex colonial origins produced a state with varying ethnic characteristics. Shiites dominate south of Baghdad, Sunnis hold sway in the western triangle zone, and Kurds are most numerous in the north, near oil-rich Kirkuk and Mosul.

Iran increasingly garners international attention. Islamic fundamentalism dramatically appeared on the political scene in 1978 as Shiite Muslim clerics overthrew the shah, an authoritarian, pro-Western ruler friendly to U.S. interests. The new leaders proclaimed an Islamic republic in which religious officials ruled both clerical and political affairs.

Today Iran supports Shiite Islamist elements throughout the region (such as Hezbollah) and has repeatedly threatened the state of Israel. The country is also seen as a threat by moderate Arab states. In 2009, the nearby United Arab Emirates, for example, invited the French to open their first new military base in decades on foreign soil, and French leaders pledged to defend against Iranian expansion. Adding uncertainty has been Iran's ongoing nuclear development program, an initiative its government claims is related to the peaceful construction of power plants (see Figure 7.31). Many in the West, however, remain unconvinced of the government's motives and demand its program be stopped or opened to more inspections before allowing for greater international cooperation with the regime. Israel, in particular, has suggested it may destroy Iranian nuclear sites before Iranian leaders have an opportunity to attack Israel with a nuclear-tipped missile. Expanded global economic sanctions against the country have also depressed the Iranian economy. Many of Iran's middle-class intellectuals, well-educated youth, and human-rights advocates are harsh critics of a government they feel is increasingly intolerant and guilty of isolating the country on the world stage.

> ## REVIEW QUESTIONS
>
> 1. Describe the role played by the French and British in shaping the modern political map of Southwest Asia and North Africa. Provide specific examples of their lasting legacy.
>
> 2. Discuss the root causes behind the Arab Spring rebellions and offer three examples of how the movement has played out politically in different ways within the region.
>
> 3. Explain how ethnic differences have shaped Iraq's political conflicts in the past 50 years.

rather than overthrowing the government. After violence escalated, troops from neighboring Saudi Arabia were called in by the Bahraini government to end the conflict. Saudi Arabia itself remains ruled by a conservative monarchy (the Saud family) unwilling to promote democratic reforms. On the surface, it supports U.S. efforts to provide stable flows of petroleum, but beneath the surface, certain elements of the regime may have financed radically anti-American groups such as Al Qaeda. The Saudi people themselves, largely Sunni Arabs, are torn among an allegiance to their royal family (and the economic stability it brings); the lure of a more democratic, open Saudi society; and an enduring distrust of foreigners, particularly Westerners. Furthermore, the Sunni majority includes Wahhabi sect members, whose radical Islamist philosophy has fostered anti-American sentiment. In addition, the large number of foreign laborers and the persisting U.S. military and economic presence within the country (one of the chief complaints of former Al Qaeda leader Osama bin Laden) create a setting ripe for political instability. In nearby Yemen, President Ali Abdullah Saleh was forced from office and elections were held in 2012. Even so, calls for democratic reforms are complicated by ongoing factionalism within the country, including pockets of Shiite militants that maintain close connections with Iran.

ECONOMIC AND SOCIAL DEVELOPMENT: LANDS OF WEALTH AND POVERTY

Southwest Asia and North Africa comprise a region of incredible wealth and discouraging poverty (Table 7.2). While some countries enjoy prosperity, due mainly to rich reserves of petroleum and natural gas, other nations are among the world's least developed. Continuing political instability contributes to the region's struggling economy. The Arab Spring rebellions, for example, while raising prospects for longer-term political reforms, also produced shorter-term challenges as economic output and tourism declined in many settings. In Iran and Syria, economic sanctions also crippled the economy. Petroleum will no doubt figure significantly in the region's future economy, but some countries in the area have also focused on increasing agricultural output, investing in new industries, and promoting tourism to broaden their economic base.

Table 7.2   DEVELOPMENT INDICATORS

Country	GNI per capita, PPP 2010	GDP Average Annual %Growth 2000–10	Human Development Index (2011)[1]	Percent Population Living Below $2 a Day	Life Expectancy (2012)[2]	Under Age 5 Mortality Rate (1990)	Under Age 5 Mortality Rate (2010)	Adult Literacy (% ages 15 and older)	Gender Inequality Index (2011)[3,1]
Algeria	8,100	3.9	.698	23.6	73	68	36	73	0.412
Bahrain	24,710	6.6	.806	—	78	17	10	91	0.288
Egypt	6,060	5.1	.644	15.4	72	94	22	66	—
Gaza and West Bank	2,900*	−0.9		<2	—	45	22	95	—
Iran	11,490	5.4	.707	8.0	70	65	26	85	0.485
Iraq	3,370	0.4	.573	21.4	70	46	39	78	0.579
Israel	27,660	3.6	.888	—	82	12	5	—	0.145
Jordan	5,800	6.7	.698	<2	73	38	22	92	0.456
Kuwait	—	8.4	.760	—	75	15	11	94	0.229
Lebanon	14,090	4.9	.793	—	72	38	22	90	0.440
Libya	16,880	5.4	.760	—	75	45	17	89	0.314
Morocco	4,600	4.9	.582	14.0	72	86	36	56	0.510
Oman	25,190	4.7	.705	—	74	47	9	87	0.309
Qatar	—	14.2	.831	—	78	21	8	95	0.549
Saudi Arabia	22,750	3.6	.770	—	74	45	18	86	0.646
Sudan	2,030	6.7	.408	44.1	60	125	103	—	0.611
Syria	5,120	5.0	.632	16.9	74	38	16	84	0.474
Tunisia	9,060	4.7	.698	8.1	75	49	16	78	0.293
Turkey	15,530	4.7	.699	4.2	73	80	18	91	0.443
United Arab Emirates	47,310	5.1	.846	—	76	22	7	90	0.234
Western Sahara	2,500*	—	—	—	67	—		—	—
Yemen	2,500	4.1	.462	46.6	65	128	77	62	0.769

[1]United Nations, *Human Development Report, 2011.*
[2]Population Reference Bureau, *World Population Data Sheet, 2012.*
[3]Gender Inequality Index—A composite measure reflecting inequality in achievements between women and men in three dimensions: reproductive health, empowerment and the labor market that ranges between 0 and 1. The higher the number, the greater the inequality.
*Additional data from the *CIA World Factbook, 2012*
Source: World Bank, *World Development Indicators, 2012*

The Geography of Fossil Fuels

The striking global geographies of oil and natural gas reveal the region's continuing importance in the world economy, as well as the extremely uneven distribution of these resources within the region (Figure 7.38). North African settings (especially Algeria and Libya), as well as the Persian Gulf region, have large sedimentary basins that contain huge reserves of oil and gas, whereas other localities (for example, Israel, Jordan, and Lebanon) lie outside zones of major fossil fuel resources. Saudi Arabia, Iran, Iraq, Kuwait, and the United Arab Emirates hold large petroleum reserves, while Iran and Qatar possess the largest regional reserves of natural gas. The distribution of fossil fuel reserves suggests that regional supplies will not be exhausted anytime soon. Overall, with only 7 percent of the world's population, the region holds a staggering 60 percent of the world's proven oil reserves. Saudi Arabia's pivotal position, both regionally and globally, is clear: Its 30 million residents live atop almost 20 percent of the planet's known oil supplies.

Regional Economic Patterns

Remarkable economic differences characterize Southwest Asia and North Africa (see Table 7.2). Some oil-rich countries have prospered greatly since the early 1970s, but in many cases fluctuating oil prices, political disruptions, and rapidly growing populations threaten future economic growth.

Higher-Income Oil Exporters   The richest countries of Southwest Asia and North Africa owe their wealth to massive oil reserves. Nations such as Saudi Arabia, Kuwait, Qatar, Bahrain, and the United Arab Emirates benefit from fossil fuel production, as well as from their relatively small populations. Large investments in transportation networks, urban centers, and other petroleum-related industries have reshaped the cultural landscape. The petroleum-processing and -shipping centers of Jubail (on the Persian Gulf) and Yanbu (on the Red Sea) are examples of this commitment to expand the Saudi economic base beyond the simple extraction of crude oil (Figure 7.39). Billions of dollars have poured into new schools, medical facilities, low-cost housing, and modernized agriculture, significantly raising the standard of living in the past 40 years.

Still, problems remain. Dependence on oil and gas revenues produces economic pain in times of falling prices, such as those seen between 2008 and 2010. Such fluctuations in world oil markets will inevitably continue in the future, disrupting construction projects, producing large layoffs of immigrant populations, and slowing investment in the region's economic and social infrastructure. In addition, countries such as Bahrain and Oman are faced with the problem of rapidly depleting their reserves over the next 20 to 30 years.

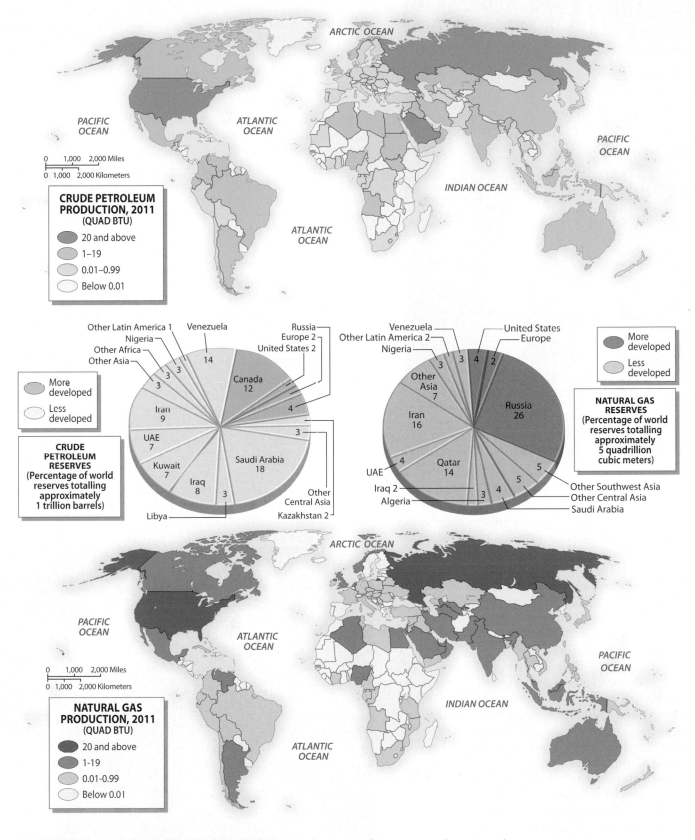

FIGURE 7.38 Crude Petroleum and Natural Gas Production and Reserves   The region plays a pivotal role in the global geography of fossil fuels. Abundant regional reserves suggest that the pattern will continue.

FIGURE 7.39 Yanbu, Saudi Arabia Pipelines facilitate the movement of oil to and from Yanbu's large coastal refineries in western Saudi Arabia.

Lower-Income Oil Exporters Some countries possess reserves of fossil fuels, but different political and economic variables have hampered sustained economic growth. In North Africa, for example, Algerian oil and natural gas overwhelmingly dominate the country's exports, but the past 20 years have also brought political instability and shortages of consumer goods. Although the country contains excellent agricultural lands in the north, the overall amount of arable land has increased little over the past 25 years, even as the country's population has grown more than 50 percent.

In Southwest Asia, Iraq also faces huge challenges. War has crippled much of Iraq's already deteriorated infrastructure, and political instability has made rebuilding its economy more difficult. Iraq suffers from high unemployment, more than 20 percent of the population remains malnourished, and only 25 percent of the country is served by dependable electricity. Oil output has increased since 2009, however, suggesting a potential for economic recovery.

The situation in Iran is also challenging. The country's oil and gas reserves are huge, but Iran is relatively poor, burdened with a stagnating standard of living. Since 1980, fundamentalist leaders have downplayed the role of international trade in consumer goods and services, fearing the import of unwanted cultural influences. Recent international sanctions on purchases of Iranian oil (related to its nuclear development program) have depressed the economy. Some economic bright spots have appeared, however: The country benefits from

energy developments in Central Asia, and literacy rates (particularly for women) have risen, reflecting a new emphasis on rural education.

Prospering Without Oil Some countries, while lacking petroleum resources, have still found paths to economic prosperity. Israel, for example, supports one of the highest standards of living in the region, even with its political challenges (see Table 7.2). Israelis and many foreigners have invested large amounts of capital to create a highly productive agricultural and industrial base. In addition, the country is a global center for high-tech computer and telecommunications products, known for its fast-paced and highly entrepreneurial business culture that resembles California's Silicon Valley (Figure 7.40). Israel also has daunting economic problems. Its struggles with the Palestinians and with neighboring states have sapped potential vitality. Defense spending absorbs a large share of total gross national income (GNI), necessitating high tax rates. Despite recent economic recovery, especially in Gaza, poverty and unemployment among Palestinians also remain unacceptably high.

Turkey also has a diversified economy, even though incomes are modest by regional standards. Lacking petroleum, Turkey produces varied agricultural and industrial goods for export. About 35 percent of the population remains in agriculture, and the country's principal commercial products include cotton, tobacco, wheat, and fruit. The industrial economy has grown since 1980, including exports of textiles, processed food, and chemicals. Turkey remains an important tourist destination in the region as well, attracting more than 6 million visitors annually in recent years.

Regional Patterns of Poverty Poorer countries of the region share the problems of the less developed world. For example, Sudan, Egypt, and Yemen each face unique economic challenges. For Sudan, continuing political problems have stood in the way of progress. Political disruptions have resulted in major food shortages. The country's transportation and communications systems have seen little new investment, and secondary school enrollments remain very low. On the other hand, Sudan's

FIGURE 7.40 Israeli High-Tech Industry These Israeli workers are employees of the Telrad factory, a high-tech division of the Koor conglomerate.

FIGURE 7.41 Cairo Slums One distinctive group of Cairo's population is the Zabbaleen, a community of Coptic Christians who collect the city's trash and live along the narrow streets of Manshiet Nasser, one of the city's most poverty-stricken neighborhoods.

fertile soils could support more farming, and its share of regional oil prospects suggests petroleum's expanding role in the economy.

Egypt's economic prospects are also unclear. Since the overthrow of the Mubarak regime in 2011, unemployment has risen, tourism has declined, and foreign investors have remained wary of the nation's uncertain political environment. Many Egyptians live in poverty, and the gap between rich and poor continues to widen (Figure 7.41). Illiteracy is widespread, and the country suffers from the **brain drain** phenomenon, as some of its brightest young people leave for better jobs in western Europe or the United States. Egypt's 80 million people already make it the region's most heavily populated state, and recent efforts to expand the nation's farmland have met with numerous environmental, economic, and political problems.

Yemen remains the poorest country on the Arabian Peninsula. Positioned far from the region's principal oil fields, Yemen's low per capita GNI puts the country on par with nations in impoverished Sub-Saharan Africa. The largely rural country relies mostly on marginally productive subsistence agriculture, and much of its mountain and desert interior lacks effective links to the outside world. Coffee, cotton, and fruits are commercial agricultural products, and modest oil exports bring in needed foreign currency. Overall, however, high unemployment remains widespread. Recent political uncertainties also cloud economic prospects.

A Woman's Changing World

The role of women in largely Islamic Southwest Asia and North Africa remains a major social issue. Female labor participation rates in the workforce are among the lowest in the world, and large gaps typically exist between levels of education for males and females. In conservative parts of the region, few women work outside the home. Even in parts of Turkey, where Western influences are widespread, it is rare to see rural women selling goods in the marketplace or driving cars in the street. More orthodox Islamic states impose legal restrictions on the activities of women. In Saudi Arabia, for example, women are not allowed to drive, although growing protests among many younger, educated Saudi females may overturn that ban. In Iran, full veiling remains mandatory in more conservative parts of the country, and police

often hassle Iranian women in more liberal dress, even in larger urban areas. Generally, Islamic women lead more private lives than men: Much of their domestic space is shielded from the world by walls and shuttered windows, and their public appearances are filtered through the use of the *niqab* (face veil) or *chador* (full-body veil).

Yet in some places, women's roles are changing, even within norms of more conservative Islamist societies. From Tunisia to Yemen, women widely participated in the Arab Spring rebellions, asserting their new political visibility in very public ways. The *local* consequences of these political and social changes are complex: In liberated Libya, young Islamic women rejoice in their freedom to wear the *niqab* in public, a practice banned under Colonel Muammar al-Qaddafi's rule. On the other hand, greater postrebellion freedoms in Egypt encouraged one young woman to pose almost naked on Facebook and Twitter, sparking controversy (and a supportive show of solidarity from a large group of Israeli women who also posed naked) and more than a million page views. Other women have been encouraged to play more active roles in political affairs, including running for office.

Algerian women demonstrate the pattern (Figure 7.42). Most studies suggest those in the younger generation are more religious than their parents and are more likely to cover their heads and drape their bodies with traditional religious clothing. At the same time, they are more likely to be educated and employed. Today 70 percent of Algeria's lawyers and 60 percent of its judges are women. A majority of university students are women, and women dominate the health-care field. These new social and economic roles help explain why birthrates are declining. The Algerian case also demonstrates how complex social processes actually unfold, rarely following simple models for how "traditional" societies evolve.

Women's roles are shifting throughout the region. In Sudan and Saudi Arabia, a growing number of women pursue high-level careers. Education may be segregated, but it is available. In Libya, more women than men graduate from the country's university system. In Western Sahara, Saharawi women play a leading role in the country's political fight for independence from Morocco, and their broader educational backgrounds and social freedoms (including the right to divorce their husbands) further separate them from Moroccan women. Women also have a more visible social position in Israel except in fundamentalist Jewish communities, where conservative social customs limit women to more traditional domestic roles.

FIGURE 7.42 Algerian Women Women now make up more than 30 percent of Algeria's newly-elected National Assembly, a higher proportion of female representation than in many Western nations.

Global Economic Relationships

Southwest Asia and North Africa share close economic ties with the rest of the world. While oil and gas remain critical commodities that dominate international economic linkages, the growth of manufacturing and tourism is also redefining the region's role in the world.

OPEC's Changing Fortunes   While OPEC does not control global oil and gas prices, it still influences the cost and availability of these pivotal products. Western Europe, the United States, Japan, China, and many less industrialized countries depend on the region's fossil fuels. In the case of Saudi Arabia, for example, petroleum and its related products comprise more than 90 percent of its exports. Many oil-producing countries (such as Saudi Arabia) form partnerships with foreign corporations, accelerating the economic integration of the region with the rest of the world. Even so, the region's major energy producers are always vulnerable to global-scale recessions, such as the downturn that hit the region hard between 2008 and 2010.

Beyond key OPEC producers, other trade flows also contribute to global economic integration. Turkey, for example, ships textiles, food products, and manufactured goods to its principal trading partners: Germany, the United States, Italy, France, and Russia. Tunisia sends more than 60 percent of its exports (mostly clothing, food products, and petroleum) to nearby France and Italy. Israeli exports emphasize the country's highly skilled workforce: Products such as cut diamonds, electronics, and machinery parts are exported to the United States, western Europe, and Japan.

Regional and International Linkages   Future interconnections between the global economy and Southwest Asia and North Africa may depend increasingly on cooperative economic initiatives far beyond OPEC. Relations with the European Union (EU) are critical. Since 1996, Turkey has enjoyed close ties with the EU, but recent attempts at full membership in the organization have failed. Other so-called Euro-Med agreements also have been signed between the EU and countries across North Africa and Southwest Asia that border the Mediterranean Sea.

Most Arab countries, however, are wary of too much European dominance. In 2005, seventeen Arab League members established the Greater Arab Free Trade Area (GAFTA), designed to eliminate all intra-regional trade barriers and spur economic cooperation. In addition, Saudi Arabia plays a pivotal role in regional economic development through organizations such as the Islamic Development Bank and the Arab Fund for Economic and Social Development.

FIGURE 7.43 Naama Bay, Egypt   At Naama Bay, scenic El Fanar beach attracts sunbathers and snorkeling enthusiasts.

Tourists are another link to the global economy. Traditional magnets such as ancient historical sites and globally significant religious localities draw millions of visitors annually. As the developed world becomes wealthier, a growing demand arises for recreational spots that offer beaches, sunshine, and novel entertainment. Indeed, many miles of the Mediterranean, Black Sea, and Red Sea coastlines are now lined with upscale, but often ticky-tacky, landscapes of resort hotels and condominiums dedicated to serving travelers' needs. More adventurous travelers seek ecotourist activities such as snorkeling in Naama Bay on Egypt's Sinai coast (Figure 7.43) or four-wheeling among the Berbers in the Moroccan backcountry. Endangered wildlife beckons photographers and poachers hoping to catch a glimpse of a South Arabian grey wolf, a Nubian ibex, or a darting Persian squirrel. All this activity means big business, and the economic impacts of tourism are likely to grow. Even in the aftermath of the disruptive Arab Spring rebellions, countries such as Tunisia and Egypt have redoubled their efforts to boost their tourist economies.

REVIEW QUESTIONS

1. Describe the basic geography of oil reserves across the region, and compare the pattern with the geography of natural gas reserves.

2. What different strategies for economic development have recently been employed by nations such as Saudi Arabia, Turkey, Israel, and Egypt? How successful have they been, and how do they relate to the theme of globalization?

Summary

- Positioned at the meeting ground of Earth's largest landmasses, Southwest Asia and North Africa have played a critical role in world history and in the processes of globalization that bind the planet together ever more tightly.

- In ancient times, the region was home to early examples of crop and livestock domestication, as well as some of the world's earliest urban centers. Three of the world's great religions—Judaism, Christianity, and Islam—also emerged beneath its desert skies.

- Despite the rich legacy of global influence and power, the peoples of Southwest Asia and North Africa are struggling early in the

21st century. Indeed, most countries are suffering from significant economic problems and political uncertainties. The region has faced difficulty and high costs in trying to expand its limited supplies of agricultural land and water resources amid fast-growing populations.

- Political conflicts have disrupted economic development across the region. Civil wars, conflicts between states, and regional tensions have worked against plans for greater cooperation and trade. Most importantly, the region must deal with the basic inconsistencies between   Western   civilization   and   more   fundamentalist interpretations of Islam, as well as with finding a lasting solution to the Israeli-Palestinian conflict.

■ As the recent Arab Spring rebellions demonstrate, future cultural and political change is difficult to anticipate. It will be guided by a complex response to Western influences, a mix of fascination and suspicion that will produce its own unique regional geography. Southwest Asia and North Africa will retain its distinctive regional identity, a character defined by its environmental setting, the rich cultural legacy of its history, the selective abundance of its natural resources, and its continuing political problems.

Key Terms

Arab League 242
Arab Spring 216
brain drain 247
choke point 220
culture hearth 217
domestication 225
exotic river 226
Fertile Crescent 225
fossil water 220
Hajj 232

hydropolitics 220
Islamic fundamentalism 218
Islamism 218
Levant 221
Maghreb 220
medina 229
monotheism 232
Organization of the Petroleum Exporting
  Countries (OPEC) 218
Ottoman Empire 233

Palestinian Authority
  (PA) 240
pastoral nomadism 225
physiological density 225
Quran 232
Shiite 233
Suez Canal 239
Sunni 233
theocratic state 232
transhumance 225

Thinking Geographically

1. How might a major project for transferring water from Turkey to the Arabian Peninsula affect the development of Saudi Arabia? What would be some of the potential political and ecological ramifications of such a project?

2. Why are birthrates declining in some countries in Southwest Asia and North Africa? Despite the cultural differences with North America, what common processes seem to be at work in both regions that have contributed to this demographic transition?

3. Imagine being the ruler of a conservative Islamist Arab state. What might be the advantages and disadvantages of opening up your country to the Internet? Divide into opposing groups and debate both sides of the argument.

4. Form groups of several students and investigate the different local causes and consequences of the Arab Spring rebellions within different countries of the region.

5. What *economic changes* could occur if Israel and the Palestinians were to reach a lasting peace? What kinds of general connections might be found between political conflicts and economic conditions throughout the region?

Mastering Geography™

Looking for additional review and test prep materials? Visit the Study Area in MasteringGeography™ to enhance your geographic literacy, spatial reasoning skills, and understanding of this chapter's content by accessing a variety of resources, including MapMaster interactive maps, videos, RSS feeds, flashcards, web links, self-study quizzes, and an eText version of *Globalization and Diversity*.

Scan to visit the author's blog for chapter updates.

http://gad4blog.
wordpress.com/
category/southwest-
asia-and-
north-africa/

Scan to visit the GeoCurrents blog.

http://geocurrents.
info/category/place/
southwest-asia-and-
north-africa

Authors' Blogs

Scan now to access the authors' blogs for up-to-date information on Southwest Asia and North Africa.

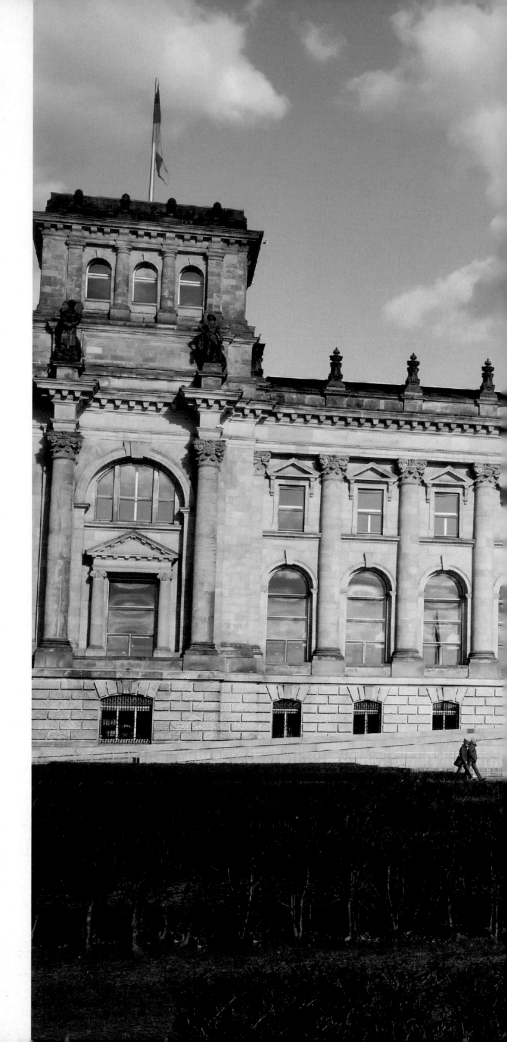

Globalization and Diversity

Historically, Europe was one of the world's earliest promoters of globalization as it spread its languages, laws, religion, politics, and economic systems throughout the world during the colonial period. Today, however, Europe struggles to adjust to the social and ethnic diversity sown by these earlier seeds as migrants from throughout the world seek a share of Europe's high standard of living.

ENVIRONMENTAL GEOGRAPHY

Europe is one of the "greenest" world regions, with strong environmental laws and regulations about recycling, energy efficiency, and pollution. Europe is also a global leader in promoting solutions to global climate change.

POPULATION AND SETTLEMENT

With low (or no) natural population growth in Europe, immigration from other world regions is a mixed blessing, providing a solution to labor shortages in some areas, but also bringing troublesome social issues to others.

CULTURAL COHERENCE AND DIVERSITY

Europe has a long history of internal cultural tensions linked to differences in language and religion. Although these differences have been largely resolved through the integrating force of the European Union (EU), other cultural tensions have surfaced associated with legal and illegal immigration to the region.

GEOPOLITICAL FRAMEWORK

After the lengthy Cold War (1945–1990) that divided Europe into two parts, east and west, the region is now experiencing political integration of former adversaries at a national level. Outbreaks of micro-nationalism, however, with agendas of autonomy, are still problematic in the Balkans, Spain, and the British Isles.

ECONOMIC AND SOCIAL DEVELOPMENT

Through the past half century, the EU has achieved remarkable success in bringing peace and prosperity to this diverse region. Nonetheless, serious economic and social challenges still remain, particularly in southern and eastern Europe.

➤ In many ways the historic Reichstag building, which today houses the German parliament, is emblematic of Europe's long and complex political geography. Built in 1894 for Germany's first democratic government, it became a symbol of Hitler's puppet government during the Nazi period, then was restored after Germany's reunification in 1990, and once again became the seat of Germany's government in 1999.

EUROPE 8

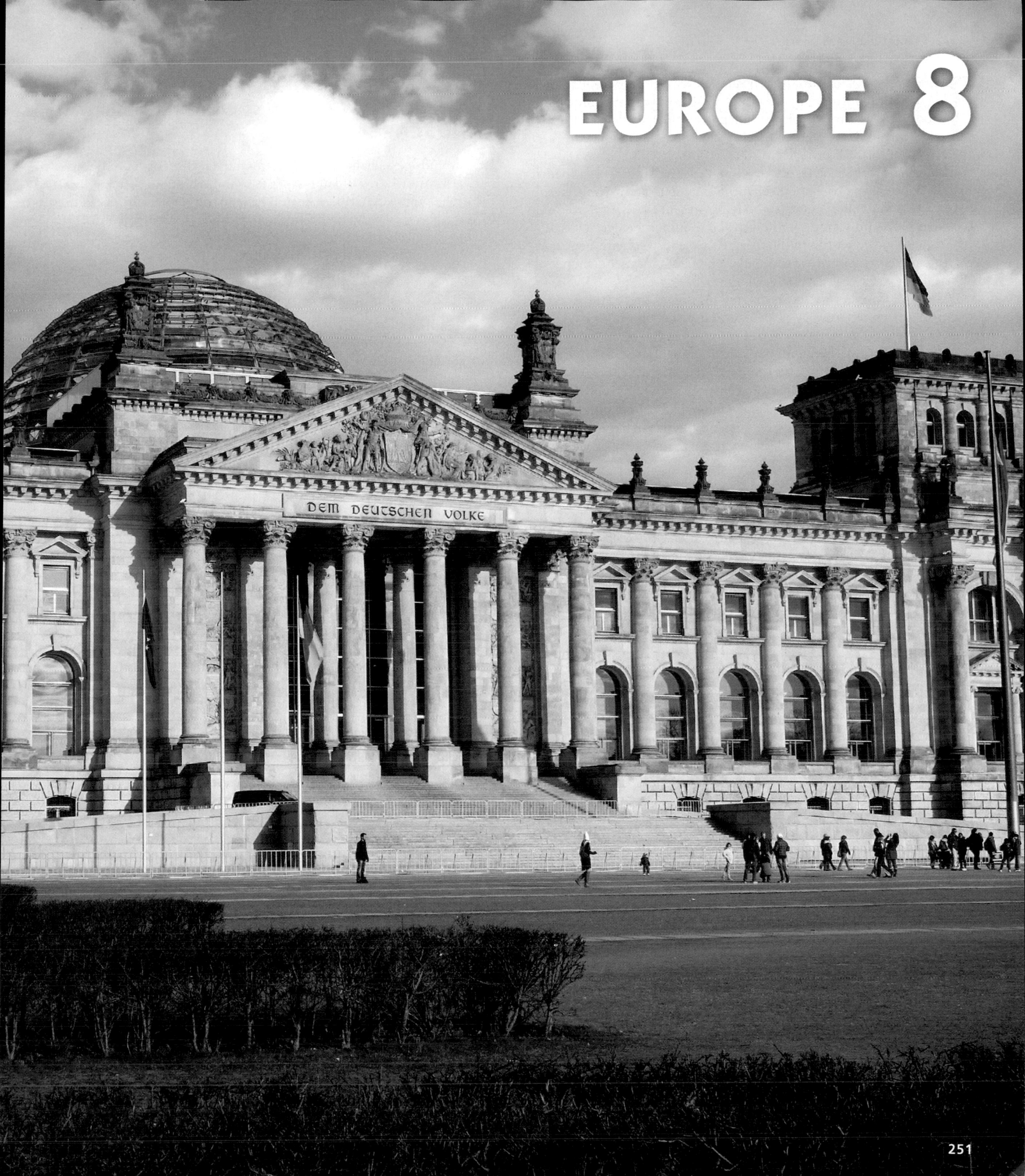

In many ways, Berlin is a poster child for Europe's complex geopolitical history of war, ruin, rebirth, division, unification, and, most recently, regional hotspot of globalization and diversity. Berlin's landscape shows it all: resplendent boulevards and palaces of a historic capital city that was then bombed into abject ruin during World War II; a struggling postwar orphan city divided by Cold War tensions embodied in the infamous Berlin Wall; heady skyscrapers and trade centers embodying present-day Berlin's role as a cosmopolitan and vibrant world city of 3.5 million people; a sustainable energy-powered capitol building that blends old with new; and, on the streets, a highly diverse population of both locals and migrants, like Europe itself.

Europe is one of the most diverse regions in the world, encompassing a wide assortment of people and places in an area considerably smaller than North America. More than half a billion people reside in this region, living in 42 countries ranging in size from giant Germany to microstates such as Andorra and Monaco (Figure 8.1).

The region's remarkable cultural diversity produces a geographic mosaic of languages, religions, and landscapes. Commonly, a day's journey finds travelers speaking two or three languages, crossing several political borders, and possibly changing money as they travel through different countries.

Although a traveler may revel in Europe's cultural and environmental diversity, these distinct regional differences are also responsible for Europe's troubled past. In the 20th century alone, Europe was the principal battleground for two world wars, followed by the 45-year **Cold War** (1945–1990), which divided the continent and the world into two hostile and highly armed camps—Europe and the United States against the former Soviet Union (now Russia) and its allies.

Today, however, Europe attempts to set aside nationalistic pride and work toward regional economic, political, and cultural integration through the **European Union (EU).** This supranational organization is made up of 27 countries, anchored by the western European states of Germany, France, Italy, and the United Kingdom, but also including eastern European countries that were former Soviet satellites. While controversial at times, the EU has unquestionably emerged as an unparalleled regional organization (Figure 8.2).

ENVIRONMENTAL GEOGRAPHY: HUMAN TRANSFORMATION OF A DIVERSE LANDSCAPE

Despite Europe's small size, its environmental diversity is extraordinary. A startling array of landscapes is found within its borders, from the Arctic tundra of northern Scandinavia to the barren hillsides of the Mediterranean islands, and from the explosive volcanoes of southern Italy to the glaciated seacoasts of Norway and Iceland.

Four factors explain this environmental diversity:

- The complex geology of this western extension of the Eurasian land mass has the newest, as well as the oldest, landscapes in the world.
- Europe's latitudinal extent from the Arctic to the Mediterranean subtropics affects climate, vegetation, and many human activities (Figure 8.3).
- This latitudinal control is modified by the moderating influences of the Atlantic Ocean, as well as the Baltic, Mediterranean, and Black Seas.
- The long history of human settlement has transformed and modified Europe's natural landscapes in fundamental ways over thousands of years.

Landform Regions

Europe can be organized into four general topographic and landform regions: the European Lowland, forming an arc from southern France to the northeast plains of Poland, but also including southeastern England; the Alpine mountain system, extending from the Pyrenees in the west to the Balkan Mountains of southeastern Europe; the Central Uplands, positioned between the Alps and the European Lowland; and the Western Highlands, which include mountains in Spain, portions of the British Isles, and the highlands of Scandinavia.

The European Lowland   This lowland, also known as the North European Plain, is the unquestioned economic focus of western

LEARNING OBJECTIVES

After reading this chapter you should be able to:

- Describe, in general terms, the topography, climate, and hydrology of Europe.
- Identify the major environmental issues in Europe, as well as the pathways being taken to resolve those problems.
- Give examples of countries with varying rates of natural growth.
- Describe the characteristics and implications of internal migration within Europe, as well as the geography of external migration to the region.

- Describe the major languages and religions of Europe.
- Summarize how the map of European states has changed in the last 100 years.
- Explain how Europe was divided during the Cold War and how Europe has changed since the end of the Cold War in 1990.
- Describe how Europe has become integrated economically and politically by the EU and its predecessors.
- Identify the major characteristics of the current economic tensions within Europe in terms of economic and social development.

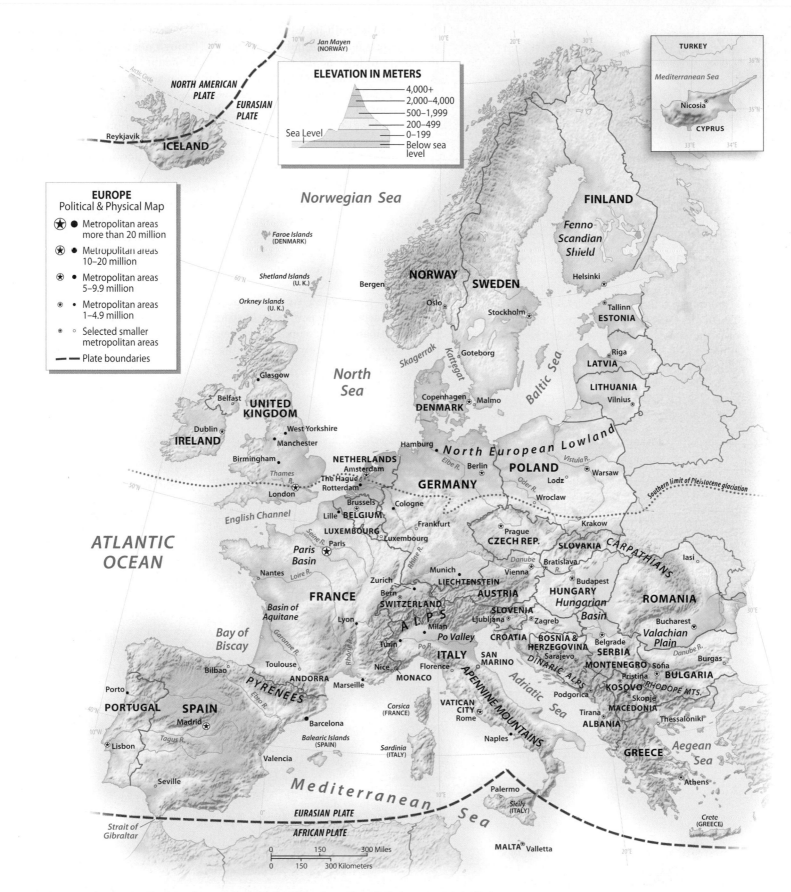

EUROPE
Political & Physical Map

- ✪ ● Metropolitan areas more than 20 million
- ✪ ● Metropolitan areas 10–20 million
- ✪ ● Metropolitan areas 5–9.9 million
- ✪ ● Metropolitan areas 1–4.9 million
- ✪ ○ Selected smaller metropolitan areas
- – – Plate boundaries

ELEVATION IN METERS

	4,000+
	2,000–4,000
	500–1,999
	200–499
Sea Level	0–199
	Below sea level

FIGURE 8.1 Europe Stretching from Iceland in the Atlantic to the Black Sea, Europe includes 42 countries, ranging in size from large states, such as France and Germany, to the microstates of Liechtenstein, Andorra, San Marino, and Monaco. Currently, the population of the region is about 531 million.

FIGURE 8.2 European Union Controversy The European Union (EU) was founded in the 1950s to facilitate greater cooperation between European countries. This it has achieved in many ways, yet in doing so it has also created controversy as the EU deliberately erodes the power of individual member countries. Here, Germans protest the 2012 EU plan to bail out weak Mediterranean economies with subsidies from fiscally stronger EU countries.

Europe, with its high population density, intensive agriculture, large cities, and major industrial regions. Though not completely flat, most of this lowland lies below 500 feet (150 meters) in elevation. In places, however, it is broken by rolling hills, plateaus, and uplands (such as in Brittany, France), where elevations exceed 1,000 feet (300 meters). Many of Europe's major rivers (the Rhine, the Loire, the Thames, and the Elbe) meander across this lowland and form broad estuaries before emptying into the Atlantic. Several of Europe's great ports, including London, Le Havre, Rotterdam, and Hamburg, are located on these strategic lowland settings.

The Rhine River delta conveniently divides the unglaciated southern European Lowland from the glaciated plains to the north, areas that were covered by a Pleistocene ice sheet until about 15,000 years ago. Because of this continental glacier, the northern European Lowland, which includes the Netherlands, Germany, Denmark, and Poland, is far less fertile than the unglaciated portion in Belgium and France (Figure 8.4). Rocky clay materials in Scandinavia were eroded and transported south by glaciers. As the glaciers retreated with a warming climate, piles of glacial debris were left on the plains of Germany and Poland. Elsewhere in the north, glacial meltwater created infertile outwash plains that have limited agricultural potential.

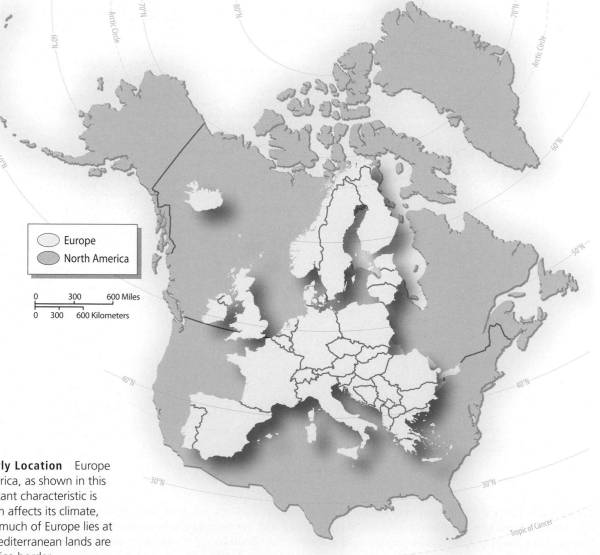

FIGURE 8.3 Europe: Size and Northerly Location Europe is about two-thirds the size of North America, as shown in this cartographic comparison. Another important characteristic is the northerly location of the region, which affects its climate, vegetation, and agriculture. As depicted, much of Europe lies at the same latitude as Canada; even the Mediterranean lands are farther north than the United States–Mexico border.

FIGURE 8.4 The European Lowland Also known as the North European Plain, this large lowland extends from southwestern France to the plains of northern Germany and into Poland. Numerous rivers drain interior Europe by crossing this region, giving rise to large port cities along the coastline.

The Alpine Mountain System The Alpine mountain system forms the topographic spine of Europe. It consists of a series of mountains running east to west from the Atlantic to the Black Sea and the southeastern Mediterranean. Though these mountain ranges carry distinct regional names, such as the Pyrenees, Alps, Apennines, Carpathians, Dinaric Alps, and Balkan Ranges, they share geologic traits. All were created more recently (about 20 million years ago) than other upland areas of Europe, and all are constructed from a complex arrangement of rock types.

The Pyrenees form the political border between Spain and France and include the microstate of Andorra. This rugged range extends almost 300 miles (480 kilometers) from the Atlantic to the Mediterranean. Within the mountain range, glaciated peaks reaching to 11,000 feet (3,350 meters) alternate with broad glacier-carved valleys. The Pyrenees are home to the Basque people in its western reaches, as well as to the distinctive Catalan-speaking minorities in the east. Both groups have strong separatist traditions.

The centerpiece of Europe's geologic system is the prototypical mountain range, the Alps, running more than 500 miles (800 kilometers) from France to eastern Austria. These impressive mountains are highest in the west, rising to more than 15,000 feet (4,600 meters) in Mt. Blanc on the French-Italian border. In Austria, to the east, the Alps are much more subdued, and few peaks exceed 10,000 feet (3,000 meters). Though easily crossed today by car or train through a system of long tunnels and valley-spanning bridges, these mountains have historically formed an important cultural divide between the Mediterranean lands to the south and central and western Europe in the north.

The Apennine Mountains are located south of the Alps; the two ranges, however, are physically connected by the hilly coastline of the French and Italian Riviera. Forming the mountainous spine of Italy, the Apennines are lower and lack the spectacular glaciated peaks and valleys of the true Alps. Farther to the south, the Apennines take on their own distinct character with the impressive and explosive volcanoes of Mt. Vesuvius (just over 4,000 feet, or 1,200 meters) outside Naples and the much higher Mt. Etna (almost 11,000 feet, or 3,350 meters), off Italy's toe on the island of Sicily. These two active volcanoes are products of the meeting between the African and Eurasian tectonic plates.

To the east, the Carpathian Mountains define the limits of the Alpine system in eastern Europe. They are a plow-shaped upland area that extends from eastern Austria to the Iron Gate gorge, a narrow passage for Danube River traffic where the borders of Romania and Serbia intersect. While about the same length as the main Alpine chain, the Carpathians are not nearly as high. The highest summits in Slovakia and southern Poland are less than 9,000 feet (2,800 meters).

Central Uplands In western Europe, a much older highland region occupies an arc between the Alps and the European Lowland in France and Germany. These mountains are much lower in elevation than the Alpine system, with their highest peaks at 6,000 feet (1,800 meters). Formed about 100 million years ago, much of this upland region is characterized by rolling landscapes about 3,000 feet (less than 1,000 meters) above sea level.

The importance of these uplands to western Europe is great because they contain the raw materials for Europe's industrial areas. In both Germany and France, for example, they have provided the iron and coal necessary for each country's steel industry. In the eastern reaches, mineral resources from the Bohemian highlands have also fueled major industrial areas in Germany, Poland, and the Czech Republic.

Western Highlands The Western Highlands define the western edge of the European subcontinent, extending from Portugal in the south, through portions of the British Isles in the northwest, to the highland backbone of Norway, Sweden, and Finland in the far north. These are Europe's oldest mountains, formed about 300 million years ago.

As with other upland areas that traverse many separate countries, specific names for these mountains differ from country to country. A portion of the Western Highlands forms the highland spine of England, Wales, and Scotland, where picturesque glaciated landscapes are found at modest elevations of 4,000 feet (1,200 meters) or less. These U-shaped glaciated valleys also appear in Norway's uplands, where they produce a spectacular coastline of **fjords**, or flooded valley inlets, similar to the coastlines of Alaska and New Zealand (Figure 8.5).

Though less elevated, the Fenno-Scandian Shield of Sweden and Finland is noteworthy because it is made up of some of the oldest rock formations in the world, dated conservatively at 600 million years. This landscape was eroded to bedrock by Pleistocene glaciers and, because of the cold climate and sparse vegetation, still has extremely thin soils that severely limit agricultural activity. Similar to other heavily glaciated areas, like the Canadian Shield of North America, numerous small lakes dot the countryside, giving evidence of the impressive erosional power of ice sheets.

Europe's Climate

Three principal climates characterize Europe (Figure 8.6). Along the Atlantic coast, a moderate and moist **maritime climate** dominates, modified by oceanic influences (see the climographs for Reykjavik,

FIGURE 8.5 Fjord in Norway During the Pleistocene epoch, continental ice sheets and glaciers carved deep U-shaped valleys along what is now Norway's coastline. As the ice sheets melted and sea level rose, these valleys were flooded by Atlantic waters, creating spectacular fjords. Many fjord settlements are accessible only by boat, linked to the outside world by Norway's extensive ferry system.

Belfast, London, and Paris). Farther inland, **continental climates** prevail, with hotter summers and colder winters (see the climographs for Oslo, Riga, Berlin, and Budapest). Finally, a dry-summer **Mediterranean climate** is found in southern Europe, from Spain to Greece (see the climographs for Barcelona, Rome, and Athens).

One of the most important climate controls is that of the Atlantic Ocean. Though most of Europe is at a relatively high latitude (London, England, for example, is slightly farther north than Vancouver, British Columbia), the mild North Atlantic current, which is a continuation of the warm Atlantic Gulf Stream, moderates coastal temperatures from Iceland and Norway to Portugal, and even inland to the western reaches of Germany. As a result, this maritime influence gives Europe a climate 5 to 10°F (3 to 6°C) warmer than comparable latitudes lacking this ocean current's effect. In the **marine west coast climate** region, no winter months average below freezing, even though cold rain, sleet, and an occasional blizzard are common winter visitors. Summers are often cloudy and overcast, with frequent drizzle and rain as moisture flows in from the ocean. Ireland, the Emerald Isle, is an appropriate icon of this maritime climate.

With increasing distance from the ocean (or where a mountain chain limits the maritime influence, as in Scandinavia), land-mass heating and cooling becomes a strong climatic control, producing hotter summers and colder winters. Indeed, all continental climates average at least one month below freezing during the winter.

In Europe, the transition between maritime and continental climates takes place close to the Rhine River border of France and Germany. Farther north, although Sweden and other nearby countries are close to the moderating influence of the Baltic Sea, a higher latitude coupled with the blocking effect of the Norwegian mountains produces cold winter temperatures characteristic of true continental climates.

The Mediterranean climate is characterized by a distinct dry season during the summer, which results from the warm-season expansion of the Atlantic (or Azores) high-pressure area. This high pressure is produced by the global circulation of air warmed in the equatorial tropics, which subsides (descends) between latitudes of 30 and 40 degrees, thus inhibiting summer rainfall. This same phenomenon also produces the Mediterranean climates of California, western Australia, parts of South Africa, and Chile. These rainless summers may attract tourists from northern Europe, but the seasonal drought can be problematic for agriculture. It is no coincidence that traditional Mediterranean cultures, such as the Arab, Moorish, Greek, and Roman, have been major innovators of irrigation technology.

Seas, Rivers, and Ports

In many ways, Europe is a maritime region with strong ties to its surrounding seas. Even landlocked countries such as Austria, Hungary, Serbia, and the Czech Republic have access to the ocean through an interconnected network of navigable rivers and canals.

Europe's Ring of Seas Four major seas encircle Europe; these water bodies are connected to each other through narrow straits with strategic importance for controlling waterborne trade and naval movement. In the north, the Baltic Sea separates Scandinavia from north-central Europe. Denmark and Sweden have long controlled the narrow Skagerrak and Kattegat straits that connect the Baltic to the North Sea. Besides its historical role as a major fishing ground, the North Sea is now well known for its rich oil and natural gas fields mined from deep-sea drilling platforms.

The English Channel (in French, *La Manche*) separates the British Isles from continental Europe. At its narrowest point, the Dover Straits, the channel is only 20 miles (32 kilometers) wide. Although England has thought of the channel as a protective moat, it has primarily been a symbolic barrier, for it deterred neither the French Normans from the continent nor the Viking raiders from the north. Since 1993, after decades of resistance by the English, the British Isles have been connected to France through the 31-mile (50-kilometer) Eurotunnel, with its high-speed rail system carrying passengers, autos, and freight.

Gibraltar guards the narrow straits between Africa and Europe at the western entrance to the Mediterranean Sea, and Britain's stewardship of this passage remains an enduring symbol of a once great sea-based empire. Finally, on Europe's southeastern flanks are the Straits of Bosporus and the Dardanelles, the narrows connecting the eastern Mediterranean with the Black Sea. Disputed for centuries, these pivotal waters are now controlled by Turkey. Though these straits are often thought of as the physical boundary between Europe and Asia, they are bridged in several places to facilitate truck and train transportation within Turkey and between Europe and Southwest Asia.

Rivers and Ports Europe is also a region of navigable rivers, connected by a system of canals and locks that allow inland barge travel from the Baltic and North seas to the Mediterranean and between western Europe and the Black Sea. Many rivers on the European Lowland—namely, the Loire, Seine, Rhine, Elbe, and Vistula—flow into Atlantic and Baltic waters. However, the Danube, Europe's longest river, flows east and south, rising in the Black Forest of Germany only a few miles from the Rhine River and running southeastward to the Black Sea, offering a connecting artery between central and eastern Europe (Figure 8.7). Similarly, the Rhône headwaters rise close to those of the Rhine in Switzerland, yet the Rhône flows southward into the Mediterranean. Both the Danube and the Rhône are connected by locks and canals with the rivers of the European Lowland, making it possible for barge traffic to travel between all of Europe's fringing seas and oceans.

As noted, major ports are found at the mouths of most western European rivers, where they serve as transshipment points for inland

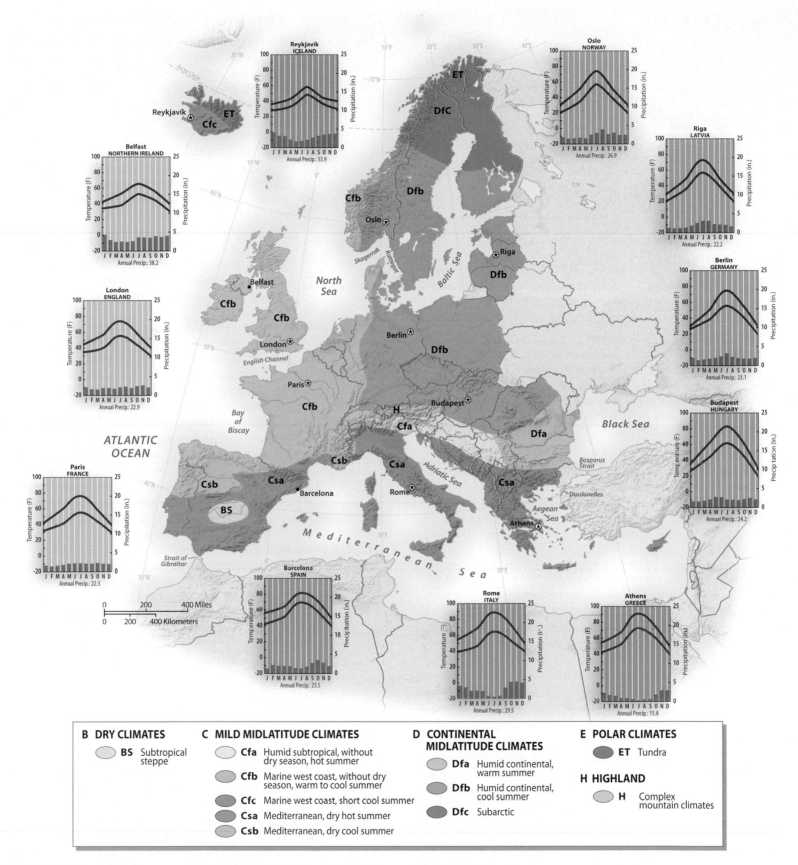

FIGURE 8.6 Climates of Europe   Three major climate zones dominate Europe. Close to the Atlantic Ocean, the marine west coast climate has cool seasons and steady rainfall throughout the year. Farther inland, continental climates have at least one month averaging below freezing, as well as hot summers, with a precipitation maximum occurring during the warm season. Southern Europe has a dry-summer Mediterranean climate.

FIGURE 8.7 Danube Barge Traffic In 1992 the Danube River, Europe's longest, was connected by canal to the Rhine River, allowing commercial barge traffic to move throughout Europe, and from the North to the Black Seas. Because Inland Water Traffic (IWT) is reportedly 80 percent cheaper than trucking, and with the increasing industrialization of eastern Europe, barge traffic has grown considerably in the last decade. In response several Danube countries are formulating plans to protect the Danube environment from increased IWT development. This photo is of a pusher tug and barge on the Danube in Serbia.

waterways and rail and truck networks. From south to north, these ports include Bordeaux at the mouth of the Garonne, Le Havre on the Seine, London on the Thames, Rotterdam (the world's largest port in terms of tonnage) at the mouth of the Rhine, Hamburg on the Elbe, and, to the east in Poland, Szcezin on the Oder and Gdansk on the Vistula.

Environmental Issues: Local and Global

Because of its long history of agriculture, resource extraction, industrial manufacturing, and urbanization, Europe has its share of environmental issues. Compounding the situation is the fact that pollution rarely stays within political boundaries. Air pollution from England, for example, creates serious acid rain problems in Sweden. Similarly, water pollution from Swiss factories on the upper Rhine River creates major problems for the Netherlands, where Rhine River water is used for urban drinking supplies (Figure 8.8). As a result of numerous trans-boundary environmental problems such as these, the EU has taken the lead in addressing the region's environmental issues. As a result, Europe is probably the "greenest" of the major world regions.

Until recently, eastern Europe was plagued by far more serious environmental problems than its western neighbors because of its history as Soviet satellite countries, where economic planning paid more attention to short-term industrial output than environmental protection. In Poland, for example, industrial effluents had reportedly wiped out all aquatic life in 90 percent of the country's rivers, with damage from air pollution affecting over half of the country's forests. Similar legacies from the Soviet period were reported in the Czech Republic, Romania, and Bulgaria. Today, however, most of these environmental issues in eastern Europe have been resolved through funding from the EU and the strengthening of national environmental laws.

Climate Change in Europe

The fingerprints of global warming are everywhere in Europe, from dwindling sea ice, melting glaciers, and sparse snow cover in arctic Scandinavia to more frequent droughts in the water-starved Mediterranean area. Furthermore, the projections for future climate change are ominous: world-class ski resorts in the Alps are forecast to have warmer winters with less snow pack, while in the lowlands, higher summer temperatures could produce more frequent heat waves that affect farmers and urban dwellers alike (Figure 8.9). In addition, rising sea levels from melting polar ice sheets will threaten the Netherlands, where much of the population lives in diked lands that are actually below sea level. Because of these threats, Europe has become a world leader in addressing global warming and has created numerous policies and programs to reduce greenhouse gas (GHG) emissions.

Europe and the Kyoto Protocol The EU entered the 1997 Kyoto negotiations with an innovative scheme, underscoring its philosophy that regional action was superior to that of individual countries. More specifically, the EU set a collective goal of an 8 percent reduction below 1990 GHG emission levels as an umbrella for its individual member states. Under this scheme, some European states were required to make large reductions in their emissions, whereas other countries were actually allowed an increase. The point was to promote growth and industrial development in the poorer countries of Spain, Greece, and Portugal, while at the same time requiring GHG emission reductions in Europe's traditional industrial core of Germany, France, and the United Kingdom. Noteworthy is that this umbrella approach has remained in place as the EU has grown from 15 members in 1997 to its current 27 member states.

The EU's Emission Trading Scheme As part of its Kyoto agreement, the EU inaugurated the world's largest carbon trading scheme in 2005 as a major pillar of its global warming policy. Under this plan, specific yearly emission caps are set for the EU's largest GHG emitters; if these emitters exceed those limits, they must compensate either by purchasing carbon equivalences from an emitter below its own cap or by buying credits from the EU carbon market. The goal of this cap-and-trade system is to make business more expensive for those that pollute, while rewarding those that stay under their carbon quota.

Currently, the plan sets caps for industries responsible for 50 percent of the EU's total GHG emissions, primarily power plants, steel factories, and cement manufacturers. Aviation will be the next sector to be added, since it is responsible for 5 percent of the region's GHG emissions.

Results to Date Despite its good intentions, the EU fell far short of its target of an 8 percent reduction of GHG emissions by 2012. Instead, the emissions reduction is closer to just 1 percent. The main reason for this failure, the EU says, is the unanticipated growth in truck transport over the past decade, coupled with high emissions from industrial development in Spain and the newer EU countries in eastern Europe. That said, it should be noted that the major industrial powers—England and Germany—reduced their emissions by over 10 percent by cutting their coal usage. Further reductions in GHG emissions have come from increased energy generation from wind and solar power, particularly in Germany.

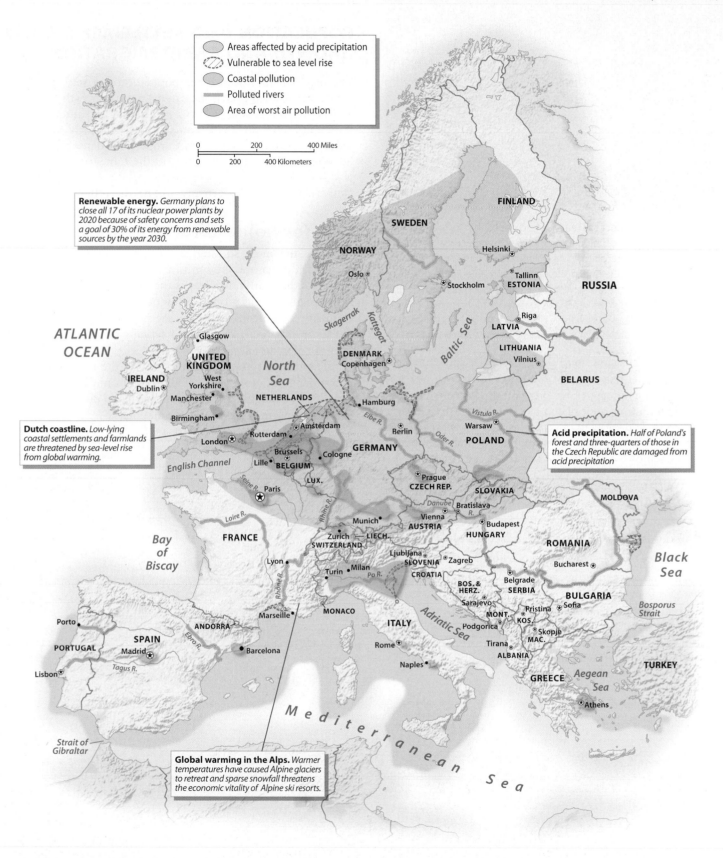

Areas affected by acid precipitation

Vulnerable to sea level rise

Coastal pollution

Polluted rivers

Area of worst air pollution

0   200   400 Miles
0   200   400 Kilometers

Renewable energy. *Germany plans to close all 17 of its nuclear power plants by 2020 because of safety concerns and sets a goal of 30% of its energy from renewable sources by the year 2030.*

Dutch coastline. *Low-lying coastal settlements and farmlands are threatened by sea-level rise from global warming.*

Acid precipitation. *Half of Poland's forest and three-quarters of those in the Czech Republic are damaged from acid precipitation*

Global warming in the Alps. *Warmer temperatures have caused Alpine glaciers to retreat and sparse snowfall threatens the economic vitality of Alpine ski resorts.*

FIGURE 8.8 Environmental Issues in Europe Although western Europe has worked energetically over the past 50 years to solve environmental problems such as air and water pollution, eastern Europe lags behind because environmental protection was not a high priority during the postwar communist period. Current efforts, however, show great promise.

FIGURE 8.9 Acid Rain and Forest Death Acid deposition (commonly called acid rain although much of the deposition takes place with snowfall) has long been a troublesome transboundary pollution problem for Europe. While much of the pollution is created in the densely populated industrial heartland of western Europe, prevailing winds carry acid-laden clouds east and northward so that much of the forest and lake damage if found in Scandinavia and eastern Europe. These dead spruce trees killed by acid deposition are in the Krkonose National Park on the Czech Republic's border with Poland.

To compensate for this slow start, the EU recently agreed to set a new target of a 20 percent reduction by 2020. To achieve that new target, Europe is promoting an aggressive program of alternative energy, drawing heavily on both wind and solar power (Figure 8.10). Currently, there are over 25,000 wind farms in Europe; this amount is expected to double by 2015, with Germany and Spain leading the way (see "Working Toward Sustainability: Germany's Renewable Energy Program").

REVIEW QUESTIONS

1. Name and locate the major lowland and mountainous areas of Europe.
2. What are the three major climate regions of Europe?
3. How would inland barge traffic get from the mouth of the Rhine to the delta of the Danube?
4. Describe the European carbon trading scheme.

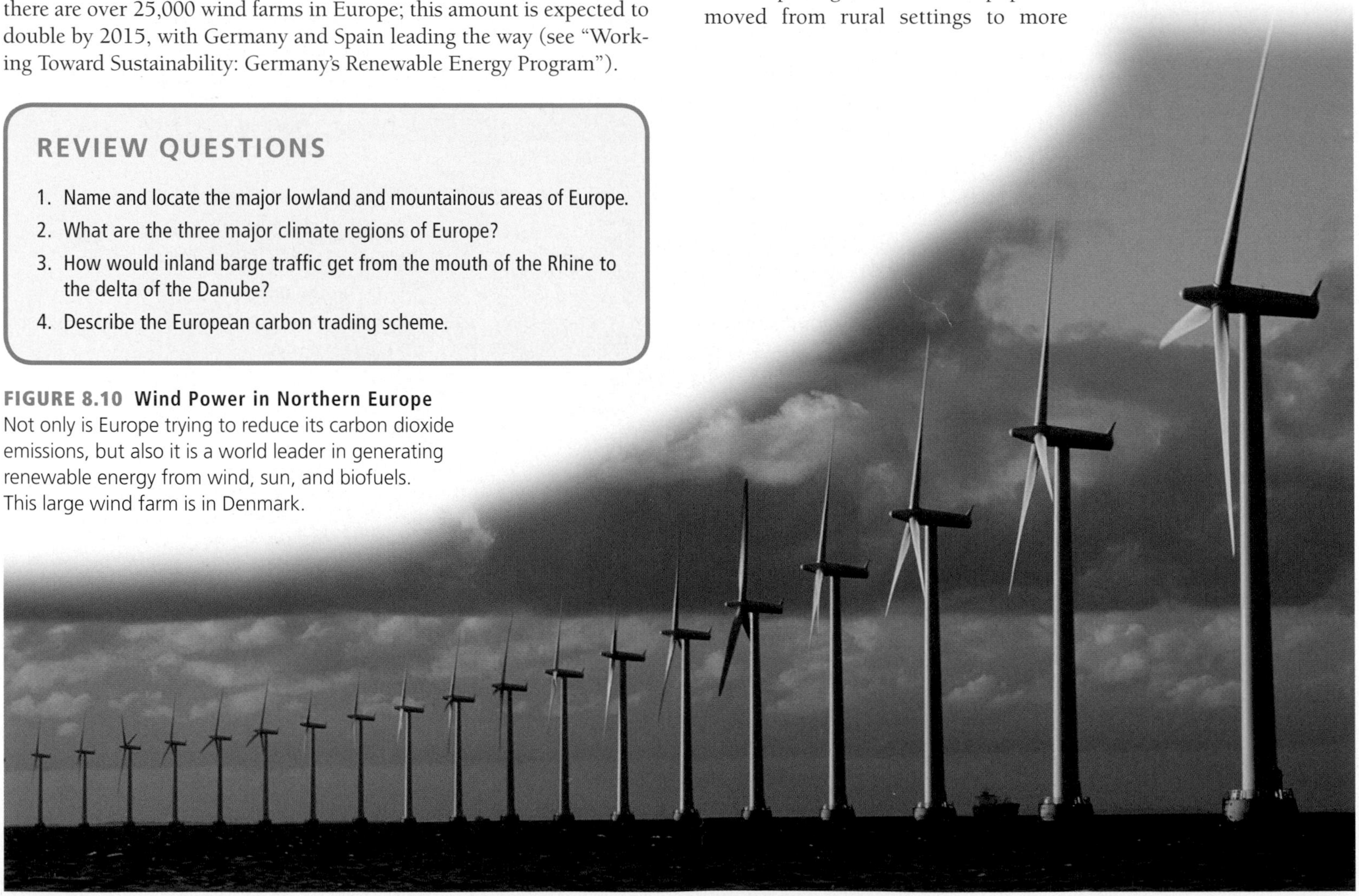

FIGURE 8.10 Wind Power in Northern Europe
Not only is Europe trying to reduce its carbon dioxide emissions, but also it is a world leader in generating renewable energy from wind, sun, and biofuels. This large wind farm is in Denmark.

POPULATION AND SETTLEMENT: SLOW GROWTH AND RAPID MIGRATION

The major themes of Europe's population and settlement geography are its very low (or even nonexistent) natural growth; the complicated and often problematic patterns of internal and international migration; and a high level of urbanization, particularly in the traditional industrial core area of western Europe. This densely settled core area includes England (the birthplace of the Industrial Revolution, which is discussed later in this chapter), northern France, Belgium, the Netherlands, western Germany, and northern Italy (Figure 8.11).

Low (or No) Natural Growth

Probably the most striking characteristic of Europe's demography is the lack of natural growth (Table 8.1). That is, in most European countries the death rate exceeds the birthrate, resulting in either very slow or no natural growth. Moreover, several countries, including Germany, are actually experiencing negative natural growth. Were it not for in-migration, they would actually decrease in population size over the next decades.

There seem to be several reasons for the lack of natural population growth in Europe. First of all, recall from Chapter 1 that the concept of the demographic transition was based on the historical change in European growth rates as populations moved from rural settings to more

Germany's Renewable Energy Program

Germany is committed to becoming the world's first major industrial economy to be powered primarily by renewable energy. More specifically, Germany's plan is to expand its wind, solar, and biomass energy facilities so that they supply more than half the country's energy needs by the year 2030. Three goals drive this new initiative: to reduce the country's global warming emissions by using clean energy, to reduce its reliance on imported gas and oil, and to become a world leader in renewable energy technology.

If successful, Germany will provide the world with a model for sustainable energy. If it fails—and everyone agrees there are serious technological and economic challenges ahead—the world's search for a clean renewable energy future could be lost.

Germany's renewable energy program was jumpstarted in 2000 with massive subsidies and incentives to expand solar and wind electricity generation. A critical component of this new plan required the country's utility companies to buy and integrate into the national power grid all of the renewable energy being produced by these new sources. As a result, solar panels blossomed everywhere: on homes, on barns, in pastures, along highways, over parking lots—anywhere there was space open to the southern sky, since there was money to be made selling solar energy to power utilities (Figure 8.1.1). Wind turbines, which are more expensive to build than rooftop solar panels, appeared on windy hilltops and along breezy coastlines,

financed by companies and investment firms, large and small.

The results from this decade-long first phase were spectacular, with renewable energy now making up over 20 percent of Germany's total energy picture. Even though solar panels are ubiquitous in the country's landscape, wind power currently produces most of the country's renewable energy (40 percent, contrasted to 8 percent for solar). In addition, Germany's renewable energy future includes plans for huge offshore wind farms.

Based upon the success of the last decade, in 2011 Germany recommitted itself to moving forward, with a target of 45 percent renewable energy by 2020, with hopes of surpassing 50 percent shortly afterward. Impetus for the country to move farther along the renewable energy pathway more quickly came from a recent decision to shut down all of Germany's 17 nuclear power plants by 2022 due to safety and fuel storage issues.

Major hurdles, however, lie ahead. For example, both wind energy and solar energy are produced by favorable weather such as sunny, windy summer days. This is not necessarily when the country's power demand is highest, which quite possibly could be on cold, windless winter evenings. How to

FIGURE 8.1.1 Solar Panels in Germany This photo of solar panels on a house in a small village near Tübingen, Germany illustrates how homeowners are taking advantage of their south-facing rooftops to generate electricity not only for their own domestic use but also to sell back to the utility company.

compensate for these daily and seasonal fluctuations remains a major issue.

Also, to integrate these new sources of renewable energy—particularly the offshore wind farms in northwest Germany—the country must completely rebuild the infrastructure supporting the national grid system. This will be costly and probably result in German industry paying more for energy than in the past. This has given rise to concerns that German products will become more expensive—and thus less competitive in the global marketplace.

Nevertheless, despite these issues, Germany seems committed to its plan for a sustainable and clean energy future.

---

urban and industrial locations. What we see today is an extension of that model's low-fertility/low-mortality fourth stage, so much so that demographers have used Europe to add a fifth, "postindustrial" stage to the demographic transition, a phase where fertility falls below replacement levels. Not only does Germany actually have a negative rate of natural increase (RNI) of –0.2, but also the large countries of England, France, and Italy have RNIs below population replacement levels. Thus, concerns have been expressed about possible labor shortages, smaller internal markets, and declining tax revenues needed to support social services (such as retirement pensions) for their aging populations (Figure 8.12).

Pro-Growth Policies   Because most European governments consider low or no population growth an economic and social liability,

programs and policies have arisen to promote population growth, ranging from attempts to ban abortion and the sale of contraceptives to what are commonly called "family-friendly" policies. In many countries, for example, family-friendly pro-growth policies include full-pay maternity and paternity leaves for both parents that include guarantees of continued employment once these leaves end, extensive child-care facilities for working parents, outright cash subsidies for having children, and free or low-cost public education and job training for children. However, even with these pervasive family-friendly policies, only two European countries—Iceland and Ireland (which is heavily Roman Catholic)—have total fertility rates (TFRs) at the replacement level of 2.1. All other European countries have lower rates of natural growth; thus, it is only because of in-migration that growth takes place.

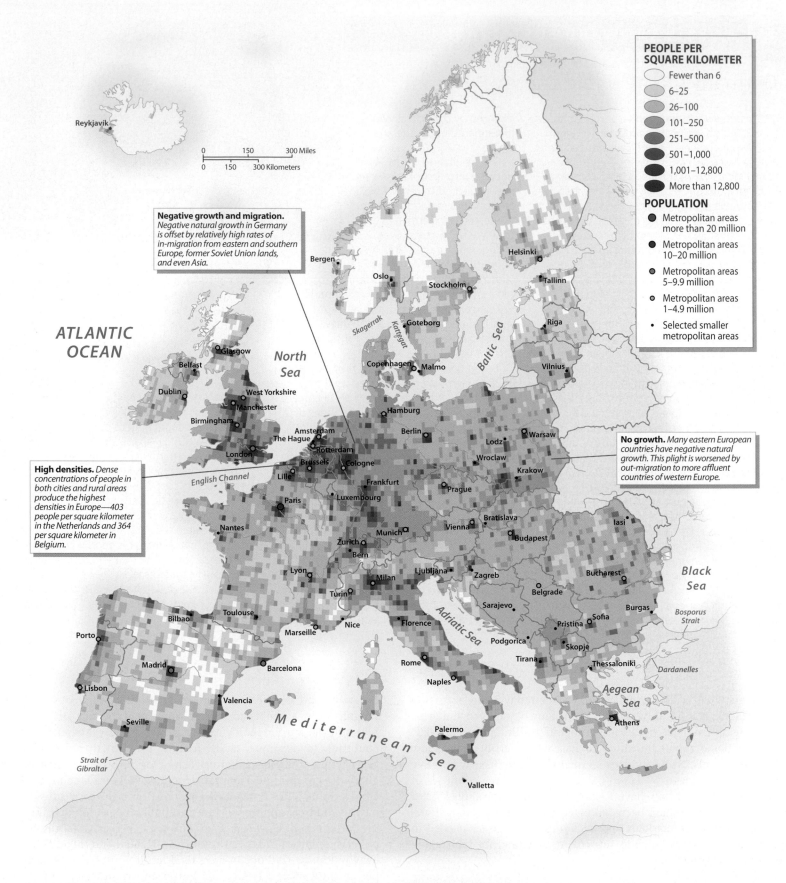

PEOPLE PER SQUARE KILOMETER
- Fewer than 6
- 6–25
- 26–100
- 101–250
- 251–500
- 501–1,000
- 1,001–12,800
- More than 12,800

POPULATION
- Metropolitan areas more than 20 million
- Metropolitan areas 10–20 million
- Metropolitan areas 5–9.9 million
- Metropolitan areas 1–4.9 million
- Selected smaller metropolitan areas

Negative growth and migration. *Negative natural growth in Germany is offset by relatively high rates of in-migration from eastern and southern Europe, former Soviet Union lands, and even Asia.*

High densities. *Dense concentrations of people in both cities and rural areas produce the highest densities in Europe—403 people per square kilometer in the Netherlands and 364 per square kilometer in Belgium.*

No growth. *Many eastern European countries have negative natural growth. This plight is worsened by out-migration to more affluent countries of western Europe.*

FIGURE 8.11 Population Map of Europe   The European region includes more than 531 million people, many of them clustered in large cities in both western and eastern Europe. As can be seen on this map, the most densely populated areas are in England, the Netherlands, Belgium, western Germany, northern France, and south across the Alps to northern Italy.

TABLE 8.1   POPULATION INDICATORS

Country	Population (millions) 2012	Population Density (per square kilometer)	Rate of Natural Increase (RNI)	Total Fertility Rate	Percent Urban	Percent <15	Percent >65	Net Migration (Rate per 1000) 2010–15[a]
Western Europe								
Austria	8.5	101	0.0	1.4	67	15	18	2.4
Belgium	11.1	364	0.2	1.8	99	17	17	1.9
France	63.6	115	0.4	2.0	78	19	17	1.6
Germany	81.8	229	−0.2	1.4	73	13	21	0.2
Liechtenstein	0.04	229	0.4	1.5	15	16	14	
Luxembourg	0.5	204	0.3	1.5	83	18	14	9.7
Monaco	0.04	36,356	0.0	—	100	13	24	
Netherlands	16.7	403	0.2	1.7	66	17	16	0.6
Switzerland	8.0	194	0.2	1.5	74	15	17	2.2
United Kingdom	63.2	260	0.4	2.0	80	18	17	3.3
Eastern Europe								
Bulgaria	7.2	65	−0.5	1.5	73	13	19	−1.4
Czech Republic	10.5	133	0.0	1.4	74	14	15	2.0
Hungary	9.9	107	−0.4	1.2	69	15	17	1.5
Poland	38.2	122	0.1	1.3	61	15	14	0.1
Romania	21.4	90	−0.4	1.3	55	15	15	−0.5
Slovakia	5.4	110	0.2	1.4	54	15	13	0.7
Southern Europe								
Albania	2.8	99	0.6	1.4	54	23	9	−3.1
Bosnia & Herzegovina	3.8	75	−0.1	1.2	46	15	14	−0.3
Croatia	4.3	76	−0.2	1.5	56	15	17	0.5
Cyprus	1.2	127	0.6	1.4	62	17	12	6.2
Greece	10.8	82	0.1	1.5	73	14	19	2.7
Italy	60.9	202	−0.1	1.4	68	14	21	3.4
Kosovo	2.3	210	1.2	2.0	—	28	7	0.0
Macedonia	2.1	80	0.2	1.5	65	17	12	0.0
Malta	0.4	1,262	0.2	1.4	100	15	16	2.4
Montenegro	0.6	45	0.2	1.6	64	19	13	−0.8
Portugal	10.6	115	−0.1	1.3	38	15	19	1.9
San Marino	0.03	530	0.3	1.2	84	15	16	
Serbia	7.1	92	−0.5	1.3	59	15	17	0.0
Slovenia	2.1	102	0.1	1.5	50	14	17	2.2
Spain	46.2	91	0.2	1.4	77	15	17	4.3
Northern Europe								
Denmark	5.6	130	0.1	1.8	72	18	17	2.2
Estonia	1.3	30	−0.0	1.5	69	16	17	0.0
Finland	5.4	16	0.2	1.8	68	16	18	1.5
Iceland	0.3	3	0.8	2.0	94	21	12	3.3
Ireland	4.7	67	1.0	2.1	60	21	12	2.2
Latvia	2.0	32	−0.5	1.1	68	14	17	−0.9
Lithuania	3.2	49	−0.2	1.5	67	15	17	−1.7
Norway	5.0	13	0.4	1.9	80	19	15	3.0
Sweden	9.5	21	0.2	1.9	84	17	19	3.3

[a]Net Migration Rate from the United Nations, Population Division, *World Population Prospects: The 2010 Revision Population Database.*
Source: Population Reference Bureau, *World Population Data Sheet, 2012.*

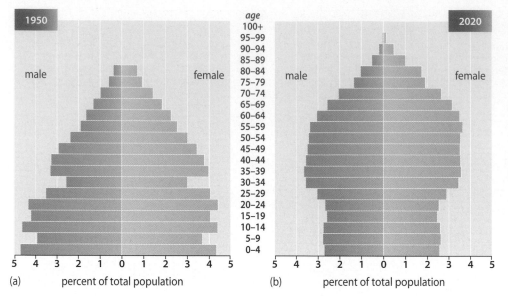

FIGURE 8.12 The Aging of Europe These two population pyramids illustrate the aging of Europe by contrasting the region's relatively youthful population of 1950 with the projected demographic makeup of Europe in the year 2020. Three components of these pyramids are noteworthy: first, the broader base of the 1950 pyramid shows a more youthful population compared to 2020; second, the bulge of older people (compared to the narrow base) of 2020 shows the aging European population; and, third, in the 1950 pyramid note the deficit of 30–34 year olds—both men and women—who one infers were killed in WW2.

Migration Within Europe

Since its birth in 1957, the EU has worked toward the goal of freer movement of both people and goods within Europe by reducing the historical economic and political barriers to cross-border travel. Today residents of the 27 EU countries can move about as they please to travel, study, and work. In fact, these new freedoms have affected not just migration patterns, but also national population growth and loss. During the past decade, for example, some 16,000 Lithuanians (out of a total population of 3.3 million) moved to Ireland to take advantage of that country's booming economy while their home country languished in the economic doldrums. More recently, when the Irish boom burst, half of these EU migrants reportedly moved back to Lithuania.

The Schengen Agreement Complementing this new European mobility is the erosion of Europe's historical national borders, where traffic used to be stopped for lengthy passport checks and vehicle inspections. Instead, today, at many former European border stations travelers can cross national boundaries without stopping or showing passports because of the **Schengen Agreement** on free travel within Europe. (This agreement was named for the city of Schengen, Luxembourg, where the declaration of intent for a border-free Europe was signed in 1985; Figure 8.13.) Although the Schengen Agreement was originally intended to facilitate movement between EU member countries, not all EU countries are members of this new "Schengenland" free travel zone, nor is membership restricted to EU member countries. The United Kingdom, for example, is an EU member, but is not a signatory to the Schengen Agreement because of concerns about protecting its historical borders. On the other hand, Norway and Iceland, both of which are not EU members, are Schengen signatories.

Despite the advantages of freer movement within Europe for residents of the EU, the Schengen Agreement has become increasingly controversial recently because the absence of internal border controls also makes it easier for illegal migrants from outside of Europe to move about relatively unhindered. Behind this controversy lie complex economic, social, and political issues, ranging from the concern of outsiders displacing European workers, to foreign terrorists traveling freely around Europe, to nationalistic resistance to local cultures being diluted by "foreigners."

International Migration to Europe

During postwar rebuilding and recovery, when western Europe's economies were booming, many countries looked to foreign migrant workers to ease labor shortages. The former West Germany, for example, depended on workers from Europe's rural and poorer periphery—Italy, the former Yugoslavia, Greece, and even Turkey—for industrial, construction, and service jobs.

FIGURE 8.13 Schengenland Border People stroll unhindered by document checks across the Poland-Germany border in the city of Görlitz. Earlier, before Poland became a member of the Schengen Agreement in late 2007, this border was a heavily policed "hard" border between Europe and the non-Schengen countries to the east where passport and visa checks were rigorously enforced.

Similarly, France and England opened their doors to migrants from former colonies in Africa and Asia, resulting in thousands of workers arriving from India, Pakistan, Hong Kong, North Africa, and the Caribbean. These immigrants formed ethnic neighborhoods in large cities throughout England, France, the Netherlands, and Spain (Figure 8.14).

Later, with the collapse of Soviet border controls in 1990, emigrants from former Soviet satellite countries poured into western Europe, seeking relief from the economic chaos in Russia and other eastern countries. This flight from the post-1989 economic and political chaos also included refugees from war-torn regions of the former Yugoslavia, particularly from Bosnia and Kosovo, as Austria, Sweden, and Germany offered amnesty to the victims of the Balkan wars.

As a result of these different migration streams, foreigners now comprise about 8 percent of the populations in Germany, France, and England, making many European countries ambivalent hosts to emerging multicultural societies, a topic discussed further in the next section of this chapter. (For comparison, 11.7 percent of the U.S. population is comprised of people classified as foreigners.)

Leaky Borders and "Fortress Europe" With high unemployment and economic stress currently widespread in Europe, the region no longer needs to import workers internationally. In fact, as noted earlier, there are fears in many European countries that the influx of foreign migrants aggravates high domestic unemployment rates as migrants work for lower wages in an off-the-books "gray" economy.

As a result, Europe is trying to stem the flow of illegal migration and also slow the amount of legal migration, consisting of colonial residents or relatives of migrants already in Europe. However, given the geography of Europe's borders, this is a vexatious task. Both Spain and Italy, for example, have extensive coastal shores accessible to seafaring illegal

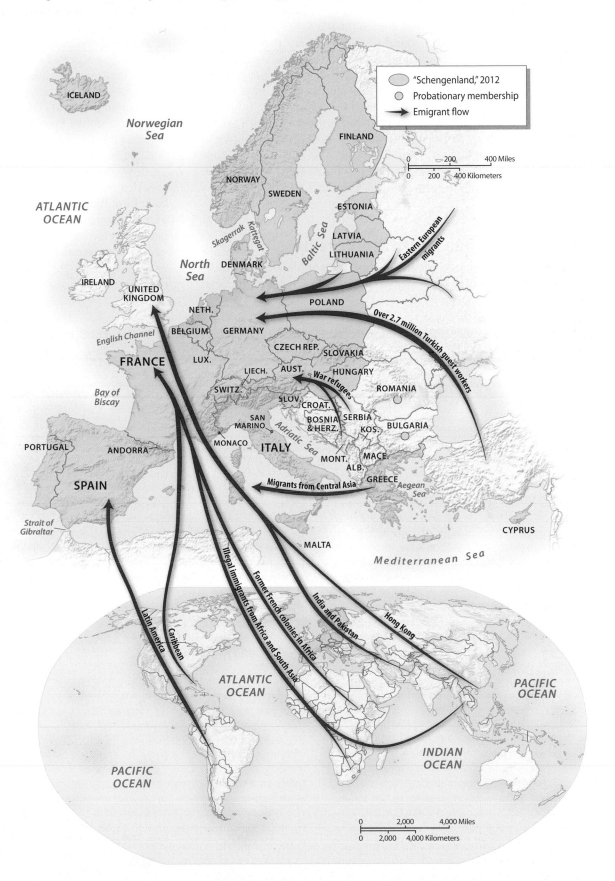

FIGURE 8.14 Migration into Europe Historically, Europe opened its doors to migrants to help solve post–World War II labor shortages. Now, however, immigration is a contentious and controversial issue, as culturally homogeneous countries such as France, England, and Germany confront tensions resulting from large numbers of immigrants, both legal and illegal.

migrants from Africa and the Near East. Greece has been accused of having a leaky border with Turkey, one that has become a favorite entry point for illegal migrants. Similar accusations have been directed at both Bulgaria and Romania, open to illegal migrants crossing the Black Sea.

Because of the soft internal borders of the Schengen countries, if illegal migrants successfully land in Spain or Italy, they can move easily into France or Austria and then freely within the heart of Europe. To help these perimeter countries police their borders, the EU has provided funds to strengthen their borders with more border guards and, in some places, with physical border fences to inhibit illegal entry. Geographically, one can think of Europe creating a system where its perimeter consists of hard borders—a "Fortress Europe," as critics (and anti-immigrant groups) call the plan—while its internal borders will be deliberately soft and porous in the spirit of the Schengen Agreement (see "Exploring Global Connections: Europe's Leaky Southern Border").

The Landscapes of Urban Europe

One of the major characteristics of Europe's population and settlement patterns is its high level of urbanization. All European countries except several small Balkan countries have more than half their populations in cities, with several western European countries more than 90 percent urbanized.

The Past in the Present   North American visitors often find European cities far more interesting than our own because of the mosaic of historical and modern landscapes, featuring medieval quarters and churches interspersed with high-rise buildings and other contemporary structures. Three historical periods dominate most European city landscapes. The medieval (900–1500), Renaissance–Baroque (1500–1800), and industrial (1800–present) periods have each left characteristic marks on the European urban scene. Learning to recognize these stages of historical growth provides visitors to Europe's cities with fascinating insights into both past and present landscapes (Figure 8.15).

The **medieval landscape** is one of narrow, winding streets, crowded with three- or four-story masonry buildings with little setback from the street. This is a dense landscape with few open spaces, except around churches and public buildings, where public squares or parks are clues to historical medieval open-air marketplaces.

As picturesque as we find medieval-era districts today, they nevertheless present challenges to contemporary inhabitants because of their narrow, congested streets and old housing. Modern plumbing

EXPLORING GLOBAL CONNECTIONS

Europe's Leaky Southern Border

Thousands of migrants are trapped in Greece, living in limbo because of their nonlegal status. They barely survive on the streets or in crowded Greek detention camps, unable (or unwilling) to return home, yet blocked by bureaucracy and red tape from reaching their desired destinations in the heart of Europe. Many are from Afghanistan, where death threats by the Taliban have forced them to flee because they worked for U.S. and NATO forces. Others are from Syria, Iraq, Pakistan, and Iran, where violence and social chaos have driven them from their homes.

Most share similar stories of paying several thousand dollars to smugglers who promised to get them into Europe through the leaky border between Turkey and Greece. Unlike rich illegal migrants, who pay $20,000 and up to fly into Europe with counterfeit passports, these migrants have walked for weeks across Iran and Turkey, moving clandestinely from village to village, sometimes led by guides from international cartels of people smugglers, other times risking it alone.

Until recently, their immediate goal was a 125-mile (200-kilometer) section of the Turkey–Greece border in the Evros district, some 200 miles (320 kilometers) directly west of Istanbul. Here some 55,000 migrants reportedly entered

Greece illegally in 2011. But recently Greece spent $13 million to block the most porous portion of their eastern border with a razor wire fence 6 miles (10 kilometers) long, 13 feet (4 meters) high, and complete with guard towers and night vision cameras (Figure 8.2.1).

Migrants successful in entering Greece illegally then have two choices: Turn themselves in to the police immediately and ask for political asylum, or try to move undetected across Greece and either northward into the Balkan countries or across the Adriatic Sea to Italy. In both cases, northward to Slovenia and across the water to Italy, the goal is enter a Schengen county with its lax border controls and travel to the heart of Europe.

Most illegal migrants, however, are caught somewhere in Greece, thus beginning their life of limbo. To achieve political amnesty, migrants must first get a court date for an amnesty hearing; this is done by navigating a bureaucratic nightmare of endless waits in long lines with thousands of other nonlegal migrants. Even with this court date (and with supporting documents that they're in danger if they return to their home country), Greece's approval rate for amnesty is notoriously low, at only 2 percent.

Aggravating the migrant bottleneck in Greece is current EU policy that migrants must apply for amnesty in the country where they

FIGURE 8.2.1   Greece's Anti-Migrant Border Fence   A Turkish observation post is visible through Greece's new 13-foot (4 meter) high border fence with Turkey in the area where reportedly hundreds of thousands of migrants had previously illegally entered Europe.

first entered Europe. This means that even if illegal migrants make it as far as Germany, France, or the United Kingdom before they are caught, they will be deported back to Greece. This EU policy, however, is currently under review and may be changed in the near future to alleviate the migrant bottleneck in Greece by allowing migrants to apply for political amnesty wherever they're arrested. The EU has also begun deportation airplane flights taking those migrants who don't qualify for political amnesty back to their home countries. As well, the EU is demanding that Turkey exercise more controls at its perimeter borders in hopes of stopping nonlegal migrants before they get closer to the southeastern EU countries.

FIGURE 8.15 Urban Landscapes This aerial view of Grosseto, Italy, shows how the historic medieval city was encircled by the Renaissance–Baroque fortifications built to protect the settlement. Today parks and public buildings are found in place of the former walls and moat.

and heating are often lacking, and rooms and hallways are small and cramped compared to present-day standards. Because these medieval districts sometimes lack modern facilities, the majority of people living in them have low or fixed incomes. In many areas, this majority is made up of the elderly, students, and ethnic migrants.

Many cities in Europe, though, are enacting legislation to upgrade and protect their historic medieval landscapes. This movement began in the late 1960s and has become increasingly popular as cultures have worked to preserve the uniqueness of special urban centers as the suburbs increasingly surrender to sprawl and shopping centers (Figure 8.16).

FIGURE 8.16 Urban Historic Preservation Salzburg, Austria was one of Europe's first cities to enact legal protection of its historic urban landscape with its 1967 law. Since then many other European cities have built upon the Salzburg approach to protect their own historic cityscapes. In this photo two historic periods dominate, the medieval dwellings and shops in the foreground and the fortress on the hill, both bracketing the 17th century Renaissance-Baroque cathedral.

In contrast to the cramped and dense medieval landscape, those areas of the city built during the **Renaissance–Baroque period** are much more open and spacious, with expansive ceremonial buildings and squares, monuments, ornamental gardens, and wide boulevards lined with palatial residences. During this period (1500–1800), a new artistic sense of urban planning arose in Europe, resulting in the restructuring of many European cities, particularly the large capitals. These changes were primarily for the benefit of the new urban elite—royalty and successful merchants. City dwellers of lesser means remained in the medieval quarters, which became increasingly crowded and cramped as more city space was devoted to the ruling classes.

During the Renaissance–Baroque period, city fortifications limited the outward spread of these growing cities, thus increasing density and crowding within. With the advent of assault artillery, European cities were forced to build an extensive system of defensive walls. Once encircled by these walls, the cities could not expand outward. Instead, as the demand for space increased within the cities, a common solution was to add several new stories to the medieval houses.

Industrialization dramatically altered the landscape of European cities. Historically, beginning in the early 19th century, factories clustered together in cities, drawn by their large markets and labor force and supplied by raw materials shipped by barge and railroad. Industrial districts of factories and worker tenements grew up around these transportation lines. In continental Europe, where many cities retained their defensive walls until the late 19th century, the new industrial districts were often located outside the former city walls, removed from the historic central city. In Paris, for example, when the railroad was constructed in the 1850s, it was not allowed to enter the city walls. As a result, terminals and train stations for the network of tracks were located beyond the original fortifications. Although the walls of Paris are long gone, this pattern of train stations lying beyond the city's historical city walls still exists today.

Not to be overlooked are the post–World War II changes to European cities as they rebuilt from the war's destruction and adapted to the political and economic demands of more recent As in North American cities, suburban sprawl has become an issue in many European countries, as people seek lower-density housing in nearby rural environments. But unlike most North American cities, European urban areas generally have well-developed public transportation systems that offer attractive alternatives to commuting by car.

REVIEW QUESTIONS

1. What European countries have the highest and lowest rates of natural population increase?

2. What European countries have the highest rates of out-migration?

3. What is the Schengen Agreement?

4. Name three stages of historical urban development still commonly found in European urban landscapes.

CULTURAL COHERENCE AND DIVERSITY: A MOSAIC OF DIFFERENCES

The rich cultural geography of Europe demands our attention for several reasons. First, the highly varied and fascinating mosaic of languages, customs, religions, ways of life, and landscapes that characterizes Europe has also strongly shaped regional identities that have all too often stoked the fires of conflict.

Second, European cultures have played leading roles in processes of globalization, as European colonialism has brought about changes in languages, religion, economies, and values in every corner of the globe. If you question this, consider cricket games in Pakistan, high tea in India, Dutch architecture in South Africa, and the millions of French-speaking inhabitants of equatorial Africa. Before modern media technologies, European culture spread across the world, changing the speech, religion, dress, and habits of millions of people on every continent.

Today, though, new waves of global culture are spreading back into Europe (Figure 8.17). Although some European cultures embrace (or simply condone) these changes, other cultures actively resist. France, for example, struggles against both U.S.-dominated popular culture and the multicultural influences of its large migrant population.

FIGURE 8.17 Global Culture in Europe U.S. popular culture is received with great ambivalence in Europe. Although many people embrace everything from fast food to Hollywood movies, at the same time they protest the loss of Europe's traditional regional cultures. This photo was taken in Prague, Czech Republic.

Geographies of Language

Language has always been an important component of nationalism and group identity in Europe (Figure 8.18). Today, although some small ethnic groups such as the Irish and the Bretons work hard to preserve their local language in order to reinforce their cultural identity, millions of Europeans are also busy learning multiple languages so they can communicate across cultural and national boundaries. The EU itself, which is primarily an integrating force in Europe, honors this linguistic mosaic by recognizing more than 20 official languages.

As their first language, 90 percent of Europe's population speaks Germanic, Romance, or Slavic languages, all of which are linguistic groups within the Indo-European family. Germanic and Romance speakers each number almost 200 million in the European region. Although Slavic languages are spoken by twice that number when Russia and its immediate neighbors are included, there are only 80 million Slavic speakers within the traditional definition of the European region.

Germanic Languages Germanic languages dominate Europe north of the Alps. Today about 90 million people speak German as their first language. It is the dominant language of Germany, Austria, Liechtenstein, Luxembourg, eastern Switzerland, and several small areas in Alpine Italy. Until recently, there were also large German-speaking minorities in Romania, Hungary, and Poland, but many of these people left eastern Europe and resettled in Germany when the Iron Curtain was lifted in 1990.

English is the second largest Germanic language, with about 60 million speakers using it as their first language. In addition, a large number of Europeans learn English as a second language, particularly in the Netherlands and Scandinavia, where many people are as fluent as native English speakers. Linguistically, English is closest to the Low German spoken along the coastline of the North Sea, which reinforces the theory that an early form of English evolved in the British Isles through contact with the coastal peoples of northern Europe. However, one of the distinctive traits of English that sets it apart from German is that almost one-third of the English vocabulary is made up of Romance words brought to England during the Norman French conquest of the 11th century.

Elsewhere in this region, Dutch (the Netherlands) and Flemish (northern Belgium) together account for another 20 million people, and roughly the same number of Scandinavians speak the closely related languages of Danish, Norwegian, and Swedish. Icelandic is a more distinctive language because of that country's geographic isolation from its Scandinavian roots.

Romance Languages Romance languages, including French, Spanish, and Italian, evolved from the vulgar (or everyday) Latin used within the Roman Empire. Today Italian is the most widely used of these Romance languages, with about 60 million Europeans speaking it as their first language. In addition to Italy, Italian is an official language of Switzerland and is also spoken on the French island of Corsica.

French is spoken in France, western Switzerland, and southern Belgium (where it is known as *Walloon*). Today there are about 55 million native French speakers in Europe. As with other languages, French also has very strong regional dialects. Linguists differentiate between two forms of French in France itself: that spoken in the

FIGURE 8.18 Language Map of Europe Ninety percent of Europeans speak an Indo-European language. These languages can be grouped into the major categories of Germanic, Romance, and Slavic languages. Ninety million Europeans speak German as a first language, which places it ahead of the 60 million who list English as their native language. However, given the large number of Europeans who speak fluent English as a second language, you could make the case that English is the dominant language of modern Europe.

north (the official form because of the dominance of Paris) and the language of the south, or *langue d'oc*. This linguistic divide expresses long-standing tensions between Paris and southern France. In the past decade, this strong regional awareness of the southwest (centered on Toulouse and the Pyrenees) has led to a rebirth of its own distinct language, *Occitanian*.

Spanish also has very strong regional variations. About 25 million people speak Castilian Spanish, the country's official language, which dominates the interior and northern areas of that large country. However, the *Catalan* form, which some argue is a completely separate language, is found along the eastern coastal fringe, centered on Barcelona, Spain's major city in terms of population and economy. This distinct language reinforces a strong sense of cultural separateness that has led to the state of Catalonia being given autonomous status within Spain.

Portuguese is spoken by another 12 million speakers in that country and in the northwestern corner of Spain, although considerably more people speak the language in Brazil, a former Portuguese colony in Latin America. Finally, Romanian represents the most eastern extent of the Romance language family; it is spoken by 24 million people in Romania. Though unquestionably a Romance language, Romanian also contains many Slavic words.

The Slavic Language Family

Slavic is the largest European subfamily of the Indo-European languages. Slavic speakers are traditionally separated into northern and southern groups, divided by the non-Slavic speakers of Hungary and Romania.

To the north, Polish has 35 million speakers, with Czech and Slovakian totaling about 15 million. These numbers pale in comparison with the number of northern Slav speakers in nearby Ukraine, Belarus, and Russia, where one can easily count more than 150 million. Southern Slav languages include three groups: 14 million Serbian and Croatian speakers (now considered separate languages because of the political differences between Serbs and Croats), 11 million Bulgarian and Macedonian speakers, and 2 million Slovenian speakers.

The use of two distinct alphabets further complicates the geography of Slavic languages (Figure 8.19). In countries with a strong Roman Catholic heritage, such as Poland and the Czech Republic, the Latin alphabet is used in writing. In contrast, countries with close ties to the Orthodox church—Bulgaria, Montenegro, Macedonia, parts of Bosnia-Herzegovina, and Serbia—use the Greek-derived **Cyrillic alphabet**.

Geographies of Religion, Past and Present

Religion is an important component of the geography of cultural coherence and diversity in Europe because many of today's ethnic tensions result from historical religious events. To illustrate, strong cultural borders in the Balkans and eastern Europe are based upon the 11th-century split of Christianity into eastern and western churches, as well as the divisions between Christianity and Islam. In Northern Ireland, blood is still shed over the tensions resulting from the 17th-century split of Christianity into Catholicism and Protestantism. Also, much of the ethnic cleansing terrorism in the former Yugoslavia resulted from the historical struggle between Christianity and Islam. In addition, over the last several decades, considerable tension has risen regarding the large Muslim migrant populations residing in England, France, and Germany. Understanding these contemporary issues involves taking a brief look at the historical geography of Europe's religions (Figure 8.20).

FIGURE 8.19 The Alphabet of Ethnic Tension   With the departure of many of Kosovo's Serbs during recent ethnic unrest, signs using the Cyrillic alphabet were taken off many stores, shops, and restaurants as Kosovars of Albanian ethnicity restated their claims to the region. Here the Albanian owner of a restaurant in the capital city of Pristina scrapes off Cyrillic letters.

The Schism Between Western and Eastern Christianity   In southeastern Europe, early Greek missionaries spread Christianity through the Balkans and into the lower reaches of the Danube. Progress was slower than in western Europe, perhaps because of continued invasions by peoples from the Asian steppes. Also, Greek missionaries refused to accept the control of Roman bishops from western Europe.

This tension with western Christianity led to an official split of the eastern church from Rome in 1054. This eastern church subsequently splintered into Orthodox sects closely linked to specific nations and states. Today, for example, we find Greek Orthodox, Bulgarian Orthodox, and Russian Orthodox churches, all with different rites and rituals.

Another factor that distinguished eastern Christianity from western was the Orthodox use of the Cyrillic alphabet instead of the Latin. Because Greek missionaries were primarily responsible for the spread of early Christianity in southeastern Europe, it is not surprising that they used an alphabet based on Greek characters. More precisely, this alphabet is attributed to the missionary work of St. Cyril in the 9th century. As a result, the division between western and eastern churches, and between the two alphabets, remains one of the most prominent and often problematic cultural boundaries in Europe.

The Protestant Revolt   Besides the division between western and eastern churches, the other great split within Christianity occurred between Catholicism and Protestantism. This division arose in Europe during the 16th century and has divided the region ever since, although with the exception of the Troubles in Northern Ireland, tensions today between these two major groups are far less problematic than in the past.

Conflicts with Islam   Both eastern and western Christian churches struggled with challenges from Islamic empires to Europe's south and east. Even though historical Islam was reasonably tolerant of Christianity in its conquered lands, Christian Europe was far less accepting of Muslim imperialism. The first crusade to reclaim Jerusalem from the Turks took place in 1095. After the Ottoman Turks conquered

FIGURE 8.20 Religions of Europe   This map shows the divide in
western Europe between the Protestant north and the Roman Catholic south.
Historically, this distinction was much more important than it is today. Note the location of the former
Jewish Pale, which was devastated by the Nazis during World War II. Today ethnic tensions with religious
overtones are found primarily in the Balkans, where adherents of Roman Catholicism, Eastern Orthodoxy,
and Islam are found in close proximity to one another.

Constantinople in 1453 and gained control over the Bosporus strait and the Black Sea, they moved rapidly to spread a Muslim empire throughout the Balkans, arriving at the gates of Vienna in the middle of the 16th century. There Christian Europe stood firm and stopped Islam from expanding into western Europe.

Ottoman control of southeastern Europe, however, lasted until the empire's end in the early 20th century. This historical presence of Islam explains the current mosaic of religions in the Balkans, with intermixed areas of Muslims, Orthodox Christians, and Roman Catholics.

A Geography of Judaism   Europe has long been a difficult home for Jews forced to leave Palestine during the reign of the Roman Empire. At that time, small Jewish settlements were located in cities throughout the Mediterranean. Later, by 900 CE, about 20 percent of the Jewish population was clustered in the Muslim lands of the Iberian Peninsula, where Islam showed greater tolerance for Judaism than had Christianity. Furthermore, Jews played an important role in trade activities both within and outside the Islamic lands. After the Christian reconquest of Iberia, however, Jews once more faced severe persecution and fled from Spain to more tolerant countries in western and central Europe.

One focus for this exodus was the area in eastern Europe that became known as the Jewish Pale. In the late Middle Ages, at the invitation of the Kingdom of Poland, Jews settled in cities and small villages in what is now eastern Poland, Belarus, western Ukraine, and northern Romania (see Figure 8.20). Jews collected in this region for several centuries in the hope of establishing a true European homeland, despite the poor natural resources of this marshy, marginal agricultural landscape.

Until emigration to North America began in the 1890s, 90 percent of the world's Jewish population lived in Europe, and most were clustered in the Pale. Even though many emigrants to the United States and Canada came from this area, the Pale remained the largest population of Jews in Europe until World War II. Tragically, Nazi Germany used this ethnic clustering by focusing its extermination activities on the Pale.

In 1939, on the eve of World War II, there were 9.5 million Jews in Europe, or about 60 percent of the world's Jewish population. During the war, German Nazis murdered some 6 million Jews in the horror of the Holocaust. Today fewer than 2 million Jews live in Europe. However, since 1990 and the lifting of quotas on Jewish emigration from Russia, Belarus, and Ukraine, more than 100,000 Jews have emigrated to Germany, giving it the fastest-growing Jewish population outside Israel (Figure 8.21).

Patterns of Contemporary Religion   Europe today has about 250 million Roman Catholics and fewer than 100 million Protestants. Generally, Catholics are found in the southern half of the region, except for significant numbers in Ireland and Poland.

Protestantism is most widespread in northern Germany, the Scandinavian countries, and England, and it is intermixed with Catholicism in the Netherlands, Belgium, and Switzerland. Because of its reaction against ornate cathedrals and statues of the Catholic Church, the landscape of Protestantism is much more sedate and subdued. Large cathedrals and religious monuments in Protestant countries are associated primarily with the Church of England, which has strong historical ties to Catholicism; St. Paul's Cathedral and Westminster Abbey in London are examples.

FIGURE 8.21 Jewish Synagogue in Berlin   Before WW II Berlin had a large and thriving Jewish population, as attested to by this synagogue built in 1866. During the Nazi period, however, the Jewish population was forcibly removed and largely exterminated. This synagogue was then used by the Nazis to store military clothing until it was heavily damaged by Allied bombing. It was restored and opened again in 1995 as a synagogue and museum.

Not to be overlooked are the estimated 13 million Muslims in Europe. Most are migrants from Africa and southwestern Asia, although some are recent converts to Islam. France has the largest population of Muslims (4.7 million), with the second largest number (4.1 million) in Germany.

European Culture in a Global Context

Europe, like all other world regions, is currently caught up in a period of profound cultural change. In fact, many would argue that the pace of cultural change in Europe has been accelerated because of the complicated interactions between globalization and Europe's internal agenda of political and economic integration. Some pundits celebrate the "New Europe" of integration and unification, but other observers refer to a more tension-filled New Europe of foreign migrants and guest workers harassed by ethnic discrimination and racism.

Migrants and Culture   Migration patterns are influencing the cultural mix in Europe. Historically, Europe spread its cultures worldwide through aggressive colonialization. Today, however, the region is experiencing a reverse flow as millions of migrants move into Europe, bringing their own distinct cultures from the far-flung countries of Africa, Asia, and Latin America. Unfortunately, in some areas of Europe, the products of this cultural exchange are highly troubling.

Ethnic clustering leading to the formation of ghettos is now common in the cities and towns of western Europe (Figure 8.22). The high-density apartment buildings of suburban Paris, for example, are home to large numbers of French-speaking Africans and Arab Muslims caught in the crossfire of high unemployment, poverty, and racial discrimination. As a result, cultural battles have emerged in many European countries. For example, in 2004, French leaders, unsettled by the country's large Muslim migrant population, attempted to speed assimilation of female high school students into French mainstream culture by banning a key symbol of conservative Muslim life: the head scarf (hijab). This rule triggered riots, demonstrations, and counterdemonstrations. As a result of these kinds of conflicts, the political landscape of many European countries now has far-right, nationalistic parties with thinly disguised platforms of excluding migrants from their countries.

Gender Issues in Europe   Despite the visibility of female political leaders and the fact that Europe is considered one of the most developed regions of the world, gender equity issues persist in government, business, and domestic life. For the EU countries as a whole, for example, male employment is 21 percent higher than for women, while women who work generally make 25 percent less than do males (Figure 8.23).

However, given the complexity of Europe with its mosaic of national, urban, rural, and migrant cultures, the nature and extent of gender issues differ widely amongst countries and regions. To illustrate, within the 27 EU countries about a quarter of the parliamentary offices are held by women. Sweden has the highest representation, with more than half of its ministers being female, while Cyprus has absolutely none. Similarly, in the business world only 11 of Europe's largest companies have women in top management, yet women make up almost a third of top management in Norway, compared to just 1 percent in Luxembourg.

FIGURE 8.23 Gender Discrimination in the EU   Emilie Turunen (at right with painted mustache), the youngest member of the European Parliament, joins with other women parliament members to protest gender discrimination in the hiring of EU workers at the Brussels headquarters.

FIGURE 8.22 Turkish Store in Germany   This Turkish store in the Kreuzberg district of Berlin, sometimes referred to as "Little Istanbul," serves not only the large Turkish population, but also German shoppers who have found this store an appealing aspect of the country's broadening ethnicity. Unfortunately, to some, Europe's migrant cultures are more problematic.

One interesting pattern is that female presence in the workforce is generally higher in the countries of eastern Europe and the Balkans. Two interrelated factors could explain this. First, women were expected to work in the communist economies of these countries from 1945 to 1990. Second, families often needed two incomes to survive during the difficult economic transition that followed the collapse of the Soviet Union in 1990 (discussed later in this chapter). Regardless of cause, the results are startling. Today Bulgaria has the highest number of female CEOs (21 percent) of any EU country, whereas Slovenia, formerly a part of socialist Yugoslavia, is the country with the least income disparity between men and women.

Women are highly represented in both government and business in the Scandinavian countries (Norway, Sweden, Denmark, Finland, and Iceland), but for very different reasons than in eastern Europe and the Balkans. It is generally agreed that the foundation of Scandinavia's gender equity comes from a combination of comprehensive child care, liberal maternity and paternity benefits that guarantee job security and career advancement after maternity and paternity leaves, and a tax code that does not punish dual-income families.

As a result, Norway and Sweden have the highest number of females in the workforce. Portugal has the third highest number at 71 percent; however, experts caution that working families in that country face significant differences, compared to Scandinavian families. In Portugal, because of its struggling economy, women usually work out of necessity rather than choice, with grandparents and other family members providing child care—unlike the government-sponsored child care common in Scandinavia.

Not to be overlooked are the extraordinarily complex gender issues within Europe's large migrant cultures and their host cultures. How France's national policies can become entangled with Muslim gender and cultural preferences was mentioned earlier. Other examples of these complexities can be found in Germany, as the state finds itself embroiled in cultural tensions that range from women's freedom beyond the family household to prosecuting Turkish honor killings where young women have paid with their life for behaviors common to German society.

REVIEW QUESTIONS

1. Describe the general location within Europe of the three major language groups: Germanic, Romance, and Slavic.

2. In general terms, describe the historic distribution within Europe of Catholicism, Protestantism, Judaism, and Islam.

3. What two countries have the largest numbers of Muslims?

GEOPOLITICAL FRAMEWORK: A DYNAMIC MAP

One of Europe's unique characteristics is its dense fabric of 42 independent states within a relatively small area; no other world region demonstrates such a mosaic of geopolitical division. Conventional wisdom says that Europe invented the nation-state. Later, these political ideas founded in Europe fueled the flames of political independence and democracy worldwide, replacing Europe's colonial rule in Asia, Africa, and the Americas.

However, Europe's geopolitical landscape has been as much problem as promise. Twice in the past century, Europe shed blood to redraw its political borders, and within the past several decades nine new states have appeared, more than half through violence and war. Today, even though some political and social animosity remains from the Balkan wars of the 1990s, most geopolitical hotspots in Europe are more about regional autonomy than creating new states (Figure 8.24).

Redrawing the Map of Europe Through War

Two world wars radically reshaped the geopolitical maps of 20th-century Europe (Figure 8.25). By the early 20th century, Europe was divided into two opposing and highly armed camps that tested each other for a decade before the outbreak of World War I in 1914. France, Britain, and Russia were allied against Germany and the Austro-Hungarian and Ottoman Empires, both of which controlled a complex assortment of ethnic groups in central Europe and the Balkans. World War I was referred to as the "war to end all wars," but it fell far short of solving Europe's geopolitical problems. Instead, according to many experts, it made another European war unavoidable.

When Germany and Austria–Hungary surrendered in 1918, the Treaty of Versailles peace process set about redrawing the map of Europe with two goals in mind: first, to punish the losers through loss of territory and severe financial reparations and, second, to recognize the nationalistic aspirations of unrepresented peoples by creating new nation-states. As a result, the new states of Czechoslovakia and Yugoslavia were born. In addition, Poland was reestablished, as were the Baltic states of Finland, Estonia, Latvia, and Lithuania.

Though the goals of the treaty were admirable, few European states were satisfied with the resulting map. New states were resentful when their ethnic citizens were left outside the new borders and became minorities in other newly shaped states. This created an epidemic of **irredentism**, or state policies for reclaiming lost territory and peoples. Examples are the large German population in the western portion of the newly created state of Czechoslovakia and the Hungarians stranded in western Romania by border changes.

This imperfect geopolitical solution was greatly aggravated by the global economic depression of the 1930s, which brought high unemployment, food shortages, and political unrest to Europe. Three competing ideologies promoted their own solutions to Europe's pressing problems: Western democracy (and capitalism); communism from the Soviet revolution to the east; and a fascist totalitarianism promoted by Mussolini in Italy and Hitler in Germany. With industrial unemployment at record rates in western Europe, public opinion fluctuated wildly between the extremist solutions of far right fascism and leftist communism and socialism. In 1936, Italy and Germany joined forces through the Rome–Berlin "axis" agreement. As in World War I, this alignment was countered with mutual protection treaties among France, Britain, and the Soviet Union. When an imperialist Japan signed a pact with Germany, the scene was set for a second global war.

Nazi Germany tested Western resolve in 1938 by first annexing Austria, the country of Hitler's birth, and then Czechoslovakia, under the pretense of providing protection for ethnic Germans located there. After signing a nonaggression pact with the Soviet Union, Hitler invaded Poland on September 1, 1939. Two days later, France and Britain declared war on Germany. Within a month, the Soviet Union moved into eastern Poland, the Baltic states, and Finland to reclaim territories lost through the peace treaties of World War I. Nazi Germany then moved westward and occupied Denmark, the Netherlands, Belgium, and France, after which it began preparations to invade England.

In 1941, the war took several startling new turns. In June, Hitler broke the nonaggression pact with the Soviet Union and, catching the Red Army by surprise, took the Baltic states and then drove deep into Soviet territory. When Japan attacked Pearl Harbor, Hawaii, in December, the United States entered the war in both the Pacific and Europe.

By early 1944, the Soviet army had recovered most of its territorial losses and moved against the Germans in eastern Europe, beginning the long communist domination in that region. By agreement with the Western powers, the Red Army stopped when it reached Berlin in April 1945. At that time, Allied forces crossed the Rhine River and

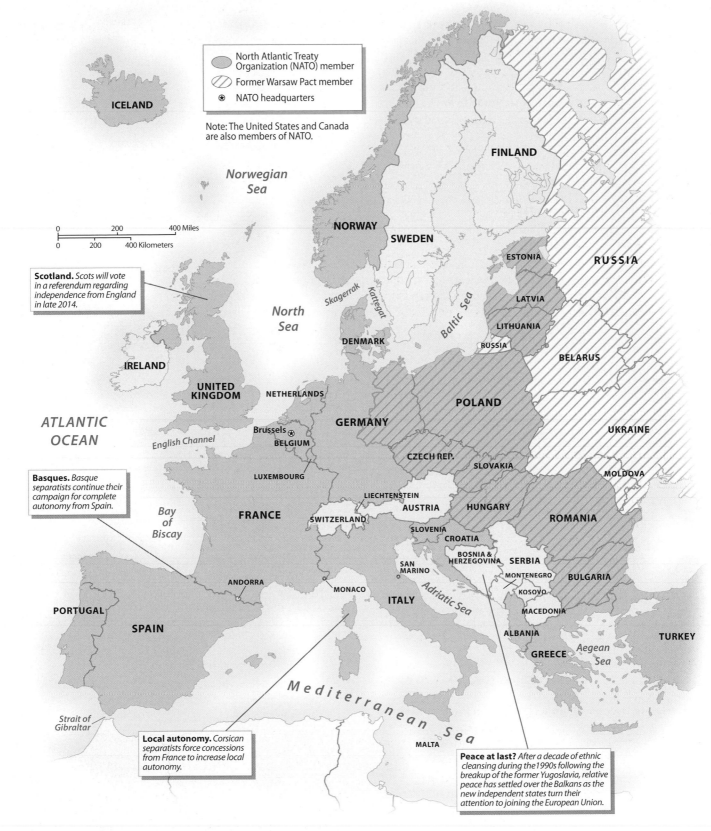

North Atlantic Treaty Organization (NATO) member

Former Warsaw Pact member

⊛ **NATO headquarters**

Note: The United States and Canada are also members of NATO.

Scotland. *Scots will vote in a referendum regarding independence from England in late 2014.*

Basques. *Basque separatists continue their campaign for complete autonomy from Spain.*

Local autonomy. *Corsican separatists force concessions from France to increase local autonomy.*

Peace at last? *After a decade of ethnic cleansing during the 1990s following the breakup of the former Yugoslavia, relative peace has settled over the Balkans as the new independent states turn their attention to joining the European Union.*

FIGURE 8.24 Geopolitical Issues in Europe Although the major geopolitical issue of the early 21st century remains the integration of eastern and western Europe into the EU, numerous issues of micro- and ethnic nationalism also engender geopolitical fragmentation. In other parts of Europe, such as Spain, France, and Great Britain, questions of local ethnic autonomy within the nation-state structure challenge central governments.

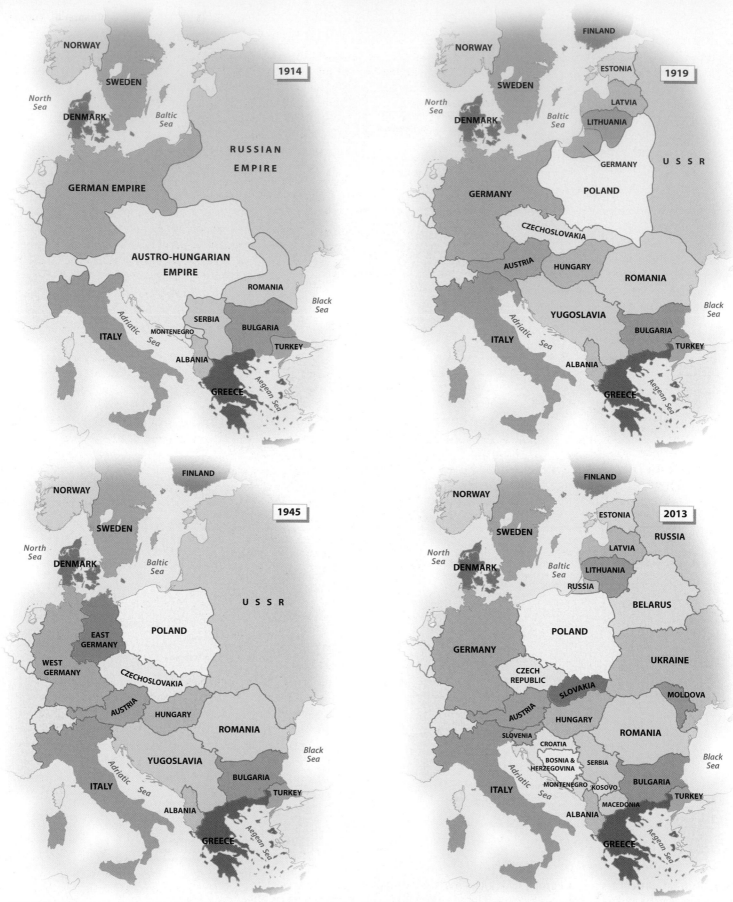

FIGURE 8.25 A Century of Geopolitical Change   At the outset of the 20th century, central Europe was dominated by the German, Austro-Hungarian (or Hapsburg), and Russian Empires. Following World War I, these empires were largely replaced by a mosaic of nation-states. More border changes followed World War II, largely as a result of the Soviet Union's turning that area into a buffer zone between itself and western Europe. With the demise of Soviet hegemony in 1989, further political change took place.

began their occupation of Germany. Immediately after Hitler's suicide, Germany signed an unconditional surrender on May 8, 1945, ending the war in Europe. But with Soviet forces firmly entrenched in the Baltic states, Poland, Czechoslovakia, Bulgaria, Romania, Hungary, Austria, and eastern Germany, the military battles of World War II were immediately replaced by an ideological Cold War between communism and democracy that lasted 45 years, until 1990.

A Divided Europe, East and West

From 1945 until 1990, Europe was divided into two geopolitical and economic blocs, east and west, separated by the infamous **Iron Curtain**, which descended shortly after the peace agreement ending World War II (Figure 8.26). East of the Iron Curtain border, the Soviet Union imposed the heavy imprint of communism on all activities—political, economic, military, and cultural. To the west, as Europe rebuilt from the destruction of the war, new alliances and institutions were created to counter the Soviet presence in Europe.

Cold War Geography   The seeds of the Cold War were planted at the Yalta Conference of February 1945, when Britain, the Soviet Union, and the United States met to plan the shape of postwar Europe. Because the Red Army was already in eastern Europe and moving quickly on Berlin, Britain and the United States agreed that the Soviet Union would occupy eastern Europe and the Western allies would occupy parts of Germany.

The larger geopolitical issue, though, was the Soviet desire for a **buffer zone** between its own territory and western Europe. This buffer zone consisted of an extensive bloc of satellite countries, dominated politically and economically by the Soviet Union, that could cushion the Soviet heartland against possible attack from western Europe. In the east, the Soviet Union took control of the Baltic states, Poland, Czechoslovakia, Hungary, Bulgaria, Romania, Albania, and briefly, Yugoslavia. Austria and Germany were divided into occupied

FIGURE 8.26 The Iron Curtain   From 1945 until 1990 the Iron Curtain divided Europe from Finland in the north to the former Yugoslavia in the south, with the most heavily fortified section along the border between East and West Germany. This photo is of a recreation of that section of the border at the Point Alpha museum near Fulda, Germany. In the background is one of the many East German observation towers located along the border.

sectors by the four (former) allied powers. In both cases, the Soviet Union dominated the eastern portion of the country, which contained the capital cities of Berlin and Vienna. Both capital cities, in turn, were divided into French, British, U.S., and Soviet sectors.

In 1955, with the creation of an independent and neutral Austria, the Soviets withdrew from their sector, effectively moving the Iron Curtain eastward to the Hungary–Austria border. However, Germany quickly evolved into two separate states, West Germany and East Germany, that remained separate until 1990.

Along the border between east and west, two hostile military forces faced each other for almost half a century. Both sides prepared for and expected an invasion by the other across the barbed wire of a divided Europe. In small military units from satellite countries, both the NATO (North Atlantic Treaty Organization) and the Warsaw Pact countries were armed with nuclear weapons, making Europe a tinderbox for a devastating third world war.

Berlin was the flashpoint that brought these forces close to a fighting war on two occasions. In winter 1948, the Soviets imposed a blockade on the city by denying Western powers access to Berlin across its East German military sector. This attempt to starve the city into submission by blocking food shipments from western Europe was thwarted by a nonstop airlift of food and coal by NATO. Then, in August 1961, the Soviets built the Berlin Wall to curb the flow of East Germans seeking political refuge in the West. The Wall became the concrete-and-mortar symbol of a firmly divided postwar Europe. For several days while the Wall was being built and the West agonized over destroying it, NATO and Warsaw Pact tanks and soldiers faced each other with loaded weapons at point-blank range. Though war was avoided, the Wall stood for 28 years, until November 1989.

The Cold War Thaw   The symbolic end of the Cold War in Europe came on November 9, 1989, when East and West Berliners joined forces to rip apart the Berlin Wall with jackhammers and hand tools (Figure 8.27).

By October 1990, East and West Germany were officially reunified into a single nation-state. During this period, all other Soviet satellite states, from the Baltic Sea to the Black Sea, also underwent major geopolitical changes that have resulted in a mixed bag of benefits and problems.

The Cold War's end came as much from a combination of problems within the Soviet Union (discussed in Chapter 9) as from rebellion in eastern Europe. By the mid-1980s, the Soviet leadership was advocating an internal economic restructuring and also recognizing the need for a more open dialogue with the West. Financial problems from supporting a huge military establishment, along with heavy losses from an unsuccessful war in Afghanistan, lessened the Soviet appetite for occupying other countries.

In August 1989, Poland elected the first noncommunist government to lead an eastern European state since World War II. Following this, with just one exception, peaceful revolutions with free elections spread throughout eastern Europe as communist governments renamed themselves and broke with doctrines of the past. In Romania, though, street fighting between citizens and military resulted in the violent overthrow and execution of communist dictator Nicolae Ceausescu.

As a result of the Cold War thaw, the map of Europe began changing once again. Germany reunified in 1990. Elsewhere,

(a)

(b)

FIGURE 8.27 The Berlin Wall In August 1961 East Germany built a concrete and barbed wire structure along the border of East Berlin to stem the flow of refugees fleeing communist rule. The Wall was the most visible symbol of the Cold War division of East and West until November 1989 when the failing Soviet Union renounced its control over Eastern Europe. a) shows the extent of the Wall zone at Brandenburg Gate (East Berlin is to the left). While b) shows Berliners of all ages physically dismantling the Wall in November 1989.

political separatism and ethnic nationalism, long suppressed by the Soviets, were unleashed in southeastern Europe as the former Yugoslavia fractured into several independent states. Similarly, on January 1, 1993, Czechoslovakia was replaced by two separate states: the Czech Republic and Slovakia.

The Balkans: Waking from a Geopolitical Nightmare

The Balkans have been a troublesome area for several centuries with their complex mixture of languages, religions, and ethnic allegiances. Throughout history, these allegiances have led to an often changing mosaic of small countries (Figure 8.28). Indeed, the term **balkanization** is used to describe the geopolitical processes of small-scale independence movements based upon ethnic fault lines. Following the fall of two competing empires in the early 20th century, much of the region was unified under the political umbrella of the former Yugoslavian state. In the 1990s, however, Yugoslavia broke apart as ethnic factionalism and nationalism produced a decade of violence, turmoil, and wars of independence, creating a geopolitical nightmare for Europe, the EU, NATO, and the world. Today, though, despite lingering tensions in several areas, there are signs that the Balkan countries are moving toward a new era of peace and stability.

Balkan Wars of Independence Following World War II, Britain, the United States, and the Soviet Union rewarded Josip Tito, the leader of an anti-Nazi guerrilla group in the Balkans, for his efforts by backing him as the leader of a new socialist Yugoslavia. The three major ethnic groups in Yugoslavia—the Serbs, Croats, and Muslims—coexisted under Tito's strong leadership by subordinating their separatist agendas to the larger goal of a communist state independent of the Soviet Union. This relative peace continued for a decade after Tito's death in 1980; however, the region's geopolitical stability unraveled in the 1990s, coincident with the lessening of Soviet control in other parts of eastern Europe.

In 1990, elections were held in Yugoslavia's different republics over the issue of secession from the mother state. Secessionist parties gained control in Slovenia and Croatia, but Serbian voters opted for continued Yugoslav unification in what observers considered to be a government-controlled election. Nonetheless, Slovenia and Croatia declared independence in 1991, Macedonia in January 1992, and Bosnia and Herzegovina in April 1992. When the Yugoslavian army attacked Slovenia, the Balkan situation got Europe's full attention, and a negotiated settlement resulted in Slovenia's independence. In Bosnia–Herzegovina, however, Serb paramilitary units waged a ferocious war of ethnic cleansing against both Muslims and Croats in a war that lasted until 1995. At that time, a complex political arrangement created a Serb republic and a Muslim–Croat federation, both ruled by the same legislature and president. Croatia also fought a devastating, but successful war of independence against Serbian nationalists.

Kosovo, in the south of Serbia (as the former Yugoslavia was now called), was another trouble spot, with long-standing tensions between Serbs and Muslims. Although Kosovo had enjoyed differing degrees of autonomy within the former Yugoslavia, this autonomy was withdrawn by Belgrade in 1990 to protect the Serb minority population. Not surprisingly, the Muslim Kosovar rebels responded by proclaiming Kosovo's independence in 1991. This act was resisted vigorously by Serbia, which responded with a violent ethnic cleansing program designed to oust the Muslims and make Kosovo a pure Serbian province (Figure 8.29). Diplomatic efforts failed to resolve this violence, and as warfare escalated in Kosovo, NATO (which included the United States) began bombing Belgrade in 1999 to force Serbia to accept a negotiated settlement. From 1999 until 2008, Kosovo was administered by the UN as a protectorate, enforced by some 50,000 peacekeepers from 30 different countries.

In 2008, the UN-sanctioned Kosovo parliament once again declared the country's independence. Although Serbia still refuses to recognize Kosovo as an independent state, over 90 countries (including the United States) have done so. At Serbia's request, the UN World Court reviewed the legality of Kosovo's independence, which it did by ruling in July 2010 that its independence was indeed legal.

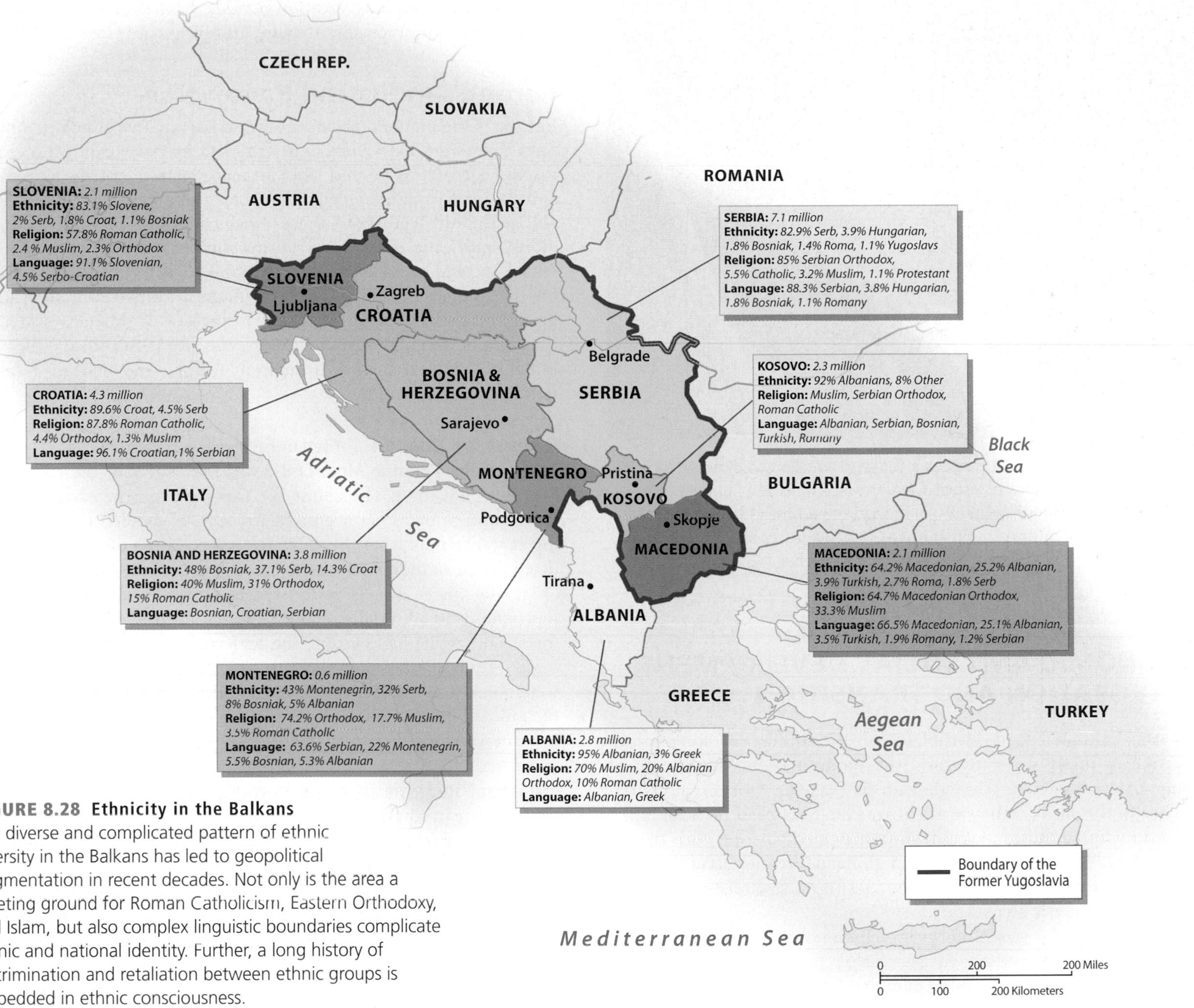
CZECH REP.
SLOVAKIA
AUSTRIA
HUNGARY
ROMANIA
SLOVENIA: 2.1 million
Ethnicity: 83.1% Slovene, 2% Serb, 1.8% Croat, 1.1% Bosniak
Religion: 57.8% Roman Catholic, 2.4 % Muslim, 2.3% Orthodox
Language: 91.1% Slovenian, 4.5% Serbo-Croatian
SERBIA: 7.1 million
Ethnicity: 82.9% Serb, 3.9% Hungarian, 1.8% Bosniak, 1.4% Roma, 1.1% Yugoslavs
Religion: 85% Serbian Orthodox, 5.5% Catholic, 3.2% Muslim, 1.1% Protestant
Language: 88.3% Serbian, 3.8% Hungarian, 1.8% Bosniak, 1.1% Romany
SLOVENIA
Ljubljana
Zagreb
CROATIA
Belgrade
BOSNIA & HERZEGOVINA
SERBIA
Sarajevo
KOSOVO: 2.3 million
Ethnicity: 92% Albanians, 8% Other
Religion: Muslim, Serbian Orthodox, Roman Catholic
Language: Albanian, Serbian, Bosnian, Turkish, Romany
CROATIA: 4.3 million
Ethnicity: 89.6% Croat, 4.5% Serb
Religion: 87.8% Roman Catholic, 4.4% Orthodox, 1.3% Muslim
Language: 96.1% Croatian, 1% Serbian
Adriatic Sea
MONTENEGRO
Pristina
KOSOVO
Black Sea
BULGARIA
ITALY
Podgorica
Skopje
MACEDONIA
BOSNIA AND HERZEGOVINA: 3.8 million
Ethnicity: 48% Bosniak, 37.1% Serb, 14.3% Croat
Religion: 40% Muslim, 31% Orthodox, 15% Roman Catholic
Language: Bosnian, Croatian, Serbian
MACEDONIA: 2.1 million
Ethnicity: 64.2% Macedonian, 25.2% Albanian, 3.9% Turkish, 2.7% Roma, 1.8% Serb
Religion: 64.7% Macedonian Orthodox, 33.3% Muslim
Language: 66.5% Macedonian, 25.1% Albanian, 3.5% Turkish, 1.9% Romany, 1.2% Serbian
Tirana
ALBANIA
MONTENEGRO: 0.6 million
Ethnicity: 43% Montenegrin, 32% Serb, 8% Bosniak, 5% Albanian
Religion: 74.2% Orthodox, 17.7% Muslim, 3.5% Roman Catholic
Language: 63.6% Serbian, 22% Montenegrin, 5.5% Bosnian, 5.3% Albanian
GREECE
Aegean Sea
TURKEY
ALBANIA: 2.8 million
Ethnicity: 95% Albanian, 3% Greek
Religion: 70% Muslim, 20% Albanian Orthodox, 10% Roman Catholic
Language: Albanian, Greek
Boundary of the Former Yugoslavia
Mediterranean Sea
0 200 200 Miles
0 100 200 Kilometers

FIGURE 8.28 Ethnicity in the Balkans
The diverse and complicated pattern of ethnic diversity in the Balkans has led to geopolitical fragmentation in recent decades. Not only is the area a meeting ground for Roman Catholicism, Eastern Orthodoxy, and Islam, but also complex linguistic boundaries complicate ethnic and national identity. Further, a long history of discrimination and retaliation between ethnic groups is embedded in ethnic consciousness.

Moving Toward Stability Although Serbia remains steadfast about reclaiming Kosovo and not recognizing its new independence, in most other matters a more moderate and less nationalistic government has led to Serbia's being reinstated in the UN, in the Council of Europe, and as an official candidate to the EU. Of the former Yugoslav republics, only Slovenia has actually achieved membership in the EU, although all Balkan countries (including Albania, which was not a part of the former Yugoslavia) have begun the lengthy and complex EU membership process.

REVIEW QUESTIONS

1. Describe briefly how the map of Europe changed with the Treaty of Versailles in 1918.
2. What European countries were considered Soviet satellites during the Cold War?
3. What countries made up the former Yugoslavia?

FIGURE 8.29 Kosovar Refugees Close to a million Muslims were expelled or fled the former Yugoslavian province of Kosovo during an ethnic cleansing campaign by Serbian forces during the late 1990s. These people are in the refugee camp at Kukës, Albania. After a long period as a NATO Protectorate, Kosovo declared its independence in 2008 in an act still challenged by Serbia.

ECONOMIC AND SOCIAL DEVELOPMENT: INTEGRATION AND TRANSITION

As the acknowledged birthplace of the Industrial Revolution, Europe in many ways invented the modern economic system of industrial capitalism. Though Europe was the world's industrial leader in the early 20th century, it was later eclipsed by both Japan and the United States as the region struggled to cope with the effects of two world wars, a decade of global depression, and the more recent Cold War. Currently (in late 2012), a fiscal crisis challenges Europe's economic structure.

In the past 50 years, however, economic integration, guided by the EU, has been increasingly successful. In fact, western Europe's success at blending national economies has given the world a new model for regional cooperation, an approach that in the near future may be imitated in Latin America and Asia. Eastern Europe, however, has not fared as well. The results of four decades of Soviet economic planning were, at best, mixed. The total collapse of that system in 1990 cast eastern Europe into a period of chaotic economic, political, and social transition that has resulted in a highly differentiated pattern of rich and poor regions. While some countries, such as the Czech Republic and Slovenia, prosper, the future prospects for Albania, Hungary, and Romania are uncertain (Table 8.2).

Accompanying western Europe's economic boom has been an unprecedented level of social development, as measured by worker benefits, health services, education, literacy, and gender equality. Though the improved social services set an admirable standard for the world, cost-cutting politicians and businesspeople now argue that these services increase the cost of business, so that European goods cannot compete in the global marketplace, especially in the recent global economic recession. As a result, many of these traditional benefits, such as job security and long vacation periods, have been eroded, causing considerable social tension in many European countries.

Europe's Industrial Revolution

Europe is the cradle of modern industrialism. Two fundamental innovations were associated with this Industrial Revolution: First, machines replaced human labor in many manufacturing processes, and, second, inanimate energy sources (water, steam, electricity, and petroleum) powered these new machines. Though we commonly apply the term *Industrial Revolution* to this transformation, which implies rapid change, it actually took more than a century for the interdependent pieces of a modern industrial system to be brought together. England was the birthplace of this new system in the years between 1730 and 1850; by the late 19th century, this new industrialism had spread throughout Europe and, within decades, to the rest of the world.

Centers of Change England's textile industry, located on the flanks of the Pennine Mountains, was the center of the earliest industrial innovation. The county of Yorkshire, on the eastern side of the Pennines, had long been a hearth area of woolen textiles, drawing raw materials from the extensive sheep herds of that region and using the clean mountain waters to wash the wool before it was spun. Originally, water wheels were used to power mechanized looms at the rapids and waterfalls of the Pennine streams, but by the 1790s the steam engine had become the preferred source of energy (Figure 8.30). Steam engines, however, needed fuel, and local wood supplies were quickly exhausted. Progress was stifled until the development of the railroad in the early 19th century, an innovation that allowed coal to be moved long distances at a reasonable cost.

Industrial Regions in Continental Europe The first industrial districts began appearing in continental Europe in the 1820s, located close to existing coalfields. The first area outside Britain was the Sambre-Meuse region, named for the two river valleys straddling the French-Belgian border. Like the English Midlands, it had a long history of cottage-based wool textile manufacturing that quickly converted to the new technology of steam-powered mechanized looms (Figure 8.31).

By 1850, the dominant industrial area in all Europe (including England) was the Ruhr district in northwestern Germany, near the Rhine River. Rich coal deposits close to the surface fueled the Ruhr's transformation from a small textile region to one of heavy industry, particularly of iron and steel manufacturing. Decades later, the Ruhr industrial region became synonymous with the industrial strength behind Nazi Germany's war machine, and thus it was bombed heavily in World War II.

Rebuilding Postwar Europe

As noted, Europe was the leader of the industrial world in 1900, when it produced 90 percent of the world's manufactured output. However, by 1945, after four decades of war and economic chaos, industrial Europe was in shambles, with many of its cities and industrial areas in ruins and much of the region's population dispirited, homeless, and hungry. Clearly, a new pathway for postwar Europe had to be forged to provide economic, political, and social security.

TABLE 8.2 DEVELOPMENT INDICATORS

Country	GNI per capita, PPP 2010	GDP Average Annual %Growth 2000–10	Human Development Index (2011)[1]	Percent Population Living Below $2 a Day	Life Expectancy (2012)[2]	Under Age 5 Mortality Rate (1990)	Under Age 5 Mortality Rate (2010)	Adult Literacy (% ages 15 and older)	Gender Inequality Index (2011)[3,1]
Western Europe									
Austria	39,790	1.8	.885	—	81	9	4	—	0.131
Belgium	38,290	1.6	.886	—	80	10	4	—	0.114
France	34,750	1.3	.884	—	82	9	4	—	0.106
Germany	38,100	1.0	.905	—	80	9	4	—	0.085
Liechtenstein	—	—	.905	—	81	—	—	—	—
Luxembourg	61,240	—	.867	—	80	—		—	0.169
Monaco	—	—	—	—	—	—		—	—
Netherlands	41,810	1.6	.910	—	81	8	4	—	0.052
Switzerland	49,960	1.9	.903	—	82	8	5	—	0.067
United Kingdom	35,840	1.8	.863	—	80	9	5	—	0.209
Eastern Europe									
Bulgaria	13,440	4.8	.771	<2	74	22	13	98	0.245
Czech Republic	22,910	3.8	.865	<2	78	14	4	—	0.136
Hungary	19,550	2.2	.816	<2	74	19	6	99	0.237
Poland	19,160	4.3	.813	<2	76	17	6	100	0.164
Romania	14,290	5.0	.781	<2	73	37	14	98	0.333
Slovakia	22,980	5.4	.834	<2	75	18	8	—	0.194
Southern Europe									
Albania	8,520	5.4	.739	4.3	75	41	18	96	0.271
Bosnia & Herzegovina	8,910	4.6	.733	<2	76	19	8	98	—
Croatia	18,890	3.2	.796	<2	77	13	6	99	0.170
Cyprus	30,300	3.1	.840	—	78	11	4	98	0.141
Greece	27,630	2.6	.861	—	80	13	4	97	0.162
Italy	31,810	0.5	.874	—	82	10	4	99	0.124
Kosovo	3,290	5.3	—	—	69	—	—	—	—
Macedonia	10,920	3.3	.728	5.9	72	39	12	97	0.151
Malta	24,660	—	.832	—	81	—		92	0.272
Montenegro	12,770	—	.771	<2	74	—		—	—
Portugal	24,590	0.7	.809	—	79	15	4	95	0.140
San Marino	—	—	—	—	84	—		—	—
Serbia	11,090	4.1	.766	<2	74	29	7	—	—
Slovenia	26,530	3.3	.884	<2	80	10	3	100	0.175
Spain	31,800	2.4	.878	—	82	11	5	98	0.117
Northern Europe									
Denmark	41,100	0.9	.895	—	79	9	4	—	0.060
Estonia	19,810	4.6	.835	<2	76	21	5	100	0.194
Finland	37,070	2.1	.882	—	80	7	3	—	0.075
Iceland	28,270	—	.898	—	82	—		—	0.099
Ireland	33,540	2.8	.908	—	79	9	4	—	0.203
Latvia	16,380	4.8	.805	<2	74	21	10	100	0.216
Lithuania	18,060	5.3	.810	<2	73	17	7	100	0.192
Norway	58,570	1.7	.943	—	81	9	3	—	0.075
Sweden	39,730	2.2	.904	—	82	7	3	—	0.049

[1]United Nations, *Human Development Report, 2011.*
[2]Population Reference Bureau, *World Population Data Sheet, 2012.*
[3]Gender Inequality Index—A composite measure reflecting inequality in achievements between women and men in three dimensions: reproductive health, empowerment and the labor market that ranges between 0 and 1. The higher the number, the greater the inequality.

Source: World Bank, *World Development Indicators, 2012.*

FIGURE 8.30 Hearth Area of the Industrial Revolution Europe's industrial revolution became on the flanks of England's Pennine Mountains where swift-running streams were used to power mechanized looms to weave cotton and wool. Later, once the railroads were developed, many of these early factories switched to coal power.

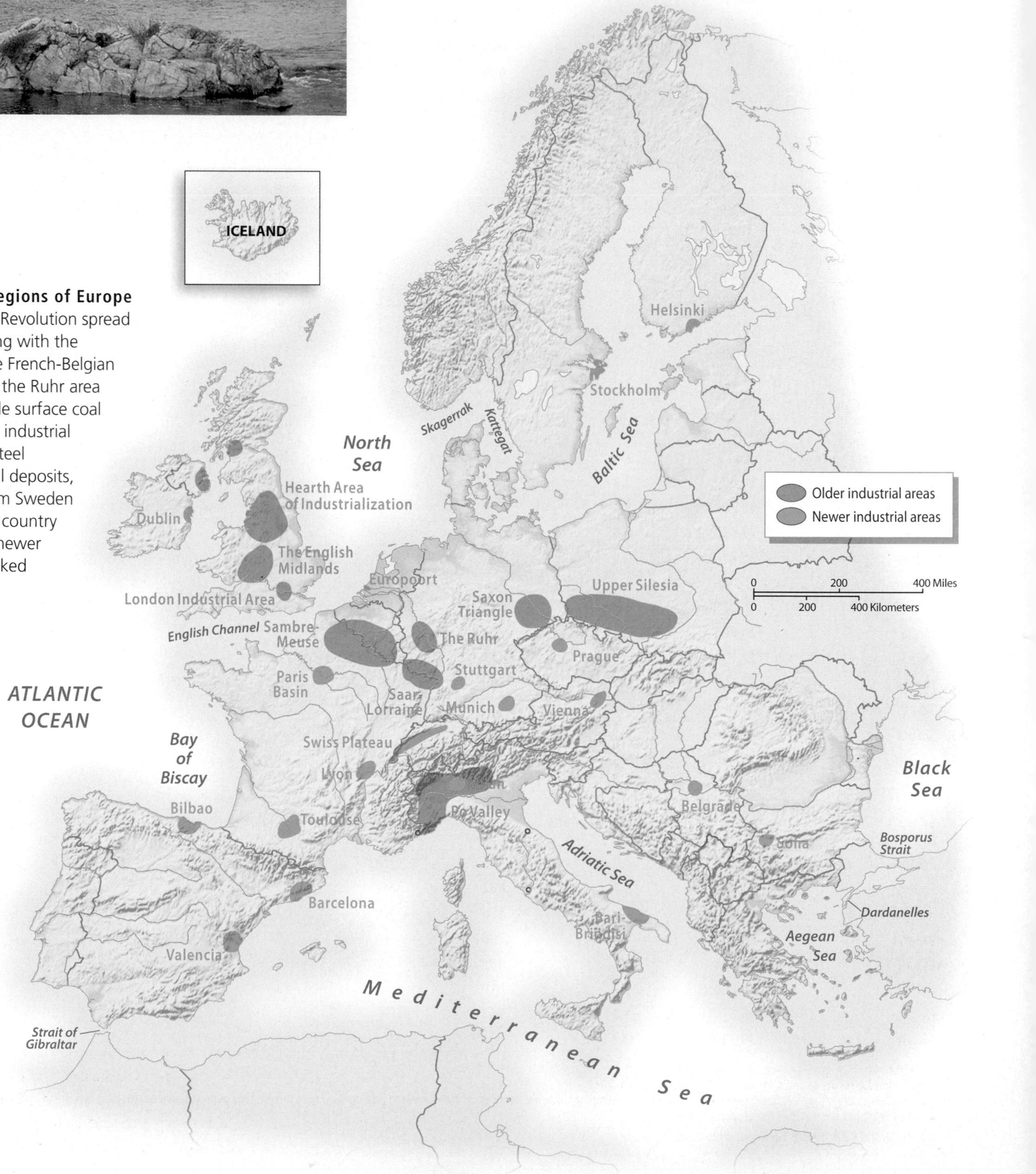

FIGURE 8.31 Industrial Regions of Europe
From England, the Industrial Revolution spread to continental Europe, starting with the Sambre-Meuse region on the French-Belgian border and then diffusing to the Ruhr area in Germany. Readily accessible surface coal deposits powered these new industrial areas. Early on, iron ore for steel manufacture came from local deposits, but later it was imported from Sweden and other areas in the shield country of Scandinavia. Most of the newer industrial areas are closely linked to urban areas.

ICELAND

Helsinki

Stockholm

North Sea

Hearth Area of Industrialization

Dublin

The English Midlands

London Industrial Area

English Channel Sambre-Meuse

ATLANTIC OCEAN

Bay of Biscay

Bilbao

Paris Basin

Saar-Lorraine

Europoort

Saxon Triangle

The Ruhr

Stuttgart

Munich

Swiss Plateau

Lyon

Toulouse

Po Valley

Upper Silesia

Prague

Vienna

Belgrade

Sofia

Black Sea

Bosporus Strait

Barcelona

Valencia

Adriatic Sea

Bari-Brindisi

Dardanelles

Aegean Sea

Skagerrak

Kattegat

Baltic Sea

Older industrial areas
Newer industrial areas

0   200   400 Miles
0   200   400 Kilometers

M e d i t e r r a n e a n   S e a

Strait of Gibraltar

The ECSC and EEC In 1950, the leaders of western Europe began discussing a new form of economic integration that would avoid the historical pattern of nationalistic independence through tariff protection, as well as the economic inefficiencies from duplication of industrial effort. Robert Schuman, France's farsighted foreign minister, proposed the radical idea that instead of rebuilding separate iron and steel facilities, Europe should share its natural resources. In May 1952, France, Germany, Italy, the Netherlands, Belgium, and Luxembourg ratified a treaty that joined them together in the European Coal and Steel Community (ECSC). Five years later, because of the resounding success of the ECSC, these six states agreed to work toward further integration by creating a larger European common market that would encourage the free movement of goods, labor, and capital (Figure 8.32). In March 1957, the Treaty of Rome was signed, establishing the European Economic Community (EEC).

From EEC to EU The EEC reinvented itself in 1965 with the Brussels Treaty, which laid the groundwork for adding a political union to the already successful economic community. In this "Second Treaty of Rome," aspirations for more than economic integration were clearly envisioned, with the creation of an EEC council, court, parliament, and political commission. At that time, the EEC became the European Community (EC) and began expanding its membership from the original six member states.

In 1991, the EC expanded its goals once again and became the European Union (EU). Although economic integration remained an underlying theme in the new constitution, with its commitment to a single currency through the European Monetary Union (discussed below), the EU moved even further into supranational affairs with discussions of common foreign policies and mutual security agreements.

In May 2004, the EU moved beyond its core membership in western Europe and added 10 new states, including a group of former Soviet-controlled communist satellites from eastern Europe. These new members—Latvia, Estonia, Lithuania, Poland, Slovakia, the

Czech Republic, Hungary, Slovenia, Malta, and Cyprus—brought the total to 25; Bulgaria and Romania were admitted in January 2007, resulting in the current 27 EU countries (Figure 8.33). As of this writing (late 2012), Iceland, Serbia, Montenegro, Macedonia, and Turkey are all formal candidates for EU membership.

The European Monetary Union As any world traveler knows, each state usually has its own monetary system. Until a decade ago, travelers in Europe used marks in Germany, francs in France, lira in Italy, and so on. Today, however, much of Europe has moved from individual state monetary systems to a common currency. This radical change began in 1999 when 11 of the then 15 EU member states joined together to form the European Monetary Union (EMU). As of that day, cross-border business and trade transactions began taking place in a new monetary unit, the euro. In 2002, euro coins and bills completely replaced the different national currencies in **Euroland**, those countries belonging to the EMU.

By adopting a common currency, Euroland members sought to increase the efficiency and competitiveness of both domestic and international business by eliminating the costs associated with business payments made in different currencies. Germany, for example, exports two-thirds of its products to other EU members. With a common currency, this business is now essentially domestic trade, protected from the fluctuations of different currencies and not subject to transaction costs. Although many traditional economists were skeptical about this new common currency system, the political goal of enhancing European unity took precedence.

Some EU member countries, however—most notably, the United Kingdom—had (and still have) considerable reservations about giving up control over their national monetary system and continue to use their traditional currency. Indeed, the early controversy over Euroland membership has been rekindled by the current European fiscal crisis. While strong economies (mainly Germany) have benefited from Euroland's common monetary policy, weaker ones (Spain, Greece, and Portugal, for example) have been hindered during recent fiscal crises by not being able to deflate their currencies to make their export goods cheaper, and thus more competitive in global markets.

Economic Disintegration and Transition in Eastern Europe

Eastern Europe has historically been less developed economically than its western counterpart. This is partially explained by the fact this region is simply not as rich in natural resources as is western Europe. But also important is that these resources have long been exploited by outside interests rather than being developed internally. This pattern began in the 19th century, when the Ottoman and Hapsburg Empires dominated eastern Europe and the Balkans. Later, it was followed by Nazi Germany's exploitation of eastern Europe, and then, most recently, the pattern continued with the Soviet Union's centralized planning from 1945 to 1991.

Even though Soviet economic planning was ostensibly an attempt to develop eastern Europe's economy by coordinating resource usage, these efforts were, in fact, meant to serve Soviet homeland interests. This system worked with mixed results for more than 40 years, but an unfortunate

FIGURE 8.32 Europe's Imported Energy As Europe uses up the last of its coal resources it becomes increasingly dependent on imported energy fuels, namely oil, natural gas, and coal. While in the past nuclear power helped reduce Europe's dependency on imported fuels, many countries are following Germany's example of phasing out nuclear power and turning instead to renewable sources such as wind and solar power. This photo is of petroleum storage facilities near Rotterdam where imported oil enters Europe.

FIGURE 8.33 The European Union The driving force behind Europe's economic and political integration has been the EU, which was formed in the 1950s as an organization with six members focused solely on rebuilding the region's coal and steel industries. As of 2012, the EU has 27 members. Besides the official applicants of Turkey, Macedonia, Croatia, and Iceland, several Balkan countries are in earlier stages of applications. Note that Norway is not a member of the EU, primarily because membership would restrict that country's fishing industry. Those same fishing restrictions may also become problematic for Iceland's membership.

result of Soviet centralized planning was that eastern European countries were plunged into economic and social chaos when the Soviet Union itself collapsed in 1991. Recovery and development since that time have been difficult, with some countries making the transition more rapidly and more fully than others, creating a geographic patchwork of both wealth and hardship throughout eastern Europe.

Transition and Change Since 1991 As mentioned, when Soviet domination over eastern Europe collapsed, so did the economic integration of the region. In place of Soviet coordination and subsidy came a painful period of economic transition that was outright chaotic in many eastern European countries. As the Soviet Union turned its attention to its own economic and political turmoil, it stopped exporting cheap natural gas and petroleum to eastern Europe. Instead, Russia sold these fuels on the open global market to gain hard currency. Without cheap energy, many eastern European industries were unable to operate and were forced to close, laying off thousands of workers (Figure 8.34). For example, in the first two years of the transition (1990–1992), industrial production fell 35 percent in Poland and 45 percent in Bulgaria. In addition, markets for eastern European products, guaranteed under Soviet planning, simply evaporated, aggravating further the collapse of eastern European economies.

To recover, eastern European countries began redirecting their economies toward western Europe. But to do this meant moving from a socialist-based economy of state ownership and control to a

FIGURE 8.35 Western Hypermarkets in Eastern Europe A customer buys fish at a new Carrefour hypermarket on the outskirts of Sofia, Bulgaria. Consumer items, including food, were historically in short supply in eastern Europe during the postwar Soviet-satellite period, with these shortages continuing through the difficult post-Soviet transition years. Today, however, there's no shortage as western super- and hypermarkets have populated eastern Europe. But whether local customers can actually afford these western goods remains unclear. Carrefour, a French firm, and the world's second-largest retailer behind Walmart, is reportedly selling its eastern Europe stores in order to expand retailing efforts in China.

capitalist economy of private ownership and free markets. Without Soviet price supports and subsidies, a completely new economic system had to be constructed, one that could compete favorably in the new global marketplace (Figure 8.35). During this transitional period, financial security was elusive, with irregular paychecks for those with jobs and uncertain welfare benefits for those without. Cronyism prevailed, as those with strong social connections profited at the expense of others who were lost in the chaos.

While some countries—namely, Poland, the Czech Republic, Slovenia, and Slovakia—made the transition quickly, others—primarily the Balkan states—are taking longer. Of considerable help in this economic and social transition was EU membership, with a complex application process that was nevertheless accompanied by an infusion of vast amounts of funds to facilitate everything from environmental clean-up to development of small businesses. Important to remember is that this recovery and transition process in eastern Europe has not ended, but instead is still going on, with greater and lesser degrees of success. The result is a patchy geographic mosaic akin in many ways to that of the world itself, with its array of more and less developed countries. This geographic diversity is not limited to eastern Europe, but can be found in all of Europe.

FIGURE 8.34 Post-1989 Hardship in Eastern Europe With the fall of communism in eastern Europe after 1989, economic subsidies and support from the Soviet Union ended. Accompanying this transition has been a high unemployment rate, resulting from the closure of many industries, such as this plant in Bulgaria.

GDI/PPP IN $

- More than 50,000
- 40,000–50,000
- 30,000–39,999
- 20,000–29,999
- Less than 20,000

5.2 Percent unemployment mid-2012

ICELAND 7.5

FINLAND 7.5

NORWAY 3.2

SWEDEN 7.3

ESTONIA 11.7

LATVIA 14.6

LITHUANIA 14.3

RUSSIA

DENMARK 8.1

IRELAND 14.5

UNITED KINGDOM 8.2

NETHERLANDS 5.0

BELGIUM 7.3

LUXEMBOURG 5.2

GERMANY 5.6

POLAND 10.1

BELARUS

UKRAINE

CZECH REP. 8.7

SLOVAKIA 13.9

MOLDOVA

LIECHTENSTEIN 2.8

AUSTRIA 4.0

HUNGARY 11.2

ROMANIA 7.5

FRANCE 10.0

SWITZERLAND 3.1

SLOVENIA 8.5

CROATIA 15.9

BOSNIA & HERZEGOVINA 43.3

SERBIA 23.7

BULGARIA 12.6

SAN MARINO

MONACO

ITALY 9.8

MONTENEGRO 11.5

KOSOVO 45.0

MACEDONIA 29.1

PORTUGAL 15.3

ANDORRA 2.9

SPAIN 24.1

ALBANIA 13.4

GREECE 21.7

MALTA 6.8

CYPRUS 10.0

ATLANTIC OCEAN

Norwegian Sea

North Sea

Baltic Sea

RUSSIA

Skagerrak

Kattegat

English Channel

Bay of Biscay

Adriatic Sea

Mediterranean Sea

Aegean Sea

Strait of Gibraltar

0   200   400 Miles
0   200   400 Kilometers

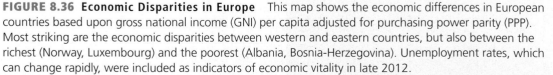

FIGURE 8.36 Economic Disparities in Europe   This map shows the economic differences in European countries based upon gross national income (GNI) per capita adjusted for purchasing power parity (PPP). Most striking are the economic disparities between western and eastern countries, but also between the richest (Norway, Luxembourg) and the poorest (Albania, Bosnia-Herzegovina). Unemployment rates, which can change rapidly, were included as indicators of economic vitality in late 2012.

Europe's Geography of Economic Disparities and Uncertainty

Despite decades of postwar economic development and integration through the EU, Europe today is a troubled region of economic and social disparities that could have global implications. Most western European and Scandinavian countries are rich, are economically viable, and enjoy the benefits of comprehensive social development. However, the peripheral countries of the Balkans and Baltic lag far, far behind by most economic and social measures. Also, and most troubling to the world economy, is the fiscal insolvency of Greece, Italy, and Spain, serious problems that challenge the very future of Europe's common currency system.

Figure 8.36 shows these economic disparities within Europe by mapping GDP per capita. It also displays data for those countries where unemployment is currently over 10 percent. On this map, one sees that the poor Balkan and Baltic countries are four times poorer than their richer cousins in western Europe and Scandinavia. Moreover, the poorest countries are three times poorer than the median GDP per capita for all of Europe. Whether those countries can ever catch up economically and socially with western Europe is an important question.

As mentioned earlier in this chapter, inhabitants of the Baltic states of Estonia, Lithuania, and Latvia, because they are EU members, can easily migrate to other EU countries with better economic opportunities. But is this migration a help or a hindrance to the economic and social development of the home countries? Opinions vary on this question.

There are fewer options for people in the non-EU countries to migrate because of visa restrictions. Most Balkan countries aspire to join the EU, which in more prosperous days would have involved economic aid to help an EU applicant country achieve economic viability before membership. However, the current fiscal crisis with Greece, Spain, and Italy has tested the limits of the willingness of richer EU countries to subsidize development in the Balkans.

Although the details of the Mediterranean economic crisis are beyond the scope of this chapter, its potential implications—both for Europe itself and for the larger global economy—are indeed relevant. Currently, Greece is the most insolvent member of the European Monetary Union, with Spain, Italy, Ireland, and France joining the list of troubled economies. Because these countries are linked through a common currency to the richer, solvent countries of Germany and the Netherlands, they share both the problem and the solution. If the richer European countries are willing to bail out the poorer ones with loans and bonds, then Euroland may survive its current crisis. At this point in time (late 2012), most analysts opine that the future of the common currency system is in doubt unless significant changes are made to its structure and management. If these changes are not made soon, then Euroland may fail, resulting in significant damage to the whole European economy.

> ### REVIEW QUESTIONS
>
> 1. What geographic factors are important in explaining the beginnings of the Industrial Revolution?
> 2. Describe the beginnings of the European Union (EU).
> 3. What factors must be considered in describing eastern Europe's postwar economic geography?
> 4. Briefly describe the current pattern of regional differences in the economic and social geography of Europe.

FIGURE 8.37 Crisis and Protest in Europe   The European economic crisis of 2012 (still unresolved as this book went to press) raised questions not just about the vitality of the European Union's common monetary system, but also about the future of the EU itself. Here, university students in Athens, Greece protest austerity measures that will raise tuition fees while reducing student services.

Summary

- In terms of environmental protection, Europe has made great progress over the last decades. Not only have individual countries enacted strong environmental legislation, but also the EU has played an important role in solving trans-boundary problems such as air and water pollution. Also important has been the EU's role in extending environmental standards to its new members in eastern Europe.

- Europe continues to face challenges related to population and migration. Although the phenomena of slow and no growth may have profound implications for the fiscal vitality of many countries, currently a more pressing problem is dealing with immigration from Asia, Africa, Latin America, and the former Soviet lands.

- Political and cultural tensions are presently high because of international migration to Europe—both legal and illegal. As the bulk of Europeans meld into a common culture, the differences increase between Europeans and migrants from other parts of the world. European countries pay lip service to building multicultural societies, but the pathways to this goal are still unclear.

- Major changes have taken place in Europe's geopolitical structure during the last few decades, with the end of the Cold War. In eastern Europe, countries formerly allied with the Soviet Union have shifted their alliances to the west through the EU, whereas in the Balkans a handful of newly independent countries has replaced the former Yugoslavia.

- Although the EU celebrated its 50th birthday in 2007 in a self-congratulatory mood, that spirit of optimism was severely challenged by the sudden economic downturn in 2008, which brought many countries to the brink of bankruptcy. Iceland, Ireland, Hungary, Spain, and Greece became unwitting poster children for that crisis. Today this ongoing financial crisis underscores both the strengths and the weaknesses of EU monetary policies.

Key Terms

balkanization 278
buffer zone 277
Cold War 252
continental climate 256
Cyrillic alphabet 270
Euroland 283

European Union (EU) 252
fjord 255
Iron Curtain 277
irredentism 274
marine west coast climate 256
maritime climate 255

medieval landscape 266
Mediterranean climate 256
Renaissance–Baroque period 267
Schengen Agreement 264

Thinking Geographically

1. To compare the scale of Europe to that of North America, draw a circle 500 miles (800 kilometers) around Frankfurt, Germany, do the same for Chicago. This distance is about a day's journey by car from each of the two cities. What does this exercise tell you about the differences in scale and social geography between the two regions?

2. Working in a small group, investigate how and why Europe has become more energy efficient than North America.

3. Map the different rates of natural increase in Europe, noting which countries will grow and which will decline in the next 20 years. Then link this map to a discussion of migration in Europe. Given these two factors, how might the population map change in 20 years?

4. Find good maps of several midsized European cities, and deduce the historical development of the cities based on their street patterns, arrangement of open spaces, boulevards, parks, and so on. Consider the differences between the medieval and Renaissance–Baroque landscapes for your interpretation.

5. Look at the ads in a French or German news magazine, and list the "globalized" English words used. Consider in your findings how different social and cultural groups may use (or not use) these foreign terms.

6. Critically examine the way boundaries were drawn in the Balkans after the demise of the former Yugoslavia in terms of religion, language, and ethnicity. With this information, provide some opinions on the Balkans' future in terms of possible problems and tensions.

7. Working in a small group, research the fiscal, political, and social changes that have taken place in a specific eastern Europe country in the last 20 years.

8. Put together a panel to debate the advantages versus the disadvantages of the European Monetary Union in times of economic and fiscal crisis.

MasteringGeography™

Looking for additional review and test prep materials? Visit the Study Area in MasteringGeography™ to enhance your geographic literacy, spatial reasoning skills, and understanding of this chapter's content by accessing a variety of resources, including **MapMaster** interactive maps, videos, RSS feeds, flashcards, web links, self-study quizzes, and an eText version of *Globalization and Diversity*.

Scan to visit the author's blog for chapter updates.

http://gad4blog.
wordpress.com/
category/europe/

Authors' Blogs

Scan now to access the authors' blogs for up-to-date information on Europe.

Scan to visit the GeoCurrents blog.

http://geocurrents.
info/category/place/
europe

Globalization and Diversity

Russia's reemergence on the global stage has been obvious since 2000. Its growing role is evident in the global energy economy, in its participation in global economic meetings such as the Group of Eight (G-8), and in its increasing geopolitical involvement in contentious settings such as North Korea.

ENVIRONMENTAL GEOGRAPHY

Many areas within the Russian domain suffered severe environmental damage during the Soviet era (1917–1991), and today air, water, toxic chemical, and nuclear pollution plagues large portions of the region.

POPULATION AND SETTLEMENT

Urban landscapes within the Russian domain reflect a fascinating mix of imperial, socialist, and postcommunist influences. Many larger urban areas within the region increasingly reflect parallel trends toward sprawl and decentralization seen elsewhere in North America and western Europe.

CULTURAL COHERENCE AND DIVERSITY

Although Slavic cultural influences dominate the region, many non-Slavic minorities shape the cultural and political geography of the domain, including varied indigenous peoples in Siberia and a complex collection of ethnic groups in the Caucasus Mountains.

GEOPOLITICAL FRAMEWORK

The post-2000 centralization of Russian political power under Vladimir Putin has contributed to economic stability in the region, but democratic freedoms have suffered amid crackdowns on the press and personal liberties. Since 2009, growing public protests have challenged Putin's authority within Russia.

ECONOMIC AND SOCIAL DEVELOPMENT

Russia's large supplies of oil and natural gas have made it a major player in the global economy, but future prosperity may increasingly hinge on unpredictable world prices for fossil fuels.

➤ This view of Moscow's growing International Business Center suggests the new connections being forged between the Russian domain and the global economy.

THE RUSSIAN 9 DOMAIN

FIGURE 9.1 The Russian Domain Russia and its neighboring states of Belarus, Ukraine, Moldova, Georgia, and Armenia make up a dynamic and unpredictable world region. Sprawling from the Baltic Sea to the Pacific Ocean, the region includes huge industrial centers, vast farmlands and thinly-settled, stretches of tundra.

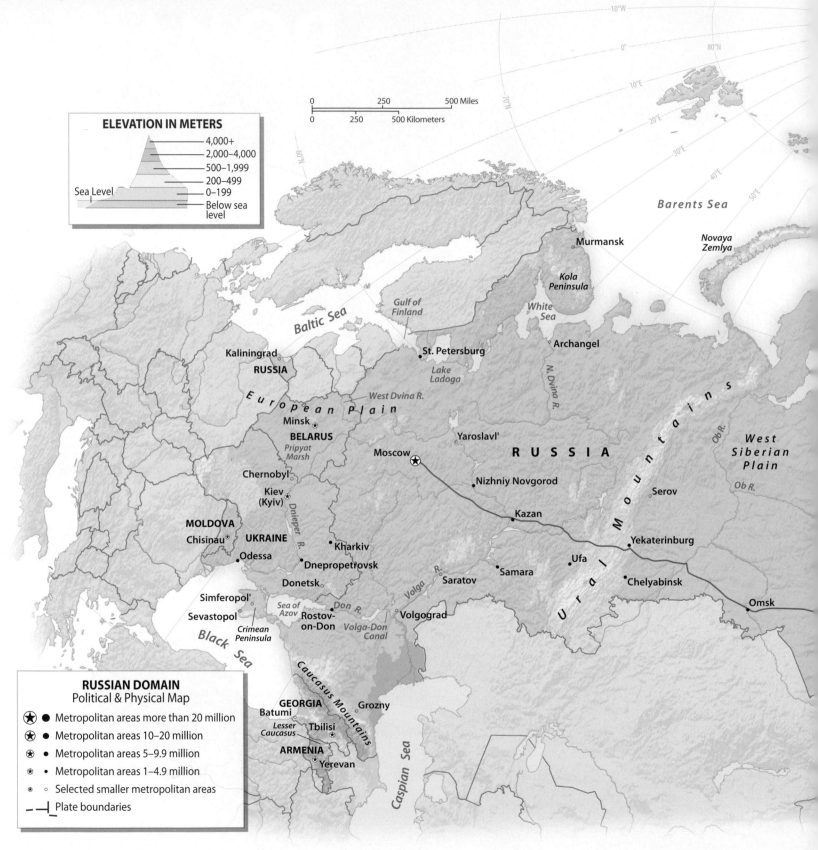

ELEVATION IN METERS

- 4,000+
- 2,000–4,000
- 500–1,999
- 200–499
- 0–199

Sea Level

Below sea level

RUSSIAN DOMAIN
Political & Physical Map

- ⭐ ● Metropolitan areas more than 20 million
- ✪ ● Metropolitan areas 10–20 million
- ✪ ● Metropolitan areas 5–9.9 million
- ✪ • Metropolitan areas 1–4.9 million
- ✪ ○ Selected smaller metropolitan areas
- —⊥ Plate boundaries

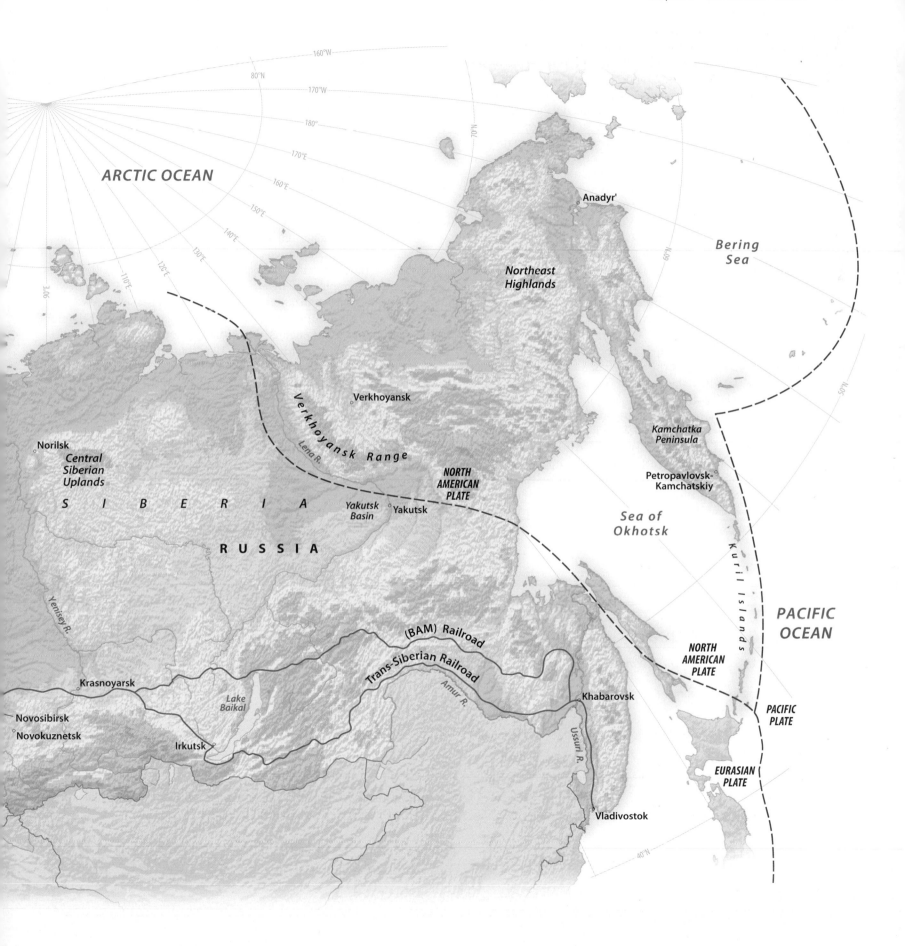

ARCTIC OCEAN

80°N

160°W

170°W

180°

170°E

160°E

150°E

140°E

130°E

120°E

110°E

100°E

90°E

70°N

60°N

50°N

40°N

Anadyr'

*Bering
Sea*

*Northeast
Highlands*

Verkhoyansk

V e r k h o y a n s k   R a n g e

Lena R.

Norilsk

*Central
Siberian
Uplands*

S I B E R I A

**NORTH
AMERICAN
PLATE**

*Yakutsk
Basin*   Yakutsk

R U S S I A

*Kamchatka
Peninsula*

Petropavlovsk-
Kamchatskiy

*Sea of
Okhotsk*

K u r i l   I s l a n d s

**PACIFIC
OCEAN**

Yenisey R.

(BAM) Railroad

Trans-Siberian Railroad

Amur R.

Krasnoyarsk

*Lake
Baikal*

Khabarovsk

**NORTH
AMERICAN
PLATE**

**PACIFIC
PLATE**

Novosibirsk

Novokuznetsk

Irkutsk

Ussuri R.

Vladivostok

**EURASIAN
PLATE**

You can see the high-rise glitz for miles. The dense assortment of towers known as the Moscow International Business Center (IBC) is located west of the Kremlin, along the banks of the Moscow River. The modern design symbolizes both the promises and the challenges of creating a new Russian economy since the end of the Soviet Union in 1991. Built on the site of decaying Soviet-era factories and an abandoned stone quarry, the IBC is becoming Russia's most dramatic, visible symbol of imaginative urban planning and a sign of growing links between the region and the global economy. Paralleling North American urban redevelopment, the project is centered on a series of mixed-use towers that offer modern offices, shopping malls, restaurants, entertainment facilities, and residential living space. Like the larger regional economy, several significant construction snags have slowed the IBC's progress. Hopes to complete the world's tallest building (the 2,000-foot [600-meter] Russia Tower) faded in 2008 when financing fell through. For a time, the site was to be a parking lot, but in 2012 new plans revealed that the tower may rise again, perhaps by 2016. Meanwhile, other towers have gradually filled in the nearby urban landscape, suggesting that the IBC is here to stay.

The IBC's uncertain rise parallels the larger performance of the Russian domain since the end of the Soviet era. Maps were suddenly redrawn late in 1991, when the once-powerful, highly centralized Soviet state was officially dissolved. In its place stood 15 former "republics" that had once been united under the Soviet Union (the Union of Soviet Socialist Republics). Now independent, each republic has tried to make its way in a post-Soviet world. The Russian Republic remained dominant in size and political influence and thus came to form the core of a new Russian domain. Today the domain sprawls across the vast northern half of Eurasia and includes not only Russia itself, but also the nations of Ukraine, Belarus, Moldova, Georgia, and Armenia (Figure 9.1). The term *domain* suggests the persistent Russian influence within the five other nations included in the region. Slavic Russia, Ukraine, and Belarus make up the region's core. Nearby Moldova and Armenia broadly remain within Russia's geopolitical orbit. Relations between Russia and Georgia remain strained, and recent conflicts in Georgia brought Russian troops into that small country.

Two significant areas that were once part of the Soviet Union have been eliminated from the domain. The mostly Muslim republics of Central Asia and the Caucasus (Kazakhstan, Uzbekistan, Kyrgyzstan, Turkmenistan, Tajikistan, and Azerbaijan) have become aligned with the Central Asia world region (see Chapter 10), while the Baltic republics (Estonia, Latvia, and Lithuania) are best grouped with Europe (see Chapter 8).

The Russian domain remains a land rich with superlatives: Endless Siberian spaces, unlimited natural resources, legends of ruthless Cossack warriors, and tales of epic wars and revolutions are all part of the region's geographic and historical mythology (Figure 9.2). Indeed, the rise of Russian civilization remarkably parallels the story of the United States. Both cultures grew from small beginnings to become imperial powers that benefited from the fur trade, gold rushes, and transcontinental railroads during the 19th century. In addition, both countries were dramatically transformed by industrialization during the 20th century.

Recently, however, the Russian domain has witnessed particularly breathtaking change. With the fall of the Soviet Union in 1991, new political and economic institutions have reshaped everyday life. Economic collapse in the late 1990s produced steep declines in living standards throughout the region. Political instability grew between neighboring states, as well as within countries. After 2000, strong and increasingly centralized leadership within Russia set the region on a different course. With the help of higher energy prices (Russia is a major exporter of oil and natural gas), real economic improvements

LEARNING OBJECTIVES

After reading this chapter you should be able to:

- Describe the major environmental issues affecting residents of the region.

- Explain the close connection among latitude, regional climates, and agricultural production.

- Identify the potential benefits and hazards of global warming within the region.

- Summarize major migration patterns, both in Soviet and in post-Soviet eras.

- Describe major urban land-use patterns in a large city such as Moscow.

- Describe the major phases of Russian expansion across Eurasia.

- Identify the key regional patterns of linguistic and religious diversity.

- Summarize the historical roots of the region's modern geopolitical system.

- Provide examples of how persistent cultural differences shape contemporary geopolitical tensions.

- Identify key ways in which natural resources, including energy, have shaped economic development in the region.

- Describe key contributions of the Soviet-era economy and list major changes that have shaped the region's economic geography since the fall of the Soviet Union in 1991.

FIGURE 9.2 Cossack   Natives of the Ukrainian and Russian steppe, the highly mobile Cossacks played a pivotal role in aiding Russian expansion into Siberia during the 16th century. Many modern descendants retain their skills of horsemanship and are proud of their distinctive ethnic heritage.

benefited most of the region. Currently, however, many of the democratic reforms in Russia and Ukraine that were welcomed with the fall of communism in 1991 are being eroded by powerful governments committed to more direct state control.

Globalization shapes the Russian domain in complex ways. The region's relationship with the rest of the world shifted during the last decade of the 20th century. Until 1991, all six countries of the Russian domain belonged to the Soviet Union, the world's most powerful communist state. Under Soviet control, the region's economy saw large increases in industrial output that made the area a major global producer of steel, weaponry, and petroleum products. The political and military reach of the Soviet Union spanned the globe, making it a superpower on a par with the United States.

Suddenly, the communist order evaporated in 1991. Russia, Ukraine, Belarus, Moldova, Georgia, and Armenia had to carve out new regional and global relationships. With the breakdown of Soviet control, the region felt the growing presence of western European and American influences. The region also became more fully exposed to both the opportunities and the competitive pressures of the global economy. The result is a world region that has seen its global linkages

redefined. Today world oil markets, foreign investment, migration patterns, and the shadowy flows of illegal drugs and Russian mafia money all demonstrate the unpredictable nature of the Russian domain's global connections.

Slavic Russia (population 140 million) dominates the region. Although only about three-quarters the size of the former Soviet Union, Russia's dimensions still make it the largest state on Earth. Its area of 6.6 million square miles (17 million square kilometers) dwarfs even Canada, and its nine time zones are a reminder that dawn in Vladivostok on the Pacific Ocean is still only evening in Moscow.

With the demise of the Soviet Union, the Russians ended almost 75 years of communist rule. After a decade of political and economic instability (1991–2000), Russia has made impressive progress in the 21st century. Russian leader Vladimir Putin built a reputation for strong leadership. Much of the economic growth has come in Russian cities, where expanding middle and professional classes are enjoying better living standards. Many rural areas, however, remain deeply mired in poverty. Another concern is that Putin's desire for wealth and power has been matched by his need for more centralized political control.

The bordering states of Ukraine, Belarus, Moldova, Georgia, and Armenia are inevitably wed to the evolution of their giant neighbor. Emerging from the shadows of Soviet dominance has been difficult. Ukraine, in particular, has the size, population, and resource base to become a major European nation, but it has struggled to create real political and economic change since independence. With 45 million people and a rich storehouse of resources, Ukraine's size of 233,000 square miles (604,000 square kilometers) is similar to that of France. Nearby Belarus is smaller (80,000 square miles [208,000 square kilometers]), and its population of almost 10 million remains closely tied economically and politically to Russia. Presently, its strikingly authoritarian and antiforeign leadership reflects many aspects of the old Soviet empire.

Moldova, with 4 million people, shares cultural links with Romania, but its economic and political connections have kept it tied to the Russian domain (Figure 9.3). South of Russia and beyond the bordering Caucasus Mountains, the Transcaucasian countries of Armenia and Georgia are similar in size to Moldova. Their populations differ culturally from that of their Slavic neighbor to the north. In addition, these two nations face significant political challenges: Armenia shares a hostile border with Azerbaijan (see Chapter 10), and Georgia's ethnic diversity and contentious relations with neighboring Russia threaten its political stability.

ENVIRONMENTAL GEOGRAPHY: A VAST AND CHALLENGING LAND

The Kamchatka Peninsula dangles dramatically into the waters of the North Pacific Ocean, far from the hectic streets of St. Petersburg or Moscow (see Figure 9.1). Unlike much of the rest of the Russian domain, which has been harmed by almost a century of reckless development and environmental exploitation, the Kamchatka region remains relatively untouched by the outside world (Figure 9.4). However, that may be changing soon, and Russia's ability to preserve the region will say a great deal about that country's future commitment to preserving and restoring its environmental health.

Six native species of Pacific salmon thrive in the region, spawning in the area's free-flowing rivers. A move to preserve the salmon habitat is growing among both Russian and global environmental organizations. The salmon are at the center of a complex environmental web focused on brown bears, seals, and Stellar's sea eagles. Salmon also remain part of the subsistence diet of the Koryak and other native peoples. In addition, the fish offer ecotourism and sports fishing opportunities found nowhere else in the world. But threats are rapidly appearing. Canadian, South Korean, and Russian oil companies hold drilling rights in the region. Poaching of salmon and salmon eggs is also on the rise. Global efforts are under way to prevent development in seven sensitive spawning areas of Kamchatka wilderness. The consolidation of several national parks on the peninsula in 2010 should make management of the region's natural resources more efficient and less costly. Such initiatives reveal a rising environmental consciousness in a part of the world that has produced some of the most toxic settings on the planet. It would also mark an important marriage between a fragile local setting and a larger collection of national and international environmental organizations dedicated to seeing the wild salmon continue to run.

A Devastated Environment

In addition to its threatened salmon, the Russian domain faces many other enormous environmental challenges. The breakup of the Soviet Union and subsequent opening of the region to international public examination revealed some of the world's most severe environmental degradation (Figure 9.5). Seven decades of intense and rapid Soviet industrialization caused environmental problems that extend across the entire region. Regional leaders are beginning to respond, however, and many observers hope greater environmental awareness will shape future political and economic decisions (see "Working Toward Sustainability: Sochi 2014 and Russia's Struggle for a 'Green' Winter Olympics").

FIGURE 9.4 Kamchatka Peninsula   The Kamchatka Peninsula offers many spectacular natural settings, as well as a habitat for a variety of Pacific salmon.

Air and Water Pollution   Poor air quality affects hundreds of cities and industrial complexes throughout the Russian domain. The Soviet policy of building large clusters of industrial processing and manufacturing plants in concentrated areas, often with minimal environmental controls, has produced a collection of polluted cities that stretches from Belarus to Russian Siberia (Figure 9.5). Growing rates of private automobile ownership have greatly increased automobile-related pollution, especially because many vehicles lack effective emission controls.

Degraded water is another hazard. Urban water supplies are vulnerable to industrial pollution, raw sewage, and demands that increasingly exceed capacity. Oil spills have harmed thousands of square miles in the tundra and taiga of the West Siberian Plain and along the Ob River. A 2011 study estimates that 1 percent of Russia's annual oil production (5 million tons) is spilled every year and the actual total could be higher. Water pollution has also affected the Volga River, the Black Sea, portions of the Caspian Sea shoreline, the waters of the Arctic Ocean off Russia's northern coast, and Siberia's Lake Baikal, the world's largest freshwater reserve (Figure 9.6).

The Nuclear Threat   The nuclear era brought added dangers to the region. The Soviet Union's nuclear weapons and energy programs often ignored issues of environmental safety. Siberia, for example, suffered regular nuclear fallout when tests were conducted in the atmosphere. Nuclear explosions also were used to move earth in dam-building projects. The once-pristine Russian Arctic has been poisoned: The area around the island of Novaya Zemlya served as an unregulated dumping ground for nuclear wastes in the Soviet era. Aging nuclear reactors also dot the region's landscape, often contaminating nearby rivers with plutonium leaks. Nuclear pollution is particularly pronounced in northern Ukraine, where the Chernobyl nuclear power plant suffered a catastrophic meltdown in 1986.

FIGURE 9.3 Chisinau, Moldova   With a population of almost 1 million people, the Moldovan capital of Chisinau is in the most economically affluent part of this small eastern European nation. Its landscape still reflects the influence of Soviet-era planning and urban design.

WORKING TOWARD SUSTAINABILITY

Sochi 2014 and Russia's Struggle for a "Green" Winter Olympics

"Sustainable development in sport is not just a trend, it's a necessity. It's not too early to talk about the Sochi 2014 legacy. Thanks to the Games, our approach to construction and environmental protection is changing and green construction standards are being introduced." These bold words from Olympic Organizing President Dmitry Chernyshenko suggest that the Russians have lofty ambitions for making the Sochi games the greenest ever. But the Russians are also struggling to live by their words and actually minimize impacts in an environmentally sensitive mountain region of southern Russia near Sochi, just to the east of the Black Sea (Figure 9.1.1). Recently, the area has been the site of what has been called the world's largest construction zone, where 55,000 workers have been building a $7+ billion venue for the 2014 Winter Games.

The Russians are sensitive to the environmental issues being raised. Russian leader Vladimir Putin (who owns a summer home in the area) has pledged that "in setting our priorities and choosing between money and the environment, we're choosing the environment." The Russians have also teamed up with the United Nations Environment Program (UNEP). In 2009, Olympic organizers signed a "Memorandum of Understanding" with UNEP, promising to make the Games the most environmentally friendly ever. Supporters have advocated recycling wastewater, replanting flora away from construction sites, and making widespread use of renewable energy technologies. Olympic organizers and Russian politicians both see Sochi as offering an overall model for how development can be more sustainable and how such efforts can attract more global investment.

Still, environmental groups have unleashed protests over the ecologically destructive nature of the project. The Sochi venues include some of the most pristine mountain settings in Eurasia, as well as potentially vulnerable coastal environments. Sochi houses the main Olympic Village and functions as the event's regional transport hub, but recently garbage dumps have swelled with untreated waste, and local rivers have run thick with brown sediment as construction projects accelerate. The nearby Krasnaya Polyana region, within the rugged and spectacular Caucasus Mountains, is the venue for alpine skiing events and for another Olympic Village (Figure 9.1.2). Critics, citing vulnerable ecological settings within the nearby Caucasus Biosphere Reserve and Sochi National Park, have already forced the relocation of the bobsled and luge runs. But one spokesperson for the World Wildlife Fund noted that "massive damage had already been caused and more is expected." Logging of endangered species, such as the Caucasian wingnut tree, also threatens to cause irreparable environmental damage.

Whether or not the actual Olympics affirm the sustainability rhetoric remains an open question. However, the controversy is a reminder of the central role that politics can play, both in the planning of international sports events and in the management of complex and fragile environments.

FIGURE 9.1.1 Sochi Olympics   The southern Russian city of Sochi plays host to the Winter Olympic Games in 2014.

FIGURE 9.1.2 Olympic Construction in the Krasnaya Polyana Region near Sochi   Environmentalists fear that major investments in Olympic infrastructure will cause permanent damage in one of Eurasia's most pristine mountain settings.

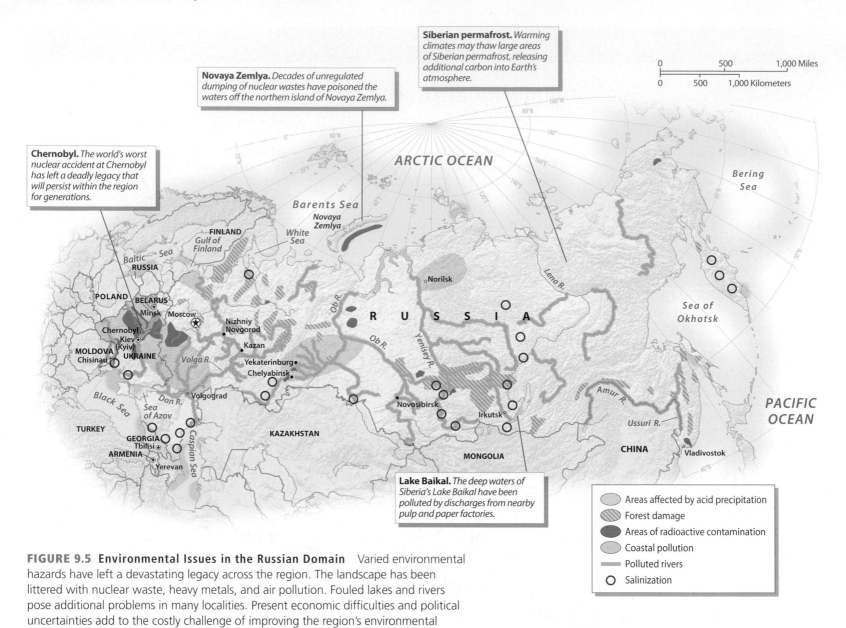

Chernobyl. *The world's worst nuclear accident at Chernobyl has left a deadly legacy that will persist within the region for generations.*

Novaya Zemlya. *Decades of unregulated dumping of nuclear wastes have poisoned the waters off the northern island of Novaya Zemlya.*

Siberian permafrost. *Warming climates may thaw large areas of Siberian permafrost, releasing additional carbon into Earth's atmosphere.*

Lake Baikal. *The deep waters of Siberia's Lake Baikal have been polluted by discharges from nearby pulp and paper factories.*

Legend:
- Areas affected by acid precipitation
- Forest damage
- Areas of radioactive contamination
- Coastal pollution
- Polluted rivers
- ○ Salinization

FIGURE 9.5 Environmental Issues in the Russian Domain   Varied environmental hazards have left a devastating legacy across the region. The landscape has been littered with nuclear waste, heavy metals, and air pollution. Fouled lakes and rivers pose additional problems in many localities. Present economic difficulties and political uncertainties add to the costly challenge of improving the region's environmental quality in the 21st century.

A Diverse Physical Setting

The Russian domain's northern latitudinal position is an important factor in shaping its basic geographies of climate, vegetation, and agriculture (see Figure 9.1). Indeed, the Russian domain provides the world's largest example of a high-latitude continental climate, where seasonal temperature extremes and short growing seasons greatly limit opportunities for human settlement (Figure 9.7).

FIGURE 9.6 Lake Baikal   Southern Siberia's Lake Baikal is one of the world's largest deep-water lakes. Industrialization devastated water quality after 1950, as pulp and paper factories poured wastes into the lake. Recent cleanup efforts have helped, but many environmental threats remain.

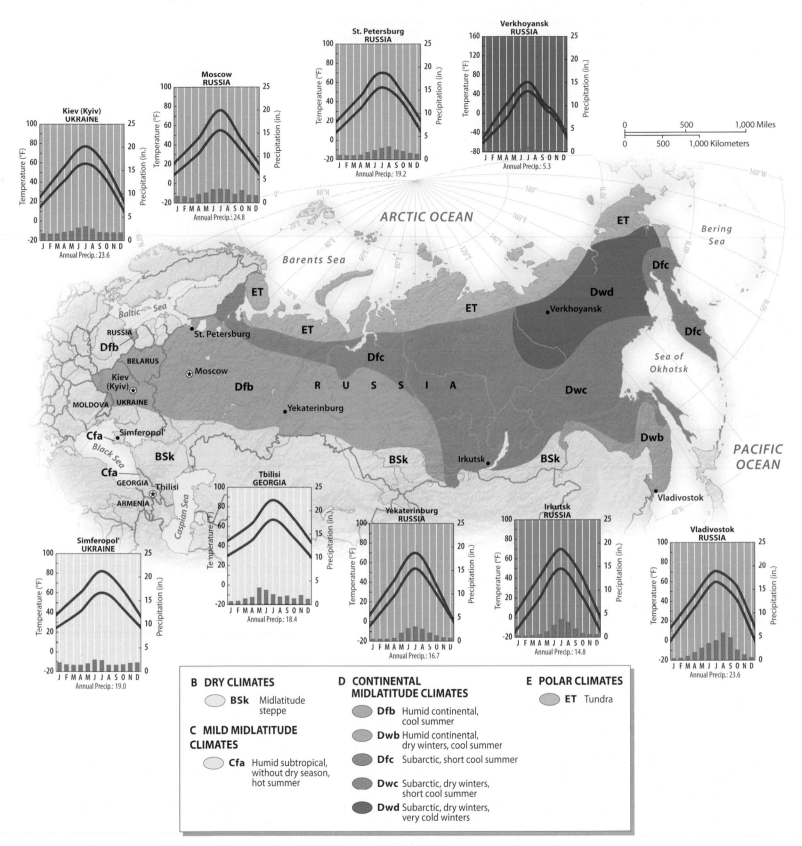

FIGURE 9.7 Climate Map of the Russian Domain   The region's northern latitude and large landmass suggest that continental climates dominate. Indeed, farming is greatly limited by short growing seasons across much of the region. Aridity imposes limits elsewhere. Only a few small zones of mild midlatitude climates are found along the temperate shores of the Black Sea.

B DRY CLIMATES

- **BSk** Midlatitude steppe

C MILD MIDLATITUDE CLIMATES

- **Cfa** Humid subtropical, without dry season, hot summer

D CONTINENTAL MIDLATITUDE CLIMATES

- **Dfb** Humid continental, cool summer
- **Dwb** Humid continental, dry winters, cool summer
- **Dfc** Subarctic, short cool summer
- **Dwc** Subarctic, dry winters, short cool summer
- **Dwd** Subarctic, dry winters, very cold winters

E POLAR CLIMATES

- **ET** Tundra

FIGURE 9.8 Agricultural Regions   Harsh climate and poor soils combine to limit agriculture across much of the Russian domain. Better farmlands are found in Ukraine and in European Russia south of Moscow. Portions of southern Siberia support wheat production, but yield marginal results. In the Russian Far East, warmer climates and better soils enable higher agricultural productivity.

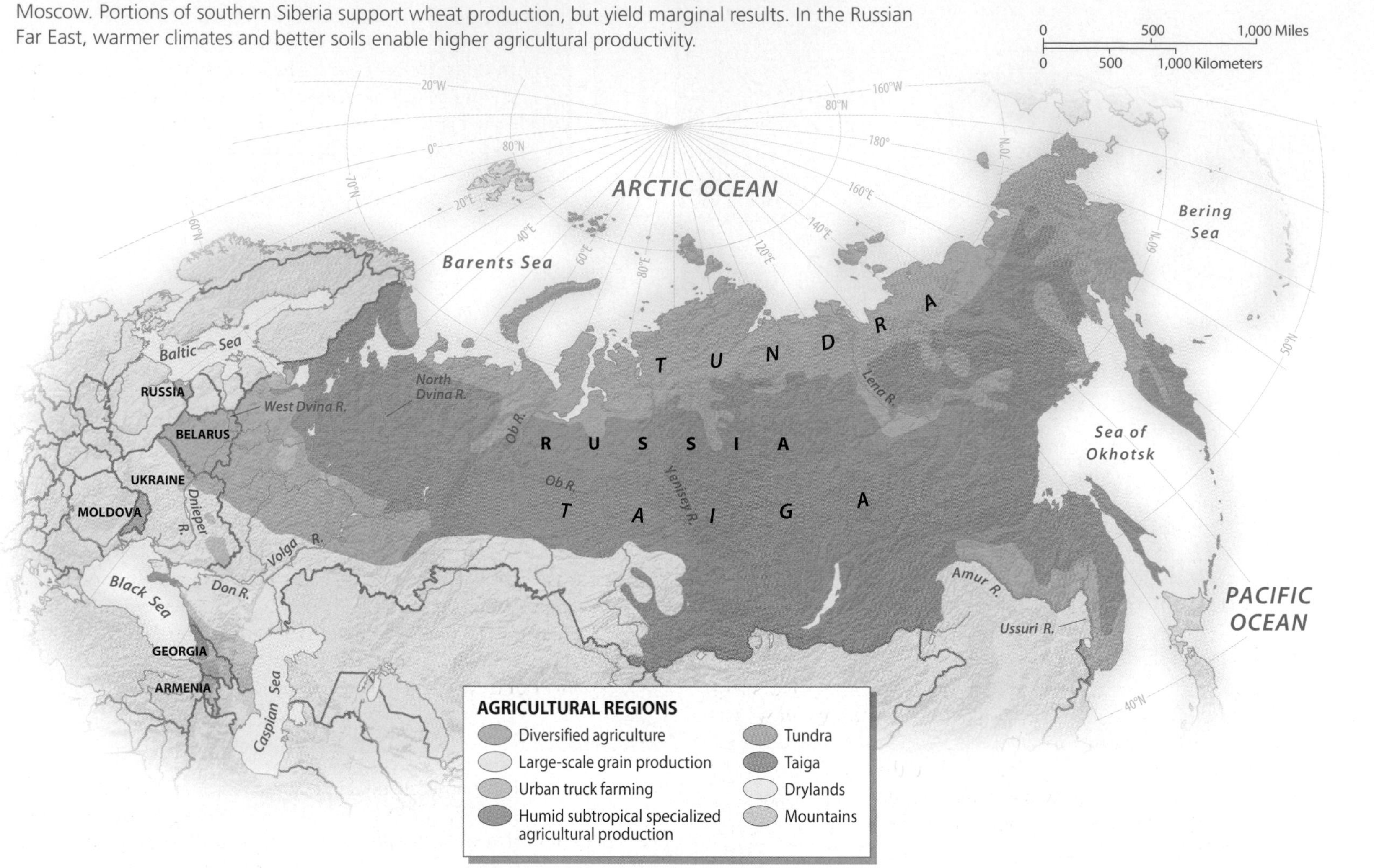

In terms of latitude, Moscow is positioned as far north as Ketchikan, Alaska, and even the Ukrainian capital of Kiev (Kyiv) sits farther north than the Great Lakes in Canada. Thus, apart from a subtropical zone near the Black Sea, the region experiences a classic continental climate with hard, cold winters and marginal agricultural potential.

The European West   An airplane flight over the western portions of the Russian domain would reveal a vast, barely changing landscape. European Russia, Belarus, and Ukraine cover the eastern portions of the vast European Plain, which runs from southwest France to the Ural Mountains. One major geographic advantage of European Russia is that different river systems, all now linked by canals, flow into four separate drainage basins. The result is that trade goods can easily flow in many directions within and beyond the region. The Dnieper and Don rivers flow into the Black Sea; the West and North Dvina rivers drain into the Baltic and White seas, respectively; and the Volga River (the longest river in Europe) runs to the Caspian Sea.

Most of European Russia experiences cold winters and cool summers by North American standards (Figure 9.7). Moscow, for example, is about as cold as Minneapolis in January, but it is not nearly as warm in July. In Ukraine, Kiev is milder, however, and Simferopol, near the Black Sea, offers wintertime temperatures that average more

than 20°F (11°C) warmer than those of Moscow. (See the climographs in Figure 9.7.)

Three distinctive environments shape agricultural potential in the European west (Figure 9.8). North of Moscow and St. Petersburg, poor soils and cold temperatures severely limit farming. The region's boreal forests have been extensively logged (Figure 9.9).

Belarus and central portions of European Russia possess longer growing seasons, but acidic **podzol soils**, typical of northern forest environments, limit output and thus the ability of the region to support a productive agricultural economy. The diversified agriculture that does occur includes grain (rye, oats, and wheat) and potato cultivation, swine and meat production, and dairying.

South of 50° latitude, agricultural conditions improve across much of southern Russia and Ukraine. Forests gradually give way to steppe environments dominated by grasslands and by fertile "black earth" **chernozem soils**. These have proven valuable for commercial wheat, corn, and sugar beet cultivation and for commercial meat production (Figure 9.10).

The Ural Mountains and Siberia   The Ural Mountains (see Figure 9.1) mark European Russia's eastern edge, separating it from Siberia. Topographically, the Urals are not a particularly impressive range.

FIGURE 9.9 Boreal Forest in Northwestern Russia The once endless boreal forests of northern Russia are quickly being harvested for commercial lumber markets in nearby western Europe. Only a small percentage of roadless forest remains today.

Still, the ancient rocks of these mountains contain valuable mineral resources, and the Urals traditionally marked European Russia's eastern cultural boundary.

East of the Urals, Siberia unfolds across the landscape for thousands of miles. The great Arctic-bound Ob, Yenisey, and Lena rivers (see Figure 9.1) drain millions of square miles of northern country, including the flat West Siberian Plain, the hills and plateaus of the Central Siberian Uplands, and the rugged and isolated Northeast Highlands. Along the Pacific, the Kamchatka Peninsula offers spectacular volcanic landscapes. Wintertime climatic conditions, however, are severe across the entire region.

Siberian vegetation and agriculture reflect the climatic setting. The north is too cold for tree growth and instead supports tundra vegetation, characterized by mosses, lichens, and a few ground-hugging flowering plants. Much of the tundra region is associated with **permafrost**, a cold-climate condition of unstable, seasonally frozen ground that limits vegetation and causes problems for railroad construction. South of the tundra, the Russian **taiga**, or coniferous forest zone, dominates a large portion of the Russian interior. With huge demand for lumber from nearby Japan and China, the eastern taiga zone has been threatened by both authorized logging and illegal timber poaching.

The Russian Far East The Russian Far East is a distinctive subregion characterized by proximity to the Pacific Ocean, more southerly latitude, and fertile river valleys, such as the Amur and the Ussuri. Located at about the same latitude as New England, the region features longer growing seasons and milder climates than those found to the west or north. Here the continental climates of the Siberian interior meet the seasonal monsoon rains of East Asia. It is a fascinating zone of ecological mixing: Conifers of the taiga mingle with Asian hardwoods, and reindeer, Siberian tigers, and leopards also find common ground.

The Caucasus and Transcaucasia In European Russia's extreme south, flat terrain rises up first to hills and then to the Caucasus Mountains, a large range stretching between the Black and Caspian seas (Figure 9.11). The Caucasus Mountains mark Russia's southern boundary and are characterized by major earthquakes. Farther south lies Transcaucasia and the distinctive natural settings of Georgia and Armenia. Patterns of both climate and terrain in the Caucasus and Transcaucasia are very complex. Rainfall is generally higher in the western zone, while the area's eastern valleys are semiarid. In areas where rainfall is adequate or where irrigation is possible, agriculture can be quite productive. Georgia in particular has long been an important producer of fruits, vegetables, flowers, and wines (Figure 9.12).

Climate Change in the Russian Domain

Given its latitude and continental climates, the Russian domain is often cited as a world region that would benefit from a warmer global climate. But such an interpretation oversimplifies the complex natural and human responses to global climate change, some of which are already occurring across the region.

FIGURE 9.10 Commercial Wheat Production This Ukrainian wheat field typifies some of the region's most productive farmland. Longer growing seasons and good soils remain enduring assets in this portion of the Russian domain.

FIGURE 9.11 Satellite Image of the Caucasus Mountains Aligned between the Black Sea (left) and the Caspian Sea (right), the rugged, snow-capped Caucasus Mountains prevent easy movement across the culturally diverse and politically contested borderlands between southern Russia and Georgia.

FIGURE 9.12 Subtropical Georgia The moderating influences of the Black Sea and a more southern latitude produce a small zone of humid subtropical agriculture in Georgia. The fertile landscapes of these tea plantations offer a sharp contrast to the colder country found north of the Caucasus.

facilitating navigation. Less severe winters might make energy development in arctic settings less costly. About 15 percent of the world's undiscovered oil reserves (and 30 percent of its undiscovered natural gas reserves) are probably located in these settings, and Russia has staked large claims to the region. In the Arctic Ocean and Barents Sea, warmer temperatures and less sea ice would translate to better commercial fishing, easier navigation, more high-latitude commerce, and more ice-free days in northern Russian ports. Indeed, by 2011, a record 34 ships negotiated the **northern sea route** along Siberia's northern coast.

Potential Hazards of Global Climate Change Even with such rosy scenarios, might long-term regional and global costs outweigh the benefits? Climate experts point to three concerns. First, rising global sea levels will hit low-lying areas of the Black and Baltic seas particularly hard. Officials in St. Petersburg, Russia's second largest city, are already contemplating significant costs associated with controlling the Baltic's rising waters. Second, changes in ecologically sensitive arctic and subarctic ecosystems are already leading to major disruptions in wildlife and indigenous human populations. Shrinking volumes of arctic sea ice habitat for polar bears have forced them to widen their search for food, bringing them into closer contact with people and disrupting seasonal hunts. Poachers have also profited, increasing illegal harvests.

Finally, the potential thawing of the Siberian permafrost has global implications. Substantial areas of northern Russia are covered with permafrost close to thawing. This same region has experienced some of the largest, most persistent global warming since 1950 (Figure 9.13).

Potential Benefits of Global Climate Change Some researchers point to benefits from warmer Eurasian climates, such as changes in agriculture (see Figure 9.8). Some models predict that the northern limit of spring cereal cultivation in northwestern Russia will shift 60–90 miles (100–150 kilometers) poleward for every 1.8°F (1°C) of warming. Elsewhere, warmer springs in Belarus might produce higher yields of maize and sunflowers. Warmer conditions might also help open Russian rivers such as the Volga and Ob to longer ice-free periods,

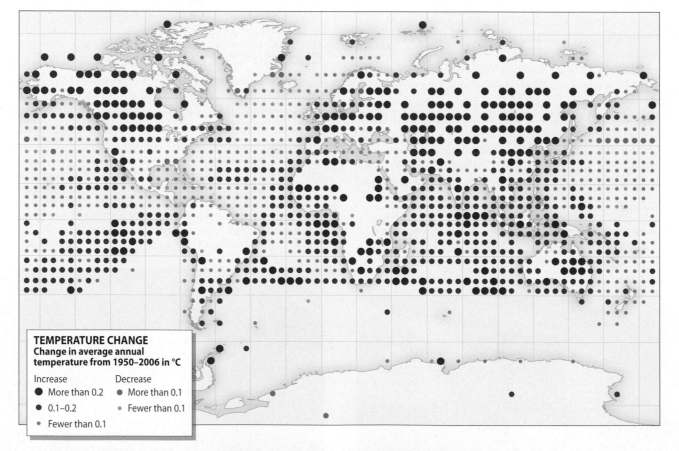

FIGURE 9.13 Change in Average Annual Temperature (°C), per Decade, 1950–2006 This map shows global temperature trends for different portions of Earth's surface between 1950 and 2006. Much of the Russian interior, including Siberia, exhibits some of the greatest examples of surface warming on the planet, portending major changes for the region's permafrost, vegetation patterns, and human population.

Permafrost contains large amounts of organic material that decomposes quickly when thawed. Minor increases in temperature could release carbon currently stored in frozen permafrost soils. Such a contribution to the world's carbon budget would likely further warm Earth, accelerating current trends.

> ### REVIEW QUESTIONS
>
> 1. Compare the climate, vegetation, and agricultural conditions of Russia's European west with those of Siberia and the Russian Far East.
> 2. Describe the high environmental costs of industrialization within the Russian domain.

POPULATION AND SETTLEMENT: AN URBAN DOMAIN

The Russian domain is home to about 200 million residents (Table 9.1). While they are widely dispersed across Eurasia, most live in cities. The region's population geography has been influenced by the distribution of natural resources and by government policies that have encouraged migration out of the traditional centers of population in the western portions of the domain.

Population Distribution

The favorable agricultural setting of the European west offered a home to more people than did the inhospitable conditions found across central and northern Siberia. Although Russian efforts over the past century have encouraged a wider dispersal of the population, it remains heavily concentrated in the west (Figure 9.14). European Russia is home to more than 100 million people, while Siberia, although far larger, holds only some 35 million. If you add the 60 million inhabitants of Belarus, Moldova, and Ukraine, the imbalance between east and west becomes even more striking.

The European Core   The region's largest cities, biggest industrial complexes, and most productive farms are located in the European Core, a subregion that includes Belarus, much of Ukraine, and Russia west of the Urals. The sprawling city of Moscow and its nearby urbanized region clearly dominate the settlement landscape, with a metropolitan area of more than 11.5 million people, the largest city in Europe (Figure 9.15).

On the shores of the Baltic Sea, St. Petersburg (earlier Leningrad, 4.9 million people) has traditionally had a great deal of contact with western Europe. Between 1712 and 1917, it served as the capital of the Russian Empire. Its handsome buildings, bridges, and canals give it an urban landscape many have compared to the great cities of western Europe (Figure 9.16).

Other urban clusters are located along the lower and middle stretches of the Volga River, including the cities of Kazan, Samara, and Volgograd. This highly commercialized river corridor, also containing important petroleum reserves, supports a diverse industrial base. Nearby, the resource-rich Ural Mountains include the gritty industrial landscapes of Yekaterinburg (1.4 million) and Chelyabinsk (1.1 million).

Other population clusters are found in Belarus and Ukraine (see Table 9.1). The Belarusian capital of Minsk (1.8 million) dominates that country, and its landscape recalls the drab Soviet-style architecture of an earlier era. In nearby Ukraine, the capital of Kiev (2.8 million) straddles the Dnieper River, and the city's old and beautiful buildings are a visual reminder of its role as a cultural and trading center within the European interior (Figure 9.17).

Siberian Hinterlands   A passenger leaving the southern Urals on a Siberia-bound train becomes aware that the land ahead is ever more sparsely settled (see Figure 9.14). The distance between cities grows, and the intervening countryside reveals a landscape shifting gradually from farms to forest. To the south, a collection of isolated, but sizable, urban centers follows the **Trans-Siberian Railroad**, a key railroad passage to the Pacific completed in 1904. The eastbound traveler encounters Omsk (1.1 million) as the rail line crosses the Irtysh River, Novosibirsk (1.5 million) at its junction with the Ob River, and Irkutsk (600,000) near Lake Baikal. The port city of Vladivostok (600,000) marks the end of the Trans-Siberian Railroad and provides access to the Pacific. To the north, a thinner sprinkling of settlements appears along the **Baikal–Amur Mainline (BAM) Railroad** (completed in 1984), which parallels the older line, but runs north of Lake Baikal to the Amur River. From the BAM line to the Arctic, the almost empty spaces of central and northern Siberia dominate the scene, interrupted only rarely by small settlements, often oriented around natural resource extraction (Figure 9.18).

Table 9.1   POPULATION INDICATORS

Country	Population (millions) 2012	Population Density (per square kilometer)	Rate of Natural Increase (RNI)	Total Fertility Rate	Percent Urban	Percent <15	Percent >65	Net Migration (Rate per 1000) 2010–15[a]
Armenia	3.3	110	0.5	1.7	64	17	10	−3.2
Belarus	9.5	46	−0.3	1.5	76	15	14	−0.2
Georgia	4.5	65	0.2	1.7	53	17	14	−5.8
Moldova	4.1	122	0.0	1.3	42	16	10	−9.0
Russia	143.2	8	−0.1	1.6	74	15	13	1.2
Ukraine	45.6	75	−0.4	1.5	69	14	15	−0.2

[a]Net Migration Rate from the United Nations, Population Division, *World Population Prospects: The 2010 Revision Population Database.*

Source: The United Nations, Population Division, World Population Prospects: The 2008 Revision Population Database.

FIGURE 9.14 Population Map of the Russian Domain Population within the region is largely clustered west of the Ural Mountains. Dense agricultural settlements, extensive industrialization, and large urban centers are found in Ukraine, in much of Belarus, and across western Russia south of St. Petersburg and Moscow.

PEOPLE PER SQUARE KILOMETER

- Fewer than 6
- 6–25
- 26–100
- 101–250
- 251–500
- 501–1,000
- 1,001–12,801
- More than 12,801

POPULATION

- Metropolitan areas more than 20 million
- Metropolitan areas 10–20 million
- Metropolitan areas 5–9.9 million
- Metropolitan areas 1–4.9 million
- Selected metropolitan areas

Regional Migration Patterns

Over the past 150 years, millions of people within the Russian domain have been on the move. These major migrations, both forced and voluntary, reveal sweeping examples of human mobility that rival the great movements from Europe and Africa or the transcontinental spread of settlement across North America.

Eastward Movement Just as settlers of European descent moved west across North America, exploiting natural resources and displacing native peoples, European Russians moved east across the Siberian frontier within Eurasia. Although these migrations into Siberia began several centuries earlier, the pace accelerated in the late 19th century after the Trans-Siberian Railroad was completed. Peasants were attracted to the region by its agricultural opportunities (in the south) and by greater political freedoms than they traditionally enjoyed under the **tsars**

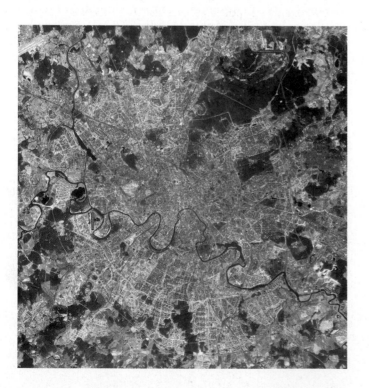

FIGURE 9.15 Metropolitan Moscow Sprawling Moscow extends more than 50 miles (80 kilometers) beyond the city center, on both sides of the Moscow River. The city is home to more than 11 million people, and the relative strength of its urban economy continues to attract migrants from elsewhere in the country, thus putting more pressure on its infrastructure.

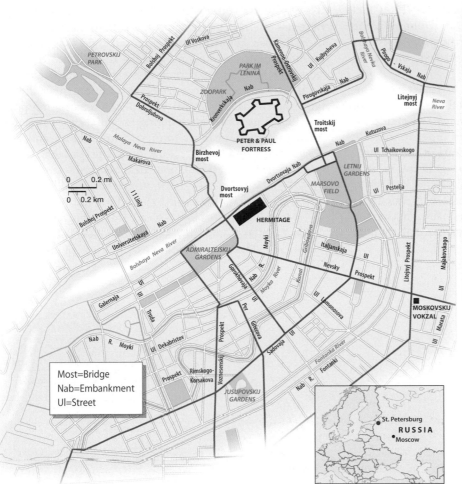

FIGURE 9.16 St. Petersburg Often beloved as Russia's most beautiful city, St. Petersburg's urban design features a varied mix of gardens, open space, waterways, and bridges. The Hermitage is one of the largest and oldest museums in the world.

(or czars; Russian for *Caesar*), the authoritarian leaders who dominated politics during the pre-1917 Russian Empire. Almost 1 million Russian settlers moved into the Siberian hinterland between 1860 and 1914.

Political Motives Political motives shaped migration patterns in the Russian domain. Particularly in the case of Russia, leaders from both the imperial and the Soviet eras forcibly relocated people to new locations, extending Russian political and economic power into the Eurasian interior. Political dissidents and troublemakers in the Soviet era were also exiled to the region's **Gulag Archipelago**, a vast collection of political prisons in which inmates often disappeared or spent years far removed from their families and

FIGURE 9.17 Kiev Kiev's historic Podil district features some of the city's most distinctive architecture, including the ornate green domes of St. Andrew's Orthodox Church (upper right).

FIGURE 9.18 Abandoned Diamond Mine, Mirny, Siberia The Siberian settlement of Mirny (population 37,000) is bordered by an abandoned diamond mine more than 1,700 feet (525 meters) deep.

communities. (The term *Gulag* is a Russian acronym for the "Chief Administration of Corrective Labor Camps and Colonies"; the term *Gulag Archipelago* is the title of a novel written by Aleksandr Solzhenitsyn, the 1970 Nobel Prize winner in literature, who wrote about these camps.)

Russification, the Soviet policy of resettling Russians into non-Russian portions of the Soviet Union, also changed the region's human geography. Millions of Russians were given economic and political incentives to move elsewhere in the Soviet Union in order to increase Russian dominance in many of the outlying portions of the country. By the end of the Soviet period, Russians made up significant minorities within former Soviet republics such as Kazakhstan (30 percent Russian), Latvia (30 percent), and Estonia (26 percent). Among its Slavic neighbors, Belarus remains 11 percent Russian and Ukraine more than 24 percent Russian, with concentrations particularly high in the eastern portions of these countries.

New International Movements In the post-Soviet era, Russification has often been reversed (Figure 9.19). Several newly independent non-Russian countries have imposed rigid language and citizenship requirements, encouraging Russian residents to leave. Often, ethnic Russians have also experienced economic and social discrimination in these non-Russian countries. In addition, since 2005, the Russian government has vigorously promoted a repatriation program for ethnic Russians worldwide, offering incentives for Russian-speaking migrants to return or move to their cultural homeland. As a result, the Central Asia and Baltic regions, once a part of the Soviet Union, have seen their Russian populations decline significantly since 1991, often by 20 to 35 percent. By 2010, more than 7 million Russians had left former Soviet republics and returned to their homeland.

Many immigrants to Russia are from ethnically non-Slavic regions. Street cleaners from Kyrgyzstan and shopkeepers from Uzbekistan are common in Moscow, where almost one-third of the country's estimated

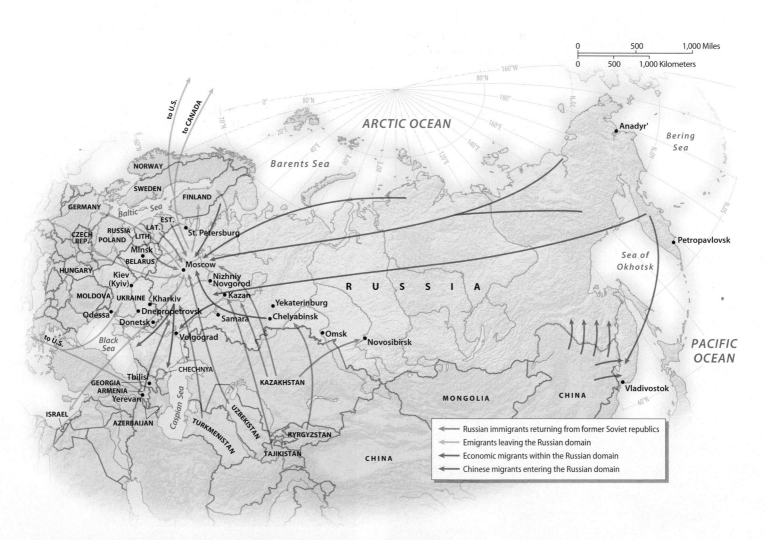

FIGURE 9.19 Recent Migration Flows in the Russian Domain Recent events are encouraging the return of ethnic Russians from former Soviet republics, while other Russians are emigrating from the domain for economic, cultural, and political reasons. Within Russia, both political and economic forces are also at work, encouraging people to be on the move. Note how most of this activity, both inflow and outflow, is centered on Moscow.

FIGURE 9.20 Chinese Immigrants in the Russian Far East Many Chinese immigrants in the Russian Far East have become small-scale entrepreneurs in the region's commercial economy.

10 million undocumented immigrants live. Elsewhere, incoming Chinese are transforming the Russian Far East, attracted by employment opportunities (a shrinking Russian workforce) in that fast-changing region (Figures 9.19 and 9.20). Between 5 and 7 million Chinese immigrants (many of them illegal) may now live in the region.

Today it is also easier to leave the region (see Figure 9.19). The job-related "brain drain" of young, well-educated, upwardly mobile Russians has been considerable. An opinion poll in 2011 revealed that 22 percent of Russia's population would prefer to leave the country for good, mostly for better economic opportunities elsewhere. Sometimes, ethnic links also influence migration patterns. For example, Russian-born ethnic Finns have moved to nearby Finland. Russia's Jewish population also continues to fall. Russians have become one of the largest new immigrant groups in the United States.

Inside the Russian City

Today most people in the Russian domain live in cities, the product of a century of urban migration and growth (see Table 9.1). Large Russian cities possess a core area, or center, that features superior transportation connections; well-stocked, upscale department stores and shops; desirable housing; and important offices (both governmental and private). In the largest urban areas, downtowns in centers such as Moscow and St. Petersburg also feature extensive public spaces and examples of monumental architecture.

Within the city, there is usually a distinctive pattern of circular land-use zones, each of which was built at a later date, moving outward from the center. Such a ring-like urban morphology is not unique. However, as a result of the power of government planners during the Soviet period, this urban form is probably more highly developed here than in most other parts of the world.

The central cores of many older cities predate the Soviet Union. Pre-1900 stone buildings often dominate older city centers. Some private mansions that were turned into government offices or subdivided into apartments during the communist period are now being privatized again. Many older buildings, however, have been leveled in rapidly growing urban settings such as downtown Moscow (Figure 9.21). Urban preservation experts estimate that since 1992 several thousand

historic buildings have been destroyed, including structures that had supposedly received official protection. Retailing malls replace many of these older structures. Nearby, nightclubs and bars are filled with pleasure seekers, where the city's professional elite mingle with foreign visitors and tourists.

Farther out from the city centers are **mikrorayons**—large, Soviet-era housing projects of the 1970s and 1980s (Figure 9.22). Mikrorayons are typically composed of massed blocks of standardized apartment buildings, ranging from 9 to 24 stories in height. The largest of these supercomplexes house up to 100,000 residents. Planners had hoped that mikrorayons would foster a sense of community, but most now serve as anonymous bedroom communities for larger metropolitan areas.

Some of Russia's most rapid urban growth has occurred on the metropolitan periphery, paralleling the North American experience. Moscow, for example, has seen its urban reach expand far beyond the city center. The surrounding administrative district (the Moscow Oblast) contains about 7 million residents and is home to more than 700 international companies. Reasons for this suburban growth include lower land prices and tax rates than in the central city, a less onerous bureaucracy, and relatively new transportation and telecommunications infrastructure. Suburban shopping malls, paralleling the North American model, are also popping up on the urban fringe.

The Demographic Crisis

Russia's Ministry of Labor and Social Development recently acknowledged that the country faces a crisis of persistently declining populations (see Table 9.1). Studies predict that Russia's population could fall by a startling 45 million by 2050 (Figure 9.23). Similar conditions are affecting other countries in the region. Declining populations, low birthrates, and relatively high mortality—particularly among middle-aged males—are all troubling symptoms of the region's population problems.

Recently, Vladimir Putin argued that the demographic decline was Russia's "most acute problem," and he has initiated a program to encourage higher birthrates. Mothers of a second child receive nontaxable cash grants, extended maternity leave, and extensive day-care subsidies. Indeed, there have been slight increases in Russian and Ukrainian birthrates, and they are now significantly higher than those of Germany, Italy, and Japan. There are long waiting lists for urban kindergarten spots in Moscow. Employers are also offering more benefits to both mothers and fathers, making parenthood more appealing for some couples. But only time will tell if the region's demographic crisis is easing.

REVIEW QUESTIONS

1. Discuss how major river and rail corridors have shaped the geography of population and economic development in the region. Provide specific examples.

2. Contrast Soviet and post-Soviet migration patterns within the Russian domain, and discuss the changing forces at work.

3. Describe some of the major land-use zones in the modern Russian city, and suggest why it is important to understand the impact of Soviet-era planning within such settings.

FIGURE 9.21 Downtown Moscow The city's downtown landscape has been rapidly transformed in the recent economic boom. It remains a complex and fascinating mix of imperial, Soviet, and post-Soviet influences.

FIGURE 9.22 Moscow Housing These drab apartment buildings in the Moscow suburbs date from the late Soviet period (1980s) and still offer housing to many middle-class residents of the city.

CULTURAL COHERENCE AND DIVERSITY: THE LEGACY OF SLAVIC DOMINANCE

For hundreds of years, Slavic peoples speaking the Russian language expanded their influence from an early homeland in central European Russia. Russian cultural patterns and social institutions spread widely during this Slavic expansion, influencing non-Russian ethnic groups that continued to live under the rule of the Russian empire. The legacy of this diffusion continues today, offering Russians a rich historical identity. It also provides a framework for examining how present-day Russians are dealing with forces of globalization and how non-Russian cultures have evolved in the region.

The Heritage of the Russian Empire

The expansion of the Russian Empire paralleled similar events in western Europe. As Spain, Portugal, France, and Britain carved out overseas empires, Russia expanded eastward and southward across Eurasia. The origin of the Russian Empire lies in the early history of the **Slavic peoples**, defined as a northern branch of the Indo-European language family. Slavic political power grew by 900 CE as these

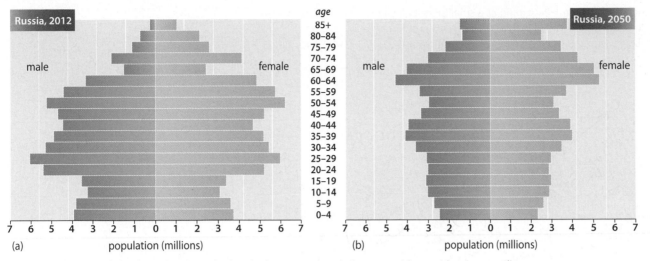

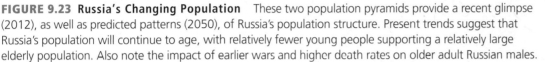

FIGURE 9.23 Russia's Changing Population   These two population pyramids provide a recent glimpse (2012), as well as predicted patterns (2050), of Russia's population structure. Present trends suggest that Russia's population will continue to age, with relatively fewer young people supporting a relatively large elderly population. Also note the impact of earlier wars and higher death rates on older adult Russian males.

people intermarried with southward-moving warriors from Sweden known as *Varangians*, or *Rus*. Within a century, the state of Rus extended from Kiev (the capital) to near the Baltic Sea. The new Kiev-Rus state interacted with the Greek Byzantine Empire, and this influence brought Christianity and the Cyrillic alphabet to the region. Even as Russians converted to **Eastern Orthodox Christianity**, which is a form of Christianity linked to eastern Europe and with church leaders in Constantinople (modern Istanbul), their Slavic neighbors to the west (Poles, Czechs, Slovaks, Slovenians, and Croatians) accepted Catholicism. This early Russian state then faltered, splitting into several principalities later ruled by invading Mongols and Tatars.

In the 14th century, northern Slavic peoples overthrew Tatar rule and established an expanding Slavic state (Figure 9.24). The new Russian Empire's core lay near the eastern fringe of the old state of Rus. Gradually, this area's language diverged from that spoken in the new core, and *Ukrainians* and Russians developed into two separate peoples. A similar development took place among northwestern Russians, who experienced several centuries of Polish rule and were transformed into a distinctive group known as the *Belarusians*.

The Russian Empire expanded remarkably in the 16th and 17th centuries. Former Tatar territories in the Volga Valley (near Kazan) were incorporated into the Russian state in the mid-1500s. The Russians also allied with the seminomadic **Cossacks**, Slavic-speaking Christians who had earlier migrated to the region seeking freedom in the ungoverned steppes (see Figure 9.2 and 9.24). This alliance smoothed the way for Russian expansion into Siberia during the 17th century. Furs and precious metals were the region's chief attractions.

Westward expansion was slow and halting. When Tsar Peter the Great (1682–1725) defeated Sweden in the early 1700s, he obtained a foothold on the Baltic Sea. There he built the new capital city of St. Petersburg, designed to give the empire better access to western Europe. Later in the 18th century, Russia defeated both the Poles and the Turks and gained all of modern-day Belarus and Ukraine. Tsarina Catherine the Great (1762–1796) was especially important in colonizing Ukraine and bringing the Russian Empire to the warm-water shores of the Black Sea.

The 19th century witnessed the Russian Empire's final expansion, mostly in Central Asia, where once-powerful Muslim states were no longer able to resist the Russian army. The mountainous Caucasus region proved a greater challenge, as residents used the rugged terrain in defending their lands. South of the Caucasus, however, Christian Armenians and Georgians accepted Russian power with little struggle because they found it preferable to rule by the Persian or the Ottoman empire.

Geographies of Language

Slavic languages dominate the region (Figure 9.25). The geographic pattern of the Belarusian people is relatively simple. Most Belarusians reside in Belarus, and most people in Belarus are Belarusians. Ukraine's situation is more complex. Russian speakers historically dominate large parts of eastern Ukraine, but make up a much smaller portion of western Ukraine's population. Similarly, the Crimean Peninsula, now part of Ukraine, has ethnic and political connections to Russia. As a result, many Ukrainian citizens in the eastern and southern parts of the country rarely speak Ukrainian. Conversely, many Ukrainians born in the west never learn Russian. Approximately 80 percent of Russia's population claims Russian linguistic identity.

Russians inhabit most of European Russia, but there are large groups of other peoples in the region. The Russian zone extends across southern Siberia to the Sea of Japan. In sparsely settled central and northern Siberia, Russians share territory with varied native peoples.

In Moldova, Romanian (a Romance language) speakers are dominant, although ethnic Russians and Ukrainians each make up about 13 percent of the country's population. Now that Russia, Ukraine, and Moldova are separate countries with a heightened sense of national distinction, this issue has emerged as a significant source of tension across the region.

The Finno-Ugric peoples include the Finnish-speaking settlers who dominate sizable portions of the non-Russian north. Altaic speakers also complicate the country's linguistic geography. They include Volga Tatars, centered on the city of Kazan. While retaining

FIGURE 9.24 Growth of the Russian Empire Beginning as a small principality in the vicinity of modern Moscow, the Russian Empire took shape between the 14th and 16th centuries. After 1600, Russian influence stretched from eastern Europe to the Pacific Ocean. Later, portions of the empire were added in the Far East, in Central Asia, and near the Baltic and Black seas.

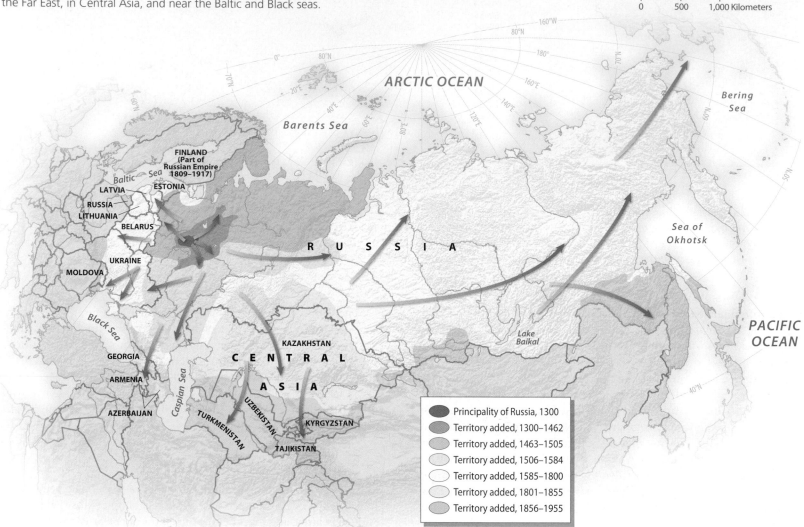

Legend:
- Principality of Russia, 1300
- Territory added, 1300–1462
- Territory added, 1463–1505
- Territory added, 1506–1584
- Territory added, 1585–1800
- Territory added, 1801–1855
- Territory added, 1856–1955

their ethnic identity, the Turkish-speaking Tatars have extensively intermarried with their Russian neighbors. Yakut and Evenki peoples of northeast Siberia also speak Turkish languages (Figure 9.26 on page 312). In the east, over 400,000 Buryats live near Lake Baikal and are tied to the cultures and history of Central Asia.

The plight of many native peoples in central and northern Siberia parallels the situation in the United States, Canada, and Australia. Poor, rural indigenous peoples remain distinct from dominant European cultures. These peoples are internally diverse, often divided into unrelated linguistic groups. Many Siberian peoples have seen traditional ways challenged by Russification, just as native peoples elsewhere have been subjected to similar pressures of cultural and political assimilation.

Transcaucasia offers a bewildering variety of languages (Figure 9.27 on page 313). From Russia, along the north slopes of the Caucasus, and to Georgia and Armenia east of the Black Sea, a complex history and physical setting combine to produce some of the world's most complex language patterns. Several language families are spoken in a region smaller than Ohio, and individual languages are represented by small, isolated cultural groups.

Geographies of Religion

Most Russians, Belarusians, and Ukrainians share a religious heritage of Eastern Orthodox Christianity. For hundreds of years, Eastern Orthodoxy served as a central cultural presence in the Russian Empire (Figure 9.28 on page 314). Indeed, church and state were tightly fused until the demise of the empire in 1917. Under the Soviet Union, however, all religion was discouraged and persecuted. More recently, with the downfall of the Soviet Union, a religious revival swept much of the Russian domain. Now about 75 million Russians are members of the Orthodox Church, including almost 500 monastic orders dispersed across the country.

Other forms of Christianity are also present. Western Ukrainians, who experienced several hundred years of Polish rule, eventually joined the Catholic Church. Eastern Ukraine, however, remained within the Orthodox framework. This religious split reinforces cultural differences between eastern and western Ukrainians. Elsewhere, Armenia has a long Christian tradition, but it differs somewhat from both Eastern Orthodox and Catholic practices. Evangelical Protestantism has also been on the rise.

FIGURE 9.25 Languages of the Russian Domain   Slavic Russians dominate the region, although many linguistic minorities are present. Siberia's diverse native peoples add cultural variety in that area. To the southwest, the Caucasus Mountains and the lands beyond contain the region's most complex linguistic geography. Ukrainians and Belarusians, while sharing a Slavic heritage with their Russian neighbors, add further variety in the west.

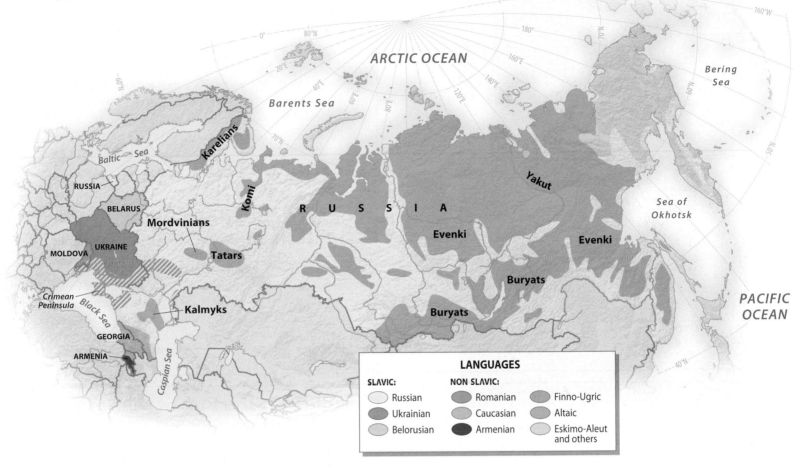

LANGUAGES

SLAVIC:	NON SLAVIC:	
Russian	Romanian	Finno-Ugric
Ukrainian	Caucasian	Altaic
Belorusian	Armenian	Eskimo-Aleut and others

Non-Christian religions also shape regional cultural geographies. Islam claims 20 million followers in the region. Most are Sunni Muslims living in the North Caucasus and the Volga Valley and near the Kazakhstan border. Growth rates among Russia's Muslim populations are three times those of non-Muslim populations. In addition, Moscow's economic boom has attracted many Muslim immigrants. Perhaps 20 percent of Moscow's population, including undocumented immigrants, is now Islamic (Figure 9.29 on page 314.). A growing Islamic political consciousness is also present, particularly in Russia. Russia, Belarus, and Ukraine are also home to more than 1 million Jews, who are especially numerous in the larger cities of the European west. In addition, Buddhists are represented by Kalmyk and Buryat peoples of the Russian interior.

Russian Culture in Global Context

Russian culture has developed its own distinctive traditions and symbols, and it has also been influenced greatly by western Europe. By the 19th century, even as Russian peasants interacted rarely with the outside world, Russian high culture had become thoroughly Westernized, and Russian composers, novelists, and playwrights gained considerable fame in Europe and the United States.

Soviet Days   During the Soviet period, new cultural influences shaped the socialist state. Initially, European-style modern art flourished in the Soviet Union, encouraged by Marxist rhetoric. By the late 1920s, however, Soviet leaders turned against modernism, viewing it as the decadent expression of the capitalist world. Increasingly, state-sponsored Soviet artistic productions centered on **socialist realism**, a style devoted to the realistic depiction of workers heroically challenging nature or struggling against capitalism. Still, traditional high arts, such as classical music and ballet, received generous state subsidies, and to this day Russian artists regularly achieve worldwide fame.

Russia's famed love affair with chess, kindled during Tsarist times, also received attention during the Soviet period. Lenin's army commander opened chess schools and promoted the game after 1921. Stalin played often, believing that the game embodied the Soviet knack for intellect and strategy. Indeed, the board game's popularity outlived the Soviet era, and 21st-century Russians still embrace the game.

Turn to the West   By the 1980s, it was clear that the attempt to fashion a new Soviet culture based on communist ideals had failed. The younger generation was more inspired by fashion and rock music from the West. American mass-consumer culture proved particularly attractive.

FIGURE 9.26 Minority Evenki Russia's indigenous Evenki population speaks an Altaic language and shares many of the economic and social problems of native peoples in North America and Australia.

After the fall of the Soviet Union, global cultural influences grew, particularly in larger cities such as Moscow. Western books and magazines flooded shops, people sought advice about home mortgages and condominium purchases, and they enjoyed the newfound pleasures of fake Chanel handbags and McDonald's hamburgers. People embraced the world that their former leaders had warned them about for generations. Cultural influences streaming into the country were not all Western in inspiration. Films from Hong Kong and Mumbai (Bombay), as well as televised romance novels (*telenovelas*) of Latin America, for example, proved even more popular in the Russian domain than in the United States.

The Music Scene Younger residents have embraced popular music, and their enthusiasm for U.S. and European performers, as well as their support of budding home-grown musicians, symbolizes the changing values of the post-Soviet generation. Today Russian MTV reaches most of the nation's younger viewers. Sony, BMG, and other labels have signed multiple Russian artists for domestic markets, boosting a home-grown pop-music culture. Increasingly, regional talent is also going global. Ukrainian singer Ruslana won the coveted Eurovision 2004 award for her song "Wild Dance." In 2009, Moscow hosted the Eurovision awards, featuring a strong performance from

Ukrainian singer, composer, and television personality Svetlana Loboda (Figure 9.30 on page 314.). She used her performance of "Be My Valentine (Anti-Crisis Girl)" and her popularity in both Russia and Ukraine as a platform to speak out against domestic violence against women, a widespread problem in her home country.

> ### REVIEW QUESTIONS
>
> 1. What were the key phases of colonial expansion during the rise of the Russian Empire, and how did each enlarge the reach of the Russian state?
>
> 2. What are some of the key ethnic minority groups (as defined by language and religion) within Russia and neighboring states?

GEOPOLITICAL FRAMEWORK: RESURGENT GLOBAL SUPERPOWER?

The geopolitical legacy of the former Soviet Union still weighs on the Russian domain. After all, the bold lettering of the "Union of Soviet Socialist Republics" dominated the Eurasian map for much of the 20th century, and the country's political influence affected every part of the world. Indeed, Russia's post-2000 political resurgence signals a return to its Soviet-era status of geopolitical dominance within the region and its increasing visibility in global affairs.

Geopolitical Structure of the Former Soviet Union

The Soviet Union rose from the ashes of the Russian Empire, which collapsed abruptly in 1917. The Russian tsars did little to modernize the country or improve the lives of the peasant population. After the tsars fell, a government representing several political groups assumed authority. Soon, however, the **Bolsheviks**, a faction of Russian communists representing industrial workers, seized power. The leader of these Russian communists was Vladimir Ilyich Ulyanov, usually known by his self-selected name, Lenin. Lenin became the architect of the Soviet Union, the state that replaced the Russian Empire. The new socialist state radically reconfigured Eurasian political and economic geography. When the Soviet Union emerged in 1917, Lenin and other communist leaders were aware that they faced a major challenge in organizing the new state.

Creating a Political Structure Soviet leaders designed a geopolitical solution that maintained their country's territorial boundaries and recognized in theory the rights of non-Russian citizens. Each major nationality received its own "union republic," provided it was situated on one of the nation's external borders (Figure 9.31 on page 315). Eventually, 15 such republics were established, creating the Soviet Union. So-called **autonomous areas** within these republics gave special recognition to smaller ethnic homelands. The massive Russian Republic sprawled over roughly three-quarters of the Soviet terrain. Even with these internal republics and autonomous areas, the Soviet Union emerged by the late 1920s as a highly centralized state, with important decisions made in the Russian capital of Moscow.

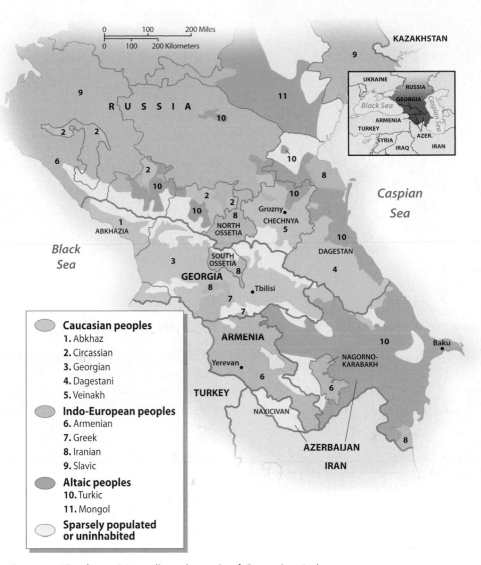

FIGURE 9.27 Languages of the Caucasus Region A complicated mosaic of Caucasian, Indo-European, and Altaic languages characterizes the Caucasus region of southern Russia and nearby Georgia and Armenia. Persisting political problems have erupted in the region as local populations struggle for more autonomy. Recent examples include independence movements in Chechnya and in nearby Dagestan.

The chief architect of this political consolidation was Joseph Stalin, who did everything he could to centralize power and assert Russian authority. The Stalin period (1922–1953) also saw the enlargement of the Soviet Union. Victorious in World War II, the country acquired Pacific islands from Japan, the Baltic republics (Estonia, Latvia, and Lithuania), and portions of eastern Europe. One strategic addition on the Baltic Sea was the northern portion of East Prussia (the port of Kaliningrad), previously part of Germany. It still forms a small, but strategic Russian **exclave**, defined as a portion of a country's territory that lies outside its contiguous land area.

After World War II, the Soviet Union expanded its influence across eastern Europe. In the words of British leader Winston Churchill, the Soviets extended an **Iron Curtain** between their eastern European allies and the more democratic nations of western Europe. As eastern Europe retreated behind the Iron Curtain, the Soviet Union and the United States became antagonists in a global **Cold War** of military competition that lasted from 1948 to 1991.

End of the Soviet System Ironically, Lenin's system of republics based on cultural differences sowed the seeds of the Soviet Union's demise. Even though the republics were never allowed real freedom, they provided a political framework that encouraged the survival of distinct cultural identities. Contrary to expectations of Soviet leaders, ethnic nationalism intensified in the post–World War II era as the Soviet system grew less repressive. When Soviet President Mikhail Gorbachev initiated his policy of **glasnost**, or greater openness, during the 1980s, several republics—most notably the Baltic states of Lithuania, Latvia, and Estonia—demanded independence. In addition, Gorbachev's policy of **perestroika**, or restructuring of the planned centralized economy, was an admission that domestic economic conditions increasingly lagged those of western Europe and the United States. A failed war in Afghanistan and increasing political protests in eastern Europe added to Gorbachev's problems.

By 1991, Gorbachev saw his authority slip away amid rising pressures for political decentralization and economic reforms. During that

FIGURE 9.28 St. Basil's Cathedral, Moscow Built in the 16th century, this famous Moscow landmark on Red Square remains a symbol of Russian culture and of the enduring power of the Russian Orthodox Church.

FIGURE 9.29 Moscow Mosque Moscow's growing immigrant population has included many Muslims from portions of the former Soviet Union, particularly from Central Asia. Today their presence in the capital is an increasingly visible element in the city's cultural landscape.

summer, Gorbachev's regime was further endangered by the popular election of reform-minded Boris Yeltsin as head of the Russian Republic and by a failed military takeover by communist hard-liners. By late December, all of the country's 15 constituent republics had become independent states, and the Soviet Union ceased to exist.

Current Geopolitical Setting

The political geography of post-Soviet Russia and the nearby independent republics has changed dramatically since the collapse of the Soviet Union in 1991 (Figure 9.32 on page 316.). All of the former republics have struggled to establish stable political relations with their neighbors.

Russia and the Former Soviet Republics For a time, it seemed that a looser political union of most of the former republics, called the **Commonwealth of Independent States (CIS)**, would emerge from the ruins of the Soviet Union. All the former republics, with the exception of the three Baltic states, soon joined the CIS (Figure 9.32). By the early 21st century, however, the CIS had developed into little more than a forum for discussion, without real economic or political power. Another complicating factor in the post-Soviet period has been the ongoing military relationships between Russia and its former republics. **Denuclearization**, the return of nuclear weapons from

FIGURE 9.30 Svetlana Loboda Ukrainian singing star Svetlana Loboda was one of the top performers at Eurovision 2009, held in Moscow. In addition to a successful career in music and television, Loboda has championed the plight of battered women throughout the region.

FIGURE 9.31 Soviet Geopolitical System   During the Soviet period, the boundaries of the country's 15 internal republics often reflected major ethnic divisions. As the Soviet empire disintegrated, the former republics became politically independent states and now form an uneasy ring of satellite nations around Russia.

outlying republics to Russian control and their partial dismantling, was completed during the 1990s. The Soviet-era nuclear arsenals of Kazakhstan, Ukraine, and Belarus were removed in the process.

Geopolitics in the South and West   In Belarus, leaders have been slow to open to political and economic opportunities. The country remains firmly within Russia's political orbit. In 2010, the two countries pledged to move toward a "union state" and expanded their common military maneuvers. Specifically, Russia and Belarus have protested the growing influence of the North Atlantic Treaty Organization (NATO) in central Europe; the two countries see that expansion as a potential military threat. In 2012, Belarusian leaders adopted one of the world's most restrictive policies on free use of the Internet within the nation, strongly discouraging the use of any "foreign" sites.

Russia's relationships with Ukraine have been less predictable. Ukraine remains highly dependent on Russian energy supplies. In addition, major Russian pipelines to central and western Europe pass through Ukraine. The result is that the two nations have often wrangled over energy prices and the availability of oil and natural gas

supplies. Early in 2009, for example, Russia briefly shut off all natural gas supplies to Ukraine amid pricing and payment disputes between Ukraine and Gazprom, the Russian energy company.

Ukraine's internal politics have also riled Russia. In 2004, Viktor Yushchenko, a reformist leader and Ukrainian nationalist, came to power after a controversial election that included an attempt to poison him. Yushchenko often criticized Russian leaders and excessive Russian interference with Ukraine's affairs. He also explored Ukraine's admission into the European Union (EU) and NATO. However, Viktor Yanukovich, one of Yushchenko's rivals, argued for closer relations with Moscow. Steep declines in Ukraine's economy in 2009 prompted widespread public protests in Kiev and calls for early elections, challenging Yushchenko. Elections in 2010 brought Yanukovich into power. Voting patterns revealed Yanukovich's popularity among Russian speakers in the country, but he also attracted disgruntled Ukrainians weary of the nation's battered economy. Results appeared to signal a tilt toward Russia. A 2012 free-trade agreement between the two countries also suggests closer ties. However political winds blow in Ukraine, Russia will continue to influence

FIGURE 9.32 Geopolitical Issues in the Russian Domain The Russian Federation Treaty of 1992 created a new internal political framework that acknowledged many of the country's ethnic minorities. Recently, however, Russian authorities have moved to centralize power and limit regional dissent. Russia's relations with several nearby states remain strained.

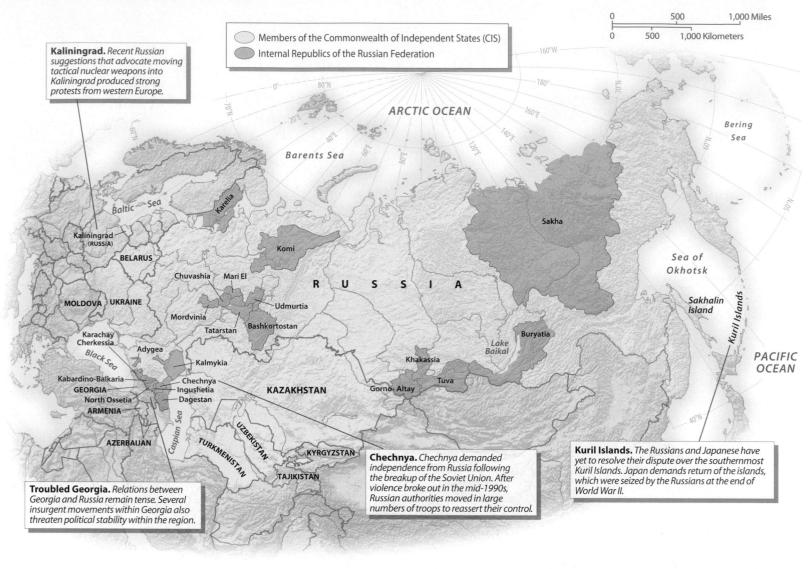

Kaliningrad. *Recent Russian suggestions that advocate moving tactical nuclear weapons into Kaliningrad produced strong protests from western Europe.*

Members of the Commonwealth of Independent States (CIS)

Internal Republics of the Russian Federation

Chechnya. *Chechnya demanded independence from Russia following the breakup of the Soviet Union. After violence broke out in the mid-1990s, Russian authorities moved in large numbers of troops to reassert their control.*

Kuril Islands. *The Russians and Japanese have yet to resolve their dispute over the southernmost Kuril Islands. Japan demands return of the islands, which were seized by the Russians at the end of World War II.*

Troubled Georgia. *Relations between Georgia and Russia remain tense. Several insurgent movements within Georgia also threaten political stability within the region.*

Ukrainian politics in any way it can to serve its own political and economic interests.

Tiny Moldova has also witnessed political tensions. Conflict has repeatedly flared in the Russian-dominated Transdniester region in the eastern part of the country, where Slavic separatists have pushed for independence. Others suggest that Moldova join with Romania and dump troublesome Transdniester in the process. Some argue that Moldova should seek a middle ground that would keep it independent, but move it out of Russia's orbit toward the potential rewards of EU, or even NATO, membership.

Transcaucasia also remains unstable. Since 2003, Georgia has moved toward closer ties with the United States, even suggesting that it might join NATO. At the same time, Georgia's own internal politics recently provoked an invasion from neighboring Russia. In 2008, Georgia attempted to reassert its control over Abkhazia and South Ossetia,

two breakaway regions that border Russia and that have wide support in that country. The Russians responded with tanks, for a time occupying large sections of Georgia. Almost 1,000 people died in the conflict, and more than 30,000 people were displaced from their homes (Figure 9.33). The situation remains tense, with Georgia claiming control over Abkhazia and South Ossetia and with Russia formally recognizing the "independence" of these microstates.

In nearby Armenia, territories of Christian Armenians and Muslim Azeris interpenetrate one another in a complex fashion. The far southwestern portion of Azerbaijan (Naxcivan) is separated from the rest of the country by Armenia, while the important Armenian-speaking district of Nagorno-Karabakh is officially an autonomous portion of Azerbaijan. After Armenia successfully occupied much of Nagorno-Karabakh in 1994, fighting between the countries diminished. No final peace treaty has been signed, however, and Azerbaijan demands

FIGURE 9.33 War in Georgia, 2008 When Georgia attempted in the summer of 2008 to assert tighter control over the breakaway regions of Abkhazia and South Ossetia, Russian troops and tanks briefly invaded the country. Tensions remain high between the two countries, and Russia has recognized the independence of both the Abkhazia and the South Ossetia regions.

the return of the territory. Meanwhile, Armenia's traditionally close connections with Russia are increasingly counterbalanced with the country's interest in building ties with the United States and the EU.

Geopolitics Within Russia Within Russia, further pressures for devolution, or more localized political control, resulted in the March 1992 signing of the Russian Federation Treaty. Theoretically, the treaty granted Russia's internal autonomous republics and its lesser administrative units greater political, economic, and cultural freedoms. Defined essentially along ethnic lines, 21 regions possess status as republics within the federation and have constitutions that often run counter to national mandates. Scattered from central Siberia to the north-facing slopes of the Caucasus, these republics reflect much of the nation's linguistic and religious diversity (see Figure 9.32).

Since 2000, Russian leaders, especially Vladimir Putin, have pushed for more centralized control in the country (Figure 9.34). Putin's prominence has been enduring: He served two terms as President (2000–2008) and one term as Prime Minister (2008–2012), and he was reelected (for a six-year term) as Russia's President in March 2012. Putin, a former internal security agent with the KGB during the Soviet period, has steadily consolidated power in the country, pushed for strong economic growth oriented around Russia's energy economy, and reasserted Russia's political and military role, both on the world stage and as a dominant regional power.

One of Putin's early political moves was to crack down on dissidents in the Russian republic of Chechnya. A strong hand, including multiple Russian military invasions of Chechnya, helped bring a halt to an independence movement in that largely Muslim region. In 2009, Russia proclaimed that the conflict had ended, bolstered by a Chechen government largely in sympathy with Moscow. Even so, instability in the region continues, particularly among Islamic separatists in Chechnya and in the nearby republics of Ingushetia and Dagestan (Figure 9.32).

FIGURE 9.34 Vladimir Putin Since 2000, Vladimir Putin's influence across the Russian domain has been immense. Within Russia, Putin (as both President and Prime Minister) has managed an impressive economic recovery while limiting civil liberties. Beyond Russia, Putin has reestablished the region's geopolitical presence on the global stage.

Russian Challenge to Civil Liberties Growing public protests since 2009 have challenged Putin's authority. Disgruntled members of the urban middle class, human rights groups, and opposition political parties have periodically protested Putin's grip on power. Protestors demand a freer press, more democracy, more open elections, and a broader commitment to economic growth. They have criticized Putin's strong-arm leadership style. They also note Putin's increasingly close ties to the **siloviki**, members of the nation's military and security forces.

Protests against the Russian central government are in part a response to the government's crackdown on civil liberties. Immediately after the fall of the Soviet Union, Russia enjoyed a genuine flowering of democratic freedoms. A multiparty political system, independent media, and a growing array of locally and regionally elected political officials signaled real change from the authoritarian legacy of the Soviet period. Since 2002, however, many hard-won civil liberties have slipped away, victims of Putin's campaign to consolidate political power, increase the authority of the central government, limit press freedoms, and silence critics who disagreed with his policies. Recently,

Russian officials have increased their surveillance and regulation of the Internet, another move to silence opposition.

The Shifting Global Setting

Since 1991, regional political tensions have continued to challenge the Russians, in both the east and the west. In East Asia, the boundary between Russia and China imposed by the Russian Empire in 1858 has never been fully accepted by Beijing. While relations between the two nations improved after a 2004 agreement clarified territorial claims along the Amur and Ussuri rivers, the potential for renewed conflict remains. In addition, Russia has played an important role in containing North Korea's nuclear ambitions in the region, pressuring their Far East neighbor to limit uranium enrichment and weapons development projects. Territorial disagreements—specifically, a dispute over the Kuril Islands—also complicate Russia's relationship with Japan. To the west, Russia has criticized the expansion of NATO. Although most Russian leaders accepted the inclusion of Poland, Hungary, and the Czech Republic in NATO, they opposed the addition of the Baltic republics (Estonia, Latvia, and Lithuania) to the increasingly powerful organization. In fact, the move provoked some Russian military leaders to recommend new troop and weapon deployments on the country's western borders.

Today Russian leaders are reasserting their nation's global political status. Russia's nuclear arsenal, while reduced in size, remains a powerful counterpoint to American, European, and Chinese interests. While Russia no longer directly challenges the United States as it did in Soviet days, it acts as a partial counterweight to the United States in international maneuverings. Russia also retains a permanent seat on the United Nations Security Council, and its inclusion in the G-8 economic meetings signifies its growing international clout.

REVIEW QUESTIONS

1. How do current geopolitical conflicts reflect long-standing cultural differences within the region?

2. Describe how Vladimir Putin has played a key role in consolidating Russia's power since 2000, both within the country and beyond.

ECONOMIC AND SOCIAL DEVELOPMENT: AN ERA OF ONGOING ADJUSTMENT

The economic future of the Russian domain remains difficult to predict. Economic declines devastated the region for much of the 1990s. Between 2002 and 2008, especially for Russia, higher oil and gas prices brought significant, but selective economic improvement. The entire region was hit hard in the 2008–2010 economic downturn. Recently, however, economic growth has picked up, particularly in Russia, where higher energy prices once again have bolstered the economy (Table 9.2).

The Legacy of the Soviet Economy

The creation of the Soviet Union in 1917 initiated radical economic changes. Under the Russian Empire, most people were peasant farmers. Following the revolution, however, the Soviet Union quickly emerged to rival many of the most powerful economies on Earth. During that era of unmatched growth, much of the present economic infrastructure was established, including new urban centers, industrial developments, and the modern network of transportation and communication linkages.

As communist leaders such as Stalin consolidated power in the 1920s and 1930s, they nationalized Russian industries and agriculture, creating a system of **centralized economic planning**, in which the state controlled production targets and industrial output. The Soviets emphasized heavy basic industries (steel, machinery, chemicals, and electricity generation), rather than consumer goods. By the late 1920s, Stalin shifted agricultural land into large collectives and state-controlled farms.

Much of the Russian domain's basic infrastructure—its roads, rail lines, canals, dams, and communications networks—originated during the Soviet period (Figure 9.35). Dam and canal construction turned many rivers into a virtual network of interconnected reservoirs. The Volga–Don Canal (completed in 1952) connected those two river systems and greatly eased the movement of raw materials and manufactured goods (Figure 9.36). The Soviets also added thousands of miles of railroad tracks in the west, and the Trans-Siberian line was modernized and complemented by the addition of the BAM

Table 9.2 DEVELOPMENT INDICATORS

Country	GNI per capita, PPP 2010	GDP Average Annual % Growth 2000–10	Human Development Index (2011)[1]	Percent Population Living Below $2 a Day	Life Expectancy (2012)[2]	Under Age 5 Mortality Rate (1990)	Under Age 5 Mortality Rate (2010)	Adult Literacy (% ages 15 and older)	Gender Inequality Index (2011)[3,1]
Armenia	5,660	9.2	.716	12.4	74	55	20	100	0.343
Belarus	13,590	8.0	.756	<2	70	17	6	100	—
Georgia	4,990	6.9	.733	32.2	74	47	22	100	0.418
Moldova	3,360	5.2	.649	4.4	69	37	19	98	0.298
Russia	19,240	5.4	.755	<2	69	27	12	100	0.338
Ukraine	6,620	4.8	.729	<2	70	21	13	100	0.335

[1]United Nations, *Human Development Report, 2011.*
[2]Population Reference Bureau, *World Population Data Sheet, 2012.*
[3]Gender Inequality Index—A composite measure reflecting inequality in achievements between women and men in three dimensions: reproductive health, empowerment and the labor market that ranges between 0 and 1. The higher the number, the greater the inequality.
Source: World Bank, *World Development Indicators, 2012.* © World Bank. License: Creative Commons Attribution CC BY 3.0.

link across central Siberia. Farther north, the Siberian Gas Pipeline was built to link the Arctic's energy-rich fields with growing demand in Europe. Overall, the postwar period produced real economic and social improvements for the Soviet people.

Despite the successes, problems increased during the 1970s and 1980s. Soviet agriculture remained inefficient, and grain imports grew. Manufacturing efficiency and quality failed to match Western standards, particularly for consumer goods. Equally troubling, the Soviet Union failed to participate fully in technological revolutions transforming the United States, Europe, and Japan. Disparities also grew between the Soviet elite and ordinary people who enjoyed few personal freedoms. By the late 1980s, the Soviet Union had reached both an economic and a political impasse.

The Post-Soviet Economy

Fundamental economic changes have shaped the Russian domain since 1991. Particularly within Russia itself, much of the highly

centralized state-controlled economy has been replaced by a mixed economy of state-run operations and private enterprise. The collapse of the communist state also meant that economic relationships between the former Soviet republics were no longer controlled by a single, centralized government. Fundamental problems of unstable currencies, corruption, and changing government policies plagued the system for much of the 1990s. Higher oil and natural gas prices have led to growth in the Russian economy, especially between 2002 and 2007 and since 2010.

Redefining Regional Economic Ties Many economic ties still link the region. Russia enjoys especially close economic ties with both Belarus and Ukraine. Russian dominance is likely to continue, as suggested by that country's leading role in the creation of a new customs union in 2010 with neighboring Belarus and Kazakhstan (and potentially Ukraine). In late 2011, Russia also signed a free-trade deal to encourage trade with Ukraine, Belarus, Moldova, and Armenia. Intriguingly, Kazakhstan, Kyrgyzstan, and Tajikistan joined the

FIGURE 9.35 Major Natural Resources and Industrial Zones   The Russian domain's varied natural resources and chief industrial zones are widely distributed. Fossil fuels are abundant, although their distance from markets often imposes special costs. In southern Siberia, rail corridors offer access to many mineral resources. In the mineral-rich Urals and eastern Ukraine, proximity to natural resources sparked industrial expansion, while Moscow's industrial might is related to its proximity to markets and capital.

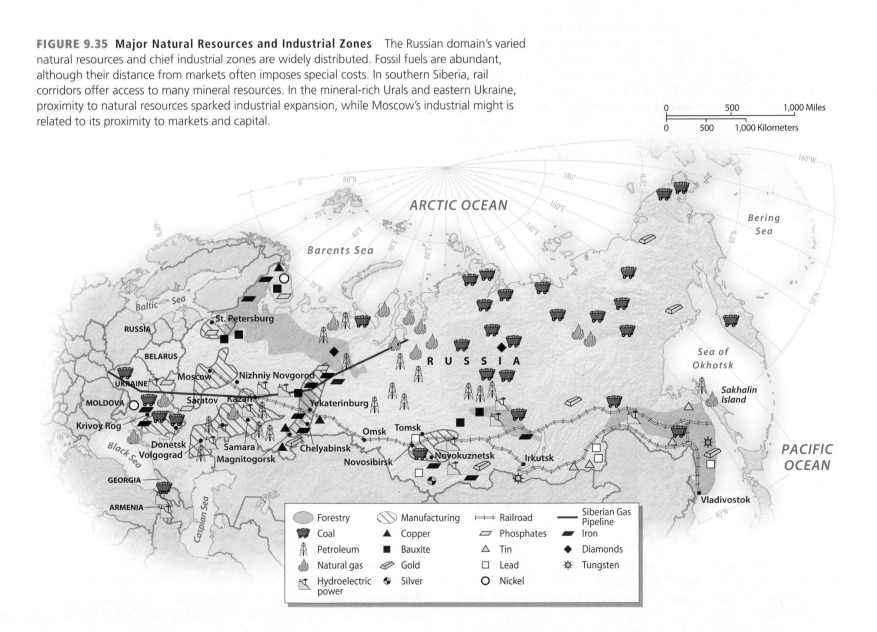

In Russia, almost 90 percent of the farmland was privatized by 2003, with many farmers forming voluntary cooperatives or joint-stock associations to work the same acreage as under the Soviet system. While crop prices have risen, costs have gone up faster, and many farmers are not skilled in dealing with the uncertainties of a market-driven economy. Overall, agriculture employs about 10 percent of Russia's workforce, but the basic distribution of crops remains little changed from the Soviet era, and short growing seasons, poor soil, and moisture deficiencies still pose challenges.

Russia encouraged more privatization in the service sector. Thousands of retailing establishments have appeared, and they now dominate that portion of the economy. In addition, the long-established "informal economy" continues to flourish. Even during the Soviet era, millions of citizens earned extra money by informally selling Western consumer goods, manufacturing food and vodka, and providing skilled services such as computer and automobile repairs. Today these barter transactions and informal cash deals form a huge part of an economy never reported to government authorities.

The natural resource and heavy industrial sectors of the economy were initially privatized in Russia, but in recent years, under Putin's management, state-run enterprises took back more control of the nation's energy assets and infrastructure. Gazprom, the huge Russian natural gas company, was privatized in 1994, but since 2005 its activities have increasingly been controlled by the state. This company is seen as a critical part of a newly emerging "state industrial policy," in which the central government is playing a more direct role in the economy. Nicknamed "Russia, Inc.," Gazprom remains one of the world's largest companies, employing more than 300,000 people and controlling huge natural gas reserves.

Especially in Russia—and particularly in that country's cities—the successes of the new economy are increasingly visible on the landscape. Luxury malls, office buildings, and more fashionable housing subdivisions are now part of the urban scene as the middle class grows in settings such as Moscow. Stroll the area near that city's Red Square and you will encounter Planet Sushi, the Moscow Maserati dealership, and trendy nightclubs. On the other hand, the gap has grown between increasing urban affluence and grinding rural poverty. In many southern Siberian villages, there are no telephones, jobs, or money. Vodka is easier to find than running water. Closed shops and a continued lack of services remain a part of everyday life in such rural settings.

FIGURE 9.36 Volga–Don Canal   This view near Volgograd suggests the enduring economic importance of the Volga–Don Canal. Built during the Soviet era, the canal remains a key commercial link that facilitates the economic integration of southern Russia.

agreement as well, prompting suggestions that Russian leaders were quietly reassembling the former Soviet Union. Georgia has followed a somewhat different path, seeking more connections with the United States and western Europe. Georgian leaders have actively explored membership in the EU. Batumi, the nation's largest Black Sea port, has become a major focus of foreign investment.

Privatization and State Control   The post-Soviet era has brought a great deal of economic uncertainty. The government initiated a massive program to privatize the Russian economy in 1993. Millions of Russians had options to buy agricultural lands and industrial companies. These initiatives opened the economy to more private initiative and investment. Unfortunately, the lack of legal and financial safeguards invited abuses and often resulted in mismanagement and corruption in the new system. Elsewhere in the region, privatization proceeded more slowly. Much of the Belarusian economy, for example, remains mired in inflexible, corrupt state-controlled companies.

The Challenge of Corruption   Throughout the Russian domain, corruption remains widespread. A 2011 study identified Russia as one of the world's most corrupt countries, where bribery was especially widespread. Doing business often means lining the pockets of government officials, company insiders, or trade union representatives. Organized crime remains pervasive in Russia. Many ties also remain between organized crime and Russian intelligence agencies. Much of the country's real wealth has been exported to foreign bank accounts. Various local and regional crime organizations divide up much of the economy. The Russian mafia has also gone global; it has been implicated in huge money-laundering schemes involving Russian, British, and U.S. banks, as well as the flow of International Monetary Fund investments into the region.

Problems of Health Care and Alcoholism   Health care is another major social problem within the Russian domain. Annual health-care

FIGURE 9.37 Health Care and Alcohol in the Russian Domain   People in the Russian domain survive on less than $700 per year for health care, compared with much higher expenditures in much of the rest of the developed world. Alcohol consumption remains highest in Russia, where it has been targeted as a key social problem by politicians and health-care experts.

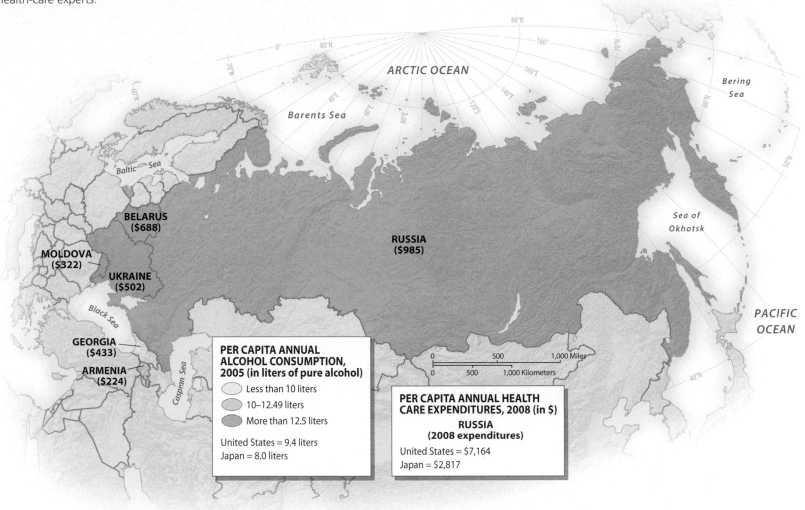

expenditures remain only a fraction of what they were during the Soviet period (Figure 9.37). To put the problem in global perspective, Ukrainians spend only about 25 percent of what Japanese spend annually on health care ($542 vs. $2,514), and Russians typically survive on barely 10 percent of what most Americans spend on health care ($638 vs. $6,714). Mortality rates for Russian men are especially grim. One in three Russian men dies before retirement. Cardiovascular disease is a key contributor to these elevated death rates.

Alcohol use in Russia (10.3 liters of pure alcohol per person annually) remains far above rates even in nearby Armenia or Georgia (see Figure 9.37). Some estimates that include illegal black market production put the Russian figure at more than 18 liters per person. For comparison, the average alcohol consumption in Japan (7.6 liters per year) is typical of many other developed-world settings, while India (0.3 liters per year) and Egypt (0.2 liters per year) exemplify settings where religious and cultural traditions discourage alcohol use. Russian leaders initiated an anti-drinking campaign in early 2010, calling their country's plight "a national disaster." Prices for vodka were raised, although bootleg liquor is cheap and widely available. An ambitious goal was set to reduce alcohol consumption by 50 percent within the next decade.

Challenges for Women   In many settings in the Russian domain, women are treated badly. Violence against women has been widely reported in the post-Soviet era. Beatings and rapes are common. A survey in Moscow suggested that one-third of divorced women had experienced domestic violence, and a women's rights group in Ukraine reported that rape was common in many villages. International organizations have sharply criticized government authorities for doing little to change the situation. In addition, **human trafficking** (a practice in which women are lured or abducted into prostitution) is a widespread problem. Armenia and Moldova are major sources of young women who become involved with prostitution in Europe and the Middle East. Within Russia, many young women from rural areas are brought to large metropolitan areas in similar illegal operations. The region is also a major global source for Internet brides, a practice that invites additional violence against women.

EXPLORING GLOBAL CONNECTIONS

Russia Exports the Nuclear Age

FIGURE 9.2.1 Soviet Nuclear-Tipped Warheads Highly enriched uranium within thousands of these Soviet-era nuclear warheads has been converted to less-enriched uranium that supplies the American nuclear power industry.

Russia's oil and gas economy plays a pivotal role in world trade, but its success in exporting the nuclear age has also created new global linkages in the post-Soviet era. After the Chernobyl nuclear disaster in 1986 (the meltdown of a nuclear power plant in Ukraine), few experts would have suggested that the Russians would emerge as leading players in the nuclear economy. The Russians argue that they have learned from their mistakes and that they are well positioned in the global marketplace to take advantage of their knowledge of nuclear energy production.

For example, Russia has captured almost 20 percent of the world's nuclear fuel market (mostly through its state-owned nuclear power company, Rosatom) and hopes to increase its share to 25 percent by 2025. Thanks to the Cold War, Russia still possesses 40 percent of the world's uranium enrichment capacity. In addition, Russia has been a key driver of the "Megatons to Megawatts" program, in which old nuclear warheads are decommissioned

and reprocessed into lower-grade nuclear fuel for export. The United States has been one of its largest customers. Ironically, the very bombs once aimed at the heartland of North America are now lighting millions of its homes and fueling its 21st-century economy (Figure 9.2.1). Experts estimate that fully 10 percent of U.S. electricity is produced from reworked nuclear material that once sat on the tips of warheads in bomb silos across Russia and Ukraine.

Russia also plays a growing role in the design, sale, and construction of new nuclear reactors around the world, mostly in developing countries such as China and India. Rosatom often underbids North American competitors such as General Electric and Westinghouse. In fact, Russia markets the special safety features in its reactors, devised after the Chernobyl meltdown. Recently, Rosatom reported that it has

15 reactors under construction and firm orders for 30 more from countries such as Turkey and Belarus. Rosatom charges $2 billion to $5 billion to build each unit and then profits further from selling its new customers the enriched uranium they need to fuel their reactors. Rosatom has set its sights on being the world's dominant player in the nuclear industry by 2030 (with annual sales targets of $50 billion), and at its current rate of growth, it may reach that goal.

Growing Economic Globalization

The relationship between the Russian domain and the world beyond has shifted greatly since the end of communism. During the Soviet era, the region was relatively isolated from the world economic system. But connections with the global economy have grown tremendously since the downfall of the Soviet Union (see "Exploring Global Connections: Russia Exports the Nuclear Age").

A More Globalized Consumer Most visibly, consumer imports now reach many residents of the Russian domain, particularly those living in larger cities. Symbols of global capitalism are visible in the heart of Moscow and, increasingly, in many other settings throughout the region. In 2010, Russia was home to 170 KFC restaurants and more than 240 McDonald's. Luxury goods from the West have also found a small, but enthusiastic market among the elite, a group noted for its devotion to BMW automobiles, Rolex watches, and other status emblems. Most Russians find such luxuries beyond their limited budgets, but they are interested in purchasing basic foods, cheap technology, and popular clothing and media from their Western neighbors or from eastern and southern Asia.

Attracting Foreign Investment Despite all of the political and economic uncertainties, most countries of the Russian domain are attracting some foreign investment. Measured by total foreign

investment, the strongest global ties by far for Russia have been with the United States, Japan, and western Europe, particularly Germany and Great Britain. The success of Russia's equity markets and the relative stability of its financial sector have been encouraging. Still, investors are put off by continuing uncertainties with Russian legal frameworks, lingering problems with slow bureaucracies and red tape, and growing concerns about the political aims of Russian leaders. In a 2009 World Bank survey measuring the business climate for foreign investors, Russia ranked in 120th place (out of 181 nations surveyed), behind Nigeria. Some of the largest investments are in Russia's oil and gas economy, but recently some of those opportunities have cooled amid the country's desire to limit foreign ownership of its energy resources and infrastructure.

Globalization and Russia's Petroleum Economy Russia's oil and gas industry remains one of the strongest economic links between the region and the global economy, and the diverse international connections it has forged suggest the increasing importance of this sector to the region's future. The statistics are impressive: Russia's energy production makes up more than one-quarter of its economic output and two-thirds of its exports. Russia has 26 percent of the world's natural gas reserves (mostly in Siberia) and is the world's largest gas exporter. As for oil, Russia is by far the world's largest non-OPEC producer, and it is the second largest oil exporter in the world (behind Saudi Arabia). It far outpaces the United States in annual output (major oilfields are

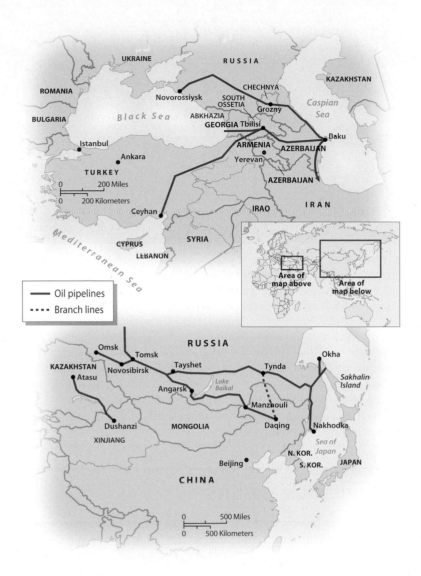

FIGURE 9.38 Russia's Expanding Pipelines These two maps show new and planned oil pipelines that are designed to expand Russia's presence in the global petroleum economy. Different pipeline projects in the Russian Far East would benefit nearby China and Japan, while projects near the Caspian Sea take pipelines through politically unstable portions of the region.

in Siberia, the Volga Valley, the Far East, and the Caspian Sea region), and it possesses more than 75 billion barrels of proven reserves.

The dynamic nature of the Russian oil and gas business exemplifies how globalization is changing the region's economy. In the Soviet era, about half of Russia's oil and gas exports went to other Soviet republics, such as Ukraine and Belarus. While these two nations still depend on Russian supplies, the primary destination for Russian petroleum products has overwhelmingly shifted to western Europe. Russia now supplies that region with more than 25 percent of its natural gas and 16 percent of its crude oil. An agreement between Russia and the

EU in 2000 aimed at the rapid expansion of these East–West linkages. The Siberian Gas Pipeline already connects distant Asian fields with western Europe via Ukraine (see Figure 9.35). Those connections are supplemented by lines through Belarus (the Yamal–Europe Pipeline) and Turkey (the Blue Stream Pipeline). Underwater pipelines (opened in 2012) beneath the Baltic (Nord Stream) and Black (South Stream) seas deliver even greater volumes of gas to northern and southern Europe. Far to the east, Russian energy companies are adding to their pipeline connections with energy-rich Sakhalin Island, and they also will construct a large new gas pipeline from Yakutia (in northern Siberia) to Vladivostok.

Expansions are also refashioning the geography of oil exports. An expanding oil port facility at Primorsk (near St. Petersburg) serves western Europe and other global markets. To the south, a large export terminal opened at Novorossiysk (on the Black Sea) in 2001, delivering Caspian Sea oil supplies to the world market via a pipeline passing through troubled Chechnya (Figure 9.38). Nearby, oil pipelines between Baku (on the Caspian Sea) and the Black and Mediterranean seas cross Azerbaijan and Georgia. In the Russian Far East, both China and Japan are lobbying hard for more pipeline projects. Russians are building a large new Siberian Pacific Pipeline to link the Siberian fields to Asian markets. The Chinese want Russian oil to flow to Daqing, where it could be refined for national and regional markets. Japan prefers a large new facility at the Pacific port of Nakhodka, well positioned to supply Japan and offering Russia easy access to global markets via the Pacific Ocean. Other links connect the system with developments on Sakhalin Island, where several major energy projects are currently being completed (Figure 9.39).

While many global companies have participated in the development of Russia's energy infrastructure, recent events suggest that state-controlled Russian companies will play a larger role in the future. Some private companies that were based in Russia, but that also utilized foreign capital, have simply been nationalized. Other North American, European, and Japanese energy companies have seen their role in Russian projects limited or eliminated. On Sakhalin Island, for example, Gazprom, the state-controlled Russian gas company, has taken control of several ventures originally dominated by companies such as Royal Dutch Shell. Ironically, Russian authorities have stripped control of these projects from foreign companies by suggesting that they failed to follow environmental regulations. Once these projects were in Russian hands, environmental concerns faded away.

FIGURE 9.39 Sakhalin Island Large-scale investments by both foreign and Russian interests have concentrated on energy-rich Sakhalin Island. The area promises to be a producer of both oil and natural gas in the years to come.

FIGURE 9.40 Novosibirsk The city of Novosibirsk, the third largest city in Russia, has become a major focus of investment and urban growth in Siberia since the disintegration of the Soviet Union.

FIGURE 9.41 Vladivostok The busy harbor of Vladivostok remains Russia's leading trade center in the Russian Far East. With easy access to markets in Japan, China, and even the United States, this Pacific port is poised to grow as Russia's economy recovers.

Local Impacts of Globalization As the discussion of the petroleum industry suggests, local impacts of globalization are highly selective. Obviously, portions of the region that are close to oil and gas wells, pipeline and refinery infrastructure, and key petroleum shipping points are greatly affected (both economically and environmentally) by the changing global energy economy. The same is true more broadly: Globalization has affected different locations in the Russian domain in very distinctive ways. In Russia, capitalism has brought its most dramatic, though selective, benefits to Moscow. Indeed, much of the country's foreign investment has flowed into the Moscow area. Beyond Moscow, St. Petersburg and the Siberian cities of Omsk and Novosibirsk (Figure 9.40) have also seen growing global investment. New oil and gas prospects in Siberia and the Russian Far East (Sakhalin Island) and near the Caspian Sea have attracted other investments. In addition, port cities such as Vladivostok are well positioned to take advantage of their accessibility to nearby markets (Figure 9.41).

Elsewhere, globalization imposes penalties. Older, less competitive locales have been hit hard. Aging steel plants, for example, no longer have the guaranteed markets for their high-cost, low-quality products that they did in the days of the planned Soviet economy. Instead, they must compete on the global market, a market that is increasingly prone to lower prices and weakening demand for many of the industrial goods that the region produces. Many of the region's extractive centers are similarly vulnerable in a global economy of rapidly changing commodity prices.

REVIEW QUESTIONS

1. Describe how centralized planning created a new economic geography across the former Soviet Union. What is its lasting impact?

2. Briefly summarize the key strengths and weaknesses of the post-Soviet Russian economy and suggest how globalization has shaped its evolution.

Summary

- The peoples of the Russian domain have endured immense challenges since 1991. Looking ahead, much will depend on Russia's success at coordinating regional economic expansion and on the political role played by the central government, both within the nation and across the region.

- Huge environmental challenges remain for the Russian domain. The legacy of the Soviet era includes polluted rivers and coastlines, poor urban air quality, and a frightening array of toxic waste and nuclear hazards.

- Declining and aging populations are part of the sobering reality for much of the region. While some localities see modest population growth related to in-migration (mostly toward expanding urban areas), many rural areas and less competitive industrial zones are likely to see continued outflows of people and very low birthrates.

- Much of the region's underlying cultural geography was formed centuries ago from the complex mix of Slavic languages, Orthodox Christianity, and numerous ethnic minorities that continue to complicate the scene today. Further changing the country are new global influences—a set of products, technologies, and attitudes that often clash with traditional cultural values.

- Much of the region's political legacy is rooted in the Russian Empire, a land-based system of colonial expansion that greatly enlarged Russian influence after 1600 and then reappeared as the Soviet Union expanded its influence. Only large remnants of that empire survive on the modern map, yet it has stamped the geopolitical character of the region in lasting ways. Beyond the region, Russia's growing visibility on the international stage signals its reemergence as a truly global political power.

- The region's future economic geography, particularly in Russia, remains tied to the fortunes of the unpredictable global energy economy.

Key Terms

autonomous area 312
Baikal–Amur Mainline (BAM) Railroad 303
Bolshevik 312
centralized economic planning 318
chernozem soil 300
Cold War 313
Commonwealth of Independent States
  (CIS) 314
Cossack 309
denuclearization 314

Eastern Orthodox Christianity 309
exclave 313
glasnost 313
Gulag Archipelago 305
human trafficking 321
Iron Curtain 313
mikrorayon 307
northern sea route 302
perestroika 313
permafrost 301

podzol soil 300
Russification 306
siloviki 317
Slavic people 308
socialist realism 311
taiga 301
Trans-Siberian Railroad 303
tsar 304

Thinking Geographically

1. How might you argue that Russia's natural environment is one of its greatest assets, as well as one of its greatest liabilities?

2. In the future, how might the forces of capitalism and free markets reshape the landscapes and land uses of cities within the Russian domain?

3. Divide into groups. On a base map of the Russian domain, have each group suggest possible political boundaries 20 years from now. Then share results and discuss.

4. What were some of the greatest strengths and weaknesses of centralized Soviet-style planning between 1917 and 1991? How were Ukraine and Belarus impacted? Why did the system ultimately fail?

5. From the perspective of a 24-year-old Russian college student and resident of Moscow, write a short essay that suggests how the economic and political changes over the past 10 to 15 years have changed your life. Exchange essays with a classmate and compare your experiences.

MasteringGeography™

Looking for additional review and test prep materials? Visit the Study Area in MasteringGeography™ to enhance your geographic literacy, spatial reasoning skills, and understanding of this chapter's content by accessing a variety of resources, including MapMaster🌐 interactive maps, videos, RSS feeds, flashcards, web links, self-study quizzes, and an eText version of *Globalization and Diversity*.

Scan to visit the author's blog for chapter updates.

http://gad4blog.
wordpress.com/
category/the-
russian-domain/

Authors' Blogs

Scan now to access the authors' blogs for up-to-date information on the Russian Domain.

Scan to visit the GeoCurrents blog.

http://geocurrents.
info/category/place/
russia-ukraine-and-
caucasus

Globalization and Diversity

For most of the past 200 years, the landlocked region of Central Asia has been geopolitically dominated by countries located in other world regions and partially cut off from the main currents of global trade. Since the downfall of the Soviet Union in 1991, however, Central Asia has emerged as a key producer of globally traded resources and as a focus of international geopolitical rivalry.

ENVIRONMENTAL GEOGRAPHY

Intensive agriculture along the rivers that flow into the deserts of Central Asia has resulted in serious water shortages, leading to the drying up of many of the region's lakes and wetlands.

POPULATION AND SETTLEMENT

Pastoral nomadism, the traditional way of life across much of Central Asia, is gradually disappearing as people settle in towns and cities.

CULTURAL COHERENCE AND DIVERSITY

In much of eastern Central Asia, the growing Han Chinese population is sometimes seen as a threat to the long-term survival of the indigenous cultures of the Tibetan and Uyghur peoples.

GEOPOLITICAL FRAMEWORK

Afghanistan and its neighbors to the north are frontline states in the struggle between radical Islamic fundamentalism and secular governments.

ECONOMIC AND SOCIAL DEVELOPMENT

Despite its abundant resources, Central Asia remains a poor region, although much of it enjoys relatively high levels of social development.

➤ Soldiers in the U.S.-allied Afghan National Army walk through a field of opium poppies as they patrol in the Taliban stronghold of Panjwai in Kandahar province in 2009.

CENTRAL ASIA 10

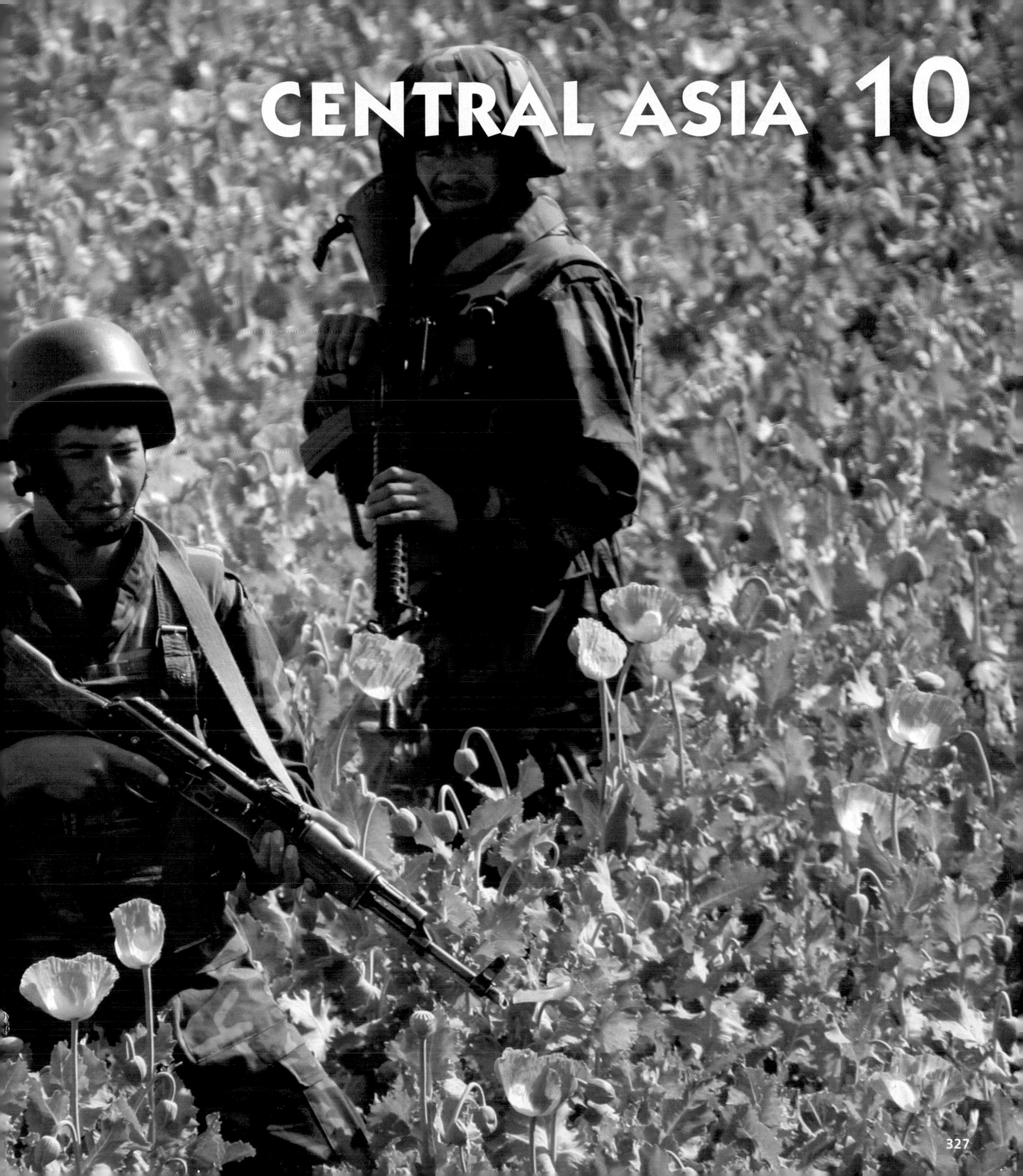

FIGURE 10.1 Central Asia   Central Asia, an extensive region in the center of the Eurasian continent, is dominated by arid plains and basins, along with high mountain ranges and plateaus. Eight independent countries—Kazakhstan, Turkmenistan, Uzbekistan, Kyrgyzstan, Tajikistan, Azerbaijan, Afghanistan, and Mongolia—form Central Asia's core. The region also includes China's lightly populated far west, which has cultural and environmental similarities to the rest of Central Asia.

Caspian Sea and Basin. *The Caspian Sea is the world's largest lake by a wide margin. It lies within the Caspian Basin, containing the world's largest area of dry land below sea level.*

The Pamir Knot. *A complex tangle of east–west and north–south trending ranges, the Pamir Knot forms the Asian highland core. Peaks reach up to 24,584 feet (7,495 meters).*

ELEVATION IN METERS

- 4000+
- 2000–4000
- 500–2000
- 200–500
- 0–200
- Below sea level

LEARNING OBJECTIVES

After reading this chapter you should be able to:

■ Identify the key environmental differences among Central Asia's desert areas, its mountain and plateau zone, and its steppe (grassland) belt and explain how these differences influence human settlement and economic development.

■ Describe the main reasons for the disappearance of the Aral Sea and outline the economic and environmental consequences of the loss of this once-massive lake.

■ Summarize the reasons why water resources are of such great importance in Central Asia and describe the ways in which people are responding to water shortages.

■ Explain why Central Asia's population is so unevenly distributed, with some areas densely settled and others essentially uninhabited.

■ Describe the differences between Central Asia's historical cities and those that have been established within the past 100 years.

■ Outline the ways in which religion divides Central Asia and describe how religious diversity has influenced the history of the region.

■ Identify how cultural globalization has affected different parts of Central Asia in distinct ways and explain why cultural globalization is controversial in much of the region.

■ Describe the geopolitical roles played in Central Asia by Russia, China, and the United States and explain why the region has been the site of pronounced geopolitical tension over the past several decades.

■ Describe how ethnic conflict has contributed to instability in Afghanistan and assess the potential of ethnic tension to destabilize the rest of the region.

■ Explain the role of oil and natural gas production in generating extremely uneven levels of economic and social development across Central Asia.

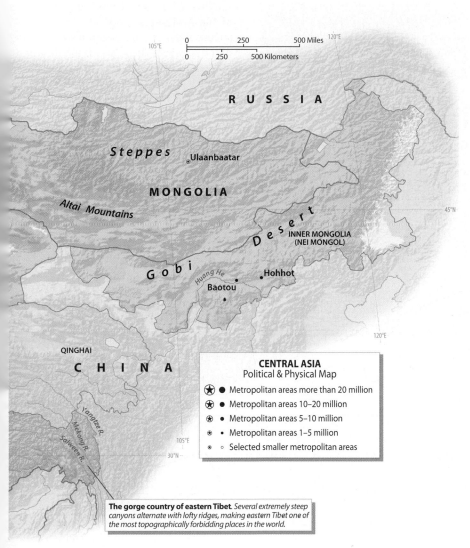

CENTRAL ASIA
Political & Physical Map
- ⊛ ● Metropolitan areas more than 20 million
- ⊛ ● Metropolitan areas 10–20 million
- ⊛ ● Metropolitan areas 5–10 million
- ⊛ • Metropolitan areas 1–5 million
- ⊛ ○ Selected smaller metropolitan areas

The gorge country of eastern Tibet. *Several extremely steep canyons alternate with lofty ridges, making eastern Tibet one of the most topographically forbidding places in the world.*

smuggling networks pass through their territories, undermining their security. But considering Afghanistan's feeble economy and meager infrastructure, few crops can compete with opium and marijuana. As a result, Afghanistan will likely remain the center of a highly globalized underground economy based on drug production.

Central Asia, located near the middle of the Eurasian supercontinent, is the only landlocked world region (Figure 10.1). As such, it is often viewed as relatively isolated from the global economic system. Such a picture, however, is now out-of-date. Afghanistan participates in the world economy mostly through the export of illegal drugs, but nearby countries have found their own global connections. Over much of the region, oil and natural gas are the main export commodities, and both Azerbaijan and Kazakhstan are now undergoing major energy booms. Elsewhere, other minerals are being mined in vast quantities. Mongolia, for example, is emerging as a major copper and coal producer. Across Central Asia, such mineral-driven engagement with the world economy brings both benefits and problems.

Defining Central Asia is not an easy matter. Most authorities agree that the region includes five former Soviet republics: Kazakhstan, Kyrgyzstan, Uzbekistan, Tajikistan, and Turkmenistan. This chapter adds another former Soviet state, Azerbaijan, as well as Mongolia and Afghanistan. In addition, the autonomous regions of western China (Tibet and Xinjiang) are counted as parts of both Central Asia and East Asia, and several other parts of western China, such as Nei Mongol (Inner Mongolia), are occasionally discussed in this chapter.

The inclusion of these additional territories in Central Asia is controversial. Azerbaijan is often classified with its neighbors in the Caucasus (Georgia and Armenia), western China fully belongs to East Asia on political grounds, and Mongolia is also often placed within East Asia because of its location and its historical connections with China. Afghanistan can also be located within either South Asia or Southwest Asia.

However, there are solid reasons for defining Central Asia as we have. The region has deep historical bonds and similar environmental and economic conditions. Azerbaijan, for example, is linked by both cultural and economic factors more to Central Asia than it is to its neighbors, Armenia and Georgia. Central Asia is also increasingly seen as a geopolitical unit, as its various countries face similar political challenges. But it is also not clear whether Central Asia will remain a coherent world region. Continuing Han Chinese migration into Tibet and especially into Xinjiang, for example, increasingly places these areas in an East Asian cultural framework.

ENVIRONMENTAL GEOGRAPHY: STEPPES, DESERTS, AND THREATENED LAKES

One of the great environmental tragedies of the late 20th century was the destruction of the Aral Sea, a vast, salty lake (until recently, larger than Lake Michigan) located on the boundary of Kazakhstan and Uzbekistan. The Aral's only sources of water are the Amu Darya and Syr Darya rivers, which flow out of the distant Pamir Mountains. Both of these rivers have been intensively used for irrigation for thousands of years, but the scale of water diversion greatly expanded after 1950. The valleys of the two rivers were some of the only areas in the Soviet Union where such warm-season crops as rice and cotton could be

P anjwai District in Afghanistan's southwestern province of Kandahar has been a difficult place for the U.S. military and its coalition allies. This arid area, where desolate stretches of desert alternate with lush irrigated fields, is regarded as the "spiritual home" of the Taliban, the radical Islamist group fighting to take over Afghanistan. Due to the Afghan surge initiated by the United States in 2009, the presence of coalition fighters in Panjwai greatly increased. In patrolling the area, troops have made some unusual discoveries. In early 2012, for example, U.S. and Afghan forces discovered a cache of 60,000 pounds (27,000 kilograms) of marijuana, one of the largest drug seizures in history. Coalition soldiers in Afghanistan are used to dealing with marijuana. In 2006, Canadian troops fought a difficult battle with the Taliban in the middle of a "forest" of 10-foot-high cannabis plants. The troops actually camouflaged one of their armored cars with marijuana leaves and branches.

Marijuana is not the only illegal drug cultivated in Afghanistan. The war-torn country is much better known for its opium growing and heroin production; as much as 90 percent of the world's opium comes from here, accounting for about a third of the total Afghan economy. Neighboring countries are angered by the situation, as drug

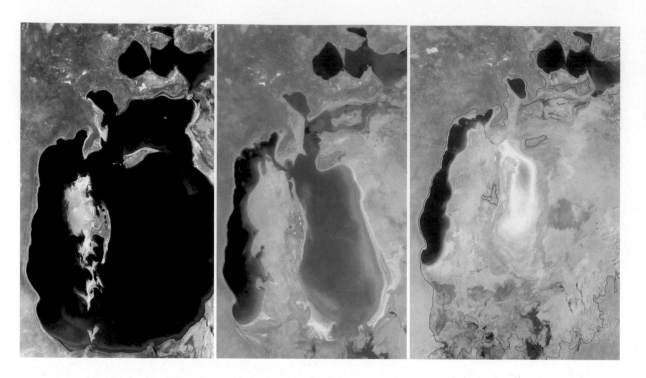

FIGURE 10.2 The Shrinking of the Aral Sea These satellite images show the steady shrinkage of the Aral Sea. The Aral was a massive lake as recently as the 1970s. It is now divided into several much smaller lakes.

grown. Soviet agricultural planners favored huge engineering projects to deliver water to arid lands to "make the deserts bloom."

Unfortunately, more water delivered to produce crops meant less freshwater for the Aral Sea. As the inflow of water was reduced, the shallow Aral began to recede. With less freshwater flowing into the lake, it also grew increasingly salty, destroying fish stocks. New islands began to emerge, and the Aral Sea split into two separate lakes in 1987 and then into three lakes in the early 2000s (Figure 10.2).

The destruction of the Aral Sea has resulted in economic and cultural damage, as well as ecological devastation. Fisheries formerly employing 40,000 workers closed down, and agriculture has suffered. The retreating lake left large salt flats on its exposed beds. Windstorms pick up the salt, along with the agricultural chemicals that had accumulated in the lake's shallows, and deposit it in nearby fields. As a result, farm yields have declined, desertification has accelerated, public health has been threatened, and the local climate has become colder in the winter and hotter in the summer.

Efforts to save what is left of the Aral Sea are currently focused on the small northern remnant. A series of dikes and dams, financed by the World Bank and the government of Kazakhstan, has managed to raise water levels over 26 feet (8 meters). Improved water quality enabled the partial revival of wildlife, and by 2009 Aral fish were once again being exported (Figure 10.3). However, the southern Aral Sea, the largest part of the basin, has seen no such recovery. By 2011, the southeastern lake had virtually disappeared.

Other Major Environmental Issues

Despite the tragedy of the Aral Sea, much of Central Asia has a relatively clean environment, largely due to its generally low population density. Industrial pollution, however, is a serious problem in the larger cities, such as Tashkent (in Uzbekistan) and Baku (in Azerbaijan). Elsewhere, the typical environmental problems of

arid environments plague the region: desertification (the spread of deserts resulting from poor land-use practices), salinization (the accumulation of salt in the soil), and desiccation (the drying up of lakes and wetlands) (Figure 10.4).

Desertification Desertification, caused by overgrazing and poor farming practices, is a major concern in Central Asia. In the eastern part of the region, the Gobi Desert has gradually spread southward, encroaching on densely settled lands in northeastern China proper. The Chinese have tried to prevent the march of desert with massive tree- and grass-planting campaigns, as the roots of such plants stabilize the soil and help to keep sand dunes from moving. Such efforts, however, have been only partially successful.

Shrinking and Expanding Lakes Western Central Asia contains several large lakes because it forms a low-lying basin, without drainage to the ocean, surrounded by mountains and other more humid areas. The world's largest lake, by a huge margin, is the Caspian Sea, located along the region's western boundary. The 4th largest was the Aral Sea, and the 15th largest is Lake Balqash, found 600 miles (960 kilometers) east of the Aral Sea in southeastern Kazakhstan. Like the Aral Sea, Lake Balqash has become smaller and saltier over the past several decades.

The story of the Caspian Sea is more complicated. The Caspian receives most of its water from the large rivers of the north—namely, the Ural and the Volga, which drain much of European Russia. Due to extensive irrigation development in the lower Volga Basin, the volume of freshwater reaching the Caspian began to decline in the second half of the 20th century. With a reduced flow of water, the water level dropped, exposing large expanses of the former lakebed. Decreased water volume and increased salinity disrupted the ecosystem and devastated fisheries. The Russian caviar industry, centered in the northern Caspian, was particularly damaged.

FIGURE 10.3 The Northern Aral Sea Due to water diversions from the Amu Darya and Syr Darya rivers, the Aral Sea has lost most of its volume and now forms two separate lakes. The government of Kazakhstan has recently built dikes to protect the water of the smaller northern Aral, which is fed by the flow of the Syr Darya River.

The water level of the Caspian Sea reached a low point in the late 1970s. At that point, it began to rise, probably because of higher-than-normal precipitation in its drainage basin. By the late 1990s, it had risen some 8.2 feet (2.5 meters). This enlargement, too, has caused problems, such as the flooding of some of the newly reclaimed farmlands in the Volga Delta. At present, the most serious environmental threat to the Caspian is pollution from the oil industry, rather than fluctuation in size.

Dam Building and Water Conflicts Roughly 80 percent of the freshwater in western Central Asia is controlled by Kyrgyzstan and Tajikistan, the two most mountainous countries of the region. Both countries are engaged in massive dam-building projects designed to control water resources and generate electricity. The Rogun Dam in Tajikistan, currently under construction with Russian aid, could be the tallest dam in the world when completed. Uzbekistan, Kazakhstan, and Turkmenistan object to the construction of these dams, fearing that Kyrgyzstan and Tajikistan will release too much water during flood periods and withhold too much during dry spells. Leaders from the five countries have met repeatedly to try to reach an agreement on water sharing, but so far they have been unable to do so. Individual countries in the region, however, have been able to enact plans that help preserve their own water resources (see "Working Toward Sustainability: The Issyk Kul Project").

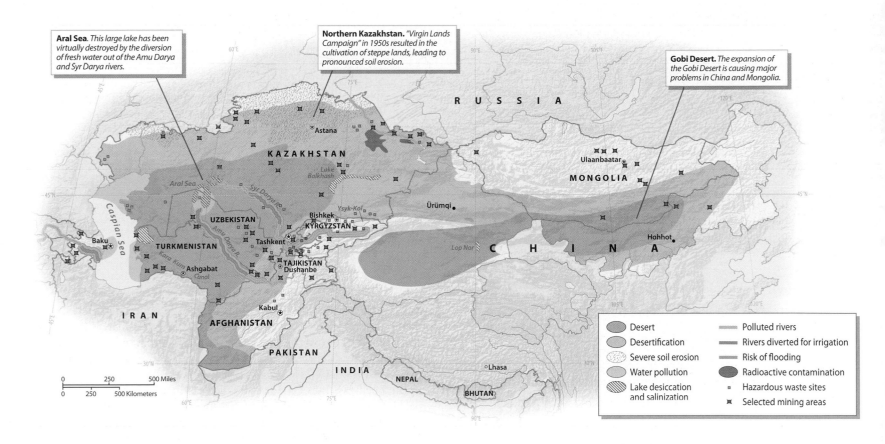

FIGURE 10.4 Environmental Issues in Central Asia Desertification is perhaps more widespread in Central Asia than in any other world region. Soil erosion and overgrazing have led to the advance of desert-like conditions in much of western China and Kazakhstan. In western Central Asia, the most serious environmental problems are associated with the diversion of river water for irrigation and the corresponding desiccation of lakes.

WORKING TOWARD SUSTAINABILITY

The Issyk Kul Project

ssyk Kul in Kyrgyzstan is the world's second largest alpine lake (after Lake Titicaca in South America), measuring 113 miles (182 km) long and 37 miles (60 km) wide (Figure 10.1.1). It dwarfs the much better-known Lake Tahoe of the United States (which measures 22 miles by 12 miles). Issyk Kul is well known in Central Asia, however, as it was once a major stop on the Silk Road and is now a significant tourist destination. (The name Issyk Kul meaning "hot lake" derives from the fact that the lake never freezes, due to its salinity.) The scenery is beautiful, the local people are hospitable, and both the lake and the adjacent mountains are rich in

FIGURE 10.1.1 Lake Issyk Kul Issyk Kul, the world's second largest alpine lake, is Kyrgyzstan's major tourist attraction. The Kyrgyz government is working hard to protect this threatened lake.

wildlife. Tourism is said to account for approximately 10 percent of Kyrgyzstan's gross domestic product, and most international tourists who visit the country head to Issyk Kul. Most of the people living along the lakeshore derive their incomes from catering to tourists.

Despite its great size, Issyk Kul is highly vulnerable to environmental degradation, which, if left unchecked, would threaten the tourist economy. Issyk Kul does not drain to the sea, so whatever flows into the lake tends to stay there. Mining and deforestation in the surrounding hills and mountains threaten water quality. Fish species introduced into the lake from elsewhere have driven several of the endemic species (those found nowhere else) to the brink of extinction. Poorly treated sewage is also a major problem, particularly near the towns where tourists congregate.

Owing to such threats, the government of Kyrgyzstan has been working hard to preserve the lake. The Issyk Kul Sustainable Development Project has been its main response. A joint project of Kyrgyzstan's Ministry of Economic Development and Trade and the Asia Development Bank, the project is focused on improving the water and sanitation infrastructure

in the main towns of the basin. The Issyk Kul area is also the site of a major biosphere reserve. Since 1995, the German government has been working with local authorities to strengthen the reserve, yet at the same time encourage sustainable tourism. Small-scale projects associated with this effort help local farmers and herders develop more sustainable practices. In the mountains, conservation efforts have focused on such threatened species as the Marco Polo sheep, the Siberian ibex, and the snow leopard.

In early 2012, a new sustainable tourism development program was announced by the Kyrgyz government: the construction of a 300-mile bicycle route all the way around the lake. A youth group called the Bai Issyk Kul Foundation came up with the idea as a way to enhance tourism, yet still protect the environment. The estimated US$1 million project will be financed by Kyrgyzstan's Ministry of Labor and Employment and Ministry of Youth Labor, although it is hoped that international investors will also contribute funds.

Admittedly, some of the recent developments in Issyk Kul will not enhance tourism or protect the environment. Since Soviet times, the lake has been used for testing naval technologies. In 2007, Russia signed an agreement with the Kyrgyz government allowing it to test sophisticated torpedoes in the lake. In 2011, India announced that it would also engage in torpedo research at the same facility.

Central Asia's Physical Regions

To understand why Central Asia experiences such great conflicts over water, it is necessary to examine the region's physical geography. In general, Central Asia is characterized by high plateaus and mountains in the south-central and southeastern areas, grassland plains (or steppes) in the northern area, and desert basins in the southwestern and central areas.

The Central Asian Highlands The highlands of Central Asia originated in one of the great geological events of Earth's history: the collision of the Indian subcontinent with the Asian mainland. This ongoing impact has created the world's highest mountains, the Himalayas, located along the boundary of South Asia and Central Asia. To the northwest, the Himalayas merge with the Karakoram Range and

then the Pamir Mountains. From the so-called Pamir Knot—a complex tangle of mountains located where Pakistan, Afghanistan, China, and Tajikistan meet—towering ranges spread outward in several directions. The Hindu Kush curves to the southwest through central Afghanistan, the Kunlun Shan extends to the east, and the Tien Shan swings out to the northeast into China's Xinjiang province. All these ranges have peaks higher than 20,000 feet (6,000 meters) in elevation.

Much more extensive than these mountain ranges is the Tibetan Plateau (Figure 10.5). This massive upland extends some 1,250 miles (2,000 kilometers) from east to west and 750 miles (1,200 kilometers) from north to south. Its elevation is as remarkable as its size; almost the entire area is more than 12,000 feet (3,700 meters) above sea level. Most of the Tibetan Plateau lies near the maximum elevation at which human life can exist. Rather than forming a flat surface, the plateau has numerous east–west mountain ranges alternating with basins.

FIGURE 10.5 Tibetan Plateau Alpine grasslands and tundra, interspersed with rugged mountains and saline lakes, dominate the Tibetan Plateau. In summer, the sparse vegetation offers forage for the herds of nomadic Tibetan pastoralists. Much of northern Tibet, however, is too high to support pastoralism and is therefore uninhabited.

Although the southeastern sections of the plateau receive adequate rainfall, most of Tibet is arid (Figure 10.6). Winters on the Tibetan Plateau are cold; and while summer afternoons can be warm, summer nights remain chilly (note the climograph for Lhasa).

The Plains and Basins Although the mountains of Central Asia are higher and more extensive than any others in the world, most of the region is characterized by plains and basins. This lower-lying zone can be divided into two main areas: a central belt of deserts and a northern strip of semiarid steppes.

The Tien Shan and Pamir Mountains divide Central Asia's desert belt into two separate segments. To the west lie the arid plains of the Caspian Sea and Aral Sea basins, located primarily in Turkmenistan, Uzbekistan, and southern Kazakhstan (Figure 10.7). The driest areas support little vegetation and contain extensive sand dunes. The climate of this region is continental: summers are dry and hot, while winter temperatures average well below freezing (see the climographs for Tashkent and Almaty in Figure 10.6). Central Asia's eastern desert belt extends for almost 2,000 miles (3,200 kilometers) from far western China at the foot of the Pamirs to the southeastern edge of Inner Mongolia. Two major deserts lie here: the Taklamakan, in the Tarim Basin of Xinjiang, and the Gobi, which runs along the border between Mongolia and the Chinese region of Inner Mongolia.

North of the desert zone, rainfall gradually increases, and desert eventually gives way to the grasslands, or steppes, of northern Central Asia. Near the region's northern boundary, trees begin to appear, outliers of the Siberian taiga (coniferous forest) of the north. Nearly continuous grasslands extend some 4,000 miles (6,400 kilometers) east to west across the entire region. Summers on the northern steppe are usually pleasant, but winters can be extremely cold.

Climate Change and Central Asia

Most climate experts expect Central Asia to be hard hit by global warming. The Tibetan Plateau has already seen marked increases in temperature, resulting in the reduction of permafrost and the retreat of mountain glaciers. The retreat of Central Asia's glaciers is especially worrisome because of their role in providing dependable flows of water for the region's rivers. In some areas, the melting of ice has resulted in the temporary flooding of lowland basins, but the long-term result will be to reduce freshwater resources in an already dry region. Climate change is also likely to reduce precipitation in the arid lowlands of western Central Asia. Prolonged and devastating droughts have recently struck Afghanistan, indicating a possible shift to a drier climate. As a result, the UN Intergovernmental Panel on Climate Change has predicted a 30 percent crop decline for Central Asia as a whole by the middle of the 21st century. A 2012 report linked warmer weather to the spread of new strains of wheat rust, which have already reportedly reduced wheat yields in parts of the region.

However, as is true elsewhere in the world, global warming will not affect all parts of Central Asia in the same way. Some areas, including the Gobi Desert and the Tibetan Plateau, could see increased precipitation. Some global climate models also indicate increased precipitation in the boreal forest zone of Siberia that extends into the mountains of northern Mongolia. But because the rivers coming out of these mountains generally flow north, their growth would provide few benefits for the rest of Central Asia.

REVIEW QUESTIONS

1. Why does Central Asia have such large lakes, and why are many of these lakes so deeply threatened?

2. How does the location of Central Asia, near the center of the world's largest landmass and at the junction of several large tectonic plates, influence the region's climate and landforms?

POPULATION AND SETTLEMENT: DENSELY SETTLED OASES AMID VACANT LANDS

Most of Central Asia is sparsely populated (Figure 10.8). Large areas are either too arid or too high to support much human life. Even many of the most favorable areas are populated only by widely scattered groups of nomadic **pastoralists** (people who raise livestock for subsistence purposes). Mongolia, which is more than twice the size of Texas, has only 2.9 million inhabitants—fewer than live in the Dallas metropolitan area. But as is common in arid environments, those few lowland settings with good soil and dependable water supplies are thickly settled.

Highland Population and Subsistence Patterns

The environment of the Tibetan Plateau is particularly harsh. Only sparse grasses and herbaceous plants can survive in this high-altitude climate, making human subsistence difficult. The only feasible way of life in this area is nomadic pastoralism based on the yak, an altitude-adapted relative of the cow. Several hundred thousand people manage to make a living in such a manner, roaming with their herds over vast distances.

Although most of the Tibetan Plateau can support only nomadic pastoralism, the majority of Tibetans are sedentary farmers. Farming in Tibet is possible only in the few locations that are *relatively* low in elevation and that have good soils and either adequate rainfall or a dependable irrigation system based on local streams. The main zone of sedentary settlement lies in the far south, where protected valleys offer these favorable conditions.

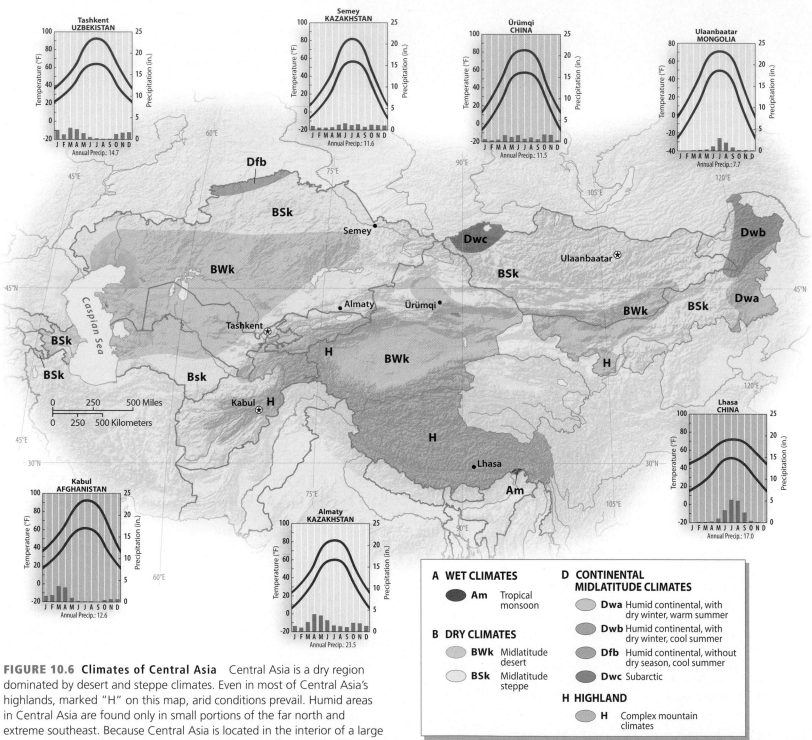

FIGURE 10.6 Climates of Central Asia   Central Asia is a dry region dominated by desert and steppe climates. Even in most of Central Asia's highlands, marked "H" on this map, arid conditions prevail. Humid areas in Central Asia are found only in small portions of the far north and extreme southeast. Because Central Asia is located in the interior of a large continent, its climate is characterized by significant differences between winter and summer temperatures.

A WET CLIMATES

Am Tropical monsoon

B DRY CLIMATES

BWk Midlatitude desert

BSk Midlatitude steppe

D CONTINENTAL MIDLATITUDE CLIMATES

Dwa Humid continental, with dry winter, warm summer

Dwb Humid continental, with dry winter, cool summer

Dfb Humid continental, without dry season, cool summer

Dwc Subarctic

H HIGHLAND

H Complex mountain climates

FIGURE 10.7 Central Asian Desert Much of Central Asia is dominated by deserts and other arid lands. The Kara Kum Desert in Turkmenistan, seen here, is especially dry, supporting little vegetation.

Throughout Central Asia, the mountainous areas are vitally important for people living in the adjacent lowlands, whether they are migratory pastoralists or settled farmers. Many herders use the highlands for summer pasture; when the lowlands are dry and hot, the high meadows provide rich grazing. The Kyrgyz (of Kyrgyzstan) are noted for their traditional economy based on **transhumance**, moving their flocks from lowland pastures in the winter to highland meadows in the summer.

Lowland Population and Subsistence Patterns

Most of the inhabitants of Central Asian deserts live in the narrow belt where the mountains meet the basins and plains. Here water supplies are adequate and soils are neither salty nor alkaline, as is often the case in the basin interiors. For example, the population distribution pattern of China's Tarim Basin forms a ring-like structure (Figure 10.9). Streams flowing out of the mountains are diverted to irrigate fields and orchards in the fertile band along the basin's edge.

The population of former Soviet Central Asia is also concentrated in the transitional zone between the highlands and the plains. A series of **alluvial fans** (fan-shaped deposits of sediments dropped by

FIGURE 10.8 Population Density in Central Asia Central Asia as a whole remains one of the world's most sparsely populated regions, although it does contain distinct clusters of higher population density. Most of Central Asia's large cities are located near the region's periphery or in its major river valleys.

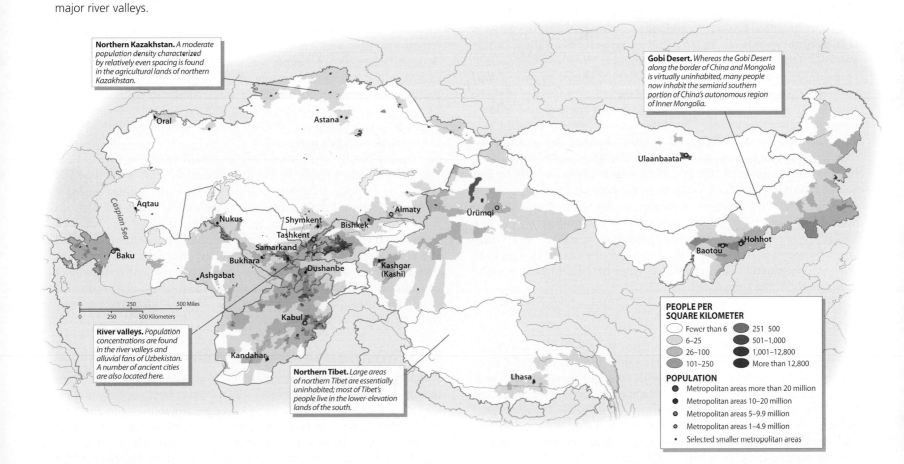

Northern Kazakhstan. *A moderate population density characterized by relatively even spacing is found in the agricultural lands of northern Kazakhstan.*

Gobi Desert. *Whereas the Gobi Desert along the border of China and Mongolia is virtually uninhabited, many people now inhabit the semiarid southern portion of China's autonomous region of Inner Mongolia.*

River valleys. *Population concentrations are found in the river valleys and alluvial fans of Uzbekistan. A number of ancient cities are also located here.*

Northern Tibet. *Large areas of northern Tibet are essentially uninhabited; most of Tibet's people live in the lower-elevation lands of the south.*

PEOPLE PER SQUARE KILOMETER

- Fewer than 6
- 6–25
- 26–100
- 101–250
- 251–500
- 501–1,000
- 1,001–12,800
- More than 12,800

POPULATION

- Metropolitan areas more than 20 million
- Metropolitan areas 10–20 million
- Metropolitan areas 5–9.9 million
- Metropolitan areas 1–4.9 million
- Selected smaller metropolitan areas

FIGURE 10.9 Population Patterns in Xinjiang's Tarim Basin The central portion of the Tarim Basin is a nearly uninhabited expanse of sand dunes and salt flats. Along the edge of the basin, however, dense agricultural and urban settlements are located where streams running out of the surrounding mountains allow for intensive irrigation. The largest of these oasis communities are found along the southwestern fringe of the basin.

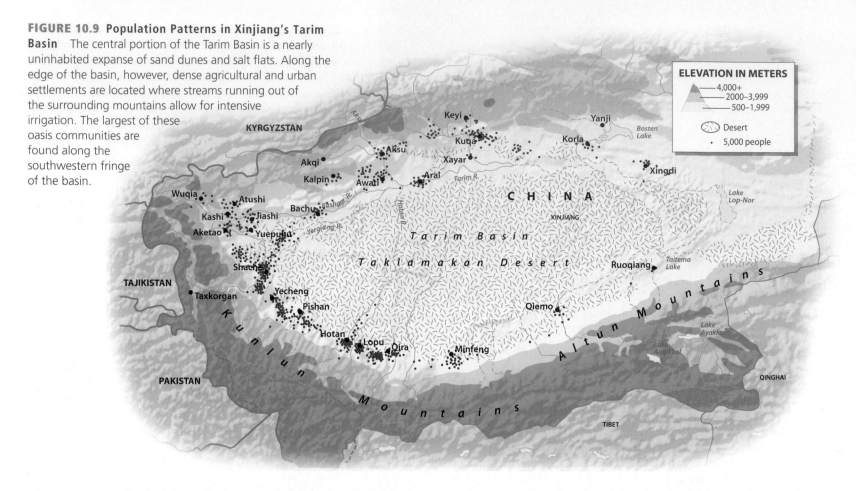

streams flowing out of the mountains) has long been devoted to intensive cultivation. **Loess**, a fertile, silty soil deposited by the wind, is also plentiful in this region. In addition, several large valleys in this area offer fertile and easily irrigated farmland. The densely populated Fergana Valley of the upper Syr Darya River is shared by three countries: Uzbekistan, Kyrgyzstan, and Tajikistan.

The steppes of northern Central Asia are the classical land of nomadic pastoralism. Until the 20th century, almost none of this area had ever been plowed and farmed. To this day, pastoralism remains a common way of life across the grasslands, particularly in Mongolia (Figure 10.10). In northwestern China and the former Soviet republics, however, many pastoral peoples have been forced to adopt sedentary lifestyles, and even in Mongolia the number of nomadic herders is steadily decreasing. In northern Kazakhstan, the Soviet regime converted the most productive pastures into wheat farms in the mid-1900s in order to increase the country's supply of grain. Consequently, northern Kazakhstan has the highest population density in the steppe region.

Population Issues

Although Central Asia has a low population density overall, some portions of it are growing at a moderately rapid pace. In western China, much of the population growth over the past 30 years has stemmed from the migration of Han Chinese into the area. In 2006, China completed the construction of a railway from Beijing to the Tibetan capital of Lhasa, which has increased the flow of migrants and tourists into the region. The trains pass through such high elevations that passengers are supplied with supplemental oxygen.

Most of the former Soviet zone of Central Asia is experiencing moderate population expansion due to its own fertility patterns. In contrast to Tibet and Xinjiang, this area has witnessed a substantial outward migration of people since 1991. At first, emigrants were mostly ethnic Russians returning to the Russian homeland. Later, as the Russian economy boomed after 2000, hundreds of thousands of men from Central Asia sought work in Russia. The economic collapse of 2008–2009 forced many of these workers back home, but by 2011 the movement had resumed.

Human fertility patterns vary substantially from one part of Central Asia to another (Table 10.1). Afghanistan, the least-developed and most male-dominated country of the region, has the highest birthrate by a wide margin. Fertility rates are in the middle range through out most of the former Soviet area. In Kazakhstan, the birthrate was well below the replacement rate as recently as 2000, but it has rebounded and is now over 2.5. Fertility here is much lower for ethnic Russians than it is for the region's indigenous peoples. Azerbaijan's fertility rate, in contrast, is only slightly above replacement level; as a result, the population pyramids of Azerbaijan and Afghanistan make an especially striking contrast (Figure 10.11).

Urbanization in Central Asia

Although the steppes of northern Central Asia had no real cities before the modern age, the river valleys have been partially urbanized for thousands of years. Such cities as Samarkand and Bukhara in Uzbekistan have long been famous for their lavish architecture (Figure 10.12). These cities contrast sharply with those built during the period of Russian/Soviet rule. Tashkent in Uzbekistan, for

FIGURE 10.10 Mongolian Grasslands   Many Mongolians still make their livings as migratory herders, roaming over large areas of the great Mongolian grasslands. Such people typically live in collapsible tents known as gers (or yurts), two of which are visible in this photograph.

example, is largely a Soviet creation and thus looks quite different from the more traditional Bukhara. Several major cities, such as Kazakhstan's former capital of Almaty, were minor settlements before Russian colonization. In cities throughout the former Soviet region, you can see the effects of centralized Soviet urban planning and design.

Parts of Central Asia have experienced substantial urbanization in recent decades. North-central Kazakhstan has recently seen the rise of a new major city, Astana, designated as the national capital due to its central location (Figure 10.13). Even Mongolia now has more people living in cities than in the countryside. Its capital city, Ulaanbaatar (Ulan Bator), is now in the middle of a major building boom. In the highlands of Central Asia, however, cities remain relatively few and far between. Only about one-quarter of the people of Tajikistan, for example, are urban residents. Tibet similarly remains a predominantly rural society, although change is occurring as Han Chinese migrants move into its cities.

Table 10.1   POPULATION INDICATORS

Country	Population (millions) 2012	Population Density (per square kilometer)	Rate of Natural Increase (RNI)	Total Fertility Rate	Percent Urban	Percent <15	Percent >65	Net Migration (Rate per 1000) 2010–15[a]
Afghanistan	33.4	51	2.8	6.2	23	46	2	4.0
Azerbaijan	9.3	107	1.3	2.3	53	22	6	0.0
Kazakhstan	16.8	6	1.4	2.6	55	25	7	−0.6
Kyrgyzstan	5.7	28	2.0	2.7	35	30	4	−6.4
Mongolla	2.9	2	1.6	2.5	63	27	4	−1.0
Tajikistan	7.1	49	2.3	3.3	26	37	3	−7.0
Turkmenistan	5.2	11	1.4	2.4	47	29	4	−1.0
Uzbekistan	29.8	29.8	1.9	2.6	51	29	4	2.9

[a]Net Migration Rate from the United Nations, Population Division, *World Population Prospects: The 2010 Revision Population Database*.
Source: Population Reference Bureau, *World Population Data Sheet, 2012*.

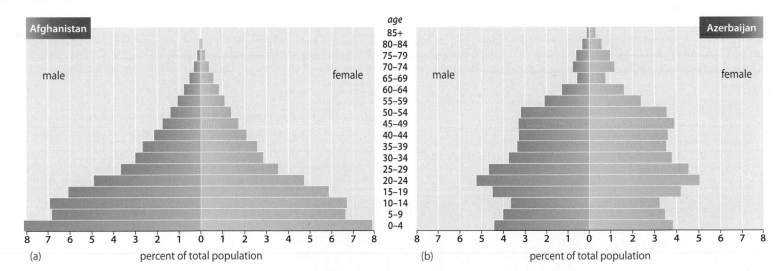

FIGURE 10.11 Steppe Pastoralism   Azerbaijan has a relative balanced population pyramid, reflecting the fact that it is close to population stability. Azerbaijan's birthrate has increased in recent year, however, as is indicated by the wider bar for ages "0-4" than for "10-14" Note also that women outlive men in Azerbaijan, which is the normal pattern. In Afghanistan, in contrast, high birthrates coupled with high death rates have created a more triangular pyramid. In this poor, male-dominated country, elderly men and elderly women are equally few in numbers.

FIGURE 10.12 Traditional Architecture in Samarkand Samarkand, Uzbekistan, is famous for its lavish Islamic architecture, some of it dating back to the 1400s. The city owes its rich architectural heritage in part to the fact that it was the capital of the empire created by the great medieval conqueror Tamerlane.

New development projects, coupled with Han migration, are transforming cities throughout Chinese Central Asia. Local controversies often follow. Beijing recently announced that the old city of Kashgar (Kashi) in Xinjiang would be largely demolished, due mainly to concerns about earthquake safety. Such actions, however, would undermine tourism, which China wants to encourage. Local activists are also struggling to retain as much of the city's traditional character as possible. An international petition drive is currently underway to have the UN Educational, Scientific and Cultural Organization declare central Kashgar a World Heritage Site in order to help preserve its old buildings.

REVIEW QUESTIONS

1. Why are large parts of Central Asia so sparsely settled, yet others have dense populations?
2. Why is the urban environment of Central Asia changing so rapidly?

FIGURE 10.13 Astana, Capital of Kazakhstan Kazakhstan's new capital city of Astana has grown rapidly in recent years, emerging as a new metropolis near the middle of the vast country. Oil wealth has supported Astana's expansion.

CULTURAL COHERENCE AND DIVERSITY: A MEETING GROUND OF DIFFERENT TRADITIONS

Although large areas of Central Asia are environmentally similar, the region's cultural traditions are somewhat more diverse. Western Central Asia is largely Muslim and is often classified as part of Southwest Asia, but in Mongolia and Tibet most people traditionally follow Tibetan Buddhism. Tibet is culturally linked to both South and East Asia, and Mongolia is historically associated with China, but neither fits easily within any other world region.

Historical Overview: Changing Languages and Populations

The river valleys and oases of Central Asia were early sites of agricultural communities. Archaeologists have discovered abundant evidence of farming villages dating back to the Neolithic period (beginning around 8000 BCE) in the Amu Darya and Syr Darya valleys. After the domestication of the horse around 4000 BCE, nomadic pastoralism, based on the raising of livestock, emerged in the steppe belt. Eventually, pastoral peoples gained power over the entire region.

The earliest recorded languages of Central Asia belonged to the Indo-European linguistic family, associated with the people who first domesticated the horse. These languages were replaced on the steppe more than 1,000 years ago by languages in the Altaic family (which includes Turkish and Mongolian). In the river valley communities as well, Turkic languages gradually began to replace the Indo-European tongues (which were closely related to Persian) as Turkic power spread through out most of Central Asia.

Contemporary Linguistic and Ethnic Geography

Today people speaking Turkic and Mongolian languages inhabit most of Central Asia (Figure 10.14). A few native Indo-European languages are confined to the southwest, while Tibetan remains the main language of the plateau. Russian is also widely spoken in the west, often as a second or third language, while Chinese is increasingly important in the east.

Tibetan Tibetan is usually placed in the Sino-Tibetan language family, suggesting a shared linguistic history between the Chinese and the Tibetan peoples. Some students of Tibetan, however, argue that no definite relationship between the two languages has ever been established. The Tibetan language is divided into several dialects that are spoken over almost the entire Tibetan Plateau. Over 90 percent of the 3 million people who live in the Tibet Autonomous Region speak Tibetan, while another 3 million or so Tibetan speakers live in the highland portions of the Chinese provinces of Qinghai and Sichuan.

Mongolian The Mongolian language includes a cluster of closely related dialects spoken by approximately 5 million people. The standard Mongolian of both the independent country of

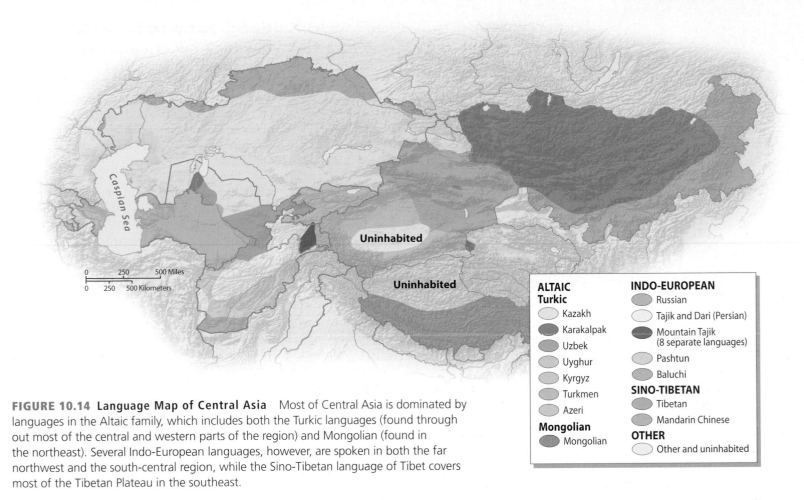

FIGURE 10.14 Language Map of Central Asia Most of Central Asia is dominated by languages in the Altaic family, which includes both the Turkic languages (found through out most of the central and western parts of the region) and Mongolian (found in the northeast). Several Indo-European languages, however, are spoken in both the far northwest and the south-central region, while the Sino-Tibetan language of Tibet covers most of the Tibetan Plateau in the southeast.

Legend:

ALTAIC
Turkic
- Kazakh
- Karakalpak
- Uzbek
- Uyghur
- Kyrgyz
- Turkmen
- Azeri

Mongolian
- Mongolian

INDO-EUROPEAN
- Russian
- Tajik and Dari (Persian)
- Mountain Tajik (8 separate languages)
- Pashtun
- Baluchi

SINO-TIBETAN
- Tibetan
- Mandarin Chinese

OTHER
- Other and uninhabited

Mongolia and China's Inner Mongolia is called *Khalkha*. Mongolian has its own distinctive script, but Mongolia itself adopted the Cyrillic alphabet of Russia in 1941. Efforts are now being made in Mongolia to revive the old script, which is still used by Mongolian speakers in northern China.

Mongolian speakers form about 90 percent of the population of Mongolia. In China's Inner Mongolia Autonomous Region, Han Chinese who migrated to the area over the past 50 years today outnumber Mongolian speakers. Currently, only about 17 percent of the 25 million residents of Inner Mongolia are Mongolian.

Turkic Languages Far more Central Asians speak Turkic languages than Mongolian and Tibetan combined. The Turkic linguistic sphere extends from Azerbaijan in the west through China's Xinjiang province in the east.

Five of the six countries of the former Soviet Central Asia—Azerbaijan, Uzbekistan, Turkmenistan, Kyrgyzstan, and Kazakhstan—are named after the Turkic languages of their dominant populations. In Azerbaijan, 90 percent of the people speak Azeri, whereas 74 percent of the people of Uzbekistan speak Uzbek, and 72 percent of the people of Turkmenistan speak Turkmen. With more than 24 million speakers, Uzbek is the most widely spoken Central Asian language. In the Amu Darya Delta in the far north of Uzbekistan, however, most people speak a different Turkish language called *Karakalpak*, while Kazakh speakers are found in the sparsely populated Uzbek deserts. Roughly 70 percent of the inhabitants of Kyrgyzstan speak Kyrgyz

as their native language, while around 63 percent of the people of Kazakhstan speak Kazakh. In Kazakhstan, more than 25 percent of the population speaks European languages, particularly Russian and Ukrainian. In general, the Kazakhs predominate in the center and south of the country, while Russians live in the agricultural districts of the north, as well as in the cities of the southeast.

Uyghur, spoken in China's Xinjiang province, dates back almost 2,000 years. The Uyghur people number about 11 million, most of whom live in Xinjiang (Figure 10.15). As recently as 1953, the Uyghur formed about 80 percent of the population of Xinjiang. By 2000, however, that number had been reduced to 45 percent, mostly due to Han Chinese migration. According to official statistics, the Han Chinese form around 40 percent of Xinjiang's population, but Uyghur activists think that the number is much larger. There are also about 1 million Kazakh speakers living in Xinjiang.

Linguistic Complexity in Tajikistan and Afghanistan The sixth republic of the former Soviet Central Asia, Tajikistan, is dominated by people who speak an Indo-European language. Tajik is so closely related to Persian that it is often considered to be a Persian (or Farsi) dialect. Roughly 80 percent of the people of Tajikistan speak Tajik as their first language. In the remote mountains of eastern Tajikistan, people speak a variety of distinctive Indo-European languages, sometimes collectively referred to as "Mountain Tajik." The remainder of Tajikistan's population speaks either Uzbek or one of a variety of minor languages.

FIGURE 10.15 **Uyghur Mosque** The Uyghur people of Xinjiang in northwestern China speak a Turkic language and follow Sunni Islam. The Id Kah Mosque in the city of Kashgar, pictured here, is the largest mosque in China, accommodating up to 10,000 people.

Geography of Religion

At one time, Central Asia was noted for its religious complexity. Major overland trading routes crossed the region, giving easy access to both merchants and missionaries. By 1500, however, the region was divided into two spiritual camps: Islam predominated in the west and center and Tibetan Buddhism in Tibet and Mongolia.

Islam in Central Asia Different Central Asian peoples are known for their different interpretations of Islam. The Pashtuns of Afghanistan are noted for their strict Islamic ideals—although critics contend that many Pashtun rules, such as never allowing women's faces to be seen in public, are based more on ethnic than religious customs (Figure 10.17). The traditionally nomadic groups of the northern steppes, such as the Kazakhs and Kyrgyz, are considered to be more relaxed about religious beliefs and practices. While most of the region's Muslims are Sunnis, Shiism is dominant among the Hazaras of central Afghanistan and the Azeris of Azerbaijan.

Under the communist regimes in China, the Soviet Union, and Mongolia, all forms of religion were discouraged. Chinese authorities and student radicals attempted to suppress Islam in Xinjiang during the Cultural Revolution of the late 1960s and early 1970s. Chinese Muslims now enjoy basic freedom of worship, but the state still

The linguistic geography of Afghanistan is even more complex than that of Tajikistan (Figure 10.16). Afghanistan became a country in the 1700s under the leadership of the Pashtun people. The rulers of Afghanistan, however, never tried to build a nation-state around Pashtun identity, in part because approximately half of the Pashtun people live in what is now Pakistan. In Afghanistan itself, estimates of the amount of people speaking Pashtun average around 40 percent. Most Pashtun speakers in Afghanistan live to the south of the Hindu Kush mountain range.

Roughly half of the people of Afghanistan speak Dari, the local form of Persian. Dari speakers are concentrated in northern Afghanistan. Two separate ethnicities divide the Dari-speaking people. Those in the west and the far north are considered to be Tajik, whereas those in the central mountains are called *Hazaras* and are said to be descendants of Mongol conquerors who arrived in the 12th century. Finally, another 10 percent or so of the people of Afghanistan speak Turkic languages, mainly Uzbek.

FIGURE 10.16 **Afghanistan's Ethnic Patchwork** Afghanistan is one of the world's most ethnically complex countries. Its largest ethnic group is the Pashtuns, who live in most of the southern portion of the country as well as in the adjoining borderlands of Pakistan. Northern Afghanistan, however, is mostly inhabited by Uzbeks, Tajiks, and Turkmens—whose main population centers are located in Uzbekistan, Tajikistan, and Turkmenistan, respectively. The Hazaras of Afghanistan's central mountains, like the Tajiks, speak a form of Persian. However, the Hazaras are considered to be a separate ethnic group in part because they, unlike other Afghans, follow Shiite rather than Sunni Islam.

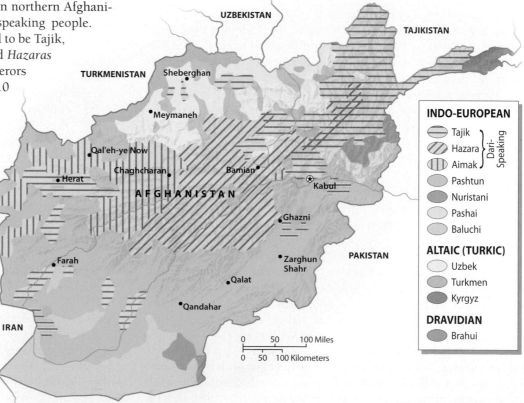

INDO-EUROPEAN
- Tajik ⎫
- Hazara ⎬ Dari-Speaking
- Aimak ⎭
- Pashtun
- Nuristani
- Pashai
- Baluchi

ALTAIC (TURKIC)
- Uzbek
- Turkmen
- Kyrgyz

DRAVIDIAN
- Brahui

FIGURE 10.17 Afghan Women in Public Especially in the Pashtun areas of Afghanistan, women have traditionally been forced to cover their entire bodies when in public areas. Extremely modest dress prevails across Afghanistan.

closely monitors religious expression out of fear that it will lead to resistance against Chinese rule. Periodic persecution of Muslims also occurred in Soviet Central Asia, and until the 1970s many observers thought that Islam would slowly disappear from the region.

Religious expression was not, however, so easily repressed. Interest in Islam began to grow in former Soviet Central Asia in the 1970s and 1980s. In the post-Soviet period, Islam continues to revive as people seek their cultural roots. In Xinjiang, Islam has served as a focus of social identity among the Uyghur people. Most Uyghur leaders, however, insist that their beliefs are not radically fundamentalist. Radical Islamic fundamentalism has emerged as a powerful political movement only in Afghanistan, parts of Tajikistan, and the Fergana Valley (mostly in Uzbekistan).

Tibetan Buddhism Mongolia and especially Tibet stand apart from the rest of Central Asia in their people's practice of Tibetan Buddhism, or Lamaism. Buddhism entered Tibet from India many centuries ago, and it merged with the native religion of the area, called *Bon*. The resulting mix is more oriented toward mysticism than are other forms of Buddhism, and it is more tightly organized. Standing at the head of Lamaist society is the Dalai Lama. Until the Chinese conquest, Tibet was essentially a **theocracy** (or religious state), with the Dalai Lama holding political as well as religious authority.

Tibetan Buddhists suffered persecution after 1959 when China invaded Tibet. Tibetan Buddhism has inspired many Tibetans to resist Chinese rule. The Dalai Lama, who fled Tibet for India in 1959, has been a powerful advocate for the Tibetan cause in international circles. During the 1960s and 1970s, an estimated 6,000 Tibetan Buddhist monasteries were destroyed; the number of active monks today is only about 5 percent of what it was before the Chinese occupation. The Chinese have more recently allowed many monasteries to reopen, but their activities remain limited and closely watched (Figure 10.18).

Central Asian Culture in International and Global Context

In cultural terms, western Central Asia is closely linked to Russia, whereas eastern Central Asia is more closely tied to China. In the east, the main cultural issue is

FIGURE 10.18 Tibetan Buddhist Monastery Tibet is well known for its large Buddhist monasteries, which at one time served as seats of political as well as religious authority. The Ganden Monastery, pictured here, is one of the three great "university monasteries" of Tibet.

FIGURE 10.19 U.S.-Style Restaurant in Baku Baku, in Azerbaijan, is a vibrant and increasingly cosmopolitan city closely linked to global economic and cultural networks. This coffee shop and wine bar caters to both local and international customers.

the migration of Han Chinese into the area, which has resulted in serious ethnic tensions. In the west, in contrast, the cultural influence of the Russians is gradually diminishing.

During the Soviet period, the Russian language spread widely through western Central Asia. Russian served both as a common language and as the language of higher education. People had to be fluent in Russian in order to reach any position of responsibility. Russian speakers settled in all the major cities. Since 1991, however, many Russian speakers have migrated back to Russia, and Russian is gradually declining in favor of local languages.

Although Central Asia is remote and poorly integrated into global cultures, it is hardly immune to the forces of globalization. The increased usage of English throughout Central Asia shows its close connections with global culture. Such influences have been especially marked in the oil cities of the Caspian Basin, such as Baku, which have seen an influx of foreign oil workers (Figure 10.19). Although

their number is low, English speakers in Central Asia, especially those with computer skills, are increasingly valued as the region strives to find a place and a voice in the global community.

REVIEW QUESTIONS

1. How do patterns of religious affiliation divide Central Asia into distinct regions, and why is religion in the region the source of increasing tensions?

2. How have the patterns of linguistic geography in Central Asia been transformed over the past two decades?

GEOPOLITICAL FRAMEWORK: POLITICAL REAWAKENING

Central Asia has played a minor role in global political affairs for the past several hundred years. Before 1991, the entire region, except Mongolia and Afghanistan, was under direct Soviet or Chinese control. Mongolia, moreover, had been a close Soviet ally, and even Afghanistan came under Soviet domination in the late 1970s. Today southeastern Central Asia is still part of China. And although the breakup of the Soviet Union saw the emergence of six new Central Asian countries, most of them retain close political ties to Russia (Figure 10.20).

Partitioning of the Steppes

Before 1500, Central Asia was a power center, a region whose mobile armies threatened the far more populous, sedentary states of Asia and Europe. The development of new weapons changed the balance of power, however, allowing the wealthier agricultural states to conquer the nomads. By the 1700s, the nomads' armies had been defeated and their lands taken. The winners in this struggle were the two largest states bordering the steppes: Russia and China.

By the mid-1700s, the Chinese empire (under the Manchu, or Qing, dynasty) stood at its greatest territorial extent, including Mongolia, Xinjiang, Tibet, and a slice of modern Kazakhstan. From the height of its power in the late 1700s, China declined rapidly. When the Manchu dynasty fell in 1912, Mongolia became independent, although China still ruled the extensive borderlands of Inner Mongolia (Nei Mongol). Tibet had earlier gained effective independence, although this status was not recognized by China.

Russia began to push south of the steppe zone in the mid-1800s, conquering most of western Central Asia by 1900. Russia advanced into this area in part to keep the British out. Britain attempted to conquer Afghanistan, but failed to do so. Subsequently, Afghanistan's position as an independent "buffer state" between the Russian Empire (later the Soviet Union) and British India remained secure.

Central Asia Under Communist Rule

Western Central Asia came under communist rule not long after the emergence of the Soviet Union in 1917. Mongolia followed in 1924. After the Chinese revolution of 1949, the communist system was also imposed on Xinjiang and Tibet.

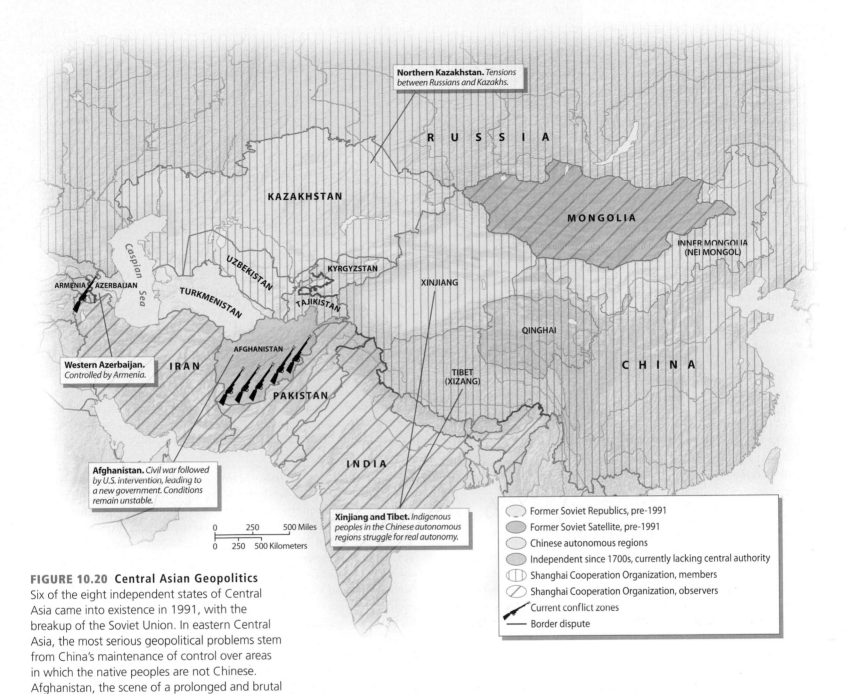

Northern Kazakhstan. *Tensions between Russians and Kazakhs.*

Western Azerbaijan. *Controlled by Armenia.*

Afghanistan. *Civil war followed by U.S. intervention, leading to a new government. Conditions remain unstable.*

Xinjiang and Tibet. *Indigenous peoples in the Chinese autonomous regions struggle for real autonomy.*

RUSSIA

KAZAKHSTAN

MONGOLIA

INNER MONGOLIA (NEI MONGOL)

Caspian Sea

ARMENIA AZERBAIJAN

UZBEKISTAN

KYRGYZSTAN

TURKMENISTAN

TAJIKISTAN

XINJIANG

IRAN

AFGHANISTAN

QINGHAI

CHINA

PAKISTAN

TIBET (XIZANG)

INDIA

0 250 500 Miles	
0 250 500 Kilometers	

Former Soviet Republics, pre-1991
Former Soviet Satellite, pre-1991
Chinese autonomous regions
Independent since 1700s, currently lacking central authority
Shanghai Cooperation Organization, members
Shanghai Cooperation Organization, observers
Current conflict zones
Border dispute

FIGURE 10.20 Central Asian Geopolitics
Six of the eight independent states of Central Asia came into existence in 1991, with the breakup of the Soviet Union. In eastern Central Asia, the most serious geopolitical problems stem from China's maintenance of control over areas in which the native peoples are not Chinese. Afghanistan, the scene of a prolonged and brutal war, has experienced the most extreme forms of geopolitical tension in the region.

Soviet Central Asia The newly established Soviet Union retained all the Central Asian territories of the Russian Empire. The new regime sought to build a Soviet society that would eventually knit together all the massive territories of the Soviet Union. Central Asia's leaders were replaced by Communist Party officials loyal to the new state, Russian immigration was encouraged, and local languages had to be written in the Cyrillic (Russian) rather than the Arabic script.

Although the early Soviet leaders thought that a new Soviet nationality would eventually emerge, they realized that local ethnic diversity would not disappear overnight. They therefore divided the Soviet Union into a series of nationally defined "union republics," in which a certain degree of cultural autonomy would be allowed. In the 1920s, Kazakhstan, Kyrgyzstan, Tajikistan, Uzbekistan, Turkmenistan, and Azerbaijan assumed their present shapes as Soviet republics.

Contrary to official intentions, the new republics encouraged the development of local national identities more than a broader Soviet identity. Also undercutting Soviet unity was the fact that cultural and economic gaps separating Central Asians from Russians did not decrease as much as planned. In addition, the region's higher birthrates led many Russians to fear that the Soviet Union risked being dominated by Turkic-speaking Muslims, contributing to the breakup of the Soviet Union.

The Chinese Geopolitical Order

After decades of political and economic chaos, China reemerged as a united country in 1949. Its new communist government quickly reclaimed most of the Central Asian territories that had slipped out of China's grasp in the early 1900s. China's new leaders promised the non-Chinese peoples a significant degree of political self-determination and cultural autonomy and thus found much local support in Xinjiang. Gaining control over Tibet was more difficult. China occupied Tibet in 1950, but the Tibetans launched a rebellion in 1959. When this was crushed, the Dalai Lama and some 100,000 followers sought refuge in India.

Loosely following the Soviet model, China established autonomous regions in areas occupied primarily by non–Han Chinese peoples, including Xinjiang, Tibet proper (called *Xizang* in Chinese), and Inner Mongolia. Such autonomy, however, often meant little, and it did not prevent the immigration of Han Chinese into these areas. Nor were all parts of Chinese Central Asia granted autonomous status. The large and historically Tibetan and Mongolian province of Qinghai, for example, remained an ordinary Chinese province.

Current Geopolitical Tensions

Although western Central Asia made the transition to independence rather smoothly, the region still suffers from several conflicts. Much of China's Central Asian territory is also unsettled, but China keeps a strong political hold on the area. Afghanistan, unfortunately, continues to suffer from a particularly brutal war.

Independence in Former Soviet Lands

The disintegration of the Soviet Union in 1991 generally proceeded peacefully in Central Asia (Figure 10.21), although a civil war that broke out in Tajikistan lasted until 1997. The six newly independent countries, however, had been dependent on the Soviet system, and it was not easy for them to chart their own courses. In most cases, authoritarian rulers, rooted in the old order, retained power and undermined opposition groups. All told, democracy has made less progress in Central Asia than in other parts of the former Soviet Union.

The newly independent countries also faced several border conflicts, due in part to the complicated political divisions established by the Soviet Union. Such problems are particularly severe in the Fergana Valley, a large, fertile, and densely populated lowland basin shared by Uzbekistan, Tajikistan, and Kyrgyzstan (Figure 10.22). Not only do boundaries here meander in a complex manner, but also the residents of many villages have maintained citizenship in a neighboring country. Partly as a result, relations between Tajikistan and Uzbekistan are tense. Tajikistan strongly objects to the fact that many ethnic Tajiks in Uzbekistan have been forced to take on an Uzbek identity.

FIGURE 10.21 The End of Soviet Rule in Central Asia An old Tajik man looks over a toppled and decapitated statue of Lenin, founder of the Soviet Union, in Dushanbe in 1991. When the Soviet Union collapsed in 1991, emblems of the old regime were immediately destroyed in many areas.

Strife in Western China

Many of the indigenous inhabitants of Tibet and Xinjiang in western China want independence or at least real political autonomy. China maintains that all its Central Asian lands are essential parts of its national territory, and it treats separatist groups harshly. In Xinjiang, a variety of groups composes the Eastern Turkistan Independence Movement. Some of these groups have a radical Islamist orientation and are classified as terrorist organizations by both China and the United States. Most, however, are secular, basing their claims on ethnicity and territory, rather than religion. Periodic violence by separatist groups typically results in quick reprisals by China. In July 2009, severe ethnic rioting broke out in Urumqi, the capital of Xinjiang. After more than 180 people were killed, China suspended Internet service and restricted cell phone coverage in the city (Figure 10.23). Order was soon reestablished, but in early 2012 China announced that it would send an additional 8,000 police officers to the region. This move angered Uyghur activists, who viewed it as an attempt to enhance the power of the Han Chinese in the region.

Anti-China protests in Tibet have also been severely repressed, but the Tibetans have brought the attention of the world to their struggle, largely through the work of the Dalai Lama. China maintains several hundred thousand troops in the region because of both Tibetan resistance and the strategic importance of the border zone with India. In March 2008, Tibetan protestors in Lhasa rioted, burning many Chinese-owned businesses. In response, China arrested hundreds of Tibetan monks and forced others to undergo "patriotic education." The Dalai Lama now agrees that Tibet should remain part of China, but Chinese leaders reject his call for true political autonomy. Tensions in the region again increased in late 2011 and early 2012, when more than 20 Tibetan monks and nuns burned themselves to death to protest Chinese policies and continuing Han Chinese migration.

FIGURE 10.22 Political Boundaries Around the Fergana Valley Some of the world's most complex political boundaries can be found in the vicinity of the Fergana Valley. The central portion of the valley belongs to Uzbekistan, which is otherwise separated from it by high mountains. The lower valley, on the other hand, is part of Tajikistan, the core area of which is likewise separated by highlands. The Fergana's upper periphery belongs to Kyrgyzstan. Note also the small enclaves of Uzbekistan within Kyrgyzstan.

FIGURE 10.23 Ethnic Tension in Xinjiang Relations between Han Chinese officials and Muslim Central Asians have grown increasingly tense in recent years. In this photo, Uyghur women are grabbing a Chinese riot police officer.

War in Afghanistan None of the conflicts in the former Soviet republics or western China compares in intensity to the struggle being waged in Afghanistan. Afghanistan's troubles began in 1978, when a Soviet-supported military "revolutionary council" seized power. The new Marxist government began to suppress religion, which led to widespread resistance. When the government was about to collapse, the Soviet Union responded with a massive invasion. Despite its power, the Soviet military was never able to control the more rugged parts of the country. Pakistan, Saudi Arabia, and the United States, moreover, ensured that the anti-Soviet forces remained well armed. When the exhausted Soviets finally withdrew their troops in 1989, brutal local warlords grabbed power across most of the country.

In 1995, a new movement called the Taliban arose in Afghanistan. Founded by young Muslim religious students, the Taliban believed in the strict enforcement of Islamic law. The Taliban model attracted large numbers of soldiers, and by September 2001 only far northeastern Afghanistan lay outside Taliban power. By 2000, however, most of Afghanistan's people were turning against the group, primarily because of the severe restrictions on daily life that the Taliban imposed. These restrictions were most pronounced for women, but even men were compelled to obey the Taliban's numerous decrees. Most forms of recreation were simply outlawed, including television, films, music—even kite flying.

The terrorist attacks of September 11, 2001, in the United States completely changed the balance of power in Afghanistan. The United States and Britain, working with Afghanistan's anti-Taliban forces,

FIGURE 10.24 War in Afghanistan Warfare continues in Afghanistan, as troops from the United States and other countries attempt to root out the remnants of the Taliban and maintain order. Tense encounters with local residents often result when foreign soldiers search for weapons and insurgent fighters.

attacked the Taliban government and soon defeated it. Although a democratic Afghan government was established within a few years, peace did not return. The new government proved corrupt and ineffective, failing to establish security in most parts of the country.

By 2004, the Taliban had regrouped, operating from safe havens in Pakistan (see Chapter 12). Afghanistan's new government has had to rely on the military power of the International Security Assistance Force (ISAF), led by the North Atlantic Treaty Organization (NATO). But despite its tens of thousands of troops, the ISAF was unable to stop the Taliban resurgence (Figure 10.24). In 2009, the United States responded by sending an additional 17,000 troops to Afghanistan. The U.S. forces have continued to target high-level Taliban leaders, often by bombing their compounds, many of which are located across the border in Pakistan. This strategy has weakened the Taliban command structure, but has also resulted in large numbers of civilian casualties, reducing local support for the war effort. In February 2012, anti-American sentiment surged across the country after it was revealed that U.S. forces had burned a number of Qurans, the holy book of Islam. Although many areas of northern Afghanistan are relatively secure, the south remains a battle-scarred war zone.

The United States and its coalition partners are planning to withdraw all troops by 2014. After that point, the government of Afghanistan will be responsible for preventing the Taliban from regaining control. Although public opinion polling shows that few Afghans want the Taliban back in power, many observers fear that the country's weak national government will have difficulties withstanding a Taliban military assault.

International Dimensions of Central Asian Tension

With the collapse of the Soviet Union in 1991, Central Asia emerged as a key arena of geopolitical tension. Several important countries, including China, Russia, Pakistan, Iran, India, and the United States, have competed for influence in the region. The political revival of Islam has also generated international geopolitical issues.

Russia's economic and military ties to Central Asia did not vanish when the Soviet Union collapsed. Russia maintains military bases in Tajikistan and Kyrgyzstan and continues to be one of the main markets for Central Asian exports (Figure 10.25). Furthermore, western Central Asia's rail links and gas and oil pipelines are mostly oriented toward Russia.

Russian and Chinese leaders became concerned about the growing influence of the United States in Central Asia after the U.S. military established bases in Uzbekistan, Kyrgyzstan, and Afghanistan following the attacks of September 11, 2001. As a result, they formed the treaty group called the **Shanghai Cooperation Organization (SCO)**, composed of China, Russia, Kazakhstan, Kyrgyzstan, Tajikistan, and Uzbekistan. The SCO seeks cooperation on such security issues as terrorism and separatism and also aims to enhance trade. The SCO works with the **Collective Security Treaty Organization (CSTO)**, a Russian-led military association that includes Belarus, Armenia, Kazakhstan, Kyrgyzstan, Tajikistan, and Uzbekistan. The two organizations work together to address military threats, crime, and drug smuggling.

FIGURE 10.25 Soldiers at the Russian Military Base in Kant, Kyrgyzstan Russia retains close interest in the former Soviet republics of Central Asia. The Russian military maintains bases in both Tajikistan and Kyrgyzstan.

Turkmenistan has not joined the SCO or the CSTO. Turkmenistan's relative isolation stems from the policies of its former president, Saparmurat Niyazov, who forced the citizens of his country to treat him as a heroic savior-figure. After Niyazov's death in 2006, many observers thought that the country would become more open, but thus far such hopes have been mostly frustrated. In 2012, Turkmenistan's president, Gurbanguli Berdymukhamedov, was reelected with 97 percent of the vote, a figure that indicates the absence of genuine democracy.

The chaotic situation in Afghanistan is Central Asia's most serious geopolitical issue, as local leaders fear that Islamic radicalism could spread from Afghanistan into their own countries. Although several Central Asian countries have expelled U.S. military bases, they have offered some support to the NATO forces operating in Afghanistan. India is also concerned about Afghanistan and has also sought to gain influence in the region. Pakistan, however, views such moves with alarm, fearing that India is trying to encircle its territory.

FIGURE 10.26 Oil Development in Azerbaijan Although oil has brought a certain amount of wealth to Azerbaijan, it has also resulted in extensive pollution. Because most of the petroleum is located either near or under the Caspian Sea, this sea—actually the world's largest lake—is becoming increasingly polluted, and its fisheries are declining.

REVIEW QUESTIONS

1. How did the collapse of the Soviet Union in 1991 change the geopolitical structure of Central Asia, and how has Russia attempted to maintain influence in the region?

2. How does the geopolitical situation of Afghanistan differ from those of the other countries of the region, and why is Afghanistan's political history so different from those of its neighbors?

ECONOMIC AND SOCIAL DEVELOPMENT: ABUNDANT RESOURCES, TROUBLED ECONOMIES

Central Asia is not a wealthy region. Afghanistan in particular stands near the bottom of almost every list of economic and social indicators. In the early years of the 21st century, however, several Central Asian countries underwent an economic boom, with annual growth rates exceeding 10 percent. The global economic crisis of 2008–2009 hit the region hard, but the energy-exporting countries of the region bounced back quickly.

Post-Communist Economies

During the communist period, economic planners sought to spread economic development widely across the Soviet Union. This effort required building large factories even in remote areas of Central Asia, regardless of the costs. Such Central Asian industries relied heavily on subsidies from the Soviet government. When those subsidies ended, the industrial base of the region collapsed, leading to a huge drop in the standard of living. As is true elsewhere in the former Soviet Union, however, some individuals became very wealthy after the fall of communism.

Since the beginning of the 21st century, Kazakhstan, Azerbaijan, and Turkmenistan have benefited from their large amounts of energy resources. Today Kazakhstan stands as the most developed Central Asian country. Its agricultural sector is productive, and it has the world's 12th largest reserves of oil and the 16th largest reserves of natural gas. The region's oldest fossil fuel industry is located in Azerbaijan (Figure 10.26), which has also attracted much international investment in recent years. Although its economy has grown rapidly, Azerbaijan remains a poor country, burdened by inadequate infrastructure. Turkmenistan, despite having the world's fourth largest reserves of natural gas, is poorer still.

Uzbekistan's fossil fuel reserves are much smaller than those of Kazakhstan and Turkmenistan, but they are still substantial. However, Uzbekistan is also a much more densely populated country, and many of its people live in overcrowded farmlands suffering from environmental stress. An authoritarian country, Uzbekistan retains many aspects of the old Soviet-style command economy (one run by governmental planners, rather than by private firms responding to the market). It is, however, the world's fifth largest cotton exporter, and its large deposits of gold and other mineral resources help keep its economy afloat.

Central Asia's two most mountainous countries, Tajikistan and Kyrgyzstan, hold most of the region's water resources, which they are developing by building massive dams. Tajikistan, however, is burdened by its remote location, rugged topography, and lack of infrastructure. With a per capita gross national income of only around $2,000, Tajikistan rates as one of Eurasia's poorest countries, with

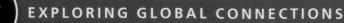

EXPLORING GLOBAL CONNECTIONS

Mongolia's Mining Boom

Mongolia has long been a poor country with relatively weak connections with the global economy. Until fairly recently, its exports were focused on animal products, such as cashmere, the soft wool of a particular kind of goat. However, Mongolia's economy, and its linkages with the rest of the world, is rapidly transforming. Mongolia is currently experiencing one of the largest mining booms the world has ever seen. Workers and foreign capital are streaming into the country, and massive quantities of minerals, particularly coal and copper, are heading out, mainly to China. More than 80 percent of Mongolia's exports are now minerals, and that figure is expected to climb to 95 percent within a few years. A single mining site, Oyu Tolgoi, in the Gobi Desert near the border with China, could account for one-third of Mongolia's total economic production by 2020 (Figure 10.2.1). As Mongolia has a small population, some observers think that it could become a wealthy county based on mineral exports alone, much like such small oil-exporting countries as Brunei and Kuwait.

Mongolia's mining boom has demographic and cultural consequences, as well as economic impact. By 2009, the country already had as many as 35,000 registered foreign workers from 108 countries; in the same year, 1,802 undocumented workers were deported,

mostly to China. The government imposes quotas on foreign workers for each sector of the economy, which range between 5 percent and 80 percent, depending on the skill levels required. Highly skilled workers from abroad demand new services, and as a result the capital city of Ulaanbaatar has seen a surge of restaurant openings. French bistros and Irish pubs are now easy to find in the booming city.

Mongolia's mineral-driven boom has generated problems, as well as opportunities. Ulaanbaatar is now reckoned by some to have the world's second worst air pollution problem. The city is growing faster than its infrastructure, generating traffic jams and worse. In some of the mineral production zones, small-scale "ninja miners" use mercury and cyanide to amalgamate gold, poisoning both themselves and local rivers in the process. In the gargantuan mining site of Oyu Tolgoi in the southern desert, water must be extracted from deep aquifers and piped in from long distances, threatening local supplies. Mining also generates huge quantities of dust in the arid climate, resulting in asthma and other respiratory disorders.

Oyu Tolgoi is not far from Mongolia's border with China, and it relies heavily on

FIGURE 10.2.1 Oyu Tolgoi Mining Area Oyu Tolgoi, in the Gobi Desert of southern Mongolia, has recently emerged as a major mining center. Investments in this area have been massive, as can be seen here.

Chinese infrastructure, Chinese workers, and the Chinese market. Mongolia has long been wary of its vast southern neighbor, which many Mongolians think could overwhelm their lightly populated country in the future. Partly as a result, Mongolia has been cultivating close ties with other Asian countries, particularly South Korea. An estimated 35,000 Mongolians now live in South Korea, many more than in any other country. Mongolia's mining industry is also highly globalized. The massive Oyu Tolgoi complex, for example, is run as a joint venture among Canada's Ivanhoe Mines, the London-based Rio Tinto Group, and the government of Mongolia.

---

two-thirds of its people living in poverty. In 2010, it was estimated that almost half of Tajikistan's labor force worked abroad, mostly in Russia, Uzbekistan, and Kazakhstan.

Kyrgyzstan also remains a very poor country, its economy heavily dependent on agriculture and gold mining. A single facility, the Kumtor Gold Mine, accounts for nearly 12 percent of Kyrgyzstan's gross domestic product and 54 percent of its industrial output; as a result, the country's economy suffered a major blow when miners at Kumtor went on strike in early 2012. Although Kyrgyzstan pushed through several market-oriented reforms in the 1990s and is the only former Soviet Central Asian country to have joined the World Trade Organization, it has experienced little sustained economic growth.

Mongolia, formerly a communist ally of the Soviet Union, also suffered an economic collapse in the 1990s, after Soviet subsidies were eliminated. Conditions worsened from 2000 until 2002, when a combination of severe winters and summer droughts devastated livestock, the traditional mainstay of the Mongolian economy. Mongolia does, however, possess vast reserves of copper, gold, and other minerals, and its economy has recently begun to boom as the mining sector expands (see "Exploring Global Connections: Mongolia's Mining Boom").

The Economy of Tibet and Xinjiang in Western China   Unlike the rest of the region, the Chinese portions of Central Asia did not experience an economic crash in the 1990s. China as a whole had one of the world's fastest-growing economies, although its centers of economic dynamism are all located in the eastern coastal zone. However, since China had been much poorer than the Soviet Union before 1991, it is not surprising that poverty in Chinese Central Asia remains widespread.

Tibet in particular remains impoverished, with its economy dominated by subsistence agriculture. China has been rapidly building roads and railroads into Tibet, greatly expanding tourism in the process. According to official statistics, the Tibetan economy's annual growth rate has been over 12 percent over the past five years. Tibetan critics, however, claim that most of the region's recent economic gains have gone to Han Chinese immigrants, rather than to indigenous Tibetans.

Xinjiang has tremendous mineral wealth, including China's largest oil reserves. Its agricultural sector is highly productive, although limited in extent. Many of the local Muslim peoples believe that the Chinese state and Han Chinese migrants are taking the region's

FIGURE 10.27 Qinghai–Tibet Railway China is engaged in a massive program to extend roads and railroads into the western half of the country. The main connection with the Tibetan Plateau is the recently completed Qinghai–Tibet Railway, shown here. This railway has resulted in a large increase in tourism in Tibet.

wealth. China is currently building major road and rail links from eastern China to Xinjiang through its official Western Development strategy (Figure 10.27). While these infrastructural projects are bringing economic benefits to the region, they are also encouraging migration from eastern China. In late 2011, the Chinese government announced that it would invest large quantities of money in state-owned enterprises in Xinjiang, focusing on petroleum, petrochemicals, and energy-related industries.

The Misery of Afghanistan Although rich in natural gas and minerals, Afghanistan is a deeply impoverished country. Foreign aid and the export of handwoven rugs have formed its legitimate economic mainstays. Although official statistics indicate that the Afghan economy expanded at the healthy pace of 7.1 percent in 2011, war, corruption, crime, and poor infrastructure have prevented most people from experiencing any real gains.

By the late 1990s, Afghanistan had emerged as the leading producer of one major commodity for the global market: illegal drugs. More than 90 percent of the world's opium, used to produce heroin, is grown in Afghanistan. In much of southern and western Afghanistan, opium is the main cash crop. Not only is it highly profitable and easy to transport, but it also uses relatively little water. As much as $100 million a year in narcotics profits have supposedly gone to the Taliban, which operates mainly in the opium-growing areas of the country. As a result, both NATO and the Afghan government have tried to convince villagers to grow alternative crops.

If Afghanistan could solve its security problems, it would probably undergo a mining boom. China is already pouring vast amounts of money into the Ainak Copper Mine, projected to become one of the largest mines in the world. When fully operational, the mine will employ an estimated 20,000 Afghans. Taliban insurgents, however, often target such foreign operations, greatly reducing international investment in the country.

Central Asian Economies in Global Context Many foreign countries are drawn to Central Asia by its natural resources (Figure 10.28).

However, if all of the region's fossil fuel fields are to be economically viable, vast pipeline systems must be constructed. The inadequate existing pipelines mostly pass through Russia, which charges high transit fees. Seeking an alternative route, an international consortium pushed through the massive Baku–Tbilisi–Ceyhan pipeline, which transports oil from the Caspian Basin to a Turkish port on the Mediterranean Sea. Opened in 2006, at a cost of $3.9 billion, it is the world's second longest oil pipeline.

More recently, energy development in Central Asia has focused more on natural gas pipelines. The United States and the European Union have supported the Nabucco plan, which would transport natural gas from the Caspian Basin through Turkey and hence into central Europe. Construction of this $10 billion project has been delayed several times, due mainly to political complications. A final decision on whether to build the pipeline is expected in mid-2013.

China is equally interested in Central Asia's natural resources, which is one reason it has been a strong supporter of the SCO. Connections to China were enhanced with the commissioning of the ambitious Central Asia–China gas pipeline, which links wells in Turkmenistan, Uzbekistan, and Kazakhstan to the China market (Figure 10.29). China has also invested heavily in regional infrastructure, as Kazakhstan in particular hopes to replace its decaying transportation system with a "new Silk Road." All told, Central Asia is poorly integrated into wider transportation networks, a problem that will require massive investments if it is to be solved.

Social Development in Central Asia

Social conditions in Central Asia vary more than economic conditions. Afghanistan falls at the bottom of the scale regarding almost every aspect of social development, but in the former Soviet territories the levels of health and education are still relatively high. Social conditions have improved recently in Tibet and Xinjiang, although they have probably not kept pace with the progress made in China proper. Mongolia is noted for its relatively high levels of social development, as well as for its fairly high level of gender equity. Women have outnumbered men in Mongolian universities since the 1980s.

Gender Issues in Former Soviet Central Asia In traditional Central Asia, the social position of women varied considerably. In general, women had more autonomy in the pastoral societies of the northern steppe zone than they did in the agricultural and urban societies located farther to the south. In Kyrgyzstan, however, women often suffered from the practice of "bride abduction," in which kidnapped women were forced to marry their abductors. Although many such kidnappings actually took place with the consent of the bride-to-be, others were true abductions. In recent years, the practice has been resurging, leading to outrage by Kyrgyz women's groups. In early 2012, the Kyrgyz parliament rejected a measure that would have helped crack down on kidnapping, largely because male legislators feared that it would also be used to prohibit polygamy. Although

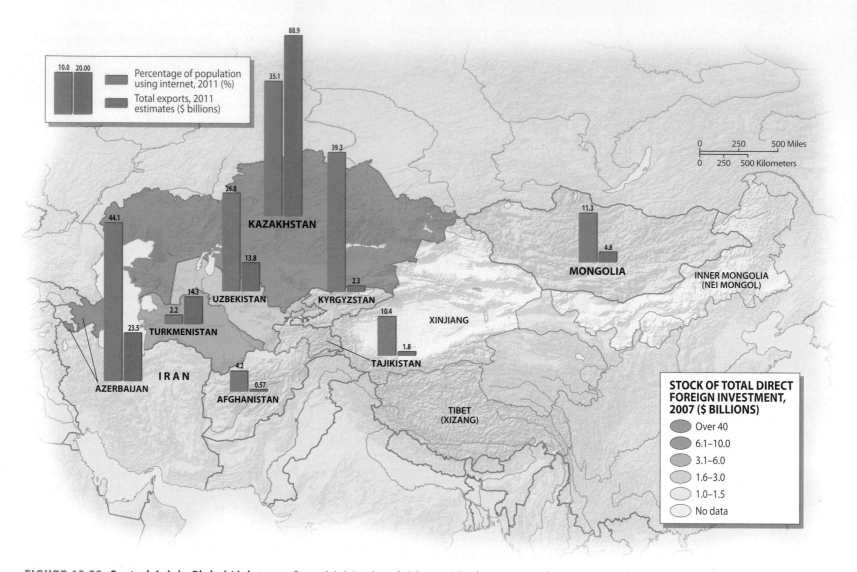

FIGURE 10.28 Central Asia's Global Linkages Central Asia's mineral-rich countries have attracted large foreign investments in recent years. Kazakhstan and Azerbaijan have especially close connections with the global economy. Countries without well-developed oil- or mining-based economies, such as Afghanistan and Tajikistan, have relatively weak global linkages, at least in regard to legal exports and foreign investment.

polygamy is technically illegal, many politically powerful Kyrgyz men still practice it.

Under communist rule in former Soviet Central Asia, efforts were made to educate women and place them in the workforce, although women often encountered limits to their career advancement. Women in these countries still have relatively high levels of education, although some fear that their position may be slipping. Central Asian countries have been attempting to address such concerns by installing quotas for women in the government. By law, one-third of the members of the Kyrgyz parliament must be female, and in Uzbekistan the figure is 30 percent. Local women's groups complain, however, that their actual power in government is limited, as the positions that they occupy have only token power.

Demographic issues in Central Asia also influence the position of women. In Tajikistan especially, so many men have relocated to

Russia for work that many households are now led by women. At the same time, local women are under pressure to follow more traditional gender roles due to the growth of more fundamentalist forms of Islam. In Kazakhstan, nationalist parties have tried to ban contraception for Kazakh women, as they hope to increase both the population of the country and the proportion of its ethnic Kazakh population.

Social Conditions and the Status of Women in Afghanistan

Afghanistan's average life expectancy is about 45 years, one of the lowest figures in the world. Infant and childhood mortality levels remain extremely high. Afghanistan endures almost constant warfare, and its rugged topography hinders the provision of basic social and medical services. Illiteracy is commonplace, especially for women. Afghanistan's 12 percent adult female literacy rate is reportedly the lowest in the world (Table 10.2).

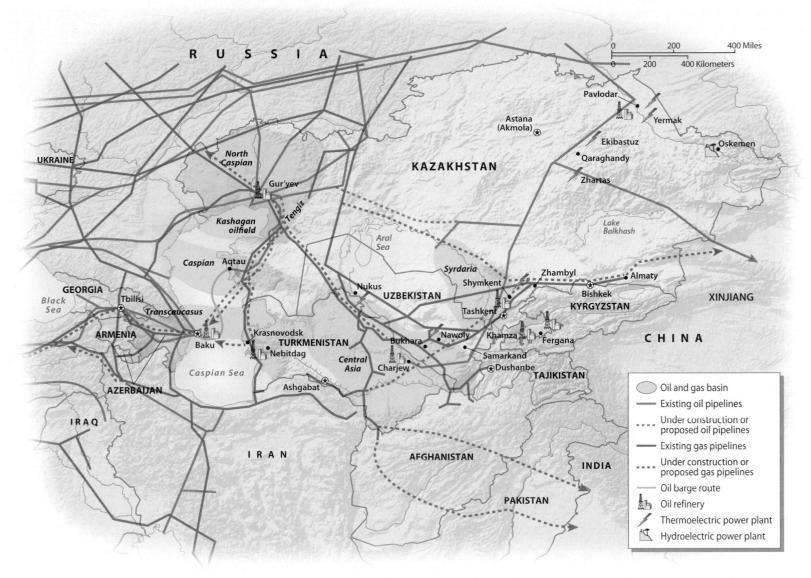

FIGURE 10.29 Oil and Gas Pipelines Central Asia has some of the world's largest oil and gas deposits and recently has emerged as a major center for drilling and exploration. Because of its landlocked location, Central Asia cannot easily export its petroleum products. Pipelines have been built to solve this problem, and several others are currently being planned. Pipeline construction is a contentious issue, however, as several of the potential pathways lie across Iran, a country that remains under U.S. sanctions.

TABLE 10.2 DEVELOPMENT INDICATORS

Country	GNI per capita, PPP 2010	GDP Average Annual %Growth 2000–10	Human Development Index (2011)[1]	Percent Population Living Below $2 a Day	Life Expectancy (2012)[2]	Under Age 5 Mortality Rate (1990)	Under Age 5 Mortality Rate (2010)	Adult Literacy (% ages 15 and older)	Gender Inequality Index (2011)[3,1]
Afghanistan	1,060	11.3	.398	—	49	209	149	28*	0.707
Azerbaijan	9,270	17.1	.700	2.8	74	93	46	100	0.314
Kazakhstan	10,770	8.3	.745	<2	69	57	33	100	0.334
Kyrgyzstan	2,070	4.4	.615	21.7	69	72	38	99	0.370
Mongolia	3,670	7.2	.653	—	68	107	32	97	0.410
Tajikistan	2,140	8.6	.607	27.7	73	116	63	100	0.347
Turkmenistan	7,490	13.6	.686	49.7	65	98	56	100	—
Uzbekistan	3,110	7.1	.641	—	68	77	52	99	—

[1]United Nations, *Human Development Report, 2011.*
[2]Population Reference Bureau, *World Population Data Sheet, 2012.*
*Additional data from the *CIA World Factbook, 2012*
[3]Gender Inequality Index—A composite measure reflecting inequality in achievements between women and men in three dimensions: reproductive health, empowerment and the labor market that ranges between 0 and 1. The higher the number, the greater the inequality.
Source: World Bank, *World Development Indicators, 2012.*

Women in traditional Afghan society—and especially in Pashtun society—have very little freedom. Restrictions on female activities intensified in the 1990s, when the Taliban gained power over most of Afghanistan. Taliban forces prevented women from working, from attending school, and often even from obtaining medical care. After the Taliban lost power, its fighters continued to target female education, destroying schools for girls throughout southern Afghanistan.

Under Afghanistan's new constitution, established in 2004, 25 percent of parliamentary seats are reserved for women (Figure 10.30). Much evidence, however, indicates that the social position of women has improved little outside the capital city of Kabul. According to one report, over 80 percent of Afghan women suffer from domestic violence. Throughout Afghanistan and especially in Kabul, many women are concerned the Taliban will again gain power and that the few gains that they have made will vanish.

FIGURE 10.30 Afghan Women in Parliament   Women in Afghanistan generally have a low social position, and they suffer from many restrictions, especially in the rural parts of the country. They do have a number of reserved seats in the country's parliament, although critics contend that female legislators have little real power.

REVIEW QUESTIONS

1. How does the distribution of fossil fuel deposits influence economic and social development in Central Asia?

2. Why do the social positions of women vary so much across Central Asia, and why is Afghanistan particularly problematic in this regard?

Summary

- Central Asia, long dominated by Russia and China, has recently reappeared on the global scene. Environmental problems, among others, have brought the region to global attention. The destruction of the Aral Sea is one of the world's worst environmental disasters. The booming oil and gas industries have also created environmental problems.

- Large movements of people in Central Asia have attracted worldwide attention. The biggest issue at present is the migration of Han Chinese into Tibet and Xinjiang, which is threatening to turn local peoples into minority groups. The migration of Russian speakers out of the former Soviet areas of Central Asia is another important demographic issue. In Afghanistan, war and continuing chaos have generated large refugee populations.

- Religious tension has recently become a cultural issue through much of western Central Asia. Radical Islamic fundamentalism remains a potent force in southern and western Afghanistan and, to a somewhat lesser extent, in the Fergana Valley. Although nonradical forms of Islam are prevalent through most of the region, Central Asian leaders have used the threat of religiously inspired violence to maintain repressive and antidemocratic policies.

- Although China maintains a firm grip on Tibet and Xinjiang, the rest of Central Asia has emerged as a key area of geopolitical competition. Russia, the United States, China, Iran, Pakistan, and India contend for influence. Central Asian countries have attempted to play these powers off against one another in order to maintain their power. Political structures throughout most of the region remain authoritarian, and in recent years several Central Asian countries have developed closer relations with China and Russia.

- The economies of Central Asia are gradually opening up to global connections, largely because of the substantial fossil fuel reserves in this region. However, Central Asia is likely to face serious economic difficulties for some time, especially because the region is not a significant participant in global trade and has attracted little foreign investment outside the oil industry. Production of drugs remains a serious problem in Afghanistan.

Key Terms

alluvial fan 335
Collective Security Treaty Organization
  (CSTO) 346

loess 336
pastoralist 333
Shanghai Cooperation Organization (SCO) 346

theocracy 341
transhumance 335

Thinking Geographically

1. To what extent will Central Asia's continental location be a disadvantage in the years to come? Is coastal access truly significant in determining a country's competitive position in the global economy?

2. Which connections with the rest of the world will prove most important for Central Asia in the near future—those based on cultural features, economic ties, or geopolitical connections?

3. Is China's control over Tibet a legitimate concern of U.S. foreign policy? Should the United States attempt to influence Chinese policy in the area?

4. How might the disappearance of Central Asia's great lakes best be addressed? Is this a topic of international concern, or should the governments of Central Asia assume full responsibility themselves?

5. What economic alternatives to drug production might be found for Afghanistan? Can the international community help the government of Afghanistan address this problem?

MasteringGeography™

Looking for additional review and test prep materials? Visit the Study Area in MasteringGeography™ to enhance your geographic literacy, spatial reasoning skills, and understanding of this chapter's content by accessing a variety of resources, including MapMaster interactive maps, videos, RSS feeds, flashcards, web links, self-study quizzes, and an eText version of *Globalization and Diversity*.

Scan to visit the author's blog for chapter updates.

http://gad4blog.
wordpress.com/
category/central-
asia/

Authors' Blogs

Scan now to access the authors' blogs for up-to-date information on Central Asia.

Scan to visit the
GeoCurrents blog.

http://geocurrents.
info/category/place/
central-asia-places

Globalization and Diversity

Japan, South Korea, Taiwan, and the coastal regions of China are among the most globalized regions of the world, whereas North Korea is probably the world's most isolated country.

ENVIRONMENTAL GEOGRAPHY

China has long experienced severe deforestation and soil erosion, but its current economic boom is generating some of the worst pollution problems in the world. Japan, South Korea, and Taiwan, however, all have extensive forests and relatively clean environments.

POPULATION AND SETTLEMENT

China is currently undergoing a major transformation as tens of millions of peasants move from impoverished villages in the interior to booming cities in the coastal region. Throughout East Asia, low birthrates and aging populations are encountered.

CULTURAL COHERENCE AND DIVERSITY

Despite the presence of several unifying cultural features, East Asia in general and China in particular are divided along several striking cultural lines. Historically, however, the entire region was tied together by Mahayana Buddhism, Confucianism, and the Chinese writing system.

GEOPOLITICAL FRAMEWORK

The growing military power of China is generating tension with other East Asian countries, while Korea remains a divided nation, with South Korea set in opposition against North Korea. As China becomes more powerful, Japan, South Korea and Taiwan are responding by strengthening their ties with the United States.

ECONOMIC AND SOCIAL DEVELOPMENT

Over the past several decades, East Asia has emerged as a core area of the world economy, with China undergoing one of the most rapid economic expansions the world has ever seen. North Korea, however, remains as desperately poor country, plagued by widespread malnutrition.

➤ Hyundai Heavy Industries Co., Ltd., is the world's largest shipbuilder, and its shipyard in Ulsan South Korea is the world's largest ship-building facility. Ulsan, with just over 1 million inhabitants, is the most economically productive city of South Korea on a per capita basis.

Up until March 11, 2011, the northeastern Japanese city of Miyako was a thriving fishing town with a population of almost 60,000. Miyako sits near one of East Asia's richest fishing grounds, owing to the meeting of the warm Kuroshio ocean current from the south and the cold Oyashio ocean current from the north. On that day, however, Miyako and nearby ports were destroyed by one of the largest and most destructive natural events of modern times, the 2011 Tohoku earthquake and tsunami. Over 15,000 lives were lost altogether, with the death toll in Miyako alone exceeding 400. Beyond the loss of lives and homes, the regional economy was devastated. Some 900 of the town's 960 fishing boats were ruined, along with the fish market, seafood-processing factories, and refueling facilities and cranes. Miyako is now struggling to rebuild, but the process will be prolonged. Few people expect that the town will ever regain its lost population or prosperity.

The repercussions of the earthquake and tsunami were enormous across Japan. The Japanese government had invested billions of dollars in anti-tsunami seawalls along almost 40 percent of its coastline, which were widely believed to provide adequate security. On March 11, 2011, however, the surge of water simply washed over the tops of the barriers, collapsing many in the process (Figure 11.1). Japan, an energy-poor country, had also invested heavily in nuclear power, which was widely regarded as relatively safe and secure. The earthquake and tsunami, however, caused severe damage to the Fukushima Daiichi nuclear power plant, releasing large quantities of radiation and forcing the evacuation of over 200,000 people. The crisis led to widespread loss of confidence in both the Japanese government and nuclear power. The Japanese people are now reconsidering their energy future, and no easy options are in sight.

East Asia is highly vulnerable to earthquakes, owing both to its geological environment and to its density of population. More people live in East Asia, which consists of China, Taiwan, Japan, South Korea, and North Korea (Figure 11.2), than any other world region. China alone, with more than 1.3 billion inhabitants, has a greater population than any other world region except South Asia.

Although East Asia is historically unified by cultural features, in the second half of the 20th century it was politically divided, with the capitalist economies of Japan, South Korea, Taiwan, and Hong Kong separated from the communist bloc of China and North Korea. Distinct

FIGURE 11.1 2011 Tohoku Tsunami   The March 11, 2011 Tohoku earthquake and tsunami was one of the wold's most devastating natural disasters of recent times. At its extreme, the tsunami wave was over 40 meters (133 feet) high, giving it enough power to wash away entire villages. More than 15,000 people died as a result.

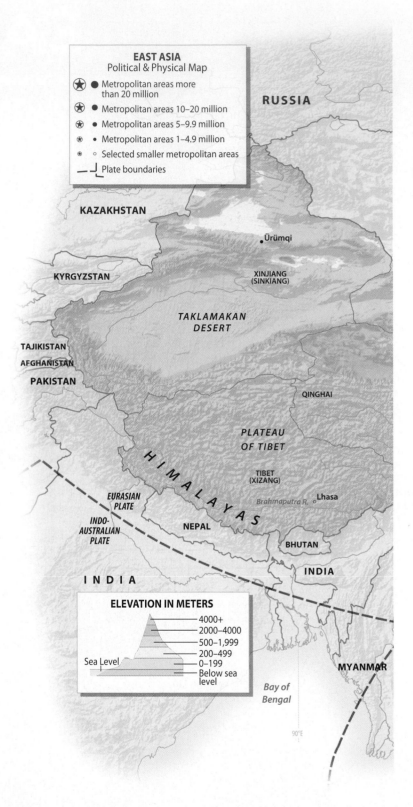

differences in levels of economic development also persisted. As Japan became a leader in the global economy, much of China remained extremely poor. By the early 21st century, however, divisions within East Asia had been reduced. Although China is still governed by the Communist Party, it is now on a path of capitalist development. Yet relations between North Korea and South Korea are still hostile, and while East Asia as a whole has seen a reduction in political tensions, the rise of China is generating increasing concerns in the rest of the region.

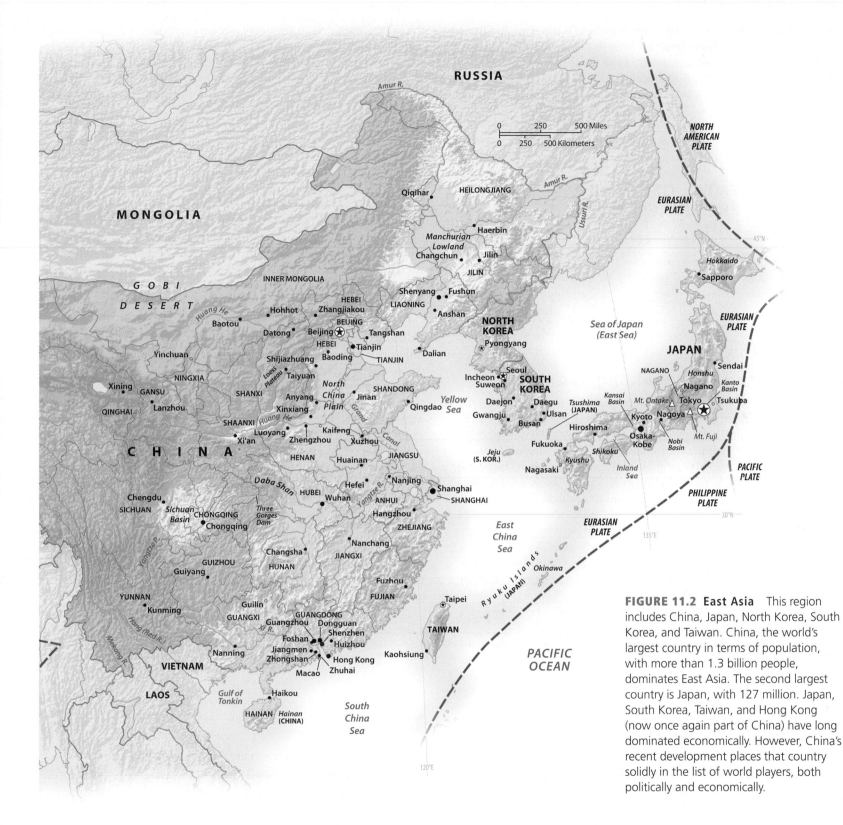

FIGURE 11.2 East Asia This region includes China, Japan, North Korea, South Korea, and Taiwan. China, the world's largest country in terms of population, with more than 1.3 billion people, dominates East Asia. The second largest country is Japan, with 127 million. Japan, South Korea, Taiwan, and Hong Kong (now once again part of China) have long dominated economically. However, China's recent development places that country solidly in the list of world players, both politically and economically.

East Asia is one of the core areas of the world economy, and it is emerging as a center of political power as well. Japan, South Korea, and Taiwan are among the world's key trading states. Tokyo, Japan's largest city, stands alongside New York and London as one of the financial centers of the globe. China has more recently emerged as one of the key global trading powers, and it now has the world's second largest economy.

In certain respects, East Asia can be easily defined by the territorial extent of its countries: China, Taiwan, North Korea, South Korea, and Japan. The political status of Taiwan, however, is ambiguous. Although Taiwan is, in effect, an independent country, China claims it as part of its own territory. As a result, Taiwan is recognized as a sovereign state by only a handful of other countries. In regard to cultural considerations, the territorial extent of East Asia is a more complicated issue, especially in regard to the western half of China. This is a huge, but lightly populated space; some 95 percent of the residents of China live in the eastern part of the country (sometimes referred to as **China proper**; see Figure 11.11). In cultural and historical terms, the indigenous peoples of western China are more closely connected to Central Asia than to the rest of East Asia. As a result, western China is covered most extensively in Chapter 10; in this chapter, we will examine this region only to the extent that it is politically part of China.

ENVIRONMENTAL GEOGRAPHY: RESOURCE PRESSURES IN A CROWDED LAND

The most serious environmental issues in East Asia are found in China. China's environmental problems stem from its large population, its rapid industrial development, and the unique features of its physical geography (Figure 11.3). One of China's most controversial environmental actions has been the building of the Three Gorges Dam on the Yangtze River.

Flooding and Dam Building in China

The Yangtze River (also called the Chang Jiang) is one of the most important physical features of East Asia. This river, the third largest (by volume) in the world, emerges from the Tibetan highlands onto the rolling lands of the Sichuan Basin, passes through a magnificent canyon in the Three Gorges area (Figure 11.4), and then meanders across the lowlands of central China before entering the sea in a large delta near the city of Shanghai. The Yangtze has historically been the main transportation corridor into the interior of China, and it has long been famous in Chinese literature for its beauty and power.

The Three Gorges Controversy   The Chinese government is trying to control the Yangtze for two main reasons: to prevent flooding and to generate electricity. To do so, it built a series of large dams,

the largest of which is a massive structure in the Three Gorges area, completed in 2006. This $39 billion structure is the largest hydroelectric dam in the world, forming a reservoir 350 miles long. It has jeopardized several endangered species (including the Yangtze River dolphin), flooded a major scenic attraction, and displaced more than 1 million people.

The Three Gorges Dam generates large amounts of electricity, supplying roughly 3 percent of China's demand. As China industrializes, its need for power is increasing rapidly. At present, most of China's energy supply comes from burning coal, which results in severe air pollution. But although the Three Gorges Dam may reduce air pollution somewhat, most environmentalists argue that the costs will exceed the benefits. Chinese government planners disagree, arguing that the Three Gorges Dam will both provide necessary electricity and reduce the threat of flooding that endangers the middle and lower stretches of the Yangtze Valley. Dam building on the Huang He River in northern China, however, has not been able to solve that region's water control problems.

Flooding in Northern China   The North China Plain has historically suffered from both drought and flooding. This area is dry most of the year; yet it often experiences heavy downpours in summer. Since ancient times, levees and canals have both controlled floods and allowed irrigation. But no matter how much effort has been put into water control, disastrous flooding has never been completely prevented (Figure 11.5).

The worst floods in northern China are caused by the Huang He, or Yellow River, which cuts across the North China Plain. As a result of upstream erosion, the Huang He carries a huge **sediment load** (the amount of suspended clay, silt, and sand in the water), making it the muddiest major river in the world. When the river enters the low-lying plain, its velocity slows, and its sediments begin to settle and accumulate in the riverbed. As a result, the level of the river gradually rises above that of the surrounding lands. Eventually, it must break free of its course to find a new route to the sea over lower-lying ground.

LEARNING OBJECTIVES

After reading this chapter you should be able to:

- Identify the key environmental differences between the island portions of East Asia (Japan and Taiwan) and the mainland.

- Describe the main environmental problems China faces today and compare them with the environmental challenges faced by Japan, South Korea, and Taiwan.

- Summarize the relationships among topography, climate, rice cultivation, and population density across East Asia.

- Explain why China's population is so unevenly distributed, with some areas densely settled and other almost uninhabited.

- Outline the distribution of major urban areas across East Asia and explain why the continued expansion of the region's largest cities is often viewed as a problem.

- Describe the ways in which religion and other systems of belief both unify and divide East Asia.

- Explain the distinction between the Han Chinese and the other ethnic groups of China, paying particular attention to language.

- Describe the geopolitical division of East Asia during the Cold War period and explain how the division of that period still influences East Asian geopolitical relations.

- Identify the main reasons behind East Asia's rapid economic growth in recent decades and discuss any possible limitations to continued expansion at such a rate.

- Describe the differences in economic and social development found across China and, more generally, across East Asia as a whole.

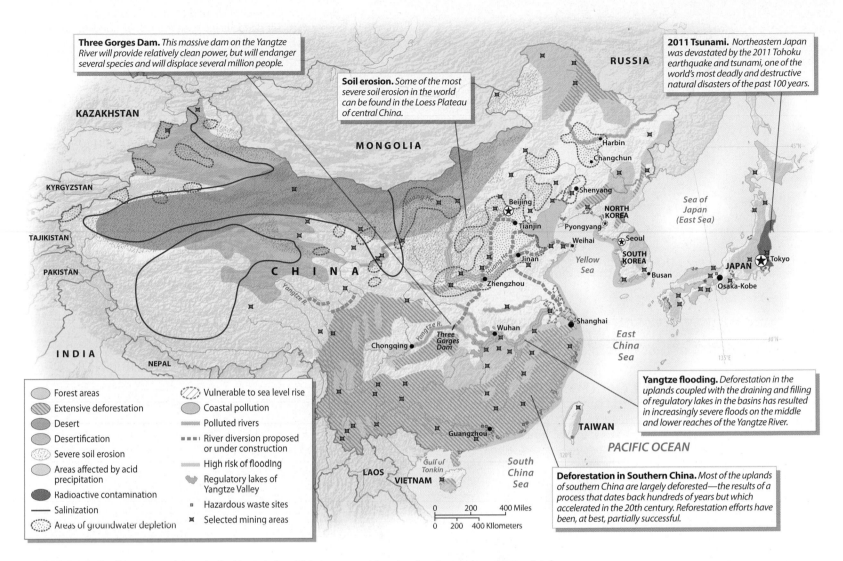

Three Gorges Dam. *This massive dam on the Yangtze River will provide relatively clean power, but will endanger several species and will displace several million people.*

Soil erosion. *Some of the most severe soil erosion in the world can be found in the Loess Plateau of central China.*

2011 Tsunami. *Northeastern Japan was devastated by the 2011 Tohoku earthquake and tsunami, one of the world's most deadly and destructive natural disasters of the past 100 years.*

Yangtze flooding. *Deforestation in the uplands coupled with the draining and filling of regulatory lakes in the basins has resulted in increasingly severe floods on the middle and lower reaches of the Yangtze River.*

Deforestation in Southern China. *Most of the uplands of southern China are largely deforested—the results of a process that dates back hundreds of years but which accelerated in the 20th century. Reforestation efforts have been, at best, partially successful.*

Legend:
- Forest areas
- Extensive deforestation
- Desert
- Desertification
- Severe soil erosion
- Areas affected by acid precipitation
- Radioactive contamination
- Salinization
- Areas of groundwater depletion
- Vulnerable to sea level rise
- Coastal pollution
- Polluted rivers
- River diversion proposed or under construction
- High risk of flooding
- Regulatory lakes of Yangtze Valley
- Hazardous waste sites
- Selected mining areas

FIGURE 11.3 Environmental Issues in East Asia This huge world region has been almost completely transformed from its natural state and continues to have serious environmental problems. In China, some of the most pressing environmental issues involve deforestation, flooding, water control, and soil erosion.

In the period of recorded history, the Huang He has changed course 26 times, usually causing much loss of human life. Since ancient times, the Chinese have attempted to keep the river within its banks by building ever-larger dikes. Eventually, however, the riverbed rises so high that the flow can no longer be contained. The resulting floods have been known to kill several million people in a single episode. While the river has not changed its course since the 1930s, most geographers agree that another change is inevitable.

Erosion on the Loess Plateau The Huang He's sediment load comes from the eroding soils of the Loess Plateau, located to the west of the North China Plain. **Loess** is a fine, windblown material that

FIGURE 11.4 The Three Gorges of the Yangtze The spectacular Three Gorges landscape of the Yangtze River is the site of a controversial dam that has displaced over 1 million people. Not only are the human costs high to those displaced, but also there may be significant ecological costs to endangered aquatic species.

was deposited on this upland area during the last ice age. In this area, loess deposits accumulated to depths of up to several hundred feet. Loess forms fertile soil, but it washes away easily when exposed to running water. Cultivation requires plowing, which leads to soil erosion. As the population of the region gradually increased, areas of woodland and grassland diminished, leading to ever-greater rates of soil loss. As the erosion process continued, great gullies cut across the plateau, steadily reducing the extent of productive land. Currently, however, major efforts are underway to stop the degradation of the Loess Plateau (see "Working Toward Sustainability: Eco-Development on the Loess Plateau").

WORKING TOWARD SUSTAINABILITY

Eco-Development on the Loess Plateau

China's Loess Plateau has the dubious distinction of suffering from the worst soil erosion problem in the world. Although it is often regarded as the cradle of Chinese civilization, the Loess Plateau is now one of the poorest parts of China. Poverty has been intensified by the loss of soil, deforestation, poor infrastructure, and periodic drought, forcing impoverished peasants to flee the plateau for cities in the more prosperous parts of China. Land rehabilitation programs undertaken in the 1950s and 1960s often failed spectacularly. At that time, the government recruited villagers to plant trees and construct both terraces and small check-dams to reduce erosion, but it provided few incentives to maintain the projects. As a result, newly planted trees often died during dry periods, and poorly constructed terraces and check-dams simply washed away.

In the 1990s, ambitious new sustainable development projects were initiated on the Loess Plateau, with backing from the government of China, the World Bank, and other international and local agencies (Figure 11.1.1). Such projects aimed both to reverse environmental degradation and to increase local wages. By the first decade of the new millennium, senior World Bank leaders were describing a particular project in the area as one of the most successful in the institution's history. Grazing was banned on many of the more vulnerable lands, and both forests and grasslands were successfully regenerated. In 2005, World Bank officials claimed that soil

erosion had been curbed on 920,000 hectares (2.3 million acres) of land and that the sediment flow in the Yellow River (Huang He) had been reduced by approximately 100 million tons a year. Equally important, living standards improved dramatically during the same period. In one case study, the per capita annual income of local farmers was shown to have grown from 585 yuan (roughly US$73) to 1624 yuan (roughly US$200). The same study indicated that the proportion of people below the poverty line dropped from 39 percent in 1999 to 12 percent in 2005. Since then, conditions have continued to improve.

Involving local villagers in the planning and implementation of the sustainable development projects in the Loess Plateau proved crucial to their success. Creating secure property rights in an equitable manner was equally important. Local communities were given the responsibility of reallocating property lines in a fair manner; after this process was completed, each family was given a secure, long-term lease to its own plot of land. As a result, any labor that local people put into reclaiming the land would benefit themselves and their families.

FIGURE 11.1.1 **Sustainable Development on the Loess Plateau** Large-scale eco-development projects funded by the World Bank and other organizations have finally begun to reduce and even reverse the environmental degradation of China's vulnerable Loess Plateau. Terracing and tree-planting efforts have been especially successful.

Similar projects elsewhere in the Loess Plateau have brought similar benefits. Several innovative agricultural practices have proved successful. Examples include *rainwater harvesting agriculture,* which involves constructing wide terraces, using plastic soil covers to reduce evaporation, and building extensive check-dams to catch eroding soil and keep it out of streams and rivers. Another project involves planting fruit trees, particularly apples, in the gullies that have been etched across the loess. The roots of the trees prevent the further entrenchment of the gullies, and the fruit provides a valuable cash crop. According to one recent study, such techniques have the potential of tripling agricultural output, while reducing soil erosion by as much as 70 percent.

FIGURE 11.6 Denuded Hillslopes in China Because of the need to clear forests for wood products and agricultural lands, many of China's mountain slopes have long been deforested. Without tree cover, soil erosion is a serious issue.

Other East Asian Environmental Problems

Although the problems associated with the Yangtze and Huang He rivers are among the most well-known environmental issues in East Asia, they are hardly the only ones. Pollution and deforestation also affect much of the region.

Forests and Deforestation Most of the uplands of China and North Korea support only grass, scrub vegetation, and stunted trees (Figure 11.6). China lacks the historical tradition of forest conservation that characterizes Japan. In much of southern China, sweet potatoes, maize, and other crops have been grown on steep and easily eroded hillsides for several hundred years. After centuries of exploitation, many upland areas have lost so much soil that they cannot easily support forests.

Although the Chinese government has started large-scale reforestation programs, few have been very successful. Today substantial forests are found only in China's far north, where a cool climate prevents fast growth, and along the eastern slopes of the Tibetan Plateau, where rugged terrain restricts commercial forestry. As a result, China suffers a severe shortage of forest resources. As its economy continues to boom, China has become a major importer of lumber, pulp, and paper.

Mounting Pollution As China's industrial base expands, other environmental problems are growing increasingly severe. The burning of high-sulfur coal has resulted in serious air pollution, which is made worse by the increasing number of automobiles. According to a 2010 World Bank study, 16 of the world's 20 most polluted cities are located in China (Figure 11.7). Air pollution results in serious respiratory problems for many Chinese city dwellers. In 2012, the president of the China Medical Association claimed that air pollution would soon become the country's biggest health threat unless the government began to take its dangers more seriously.

Environmental Issues in Japan Considering its large population and intensive industrialization, Japan's environment is relatively clean. Actually, the very density of its population confers certain environmental

advantages. In crowded cities, for example, travel by private automobile is often slower than travel on subways. In the 1950s and 1960s—Japan's most intensive period of industrial growth—the country suffered from some of the world's worst water and air pollution. Soon afterward, the Japanese government passed strict environmental laws.

Japan's environmental cleanup was aided by its location because winds usually carry smog-forming chemicals out to sea. Equally important has been the phenomenon of **pollution exporting**. Because of Japan's high cost of production and its strict environmental laws, many Japanese companies have relocated their dirty factories to other areas, especially China and Southeast Asia. This practice, which is also followed by the United States and western Europe, means that Japan's pollution is partially displaced to poorer countries.

Climate Change and East Asia

East Asia has come to occupy a central position in global warming debates, largely because of China's extremely rapid increase in carbon emissions. From levels only about half those of the United States in 2000, China's total production of greenhouse gases surpassed those of the United States in 2007. This staggering rise has been caused both by China's explosive economic growth and by the fact that it relies on burning coal to generate most of its electricity.

The potential effects of global warming in China have serious implications globally. According to one report, the country's production of wheat, corn, and rice could fall by as much as 37 percent if average temperatures increase by 3.5 to 5.5°F (2 or 3°C) over the next 50 to 80 years. Increased evaporation rates could greatly intensify the water shortages that already plague much of northern China. In the wet zones of southern China, global warming concerns center on the probability of more intense storms in this already flood-prone region.

In June 2007, China released its first national plan on climate change, which called for major gains in energy efficiency, as well as a partial transition to renewable sources of energy. As a result, its government has been heavily subsidizing the manufacturing of solar panels for generating electricity, which China sees as a key technology of the 21st century. China's global warming strategy also includes

FIGURE 11.7 Air Pollution in China Major Chinese cities suffer from some of the worst air pollution in the world. As can be seen in this photo of Shanghai, the resulting haze can make it difficult to see for more than a few city blocks.

a major expansion of nuclear power and the continuation of ambitious reforestation efforts. Current plans call for non-fossil energy to account for 15 percent of China's energy mix by 2020 and about 40 percent by 2050.

China's contribution to the causes of global warming has tended to overshadow those of other East Asian countries. Japan, South Korea, and Taiwan are all major emitters of greenhouse gases, but all also have energy-efficient economies, releasing far less carbon than the United States on a per capita basis. Japanese high-tech companies, like those of South Korea and Taiwan, also realize that it could be very good business to develop new energy technologies to reduce the threat of climate change.

East Asia's Physical Geography

An examination of the physical geography of East Asia helps shed light on the environmental issues it faces. East Asia is situated in the same general latitudinal range as the United States, although it extends considerably farther north and south. The northernmost tip of China lies as far north as central Quebec, while China's southernmost point is at the same latitude as Mexico City. The climate of southern China is thus roughly comparable to that of the Caribbean, while the climate of northern China is similar to that of south-central Canada (Figure 11.8; note particularly the climographs for Hong Kong, Chongqing, Beijing, and Shenyang).

FIGURE 11.8 Climate Map of East Asia East Asia is located in roughly the same latitudinal zone as North America, so there are climatic parallels between the two world regions. The northernmost tip of China lies at about the same latitude as Quebec and shares a similar climate, whereas southern China approximates the climate of Florida. In Japan, maritime influences produce a milder climate.

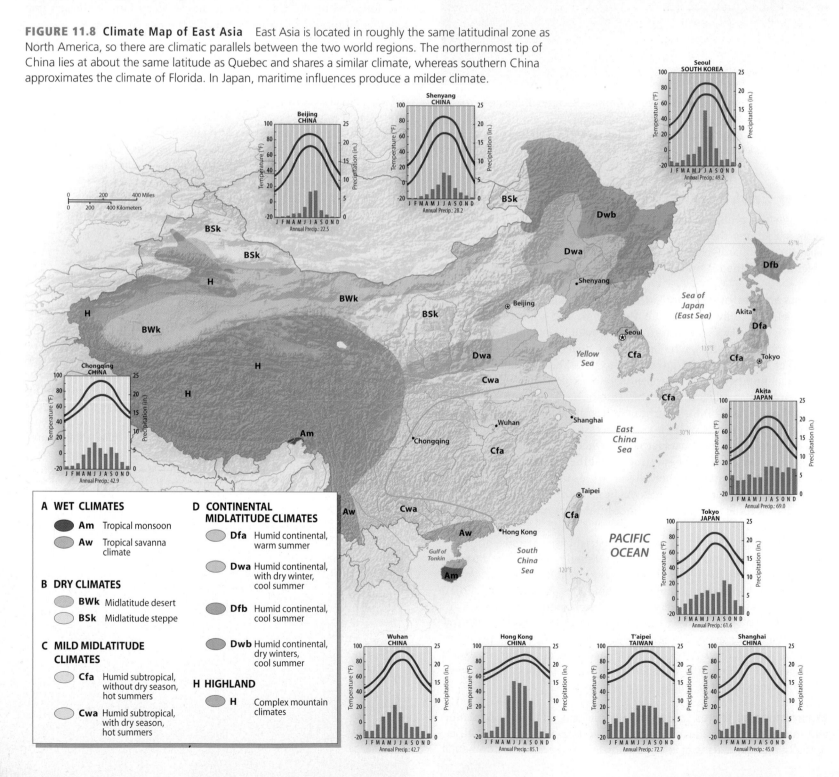

Much of East Asia is geologically active, with earthquakes and volcanic eruptions occurring frequently. Such geological hazards are most common in Japan, but are often more destructive in China, due to lax construction standards. The May 2008 Sichuan earthquake in central China resulted in roughly 70,000 deaths and left over 4 million people homeless. Many children died when poorly built schools collapsed.

Japan's Physical Environment Although slightly smaller than California, Japan is more elongated, extending farther north and south. As a result, Japan's extreme south, in southern Kyushu and the Ryukyu Archipelago, is subtropical, while northern Hokkaido is almost subarctic. Most of the country, however, is distinctly temperate. The climate of Tokyo is similar to that of Washington, DC—although Tokyo receives significantly more rain (see the climograph in Figure 11.8).

The Pacific coast of Japan is separated from the Sea of Japan coast by a series of mountain ranges (Figure 11.9). Japan is one of the world's most rugged countries, with mountainous terrain covering some 85 percent of its territory. Most of these uplands are heavily forested (Figure 11.10). Japan owes its lush forests both to its mild, rainy climate and to its long history of conservation. For hundreds of years, both the Japanese state and its village communities have enforced strict forest-conservation rules, ensuring that timber and firewood extraction would be balanced by tree growth.

Small alluvial plains are located along parts of Japan's coastline and are interspersed among its mountains. These areas have long been cleared and drained for intensive agriculture. The largest Japanese lowland is the Kanto Plain to the north of Tokyo, but even it is only some 80 miles wide and 100 miles long (130 kilometers wide and 160 kilometers long). The country's other main lowland basins are the Kansai, located around Osaka, and the Nobi, centered on Nagoya.

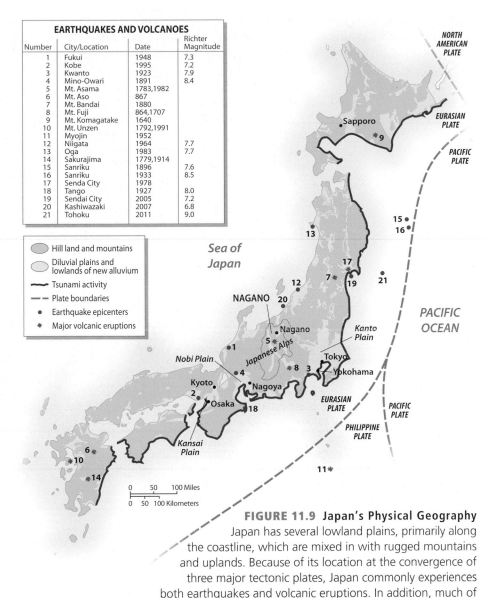

FIGURE 11.9 Japan's Physical Geography Japan has several lowland plains, primarily along the coastline, which are mixed in with rugged mountains and uplands. Because of its location at the convergence of three major tectonic plates, Japan commonly experiences both earthquakes and volcanic eruptions. In addition, much of Japan's coast is vulnerable to devastating tsunamis (tidal waves) caused by earthquakes in the Pacific Basin.

EARTHQUAKES AND VOLCANOES			
Number	City/Location	Date	Richter Magnitude
1	Fukui	1948	7.3
2	Kobe	1995	7.2
3	Kwanto	1923	7.9
4	Mino-Owari	1891	8.4
5	Mt. Asama	1783,1982	
6	Mt. Aso	867	
7	Mt. Bandai	1880	
8	Mt. Fuji	864,1707	
9	Mt. Komagatake	1640	
10	Mt. Unzen	1792,1991	
11	Myojin	1952	
12	Niigata	1964	7.7
13	Oga	1983	7.7
14	Sakurajima	1779,1914	
15	Sanriku	1896	7.6
16	Sanriku	1933	8.5
17	Senda City	1978	
18	Tango	1927	8.0
19	Sendai City	2005	7.2
20	Kashiwazaki	2007	6.8
21	Tohoku	2011	9.0

FIGURE 11.10 Forested Landscapes of Japan Although much of Japan is heavily forested and supports a wood products industry, the yield is not large enough to satisfy demand. As a result, Japan imports timber extensively from North America, Southeast Asia, and Latin America. Densely populated urban areas in Japan often end abruptly at the base of nearby mountains, which in most cases are densely forested, as can be seen in this satellite image of Kyoto.

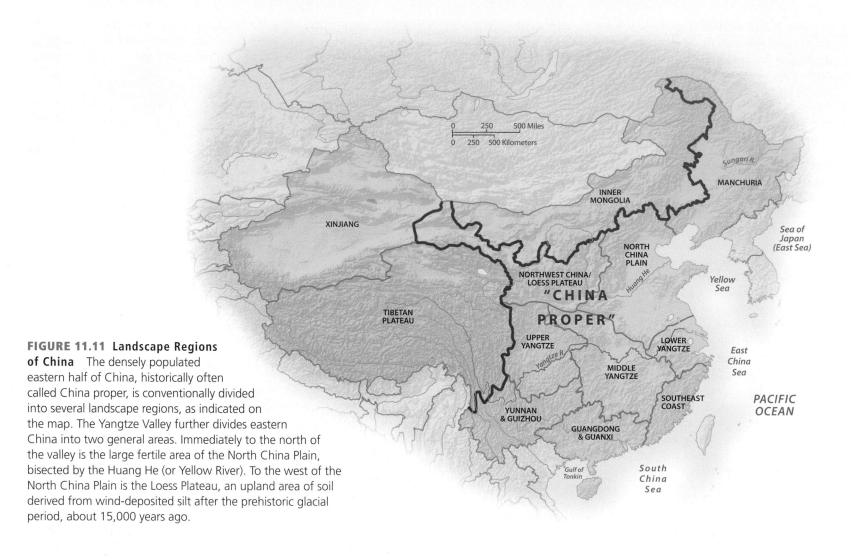

FIGURE 11.11 Landscape Regions of China The densely populated eastern half of China, historically often called China proper, is conventionally divided into several landscape regions, as indicated on the map. The Yangtze Valley further divides eastern China into two general areas. Immediately to the north of the valley is the large fertile area of the North China Plain, bisected by the Huang He (or Yellow River). To the west of the North China Plain is the Loess Plateau, an upland area of soil derived from wind-deposited silt after the prehistoric glacial period, about 15,000 years ago.

Taiwan's Environment Taiwan, an island about the size of Maryland, sits at the edge of the continental landmass. To the west, the Taiwan Strait is only about 200 feet (60 meters) deep; to the east, ocean depths of many thousands of feet are found 10 to 20 miles (15 to 30 kilometers) offshore.

Taiwan's central and eastern regions are rugged and mountainous, while the west is mainly a lowland alluvial plain. Bisected by the Tropic of Cancer, Taiwan has a mild winter climate, but it is often hit by typhoons in the early autumn. Unlike nearby areas of China proper, Taiwan still has extensive forests, especially in its remote central and eastern uplands.

Chinese Environments Even if you exclude China's far western provinces, China is a vast country with diverse environmental regions. For the sake of convenience, it can be divided into two main areas: one lying to the north of the Yangtze (or Chang Jiang) River Valley and the other including the Yangtze and all areas to the south. As Figure 11.11 shows, each of these areas can be subdivided into several distinct regions.

Large valleys and moderate-elevation plateaus are found in the far south, where the climate is tropical. The coastal areas in southeast China are rugged and offer limited agricultural opportunities (Figure 11.12). North of the Yangtze Valley, the climate is both colder

FIGURE 11.12 The Fujian Coast of China The rugged coastal province of Fujian lies in southeast China. Here the coastal plain is narrow and the shoreline deeply indented, producing a striking landscape. Because of limited agricultural opportunities along this rugged coastline, many residents of Fujian work in maritime activities.

North Korea, Russia, and Mongolia. Although winters here can be brutally cold, summers are usually warm and moist. Manchuria's upland areas are home to some of China's best-preserved forests and wildlife refuges.

Korean Landscapes   Korea forms a well-defined peninsula, partially cut off from Northeast China by rugged mountains and sizable rivers. The far north, which just touches Russia's Far East, has a climate similar to that of Maine, whereas the southern tip is more like the Carolinas. Korea, like Japan, is a mountainous land with scattered alluvial basins. The lowlands of the southern portion of the peninsula are more extensive than those of the north, giving South Korea a distinct agricultural advantage over North Korea. The north, however, has far more abundant natural resources. The uplands of North Korea have been heavily deforested, whereas those of South Korea have seen extensive reforestation since the end of World War II.

FIGURE 11.13 Drought in Northern China   The Yellow River (or Huang He) is sometimes called the "cradle of Chinese civilization" owing to its historical importance. However, due to the increasing extraction of water for agriculture and industry, the river now often runs dry in its lower reaches. In this photo, two children are playing on a boat in the dried-up riverbed in China's Henan Province.

REVIEW QUESTIONS

1. Why has China become the world's largest emitter of greenhouse gases, and what is its government doing about this problem?

2. Why is Japan so much more heavily forested than China?

---

and drier. Summer rainfall is generally abundant, but the other seasons are usually dry. The North China Plain, a large, flat area of fertile soil crossed by the Huang He (or Yellow River), is cold and dry in winter and hot and humid in summer. Overall precipitation is somewhat low and unpredictable, and much of the North China Plain is in danger of **desertification** (the spread of desert conditions), which often results in sand- and dust-storms. Seasonal water shortages are growing increasingly severe through much of the region as withdrawals for irrigation and industry increase (Figure 11.13).

China's northeastern region is usually referred to as Manchuria in English, but Chinese writers generally call it simply Northeast China. It is dominated by a broad, fertile lowland sandwiched between mountains and uplands stretching along China's borders with

POPULATION AND SETTLEMENT: A REALM OF CROWDED LOWLAND BASINS

East Asia is one of the most densely populated regions of the world (Table 11.1). The lowlands of Japan, Korea, and China are among the most intensely used portions of Earth, containing not only the major cities, but also most of the agricultural lands of these countries. Although the density of East Asia is extremely high, the region's population growth rate has declined dramatically (Figure 11.14). In Japan, the current concern is population loss. Although China's population is still expanding, its rate of growth is low by global standards.

TABLE 11.1   POPULATION INDICATORS

Country	Population (millions) 2012	Population Density (per square kilometer)	Rate of Natural Increase (RNI)	Total Fertility Rate	Percent Urban	Percent <15	Percent >65	Net Migration (Rate per 1000) 2010–12[a]
China	1,350.4	141	0.5	1.5	51	16	9	−0.3
Hong Kong	7.1	6,487	0.8	1.2	100	12	14	7.9
Japan	127.6	338	−0.2	1.4	86	13	24	0.4
North Korea	24.6	204	0.5	2.0	60	23	9	0.0
South Korea	48.9	491	0.4	1.2	82	16	11	−0.1
Taiwan	23.3	646	0.2	1.1	78	15	11	

[a]Net Migration Rate from the United Nations, Population Division, *World Population Prospects: The 2010 Revision Population Database.*
Source: Population Reference Bureau, *World Population Data Sheet, 2012.*

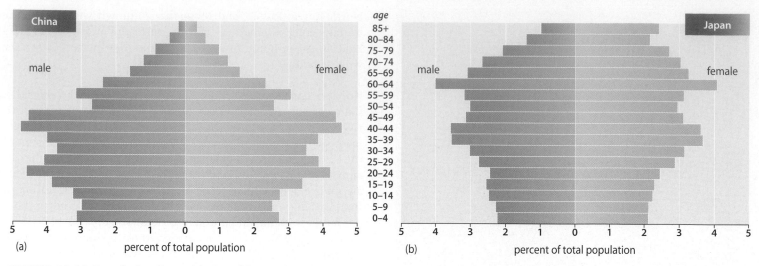

(a)   percent of total population   (b)   percent of total population

FIGURE 11.14 Population Pyramids for China and Japan   Japan has one of the world's oldest, and most rapidly aging, populations, due both to its low birthrate and its high life-expectancy figure. The large Japanese population in the over-60 category places a large burden on the country's economy. China has a more balanced demographic profile, but it also has a low birth-rate, which could result in similar problems arising in another 10 to 20 years.

Japanese Settlement and Agricultural Patterns

Japan is a highly urbanized country, supporting two of the largest metropolitan areas in the world: Tokyo and Osaka. It is also one of the world's most mountainous countries, and its uplands are lightly populated. Agriculture must therefore share the limited lowlands with cities and suburbs, resulting in extremely intensive farming practices.

Japan's Agricultural Lands   Japanese agriculture is largely limited to the country's coastal plains and interior basins. Rice is Japan's major crop, and irrigated rice demands flat land. Japanese rice farming has long been one of the most productive forms of agriculture in the world, helping to support a large population—some 127 million people—on a relatively small and rugged land. Although rice is grown in almost all Japanese lowlands, the country's premier rice-growing districts lie along the Sea of Japan coast of central and northern Honshu.

Vegetables are also grown intensively in all the lowland basins, even on tiny patches within urban neighborhoods (Figure 11.15). The valleys of central and northern Honshu are famous for their temperate-climate fruit, while citrus comes from the milder southwestern reaches of the country. Crops that thrive in a cooler climate, such as potatoes, are produced mainly in Hokkaido and northern Honshu.

Settlement Patterns   All Japanese cities—and the vast majority of the Japanese people—are located in the same lowlands that support the country's agriculture. Many of Japan's more remote areas, moreover, are currently experiencing population loss as younger people increasingly move to larger cities. Not surprisingly, the three largest metropolitan areas—Tokyo, Osaka, and Nagoya—sit near the centers of the three largest plains.

Overall, Japan is a densely populated country (Figure 11.16). The fact that its settlements are largely restricted to the lowlands means that Japan's effective population density—the actual crowding that the country experiences—is one of the highest in the world. This is especially true in the main industrial belt, which extends from Tokyo south and west through Nagoya and Osaka to the northern coast of Kyushu. Due to such space limitations, all Japanese cities are characterized by dense settlement patterns. In the major urban areas, the amount of available living space is highly restricted for all but the

FIGURE 11.15 Japanese Urban Farm   Japanese landscapes often combine dense urban settlement with small patches of intensively farmed land. Here a small farm is immediately adjacent to an urban neighborhood.

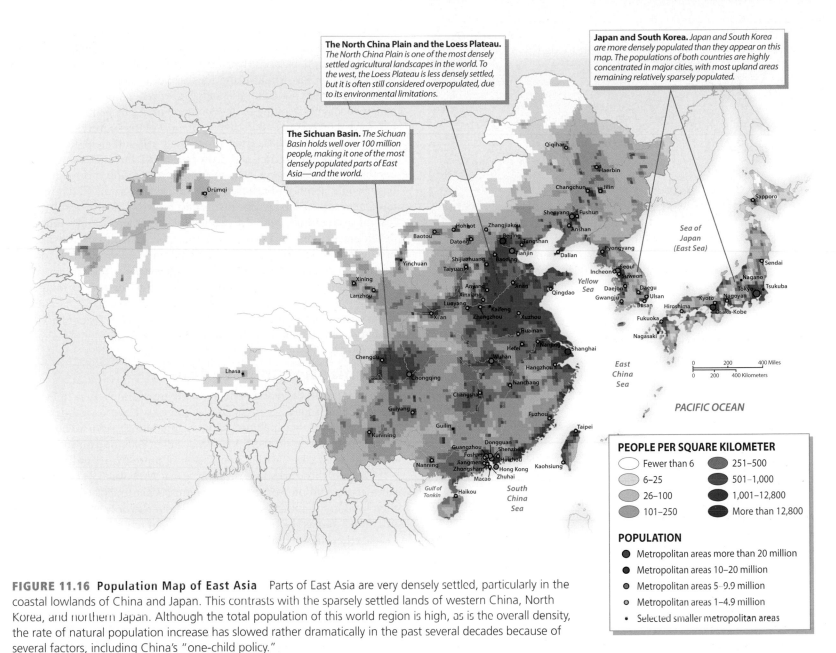

The North China Plain and the Loess Plateau. *The North China Plain is one of the most densely settled agricultural landscapes in the world. To the west, the Loess Plateau is less densely settled, but it is often still considered overpopulated, due to its environmental limitations.*

The Sichuan Basin. *The Sichuan Basin holds well over 100 million people, making it one of the most densely populated parts of East Asia—and the world.*

Japan and South Korea. *Japan and South Korea are more densely populated than they appear on this map. The populations of both countries are highly concentrated in major cities, with most upland areas remaining relatively sparsely populated.*

PEOPLE PER SQUARE KILOMETER

- Fewer than 6
- 6–25
- 26–100
- 101–250
- 251–500
- 501–1,000
- 1,001–12,800
- More than 12,800

POPULATION

- Metropolitan areas more than 20 million
- Metropolitan areas 10–20 million
- Metropolitan areas 5–9.9 million
- Metropolitan areas 1–4.9 million
- Selected smaller metropolitan areas

FIGURE 11.16 Population Map of East Asia   Parts of East Asia are very densely settled, particularly in the coastal lowlands of China and Japan. This contrasts with the sparsely settled lands of western China, North Korea, and northern Japan. Although the total population of this world region is high, as is the overall density, the rate of natural population increase has slowed rather dramatically in the past several decades because of several factors, including China's "one-child policy."

most affluent families. Many observers argue that Japan should allow its cities and suburbs to expand into nearby rural areas. However, because most uplands are too steep for residential use, such expansion would have to occur at the expense of agricultural land.

Settlement and Agricultural Patterns in China, Taiwan, and Korea

Like Japan, Taiwan and South Korea are essentially urban. China, however, remains largely rural, with only about half its population living in cities. Almost all Chinese cities are growing at a rapid pace, however, despite government efforts to keep their size under control.

China's Agricultural Regions   A line drawn just to the north of the Yangtze Valley divides China into two main agricultural regions. To the south, rice is the dominant crop. To the north, wheat, millet, and sorghum are the most common.

In southern and central China, population is highly concentrated in the broad lowlands, which are famous for their fertile soil and intensive agriculture. Planting and harvesting of crops occurs year-round in most of southern and central China; summer rice alternates with winter barley or vegetables, and in many areas two rice crops as well as one winter crop are grown. Southern China also produces a wide variety of tropical and subtropical crops, and moderate slopes throughout the area produce sweet potatoes, corn, and other upland crops.

In northern China, population distribution is more variable. The North China Plain has long been one of the most thoroughly **anthropogenic landscapes** in the world (that is, a landscape that has been heavily transformed by human activities). Virtually its entire extent is either cultivated or occupied by houses, factories, and other structures of human society. Manchuria, on the other hand, was a lightly populated frontier zone as recently as the mid-1800s. Today, with a population of more than 100 million, its central plain is thoroughly settled. The Loess Plateau is also relatively crowded, despite its harsh environment (Figure 11.17).

FIGURE 11.17 Loess Settlement The Loess Plateau is densely settled, considering its harsh environment. Over much of the plateau, subterranean dwellings are carved out of the soft sediments. This area is prone to major earthquakes that exact a high toll on the local population because of dwelling collapses.

Settlement and Agricultural Patterns in Korea and Taiwan

Korea is also densely populated. It contains some 73 million people (24.6 million in North Korea and 28.9 million in South Korea) in an area smaller than Minnesota. South Korea's population density is actually significantly higher than that of Japan. Most Koreans are crowded into the alluvial plains and basins of the west and south. The highland spine, extending from the far north to northeastern South Korea, remains relatively sparsely settled. South Korean agriculture is dominated by rice. North Korea relies heavily on corn and other upland crops that do not require irrigation.

Taiwan is the most densely populated state in East Asia. Roughly the size of the Netherlands, it contains more than 23 million inhabitants. Its overall population density is one of the highest in the world. Because mountains cover most of central and eastern Taiwan, virtually the entire population is concentrated in the narrow lowland belt in the north and west. In this area, large cities and numerous factories are scattered amid lush farmlands.

East Asian Agriculture and Resources in Global Context

Although East Asian agriculture is highly productive, it cannot feed the huge number of people who live in the region. Japan, Taiwan, and South Korea are major food importers, and China has recently moved in the same direction. Other resources are also being drawn in from all quarters of the world by the powerful economies of East Asia.

Japan may be self-sufficient in rice, but it is still one of the world's largest food importers. Japan imports much of its meat and the feed used in its domestic livestock industry from the United States, Brazil, Canada, and Australia. These same countries supply soybeans and wheat. Japan has one of the highest rates of fish consumption in the world, and the Japanese fishing fleets must scour the world's oceans to meet the demand. Japan also depends on imports to supply its demand for forest resources. Although its own forests produce high-quality cedar and cypress logs, Japan buys most of its construction lumber and pulp (for papermaking) from western North America and Southeast Asia.

South Korea has followed Japan in obtaining food and forestry resources from abroad. In 2008, as global grain prices increased rapidly, large South Korean companies began to negotiate long-term leases for vast tracts of farmland in poor tropical countries. South Korean companies are also investing heavily in natural resource extraction in other parts of the world, particularly in central Asia. Unlike South Korea, North Korea relies on its own farmland for most of it food production. North Korea's inefficient agricultural system, however, is unable to supply adequate foodstuff, resulting in devastating famines in years of drought or excessive rain, as well as the stunting of most of the country's population.

Through the 1980s, China remained self-sufficient in food, despite its huge population and crowded lands. But increasing wealth, combined with changing diets and the introduction of foreign foods, has brought about an increased consumption of meat, which requires large amounts of feed grain. Economic growth has also resulted in a loss of agricultural lands to residential, commercial, and industrial development. China's current demands for agricultural and mineral products are reordering patterns of global trade. Many Latin American, African, and Asian countries now rely heavily on the Chinese market. China continues to invest heavily in infrastructure, farming, and mining projects in Africa, Latin America, and elsewhere. As a result, some observers fear that certain parts of Africa are at risk of becoming virtual economic colonies of China (Figure 11.18).

Urbanization in East Asia

China has one of the world's oldest urban foundations. In medieval and early modern times, East Asia as a whole possessed a well-developed system of cities that included some of the largest settlements on the planet. In the early 1700s, Tokyo, then called Edo, probably overshadowed all other cities, with a population of more than 1 million.

Despite this early urbanization, East Asia was overwhelmingly rural at the end of World War II. Some 90 percent of China's people then lived in the countryside, and even Japan was only about 50 percent urbanized. But as the region's economy began to grow after the war, so did its cities. Japan, Taiwan, and South Korea are now between 78 and 86 percent urban, which is typical for advanced industrial countries.

FIGURE 11.18 Chinese Overseas Investments China is now investing vast sums of money in many of the world's underdeveloped countries. Critics contend that Chinese investment is aimed merely at securing natural resources, whereas supporters argue that it brings many local benefits, both economic and cultural. Here Senegalese and Chinese workers observe a ceremony at the site of a new national theater in Dakar, Senegal, financed by the Chinese government.

A little more than half of the people of China now live in cities, rising from a figure of only 26 percent in 1990. The current movement of Chinese people from the countryside to urban areas is one of the largest population transfers that the world has ever seen.

Chinese Cities Traditional Chinese cities were clearly separated from the countryside by defensive walls. Most were planned in accordance with strict geometric principles, with straight streets meeting at right angles. The old-style Chinese city was dominated by low buildings. Houses were typically built around courtyards, and narrow alleyways served both commercial and residential functions.

China's cities began to change as Europeans started to gain power in the region in the 1800s. Several port cities were taken over by European interests, which proceeded to build Western-style buildings and modern business districts. By far the most important of these semicolonial cities was Shanghai, built near the mouth of the Yangtze River, the main gateway to interior China. Although Shanghai declined after the Communist Party came to power in 1949, it has recently experienced a major revival. Migrants are now pouring into Shanghai, and building cranes crowd the skyline. Official statistics put the population of the metropolitan area at 23 million.

Beijing was China's capital during the Manchu period (1644–1912), a status it regained in 1949. Under communist rule, Beijing was radically transformed; old buildings were razed, and broad avenues were plowed through old neighborhoods. Crowded residential districts gave way to large blocks of apartment buildings and massive government offices. Some historically significant buildings were saved; the buildings of the Forbidden City, for example, where the Manchu rulers once lived, survived as a complex of museums.

In the 1990s, Beijing and Shanghai vied for primary position among Chinese cities, with Tianjin, serving as Beijing's port, coming in a close third. All three of these cities have historically been removed from the regular provincial structure of the country and granted their own metropolitan governments. In 1997, another major city, Hong Kong, passed from British to Chinese control and was granted a distinctive status as a self-governing "special administrative region." The greater metropolitan area of the Xi Delta—composed of Hong Kong, Shenzhen, and Guangzhou (called Canton in the West)—is now one of China's premier urban areas.

City Systems of Japan and South Korea The urban structures of Japan and South Korea are quite different from those of China. South Korea is noted for its pronounced **urban primacy** (the concentration of total urban population in a single city), whereas Japan is the center of a new urban phenomenon—the **superconurbation**, also called a **megalopolis** (a huge zone of coalesced metropolitan areas). (We saw another example of a megalopolis in Chapter 3: the Boston/New York/Philadelphia/Washington, DC corridor.)

Seoul, the capital of South Korea, is by far the largest city in the country. Built along the banks of the Han River, the city is home to more than 10 million people, and its greater metropolitan area contains some 40 percent of South Korea's total population (Figure 11.19). All of South Korea's major governmental, economic, and cultural institutions are concentrated there. However, Seoul's explosive and generally unplanned growth has resulted in serious congestion.

FIGURE 11.19 Contemporary Seoul This vibrant metropolis of more than 10 million inhabitants is the primate city of South Korea. Seoul is nestled among forested mountains, a location that is considered very favorable according to the principles of geomancy (a landscape-based system of divination common in East Asia).

Japan has traditionally been characterized by urban "bipolarity," rather than urban primacy. Until the 1960s, Tokyo, the capital and main business and educational center, together with the neighboring port of Yokohama, was balanced by the trading center of Osaka and its port, Kobe (Figure 11.20). Kyoto, the former imperial capital and the

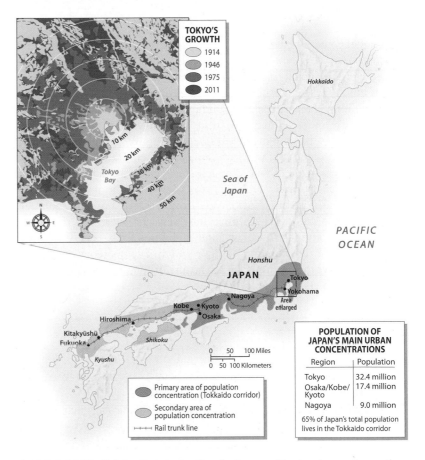

TOKYO'S GROWTH
- 1914
- 1946
- 1975
- 2011

POPULATION OF JAPAN'S MAIN URBAN CONCENTRATIONS

Region	Population
Tokyo	32.4 million
Osaka/Kobe/Kyoto	17.4 million
Nagoya	9.0 million

65% of Japan's total population lives in the Tokaido corridor

- Primary area of population concentration (Tokkaido corridor)
- Secondary area of population concentration
- Rail trunk line

FIGURE 11.20 Urban Concentration in Japan The inset map shows the rapid expansion of Tokyo in the postwar decades. Today the greater Tokyo metropolitan area is home to almost 30 million people. The larger map shows the cluster of urban settlements along Japan's southeastern coast. The major area of urban concentration is between Tokyo and Osaka, a distance of some 300 miles, known as the *Tokaido corridor*. Roughly 65 percent of Japan's population lives in this area. The Osaka–Kobe metropolitan area ranks second to Tokyo, with approximately 14 million inhabitants.

FIGURE 11.21 Tokyo Apartments The extremely high population density of Tokyo and other large Japanese cities forces most people to live in crowded apartment blocks.

traditional center of elite culture, is also located in the Osaka region. As Japan's economy boomed in the 1960s through the 1980s, however, so did Tokyo: The capital city outpaced all other urban areas in almost every urban function. The Greater Tokyo metropolitan area today contains 25 to 35 million persons, depending on how its boundaries are defined.

Japanese cities sometimes strike foreign visitors as rather gray and dull places, lacking historical interest (Figure 11.21). Little of the country's premodern architecture remains intact. Traditional Japanese buildings were made of wood, which survives earthquakes much better than stone or brick. However, fires have therefore been a long-standing hazard, and in World War II the U.S. Air Force firebombed most Japanese cities. (Hiroshima and Nagasaki were, on the other hand, completely destroyed by nuclear bombs.) The one exception was Kyoto, the old imperial capital, which was spared devastation. As a result, Kyoto is famous for its beautiful temples, which ring the basin in which central Kyoto lies.

REVIEW QUESTIONS

1. Why does East Asia get so much of its food and so many of its natural resources from other parts of the world?

2. How is the urban landscape of East Asia, and particularly that of China, currently changing?

CULTURAL COHERENCE AND DIVERSITY: A CONFUCIAN REALM?

East Asia is in some respects one of the world's most culturally unified regions. Although different parts of East Asia have their own particular cultures, the entire region shares certain historically rooted ways of life and systems of ideas. Most of these common features can be traced back to ancient Chinese civilization. Chinese culture emerged roughly 4,000 years ago, largely in isolation from the eastern hemisphere's other early centers of civilization in the valleys of the Indus, Tigris-Euphrates, and Nile rivers.

Unifying Cultural Characteristics

The most important unifying cultural characteristics of East Asia are related to religious and philosophical beliefs. Throughout the region, Buddhism and especially Confucianism have shaped both individual beliefs and social and political structures. Although the role of traditional belief systems has been seriously challenged in recent decades, especially in China, traditional cultural patterns remain.

The Chinese Writing System The clearest distinction between East Asia and the world's other cultural regions is found in written language. Writing systems elsewhere in the modern world are based on the alphabetic principle, in which each symbol represents a distinct sound. East Asia evolved an entirely different system of **ideographic writing**, in which each symbol (or ideograph—often called *character*) usually represents an idea, rather than a sound.

The East Asian writing system can be traced to the dawn of Chinese civilization. As the Chinese Empire expanded, the Chinese writing system spread. Japan, Korea, and Vietnam all came to use the same system, although in Japan it was substantially modified, while in Korea it was later largely replaced by an alphabetic system. A major disadvantage of Chinese writing is that it is difficult to learn; to be literate, a person must memorize thousands of characters. Its main benefit is that two literate persons do not have to speak the same language to be able to communicate because the written symbols they use to express their ideas are the same.

The writing system of Japan is particularly complex (Figure 11.22). Initially, the Japanese simply borrowed Chinese characters, referred to in Japanese as **kanji**. However, because of the grammatical differences between spoken Japanese and Chinese, the exclusive use of *kanji* resulted in awkward sentences. The Japanese solved this problem by developing **hiragana**, a writing system in which each symbol represents a distinct syllable, or combination of a consonant and a vowel sound. A parallel system, called *katakana,* was devised for spelling words of foreign origin. Written Japanese employs a complex mixture of all three of these symbolic systems.

The Confucian Legacy Just as the use of a common writing system helped build cultural linkages throughout East Asia, so, too, the idea system of **Confucianism** (the philosophy developed by Confucius) came to occupy a significant position in the region. Indeed, so strong is the heritage of Confucius that some writers refer to East Asia as the "Confucian world."

The premier philosopher of China, Confucius (or Kung Fu Zi, in Mandarin Chinese) was born in 551 BCE, a period of political instability. Confucius's goal was to create a philosophy that could generate social stability. While Confucianism is sometimes considered to be a religion, Confucius himself was far more interested in how to lead a correct life and organize a proper society. He stressed obedience to authority, but thought that those in power must act in a caring manner. Confucian philosophy also emphasizes education. The most basic level of the traditional Confucian moral order is the family unit,

FIGURE 11.22 Japanese Writing The writing system of Japan was originally based on Chinese characters, known in Japan as *kanji*. Because of grammatical differences, however, the Japanese developed two unique "alphabets" of syllables, known as *katakana* and *hiragana*. *Kanji* and *katakana* symbols are visible in this photo.

considered the bedrock of society. The ideal family structure is patriarchal (male-dominated), and children are taught to obey and respect their parents.

The significance of Confucianism in East Asian development has long been debated. In the early 1900s, many observers believed that this conservative philosophy, based on respect for tradition and authority, was responsible for the economically backward position of China and Korea. But because East Asia has more recently enjoyed the world's fastest rates of economic growth, such a position is no longer supportable. New arguments claim that Confucianism's respect for education and the social stability that it generates give East Asia an advantage in international competition. It must also be recognized, however, that Confucianism has lost much of the hold that it once had on public morality throughout East Asia.

Religious Unity and Diversity in East Asia

Certain religious beliefs have worked alongside Confucianism to unite the East Asian region. The most important culturally unifying beliefs are associated with Mahayana Buddhism. Other religious practices, however, have tended to challenge the cultural unity of the region.

Mahayana Buddhism Buddhism, a religion that stresses the quest to escape an endless cycle of rebirths and reach union with the cosmos (a state called nirvana), originated in India in the 6th century BCE. By the 2nd century CE, Buddhism had reached China, and within a few hundred years it had spread throughout East Asia. Today Buddhism remains widespread everywhere in the region, although it is far less significant here than it is in mainland Southeast Asia, Sri Lanka, and Tibet (Figure 11.23).

The variety of Buddhism practiced in East Asia—Mahayana, or Great Vehicle—is distinct from the Therevada Buddhism of Southeast Asia. Most importantly, Mahayana Buddhism simplifies the quest for nirvana, in part by putting forward the existence of beings who refuse divine union for themselves in order to help others spiritually. Mahayana Buddhism also allows its followers to practice other religions. Thus, many Japanese are both Buddhists and Shintoists, while many Chinese consider themselves to be both Buddhists and Taoists (as well as Confucianists).

Shinto The Shinto religion is so closely bound to the idea of Japanese nationality that it is questionable whether a non-Japanese person can follow it. Shinto began as the worship of nature spirits, but it was gradually refined into a subtle set of beliefs about the harmony of nature and its connections with human existence. Shinto is still a place- and nature-centered religion. Certain mountains, particularly Mount Fuji, are considered sacred and are thus climbed by large numbers of people. Major Shinto shrines, often located in scenic places, attract numerous religious pilgrims. The most notable of these is the Ise Shrine south of Nagoya, devoted to the emperor of Japan (Figure 11.24).

Taoism and Other Chinese Belief Systems Similar to Shinto, the Chinese religion Taoism (or Daoism) is rooted in nature worship. Also like Shinto, it stresses spiritual harmony. Taoism is indirectly associated with *feng shui*, which is the Chinese and Korean practice

FIGURE 11.23 The Buddhist Landscape Mahayana Buddhism has traditionally been practiced throughout East Asia. This Golden Buddha statue is located in Baomo Park in Chi Lei Village, Guangdong Province, China.

however, rigid Marxist beliefs are still strictly enforced. North Korea is also noted for its official ideology of *juche*, or "self-reliance." Ironically, juche demands absolute loyalty to the country's repressive political leaders.

Linguistic and Ethnic Diversity in East Asia

Written languages may have helped unify East Asia, but the same cannot be said for spoken languages (Figure 11.25). Japanese and Mandarin Chinese may partially share a system of writing, but the two languages have no direct relationship. Like Korean, however, Japanese has adopted many words of Chinese origin.

FIGURE 11.24 Japan's Ise Grand Shrine   The Ise shrine complex in central Japan is often regarded as the central place of the Japanese religion of Shinto. The Ise Grand Shrine, shown here, is dedicated to the goddess Amaterasu-omikami.

of designing buildings in accordance with the spiritual powers that supposedly flow through the local topography. Even in hypermodern Hong Kong, skyscrapers worth millions of dollars have occasionally gone unoccupied because their construction failed to follow feng shui principles.

Minority Religions   Followers of virtually all world religions can be found in East Asia. More than 1 million Japanese, for example, belong to Christian churches. Christianity has spread much more extensively in South Korea, where it is now followed by between 25 and 35 percent of the population. South Korea now sends more Christian missionaries abroad than any other country except the United States. Christianity is also spreading rapidly in China, causing Beijing's communist leadership some concern. In 2008, official Chinese statistics listed some 20 million Protestants and 10 million Roman Catholics, but some independent sources argue that the Christian population could be much higher. A 2011 report by the Pew Forum on Religion and Public Life put the figure at 67 million.

China's Muslim community is much more deeply rooted than its Christian population. Roughly 10 million Chinese-speaking Muslims, called Hui, are concentrated in Gansu and Ningxia in the northwest and in Yunnan Province along the south-central border. Smaller clusters of Hui, often separated in their own villages, live in almost every province of China.

Secularism in East Asia   Despite all its varied forms of religious expression, East Asia is one of the most secular regions of the world. In Japan, although a small section of the population is highly religious, most people only occasionally observe Shinto or Buddhist rituals. Japan also has several "new religions," a few of which are noted for their strong beliefs. But for Japanese society as a whole, religion is not particularly important.

After the communist regime took power in China in 1949, all forms of religion and traditional philosophy—including Confucianism—were severely discouraged. Under the new regime, **Marxism** (or communism), an atheistic philosophy, became the official belief system. With the easing of Marxism during the 1980s and 1990s, however, many forms of religious expression began to return. In North Korea,

Language and National Identity in Japan   Japanese, according to most linguists, is not related to any other language. Korean is also usually classified as the only member of its language family. Some linguists, however, think that Japanese and Korean should be classified together because they share many basic grammatical features.

From many perspectives, the Japanese form one of the world's most homogeneous peoples. In earlier centuries, however, the Japanese islands were divided between two very different peoples: the Japanese living to the south and the Ainu inhabiting the north. The Ainu are physically distinct from the Japanese and possess their own language (Figure 11.26). For centuries, the two groups competed for land, and by the 10th century CE the Ainu were mostly restricted to the northern island of Hokkaido. Today only about 25,000 Ainu remain, and the Ainu language is almost extinct.

The Japanese language is divided into several dialects, but only in the Ryukyu Islands is a variant of Japanese spoken that is so distinct it might be considered a separate language. Many Ryukyu people believe that they have not been considered full members of the Japanese nation, and they have suffered a certain amount of discrimination.

Approximately 600,000 people of Korean descent living in Japan today have also felt discrimination. Most of them were born in Japan and speak Japanese, rather than Korean. Despite their deep bonds to Japan, however, such individuals are not easily able to obtain Japanese citizenship. Perhaps as a result of such treatment, a substantial minority of Japanese Koreans hold radical political views and have tended to support North Korea.

Starting in the 1980s, other immigrants began to arrive in Japan, mostly from the poorer countries of Asia. Because Japan severely restricts the flow of immigrants, many do not have legal status. Men from China and southern Asia typically work in construction; women from Thailand and the Philippines often work as entertainers or prostitutes. Overall, immigration is less pronounced in Japan than in most other wealthy countries, and relatively few migrants acquire permanent residency, let alone citizenship.

Language and Identity in Korea   The Koreans are also a relatively homogeneous people. The vast majority of people in both North and South Korea speak Korean and consider themselves to be members of the Korean nation. However, a strong sense of regional identity persists, which can be traced back to the medieval period, when the peninsula was divided into three separate kingdoms.

Not all Koreans live in Korea. Several million reside directly across the border in northern China. Desperately poor North Koreans

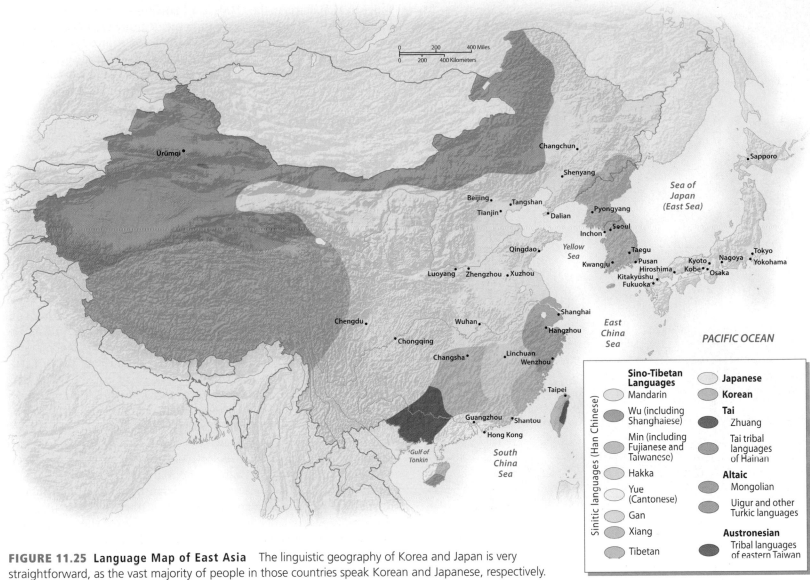

FIGURE 11.25 Language Map of East Asia   The linguistic geography of Korea and Japan is very straightforward, as the vast majority of people in those countries speak Korean and Japanese, respectively. In China, the dominant Han Chinese speak a variety of closely related *Sinitic* languages, the most important of which is Mandarin Chinese. In the peripheral regions of China, a large number of languages—belonging to several different linguistic families—are spoken.

often try to sneak across the border to join these Korean-speaking Chinese communities, but the Chinese government regards such migrants as a security threat and thus returns them to North Korea when it can. A more recent Korean **diaspora** (a scattering of a particular group of people over a vast geographical area) has brought hundreds of thousands of people to the United States, Canada, Australia, New Zealand, the Philippines, and other countries. Most of the estimated 50 million people of Chinese background who live in other countries trace their ancestries to these non-Mandarin-speaking areas of southeastern China.

Language and Ethnicity Among the Han Chinese   The geography of language and ethnicity in China is more complex than that of Korea or Japan. This is true even if we consider only the eastern half of the country, so-called *China proper*. The most important distinction is that separating the Han Chinese from the non-Han peoples. The Han, who form the vast majority, are those people who have historically been incorporated into the Chinese cultural and political systems and

whose languages are expressed in the Chinese writing system. They do not, however, all speak the same language.

The area of northern, central, and southwestern China—a vast area extending from Manchuria through the middle and upper Yangtze Valley to the valleys of Yunnan in the far south—constitutes a single linguistic zone. The spoken language here is called Mandarin Chinese in English. In China today, Mandarin—locally called *Putonghua*, or "common language"—is the national tongue.

In southeastern China, from the Yangtze Delta to China's border with Vietnam, several separate, but related languages are spoken. Traveling from south to north, we encounter Cantonese (or Yue), spoken in Guangdong Fujianese (alternatively Hokkienese, or Min locally), spoken in Fujian and Shanghaiese (or Wu), spoken in and around Shanghai. Linguistically speaking, these are true languages because they are not mutually intelligible. They are usually called dialects, however, because they have no distinctive written form.

Despite their many differences, all the languages of the Han Chinese are closely related to each other, belonging to the Sinitic language

FIGURE 11.26 Ainu Men The indigenous Ainu people of northern Japan are much reduced in population, but they still maintain some of their cultural traditions. Here Ainu men participate in the Marimo Festival on the northern Japanese island of Hokkaido.

subfamily. Because their basic grammars and sound systems are similar, it is not difficult for a person speaking one of these languages to learn another. All Sinitic languages are **tonal**, so the meaning of each basic syllable changes according to the pitch in which it is uttered, and most of their words are composed of a single syllable (although compound words can be formed from several syllables).

The Non-Han Peoples Many of the remote upland districts of China proper are inhabited by various groups of non-Han peoples speaking non-Sinitic languages. Such peoples are often classified as **tribal**, implying that they have a traditional social order based on self-governing village communities. Such a view is not entirely accurate, however, because some of these groups once had their own kingdoms and all are now subject to the Chinese state (Figure 11.27).

As many as 11 million Manchus live in Manchuria. Few Manchus, however, speak their own language, having abandoned it for Mandarin Chinese. This is an ironic situation because the Manchus ruled the entire Chinese Empire from 1644 to 1912. Until the later part of this period, the Manchus prevented the Han from settling in central and northern Manchuria. Once Han Chinese were allowed to

move to Manchuria in the 1800s, the Manchus soon found themselves vastly outnumbered. As a result, their own culture began to disappear.

Much larger communities of non-Han peoples are found in south-central China, especially in Guangxi, Guizhou, and Yunnan. Although most of the residents of Yunnan are Han Chinese, the remote areas of the province are inhabited by a wide array of indigenous peoples (Figure 11.28). Because most of its inhabitants in the uplands and remote valleys speak languages of the Tai family, Guangxi has been designated an **autonomous region**. Critics contend, however, that little real autonomy has ever existed. (In addition to Guangxi, there are four other autonomous regions in China. Three of these—Xizang [Tibet], Nei Mongol [Inner Mongolia], and Xinjiang—are located in Central Asia and are thus discussed at length in Chapter 10. The final autonomous region, Ningxia, located in northwestern China, is distinguished by its large concentration of Hui [Mandarin-speaking Muslims].)

Language and Ethnicity in Taiwan Taiwan is noted for its linguistic and ethnic complexity. In the island's mountainous eastern region, a few small groups of "tribal" peoples speak Austronesian languages related to those of Indonesia. These peoples resided throughout Taiwan before the 16th century. At that time, however, Han migrants began to arrive in large numbers. Most of the newcomers spoke Fujianese dialects, which eventually evolved into the distinctive language called Taiwanese.

Taiwan was transformed almost overnight in 1949, when China's nationalist forces, defeated by the communists, sought refuge on the island. Most of the nationalist leaders spoke Mandarin, which they made the official language. Taiwan's new leadership discouraged Taiwanese, viewing it as a local dialect. As a result, considerable tension developed between the Taiwanese and Mandarin communities. Only in the 1990s did Taiwanese speakers begin to reassert their linguistic identity. At present, many proponents of the Taiwanese language advocate formal independence from China, whereas those who favor Mandarin more often hope for eventual reunification.

FIGURE 11.27 Tribal Villages in South China Non-Han people are usually classified as "tribal" in China, which assumes they have a traditional social order based on self-governing village communities. Shown are Yi people at an open-air market in the village of Xhanghe in Yunnan Province.

FIGURE 11.28 Language Groups in Yunnan China's Yunnan Province is the most linguistically complex area in East Asia. In Yunnan's broad valleys and relatively level plateau areas, as well as in its cities, most people speak Mandarin Chinese. In the hills, mountains, and steep-sided valleys, however, a wide variety of tribal languages, falling into several linguistic families, is spoken. In certain areas, several different languages can be found in very close proximity.

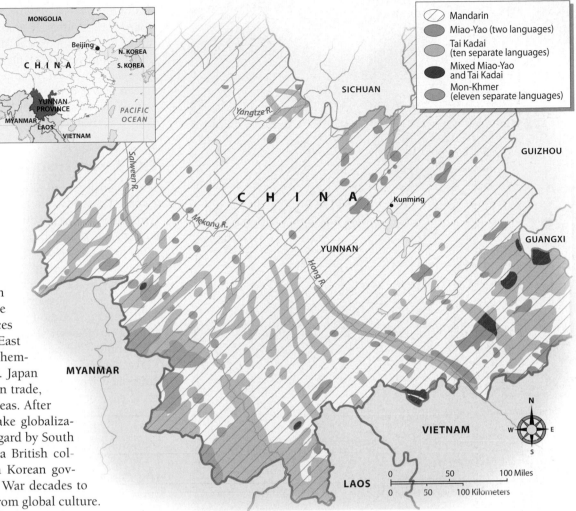

Legend:
- Mandarin
- Miao-Yao (two languages)
- Tai Kadai (ten separate languages)
- Mixed Miao-Yao and Tai Kadai
- Mon-Khmer (eleven separate languages)

East Asian Cultures in Global Context

East Asia has long been torn between separating itself from the rest of the world and welcoming foreign influences and practices. Until the mid-1800s, all East Asian countries attempted to insulate themselves from Western cultural influences. Japan subsequently opened its ports to Western trade, but remained uncertain about foreign ideas. After its defeat in 1945, Japan decided to make globalization a priority. It was followed in this regard by South Korea, Taiwan, and Hong Kong (then a British colony). However, the Chinese and North Korean governments sought during the early Cold War decades to isolate themselves as much as possible from global culture. Such a stance is still maintained in North Korea.

The Globalized Fringe The capitalist countries of East Asia are characterized to some extent by a cultural internationalism, especially in the large cities. Virtually all Japanese, for example, study English for 6 to 10 years, and although relatively few learn to speak it fluently, most can read and understand a good deal. Business meetings among Japanese, Chinese, and Korean firms are often conducted in English.

The current cultural flow is not merely from a globalist West to a previously isolated East Asia. Instead, the exchange has become reciprocal. Hong Kong's action films are popular throughout most of the world and have come to influence filmmaking techniques in Hollywood. Japan nearly dominates the world market in video games, and its *anime* style of animated film and television programming is now following karaoke bars in their overseas movement. Employment in Japan's anime industry, however, peaked around 2005 and has been slowly declining ever since. Critics claim that the industry is too focused on niche audiences, noting as well that it is suffering from competition by firms in South Korea and other Asian countries.

South Korea's popular culture industry, on the other hand, is thriving, as Korean music, movies, and television shows have become extremely popular across much of the world. K-op, a South Korean musical genre based on rock, hip-hop, and electropop styles, has been especially fashionable abroad, its global reach enhanced by its fans' use of Facebook, Twitter, and YouTube to publicize their favorite artists (Figure 11.29). In 2012, the music video "Gangnam Style" by the South Korean artist PSY became with third most viewed video in YouTube history, with more than 590 million hits as of October 29, 2012.

The Chinese Heartland In one sense, Japan is more culturally predisposed to cosmopolitanism than is China. Before the modern era, the Japanese borrowed heavily from other cultures (particularly from China itself), whereas the Chinese have historically been more self-sufficient. The southern coastal Chinese, however, have long had a stronger orientation toward foreign lands.

In most periods of Chinese history, the internal orientation of the center prevailed over the external orientation of the southern coast. After the communist victory of 1949, only the small British enclave of Hong Kong was able to maintain international cultural connections. In the rest of the country, a grim and puritanical cultural order was rigidly enforced. After China began to liberalize its economy and open its doors to foreign influences in the late 20th century, however, the southern coastal region suddenly assumed a new prominence. Through its coastal cities, global cultural patterns began to penetrate the rest of the country. The result has been the emergence of a vibrant and somewhat flashy urban popular culture in China that contains such global features as nightclubs, karaoke bars, fast-food franchises, and theme parks (Figure 11.30). In 2008, China demonstrated its openness, as well as its own cultural power, to the international community by successfully hosting the Summer Olympics in Beijing.

FIGURE 11.29 Korean Pop Star Se7en Korea has recently become noted for its export of popular culture. South Korean music, movies, and television shows are very popular in the rest of Asia and in other parts of the world as well. Stars of the K-Pop musical genre, such as Se7en, are idolized by many.

REVIEW QUESTIONS

1. What features mark the Han Chinese as an ethnic group?

2. How has the geography of religion changed in East Asia since the end of World War II?

GEOPOLITICAL FRAMEWORK: THE IMPERIAL LEGACIES OF CHINA AND JAPAN

Much of the political history of East Asia revolves around the centrality of China and the ability of Japan to remain outside China's grasp. The traditional Chinese conception of geopolitics was based on the idea of a universal empire: All territories were either supposed to be a part of the Chinese Empire, pay tribute to it, and acknowledge its supremacy or supposed to stand outside the system altogether. When China could no longer maintain its power in the face of European aggression, the East Asian political system began to fall apart. As European power declined in the 1900s, China and Japan competed for regional leadership. After World War II, East Asia was split by larger Cold War rivalries (Figure 11.31).

The Evolution of China

The original core of Chinese civilization was the North China Plain and the Loess Plateau. For many centuries, periods of unification alternated with times of division into competing states. The most important episode of unification occurred in the 3rd century BCE, under the short-lived Qin Dynasty. The Han Dynasty quickly followed the Qin, ruling from 206 BCE to 220 CE. During this period, China expanded vigorously to the south of the Yangtze Valley. Subsequently, the ideal of the unity of China triumphed, helping to join the Han Chinese into a single people. Although periods of disunity followed the collapse of China's various dynasties, reunification always followed

Various Chinese dynasties attempted to conquer Korea, but the Koreans resisted. Eventually, China and Korea worked out an arrangement whereby Korea paid token tribute and acknowledged the supremacy of the Chinese Empire. In return, Korea received trading privileges and retained independence. When foreign armies invaded Korea—as did those of Japan in the late 1500s—China sent troops to support its "vassal kingdom."

The Qing Dynasty of the Manchus The most significant conquest of China occurred in 1644, when the Manchus toppled the Ming Dynasty and replaced it with the Qing (also spelled Ch'ing) Dynasty. As earlier conquerors did, the Manchus retained the Chinese bureaucracy and made few institutional changes. Their strategy was to adapt themselves to Chinese culture, while at the same time preserving their own identity as an elite military group. This system functioned well until the mid-19th century, when the

FIGURE 11.30 Chinese Theme Park As China's economy grows, its people are spending increasing amounts of money on entertainment. Theme parks, which now number over 2,000, are particularly popular.

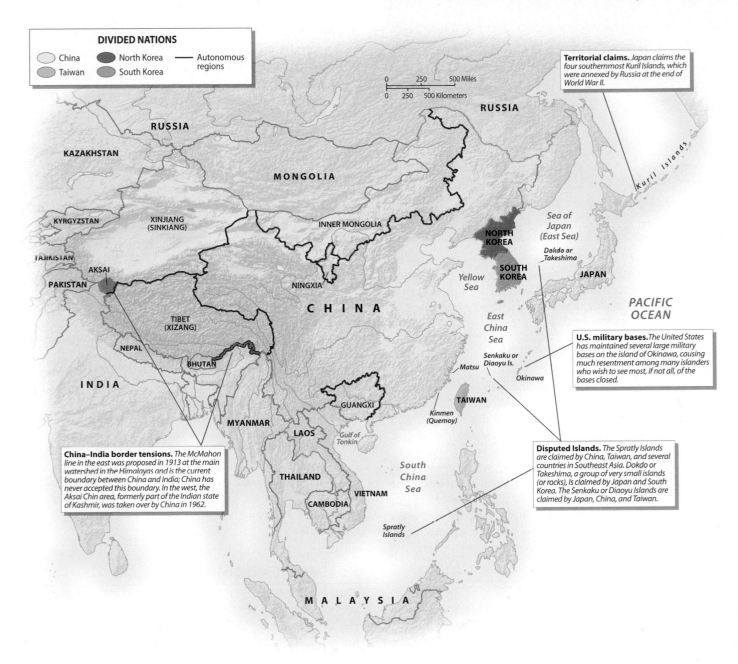

FIGURE 11.31 Geopolitical Issues in East Asia   East Asia remains one of the world's geopolitical hot spots. Tensions are particularly high between capitalist, democratic South Korea and the isolated communist regime of North Korea, as well as between China and Taiwan. China has had several border disputes, one of which involves a group of small islands in the South China Sea. Japan and Russia have not been able to resolve their quarrel over the southern Kuril Islands.

Chinese Empire began to crumble at the hands of European and, later, Japanese power.

The Modern Era   From its height in the 1700s, the Chinese Empire declined rapidly in the 1800s, as it failed to keep pace with the technological progress of Europe. Threats to the empire had always come from the north, and imperial officials saw little danger from European merchants operating along their coastline. But the Europeans were distressed by the amount of silver needed to obtain Chinese silk, tea, and other products. In response to their lack of alternative goods, the British began to sell opium, which Chinese authorities rightfully

viewed as a threat. When the imperial government tried to suppress the opium trade in the 1840s, Britain attacked and quickly prevailed (Figure 11.32).

This first "opium war" introduced a century of political and economic chaos in China. The British demanded and received trade privileges in selected Chinese ports. As European businesses penetrated China and weakened local economic interests, anti-Manchu rebellions broke out. At first, all such uprisings were crushed—but not before causing tremendous destruction. Meanwhile, European power continued to advance. In 1858, Russia annexed the northernmost reaches of Manchuria, and by 1900 China had been divided into

FIGURE 11.32 Opium War Great Britain humiliated China in two "opium wars" in the early 1800s, forcing the much larger country to open its economy to foreign trade and to grant Europeans extraordinary privileges. This image shows the East India Company steamer *Nemesis* destroying Chinese war junks in January 1841.

separate **spheres of influence** (Figure 11.33). (In a sphere of influence, the colonial power has no formal political authority, but does have informal influence and tremendous economic clout.)

A successful rebellion in 1911 finally toppled the Manchus and destroyed the empire, but subsequent efforts to establish a unified Chinese Republic were not successful. In many parts of the country, local military leaders ("warlords") grabbed power for themselves. By the 1920s, it appeared that China might be completely torn apart. The Tibetans had gained autonomy; Xinjiang was under Russian influence; and in China proper, Europeans and local warlords vied with the weak Chinese Republic for power. Japan was also increasing its demands and seeking to expand its territory.

The Rise of Japan

Japan did not emerge as a unified state until the 7th century, some 2,000 years later than China. From its earliest days, Japan looked to China for intellectual and political models. Its offshore location, however, insulated Japan from Chinese rule. Between 1000 and 1580, Japan had no real unity, being divided into several small warring states.

The Closing and Opening of Japan By the early 1600s, Japan had been reunited by the armies of the Tokugawa **shogunate** (a **shogun** is a military leader who in theory remains under the emperor; in practice, the shoguns ruled the country). At this time, Japan attempted to isolate itself from the rest of the world. Until the 1850s, Japan traded with China mostly through the Ryukyu islanders and with Russia through Ainu go-betweens. The only Westerners allowed to trade in Japan were the Dutch, and their activities were strictly limited.

Japan remained largely closed to foreign commerce and influence until U.S. gunboats sailed into Tokyo Bay in 1853 to demand trade access. Aware that China was losing power, Japanese leaders set about modernizing their economic, administrative, and military systems. This effort accelerated when the Tokugawa shogunate was toppled in 1868 by the Meiji Restoration. (It is called a *restoration* because it was

carried out in the emperor's name, but it did not give the emperor any real power.) Unlike China, Japan successfully strengthened its government and economy.

The Japanese Empire Japan's new rulers realized that their country remained threatened by European imperial powers. They decided that the only way to meet the challenge was to expand their own territory. Japan soon took control over Hokkaido and began to move farther north into the Kuril Islands and Sakhalin.

In 1895, the Japanese government tested its newly modernized army against China, winning a quick victory that gave it control of Taiwan. Tensions then mounted with Russia as the two countries competed for power in Manchuria and Korea. The Japanese defeated the Russians in 1905, giving Japan considerable influence in northern China. With no strong rival in the area, Japan annexed Korea in 1910. Alliance with Britain, France, and the United States during World War I brought further gains, as Japan was awarded Germany's island colonies in Micronesia.

The 1930s brought a global depression, greatly reducing world trade and putting a resource-dependent Japan in a difficult situation. The country's leaders sought a military solution, and in 1931 Japan conquered Manchuria. In 1937, Japanese armies moved south, occupying the North China Plain and the coastal cities of southern China. During this period, Japan's relations with the United States deteriorated. When the United States cut off the export of scrap iron, Japan began to experience a resource crunch.

In 1941, Japan's leaders decided to destroy the American Pacific fleet in order to clear the way for the conquest of resource-rich Southeast Asia. Their grand strategy was to unite East and Southeast Asia into a "Greater East Asia Co-Prosperity Sphere." This "sphere" was to be ruled by Japan and was designed to keep the Americans and Europeans out.

Postwar Geopolitics

With the defeat of Japan at the end of World War II in 1945, East Asia became dominated by rivalry between the United States and the Soviet Union. Initially, the American interests prevailed in Japan, South Korea, and Taiwan, while Soviet interests advanced on the mainland. Soon, however, East Asia began to experience a revival.

Japan's Revival Japan lost its colonial empire when it lost World War II. Its territory was reduced to the four main islands plus the Ryukyu Archipelago. In general, the Japanese government agreed to this loss of land. The only remaining territorial conflict concerns the four southernmost islands of the Kuril chain, which were taken by the Soviet Union in 1945. Japan still claims these islands, and Russia refuses to discuss the issue.

Japan's military power was limited by the constitution imposed on it by the United States, forcing Japan to rely in part on the U.S. military for its defense needs. Many Japanese citizens, however, believe that their country ought to provide its own defense. Slowly but steadily, meanwhile, Japan's military has emerged as a strong regional force, despite the limits imposed on it. North Korean nuclear bomb making and missile testing have raised security concerns in Japan. China's growing power is also a concern. New military guidelines announced in 2010 call for the Japanese military to focus more on the potential threat posed by China.

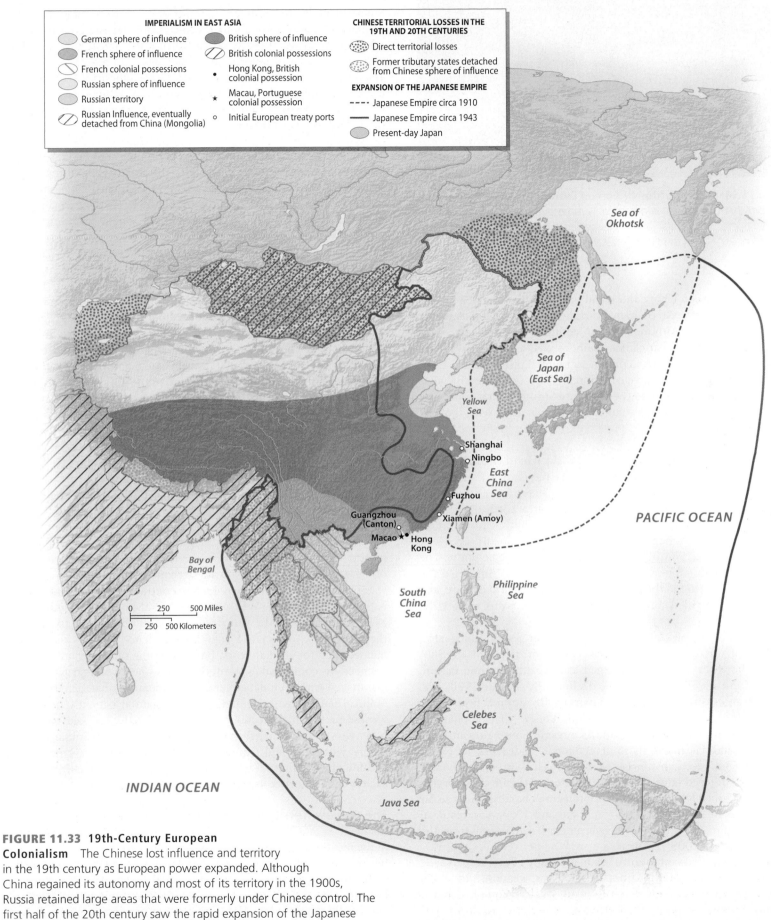

IMPERIALISM IN EAST ASIA

German sphere of influence
French sphere of influence
French colonial possessions
Russian sphere of influence
Russian territory
Russian Influence, eventually detached from China (Mongolia)

British sphere of influence
British colonial possessions
Hong Kong, British colonial possession
Macau, Portuguese colonial possession
Initial European treaty ports

CHINESE TERRITORIAL LOSSES IN THE 19TH AND 20TH CENTURIES

Direct territorial losses
Former tributary states detached from Chinese sphere of influence

EXPANSION OF THE JAPANESE EMPIRE

Japanese Empire circa 1910
Japanese Empire circa 1943
Present-day Japan

Sea of Okhotsk

Sea of Japan (East Sea)

Yellow Sea

East China Sea

Shanghai
Ningbo
Fuzhou
Guangzhou (Canton)
Xiamen (Amoy)
Macao
Hong Kong

PACIFIC OCEAN

Bay of Bengal

0   250   500 Miles
0   250   500 Kilometers

South China Sea

Philippine Sea

INDIAN OCEAN

Celebes Sea

Java Sea

FIGURE 11.33 19th-Century European Colonialism   The Chinese lost influence and territory in the 19th century as European power expanded. Although China regained its autonomy and most of its territory in the 1900s, Russia retained large areas that were formerly under Chinese control. The first half of the 20th century saw the rapid expansion of the Japanese Empire, which ended with the defeat of Japan in World War II.

FIGURE 11.34 Funeral of North Korean Leader Kim Jong-il in December 2011 North Korea is a highly repressive state that has cultivated "personality cults" around its leaders. The funeral of Kim Jong-il in December 2011 thus entailed a massive public ceremony.

Tensions between China and Japan have grown over the Senkaku Islands to the northeast of Taiwan (called the Diaoyu Islands in Mandarin Chinese). Although Japan controls these small, uninhabited islands, China also claims them, as well as the surrounding oceanic territory—which may contain substantial oil resources. Anti-Japanese feelings in China have occasionally been reinforced by visits of Japanese prime ministers to the Yasukuni Shrine, which contains a military cemetery in which several war criminals from World War II are buried.

The Division of Korea The end of World War II brought much greater changes to Korea than to Japan. As the end of the war approached, the Soviet Union and the United States agreed to divide the country; Soviet forces were to occupy the area north of the 38th parallel, whereas U.S. troops would occupy the south. This soon resulted in the establishment of two separate governments. In 1950, North Korea invaded South Korea, seeking to reunify the country. The United States, with support from the United Nations, supported the south, while China aided the north. The war ended in a stalemate, and Korea remains a divided country, its two governments still technically at war.

Large numbers of U.S. troops remained in South Korea after the war, as it was a poor country that could not defend itself. Subsequently, however, the south emerged as a wealthy trading nation, while the fortunes of the north have declined. By the late 1990s, the South Korean government came to favor a softer approach to North Korea. In 1998, it established a "Sunshine Policy" that emphasized peaceful cooperation with the north. As a result, South Korean firms invested substantial funds in joint economic endeavors in North Korea, and South Korean tourists were allowed to cross the border for closely monitored trips to famous locations. Despite these peaceful moves by South Korea, North Korea remained hostile, going so far as to detonate small nuclear bombs in 2006 and 2009. As a result of such provocations, South Korea's government backed away from the Sunshine Policy in 2008.

The international community has made many efforts to persuade North Korea to abandon its quest for nuclear weapons. In 2008, North Korea agreed to shut down its main nuclear complex, but it continued its nuclear program at other more secretive facilities. The death of North Korea's leader Kim Jong-il in 2011, followed by his

replacement by his untested 28-year-old son Kim Jong-un, led to heightened uncertainty about the direction that the country would take (Figure 11.34). In early 2012, North Korea agreed to a partial nuclear freeze and a suspension of missile testing in return for U.S. food aid. A few weeks later, however, North Korea announced that it would fire a long-range rocket in order to launch a satellite. Because such a rocket could be used to deliver a nuclear weapon, the United States responded by suspending food aid.

The Division of China World War II brought tremendous destruction and loss of life to China. Before the war began, China had already been engaged in a civil war between nationalist and communist forces. After Japan invaded China proper in 1937, the two camps cooperated; but as soon as Japan was defeated, China again found itself in a civil war. In 1949, the communists proved victorious, forcing the nationalists to retreat to Taiwan. The mainland was then renamed the People's Republic of China, while the nationalist government on Taiwan retained the name the Republic of China.

A dormant state of war between China and Taiwan persisted for decades after 1949. The Beijing government still claims Taiwan as an integral part of China and vows that it will eventually reclaim it. The nationalists in Taiwan long insisted that they represented the true government of China and that Taiwan was merely one province of a temporarily divided country. By the end of the 20th century, however, almost all Taiwanese had given up on the idea of taking over China itself, and many began to press openly for formal independence of the island.

The idea of Chinese unity continues to be influential both in China and abroad. In the 1950s and 1960s, the United States recognized Taiwan as the only legitimate government of China, but its policy changed after U.S. leaders decided that it would be more useful to recognize mainland China. Soon, China entered the United Nations, and Taiwan found itself diplomatically isolated. Taiwan continues to be recognized, however, as the legitimate government of China by several small countries in Africa, the Americas, and the Pacific. Most of these countries receive Taiwanese economic aid in return.

The geopolitical status of Taiwan continued to be a controversial issue in Taiwan itself. When a supporter of political separation was elected president of Taiwan in 2000, China threatened to invade if the island were to declare formal independence. Military tensions remained high for several years, but during the same period economic connections continued to strengthen. Taiwanese voters began to grow dissatisfied with an independence movement that had generated geopolitical tension without delivering substantial benefits. In the Taiwanese presidential elections of 2008 and 2012, the old nationalist party won clear victories, in part by promising to maintain good relations with mainland China. The policy of both governments then became one of "mutual non-denial," based on unofficially accepting each other's existence as separate states.

The Chinese Territorial Domain Despite the fact that it has been unable to regain Taiwan, China has been successful in retaining most of the territories that the Manchus controlled. In the case of Tibet, this has required considerable force. Resistance by the Tibetans compelled China to launch a full-scale invasion in 1959. The Tibetans, however, have continued to struggle for real autonomy, if not actual independence, as they fear that the Han Chinese now moving to Tibet will eventually outnumber them and undermine their culture (Figure 11.35).

The postwar Chinese government also retained control over Xinjiang in the northwest, as well as Inner Mongolia (or Nei Mongol), a vast territory stretching along the Mongolian border. The native peoples of Xinjiang prefer to call the region Eastern Turkestan to emphasize its Turkic heritage. The Han Chinese, however, reject this term because it challenges the unity of China. Like Tibet, Nei Mongol and Xinjiang are classified as autonomous regions. The peoples of Xinjiang are asserting their religious and ethnic identities, and separatist attitudes are common. Most Han Chinese, however, regard Nei Mongol and Xinjiang as integral parts of their country, and they regard any talk of independence as treasonous.

One territorial issue was finally resolved in 1997, when China reclaimed Hong Kong from Britain. In the isolationist 1950s, 1960s, and 1970s, Hong Kong acted as China's window on the outside world, and it grew wealthy as a capitalist city. As Chinese relations with the outer world opened in the 1980s, Britain decided to honor its treaty provisions and return Hong Kong to China. China, in turn, promised that Hong Kong would become a **special administrative region**, retaining its fully capitalist economic system for at least 50 years. Civil liberties not enjoyed in China itself also remain protected in Hong Kong.

In 1999, Macao, the last colonial territory in East Asia, was returned to China, becoming the country's second special administrative region. This small former Portuguese enclave, located across the estuary from Hong Kong, has functioned largely as a gambling refuge. In 2008, gambling revenues in Macao surpassed those of Las Vegas, making it the world capital of commercial wagering (Figure 11.36).

The Global Dimension of East Asian Geopolitics

In the early 1950s, East Asia was divided into two hostile Cold War camps: China and North Korea were allied with the Soviet Union, while Japan, Taiwan, and South Korea were linked to the United States. The China–Soviet Union alliance soon deteriorated into

FIGURE 11.36 Casino in Macao Macao, reclaimed by China from Portugal in 1999, retains a unique mixture of Chinese and Portuguese cultural influences. Its economic mainstay remains gambling, which is prohibited in the rest of China.

FIGURE 11.35 Chinese Soldiers in Tibet Following a full-scale invasion of Tibet in 1959, China continues to increase its presence through its military forces, the relocation of migrants into the area from other parts of China, and rebuilding programs that mask the traditional Tibetan landscape. Here a Chinese soldier oversees Tibentan pilgrims at a festival.

mutual hostility, however, and in the 1970s China and the United States found that they could work with each other, sharing as they did a common enemy in the Soviet Union.

The end of the Cold War, coupled with the rapid economic growth of China, again altered the balance of power in East Asia. The United States no longer needed China to offset the Soviet Union, and the U.S. military has become increasingly worried about the growing power of the rapidly modernizing Chinese army. Through the first decade of the new century, China's military budget grew at an average annual rate of about 10 percent. Several of China's neighbors have become concerned about its growing strength. As a result, South Korea and especially Japan have been eager to maintain close military ties with the United States. Currently, some 35,000 U.S. troops are stationed in Japan, and another 28,000 are stationed in South Korea.

China is thus coming of age as a major force in global politics. Whether it is a force to be feared by other countries is a matter of considerable debate. Chinese leaders insist that they have no intention of interfering in the internal affairs of other countries. They do, however, regard concerns expressed by the United States and other countries about their human rights record, as well as about their activities in Tibet, as excessive meddling in their internal affairs.

REVIEW QUESTIONS

1. How have geopolitical issues in East Asia changed since the end of the Cold War?

2. How did the decline of China during the 1800s affect the geopolitical structure of East Asia?

ECONOMIC AND SOCIAL DEVELOPMENT: A CORE REGION OF THE GLOBAL ECONOMY

East Asia shows vast disparities in economic and social well-being (Table 11.2). Japan's urban belt has one of the world's greatest concentrations of wealth, whereas many interior districts of China have experienced little development in decades. Overall, East Asia has experienced rapid economic growth since the 1970s, but again growth has not been evenly distributed. North Korea, for example, has seen its living standards decline sharply over the past two decades.

Japan's Economy and Society

Japan was the pacesetter of the world economy in the 1960s, 1970s, and 1980s. In the early 1990s, however, the Japanese economy experienced a major setback, and growth has remained slow ever since. Despite its recent problems, Japan is still by some measurements the world's second largest economic power.

Japan's Boom and Bust Although Japan's heavy industrialization began in the late 1800s, most of its people remained poor. The 1950s, however, saw the beginnings of the Japanese "economic miracle." With its empire gone, Japan was forced to export manufactured materials. Beginning with inexpensive consumer goods, Japanese industry moved to more sophisticated products, including automobiles, cameras, electronics, machine tools, and computer equipment. By the 1980s, Japan was the leader in many segments of the global high-tech economy.

The early 1990s saw the collapse of Japan's inflated real estate market, leading to a banking crisis. At the same time, many Japanese companies relocated factories to Southeast Asia and China. As a result, Japan's economy stagnated for several years. The Japanese government tried to revitalize the economy through massive state spending, resulting in large public deficits.

Despite its economic problems, Japan remains a core country of the global economic system. Its economic influence spans the globe, as Japanese multinational firms invest heavily in production facilities in North America and Europe, as well as in developing countries. Japan remains a world leader in a range of high-tech fields, including robotics, optics, and machine tools for the semiconductor industry (Figure 11.37).

The global economic crisis of 2008–2009 caused much damage in Japan, as export markets collapsed. Just as Japan's economy was beginning to recover, the country was hit by the massive Tohoku earthquake and tsunami of 2011. As a result of this disaster, the Japanese economy shrank by roughly 1 percent in 2011.

Living Standards and Social Conditions in Japan Despite high levels of economic development, Japanese living standards remain somewhat lower than those of the United States. Housing, food, transportation, and services are particularly expensive in Japan. Certain amenities that are standard in the United States, such as central heating, remain relatively rare in Japan.

Although the Japanese may live in cramped quarters and pay high prices for basic products, they also enjoy many benefits unknown in the United States. Unemployment remains lower than in the United States, health care is provided by the government, and crime rates are extremely low. By such social measures as literacy, infant mortality, and average life expectancy, Japan surpasses the United States by a comfortable margin. Japan also lacks the extreme poverty found in certain pockets of U.S. society. Critics contend, however, that Japan suffers from a significant degree of hidden poverty, which largely remains outside of public view.

TABLE 11.2 DEVELOPMENT INDICATORS

Country	GNI per capita, PPP 2010	GDP Average Annual % Growth 2000–10	Human Development Index (2011)[1]	Percent Population Living Below $2 a Day	Life Expectancy (2012)[2]	Under Age 5 Mortality Rate (1990)	Under Age 5 Mortality Rate (2010)	Adult Literacy (% ages 15 and older)	Gender Inequality Index (2011)[3,1]
China	7,640	10.8	.687	29.8	75	48	18	94	0.209
Hong Kong	47,480	4.6	.898	—	83	—	—	—	—
Japan	34,610	0.9	.901	—	83	6	3	—	0.123
North Korea	1,800*	—	—	—	69	45	33	100	—
South Korea	29,110	4.1	.897	—	81	8	5	—	0.111
Taiwan	38,200*	4.0*	—	—	79	—	—	96*	—

[1]United Nations, *Human Development Report, 2011.*

[2]Population Reference Bureau, *World Population Data Sheet, 2012.*

*Additional data from the *CIA World Factbook, 2012.*

[3]Gender Inequality Index—A composite measure reflecting inequality in achievements between women and men in three dimensions: reproductive health, empowerment and the labor market that ranges between 0 and 1. The higher the number, the greater the inequality.

Source: World Bank, *World Development Indicators, 2012.*

FIGURE 11.37 Automated Japanese Auto Factory Part of Japan's economic success has resulted from the automation of its factory assembly lines. Here Mazda automobiles are assembled in a Hiroshima plant. These cars are destined for the East Coast of the United States.

Women in Japanese Society Critics often point out that Japanese women have not shared the benefits of their country's success. Advanced career opportunities remain limited for women, especially those who marry and have children. The expectation remains that mothers will devote themselves to their families and to their children's education. Japanese businessmen often work or socialize with their coworkers until late every evening and thus contribute little to child care.

One response to the difficult conditions faced by Japanese women has been a drop in the marriage rate. Japan has seen an even more dramatic decline in its fertility rate. Whether this is due to the domestic difficulties faced by Japanese women or is merely a result of the pressures of a postindustrial society is an open question. Fertility rates have, after all, dropped even lower in many parts of Europe. But regardless of the cause, a shrinking population means an aging population, and increasing numbers of Japanese retirees will have to be supported by smaller numbers of workers. Japan's population seems to have peaked in 2008 and is now experiencing a slow decrease.

The Newly Industrialized Countries

The Japanese path to development was successfully followed by its former colonies, South Korea and Taiwan. Hong Kong also emerged in the 1960s and 1970s as a newly industrialized economy, although its economic and political systems are different.

The Rise of South Korea The postwar rise of South Korea was even more remarkable than that of Japan. During the period of Japanese occupation, Korean industrial development was concentrated in the north, which is rich in natural resources. The south, in contrast, remained a densely populated, poor, agrarian region.

In the 1960s, the South Korean government began a program of export-led economic growth. It guided the economy with a heavy hand and denied basic political freedom to the Korean people. By the 1970s, such policies had proved highly successful in the economic realm. Huge Korean industrial conglomerates, known as *chaebol,* moved from exporting inexpensive consumer goods to heavy

industrial products and then to high-tech equipment. As the South Korean middle class expanded, pressure for political reform increased, and in the 1980s South Korea made the transition from an authoritarian to a democratic country.

Through the 1980s, South Korean firms remained dependent on the United States and Japan for basic technology. By the 1990s, however, this was no longer the case, and South Korea emerged as one of the world's main producers of semiconductors. South Korean wages have also risen at a rapid rate. The country has invested heavily in education, which has served it well in the global high-tech economy. Increasingly, South Korean companies are themselves becoming multinational, building new factories in low-wage countries elsewhere in Asia and Latin America, as well as in the United States and Europe (see "Exploring Global Connections: South Korean Investments in Central Asia").

The political and social development of South Korea has not been as smooth as its economic progress. Issues of economic globalization often provoke serious political conflicts. In 2008, for example, Seoul was temporarily paralyzed when up to 40,000 demonstrators took to the streets to protest the government's decision to allow the import of beef from the United States, which had previously been banned due to concerns about mad cow disease. A free trade agreement between South Korea and the United States, which was ratified by both countries in late 2011, similarly resulted in heated protests in Seoul (Figure 11.38).

Like other trade-dependent countries, South Korea experienced a major recession in 2008–2009, with its stock market declining by 40 percent and its currency dropping 26 percent. Economic recovery in South Korea, however, occurred more rapidly than in most other industrialized countries. By 2011, its economy was expanding at a healthy rate of 3.6 percent.

Taiwan and Hong Kong Taiwan and Hong Kong have experienced rapid economic growth since the 1960s. The Taiwanese government, like the governments of South Korea and Japan, has guided the economic development of the country.

FIGURE 11.38 Protests in South Korea Massive political protests are common in South Korea. In October 2011, heated protests accompanied the ratification of the United States–South Korea Free Trade Agreement.

EXPLORING GLOBAL CONNECTIONS

South Korean Investments in Central Asia

South Korea is currently investing heavily in Central Asian economic initiatives. Such connections were highlighted in the summer of 2011, when South Korean President Lee Myun-bak undertook a tour of Mongolia, Uzbekistan, and Kazakhstan, and again in March 2012, when he revisited Kazakhstan (Figure 11.2.1). In Mongolia, Lee signed a "memorandum of understanding" with the Mongolian government focused on Korean investment in mining, the electricity sector, and infrastructure. In Uzbekistan, he made economic deals valued at $4.2 billion, centered on the development of the Surgil natural gas field near the remnants of the Aral Sea. It was also agreed that South Korea would transfer information technology to Uzbekistan and help modernize its stock exchange system. In Kazakhstan, Lee signed deals worth over $7 billion. Included here were agreements for South Korean firms to construct two power plants in the city of Balkhash, which will eventually supply more than 7 percent of Kazakhstan's electricity, and to build a major petrochemical complex near the Caspian Sea.

Kazakhstan's government foresees deepening economic ties with South Korea. Kazakh President Nursultan has long regarded South Korea as a model for economic development. In the early 1990s, a South Korean economist, Chan Young Bang, drafted the original design for Kazakhstan's economic privatization program. In 2011, Kazakhstan's Deputy Prime Minister Asset Issekeshev reportedly commented, "After signing the agreements, South Korea will become the number one investor in Kazakhstan within the process of industrialization of the country." (*Source: South Korea deepens role in Central Asia* By Robert M Cutler. ASIA TIMES ONLINE: http://www.atimes.com/atimes/Central_Asia/MI02Ag01.html)

Critics contend that South Korea's growing economic ties with Central Asia have troubling political implications. In early 2012, South Korea deported several Uzbek refugees, even though it was suspected that they would be subjected to torture in Uzbekistan. According to U.S. embassy cables made public by WikiLeaks, Uzbekistan and South Korea agreed in 2006 to root out illegal Uzbek laborers in South Korea. Several thousand workers from Uzbekistan are still believed to live in South Korea, but the number had been significantly larger in the early 2000s.

South Korean economic ties with both Uzbekistan and Kazakhstan are facilitated by the large number of ethnic Koreans living in those countries. As many as 200,000 Koryo-sarams—as ethnic Koreans living in Central Asia are called—reside in Uzbekistan, with another 100,000 in Kazakhstan. The Koryo-sarams are the descendants of Koreans who were deported by the Soviet Union from the Russian Far East to Central Asia in the 1930s. The deportations were brutal, resulting in an estimated 40,000 deaths. Eventually, however, many of the ethnic Koreans in the region acquired high levels of education and reached positions of economic responsibility.

By 1989, the majority of the ethnic Koreans in Central Asia considered Russian—rather than Korean, Uzbek, or Kazakh—to be their first language. Many members of the community, however, are trilingual to some extent, speaking Russian and Korean in addition to either Uzbek or Kazakh. Such widespread linguistic capabilities help smooth the progress of South Korean firms investing in Central Asia.

FIGURE 11.2.1 South Korea's Central Asia Connection   South Korea has recently made agreements for massive investments in Central Asian industry and infrastructure. In March 2012, South Korean President Lee Myung-bak and Kazakh President Nursultan Nazarbayev signed deals worth US$7 billion.

Hong Kong, unlike its neighbors, has been characterized by one of the most **laissez-faire** economic systems in the world (laissez-faire refers to market freedom with little governmental control). Governmental involvement has been minimal, which is one reason members of the city's business elite were nervous about the transition to Chinese rule. Hong Kong traditionally functioned as a trading center, but in the 1960s and 1970s it became a major producer of textiles, toys, and other consumer goods. By the 1980s, however, such cheap products could no longer be made in such an expensive city. Hong Kong industrialists subsequently began to move their plants to southern China, while Hong Kong itself increasingly specialized in business services, banking, telecommunications, and entertainment.

Both Taiwan and Hong Kong have close overseas economic connections. Linkages are particularly tight with Chinese-owned firms located in Southeast Asia and North America. Taiwan's high-technology businesses are also intertwined with those of Silicon Valley in the United States. Hong Kong's economy is also closely bound with that of the United States (as well as those of Canada and Britain), but its closest connections are with the rest of China.

Such high levels of globalization resulted in major recessions in both economies during the global economic crisis of 2008–2009. However, as was the case in South Korea, the economy of Taiwan was able to bounce back quite rapidly once the crisis had passed. Growth was bolstered by a 2010 agreement with China that allowed

Taiwanese financial firms more freedom to operate on the mainland, just as it allowed mainland firms to invest more freely in Taiwan.

Chinese Development

China dwarfs all of the rest of East Asia in both physical size and population. Its economic takeoff is thus reshaping the economy of the entire world. Despite its recent growth, however, China's economy has several weaknesses. For example, much of the vast interior remains trapped in poverty, and many of its largest industries are not competitive. The future of the Chinese economy is one of the biggest uncertainties facing both East Asia and the world economy as a whole.

China Under Communism   More than a century of war, invasion, and near chaos in China ended in 1949, when the communist forces led by Mao Zedong seized power. The new government, inheriting a weak economy, set about nationalizing private firms and building heavy industries. These plans were most successful in Manchuria, where a large amount of heavy industrial equipment had been left by the Japanese.

In the late 1950s and 1960s, however, China experienced two economic disasters. The first, ironically called the "Great Leap Forward," entailed the idea that small-scale village workshops could produce the large quantities of iron needed for sustained industrial growth. Communist Party officials demanded that these inefficient workshops meet unreasonably high production quotas. The result was a horrific famine that may have killed 20 million persons. The early 1960s saw a return to more practical policies, but toward the end of the decade a new wave of radicalism swept through China. This "Cultural Revolution" aimed at mobilizing young people to rid the country of "undesirable" traditional social values and replace them with communist ideology. Thousands of experienced industrial managers and college professors were expelled from their positions. Many were sent to villages to be "reeducated" through hard physical labor; others were simply killed. The economic consequences were devastating.

Toward a Postcommunist Economy   When Mao Zedong, who had been revered as an almost superhuman being, died in 1976, China faced a crucial turning point. Its economy was nearly stagnant and its people desperately poor. However, the economy of Taiwan, its rival, was booming. This led to a political struggle between pragmatists hoping for change and dedicated communists. The pragmatists emerged victorious, and by the late 1970s it was clear that China would embark on a different economic path. The new China would seek closer connections with the world economy and take a modified capitalist road to development.

China did not, however, transform itself into a fully capitalist country. The state continued to run most heavy industries, and the Communist Party kept a monopoly on political power. Instead of suddenly abandoning the communist model, as the former Soviet Union did, China allowed cracks to appear in which capitalist businesses could take root and thrive.

Industrial Reform   An important early industrial reform involved opening **Special Economic Zones (SEZs)**, in which foreign investment was welcome and state interference was minimal. The Shenzhen SEZ, adjacent to Hong Kong, proved particularly successful (Figure 11.39). Additional SEZs were soon opened, mostly in the coastal region. The

FIGURE 11.39 Shenzhen   The city of Shenzhen, adjacent to Hong Kong, was one of China's first Special Economic Zones. It has recently emerged as a major city in its own right.

basic strategy was to attract foreign investment that could generate exports, the income from which could supply China with the capital it needed to build its infrastructure (roads, electrical and water systems, telephone exchanges, and the like). China is now trying to use the SEZ model to spread economic development into the interior portion of the country. In 2010, for example, the entire city of Kashgar in western Xinjiang was declared a Special Economic Zone.

In the 1980s and 1990s, other capitalistic reforms were also enacted. Former agricultural cooperatives were allowed to produce for the market. Many of these "township and village enterprises" proved highly successful. By the early 1990s, the Chinese economy was growing at roughly 10 percent a year, perhaps the fastest rate of expansion the world has ever seen. China emerged as a major trading nation, and by the mid-1990s it had huge trade surpluses, especially with the United States. Seeking to strengthen its connections with the global economic system, in 2001 China joined the World Trade Organization (WTO), a body designed to facilitate free trade and provide ground rules for international economic exchange.

Despite China's reliance on trade, it weathered the global economic crisis of 2008–2009 better than most other countries. Early in the crisis, China's government announced a massive stimulus program, pledging to inject $586 billion into the economy over a two-year period. Although China's economic growth did slow down, the country did not go into recession. In 2011, the Chinese economy was again growing at a sizzling rate of more than 9 percent. Some critics contend that the Chinese economy, and especially its real estate market, is overheated and will eventually suffer a severe recession, and economic growth did slow significantly in 2012. The Chinese government is concerned, however, and it is now trying to increase domestic consumption in order to decrease reliance on exports.

Criticism of the Chinese Economic Model   China's economic expansion has created tensions with the United States. China exports far more to the United States than it imports, creating economic imbalances. Other foreign critics accuse China of unfairly keeping the price of labor low in order to enhance exports. China's large and growing holdings of U.S. Treasury bonds, however, make it difficult for the United States to exert much pressure on the Chinese economy. This topic became an issue in the U.S. presidential election of 2012, when several American politicians accused China of undervaluing its currency to keep its exports cheap and thus take advantage of the U.S. market.

Critics of China also focus on the mistreatment of labor in factories, especially those producing electronics goods for export. Employees in such plants are often exposed to hazardous chemicals and are often forced to work overtime under harsh condition without additional pay. In 2011 and 2012, the global labor-rights movement led major protests against Foxconn, a Taiwanese electronics firm that manufactures iPads, iPhones, Kindles, and other U.S.-designed devices in Chinese factories. Subsequent investigations showed that some of the charges against Foxconn were exaggerated; however, it is difficult to deny that conditions in many Chinese factories are brutal. The processing of electronics waste from other countries is an especially hazardous economic sector, and it reportedly employs many child laborers.

Critics also point to the fact that political freedom has not accompanied China's creation of a more open, market-driven economy. Critics of the state still face imprisonment, and freedom of the press is highly limited. Reporters Without Borders, a global organization that advocates freedom of the press, rates China as one of the world's 12 "enemies of the Internet." In 2012, the same organization ranked China in the 174th position, out of 179, in regard to freedom of the press in general.

Social and Regional Differentiation The Chinese economic surge brought about by the reforms of the late 1970s and 1980s resulted in growing **social and regional differentiation**. In other words, certain groups of people—and certain portions of the country—prospered, while others did not. Despite its socialist government, the Chinese state encouraged the formation of an economic elite, having concluded that only wealthy individuals can adequately transform the economy. The least-fortunate Chinese citizens were sometimes left without work, and many millions migrated from rural villages to seek employment in the booming coastal cities. The government attempted to control the transfer of population, but with only partial success. Shantytowns and homeless populations began to emerge around some of China's cities.

Although China is still officially a communist country, it now has a much more unequal distribution of wealth than its officially capitalist neighbors—Japan, South Korea, and Taiwan. According to some statistical measurements, wealth is now more evenly distributed in the United States than in China.

China's Booming Coastal Region Most of the benefits from China's economic transformation have flowed to the coastal region and to the capital city of Beijing. The southern provinces of Guangdong and Fujian were the first to benefit. These provinces have profited from their close connections with the overseas Chinese communities of Southeast Asia and North America. (The vast majority of overseas Chinese emigrants came from these provinces.) Their location close to Taiwan and especially Hong Kong also proved helpful.

By the 1990s, the Yangtze Delta, centered on Shanghai, reemerged as the economic core of China. The Chinese government has encouraged the development of huge industrial, commercial, and residential complexes, hoping to take advantage of the region's vitality. The Suzhou Industrial Park is now a hypermodern city of more than a half million people, thanks largely to a $20 billion investment, most of it from Singapore. Shanghai's Pudong industrial development zone has attracted $10 billion, much of it going to the construction of a new airport and subway system. Shanghai now has the world's largest container port.

The Beijing–Tianjin region has also played a major role in China's economic boom. Its main advantages are its proximity to political power and its position as the gateway to northern China. The other coastal provinces of northern China have also done relatively well.

Interior and Northern China The interior and northern parts of China, in contrast, have seen much less economic expansion than the rest of the country. Manchuria remains relatively well-off as a result of fertile soil and early industrialization, but it has not participated much in the recent boom. Many of the state-owned heavy industries of the Manchurian **rust belt**, or zone of decaying factories, are relatively inefficient.

Most of the interior provinces of China have also missed the recent wave of growth. In many areas, rural populations are hurt by environmental degradation. One consequence is high levels of underemployment and out-migration. As a result of such regional differences, China is building roads and railway lines and undertaking other projects to encourage development in the interior. But by most measures, poverty still increases with distance from the coast (Figure 11.40).

Social Conditions in China

Despite its pockets of persistent poverty, China has made significant progress in social development. Since coming to power in 1949, the communist government has made large investments in medical care and education, and today China has impressive health and longevity figures. The literacy rate remains lower than those of Japan, South Korea, and Taiwan, but because almost all children attend elementary school, it will probably rise substantially in the coming years.

Not surprisingly, human well-being in China varies from region to region. The literacy rate remains relatively low in many of the poorer parts of China, including the uplands of Yunnan and Guizhou. Such regional disparities may increase as the gap between the wealthy and the poor grows. Some evidence also indicates that the social well-being of undocumented migrant workers in Chinese cities is declining.

China's Population Quandary Population policy remains an unsettling issue for China. With more than 1.6 billion people highly concentrated in less than half of its territory, eastern China is very densely populated. In 1978, its government had become so concerned that it instituted the infamous "one-child policy." Under this plan, couples in normal circumstances are expected to have only a single offspring and can suffer financial penalties if they do not comply (Figure 11.41). This strategy has been successful; the total fertility level is now only 1.5. China will, however, probably reach more than 1.5 billion people before its population stabilizes. Some scholars think that, if the one-child policy had not been instituted, China would have as many as 300 million more people than it does today.

Fertility levels in China, as might be expected, vary from province to province. Birth rates are relatively low in most large cities, the Yangtze Delta, parts of the Sichuan Basin, and parts of the North China Plain; they are relatively high in many upland areas of south China, the Loess Plateau, and northwestern China. Higher levels of fertility in these poorer and more rural areas will probably lead to increased migration to the booming coastal cities.

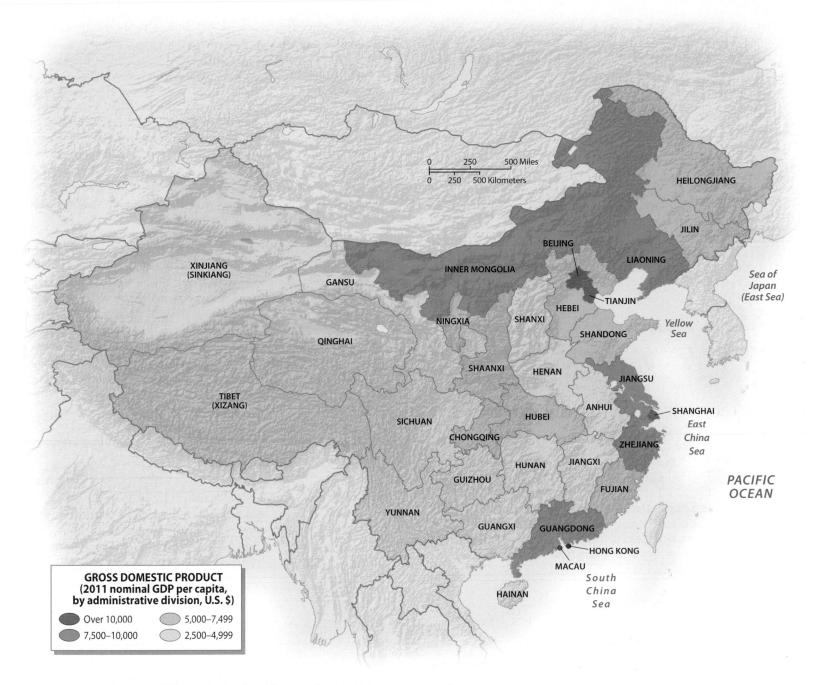

FIGURE 11.40 Economic Differentiation in China   Although China has seen rapid economic expansion since the late 1970s, the benefits of growth have not been evenly distributed throughout the country. Economic prosperity and social development are concentrated on the coast, especially in Shanghai, Beijing, and Tianjin. Most of the interior remains mired in poverty. The poorest part of China is the upland region of Guizhou in the south-central part of the country.

While China's policy has reduced its population growth rate, it has also generated social tensions and human-rights abuses. Particularly troubling is the growing gender imbalance. In 2010, 119 boys were born for every 100 girls. This gap reflects the practice of honoring a family's ancestors; because family lines are traced through male offspring, only a male heir can maintain the family's lineage. Some couples decide to bear more than one child, regardless of the penalties they may face. Another option is gender-selective abortion; if ultrasound reveals a female fetus, the pregnancy is sometimes terminated. In 2004, the Chinese government banned this practice, but much

evidence suggests that it continues to occur. Poor couples not uncommonly abandon baby girls, and young boys are occasionally kidnapped and sold to wealthy couples without a son.

The Chinese government is currently attempting to find solutions for the problems generated by its population policies and demographic imbalance. In 2011, the country's chief of population and family planning was pleased to announce that the sex ratio at birth had declined to 118 boys per 100 girls, a drop that he attributed to a crackdown on sex-selective abortions. In the same year, the government also announced that it would soon review the one-child policy and was considering

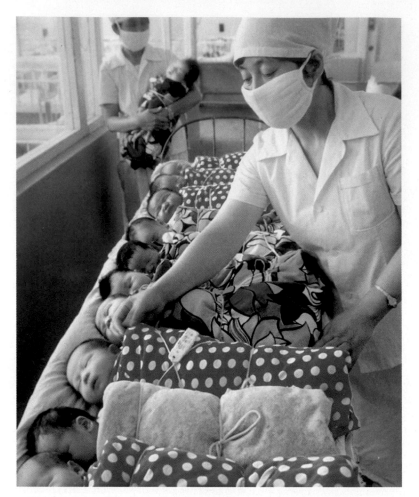

FIGURE 11.41 China's Population Policies One aspect of China's population policies is the expansion of child-care facilities so that mothers can be near their children while at work. This enables women to resume participating in the workforce soon after giving birth. This photo shows a typical day-care center attached to an industrial plant in Guangdong Province in coastal China.

allowing all couples to have a second child. Already, the policy is being relaxed and is no longer applied to most couples.

The Position of Women Women have historically had a relatively low position in Chinese society, as is true in most other civilizations. One traditional expression of this was the practice of foot binding: The feet of almost half the girls of China in the 1800s were deformed through breaking and binding in order to produce a dainty appearance. This crippling and painful practice was eliminated only in the 20th century. In certain areas of southern China, it was also common in traditional times for girls to be married, and hence to leave their own families, when they were mere toddlers (such marriages, of course, would not be consummated for many years).

Not all women suffered such disabilities in premodern China. Some individuals achieved fame and fortune—a few even through military service. Both the nationalist and the communist governments have, moreover, sought to begin equalizing the relations between the sexes. Many of their measures have been successful, and women now have a relatively high level of participation in the Chinese workforce. But it is still true that throughout East Asia—in Japan no less than in China—few women have achieved positions of power in either business or government. Chinese women also suffer from governmental restrictions on reproductive freedom, and forced prostitution remains common in some areas. Yet as China modernizes and its urban economy grows, the position of its women will probably improve.

REVIEW QUESTIONS

1. How has the economic development of Japan, South Korea, Taiwan, and China been similar since the end of World War II, and how has it been different in the different countries?

2. Why do levels of social and economic development vary so extensively from the coastal region of China to the interior portions of the country?

Summary

- The economic success of East Asia has been accompanied by severe environmental degradation. Japan, South Korea, and Taiwan managed to overcome environmental crises by enacting strict protective legislation and by moving many of their most polluting industries overseas. In China, however, pollution in cities is so serious that it has had major negative effects on human health, and much of the Chinese countryside suffers from such problems as soil erosion and desertification.

- Although East Asia is a very densely populated region, it has seen its birth rates plummet in recent decades. Japan is now facing population decline, which could become quite serious within a few decades. In China, the biggest demographic challenge results from the massive movement of people from the interior to the coast and from rural villages to the rapidly expanding cities.

- East Asia is united by deep cultural and historical bonds. China has had a particularly large influence because, at one time or another, it covered nearly the entire region. Overall, the most prosperous

parts of East Asia have seen the striking development of cultural globalization over the past several decades. Cultural trends from Europe and North America have been enthusiastically adopted, just as East Asian cultural ideas and practices have spread to many other parts of the world.

- Geopolitically, East Asia remains a region characterized by strife. China and Korea are still suspicious of Japan, and they worry that it might rebuild a strong military force. Japan, for its part, is concerned about the growing military power of China and especially about the nuclear arms and missiles of North Korea. Relations between North and South Korea deteriorated after North Korea renewed its nuclear weapons program in 2006. North Korean weapons development, moreover, is a global concern that deeply involves the United States.

- With the notable exception of North Korea, all East Asian countries have experienced major economic growth since the end of World War II. Such growth has had large global consequences, as East Asia massively exports to and imports from all other major areas of the global economy. The most important story of the global economy itself over the past two decades has probably been the rapid rise of China. That rise, however, has generated many problems, both domestically and abroad.

Key Terms

anthropogenic landscape 367
autonomous region 374
China proper 358
Confucianism 370
desertification 365
diaspora 373
hiragana 370
ideographic writing 370

kanji 370
laissez-faire 384
loess 359
Marxism 372
pollution exporting 361
rust belt 386
sediment load 358
shogun, shogunate 378

social and regional differentiation 386
special administrative region 381
Special Economic Zone (SEZ) 385
sphere of influence 378
superconurbation (megalopolis) 369
tonal 374
tribal 374
urban primacy 369

Thinking Geographically

1. What are the advantages and disadvantages of China's dam-building policy? Is it wise for the Chinese government to emphasize dam construction?

2. What would be the main results, both positive and negative, if Japan were to allow the importation of rice and open its agricultural lands to urban development?

3. What might be the main consequences if China were to grant true autonomy to the Tibetans and other non-Han peoples? Is there any chance that Tibet might eventually gain independence? Could Taiwan possibly gain formal independence?

4. Discuss the potential ramifications of the United States' restricting the importation of Chinese goods in order to put pressure on the Chinese government for human-rights reforms.

5. What is the likelihood of East Asia emerging as the center of the world economy over the next 50 years? What would be some of the main consequences of such a development?

MasteringGeography™

Looking for additional review and test prep materials? Visit the Study Area in MasteringGeography™ to enhance your geographic literacy, spatial reasoning skills, and understanding of this chapter's content by accessing a variety of resources, including **MapMaster** interactive maps, videos, RSS feeds, flashcards, web links, self-study quizzes, and an eText version of *Globalization and Diversity*.

Scan to visit the author's blog for chapter updates.

http://gad4blog.
wordpress.com/
category/east-asia/

Authors' Blogs

Scan now to access the authors' blogs for up-to-date information on East Asia.

Scan to visit the GeoCurrents blog.

http://geocurrents.
info/category/place/
east-asia

Globalization and Diversity

Although several cities in western and southern India have recently emerged as centers of the global high-tech economy, much of South Asia is relatively isolated from the world economy. Resistance to globalization, moreover, remains widespread across much of the region.

ENVIRONMENTAL GEOGRAPHY

The arid parts of South Asia suffer from water shortages and salinization of the soil, whereas the humid areas often experience devastating monsoon floods.

POPULATION AND SETTLEMENT

South Asia will soon become the most populous region of the world. Birth rates have, however, decreased substantially in recent years.

CULTURAL COHERENCE AND DIVERSITY

The South Asia region is one of the most culturally diverse parts of the world, with India alone having more than a dozen official languages, as well as numerous adherents of most major religions.

GEOPOLITICAL FRAMEWORK

South Asia is burdened not only by several violent secession movements, but also by the ongoing struggle between India and Pakistan, which are both armed with nuclear weapons.

ECONOMIC AND SOCIAL DEVELOPMENT

Although South Asia is one of the poorest regions of the world, certain areas are experiencing rapid economic growth and technological development.

➤ The Ganges, a holy river in the Hindu religion, runs trough the impoverished Indian state of Bihar. In this photo we see people performing acts of religious devotion on the banks of the river near Patna, the capital of Bihar, in 2010.

The state of Bihar in northern India has long been noted for its fertile soils, large rivers, and lush rural farmlands—and also for its intense poverty, lack of development, and corrupt governance. With more than 100 million people living in an area just a little larger than Ireland or South Carolina, Bihar is one of the most densely populated places in the world. Until recently, economic and social conditions in Bihar were so poor that many people had given up on the state, which was best noted in India for ranking last in almost every developmental indicator. However, over the past five years, Bihar has achieved a remarkable turnaround. Government campaigns against corruption proved so successful that by the end of 2011, Bihar was shockingly declared to be the least corrupt state in India. In the same year, Bihar's economy grew at the blazing rate of 11 percent. Suddenly, the standard of living began to rise rapidly, even in the state's rural areas. Motorcycles, tractors, cell phones, and televisions are now common in villages that had no electricity or motor vehicles a generation ago.

Although Bihar has made great progress, its gains should not be exaggerated. Most of the state's inhabitants are still poor, malnutrition is still widespread, and corruption remains a problem (Figure 12.1). In Bihar's poorest districts, moreover, insurgent fighters continue to struggle against the Indian police, seeking to overthrow the government of the entire country. Even Bihar's gains come at a significant cost. The rapid increase in car and motorcycle ownership, for example, is straining the state's inadequate transportation system and is generating pollution that contributes to global climate change.

Over much of South Asia, similar developments are unfolding. Most of the region has experienced substantial economic and social development over the past several decades, and some areas have emerged as world-class business centers, often specializing in high-tech fields. But other areas have failed to make such gains, and across the region rampant environmental degradation threatens both human communities and natural ecosystems. Population growth adds further pressure; given its current rate of expansion, South Asia will soon surpass East Asia as the world's most populous region. Geopolitical tensions also persist both

FIGURE 12.1 Rural Poverty in Bihar, India Bihar is Indian least developed state, noted for its rural poverty, as can be seen in this photograph. Over the past few years, however, Bihar has begun to make substantial economic progress.

within and among the various countries of the region. Since gaining independence from Britain in 1947, the two largest countries in the region, India and Pakistan, have fought several wars and remain locked in conflict. Political tensions have reached such heights that many experts consider South Asia the leading candidate for the site of nuclear war. Religious divisions contribute to the geopolitical turmoil, for India is primarily a Hindu country (with a large Muslim minority), while neighboring Pakistan and Bangladesh are both predominantly Muslim.

South Asia as a whole forms a distinct landmass separated from the rest of the Eurasian continent by a series of sweeping mountain ranges, including the Himalayas—the highest mountains in the world. It is often called the **Indian subcontinent**, in reference to its largest country. South Asia also includes several islands in the Indian Ocean, including the countries of Sri Lanka and the Maldives, as well as the Indian territories of the Lakshadweep, Andaman, and Nicobar islands (Figure 12.2).

LEARNING OBJECTIVES

After reading this chapter you should be able to:

- Explain how the monsoon is generated and describe its importance for South Asia.

- Describe the geological relationship between the Himalayas and other high mountains of northern South Asia and the flat, fertile plain of the Indus and Ganges river valleys.

- Outline the ways in which the patterns of human population growth in South Asia have changed over the past several decades and explain why they vary so strikingly from one part of the region to another.

- Identify the causes of the explosive growth of South Asia's major cities and describe both the benefits and the problems that result from the emergence of such large cities.

- Compare and contrast the ways in which India and Pakistan have dealt with the problems of building national cohesion, considering the fact that both countries contain numerous, distinctive language groups.

- Summarize the historical relationship between Hinduism and Islam in South Asia and explain why so much tension exists between the two religious communities today.

- Explain why South Asia was politically partitioned at the end of the period of British rule and show how the legacies of partition have continued to generate political and economic difficulties in the region.

- Describe the various challenges that India, Pakistan, and Sri Lanka have faced from insurgency movements that seek to carve out new independent states from their territories.

- Explain why European merchants were so eager to trade in South Asia in the 16th, 17th, and 18th centuries and describe how their activities influenced the region's later economic development.

- Summarize the ways in which economic and social development varies across the different regions of South Asia and explain why such variability is so pronounced.

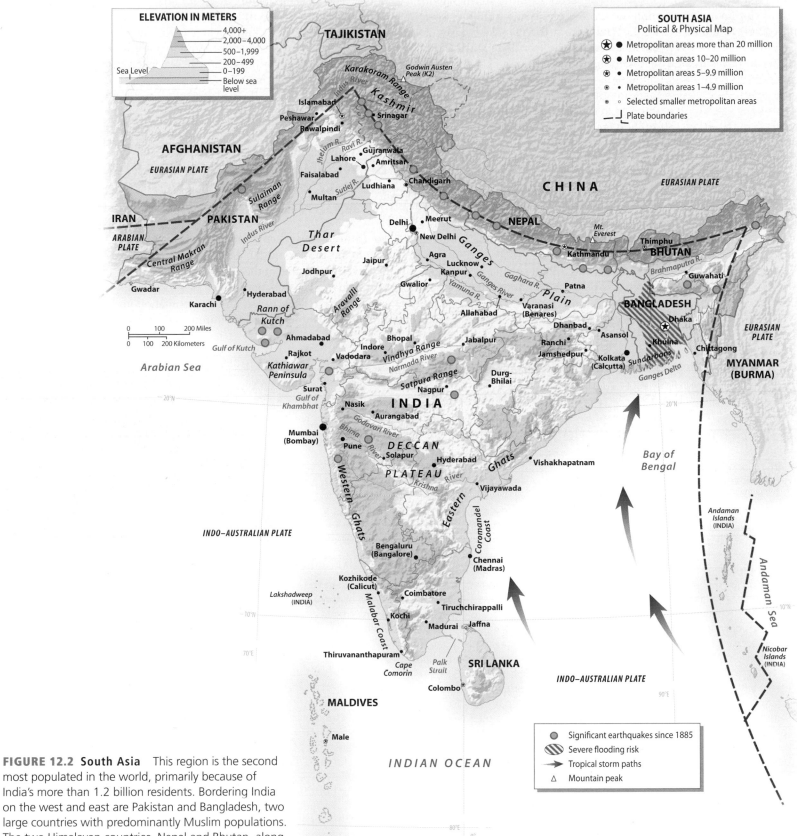

FIGURE 12.2 South Asia   This region is the second most populated in the world, primarily because of India's more than 1.2 billion residents. Bordering India on the west and east are Pakistan and Bangladesh, two large countries with predominantly Muslim populations. The two Himalayan countries, Nepal and Bhutan, along with the island nations Sri Lanka and the Maldives, round out the region. Although South Asia is well known for its poverty, it is also home to many thriving high-technology firms.

Three countries—India, Pakistan, and Bangladesh—dominate the South Asia landmass. India is by far the largest country in the region, both in size and in population. Covering more than 1 million square miles (2.60 million square kilometers) from the Himalayan crest to the southern tip of the peninsula, India is the world's seventh largest country in terms of area and, with more than 1.2 billion inhabitants, second only to China in population. Pakistan, the next largest country in South Asia, is less than one-third the size of India. Stretching from the high northern mountains to the arid coastline on the Arabian Sea, its population of 180 million is only about 15 percent of India's. Bangladesh, on India's eastern shoulder, was originally created as East Pakistan in the hurried division of India in 1947 and achieved independence after a brief civil war in 1971. Although a small country in area (54,000 square miles [140,000 square kilometers]), Bangladesh is one of the world's most densely populated places, with 153 million people living in an area about the size of Wisconsin. Bangladesh has a short border with Burma (Myanmar), but it is otherwise bordered only by India.

The other countries of South Asia have much smaller populations than the three demographic giants. Nepal and Bhutan are both located in the Himalayan Mountains, sandwiched between India and the Tibetan Plateau of China. Nepal, with some 31 million people, is much larger than Bhutan, which has fewer than 1 million inhabitants. The two island countries of Sri Lanka (formerly Ceylon) and the Maldives round out South Asia. Sri Lanka is a large island with more than 21 million inhabitants, whereas the Maldives is a collection of tiny atoll islands that together support only about 300,000 people. If global warming continues and if sea levels rise as predicted, the Maldives—where the highest point is only 6 feet (2 meters) above sea level—will be entirely flooded, and its inhabitants will have to seek higher ground elsewhere.

ENVIRONMENTAL GEOGRAPHY: DIVERSE LANDSCAPES, FROM TROPICAL ISLANDS TO MOUNTAIN RIM

South Asia's diverse environmental geography ranges from the highest mountains in the world to densely populated delta islands barely above sea level; from some of the wettest places on Earth to scorching deserts; and from tropical rain forests to eroded scrublands (Figure 12.2). Across most of the region, dense human population and rapid industrialization are generating severe environmental problems. The rapid increase

WORKING TOWARD SUSTAINABILITY

The Bangladeshi Textile Industry Comes Clean?

The textile industry, focused on ready-made garments, is the lifeblood of the Bangladeshi economy, accounting for 80 percent of the country's exports. Bangladesh is now the world's second largest textile exporter, trailing only China. The rapid expansion of the industry over the past several decades has been responsible for much of the country's impressive economic growth. Some 3.5 million people now work in the Bangladeshi textile and garment factories, 85 percent of whom are women.

This highly profitable textile industry, however, has plenty of problems. Wages are reportedly the lowest in the world, at less than $2 a day, and working conditions are harsh and often dangerous. In 2010, a fire on the 11th floor of a clothing factory resulted in 29 deaths, mostly from panicking workers stampeding toward the exits. Labor organizers seeking better conditions and pay have sometimes been attacked and occasionally killed. Despite such reprisals, protests and strikes by workers are becoming increasingly common. Some observers now doubt whether the Bangladeshi textile industry can continue to thrive based merely on its extremely low cost structure.

Textile production is also environmentally damaging. An estimated 56 million tons of untreated water, used for washing and finishing garments, are dumped into Bangladeshi rivers every year (Figure 12.1.1). A larger problem is the discharge of excess dye and other chemical substances. Chemicals used for dyeing clothing are often particularly toxic, resulting in fish kills and contaminating the drinking water of millions of impoverished citizens.

Bangladesh, however, is responding to the environmental threats posed by the industry, seeking to create a sustainable textile production system. In early 2010, for example, the Bangladesh Department of Environment imposed a major fine on a textile dyeing unit that was pouring untreated waste into the Turag River. The department further insisted that the factory cease operations until it installed an effluent treatment plant.

Foreign countries and companies that import clothing from Bangladesh are also seeking to enhance the sustainability of the textile industry. The government of the Netherlands, for example, recently pledged US$6.5 million to help factories implement cleaner production practices and install new water-saving technologies. It is hoped that such changes will result in both reduced pollution and enhanced profitability. The

FIGURE 12.1.1 Pollution in the Bangladeshi Textile Industry Bangladesh's thriving textile industry results in extensive water pollution. The Buriganga River, seen here, suffers from particularly high levels of contamination, mostly from dyes.

global company Huntsman Textile Effects has recently partnered with the Bangladeshi firm DBL Group in a project that will supposedly reduce water usage by roughly 50 percent and cut energy consumption by some 30 percent, while simultaneously increasing production by 30 percent. Other global companies are also turning their attention to this issue. A program announced in 2012 involves the participation of such leading retailers as H&M, Levi Strauss, Carrefour, and Tesco in cleaning up the Bangladeshi clothing industry. If the project works as hoped, it could result in an annual saving of 10.5 billion liters of water, as well as cash savings of US$70 million.

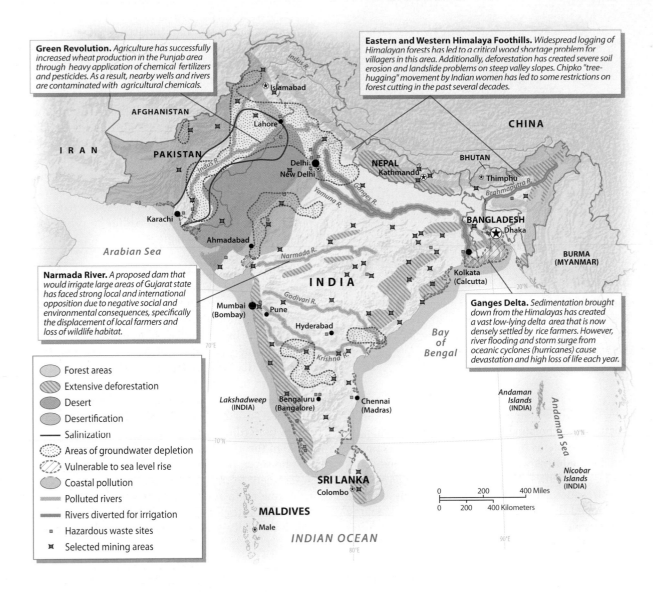

Green Revolution. *Agriculture has successfully increased wheat production in the Punjab area through heavy application of chemical fertilizers and pesticides. As a result, nearby wells and rivers are contaminated with agricultural chemicals.*

Eastern and Western Himalaya Foothills. *Widespread logging of Himalayan forests has led to a critical wood shortage problem for villagers in this area. Additionally, deforestation has created severe soil erosion and landslide problems on steep valley slopes. Chipko "tree-hugging" movement by Indian women has led to some restrictions on forest cutting in the past several decades.*

Narmada River. *A proposed dam that would irrigate large areas of Gujarat state has faced strong local and international opposition due to negative social and environmental consequences, specifically the displacement of local farmers and loss of wildlife habitat.*

Ganges Delta. *Sedimentation brought down from the Himalayas has created a vast low-lying delta area that is now densely settled by rice farmers. However, river flooding and storm surge from oceanic cyclones (hurricanes) cause devastation and high loss of life each year.*

Legend:
- Forest areas
- Extensive deforestation
- Desert
- Desertification
- — Salinization
- Areas of groundwater depletion
- Vulnerable to sea level rise
- Coastal pollution
- Polluted rivers
- Rivers diverted for irrigation
- ▫ Hazardous waste sites
- ✳ Selected mining areas

FIGURE 12.3 Environmental Issues in South Asia   As might be expected in a highly diverse and densely populated region, there are a wide range of environmental problems in South Asia. These range from salinization of irrigated areas in the dry lands of Pakistan and western India to groundwater pollution from Green Revolution fertilizers and pesticides. In addition, deforestation and erosion are widespread in upland areas.

in automobile use in particular is creating a serious smog problem in the region's urban areas. Air pollution in Delhi, one of India's largest cities, is now periodically worse than that of Beijing in China. South Asia has also suffered from some of the world's worst environmental disasters. The 1984 gas leak of a fertilizer plant in Bhopal, India, for example, killed more than 2,500 persons. Today, across the region, environmental activists, government agencies, and industrial producers are responding to the crisis, often in innovative ways. (See "Working Toward Sustainability: The Bangladeshi Textile Industry Comes Clean?")

Natural Hazards and Landscape Change in South Asia

Owing to specific features of its physical geography, South Asia suffers from some severe natural hazards and environmental problems (Figure 12.3). Particular problems include flooding in the region's large river deltas, deforestation in the uplands, and desertification in the northwest. Compounding all these problems are the immense numbers of new people added each year through natural population growth.

The Precarious Situation of Bangladesh   The link between population pressure and environmental problems is nowhere clearer than in the delta area of Bangladesh, where the search for fertile farmland has driven people into hazardous areas, putting millions at risk from seasonal flooding, as well as from the powerful cyclones (tropical storms) that form over the Bay of Bengal. For thousands of years, drenching monsoon rains have eroded huge quantities of sediment from the Himalayan slopes, which is then transported to the sea by the Ganges and Brahmaputra rivers, gradually building this low-lying delta environment.

FIGURE 12.4 Flooding in Bangladesh Devastating floods are common in the low-lying delta lands of Bangladesh. Heavy rains come with the southwest monsoon, especially to the Himalayas, and powerful cyclones often develop over the Bay of Bengal.

Although periodic flooding is a natural, even beneficial, phenomenon that enlarges deltas by depositing fertile river-borne sediment, huge flooding is a serious problem. In September 1998, for example, more than 22 million Bangladeshis were made homeless when water covered two-thirds of the country (Figure 12.4). Equally serious flooding in August 2007 caused much less damage, largely because of an internationally funded development program that was able to elevate hundreds of thousands of houses above the 1998 high-water mark. With Bangladesh's population growing rapidly, however, there is a strong possibility that flooding will take higher tolls in the coming decades as desperate farmers continue to relocate into the hazardous lower floodplains. Continuing deforestation of the Ganges and Brahmaputra headwaters magnifies the problem.

The summer of 2010 saw heavy rains across northern South Asia, resulting in renewed flooding in Bangladesh, which again left thousands of people homeless. During this event, however, damage in Pakistan was much worse. With damages estimated at US$43 billion, more than 2,000 people dead, and more than 20 million people affected, the 2010 Pakistan flood is commonly regarded as the worst natural disaster that the country has ever experienced.

Forests and Deforestation Many scholars link the catastrophic floods periodically experienced in much of South Asia with deforestation in the region's uplands. As trees are cleared away from the mountains and foothills of the Himalayas, runoff increases in the adjacent lowlands of Bangladesh, Pakistan, and northern India. As a result, efforts to protect remaining forest coverage in the region have intensified.

Tropical monsoon forests and savanna woodlands once covered most of South Asia, except for the desert areas in the northwest. In most areas, however, tree cover has vanished as a result of human activities. The Ganges River Valley and coastal plains of India, for example, were largely deforested hundreds of years ago to make room for agriculture. Elsewhere, forests were cleared more gradually for

agricultural, urban, and industrial expansion. More recently, hill slopes in the rugged landscapes of eastern and northern South Asia have been logged for commercial purposes. Extensive forests can still be found, however, in some areas.

As a result of deforestation, many South Asian villages suffer from a shortage of fuel wood for household cooking, forcing people to burn dung cakes from cattle. Although this low-grade fuel provides adequate heat, it prevents manure from being used as fertilizer. Where wood is available, collecting it may involve many hours of female labor because the remaining sources of wood are often far from the villages. In many areas, extensive eucalyptus stands have been planted to supply fuel wood and timber. These nonnative Australian trees support little or no wildlife, thus adding to problems of declining biodiversity throughout the region.

Wildlife Although the overall environmental situation in South Asia is rather bleak, wildlife protection inspires some optimism. The region has managed to retain a diverse assemblage of wildlife despite heavy population pressure and intense poverty. The only remaining Asiatic lions live in India's state of Gujarat, and even Bangladesh retains a viable population of tigers in the Sundarbans, the mangrove forests of the Ganges Delta. Wild elephants still roam several large reserves in India, Sri Lanka, and Nepal.

The protection of wildlife in India far exceeds that in most other parts of Asia. The official Project Tiger, which currently operates 42 preserves, is credited with increasing the country's tiger population from 1,200 in the 1970s to roughly 3,500 by the 1990s. A 2008 wildlife census, however, revealed a sharp drop in numbers, prompting the Indian government to pledge $153 million in further funding and to consider relocating up to 200,000 villagers to make more room for tiger populations. Advocates for the rural poor of India are not pleased by such plans, which they believe place the needs of wildlife above those of people.

South Asia's Monsoon Climates

The dominant climatic factor for most of South Asia is the **monsoon**, the seasonal change of wind direction that corresponds to wet and dry periods (Figure 12.5). During the winter, a large high-pressure system forms over the cold Asian landmass. Because winds flow from high pressure to low, cool, dry winds flow outward from the continental interior across South Asia. The resulting dry season extends from November until February. As winter turns to spring, these winds diminish, resulting in the hot, dry season of March through May. Eventually, the buildup of heat over South Asia and Southwest Asia produces a large low-pressure cell. By early June, the low-pressure cell is strong enough to cause a shift in wind direction so that warm, moist air from the Indian Ocean moves toward the continental interior. This signals the onset of the warm and rainy season of the southwest monsoon, which lasts from June through October (Figure 12.6).

Orographic rainfall is caused by the uplifting and cooling of moist monsoon winds over the Western Ghats and the Himalayan foothills. As a result, some areas receive more than 200 inches (508 centimeters) of rain during the four-month wet season (Figure 12.7). Cherrapunji, in northeastern India, is one of the world's wettest places, with an average rainfall of 450 inches (1,130 centimeters); see the climograph in Figure 12.7. On the Deccan Plateau, however, rainfall is dramatically reduced by a strong **rain-shadow effect**. A *rain shadow* is the area of low rainfall found on the downwind side of a mountain

FIGURE 12.5 The Summer and Winter Monsoons   Low pressure centered over South and Southwest Asia draws in warm, moist air masses during the summer that bring heavy monsoon rains to most of the region. Usually, these rains begin in June and last for several months. During the winter, high pressure forms over northern Asia. As a result, winds are reversed from those of the summer. During this season, only a few coastal locations in eastern India and Sri Lanka receive substantial rain.

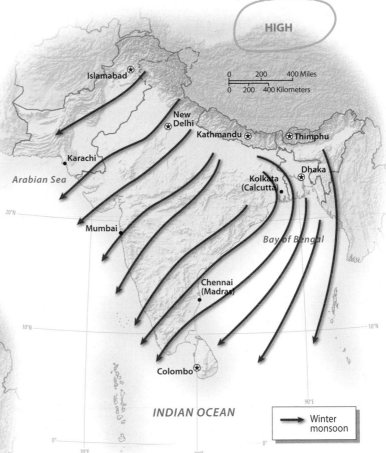

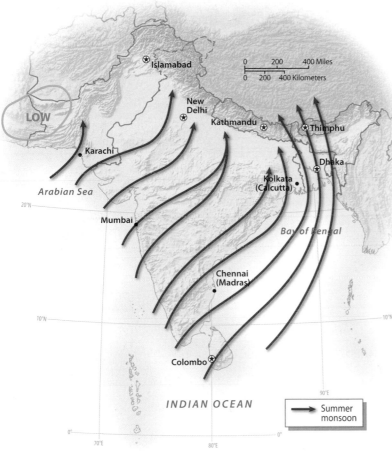

range (see Chapter 2). As winds move downslope, the air becomes warmer, and dry conditions usually prevail; see the climographs for Hyderabad, Delhi, and Karachi in Figure 12.7.

Climate Change and South Asia

Due to its climate and landforms, much of South Asia is highly vulnerable to global warming. Even a minor rise in sea level will inundate large areas of the Ganges–Brahmaputra Delta in Bangladesh. Already, more than 18,500 acres (7,500 hectares) of swampland in the Sundarbans region have been submerged. A 2007 report from the Indian government suggests that up to 7 million people could be displaced from coastal areas by the end of the century, due to a predicted 3.3-foot (1-meter)

FIGURE 12.6 Monsoon Rain   During the summer monsoon, some Indian cities, such as Mumbai (Bombay), receive more than 70 inches (178 centimeters) of rain in just three months. These daily torrents cause floods, power outages, and daily inconvenience. When the monsoon downpours finally arrive, people often react with joy, as can be seen in this photograph.

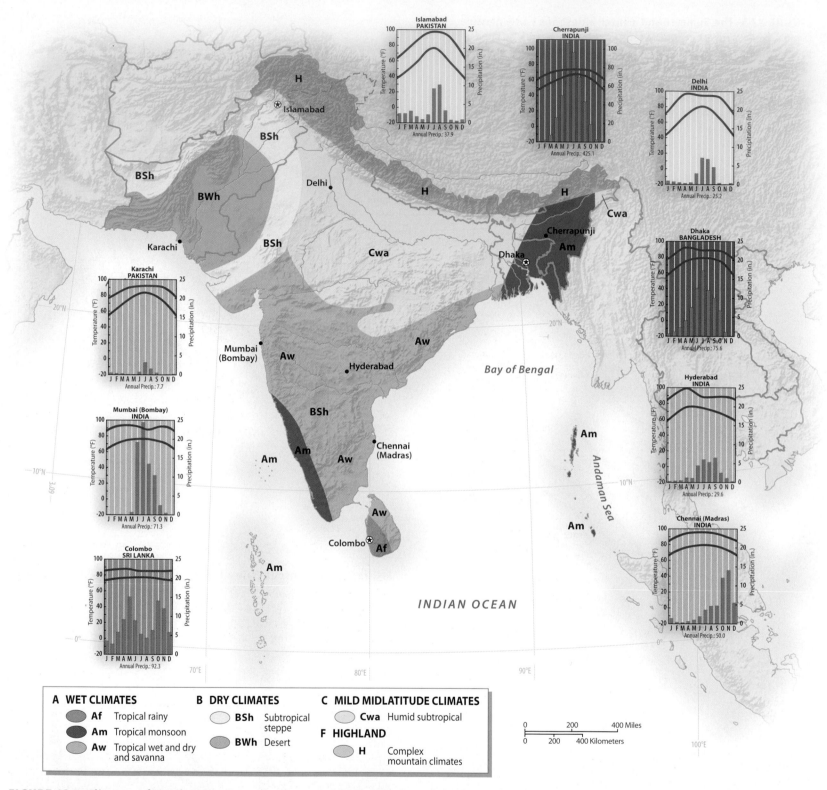

FIGURE 12.7 Climates of South Asia Except for the extensive Himalayas, South Asia is dominated by tropical and subtropical climates. Many of these climates show a distinct summer rainfall season that is associated with the southwest monsoon. The climographs for Mumbai (Bombay) and Delhi are excellent illustrations.

rise in sea level. If the most severe sea-level forecasts are realized, the atoll nation of the Maldives will simply vanish beneath the waves.

South Asian agriculture is likely to suffer from several problems linked to climate change. Many Himalayan glaciers are retreating, threatening the dry-season water supplies of the Indus–Ganges Plain, an area that already suffers from overuse of groundwater resources.

Increased winter temperatures of up to 6.4°F (3°C) could undermine the vital wheat crop of Pakistan and northwestern India. In parts of South Asia, climate change could result in increased rainfall due to an intensification of the summer monsoon. Unfortunately, much of this rainfall would come from intense cloudbursts, and as a result flooding and soil erosion would likely increase.

India signed the Kyoto Protocol in 2002, but as a developing country, it does not have to follow the main provisions of the treaty. With its poor and largely non-industrial economies, South Asia still has a low per capita output of greenhouse gases. However, India's economy in particular not only is growing rapidly, but also depends heavily on burning coal to generate electricity. According to official estimates, if India is to maintain an economic growth rate of 8 percent over the next quarter century, it will have to triple or even quadruple its primary energy supply.

Climate change may put Pakistan in a particular bind. As much as 90 percent of the irrigation water for this mostly desert country comes from the mountains of Kashmir, a contested region divided between Pakistan and India. According to estimates of the Intergovernmental Panel on Climate Change, the glaciers of Kashmir that feed Pakistan's rivers could be mostly gone by 2035. Because Pakistan already experiences periodic water shortages, many experts believe that it must cooperate with India to develop and conserve the water resources of the Kashmir region. However, considering the geopolitical tension between the two countries, which is focused on Kashmir, any such agreement seems unlikely.

Efforts are being made to prepare South Asia for possible climate change. In 2011, the global organization called CGIAR instituted a program aimed at creating "climate-smart villages" across the region. The program encourages tree planting, rainwater harvesting, careful water management, and soil conservation. It also facilitates the sharing of information on weather and climate, as well as on market conditions, though smart-phone networks.

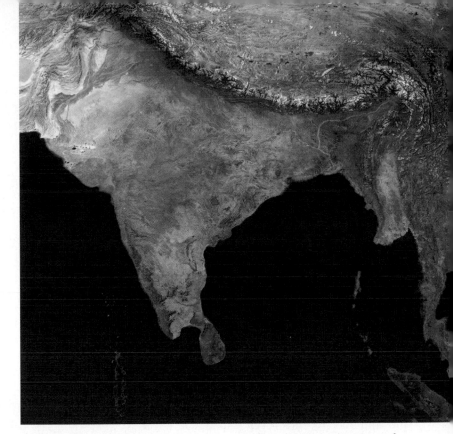

FIGURE 12.8 South Asia from Space The four physical subregions of South Asia are clearly seen in this satellite photograph, from the snow-clad Himalaya Mountains in the north to the islands of the south. The irrigated lands of the Indus River Valley in Pakistan are clearly visible in the upper left.

Physical Subregions of South Asia

To better understand environmental conditions in this diverse region, we can divide South Asia into four physical subregions, starting with the high mountain ranges of its northern edge and extending to the tropical islands of the far south. Lying south of the mountains are the extensive river lowlands that form the heartland of both India and Pakistan. Between these river lowlands and the island countries is the vast area of peninsular India, extending more than 1,000 miles (1,600 kilometers) from north to south (Figure 12.8).

Mountains of the North South Asia's northern rim of mountains is dominated by the great Himalayan Range, forming the northern borders of India, Nepal, and Bhutan. More than two dozen peaks exceed 25,000 feet (7,600 meters), including the world's highest mountain, Everest, on the Nepal–China (Tibet) border. To the east are the lower Arakan Yoma Mountains, forming the border between India and Burma (Myanmar) and separating South Asia from Southeast Asia.

These mountain ranges are a result of the dramatic collision of northward-moving peninsular India with the Asian landmass. The entire region is still geologically active, putting all of northern South Asia in serious earthquake danger. A massive earthquake in the Pakistani-controlled section of Kashmir on October 8, 2005, for example, resulted in roughly 100,000 deaths and left more than 3 million people homeless. Another huge quake hit Pakistan's Balochistan Province in 2011, but as it was centered in a sparsely settled area, the death toll was relatively low.

Although most of South Asia's northern mountains are too rugged and high to support dense human settlement, major population clusters are found in the Kathmandu Valley of Nepal, situated at 4,400 feet (1,300 meters), and the Valley, or Vale, of Kashmir in northern India, at 5,200 feet (1,600 meters).

Indus–Ganges–Brahmaputra Lowlands South of the northern mountains lie large lowlands created by three major river systems, which have deposited sediments to build huge alluvial plains of fertile and easily farmed soils. These densely settled lowlands constitute the core population areas of Pakistan, India, and Bangladesh.

The Indus River, flowing from the Himalayas through Pakistan to the Arabian Sea, provides much-needed irrigation for Pakistan's southern deserts. More famous, however, is the Ganges, which flows southeasterly some 1,500 miles (2,400 kilometers) and empties into the Bay of Bengal. The Ganges has provided the fertile alluvial soil that has made northern India one of the world's most densely settled areas. Given the central role of this important river throughout Indian history, it is understandable why Hindus consider the Ganges sacred. Finally, the Brahmaputra River, which rises on the Tibetan Plateau, flows more than 1,700 miles (2,700 kilometers) before joining the Ganges in central Bangladesh and spreading out over the world's largest river delta.

Peninsular India Extending southward from the river lowlands is peninsular India, made up primarily of the Deccan Plateau, which is bordered on each side by narrow coastal plains backed by north–south mountain ranges. On the west are the higher Western Ghats, which are generally about 5,000 feet (1,500 meters) in elevation; to the east, the Eastern Ghats are lower and less continuous. On both coastal plains, fertile soils and an adequate water supply support population densities comparable to those of the Ganges lowland to the north.

Soil quality ranges from fair to poor over much of the Deccan Plateau, but in the state of Maharashtra lava flows have produced particularly fertile black soils. Unfortunately, much of the area does not have a reliable water supply for agriculture. The western portion of the plateau lies in the rain shadow of the Western Ghats, giving it a semiarid climate. For centuries, small reservoirs have collected monsoon rainfall for use during the dry season. More recently, deep wells and powerful pumps have allowed groundwater development to support more widespread irrigation.

Partly because of the overuse of groundwater resources, the Indian government is building a series of large dams to provide for irrigation. Dam building, however, is controversial because the resulting reservoirs displace large numbers of rural residents. A case in point is the Sardar Sarovar Dam project on the Narmada River in the state of Madhya Pradesh, which has dislodged more than 100,000 people and has recently been expanded. Local residents and activists throughout India have joined forces in opposition, but farmers in neighboring Gujarat, who are reaping the benefits, strongly support the project.

The Southern Islands   At the southern tip of peninsular India lies the island country of Sri Lanka. Sri Lanka is ringed by extensive coastal plains and low hills, but mountains reaching more than 8,000 feet (2,400 meters) occupy the southern interior, providing a cool, moist climate. Because the main monsoon winds arrive from the southwest, that portion of the island is much wetter than the rain-shadow areas of the north and east.

The Maldives, a chain of more than 1,200 islands off the southwestern tip of India, forms a separate country. The combined land area of these islands is only about 116 square miles (290 square kilometers), and only one-quarter of the islands are inhabited. Like many islands of the South Pacific, the islands of the Maldives are low coral atolls, with a maximum elevation of just over 6 feet (2 meters) above sea level.

POPULATION AND SETTLEMENT: THE DEMOGRAPHIC DILEMMA

South Asia will soon surpass East Asia as the world's most populous region (Figure 12.9). India alone is home to over 1.2 billion people, while Pakistan and Bangladesh, with 180 and 153 million residents, respectively, rank among the world's 10 most populous countries

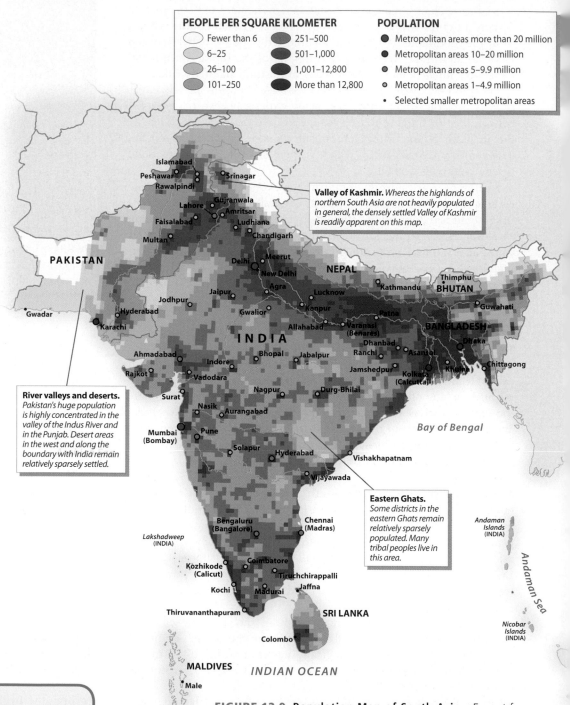

FIGURE 12.9 Population Map of South Asia   Except for the desert areas of the west and the high mountains of the north, South Asia is a densely populated region. Particularly high densities of people are found on the fertile plains along the Indus and Ganges rivers and in India's coastal lowlands. In rural areas, the population is typically clustered in villages, often located near water sources, such as streams, wells, canals, or small tanks that store water between monsoon rains.

REVIEW QUESTIONS

1. Why is the monsoon so crucial to life in South Asia?

2. Why is flooding such an important environmental issue in Bangladesh and adjacent areas of northeastern India?

TABLE 12.1 POPULATION INDICATORS

Country	Population (millions) 2010	Population Density (per square kilometer)	Rate of Natural Increase (RNI)	Total Fertility Rate	Percent Urban	Percent <15	Percent >65	Net Migration (rate per 1000) 2005–10[a]
Bangladesh	152.9	1,062	1.6	2.3	25	31	5	−1.0
Bhutan	0.7	15	1.3	2.6	35	30	5	2.6
India	1,259.7	383	1.5	2.5	31	31	5	−0.2
Maldives	0.3	1,110	1.9	2.3	35	27	5	−0.0
Nepal	30.9	210	1.8	2.6	17	36	4	−0.6
Pakistan	180.4	227	2.1	3.6	35	35	4	−1.4
Sri Lanka	21.2	323	1.2	2.2	15	25	8	−2.3

[a]Net Migration Rate from the United Nations, Population Division, *World Population Prospects: The 2008 Revision Population Database.*

Source: Population Reference Bureau, *World Population Data Sheet, 2010.*

(Table 12.1). Furthermore, much of South Asia is still experiencing rapid population growth. Although South Asia has made remarkable agricultural gains over the past several decades, widespread concern still persists about its ability to feed itself. The threat of crop failure, although much reduced, remains, in part because much South Asian farming is vulnerable to the unpredictable monsoon rains.

India's total fertility rate (TFR) has dropped rapidly, falling from 6 in the 1950s to the current rate of 2.5. In western and southern India, fertility rates are now generally at or below replacement levels. In much of northern India, however, birthrates remain high; the average woman in the poor state of Bihar gives birth to 3.9 children. A distinct cultural preference for male children is found in most of South Asia, a tradition that further complicates family planning. Although performing sex-selective abortions is illegal in India, the practice persists. In much of northern India, only about 8 girls are born for every 10 boys. In southern India and Sri Lanka, where women generally have a higher social position, sex ratios are balanced, and birthrates are much lower.

Pakistan has seen a rapid reduction in its TFR in recent years, but at 3.6, it is still well above the replacement level. As a result, Pakistan's population will probably be over 250 million by 2050. This is a worryingly high number, considering the country's arid environment, underdeveloped economy, and political instability (Figure 12.10). Bangladesh

FIGURE 12.11 Family Planning in Bangladesh Bangladesh has been one of the most successful nations in South Asia in reducing its fertility rate through family planning. The government of Bangladesh advertises family planning services heavily, even on its postage stamps.

has been more successful than Pakistan in reducing its birthrate. As recently as 1975, its TFR was 6.3, but it dropped to 2.3 by 2012. The success of family planning can be partly attributed to strong support from Bangladesh's government, advertised through radio, billboards, and even postage stamps (Figure 12.11).

Migration and the Settlement Landscape

South Asia is one of the least urbanized regions in the world, with around one-third of its people living in cities. Most South Asians reside in compact rural villages. Rapid migration from villages to large cities, however, is occurring. This often results as much from desperate conditions in the countryside as it does from employment opportunities in the city. Increased mechanization of agriculture, along with the

FIGURE 12.10 Population Pyramids of Pakistan and Sri Lanka Sri Lanka has had a relatively low birth rate for decades, as well as relatively high longevity figures, and thus has a well-balanced population pyramid. Pakistan, in contrast, has a much higher birth rate as well as a lower average life-span, and thus has a bottom-heavy pyramid.

expansion of large farms at the expense of subsistence cultivation, pushes many people to the region's rapidly growing urban areas.

The most densely settled areas of South Asia are those with fertile soils and dependable water supplies. The highest rural population densities are found in the core area of the Ganges and Indus river valleys and on the coastal plains of India. Settlement is less dense on the Deccan Plateau and is relatively sparse in the highlands of the far north and the arid lands of the northwest.

Many South Asians have migrated in recent years from poor and densely populated areas to less densely populated or wealthier areas. Migrants are often attracted to large cities such as Mumbai (Bombay), but those from Bangladesh are settling in large numbers in rural portions of northeastern India, creating ethnic tensions. Sometimes migrants are forced out by war; a large number of both Hindus and Muslims from Kashmir, for example, have sought security away from their battle-scarred homeland.

Many experts are concerned about internal migration in South Asia, which results in huge shantytowns and soaring homeless populations in the country's largest cities. A recent World Bank report advises India to encourage the development of its cities and to invest more in urban infrastructure, noting that cities produce two-thirds of India's gross domestic product and over 90 percent of its government revenues.

Agricultural Regions and Activities

South Asian agriculture has historically been relatively unproductive, especially compared with that of East Asia. Since the 1970s, however, agricultural production has grown rapidly. Unfortunately, many South Asian farmers have gone deeply in debt, threatening future agricultural gains.

Crop Zones   South Asia can be divided into several distinct agricultural regions, all with different problems and potentials. These regions are based on the production of three subsistence crops—rice, wheat, and millet.

Rice is the main crop and foodstuff in the lower Ganges Valley, along the lowlands of India's eastern and western coasts, in the delta lands of Bangladesh, along Pakistan's lower Indus Valley, and in Sri Lanka (Figure 12.12). This distribution reflects the large volume of

FIGURE 12.12 Rice Cultivation   A large amount of irrigation water is needed to grow rice, as is apparent from this photo. Rice is also the main crop in the lower Ganges Valley and Delta, along the lower Indus River of Pakistan, and in India's coastal plains. This photo shows farmers transplanting rice seedlings in Sri Lanka.

FIGURE 12.13 Green Revolution Farming   Because of "miracle" wheat strains that have increased yields in the Punjab area, this region has become the breadbasket of South Asia. India has more than doubled its wheat production in the past 25 years and has moved from continual food shortages to self-sufficiency. This photo shows a man scattering fertilizer in a green-revolution rice field in India.

irrigation water needed to grow the crop. The amount of rice grown in South Asia is impressive: India ranks behind only China in world rice production, and Bangladesh is the fourth largest producer.

Wheat is the principal crop of the northern Indus Valley and in the western half of India's Ganges Valley. South Asia's "breadbasket" is the northwestern Indian state of Punjab and adjacent areas in Pakistan. Here the so-called Green Revolution has been particularly successful in increasing grain yields. In the less fertile areas of central India, millet and sorghum are the main crops, along with root crops such as manioc. In general, wheat and rice are the preferred staples throughout South Asia, but poorer people must often subsist on millet and root crops.

The Green Revolution   The main reason South Asian agricultural growth has kept up with population growth is the **Green Revolution**, which originated during the 1960s in agricultural research stations established by international development agencies. By the 1970s, efforts to breed high-yield varieties of rice and wheat had reached their initial goals. As a result, South Asia was transformed from a region of chronic food deficiency to one of self-sufficiency. India more than doubled its annual grain production between 1970 and the mid-1990s (Figure 12.13).

Although the Green Revolution was an agricultural success, many experts highlight its ecological and social costs. Serious environmental problems result from the chemical dependency of the new crop strains. Not only do these crops typically need large quantities of industrial fertilizer, which is both expensive and polluting, but also they require frequent pesticide applications because they lack natural resistance to plant diseases and insects.

Social problems have also followed the Green Revolution. In many areas, only the more prosperous farmers are able to afford the new seed strains, irrigation equipment, farm machinery, fertilizers, and pesticides. As a result, poorer farmers have often been forced from their lands, becoming wage laborers for their more successful neighbors or migrating to crowded cities. To purchase the necessary inputs, moreover, most farmers have had to borrow large amounts of money. As lagging crop prices prevented many from repaying their debts, a wave of suicides occurred. According to a 2011 report published by the Centre of Human Rights and Justice, an Indian farmer commits suicide every 30 minutes on average, creating the largest wave of suicides in world history.

The Green Revolution has fed South Asia's expanding population over the past several decades, but whether it will be able to continue doing so remains unclear. An alternative option is to expand water delivery systems (through either canals or wells), as many fields are not irrigated. Irrigation, however, brings its own problems. In much of Pakistan and northwestern India, where irrigation has been practiced for generations, soil **salinization**, or the buildup of salt in fields, is already a major problem. In addition, groundwater is being depleted, especially in Punjab, India's breadbasket. On the other hand, optimists point out that India's production continues to grow and that famines, which were once common, are now a thing of the past.

Urban South Asia

Although South Asia remains a largely rural society, many of its cities are large and growing quickly. India alone has more than 40 metropolitan areas with over 1 million inhabitants. According to a 2010 report, India's urban population will increase from 340 million in 2008 to 590 million in 2030. Because of this rapid growth, South Asian cities have serious problems with homelessness, poverty, congestion, water shortages, air pollution, and sewage disposal. Throughout South Asia, sprawling squatter settlements, or **bustees**, exist in and around urban areas, providing meager shelter for many migrants.

Mumbai (Bombay)   The largest city in South Asia, Mumbai (often called by its former name, Bombay) is India's financial, industrial, and commercial center. Mumbai itself contains roughly 14 million people, while its metropolitan area is home to more than 22 million. Mumbai is responsible for much of India's foreign trade, has long been a manufacturing center, and is the focus of India's film industry—the world's largest. Mumbai's economic vitality draws people from all over India, resulting in simmering ethnic tensions.

Because of the city's restricted space, most of Mumbai's growth has taken place to the north and east of the historic city. Building restrictions in the downtown area have resulted in skyrocketing commercial and residential rents, which are some of the highest in the world. Even members of the city's thriving middle class have difficulty finding adequate housing. Hundreds of thousands of less-fortunate immigrants live in "hutments," crude shelters built on formerly busy sidewalks (Figure 12.14). The least fortunate sleep on the street or in simple plastic tents, often placed along busy roadways.

Mumbai's notorious road congestion eased somewhat in 2009, after the completion of a massive eight-lane, $340 million bridge linking the central city to its northern suburbs. The Mumbai Metro, an ambitious rapid transit system, is scheduled to be completed in 2021, promising further improvements.

FIGURE 12.14 Mumbai Hutments   Hundreds of thousands of people in Mumbai live in crude hutments, with no sanitary facilities, built on formerly busy sidewalks. Hutment construction is forbidden in many areas, but wherever it is allowed, sidewalks quickly disappear.

Kolkata (Calcutta)   To many, Kolkata—more often called by its old name, Calcutta—symbolizes the problems faced by rapidly growing cities in developing countries. Approximately 1 million people here sleep on the streets every night. And with approximately 15 million people in its metropolitan area, Kolkata falls far short of supplying the rest of its residents with water, power, and sewage treatment. Electrical power is woefully inadequate, and during the wet season many streets are routinely flooded.

With rapid growth as migrants pour in from the countryside, a mixed Hindu–Muslim population that generates ethnic tension, a decayed economic base, and an overloaded infrastructure, Kolkata faces a troubled future. Yet it remains a culturally vibrant city, noted for its fine educational institutions, theaters, and publishing firms. Kolkata is currently trying to nurture an information technology industry, but it remains to be seen whether it will prove successful.

Karachi   Pakistan's largest urban area and commercial core, Karachi is one of the world's fastest-growing cities. Its metropolitan population, already somewhere between 12 and 18 million, is expanding at about 5 percent per year (Figure 12.15). Karachi served as Pakistan's

FIGURE 12.15 Karachi Street Scene   Karachi, Pakistan's largest city and main port, is noted for both its economic power and its ethnic violence, as well as for its congested and colorful streets. British influences are evident in this photo of the Karachi Empress Market and bus station.

capital until 1963, when the new city of Islamabad was created in the northeast. Karachi suffered relatively little from the departure of government functions; it is still the most cosmopolitan city in Pakistan, its main streets lined with businesses and high-rise buildings.

Karachi also suffers from political and ethnic tensions that have periodically turned parts of the city into armed camps and even battlegrounds. In the early decades of Pakistan's independence, Karachi's main conflict was between the Sindis, the region's native inhabitants, and the Muhajirs, the Muslim refugees from India who settled in the city after division from India in 1947. More recently, clashes between Sunni and Shiite Muslims have intensified, as have those between Pashtun migrants from northwestern Pakistan and other residents. In 2011 alone, more than 800 people were killed in politically motivated violence in the city.

REVIEW QUESTIONS

1. Why has the Green Revolution been so controversial, considering the fact that it has greatly increased South Asia's food supply?

2. What are the major advantages and disadvantages of the growth of South Asia's mega-cities?

CULTURAL COHERENCE AND DIVERSITY: A COMMON HERITAGE UNDERMINED BY RELIGIOUS RIVALRIES

Historically, South Asia is a well-defined cultural region. A thousand years ago, virtually the entire area was united by the religion of Hinduism. The subsequent arrival of Islam added a new religious element, but did not undermine the region's cultural unity. British imperialism later added other cultural features to the region, from the widespread use of English to a passion for cricket. Since the mid-20th century, however, religious strife has intensified, leading some to question whether South Asia can still be considered a culturally unified region.

India has been a secular state since its creation. Since the 1980s, this political tradition has come under pressure from the growth of **Hindu nationalism**, which promotes Hindu values as the foundation of Indian society. Hindu nationalists have gained considerable political power through the Bharatiya Janata Party (BJP). In several high-profile instances, Hindu mobs demolished Muslim mosques that had allegedly been built on the sites of ancient Hindu temples. Since 2000, however, the Hindu nationalist movement has declined, and the more secular political parties have won most of India's recent elections. Although bitter divisions persist, many efforts are being made to promote religious understanding (Figure 12.16).

In Pakistan, Islamic fundamentalism has been a divisive issue. Powerful fundamentalist leaders want to make Pakistan a religious state under Islamic law, a plan rejected by the country's secular intellectuals and international businesspeople. The government has attempted to mediate between the two groups, but with little success. Substantial areas of the country have come under the control of Islamist insurgents, leading some observers to question whether Pakistan can hold together as a country.

FIGURE 12.16 The Controversy of the Ayodhya Mosque The dismantling of the Babri Mosque on Ayodhya by Hindu nationalists in 1992 resulted in intense religious conflicts in many parts of India. More recently Hindu-Muslim brotherhood ("Bhai Bhai") groups have emerged to try to enhance understanding and mutual respect. A rally of one such group is depicted here.

Origins of South Asian Civilizations

Many scholars think that the roots of South Asian culture extend back to the Indus Valley civilization, which flourished 4,500 years ago in what is now Pakistan. This remarkable urban-oriented society vanished almost entirely around 1800 BCE. By 800 BCE, however, a new focus of civilization had emerged in the middle Ganges Valley.

Hindu Civilization The religion that emerged out of the early Ganges Valley civilization was Hinduism, a complicated faith that lacks a single system of belief. Certain deities are recognized, however, by all believers, as is the notion that these various gods are all expressions of a single divine entity (Figure 12.17). Hindus also share a common set of epic stories, usually written in **Sanskrit**, the sacred language of their religion. Hinduism is noted for its mystical tendencies, which have long inspired many to seek an ascetic lifestyle,

FIGURE 12.17 Hindu Temple Hindu temples are a prominent feature of the Indian landscape, and as India's economic grows, lavish new temples continue to be built. The Akshardham temple complex in Delhi, part of which is seen here, opened in 2005.

renouncing property and sometimes all regular human relations. One of its hallmarks is a belief in the transmigration of souls from being to being through reincarnation. Hinduism is also associated with India's **caste system**, the strict division of society into hereditary groups that are ranked as ritually superior or inferior to one another.

Buddhism Ancient India's caste system was challenged from within by Buddhism. Siddhartha Gautama, the Buddha, was born in 563 BCE in an elite caste. He rejected the life of wealth and power, however, and sought instead to attain enlightenment, or mystical union with the universe. He preached that the path to such enlightenment (or *nirvana*) was open to all, regardless of social position. His followers eventually established Buddhism as a new religion. Buddhism spread throughout South Asia and later expanded into East, Southeast, and Central Asia. But it never fully replaced Hinduism in India, and by 500 CE Buddhism was disappearing from most of South Asia.

Arrival of Islam The next major challenge to Hindu society—Islam—came from the outside. Around the year 1000, Turkic-speaking Muslims began to invade from Central Asia. By the 1300s, most of South Asia lay under Muslim power, although Hindu kingdoms persisted in southern India. During the 16th and 17th centuries, the **Mughal (or Mogul) Empire**, the most powerful of the Muslim states, dominated much of the region from its power center in the upper Indus–Ganges Basin (Figure 12.18).

At first, Muslims formed a small ruling elite, but over time increasing numbers of Hindus converted to the new faith. Conversions were most pronounced in the northwest and northeast, with the areas now known as Pakistan and Bangladesh becoming predominantly Muslim.

The Caste System Caste is one of the historically unifying features of South Asia, as certain aspects of caste organization are found even among the Muslim and Christian populations of the region. Islam, however, has gradually reduced its significance, and even in India caste is now being de-emphasized, especially among the more educated social groups. But caste remains significant, especially in rural India. Marriage across caste lines is still relatively rare, and caste issues figure heavily in Indian politics. In 2006, a major controversy erupted when the Indian government increased the number of positions in public universities

reserved for members of the lower castes. Proponents claimed that such a measure was necessary to address discrimination, whereas opponents claimed that it would result in lower educational standards. Further controversies emerged in 2011, when the government established a separate quota for government employment for members of religious minorities within the "backward caste" category.

Caste is actually a rather clumsy term that refers to the complex social order of the Hindu world. It combines two distinct local concepts: *varna* and *jati*. *Varna* refers to the ancient fourfold social hierarchy of the Hindu world, which distinguishes the Brahmins (priests), Kshatriyas (warriors), Vaishyas (merchants), and Sundras (farmers and craftsmen), in declining order of ritual purity. Standing outside this traditional order are the so-called untouchables, now usually called **Dalits**, whose ancestors held "impure" jobs, such as those associated with leather working or trash collection. *Jati*, on the other hand, refers to the hundreds of local endogamous ("marrying within") groups that exist at each *varna* level. Different *jati* groups are often called *subcastes*.

Contemporary Geographies of Religion

In the simplest terms, South Asia has a Hindu heritage overlain by a significant Muslim presence. Such a picture fails, however, to capture the enormous diversity of religion in contemporary South Asia (Figure 12.19).

Hinduism Fewer than 1 percent of the people of Pakistan are Hindu, and in Bangladesh and Sri Lanka Hinduism is a minority religion. However, in India and Nepal, Hinduism is clearly the majority faith. In most of central India, more than 90 percent of the population is Hindu. Hinduism is itself a geographically complicated religion, with different aspects of faith varying across different parts of India.

Islam Islam may be a minority religion for South Asia as a whole, but it is still very widespread, counting more than 500 million followers. Bangladesh and especially Pakistan are overwhelmingly Muslim. India's Muslim community, although constituting only some 15 percent of the country's population, is still roughly 175 million strong.

Muslims live in almost every part of India. They are, however, concentrated in four main areas: in most large cities; in Kashmir, in the far north, particularly in the densely populated Vale of Kashmir (more than 80 percent of the population here follows Islam); in the central Ganges Plain, where Muslims constitute 15 to 20 percent of the population; and in the southwestern state of Kerala, which is approximately 25 percent Muslim.

Interestingly, Kerala was one of the few parts of India that never experienced prolonged Muslim rule. Islam in Kerala was historically connected to trade across the Arabian Sea. Kerala's Malabar Coast historically supplied spices and other luxury products to Southwest Asia, encouraging many Arab traders to settle there. Gradually, many of Kerala's native residents converted to the new religion as well. The same trade routes brought Islam to Sri Lanka, which is approximately 9 percent Muslim, and to the Maldives, which is almost entirely Muslim.

FIGURE 12.18 The Red Fort The Red Fort of Delhi, completed in 1648, was the power center of the Mughal Empire. Today this massive fortification, one of the largest in the world, is a major tourist destination.

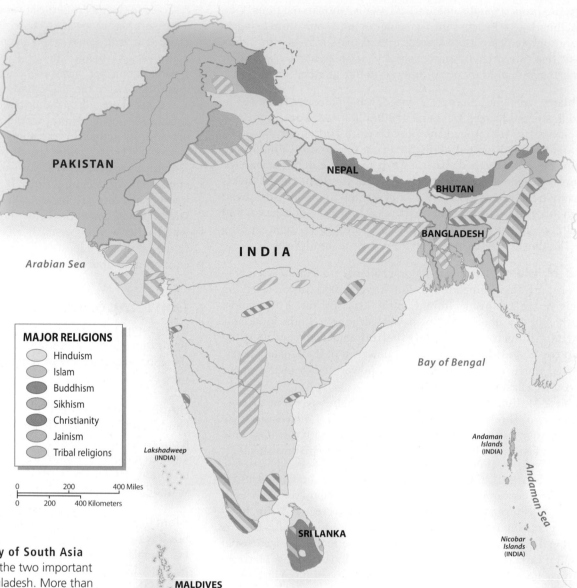

FIGURE 12.19 Religious Geography of South Asia
Hindu-dominated India is bordered by the two important
Muslim countries of Pakistan and Bangladesh. More than
150 million Muslims, however, live within India, making up
roughly 15 percent of the total population. Of particular note
are the Muslims in northwest Kashmir and in the Ganges Valley.
Sikhs form the majority population in India's state of Punjab.
Also note the Buddhist populations in Sri Lanka, Bhutan, and
northern Nepal; the areas of tribal religion in the east; and the
centers of Christianity in the southwest.

FIGURE 12.20 Sikh Soldiers Members of India's Sikh religion
developed military traditions when their faith was persecuted in earlier
centuries, Today, many Sikh men serve in the India military. This photo
shows members of a Sikh regiment marching in Delhi's in celebration of
India's 51st anniversary of its transition to a republic in 2001.

Sikhism The tension between Hinduism and Islam in northern
South Asia gave rise to a new religion called **Sikhism**. Sikhism originated in the 1400s in the Punjab, near the modern boundary between
India and Pakistan. The Punjab was the site of intense religious competition at the time; Islam was gaining converts, and Hinduism was
on the defensive. The new faith combined elements of both religions.
Many orthodox Muslims viewed Sikhism as dangerous because it incorporated elements of their own religion in a manner contrary to accepted beliefs. Periodic persecution led the Sikhs to adopt a militantly
defensive stance. Even today many Sikh men work as soldiers and
bodyguards, both in India and abroad (Figure 12.20).

At present, the Indian state of Punjab is approximately 60 percent Sikh. Small, but often influential groups of Sikhs are scattered across the rest of India. Devout Sikh men are immediately visible because they do not cut their hair or their beards. Instead, they wear their hair wrapped in a turban and often tie their beards close to their faces.

Buddhism and Jainism Although Buddhism virtually disappeared from India in medieval times, it persisted in Sri Lanka. Among the island's dominant Sinhalese people, Theravada Buddhism developed into a national religion. In the high valleys of the Himalayas, the Tibetan form of Buddhism emerged as the majority faith. The town of Dharamsala in the northern Indian state of Himachal Pradesh is the seat of Tibet's government-in-exile and of its spiritual leader, the Dalai Lama, who fled Tibet in 1959 after an unsuccessful revolt.

At roughly the same time as the birth of Buddhism (circa 500 BCE), another religion emerged in northern India: **Jainism**. This religion also stressed nonviolence, taking this creed to its ultimate extreme. Jains are forbidden to kill any living creatures, and as a result the most devoted members of the community wear gauze masks to prevent them from inhaling small insects. Agriculture is forbidden to Jains because plowing can kill small creatures. As a result, most members of the faith have looked to trade for their livelihoods. Today the Jains form a relatively prosperous community concentrated in northwestern India.

Other Religious Groups The Parsis, concentrated in Mumbai, form a small, but influential religious group. Followers of Zoroastrianism, the ancient faith of Iran, Parsi refugees fled to India in the 7th century (Figure 12.21). The Parsis prospered under British rule, forming some of India's first modern industrial companies, such as the Tata Group of industrial companies. Intermarriage and low fertility, however, now threaten the survival of this small community.

Indian Christians are more numerous than either Parsis or Jains. Their religion arrived some 1,700 years ago, as missionaries from Southwest Asia brought Christianity to India's southwestern coast. Today roughly 20 percent of the people of Kerala follow Christianity. Several Christian sects are represented, but the largest are affiliated with the Syrian Christian Church of Southwest Asia. Another stronghold of Christianity is the small Indian state of Goa, a former Portuguese colony. Here Roman Catholics make up roughly half of the population.

During the colonial period, British missionaries went to great efforts to convert South Asians to Christianity. They had little success, however, in Hindu, Muslim, and Buddhist communities. The remote tribal districts of British India proved to be more receptive to missionary activity, especially those in the northeast. The Indian states of Nagaland, Meghalaya, and Mizoram now have Christian majorities, with more than 75 percent of the people of Nagaland belonging to the Baptist Church.

Geographies of Language

South Asia's linguistic diversity rivals its religious diversity. In northern South Asia, most languages belong to the Indo-European family, the world's largest. The languages of southern India, on the other hand, belong to the **Dravidian language** family, which is found only in South Asia. Along the mountainous northern rim of the region, a third linguistic family, Tibeto-Burman, dominates. Within these broad divisions are many different languages, each associated with a distinct culture. In many parts of South Asia, several languages are spoken within the same area, and the ability to speak several languages is common everywhere (Figure 12.22).

Each of the major languages of India is associated with an Indian state, as the country deliberately structured its political subdivisions along linguistic lines a decade after attaining independence. As a result, the Gujarati language is spoken in Gujarat, Marathi in Maharashtra, Oriya in Orissa, and so on. Two of these languages, Punjabi and Bengali, extend into Pakistan and Bangladesh, respectively, as the political borders here were established on religious rather than linguistic lines. Nepali, the national language of Nepal, is also spoken in many of the mountainous areas of northern India. Minor dialects and languages abound in many of the more remote areas.

The Indo-European North The most widely spoken language of South Asia is **Hindi** (not to be confused with the Hindu religion). With more than 500 million native speakers, Hindi is by some measurements the world's second most widely spoken language. It plays a prominent role in present-day India, both because so many people speak it and because it is the main language of the Ganges Valley. Hindi is an official language in 10 Indian states, and it is widely studied throughout the country.

Bengali, the second most widely spoken language in South Asia, is the official language of Bangladesh and the Indian state of West Bengal. Spoken by roughly 200 million people, Bengali is the world's ninth most widely spoken language. It also has an extensive literature, as West Bengal (and particularly its capital city, Kolkata [Calcutta])

FIGURE 12.21 Parsi Temple India's small but influential Parsi community follows the ancient Iranian religion of Zoroastrianism. This photo shows decorations in the style of ancient Iran on a Parsi temple in Mumbai, India.

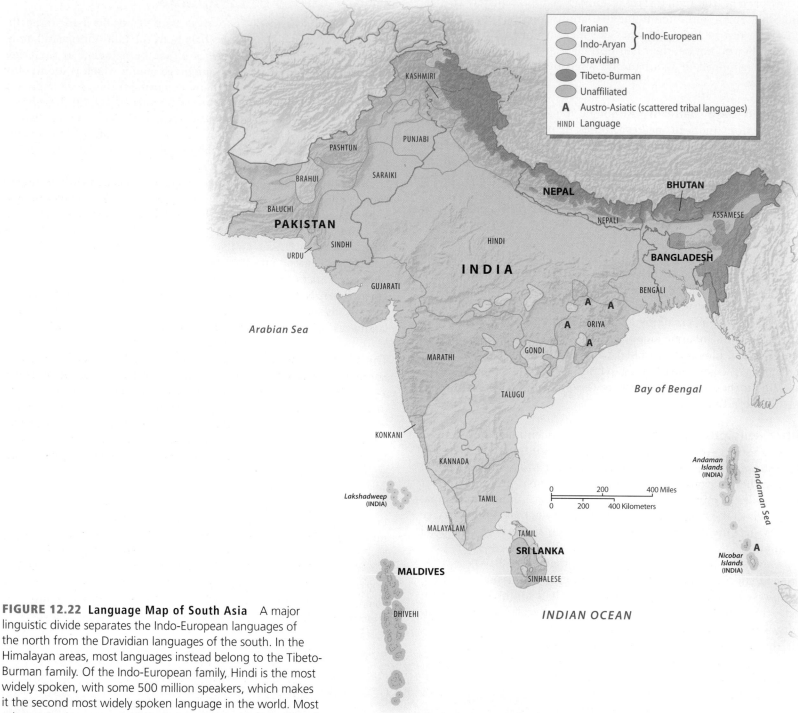

FIGURE 12.22 Language Map of South Asia A major
linguistic divide separates the Indo-European languages of
the north from the Dravidian languages of the south. In the
Himalayan areas, most languages instead belong to the Tibeto-
Burman family. Of the Indo-European family, Hindi is the most
widely spoken, with some 500 million speakers, which makes
it the second most widely spoken language in the world. Most
other major languages are closely associated with states in India.

has long been one of South Asia's leading literary and intellectual
centers (Figure 12.23).

The Punjabi-speaking zone in the west was split at the time of independence between Pakistan and the Indian state of Punjab. While
almost 100 million people speak Punjabi, this language does not
have the significance of Bengali. Punjabi did not become the national
language of Pakistan, even though it is the day-to-day language of
almost half of the country's population. Instead, that distinction was
given to Urdu.

Urdu, like Hindi, originated on the plains of northern India. The
difference between the two was largely one of religion: Hindi was the

language of the Hindu majority, Urdu that of the Muslim minority.
Because of this distinction, Hindi and Urdu are written differently—
Hindi in the Devanagari script (derived from Sanskrit) and Urdu in
the Arabic script. Although Urdu contains many words borrowed
from Persian, its basic grammar and vocabulary are almost identical to
those of Hindi. With independence in 1947, millions of Urdu-speaking
Muslims from the Ganges Valley fled to Pakistan. Because Urdu had a
higher status than Pakistan's native tongues, it was quickly established
as the new country's official language. Although only about 8 percent
of the people of Pakistan learn Urdu as their first language, more than
90 percent are able to speak and understand it.

FIGURE 12.23 Kolkata (Calcutta) Bookstore Although Kolkata (Calcutta) is noted in the West mostly for its abject poverty, the city is also known in India for its vibrant cultural and intellectual life, illustrated by its large number of bookstores, theaters, and publishing firms.

FIGURE 12.24 Multilingualism This four-language sign, in Malayalam, Tamil, Kannada, and English, shows the multilingual nature of contemporary South Asia. In many ways, English, the colonial language of British rule, still serves to bridge the gap between the many different languages of the region.

Languages of the South The four main Dravidian languages are confined to southern India and northern Sri Lanka. As in the north, each language is closely associated with an Indian state: Kannada in Karnataka, Malayalam in Kerala, Telugu in Andhra Pradesh, and Tamil in Tamil Nadu. Tamil is usually considered the most important member of the family because it has the longest history and the largest literature. Tamil poetry dates back to the 1st century CE, making it one of the world's oldest written languages.

Although Tamil is spoken in northern Sri Lanka, the country's majority population, the Sinhalese, speak an Indo-European language. Apparently, the Sinhalese migrated from northern South Asia several thousand years ago, settling primarily on the island's fertile southwestern coast and central highlands. These same people also migrated to the Maldives, where the national language, Dhivehi, is essentially a Sinhalese dialect. The drier north and east of Sri Lanka, on the other hand, were settled mainly by Tamils from southern India. Some Tamils later moved to the central highlands, where they were employed as tea-pickers on British-owned estates.

Linguistic Dilemmas The multilingual countries of Sri Lanka, Pakistan, and India are all troubled by linguistic conflicts. Such problems are most complex in India, simply because India is so large and has so many different languages (Figure 12.24).

Indian nationalists have long dreamed of a national language that could help unify their country. But **linguistic nationalism**, or the linking of a specific language with political goals, often faces the resistance of many people. The obvious choice for a national language would be Hindi, and Hindi was indeed declared as such in 1947. Raising Hindi to this position, however, angered many non-Hindi speakers, especially in the Dravidian south. It was eventually decided that both Hindi and English would serve as official languages of India as a whole, but that each Indian state could select its own official language. As a result, 22 separate Indian languages now have such status.

Regardless of opposition, Hindi is expanding, especially in the Indo-European north. Here local languages are closely related to Hindi, which can therefore be learned fairly easily. Hindi is spreading through education and, more significantly, through television and motion pictures. Films and television programs are made in several

northern languages, but Hindi remains primary. In a poor, but modernizing country such as India, where many people experience the wider world largely through moving images, the influence of a national film and television culture can be substantial.

Despite its spread, Hindi remains foreign to much of India. National-level communication is thus conducted mainly in English. Although many Indians want to de-emphasize English, others advocate it as a neutral national language because all parts of the country have an equal stake in it. Furthermore, English gives substantial international benefits. English-medium schools abound throughout South Asia, and many children of the elite learn this global language well before they begin school.

South Asia in Global Cultural Context

The widespread use of English in South Asia not only helps global culture spread throughout the region, but also has helped South Asians' cultural production reach a global audience. The global spread of South Asian literature, however, is nothing new. As early as the turn of the 20th century, Rabindranath Tagore gained international acclaim for his poetry and fiction, earning the Nobel Prize for Literature in 1913. Not long afterward, the "Bollywood" films produced in Mumbai (Bombay) gained great popularity across much of the developing world. Bollywood continues to have global appeal: In 2010, the Indian film *My Name Is Khan* earned more than US$23 million in foreign markets alone.

The expansion of South Asian culture abroad has been accompanied by the spread of South Asians themselves. Migration from South Asia during the time of the British Empire led to the establishment of large communities in such distant places as eastern Africa, Fiji, and the southern Caribbean (Figure 12.25). Subsequent migration has been aimed more at the developed world; there are now several million people of South Asian origin living in Britain, and a similar number live in North America. Many present-day migrants to the United States are doctors, software engineers, and members of other professions.

In South Asia itself, the globalization of culture has brought tensions as severe as those felt anywhere else in the world. Traditional Hindu and Muslim religious norms frown on any overt display of sexuality—a staple feature of global popular culture. Religious leaders thus often criticize Western films and television shows as being

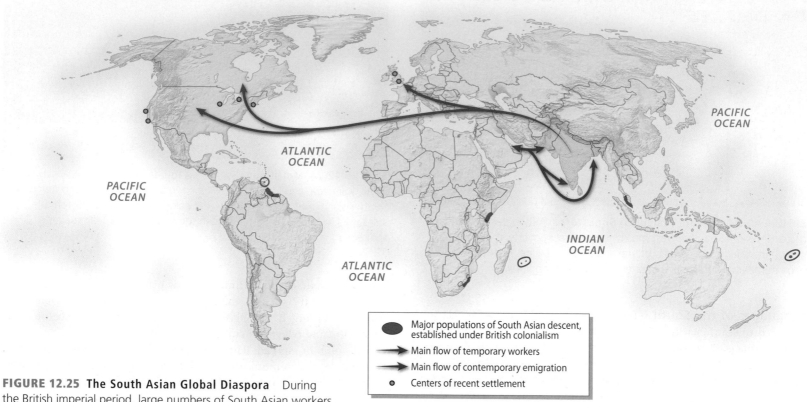

Major populations of South Asian descent, established under British colonialism

→ Main flow of temporary workers

→ Main flow of contemporary emigration

○ Centers of recent settlement

FIGURE 12.25 The South Asian Global Diaspora   During the British imperial period, large numbers of South Asian workers settled in other colonies. Today, roughly 50 percent of the population of such places as Fiji and Mauritius is of South Asian descent. More recently, large numbers have settled, and are still settling, in Europe (particularly Britain) and North America. Large numbers of temporary workers, both laborers and professionals, are employed in the wealthy oil-producing countries of the Persian Gulf.

immoral. Although India is a relatively free country, both the national government and the state governments periodically ban films and books that are considered too sexual. In 2012, for example, *The Girl with the Dragon Tattoo* was banned because of its adult scenes. Other films have been barred from India for other reasons. *Indiana Jones and the Temple of Doom*, for example, was banned in 1984 for its racist portrayals of Indians.

Still, the pressures of internationalization are hard to resist. In the tourism-oriented Indian state of Goa, such tensions are on full display. There German and British sun worshipers often wear nothing but skimpy swimwear, whereas Indian women tourists go into the ocean fully clothed. Young Indian men, for their part, often simply walk the beach and gawk, good-naturedly, at the semi-naked foreigners (Figure 12.26).

REVIEW QUESTIONS

1. Why has religion become such a contentious issue in South Asia over the past several decades?

2. How have India and Pakistan, the two largest countries of South Asia, tried to foster national unity in the face of ethnic and linguistic fragmentation?

FIGURE 12.26 Goa Beach Scene   The liberal Indian state of Goa, formerly a Portuguese colony, is now a major destination for tourists, both from within India and from Europe and Israel. European tourists come in the winter for sunbathing and for "Goan rave parties," where ecstasy and other drugs are widely available. Indian tourists typically find the scantily clad foreigners unusual, if not bizarre.

GEOPOLITICAL FRAMEWORK: A DEEPLY DIVIDED REGION

Before the coming of British imperialism, South Asia had never been politically united. At times, a few empires ruled most of the subcontinent, but none covered its entire extent. The British, however, brought the entire region under their power by the middle of the 19th century. Independence in 1947 brought the separation of Pakistan from India; in 1971, Pakistan itself was divided, with the independence of Bangladesh, formerly East Pakistan. Today serious geopolitical issues continue to plague the region (Figure 12.27).

South Asia Before and After Independence in 1947

During the 1500s, when Europeans first arrived, most of northern South Asia was ruled by the Muslim Mughal Empire (Figure 12.28), while southern India remained under the control of the Hindu kingdom of *Vijayanagara*. European merchants, eager to obtain spices, textiles, and other Indian products, established several coastal trading posts. The Portuguese carved out an enclave in Goa, while the Dutch gained control over much of Sri Lanka, but neither was a significant threat to the Mughals. In the early 1700s, however, the Mughal Empire weakened rapidly, with competing states emerging in its former territories.

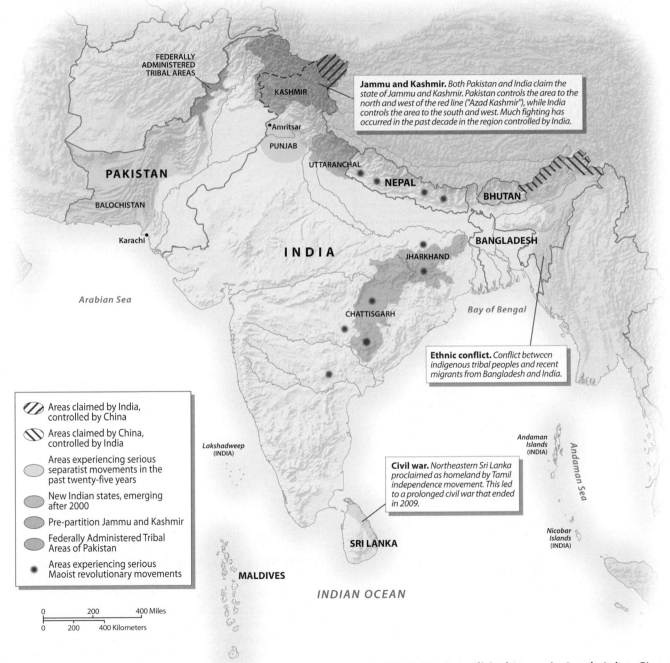

Jammu and Kashmir. *Both Pakistan and India claim the state of Jammu and Kashmir. Pakistan controls the area to the north and west of the red line ("Azad Kashmir"), while India controls the area to the south and west. Much fighting has occurred in the past decade in the region controlled by India.*

Ethnic conflict. *Conflict between indigenous tribal peoples and recent migrants from Bangladesh and India.*

Civil war. *Northeastern Sri Lanka proclaimed as homeland by Tamil independence movement. This led to a prolonged civil war that ended in 2009.*

Areas claimed by India, controlled by China

Areas claimed by China, controlled by India

Areas experiencing serious separatist movements in the past twenty-five years

New Indian states, emerging after 2000

Pre-partition Jammu and Kashmir

Federally Administered Tribal Areas of Pakistan

Areas experiencing serious Maoist revolutionary movements

FIGURE 12.27 Geopolitical Issues in South Asia Given the cultural mosaic of South Asia, it is not surprising that ethnic tensions have created numerous geopolitical problems in the region. Particularly troubling are ethnic tensions in Sri Lanka, Kashmir, and northeastern India.

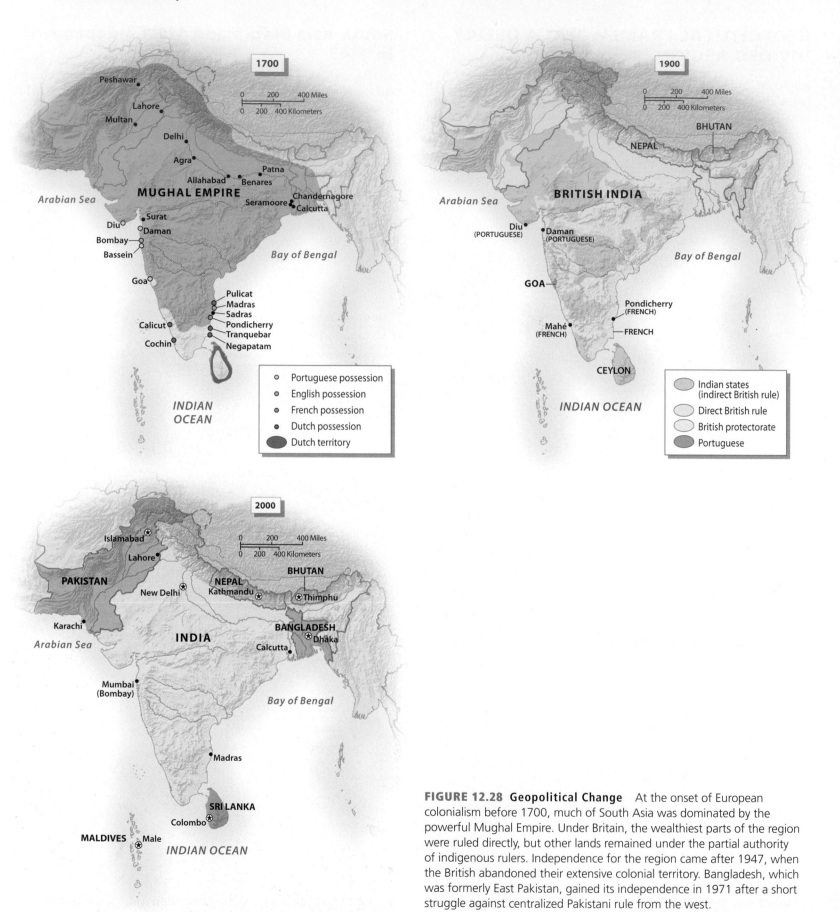

FIGURE 12.28 Geopolitical Change At the onset of European colonialism before 1700, much of South Asia was dominated by the powerful Mughal Empire. Under Britain, the wealthiest parts of the region were ruled directly, but other lands remained under the partial authority of indigenous rulers. Independence for the region came after 1947, when the British abandoned their extensive colonial territory. Bangladesh, which was formerly East Pakistan, gained its independence in 1971 after a short struggle against centralized Pakistani rule from the west.

FIGURE 12.29 Former British Hill Station   Britain's imperial officials in India often longed for the cool weather and lush gardens of their homeland. As a result, Britain built many high-elevation "hill stations" as resorts. Today these one-time symbols of imperialism are popular tourist destinations for the Indian elite. Part of the former hill station of Darjeerling in West Bengal, India is visible in this photo.

The British Conquest

The unsettled conditions of the 1700s provided an opening for European imperialism. The British and French, having largely displaced the Dutch and Portuguese, competed for trading posts. Before the Industrial Revolution, Indian cotton textiles were considered the finest in the world, and European merchants needed large quantities for their global trading networks. After Britain's victory over France in the Seven Years' War (1756–1763), the French retained only a few minor coastal cities. Elsewhere, the **British East India Company**, the private organization that acted as an arm of the British government, was free to carve out its own South Asian empire. By the 1840s, British control over South Asia was essentially completed. Valuable local allies, however, were allowed to remain in power, provided that they did not threaten British interests. The territories of these indigenous (or "princely") states were gradually reduced, while British advisors increasingly dictated their policies.

The continual expansion of British power led to a rebellion in 1856 across much of South Asia. When this uprising (often called the *Sepoy Mutiny*) was finally crushed, a new political order was implemented. South Asia was now under the authority of the British government, with the Queen of England acting as its head of state (Figure 12.29). Britain enjoyed direct control over the region's most productive and densely populated areas, including almost the entire Indus–Ganges Valley and coastal plains. In more remote areas, Britain ruled indirectly, with native "princes" continuing to occupy their thrones.

British officials were always concerned about threats to their immensely profitable Indian colony, particularly from the Russians advancing across Central Asia. In response, they attempted to secure their boundaries. In some cases, this merely required making alliances with local rulers. In such a manner, Nepal and Bhutan retained their independence. In the extreme northeast, some small states and tribal territories, most of which had never been part of the South Asian cultural sphere, were taken over by the British Empire. A similar policy was conducted on the vulnerable northwestern frontier.

Independence and Partition   The framework of British India began to unravel in the early 20th century, as the people of South Asia increasingly demanded independence. The British, however, were determined to stay, and by the 1920s South Asia was caught up in massive political protests.

The leaders of the rising nationalist movement faced a dilemma in attempting to organize an independent country. Many leaders, including Mohandas Gandhi—the father-figure of Indian independence—favored a unified state that would include all British territories in mainland South Asia. Most Muslim leaders, however, feared that a unified India would leave their people in a vulnerable position. They therefore argued for the division of British India into two new countries: a Hindu-majority India and a Muslim-majority Pakistan. In several parts of northern South Asia, however, Muslims and Hindus were settled in roughly equal numbers. Another problem was the location of areas of clear Muslim majority on opposite east–west sides of the subcontinent, in present-day Pakistan and Bangladesh.

When the British finally withdrew in 1947, South Asia was indeed divided into India and Pakistan. Partition was a horrific event. Not only were some 14 million people displaced, but also roughly 1 million were killed. Hindus and Sikhs fled from Pakistan, to be replaced by Muslims fleeing India (Figure 12.30).

FIGURE 12.30 Partition, 1947   Following Britain's decision to leave South Asia, violence and bloodshed broke out between Hindus and Muslims in much of the region. With the partition of British India into India and Pakistan, millions of people were forced to flee their homes, often in overcrowded trains.

The Pakistan that emerged from partition was for several decades a clumsy two-part country, with its western section in the Indus Valley and its eastern portion in the Ganges Delta. The Bengalis, occupying the poorer eastern section, complained that they were treated as second-class citizens. In 1971, they launched a rebellion and, with the help of India, quickly prevailed. Bangladesh then emerged as a new country. This second partition did not solve Pakistan's problems, however, as the country remained politically unstable and prone to military rule. Pakistan retained the British policy of allowing almost full autonomy to the Pashtun tribes living along its border with Afghanistan, a relatively lawless area marked by clan fighting. This area would later lend much support to Afghanistan's Taliban regime and to Osama bin Laden's Al Qaeda organization.

Ethnic Conflicts in South Asia

After India and Pakistan gained independence, ethnic and religious tensions continued to plague many parts of South Asia. The region's most complex—and perilous—struggle is that in Kashmir, which involves both India and Pakistan.

Kashmir   Relations between India and Pakistan were hostile from the start, and the situation in Kashmir has kept the conflict burning (Figure 12.31). During the British period, Kashmir was a large princely state with a primarily Muslim core joined to a Hindu district in the south (Jammu) and a Tibetan Buddhist district in the northeast (Ladakh). Kashmir was ruled by a Hindu **maharaja**, a local king subject to British advisors. During partition, Kashmir came under severe pressure from both India and Pakistan. After troops from Pakistan gained control of western Kashmir, the maharaja decided to join India. But neither Pakistan nor India would accept the other's control over any portion of Kashmir, and as a result they have since fought several wars over the issue.

Although the Indo-Pakistani boundary has remained fixed, fighting in Kashmir has continued, reaching a peak in the 1990s. As recently as 2010, massive demonstrations against the Indian military presence in Kashmir Valley resulted in 112 deaths. Many Muslim Kashmiris would like to join their homeland to Pakistan, others prefer that it remain a part of India, and a large portion would rather see it become an independent country. Indian nationalists are determined that Kashmir remain part of India. Militants from Pakistan continue to cross the border to fight the Indian army, ensuring that tensions between the two countries remain high. The result has been a low-level but periodically brutal war. The Vale of Kashmir, with its lush fields and orchards nestled among some of the world's most spectacular mountains, was once one of South Asia's premier tourist destinations. By 2012, tourism in the valley was beginning to revive, but as the underlying tensions persist, the tourist trade remains highly vulnerable.

The Northeast Fringe   A relatively obscure series of ethnic conflicts emerged in the 1980s in the uplands of India's extreme northeast. Much of this area has never really been part of the South Asian cultural sphere, and many of its peoples want autonomy, if not actual independence. Northeastern India is still relatively lightly populated and as a result has attracted millions of migrants from Bangladesh and northern India. Many local people view this movement as a threat to their lands and culture. On several occasions, local guerillas have attacked newcomer villagers and, in turn, have suffered reprisals from the Indian military. This is a remote area, however, and relatively little information from it reaches the outside world. According to the South

FIGURE 12.31 Conflict in Kashmir
Unrest in Kashmir maintains hostility between the two nuclear powers of India and Pakistan. Under the British, this region of predominantly Muslim population was ruled by a Hindu maharaja, who managed to join the province to India upon partition. Today many Kashmiris wish to join Pakistan, while many others argue for an independent state.

FIGURE 12.32 India–Bangladesh Fence   India began building a border fence between its territory and that of Bangladesh in 2003 in order to reduce illegal immigration and to stop the influx of militants. Members of the Indian Border Security Force visible in this photograph are patrolling a segment of the border fence.

Asia Terrorism Portal, fighting in northeastern India resulted in almost 5,000 fatalities between 2005 and June 2012.

Tensions in the northeast have complicated India's relations with Bangladesh. India accuses Bangladesh of allowing separatists sanctuary on its side of the border and objects as well to continuing Bangladeshi emigration. As a result, India is currently building a 2,500-mile (4,000-kilometer), $1.2 billion fence along the border between the two countries (Figure 12.32).

Sri Lanka   Ethnic violence in Sri Lanka has been especially severe. Here the conflict stems from both religious and linguistic differences. Northern Sri Lanka is dominated by Hindu Tamils, whereas the island's majority group is Buddhist in religion and Sinhalese in language. Relations between the two communities have historically been fairly good, but tensions mounted soon after independence (Figure 12.33). Sinhalese nationalists have favored a centralized government, some of them calling for an officially Buddhist state. Most Tamils want political and cultural autonomy, and they have accused the government of discriminating against them.

In 1983, war erupted when the rebel force known as the *Liberation Tigers of Tamil Eelam*, or "Tamil Tigers," attacked the Sri Lankan army. By the 1990s, most of northern Sri Lanka was under the control of the Tamil Tigers. A Norwegian-brokered cease-fire in 2000 brought

some hope that the conflict was winding down, but fighting again intensified in 2006. In 2007, the Sri Lankan government abandoned negotiations, instead launching an all-out offensive. In May 2009, government forces crushingly defeated the Tamil Tiger army, killing the organization's leaders. A United Nations report released in 2011 found evidence that both the Sri Lankan army and the Tamil Tigers had engaged in war crimes and crimes against humanity. Although Sri Lanka is currently at peace, many of its Tamil citizens remain bitterly opposed to the government.

The Maoist Challenge

Not all of South Asia's conflicts are rooted in ethnic or religious differences. Poverty, inequality, and environmental degradation in east-central India, for example, have generated a revolutionary movement inspired by former Chinese communist leader Mao Zedong. Here the fighting is sporadic, but persistent. An estimated 1,100 people lost their lives due to this struggle in 2009 alone, prompting Manmohan Singh, India's prime minister, to refer to the Maoist insurgency as "the single biggest internal security challenge ever faced by our country." By 2011, however, the level of violence had dropped by about 50 percent, and the number of affected districts had declined as well.

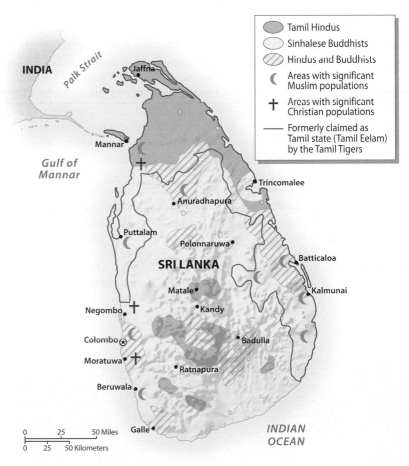

FIGURE 12.33 Civil War in Sri Lanka   The majority of Sri Lankans are Sinhalese Buddhists, many of whom maintain that their country should be a Buddhist state. A Tamil-speaking Hindu minority in the northeast strongly resists this idea. Tamil militants waged war against the Sri Lankan government for several decades, hoping to create an independent country their northern and eastern homeland. They were decisively defeated by the Sri Lankan military, however, in 2009.

Maoism has been an even greater challenge to the government of Nepal. Nepalese Maoists, frustrated by the lack of development in rural areas, emerged as a significant force in the 1990s. By 2005, they controlled over 70 percent of the country. At the same time, Nepal's urban population turned against the country's monarchy, launching massive protests. In 2006, Nepal's king, Gyanendra Shah, agreed to restore democratic rule, while at the same time the Maoist rebels announced that they would quit fighting and enter the democratic political process. In 2008, the king stepped down, and Nepal became a republic, with the leader of the former Maoist rebels serving as prime minister.

The end of the monarchy has not brought stability to Nepal. Several governments have been formed and then disbanded since the end of the monarchy. A new crisis emerged in 2012 when the country's bickering political parties failed to create a new constitution designed to restore stability. At the same time, the indigenous people of Nepal's southern lowlands, distressed by the migration of settlers from the more densely populated hill country, have been pushing for greater representation and threatening to rebel if their demands are not met.

International Geopolitics

As the case of Kashmir shows, South Asia's major international geopolitical problem is the continuing cold war between India and Pakistan (Figure 12.34). Since India gained independence, the two countries have regarded each other as enemies. Today the stakes are extremely high, as both India and Pakistan have numerous nuclear weapons.

During the global Cold War, Pakistan allied itself with the United States, and India leaned slightly toward the Soviet Union. Such alliances fell apart with the end of the superpower conflict in the early 1990s. Since then, Pakistan has forged an informal alliance with China, which has long been in military competition with India. India, meanwhile, has gradually been moving into an informal alliance with the United States.

The conflict between India and Pakistan became more complex after the attacks of September 11, 2001, on the United States. Until that time, Pakistan had been supporting Afghanistan's Taliban regime. After

FIGURE 12.34 Border Tensions An Indian officer looks through binoculars in war-torn Kashmir. Relationships between India and Pakistan have been extremely tense since India gained independence in 1947. Moreover, with both countries now nuclear powers, the fear that border hostilities will escalate into wider warfare has become a nightmarish possibility.

Osama bin Laden's attacks on the World Trade Center and Pentagon, the United States gave Pakistan a stark choice: Either Pakistan would assist the United States in its fight against the Taliban and receive financial aid in return, or it would lose favor with the U.S. government. Pakistan agreed to help, offering valuable intelligence to the U.S. military.

Pakistan's decision to help the United States came with large risks. Both Osama bin Laden and the Taliban enjoyed substantial support among the Pashtun people of Pakistan's wild northwestern region. After suffering several military reversals, Pakistan decided to negotiate with radical Islamist groups operating in the Pashtun region, and has allowed them to maintain virtual control over sizable areas. From these bases, militants have launched numerous attacks on U.S. forces in Afghanistan and have attempted to gain control over broader swaths of Pakistan's territory. The United States has responded mainly by using unmanned drones to attack insurgent leaders, a tactic that has resulted in large numbers of civilian casualties and pronounced anti-American sentiment throughout Pakistan.

The security crisis in Pakistan has intensified in recent years. In November 2008, Pakistan's president declared a state of emergency, and in the following month Benazir Bhutto, Pakistan's former prime minister, was assassinated by radicals as she campaigned for the 2008 election. Although the elections went on relatively smoothly, the new government faced severe challenges as Islamists expanded their control over remote areas. Pakistan estimates that it has spent almost $70 billion combating radical Islamists, a struggle that has resulted in the displacement of some 3 million people. Relations with the United States deteriorated in May 2011 after U.S. forces launched a raid deep in Pakistan's territory, without informing Pakistani officials, and killed Osama bin Laden. Later that year, Pakistan shut down NATO supply lines extending across its territory into Afghanistan in retaliation for a U.S.-led attack on two Pakistan military outposts near the Afghan border that resulted in the deaths of 24 Pakistani soldiers. Although Pakistan later agreed to reopen the supply lines after the U.S. apologized for the attack, anti-American attitudes remain pronounced in Pakistan, just as anti-Pakistani attitudes remain strong in the United States.

Pakistan is a troubled country, facing not only a deep Islamist insurgency and a seemingly unsolvable geopolitical conflict with India, but also a low-level ethnic rebellion in its southwestern province (Baluchistan). Its relations with India, moreover, deteriorated in November 2008 when terrorists operating from Pakistan launched a series of coordinated attacks on tourist facilities and public places in Mumbai, killing 173 people. Pakistan responded by investigating the event and arresting several alleged plotters, but many Indians suspect that the attack had the support of certain elements of Pakistan's government and military. Both Indian and Pakistani leaders, however, decided that it would be in their own best interests to reduce tensions. Talks between the two countries resumed in 2011, and in 2012 India agreed to remove restrictions on investments from Pakistan in Indian companies.

REVIEW QUESTIONS

1. How have relations between India and Pakistan influenced South Asian geopolitical developments over the past several decades?

2. Why has South Asia experienced so many insurgencies since the end of British rule in 1947?

ECONOMIC AND SOCIAL DEVELOPMENT: RAPID GROWTH AND RAMPANT POVERTY

South Asia is one of the poorest regions of the world, yet it is also the site of great wealth. Many of South Asia's scientific and technological accomplishments are world-class, but the area also has some of the world's highest illiteracy rates. Although South Asia's high-tech businesses are closely integrated with the global economy, the South Asian economy as a whole was until recently one of the world's most isolated.

South Asian Poverty

One of the clearest measures of human well-being is nutrition, and by this measure South Asia ranks very low. No other region has so many chronically undernourished people. According to the United Nations Children's Fund, one-third of the world's malnourished children live in India. Roughly two-thirds of the people of India live on less than $2 a day, and Bangladesh is still poorer (Table 12.2) (Figure 12.35).

Despite such deep and widespread poverty, South Asia should not be regarded as a zone of misery. More than 300 million Indians are now rated by local standards as members of the "middle class" who can buy such modern goods as televisions, motor scooters, and washing machines. This large market has begun to interest corporate executives worldwide. By the early years of the 21st century, India's economy was growing at the extremely rapid rate of 7 to 8 percent a year. The global economic crisis of 2008–2009 reduced India's economic growth, but by 2010 it was again expanding at an annual rate of more than 8 percent. Growth cooled off again in 2012, but by global standards the Indian economy remains healthy.

Geographies of Economic Development

After gaining independence, the governments of South Asia attempted to build new economic systems that would benefit their own people, rather than foreign countries or corporations. Planners initially stressed heavy industry and economic self-sufficiency. Some gains were realized, but the overall pace of development remained slow.

FIGURE 12.35 Poverty in India India's rampant poverty results in a significant amount of child labor. In this photo, a 10-year-old boy is moving a large burden of plastic waste by bicycle.

Since the 1990s, however, governments in the region have gradually opened their economies to the global economic system. In the process, core areas of development and social progress have emerged, surrounded by large peripheral zones that have lagged behind.

TABLE 12.2 DEVELOPMENT INDICATORS

Country	GNI per capita, PPP 2010	GDP Average Annual % Growth 2000–10	Human Development Index (2011)[1]	Percent Population Living Below $2 a Day	Life Expectancy (2012)[2]	Under Age 5 Mortality Rate (1990)	Under Age 5 Mortality Rate (2010)	Adult Literacy (% ages 15 and older)	Gender Equity (2011)[3,1]
Bangladesh	1,810	5.9	.500	76.5	69	143	48	56	0.550
Bhutan	4,990	—	.522	29.8	69	—		53	0.495
India	3,400	8.0	.547	68.7	65	115	63	63	0.617
Maldives	8,110	—	.661	—	74	—		98	0.320
Nepal	1,210	3.8	.458	57.3	68	141	50	59	0.558
Pakistan	2,790	5.1	.504	60.2	65	124	87	56	0.573
Sri Lanka	5,010	5.6	.691	29.1	75	32	17	91	0.419

[1]United Nations, *Human Development Report, 2011.*

[2]Population Reference Bureau, *World Population Data Sheet,* 2012.

[3]Gender Inequality Index—A composite measure reflecting inequality in achievements between women and men in three dimensions: reproductive health, empowerment, and the labor market, that ranges between 0 and 1. The higher the number, the greater the inequality.

Source: World Bank, *World Development Indicators, 2012.*

FIGURE 12.36 Tourism in Nepal Nepal has long been one of the world's main destinations for adventure tourism, although business has suffered greatly in recent years due to the country's Maoist insurgency. Many tourists in Nepal stay in inexpensive lodges. Advertisement for several such lodges are visible in this photo.

The Himalayan Countries Both Nepal and Bhutan are disadvantaged by their rugged terrain and remote locations and by the fact that they have been relatively isolated from modern technology and infrastructure. However, such measurements can be misleading, especially for Bhutan, because many areas in the Himalayas are still largely subsistence oriented.

Until recently, Bhutan remained purposely disconnected from the modern world economy, allowing its small population to live in a relatively pristine natural environment. Although Bhutan now allows direct international flights, cable television, and the Internet, it still charts its own course, emphasizing "gross national happiness" over economic growth. Nepal, on the other hand, is more heavily populated and suffers much more severe environmental degradation. It is also more closely integrated with the Indian economy. Nepal has relied heavily on international tourism, but its tourist industry contracted sharply as its political crisis deepened after 2002 (Figure 12.36).

Bangladesh By several measures, Bangladesh is the poorest South Asian country. Environmental degradation and colonialism have contributed to Bangladesh's poverty, as did the partition of 1947. Most of pre-partition Bengal's businesses were located in the west, which went

FIGURE 12.37 Grameen Bank The internationally famous Grameen Bank supplies low-interest micro-loans to the women of Bangladesh. The photo shows funds being dispensed a Grameen branch branch meeting.

to India. Slow economic growth and a rising population in the first several decades after independence meant that Bangladesh remained a poor country. Almost 40 percent of Bangladeshis currently live on less than $1 a day.

Economic conditions in Bangladesh have, however, improved in recent years. Bangladesh is internationally competitive in textile and clothing manufacture, in part because its wage rate is so low. Low-interest credit provided by the internationally acclaimed Grameen Bank has given hope to many poor women in the country, allowing the emergence of several small-scale enterprises (Figure 12.37). Bangladesh has also discovered substantial reserves of natural gas, although it has been slow to develop them. Political instability and environmental degradation, however, cloud the country's economic future.

Pakistan Like Bangladesh, Pakistan suffered deeply from partition in 1947. Even so, for several decades after independence Pakistan maintained a more productive economy than did India. The country has a strong agricultural sector, as it shares the fertile Punjab with India. Pakistan also boasts a large textile industry, based on its huge cotton crop. Pakistan's economy, however, is less dynamic than that of India, with a lower potential for growth. Pakistan is burdened by very high levels of defense spending. In addition, a small, but powerful landlord class controls much of its best agricultural lands, yet pays virtually no taxes. Unlike India, Pakistan has not been able to develop a successful high-tech industry.

Pakistan's economic growth accelerated in the early 2000s after the country was granted concessions by the international community for its role in combating global terrorism. In 2008, however, its inflation rate had risen to over 20 percent, and huge budget deficits required a bailout from the International Monetary Fund. In 2011, Pakistan's economic growth rate was only about half that of India. Electricity shortages, a lack of foreign investment, insurgency, and general political instability contribute to Pakistan's unsettled economic future.

Sri Lanka and the Maldives Sri Lanka's economy is by several measures the most highly developed in South Asia. The country's exports are concentrated in textiles and agricultural products, such as rubber and tea. By global standards, however, Sri Lanka is still a very poor country, its progress undermined by its prolonged civil war. As the conflict comes to an end, Sri Lanka will probably benefit from its high levels of education, its abundant natural resources, and its tremendous tourist potential. China has invested heavily in the country in recent years.

The Maldives is the most prosperous South Asian country based on per capita economic output, but its total economy, like its population, is very small. Most of the country's revenues are gained from fishing and international tourism. Critics claim that most of the revenues from the tourist economy go to a very small segment of the population. Tourism revenue declined sharply in 2012, moreover, due mainly to political instability.

India's Less Developed Areas

Although India's per capita gross national income is similar to that of Pakistan, its total economy is much larger. As the region's largest country, India has far more internal variation in economic development (Figure 12.38). The most basic economic division is that between its more prosperous southern and western areas and its poorer districts in the north and east.

India's least developed and most corrupt area has long been Bihar, a state of 103 million people located in the lower Ganges Valley. Bihar's per capita level of economic production is less than one-third that of India as a whole. Neighboring Uttar Pradesh, India's most populous state, is also extremely poor. It, too, is densely settled and has experienced little industrial development. Both states are also noted for their socially conservative outlooks and caste tensions. Other states of north-central India, such as Madhya Pradesh, Jharkhand, Chhattisgarh, and Orissa, have also experienced relatively little economic development. Over the past few years, however, economic growth has been picking up in most of these regions, leading to renewed hope for genuine development.

India's Centers of Economic Growth

The west-central states of Gujarat and Maharashtra are noted for their industrial and financial power, as well as for their agricultural productivity. Gujarat was one of the first parts of South Asia to experience industrialization, and its textile mills are still among the most productive in the region (Figure 12.39). Gujaratis are well known as merchants and overseas traders, and they are heavily represented in the **Indian diaspora**, the migration of Indians to foreign countries. Cash remittances sent home from these emigrants help the state's economy. Gujarat is also considered to be a well-governed state, although it also has some of the worst Hindu–Muslim relations in India.

Maharashtra is usually viewed as India's economic pacesetter. Its huge city of Mumbai (Bombay) has long been the financial center and media capital of India. Major industrial zones are located around Mumbai and in several other parts of Maharashtra. In recent years, Maharashtra's economy has grown more quickly than those of most other Indian states. Its per capita level of economic production is now roughly 50 percent greater than that of India as a whole.

In the northwestern Indian states of Punjab and Haryana, showcases of the Green Revolution, per capita levels of economic output are similar to those of Maharashtra and Gujarat. Their economies have rested largely on agriculture, but investments have been made recently in food processing and other industries. On Haryana's eastern border lies the National Capital Territory of Delhi, where much of India's political power and wealth are concentrated. This territory, in turn, is divided into

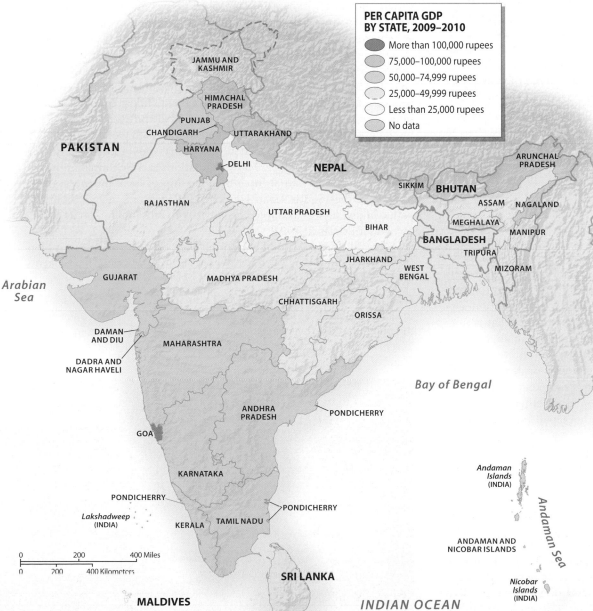

PER CAPITA GDP BY STATE, 2009–2010
- More than 100,000 rupees
- 75,000–100,000 rupees
- 50,000–74,999 rupees
- 25,000–49,999 rupees
- Less than 25,000 rupees
- No data

FIGURE 12.38 India's Economic Disparities Levels of economic development vary profoundly from one part of India to another, as can be seen in this map of per capita gross domestic product. Per capita economic production in Delhi is roughly seven times greater than that of Bihar.

FIGURE 12.39 Gujarat Factory The western Indian state of Gujarat in one of India's main manufacturing centers. This photograph shows a modern cotton mill in the city of Ahmedabad.

FIGURE 12.40 India's Silicon Plateau The Bangalore suburb of Whitefield is usually considered to be the heartland of India's high-tech sector. This photograph shows Whitefield's International Technical Park.

nine districts, one of which—New Delhi—is the official capital city of India. New Delhi was established by the British in 1911, adjacent to the old Mughal capital, which is now called Old Delhi or Central Delhi.

The center of India's fast-growing high-technology sector lies farther to the south, especially in Bangalore and Hyderabad. The Indian government selected the upland Bangalore area, which is noted for its pleasant climate, for technological investments in the 1950s. Other businesses soon followed. In the 1980s and 1990s, a quickly growing computer software and hardware industry emerged, earning Bangalore the label "Silicon Plateau" (Figure 12.40). By 2000, large numbers of American software, accounting, and data-processing jobs were being transferred, or "outsourced," to Bangalore and other Indian cities. In particular, call centers—which international customers phone to receive information, make purchases, or get advice on technical issues—have often been transferred from the United States and Europe to cities in India. The southern Indian states of Tamil Nadu and Kerala have also seen rapid growth in recent years, due in part to booming information technology industries.

India has proved especially competitive in software because software development does not require a sophisticated infrastructure. Computer code can be exported by wireless telecommunication systems without the use of modern roads or port facilities. What is necessary, of course, is technical talent, and this India has in abundance. Many Indian social groups have long been highly committed to education, and India has been a major scientific power for decades. With the growth of the software industry, India's brainpower has finally begun to create economic gains (Figure 12.41). Whether such developments can spread benefits beyond the rather small high-tech areas they presently occupy remains to be seen.

Globalization and South Asia's Economic Future

Throughout most of the second half of the 20th century, South Asia was relatively isolated from the world economy. Even today, especially compared with East Asia or Southeast Asia, the region's volume of foreign trade and its influx of foreign direct investment are still relatively small. But globalization is advancing rapidly, especially in India.

To understand South Asia's recent low level of globalization, it is necessary to examine its economic history. After independence, India's

economic policy was based on widespread private ownership, combined with government control of planning, resource allocation, and certain industries. India also established high trade barriers to protect its economy from global competition. This mixed socialist-capitalist system encouraged the development of heavy industry and allowed India to become nearly self-sufficient, even in the most technologically sophisticated goods. By the 1980s, however, problems with this model were becoming apparent. Slow economic growth meant that the percentage of Indians living in poverty remained almost constant. At the same time, countries such as China and Thailand were experiencing rapid development after opening their economies to globalization.

In response to these difficulties, India's government began to open its economy in 1991. Many regulations were eliminated, tariffs were reduced, and partial foreign ownership of local businesses was allowed. Other South Asian countries followed a somewhat similar path. Pakistan, for example, began to privatize many of its state-owned industries in 1994.

Overall, India's economic reforms have proved successful. Indian information technology firms are world-class, and many have begun

FIGURE 12.41 Indian Institutes of Technology The 16 campuses of the Indian Institutes of Technology provide world-class technical training, In this photograph, software engineers pose with new Aakash-2 computing tablets in their lab at the Indian Institute of Technology (IIT) campus in Mumbai in 2012.

to expand globally (see "Exploring Global Connections: Indian Companies Invest in Europe"). Growth has been so rapid in this sector that some companies are now having difficulty finding and retaining qualified workers; as a result, wages are increasing rapidly. Recent growth has also demonstrated the need for India to improve its infrastructure, but it is not clear how India will be able to afford the necessary investments in roads, railroads, and facilities for electricity generation and transmission.

The gradual internationalization and deregulation of the Indian economy has generated substantial opposition. Foreign competitors are now seriously challenging some domestic firms. Cheap manufactured goods from China are seen as an especially serious threat. Moreover, hundreds of millions of Indian peasants and slum-dwellers have seen few benefits from their country's rapid economic growth.

Opposition has also grown in the United States and Europe to the outsourcing of jobs to India. Due both to such opposition and to rising wages in India's high-tech centers, Indian companies are now beginning to turn the tables by relocating some of their operations back to North America. Mumbai-based Aegis Communications, for example, recently opened a call center in New York, which focuses on providing medical information.

Although India gets most of the media attention, other South Asian countries have also experienced significant economic globalization in recent years. Besides exporting textiles and other consumer goods, Bangladesh, Pakistan, and Sri Lanka send large numbers of their citizens to work abroad, particularly in the Persian Gulf countries. Remittances from foreign workers are Bangladesh's second largest source of income, reaching more than $11 billion a year in 2012. On a per capita basis, remittances are even more important for Sri Lanka. Out of a total population of 21 million, roughly 1.5 million Sri Lankans work abroad, 90 percent of them in the Middle East.

Social Development

South Asia has relatively low levels of health and education, which is not surprising considering its poverty. As might be expected, people in the more developed areas of western and southern India are healthier, live longer, and are better educated, on average, than people in the poorer areas, such as the lower Ganges Valley. Bihar, with a literacy rate of only about 50 percent, is at the bottom of most social-development rankings, while Kerala, Punjab, Gujarat, and Maharashtra are near the top. Several key measurements of social welfare are higher in India than in Pakistan. Pakistan has done a poor job of educating its people, which is one reason extremist Islamic organizations have been able to recruit effectively in much of the country.

Several oddities stand out when comparing South Asia's map of economic development with its map of social well-being. Portions of India's extreme northeast, for example, show relatively high literacy rates despite their poverty, due largely to the educational efforts of Christian missionaries. In Mizoram, for example, the literacy rate is over 90 percent, the highest in India. In overall terms, however, southern South Asia outpaces other parts of the region in regard to social development.

The Educated South Southern South Asia's relatively high levels of social welfare are clearly visible in Sri Lanka. Considering its meager economy and prolonged civil war, Sri Lanka must be considered a social developmental success. Its average life expectancy is 75 years, and its literacy rate is 92 percent. The Sri Lankan government has

FIGURE 12.42 Education in Kerala India's southwestern state of Kerala, which has virtually eliminated illiteracy, is South Asia's most highly educated region. It also has the lowest fertility rate in South Asia. Because of this, many argue that women's education and empowerment is the best and most enduring form of contraception.

achieved these results through universal primary education and inexpensive medical clinics.

On the mainland, Kerala in southwestern India has achieved even more impressive results. Kerala is extremely crowded and has long had difficulty feeding its population. Although Kerala's economy has grown strongly in recent years, its per capita economic output is only a little above average for India. Kerala's level of social development, however, is the highest in India (Figure 12.42): 90 percent of the people of Kerala are literate, the state's average life expectancy is 75 years, and several diseases, such as malaria, have been essentially eliminated.

Some observers attribute Kerala's social successes to its state policies. Kerala has often been led by a socialist party that has stressed education and community health care. While this has no doubt been an important factor, it does not seem to offer a complete explanation. Some researchers suggest that one of the key factors is the relatively high social position of women in Kerala.

Gender Relations and the Status of Women It is often argued that South Asian women have a very low social position in both the Hindu and the Muslim traditions. Throughout most of India, women traditionally leave their own families shortly after puberty to join those of their husbands. As outsiders, often in distant villages, young brides have little freedom and few opportunities. In Pakistan, Bangladesh, and such Indian states as Rajasthan, Bihar, and Uttar Pradesh, female literacy lags far behind male literacy.

An even more disturbing statistic is that of gender ratios, the relative proportion of males and females in the population. A 2011 study found that for India as a whole, only 914 girls are born for every 1,000 boys and that in parts of northern India the ratio is as low as 824 to 1,000. An imbalance of males over females often results from differences in care. In poor families, boys typically receive better nutrition and medical care than do girls, which results in higher rates of survival. An estimated 10 million girls, moreover, have supposedly been lost in northern India due to sex-selective abortion over the past 20 years. Economics play a major role in this situation. In rural households, boys are usually viewed as a blessing because they typically remain with and work for the well-being of their families. In the poorest groups, elderly people (especially widows) subsist largely on what their sons provide. Girls, on the other hand, marry out of their families at an early age and must be provided with a dowry. They are thus seen as an economic liability.

EXPLORING GLOBAL CONNECTIONS

Indian Companies Invest in Europe

Most large multinational corporations, which produce goods and services in many different countries, are based in the core areas of the world economy. Such companies generally rely on the financial and technological resources of wealthy cities, yet often seek production sites in poor countries with low wages and few regulations. As a result, multinational corporations have been criticized for their role in maintaining a stark global division between the developed and the underdeveloped worlds. As globalization proceeds, however, such a view is becoming outmoded. Increasing numbers of powerful multinational corporations are based in relatively poor parts of the world, and several are now investing heavily in wealthy countries.

Indian companies in particular are noted for their investments in Europe in recent years. Several well-known European brands are now owned by Indian firms. The famous British automobile makers Jaguar and Land Rover, for example, were purchased by India's Tata Motors, part of the globe-spanning Tata Group, in 2008 (Figure 12.2.1). The Tata Group also controls Tetley Tea, a historical British firm.

Indian high-tech companies have been especially active in Europe. Bangalore-based Infosys—which specializes in business consulting and outsourcing services, as well as computer hardware and software—has continued to invest heavily despite the recent European economic crisis. As of early 2012, Infosys employed 5,500 people in Europe, with its European operations accounting for almost a quarter of its revenues. Pharmaceutical firms have also been active players. The Indian firm Ranbaxy Laboratories Limited, which is itself owned in part by the Japanese pharmaceutical giant Daiichi Sankyo, has major operations in the United Kingdom, Germany, and France and is currently expanding in Italy and Spain.

European countries and firms often welcome Indian investment. A publication by the German Federal Ministry of Transport, Building and Urban Affairs, for example, advocates "Eastern Germany as a location for Indian Direct Investment." The publication showcases several successful Indian ventures in the region. Suzlon Energy, the world's fifth largest producer of wind turbines, is now designing machines in the eastern German cities of Berlin and Rostock, where it cooperates with local universities and technical schools and takes advantage of the region's well-trained engineers. Suzlon has operated much longer in the western German city of Hamburg, but is currently reorienting its operations to the eastern part of the country, due mostly to lower labor and living costs.

FIGURE 12.2.1 Indian Investments in Europe   India's Tata Group is one of the world's biggest industrial companies. Many familiar Western firms, such as the auto manufactures Jaguar and Land Rover, are owned by the Tata Group.

Some Indian investments in Europe have proved much more controversial, particularly those that involve professional sports teams. In 2010, the Indian poultry firm Venkey's purchased the Blackburn Rovers Football Club, a soccer team that dates back to 1875. The new management began to change the coaching staff, resulting in protests by fans. In May 2011, a disgruntled fan released a chicken on the playing field wrapped in a Blackburn flag emblazoned with the word "out," intended as a message for the club's owners. In 2001, the Bahrain-based Indian business tycoon Ahsan Ali Syed purchased the Spanish soccer club Racing de Santander. Controversy again resulted when the club went through three different mangers in the 2011–2012 season. Fans claim that Syed knows little about the sport and is therefore not able to run it competently.

The social position of women is improving in many parts of South Asia, especially in the more prosperous areas where employment opportunities outside the family are emerging. But even in many of the region's middle-class households, women still experience deep discrimination. Indeed, dowry demands are increasing in some areas, and several well-publicized murders of young brides have occurred whose families failed to deliver an adequate supply of goods. Although the social bias against women across the north is striking, it is much less evident in southern South Asia, especially Kerala and Sri Lanka.

REVIEW QUESTIONS

1. How has the economy of India been transformed since the reforms of 1991?

2. Why do levels of social and economic development vary so much across South Asia?

Summary

- South Asia is a large, complex, and densely populated area that in many ways has been overshadowed by neighboring world regions: by the uneven globalization of Southeast Asia, by the size and political weight of East Asia, and by the geopolitical tensions of Southwest Asia. Much of that is changing, however, as South Asia now figures prominently in discussions of world problems and issues.

- Environmental degradation and instability pose particular problems for South Asia. Due to its monsoon climate, both floods and droughts tend to be more problematic here than in most other world regions. Global climate change directly threatens the low-lying Maldives and may play havoc with the monsoon-dependent agricultural systems of India, Pakistan, and Bangladesh.

- Continuing population growth in this already densely populated region demands attention. Although fertility rates have declined in recent years, Pakistan, northern India, and Bangladesh cannot easily meet the demands imposed by their expanding populations.

Increasing social and political instability, as well as environmental degradation, may result as cities mushroom in size and rural areas grow more crowded.

- South Asia's diverse cultural heritage, shaped by peoples speaking several dozen languages and following several major religions, makes for a particularly rich social environment. Unfortunately, cultural differences have often translated into political conflicts. Ethnically- or religiously-based separatist movements have severely challenged the governments of Pakistan, India, and Sri Lanka. In India, moreover, religious strife between Hindus and Muslims persists.

- Geopolitical tensions within South Asia are particularly severe, again demanding global attention. The long-standing feud between Pakistan and India escalated dangerously in the late 1990s, leading many observers to conclude that this was the part of the world most likely to experience a nuclear war. Although tensions between the two countries have more recently been reduced, the underlying sources of conflict—particularly the struggle in Kashmir—remain unresolved.

- Although South Asia remains one of the poorest parts of the world, much of the region has seen rapid economic expansion in recent years. Many argue that India in particular is well positioned to take advantage of economic globalization. Large segments of its huge labor force are well educated and speak excellent English, the major language of global commerce. But will these global connections help the vast numbers of India's poor or only the small number of its economic elite? Advocates of free markets and globalization tend to see a bright future, whereas skeptics more often see growing problems.

Key Terms

British East India Company 413
bustee 403
caste system 405
Dalit 405
Dravidian language 407
Green Revolution 402
Hindi 407

Hindu nationalism 404
Indian diaspora 419
Indian subcontinent 392
Jainism 407
linguistic nationalism 409
maharaja 414
monsoon 396

Mughal (or Mogul) Empire 405
orographic rainfall 396
rain-shadow effect 396
salinization 403
Sanskrit 404
Sikhism 406
Urdu 408

Thinking Geographically

1. How might the vulnerable countries of Bangladesh and the Maldives best respond to the challenges posed by the threat of global climate change? What responsibilities do wealthy countries such as the United States have in regard to such issues?

2. What are the pros and cons of the Green Revolution as a means of increasing South Asia's food supplies? What are the advantages and disadvantages of expanding irrigated agriculture in India and Pakistan? What is the outlook for the next decade?

3. What are the drawbacks and benefits of using English as a national language in India? Might it help or hinder unity? Would it increase or decrease India's links to the contemporary world?

4. How could Pakistan best respond to the violence that currently plagues its Pashtun-speaking areas? Should the government of Pakistan try to establish control over the Federally Administered Tribal Areas?

5. Does the information technology industry offer a way for South Asia to achieve economic and social development, or are the benefits of the industry likely to help only a small segment of the population? What kinds of policies might the governments of the region enact in order to gain greater benefits from the emerging high-tech economy?

MasteringGeography™

Looking for additional review and test prep materials? Visit the Study Area in MasteringGeography™ to enhance your geographic literacy, spatial reasoning skills, and understanding of this chapter's content by accessing a variety of resources, including MapMaster interactive maps, videos, RSS feeds, flashcards, web links, self-study quizzes, and an eText version of *Globalization and Diversity*.

Scan to visit the author's blog for chapter updates.

Scan to visit the GeoCurrents blog.

Authors' Blogs

Scan now to access the authors' blogs for up-to-date information on South Asia.

http://gad4blog.
wordpress.com/
category/south-asia/

http://geocurrents.
info/category/place/
south-asia

Globalization and Diversity

Southeast Asia has been a key player in global commerce for hundreds of years, and it has long been open to cultural influences from other parts of the world. Today Southeast Asian leaders are trying to cope with globalization on their own terms, working to create a more economically and politically united world region.

ENVIRONMENTAL GEOGRAPHY

The rain forests of Southeast Asia are vital centers of biological diversity, but they are much diminished due to commercial logging and agricultural expansion.

SETTLEMENT AND POPULATION

As is true for many other regions, Southeast Asia's river valleys, deltas, and areas of fertile volcanic soil tend to be densely populated, whereas most of its upland areas are lightly settled.

CULTURAL COHERENCE AND DIVERSITY

This region is noted for both its linguistic and its religious diversity. Many areas of Southeast Asia, however, are plagued by ethnic conflicts and religious tensions.

GEOPOLITICAL FRAMEWORK

Southeast Asia is one of the most geopolitically united regions of the world, with all but one of its countries belonging to the Association of Southeast Asian Nations (ASEAN).

ECONOMIC AND SOCIAL DEVELOPMENT

Southeast Asia is a region of startling economic contrasts. It contains some of the world's most globalized and dynamic economies, as well as some of the most isolated and impoverished.

➤ **The Irrawaddy River, shown in this photo, is often seen as a symbol of Burma. With major Chinese backing, Burma had begun construction of the massive Myitsone Dam project on the river in the northern part of the country. Due to local opposition and concerns about Chinese influence, Burma abruptly canceled the project in late 2011.**

SOUTHEAST ASIA 13

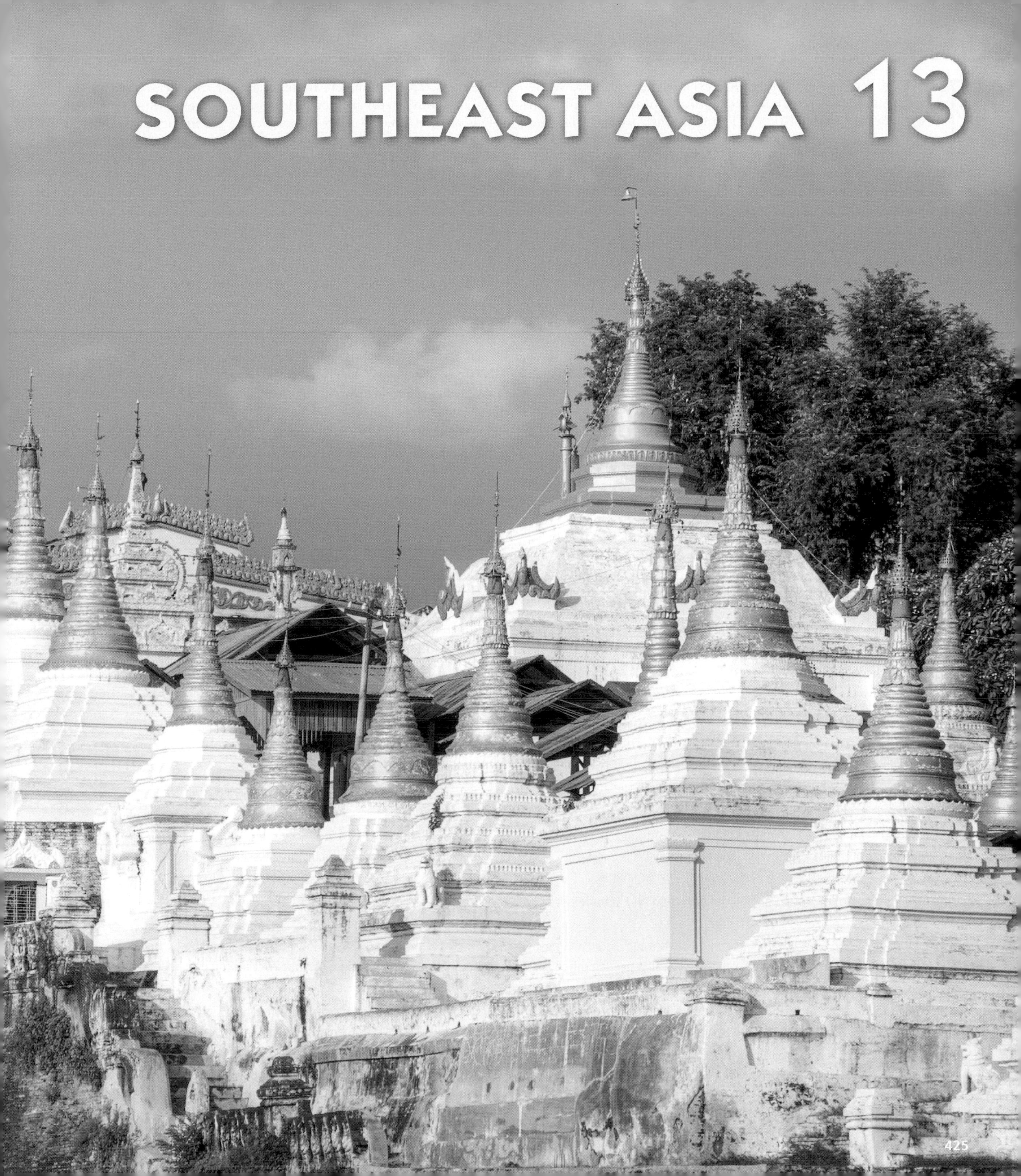

Northern Burma (Myanmar), where the broad Irrawaddy Valley gives way to rugged highlands, was a tense place in the summer of 2011. Preliminary construction had begun on the huge, Chinese-financed Myitsone Dam, infuriating local residents who would lose their lands to the new reservoir. Local resistance was organized by a well-armed ethnic militia, the Kachin Independence Army (KIA), but the Burmese army was driving the KIA troops deeper into the mountains.

Then suddenly everything began to change. The government of Burma abruptly announced the cancellation of the dam, angering its Chinese backers. Burma also promised to free hundreds of political prisoners and more generally to open up to the global community. Shortly afterward, U.S. Secretary of State Hillary Clinton visited the country. In January 2012, Burma signed a peace treaty with one of the main ethnic militias that it had been battling for decades. A few days later, the United States agreed to exchange ambassadors, reestablishing diplomatic relations, which had been suspended since 1990. Burma, in short, appeared to be changing from a repressive country relatively isolated from most of the rest of the world to a much more open and globalized state. (While the government of Burma insists that the country be called Myanmar, both the country's democratic opposition and the United States favor the old name of Burma, which we use in this chapter.)

In opting to engage with the international community, Burma was following the lead of its major Southeast Asian neighbors. Over the past 30 years, Southeast Asia has been one of the most globally connected parts of the developing world, demonstrating both the promises and the perils of globalization. As foreign investments have flowed in, the region has experienced both economic booms and severe recessions.

Southeast Asia's involvement with the larger world is not new (Figure 13.1). Chinese and Indian connections date back many centuries. Later, commercial ties with the Southeast Asia opened the doors to Islam, and today Indonesia is the world's most populous Muslim country. More recently came the impact of the West, as Britain, France, the Netherlands, and the United States gained control of large Southeast Asian colonies. Southeast Asia's resources and its strategic location made it a major battlefield during World War II. Yet long after peace was restored in 1945, warfare of a different sort continued in this region. As colonial powers withdrew and were replaced by newly independent countries, Southeast Asia became a battleground for world powers and their competing economic systems.

Although communism eventually prevailed in Vietnam, Laos, and Cambodia, all of these countries later opened their economies to the global market. Today the struggle between capitalism and communism has taken a back seat to the economic and ethnic problems confronting Southeast Asia. Relations among the various countries of the region are now generally good. The **Association of Southeast Asian Nations (ASEAN)**, which includes every country in the region except East Timor (Timor Leste), has created a generally effective system of regional cooperation.

Southeast Asia is a clearly defined region consisting of 11 countries that vary widely in spatial extent, population, cultural traits, and levels of economic and social development. Geographically, these countries are commonly divided into those on the Asian continent (mainland Southeast Asia) and those on islands (insular Southeast Asia). The mainland includes Burma, Thailand, Cambodia, Laos, and Vietnam. Although Burma has the largest mainland territory, Vietnam has the largest population of the mainland states, with 89 million people.

Insular Southeast Asia includes the large countries of Indonesia, the Philippines, and Malaysia, as well as the small countries of Singapore, Brunei, and East Timor. Although classified as part of the insular realm because of its cultural and historical background, Malaysia actually splits the difference between the continent and the

LEARNING OBJECTIVES

After reading this chapter you should be able to:

- Identify the key environmental differences between the equatorial belt of insular Southeast Asia and the higher-latitude zone of mainland Southeast Asia and explain how these differences influence human settlement and economic development.

- Describe the driving forces behind deforestation and habitat loss in the different regions of Southeast Asia.

- Explain the ways in which the interaction of tectonic plates and the resulting volcanism and seismic activity have influenced Southeast Asian history and development.

- Show how the differences among plantation agriculture, rice growing, and swidden cultivation in Southeast Asia have molded settlement patterns.

- Describe the role of primate cities and other massive urban centers in the development of Southeast Asia.

- Outline the ways in which religions from other parts of the world have spread through Southeast Asia, including how religious diversity has influenced the history of the region.

- Identify the controversies surrounding cultural globalization in Southeast Asia, explaining why some people in the region welcome the process, whereas others resist it.

- Trace the origin and spread of ASEAN and explain how this organization has influenced geopolitical relations in the region.

- Describe the major ethnic conflicts in Southeast Asia, showing why certain countries in the region have such deep problems in this regard.

- Explain why levels of economic and social development vary so widely across the Southeast Asian region.

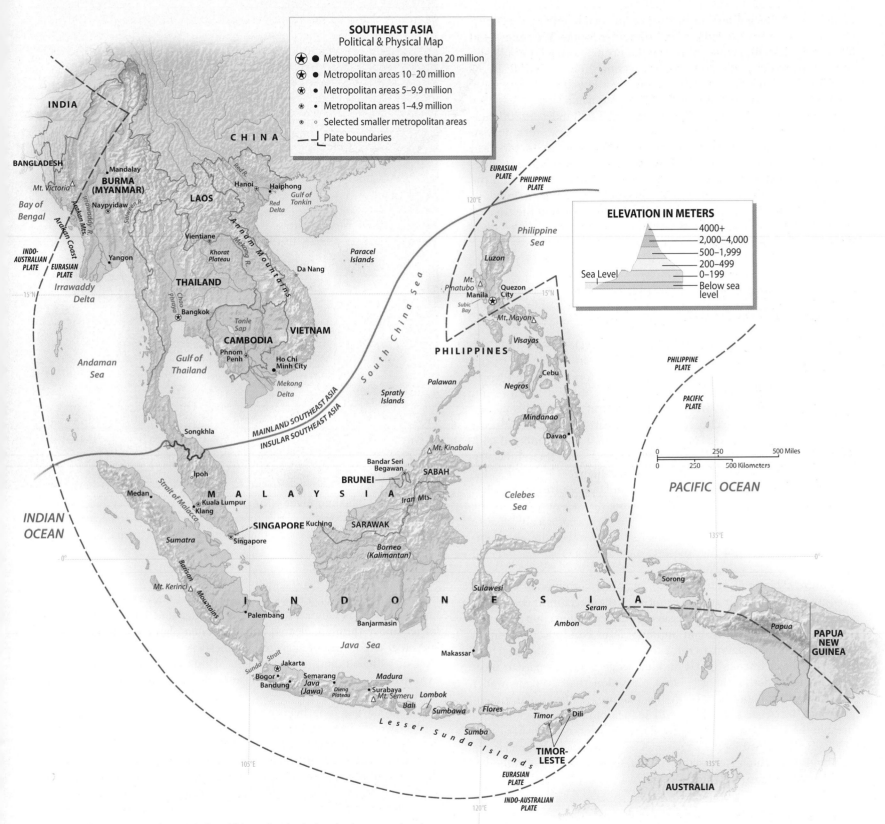

SOUTHEAST ASIA
Political & Physical Map

★● Metropolitan areas more than 20 million

★● Metropolitan areas 10–20 million

★● Metropolitan areas 5–9.9 million

⊛● Metropolitan areas 1–4.9 million

⊚○ Selected smaller metropolitan areas

⌐┘ Plate boundaries

ELEVATION IN METERS

4000+
2,000–4,000
500–1,999
200–499
Sea Level ┤ 0–199
Below sea
level

FIGURE 13.1 Southeast Asia This region includes the large peninsula
in the southeastern corner of Asia, as well as a large number of islands
scattered to the south and east. It is commonly divided into two subregions:
mainland Southeast Asia, which includes Burma, Thailand, Laos, Cambodia,
and Vietnam; and insular (or island) Southeast Asia, which includes
Indonesia, the Philippines, Malaysia, Brunei, Singapore, and East Timor.
Malaysia consists of the tip of the mainland peninsula and most of the
northern part of the island of Borneo.

islands. Part of its national territory is on the mainland's Malay Peninsula and part is on the large island of Borneo, some 300 miles (480 kilometers) distant. Borneo also includes Brunei, a small, but oil-rich country, of roughly 400,000 people covering an area slightly larger than Rhode Island. Singapore is essentially a city-state, occupying a small island just to the south of the Malay Peninsula.

Indonesia is an island nation, stretching 3,000 miles (4,800 kilometers), or about the same distance as from New York to San Francisco. From Sumatra in the west to New Guinea in the east, Indonesia contains more than 13,000 separate islands. Not only does it dwarf all other Southeast Asian states in size, but also it is by far the largest in population. With roughly 241 million people, it ranks as the world's fourth most populated country. Lying north of the equator is the Philippines, a country of 96 million people spread over some 7,000 islands, both large and small.

ENVIRONMENTAL GEOGRAPHY: A ONCE-FORESTED REGION

Southeast Asia is an almost entirely tropical region, with only the northern portion of Burma extending north of the Tropic of Cancer. Most of insular Southeast Asia lies in the equatorial zone and hence is characterized by heavy rainfall over most of the year. As a result, historically, tropical rain forests covered most of the island zone. Although rain forests were cleared out of a few parts of insular Southeast Asia hundreds of years ago, most of the region was still covered with thick forests in the middle of the 20th century. Since then, however, deforestation has resulted in rapid landscape change.

Unlike the insular zone, most of mainland Southeast Asia lies outside of the equatorial zone, where a long dry season prevents the establishment of rain-forest vegetation. The less ecologically diverse tropical "wet and dry" forests of this region, however, are also valuable, containing several tree species, such as teak, that are targeted by loggers.

Forest clearing is not the only major environmental concern in Southeast Asia. As is true over most of the developing world, air and water pollution are major issues. Another major environmental issue in Southeast Asia is dam building. Dam building has been an especially controversial matter in Laos, Burma, and Cambodia, where China has been financing and building large dams, in part to meet its own energy needs. In early 2012, Laos alone had 10 dams under construction and another 25 in the planning stage. Environmental groups fear that such dams will reduce biological diversity and threaten the livelihoods of many local people. Freshwater fisheries are especially threatened by the large dams being planned for the massive Mekong River, which is shared by China, Laos, Thailand, Cambodia, and Vietnam (Figure 13.2).

Patterns of Deforestation

Deforestation and related environmental problems are major issues throughout most of Southeast Asia (Figure 13.3). Although countries such as Indonesia have transformed large areas of forests into croplands in order to help feed their expanding populations, population growth is not the main cause of deforestation. Most forests are cut so that the wood products can be exported to other parts of the

FIGURE 13.2 Mekong River The Mekong River, which flows through mainland Southeast Asia, is one of the world's largest and most ecologically productive rivers. It is increasingly threatened by dam construction.

world. Initially Japan, Europe, and the United States were the main importers, but as China industrializes, its demand has grown. After loggers move through, cleared lands are often planted with oil palms and other export-oriented crops. Indonesia is the world's largest supplier of palm oil, and its production of this valuable commodity is expected to increase by 6 percent annually over the next few years (Figure 13.4).

Malaysia has long been one of the leading exporters of tropical hardwoods from Southeast Asia. Mainland Malaysia was largely deforested by 1985, when a cutting ban was imposed. Since then, logging has been concentrated in the states of Sarawak and Sabah on the island of Borneo. In these areas, the granting of logging concessions to Malaysian and foreign firms has caused considerable problems with local tribal people by disrupting their traditional resource base.

Thailand cut down more than 50 percent of its forests between 1960 and 1980. This loss was followed by a series of logging bans that virtually eliminated commercial forestry by 1995. Damage to the landscape, however, was severe; flooding increased in lowland areas, and erosion on hillslopes led to such problems as the accumulation of silt in irrigation works and hydroelectric facilities. Many of

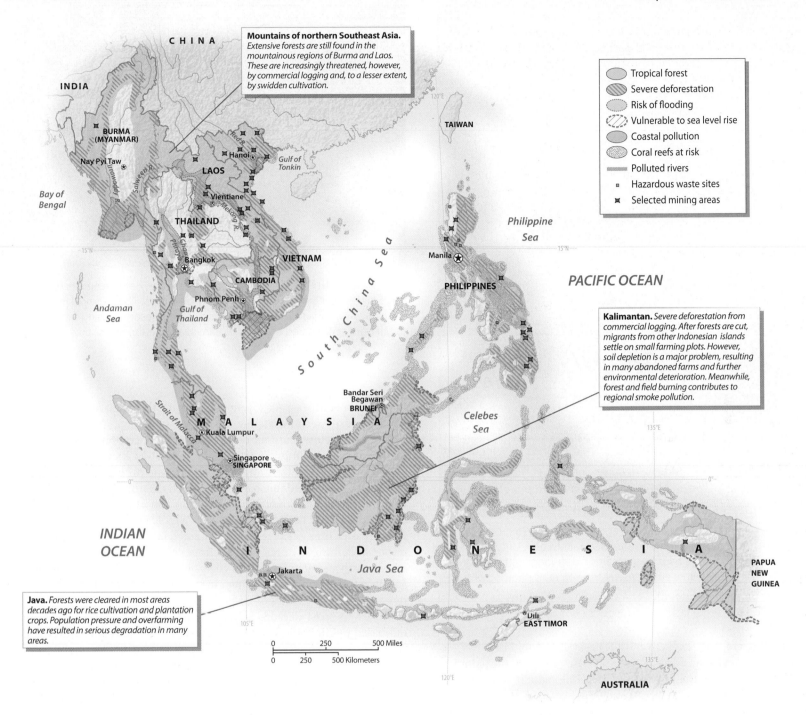

Mountains of northern Southeast Asia. *Extensive forests are still found in the mountainous regions of Burma and Laos. These are increasingly threatened, however, by commercial logging and, to a lesser extent, by swidden cultivation.*

Tropical forest
Severe deforestation
Risk of flooding
Vulnerable to sea level rise
Coastal pollution
Coral reefs at risk
Polluted rivers
Hazardous waste sites
Selected mining areas

Kalimantan. *Severe deforestation from commercial logging. After forests are cut, migrants from other Indonesian islands settle on small farming plots. However, soil depletion is a major problem, resulting in many abandoned farms and further environmental deterioration. Meanwhile, forest and field burning contributes to regional smoke pollution.*

Java. *Forests were cleared in most areas decades ago for rice cultivation and plantation crops. Population pressure and overfarming have resulted in serious degradation in many areas.*

FIGURE 13.3 Environmental Issues in Southeast Asia   Southeast Asia was once one of the most heavily forested regions of the world. However, most of the tropical forests of Thailand, the Philippines, peninsular Malaysia, Sumatra, and Java have been destroyed by a combination of commercial logging and agricultural settlement. The forests of Kalimantan (Borneo), Burma, Laos, and Vietnam, moreover, are now being rapidly cleared. Water and urban air pollution, as well as soil erosion, is also widespread in Southeast Asia.

these cutover lands are being reforested with fast-growing Australian eucalyptus trees, which do not support local wildlife. The Thai forestry ban, moreover, has resulted in increased logging in the remote areas of Laos and Cambodia, as well as Burma, much of which is done illegally.

Indonesia, the largest country in Southeast Asia, has fully two-thirds of the region's forest area, including about 10 percent of the world's true tropical rain forests. Most of Sumatra's forests have been cut, however, and those of Kalimantan (Borneo) are rapidly receding. Indonesia's last major forestry frontier is on the island of New Guinea,

FIGURE 13.4 Forest Clearance Insular Southeast Asia supports some of the world's most ecologically diverse tropical rain forests. Unfortunately, many of the region's forests have been cleared. After logging, forests are often replaced with oil palm plantations, as shown in this photograph, taken in Indonesia's Papua province.

FIGURE 13.5 Urban Air Pollution Air pollution has reached a crisis stage in the rapidly industrializing cities of Southeast Asia, particularly in Bangkok, Manila, and Jakarta. People sometimes resort to using face masks to filter out soot and other forms of particulate matter. Forest fires, which often follow logging, add to the problem.

where forests are still extensive. Indonesia is trying, however, to preserve the remaining areas of habitat elsewhere in the country. Kutai National Park in the province of East Kalimantan, for example, covers more than 741,000 acres (300,000 hectares). Conservation officials hope that this and other protected areas will allow animals a chance to survive in the wild, including the orangutan, which now lives only in a small portion of northern Sumatra and a somewhat larger area of Kalimantan.

Smoke and Air Pollution

Until recently, most of Southeast Asia's residents seemed unconcerned about the widespread air pollution created by a combination of urban smog and smoke from forest clearing. Then, late in the 1990s, the region suffered from two consecutive years of disastrous air pollution that served as a wake-up call (Figure 13.5). Although the situation later improved, Malaysian officials had to declare another emergency in 2005 because of widespread smoke from forest fires. In 2010, fires

in northern Southeast Asia were so extensive that their smoke plumes could be easily seen in satellite images.

Several factors, both natural and economic, combined to produce the region's air pollution disaster of the late 1990s. First, large portions of Southeast Asia suffer from periodic extreme drought, often caused by El Niño (discussed in Chapter 4), which can turn the normally wet tropical forests into tinderboxes. Drought can also dry out the widespread peat bogs of coastal wetlands, which continued to burn for months after fires began. Second, commercial forest cutting has been responsible for many fires, as the leftover slash (branches, small trees, and so forth) is often burned to clear out the land. The third factor is Southeast Asia's rapidly growing cities, where cars, trucks, and factories emit huge quantities of air pollutants.

Over the past decades, several Southeast Asian cities have built rail-based public transportation systems in order to reduce traffic and vehicular emissions. Bangkok in particular has made substantial gains in this area, substantially reducing its levels of ozone and sulfur dioxide. In the Philippines, Manila is still suffering from deteriorating air quality, but new initiatives promise to improve the situation (see "Working Toward Sustainability: Makati's Green City Initiative and the Rise of the Ejeepney").

Climate Change and Southeast Asia

As most of Southeast Asia's people live in coastal environments, the region is highly vulnerable to the rise in sea level associated with global warming. Periodic flooding is already a major problem in many of the region's low-lying cities, particularly Bangkok and Manila. Southeast Asian farmland is also concentrated in delta environments and thus could suffer from saltwater intrusion and higher storm surges. Environmentalists also fear that aside from rising sea levels, higher temperatures themselves could reduce rice yields through much of the region.

Makati's Green City Initiative and the Rise of the Ejeepney

The jeepney is not only the main form of urban transportation in the Philippines, but also a symbol of the country. These colorful vehicles are instantly recognizable, noted as much for their lavish decoration as for their crowded seating. Most jeepneys have their own names, and many feature small statues on the hoods, as well as assorted painted slogans.

The jeepney—a name that combined *jeep* and *jitney*—originated immediately after World War II. At that time, the public transportation infrastructure of the Philippines was largely destroyed, but the country had a huge stock of jeeps left there by the U.S. military. Local entrepreneurs soon began turning these vehicles into people carriers, usually by elongating the beds and adding bench seating and sunshades. Soon the Philippine government was regulating routes and licensing drivers. Eventually, an entire industry emerged around the vehicle.

Jeepneys have provided the cities of the Philippines with inexpensive and reliable transportation for decades, but they have also generated problems. Seating is cramped, fuel efficiency is low, and emissions are high. Older, diesel-burning jeepneys are especially polluting. A recent study found that a typical jeepney seating 16 passengers uses as much fuel as a bus that can hold 54. As environmental pressure against the vehicle has grown, many manufacturers have gone out of business, threatening the future of the Philippine transportation system.

Over the past few years, however, a new kind of vehicle has appeared that promises a revival of the system: the ejeepney (or electric jeepney). Ejeepneys, most of which are made in the Philippines, are not capable of going more than 40 kilometers per hour. But in the crowded streets of most Philippines cities, such a speed is seldom surpassed. The new ejeepneys are relatively clean and energy-efficient, capable of carrying 17 passengers for 120 kilometers on an eight-hour electric charge (Figure 13.1.1).

Ejeepneys have made the greatest impact in Makati, a city of 600,000 residents near the heart of the Manila metropolitan area. Makati is the business capital of the Philippines and a relatively prosperous city, allowing it to undertake a number of forward-looking urban initiatives. Working with environmental organizations, Makati's civic leaders have been pushing the ejeepney since 2008. As a result of this and other environmental projects, Makati has been labeled by Greenpeace as Southeast Asia's first "Energy-Efficient City."

Makati's environmental program focuses on transportation and energy use. Urban walkways covering 30.4 kilometers now span the central business district and link it with mass transportation routes. Energy efficiency is also stressed, especially for city-owned

FIGURE 13.1.1 Ejeepneys in the Philippines The public transportation system in the Philippines has long been dominated by the colorful jeepney. The cramped vehicles, however, are highly polluting and not very fuel-efficient. The new ejeepneys are not as elaborately decorated, but they are much more environmentally friendly.

buildings and street lighting. Environmentally sensitive solid waste management programs and rainwater harvesting systems have also been put in place. Makati's mayor promises that the city will remain committed to building an environmentally sustainable urban infrastructure.

Considering its relative prosperity, Makati can pursue such ambitious environmental projects much more easily than any other Philippine city. But as energy efficiency results in economic savings, many experts think that the gains made by Makati will inspire the leaders of other urban areas. Ejeepneys, at the very least, are spreading quickly, offering reduced air pollution and enhanced energy efficiency for the Philippines as a whole.

Changes in precipitation across Southeast Asia, brought about by global warming, remain uncertain. Many experts foresee an intensification of the monsoon pattern, which could bring increased rainfall to much of the mainland. While enhanced precipitation would likely result in more destructive floods, it could bring some agricultural benefits to dry areas such as Burma's central Irrawaddy Valley. Others think that both dry spells and flooding could be intensified. In 2010, northern mainland Southeast Asia saw one of the worst droughts in recorded history. Then, in the summer of 2011, Thailand experienced massive flooding, which affected almost 13 million people, caused US$45 billion in damge, and innundated 14.8 mllion acres (6 million hectares) of land.

Southeast Asia's overall carbon emissions from conventional sources remain low by global standards. When greenhouse gas output is associated with deforestation is factored in, however, Southeast Asia's role in global climate change is revealed to be much larger. By some estimates, Indonesia is the world's third largest contributor to the problem, after China and the United States. Much of Indonesia's carbon emissions stem from the burning and oxidation of peat in deforested swamps. (Peat is the partially decayed organic matter that accumulates in saturated soils.) During dry periods, peat soils sometimes burn, releasing the stored carbon (Figure 13.6). Both Indonesia and Malaysia, moreover, are actively draining coastal wetlands to make room for agricultural expansion, a process that results in the gradual oxidation of the peat. Ironically, many of these swamps are being destroyed in order to produce biodiesel from palm oil, a supposedly ecofriendly form of energy.

FIGURE 13.6 Burning Peatlands   In Southeast Asia's coastal wetlands, organic peat soils often burn after they have been dried out by drought or drainage, contributing to air pollution and climate change. This photograph shows burning trees and peatlands in Indonesia's Riau Province.

Patterns of Physical Geography

The different kinds of forest found in insular and mainland Southeast Asia mostly reflect differences in climate. The two subregions, however, are also marked by striking variations in landforms and other aspects of the physical environment. As a result, the physical geography of each is discussed separately below.

Mainland Environments   Mainland Southeast Asia is an area of rugged uplands mixed with broad lowlands associated with large rivers. The region's northern boundary lies in a cluster of mountains connected to the highlands of western Tibet and south-central China. In the far north of Burma, peaks reach 18,000 feet (5,500 meters). From this point, a series of distinct mountain ranges spreads out, extending through western Burma, along the Burma–Thailand border, and through Laos into southern Vietnam.

Several large rivers flow southward out of Tibet and its adjacent highlands into mainland Southeast Asia. The valleys and deltas of these rivers are the centers of both population and agriculture in mainland Southeast Asia. The longest river is the Mekong, which flows through Laos and Thailand and then across Cambodia before entering the South China Sea through a large delta in southern Vietnam. Second longest is the Irrawaddy, which flows through Burma's central plain before reaching the Bay of Bengal. Two smaller rivers are equally significant: the Red River, which forms a heavily settled delta in northern Vietnam, and the Chao Phraya, which has created the fertile alluvial plain of central Thailand. The lower Chao Phraya Valley is easily seen in satellite images, especially when the river is flooding (Figure 13.7).

The centermost area of mainland Southeast Asia is Thailand's Khorat Plateau, which is neither a rugged upland nor a fertile river valley. This low sandstone plateau averages about 500 feet (175 meters) in height and is noted for its thin, poor soils. Water shortages and periodic droughts pose challenges throughout this extensive area.

The Influence of the Monsoon   Almost all of mainland Southeast Asia is affected by the seasonally shifting winds known as the *monsoon*. The climate of this area is characterized by a distinct hot

and rainy season from May to October. This is followed by dry but still generally hot conditions from November to April (Figure 13.8). Only the central highlands of Vietnam and a few coastal areas receive significant rainfall during this period.

Two types of tropical climates are dominant in mainland Southeast Asia. Although both are affected by the monsoon, they differ in the total amount of rainfall received during the year. Along the coasts and in the highlands, the tropical monsoon climate (Am) dominates. Rainfall totals for this climate zone usually average more than 100 inches (250 centimeters) each year (see the climographs for Rangoon and Da Nang). The greater portion of the mainland falls into the tropical savanna (Aw) climate type, where annual rainfall totals are about half those of the Am region (see the climographs for Bangkok and Vientiane). In the so-called dry zone of central Burma, however, rainfall is distinctly lower and droughts are common.

Insular Environments   The main feature of insular Southeast Asia is its island environment. Although Indonesia contains thousands of islands, it is dominated by the four large landmasses of Sumatra, Kalimantan (or Borneo), Java, and Sulawesi. A string of active volcanoes extends along the length of eastern Sumatra across Java and into the Lesser Sunda Islands. In the Philippines, the two largest and most important islands are Luzon (about the size of Ohio) in the north and Mindanao (the size of South Carolina) in the south. Sandwiched between them are the Visayan Islands, which number roughly a dozen. The topography of the Philippines includes rugged upland landscapes, as well as numerous volcanoes.

Insular Southeast Asia has so many volcanoes because it lies along the boundaries of several tectonic plates (Figure 13.9). This geologically active situation generates several natural hazards, including earthquakes, toxic mud volcanoes, and **tsunamis** (sometimes

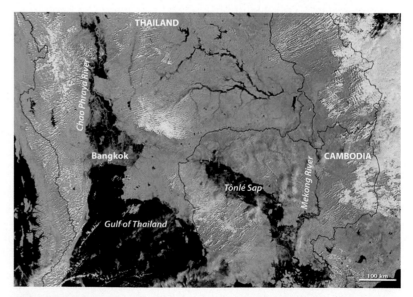

FIGURE 13.7 Central Thailand   The flat, fertile, and well-watered delta and lower valley of the Chao Phraya River in central Thailand are visible in this satellite image, taken when the river was flooding in late 2011. Cambodia's Tonlé Sap (Great Lake), swollen from the same floods, is also clearly evident.

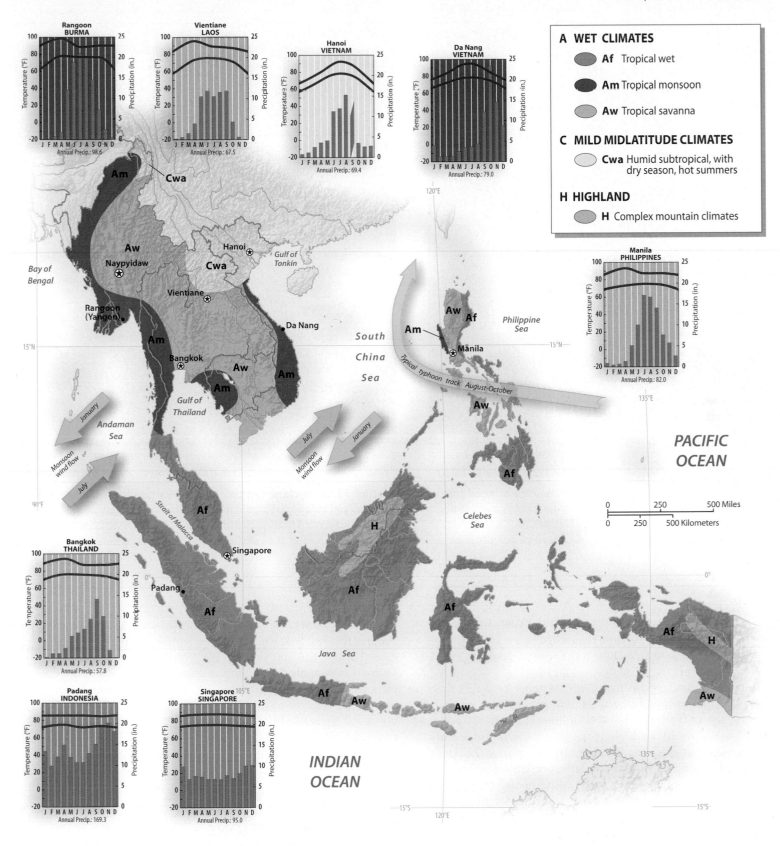

FIGURE 13.8 Climate Map of Southeast Asia Most of insular Southeast Asia is characterized by the constantly hot and humid climates of the equatorial zone. Mainland Southeast Asia, on the other hand, has the seasonally wet and dry climates of the tropical monsoon and tropical savanna types. Only the far north features subtropical climates, with relatively cool winters. The northern half of the region is strongly influenced by the seasonally shifting monsoon winds. Northeastern Southeast Asia—and especially the Philippines—often experiences typhoons from August to October.

FIGURE 13.9 Bromo Volcano Insular Southeast Asia is noted for its widespread volcanism, which has created fertile soils in many areas, while also generating many natural hazards. This photograph shows the eruption of Bromo Volcano on the island of Java in Indonesia on January 22, 2011.

FIGURE 13.10 Cyclone Nargis Cyclone Nargis slammed into southern Burma in early May 2008, resulting in the country's worst natural disaster in recorded history. This photograph shows devastated families waiting for relief near their destroyed homes on Haing Guy Island in southwestern Burma.

called *tidal waves*). An undersea earthquake off the coast of Sumatra in December 2004 generated a tsunami that killed some 230,000 people in Southeast and South Asia, making it the second deadliest earthquake in recorded history.

The climates of insular Southeast Asia are more complex than those of the mainland, largely because the insular belt extends across a greater span of latitude. Most of Indonesia lies in the equatorial zone, which results in high levels of precipitation evenly distributed throughout the year (see the climograph for Padang). Southeastern Indonesia and East Timor, however, experience a prolonged dry season from June to October. Most of the Philippines experiences dry conditions from November to April (see the climograph for Manila).

The Typhoon Threat The coastal areas of mainland Southeast Asia and the Philippines are highly vulnerable to tropical cyclones, or **typhoons** as they are called in the western Pacific. These strong storms bring devastating winds and torrential rain. Each year several typhoons hit Southeast Asia, causing heavy damage and loss of life through flooding and landslides. Deforestation and farming on steep hillsides make the problem much worse.

In early May 2008, southern Burma was slammed by Cyclone Nargis, the worst storm in recent Southeast Asian history. Nargis caused some 150,000 deaths, left 1 million people homeless, and resulted in roughly $10 billion in damages (Figure 13.10). Casualty rates were particularly high because Burma's repressive government reacted very slowly to the crisis, restricting the activities of international aid agencies. In December 2011, Typhoon Washi swept away entire neighborhoods in the southern Philippines, affecting more than 330,000 people.

REVIEW QUESTIONS

1. Why do the mainland and insular regions of Southeast Asia have such distinctive climates and landforms, and how have these differences impacted the human communities of these two regions?

2. How have people changed the physical landscapes of Southeast Asia over the past 50 years?

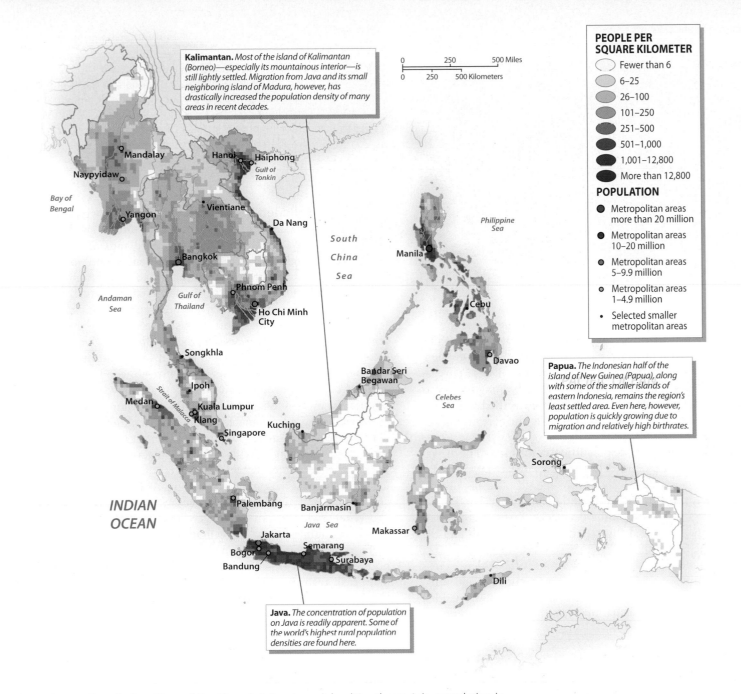

Kalimantan. *Most of the island of Kalimantan (Borneo)—especially its mountainous interior—is still lightly settled. Migration from Java and its small neighboring island of Madura, however, has drastically increased the population density of many areas in recent decades.*

PEOPLE PER SQUARE KILOMETER
- Fewer than 6
- 6–25
- 26–100
- 101–250
- 251–500
- 501–1,000
- 1,001–12,800
- More than 12,800

POPULATION
- Metropolitan areas more than 20 million
- Metropolitan areas 10–20 million
- Metropolitan areas 5–9.9 million
- Metropolitan areas 1–4.9 million
- Selected smaller metropolitan areas

Papua. *The Indonesian half of the island of New Guinea (Papua), along with some of the smaller islands of eastern Indonesia, remains the region's least settled area. Even here, however, population is quickly growing due to migration and relatively high birthrates.*

Java. *The concentration of population on Java is readily apparent. Some of the world's highest rural population densities are found here.*

FIGURE 13.11 Population Map of Southeast Asia In mainland Southeast Asia, population is concentrated in the valleys and deltas of the region's large rivers. In the uplands, population density remains relatively low. In Indonesia, density is extremely high on Java, an island noted for its fertile soil and large cities. Some of Indonesia's outer islands, especially those of the east, remain lightly settled. Overall, population density is high in the Philippines, especially in central Luzon.

POPULATION AND SETTLEMENT: SUBSISTENCE, MIGRATION, AND CITIES

Southeast Asia's population issue is quite different in scale from those of East Asia and South Asia. With almost 600 million people, Southeast Asia is still *relatively* sparsely settled. Over most of the region, extensive tracts of land with infertile soil and rugged topography remain thinly inhabited. In contrast, dense populations live in the region's deltas, coastal areas, and zones of fertile volcanic soil (Figure 13.11). Southeast Asia experienced rapid population growth in the second half of the 20th century. More recently, birthrates have dropped quickly, especially in the wealthier countries of the region.

Settlement and Agriculture

Much of insular Southeast Asia has relatively infertile soil, which cannot easily support intensive agriculture and high rural population densities. Although the island forests are lush and biologically rich, plant nutrients are locked up in the vegetation itself, rather than being stored in the soil where they would easily benefit agriculture. Furthermore, the constant rain of the equatorial zone tends to wash nutrients away. Agriculture must thus rely on constant field rotation or the application of large amounts of fertilizer.

There are some notable exceptions to this pattern. Unusually rich soils connected to volcanic activity are scattered throughout much of the region, but are particularly widespread on the island of Java. With

FIGURE 13.12 Swidden Agriculture In the uplands of Southeast Asia, swidden (or slash-and-burn) agriculture is widely practiced. When done by tribal peoples with low population densities, swidden is not environmentally harmful. This photo shows a small swidden plot that has recently been burned.

Swidden in the Uplands Also known as *shifting cultivation* or *slash-and-burn* agriculture, swidden is practiced throughout the rugged uplands of Southeast Asia (Figure 13.12). In the **swidden** system, small plots of several acres of forest or brush are periodically cut by hand. The fallen vegetation is then burned, and the ash naturally spreads over the ground. This practice transfers nutrients to the soil before subsistence crops are planted. Yields remain high for several years and then drop off as the soil nutrients are exhausted and insect pests and plant diseases multiply. These plots are abandoned after a few years and return to woody vegetation. The cycle of cutting, burning, and planting then moves to another small plot not far away—thus the term *shifting cultivation*.

Swidden is a sustainable form of agriculture when population densities remain relatively low. Today, however, the swidden system is increasingly threatened. With higher population densities, the rotation period must be shortened, which damages soil resources. Swidden farming is also harmed by commercial logging, which both displaces farmers and removes soil nutrients from the ecosystem as logs are exported.

When swidden can no longer support the population, upland people sometimes adapt by switching to cash crops that will allow them to participate in the commercial economy. In the mountains of northern Southeast Asia, also called the **Golden Triangle**, one of the main cash crops has historically been opium, grown by local farmers for the global drug trade. Intensive efforts by national governments and the United Nations virtually eliminated opium growing in Laos and northern Thailand by 2006, and even Burma saw a huge reduction

more than 50 volcanoes, Java is a fertile island that supports a wide range of tropical crops and a very high population density. Roughly 138 million people live on Java, an island smaller than Iowa, giving it one of the world's highest rural population densities. Dense populations are also found in pockets of fertile alluvial soils along the coasts of insular Southeast Asia. One particularly densely settled area is the central lowlands of Luzon near the city of Manila, the core area of the Philippines.

In mainland Southeast Asia, population is concentrated in the agriculturally intensive valleys and deltas of the large rivers, whereas most upland areas remain relatively lightly settled. The population core of Thailand is formed by the valley and delta of the Chao Phraya River, just as Burma's is focused on the Irrawaddy River. Vietnam has two distinct core areas: the Red River Delta in the far north and the Mekong Delta in the far south.

Agricultural practices and settlement forms vary widely across the complex environments of Southeast Asia. Generally speaking, however, three farming and settlement patterns are apparent: swidden in the upland areas and both plantation agriculture and rice cultivation in the lowland areas.

FIGURE 13.13 Tea Harvesting in Indonesia Plantation crops, such as tea, are major sources of exports for several Southeast Asian countries. Coconut, rubber, palm oil, and coffee are other major cash crops. Many of these crops require large amounts of labor, particularly at harvest time.

in the crop. But high prices have lured many farmers back to opium, with total production doubling between 2006 and 2012.

Plantation Agriculture With European colonization, Southeast Asia became a focus for plantation agriculture, growing high-value specialty crops ranging from coconuts to rubber. Even in the 19th century, Southeast Asia was linked to world trade flow through the plantation system. Forests were cleared and swamps drained to make room for commercial farms; labor was supplied by native people or by workers brought in from India or China (Figure 13.13).

Plantations are still an important part of Southeast Asia's geography and economy. Most of the world's natural rubber is produced in Malaysia, Indonesia, and Thailand. Cane sugar has long been a major crop of the Philippines and parts of Indonesia, although it is no longer very profitable; as a result, sugar areas in the Philippines are associated with intense rural poverty. Indonesia is the region's leading producer of tea, and Vietnam dominates the production of coffee. In recent years, oil palm plantations have been spreading through much of the region, often at the expense of tropical rain forests. Coconuts are widely grown in the Philippines, Indonesia, and elsewhere.

Rice in the Lowlands The lowland basins of mainland Southeast Asia are largely devoted to intensive rice cultivation. Throughout almost all of Southeast Asia, rice is the preferred staple food. Rice harvests are increasingly traded to meet the needs of expanding urban markets throughout the world. Three delta areas have been the focus for commercial rice cultivation: the Irrawaddy in Burma, the Chao Phraya in Thailand, and the Mekong in Vietnam. The use of agricultural chemicals and high-yield crop varieties has allowed production to keep pace with population growth, although at the cost of significant environmental damage.

As of 2012, the world's two largest rice exporters were Thailand and Vietnam. In 2008, these two countries, along with Burma, Cambodia, and Laos, joined together to form the Organization of Rice Exporting Countries (OREC), which seeks to maintain high and even prices. The Philippines has strongly criticized the organization, as it is the world's top rice-importing country. Although agriculture in much of the Philippines is dominated by rice, farmers in the country have been unable to keep pace with the country's rapidly growing population.

Recent Demographic Changes

Because Southeast Asia is not facing the same kind of population pressure as East or South Asia, a wide range of government population policies exists. While several countries are concerned about rapid growth and thus have strong family-planning programs, others believe that their populations are too small (Figure 13.14). In those countries facing rapid demographic expansion, internal relocation of people away from densely populated areas to outlying districts is a common outcome.

Population Contrasts The Philippines, the second most populous country in Southeast Asia, has a relatively high birthrate, with a total fertility rate over 3.0 (Table 13.1). Effective family planning here has been difficult to establish. For example, when a popular democratic government replaced a dictatorship in the 1980s, the Philippine Roman Catholic Church, which played an active role in the peaceful revolution, pressured the new government to cut funding for family-planning programs. Due to a combination of rapid growth and economic stagnation, many Filipinos have been forced to migrate (Figure 13.15). An estimated 11 million Filipinos now live and work abroad, with over 1 million in Saudi Arabia alone.

The highest total fertility rates (TFRs) in mainland Southeast Asia are found in Laos and Cambodia, two countries of Buddhist religious tradition. Here the relatively high birthrates are best explained by the country's low level of economic and social development. Still, both countries have experienced major fertility declines in recent years, with the TFR of Laos dropping from over 6 in 1985 to roughly 3.5Z in 2011. In Thailand, which shares cultural traditions with Laos yet is considerably more developed, the birthrate is now at 1.6, well below the replacement level. The fertility rate in Vietnam is also below the replacement level, while that of Burma is close to it.

Indonesia, with the region's largest population at 241 million, has also seen a dramatic decline in fertility in recent decades, although its fertility rate remains slightly above the replacement level. If the present trend continues, however, Indonesia will reach population stability before most other large developing countries. As with Thailand, this drop in fertility seems to have resulted from government family-planning efforts coupled with urbanization and improvements in education.

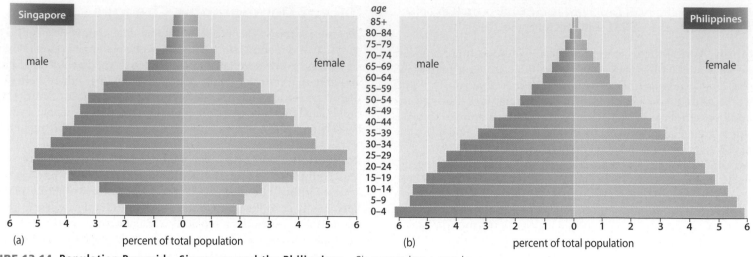

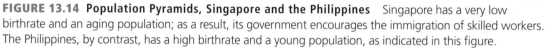

FIGURE 13.14 Population Pyramids, Singapore and the Philippines Singapore has a very low
birthrate and an aging population; as a result, its government encourages the immigration of skilled workers.
The Philippines, by contrast, has a high birthrate and a young population, as indicated in this figure.

The city-state of Singapore stands out on the demographic charts, as its fertility rate dropped below the replacement level in the mid-1970s and is now one of the lowest in the world. Its government is concerned about this situation and is actively promoting marriage and childbearing, particularly among the most highly educated segment of its population. In Singapore's "baby bonus program," introduced in 2001 and enhanced in 2008, couples can get a cash gift of up to US$4,000 for their first and second babies and up to US$6,000 for their third and fourth. Despite this program, the fertility rate of Singapore continues to decline. Yet because of migration, the country's population has been steadily increasing. The government has been especially eager to encourage highly skilled people to relocate to the city-state, offering high wages and other benefits.

Growth and Migration Indonesia has the most explicit policy of **transmigration**, or relocation of its population from one region to another within its national territory. Primarily because of migration from densely populated Java, the population of the outer islands of Indonesia has grown rapidly since the 1970s. The province of East Kalimantan, for example, experienced a growth rate of 30 percent per year during the last two decades of the 1900s.

Table 13.1 POPULATION INDICATORS

Country	Population (millions) 2010	Population Density (per square kilometer)	Rate of Natural Increase (RNI)	Total Fertility Rate	Percent Urban	Percent <15	Percent >65	Net Migration (Rate per 1000) 2005–10[a]
Burma (Myanmar)	54.6	81	1.1	2.3	31	28	5	−0.4
Brunei	0.4	72	1.6	2.0	72	26	4	1.7
Cambodia	15.0	83	1.7	3.0	21	33	4	−1.8
East Timor	1.1	76	2.6	5.7	30	42	3	−0.8
Indonesia	241	127	1.3	2.3	43	27	6	−0.8
Laos	6.5	28	2.0	3.5	27	38	4	−2.3
Malaysia	29.0	88	1.5	2.6	63	27	5	0.6
Philippines	96.2	321	1.9	3.2	63	35	4	−2.1
Singapore	5.3	7,751	0.5	1.2	100	17	9	6.6
Thailand	69.9	136	0.5	1.6	34	21	9	1.1
Vietnam	89.0	268	1.0	2.0	31	24	7	−0.5

[a]Net Migration Rate from the United Nations, Population Division, *World Population Prospects: The 2008 Revision Population Database.*
Source: Population Reference Bureau, *World Population Data Sheet, 2010.*

FIGURE 13.15 Filipino Overseas Employment Fair Due to a combination of poor economic conditions at home, rapid population growth, and good English-language skills, many Filipinos seek employment abroad. Here job seekers crowd an overseas employment fair.

High social and environmental costs accompany these relocation programs. Javanese peasants, accustomed to working the fertile soils of their home island, often fail in their attempts to grow rice in the former rain forest of Kalimantan. In some areas, farmers have little choice but to adopt a semi-swidden form of cultivation, a process associated with further deforestation, as well as conflicts with indigenous peoples. Partly because of these problems, the Indonesian government significantly reduced its official transmigration program in 2000. The government still helps around 60,000 people migrate to the outer islands every year, however, and many others borrow money or use their personal savings to do the same (Figure 13.16).

FIGURE 13.16 Transmigration in Indonesia Migration from densely settled to sparsely settled areas of Southeast Asia has resulted in the creation of thousands of new communities. Many of these communities have minimal transportation and communication facilities and receive few governmental services. Shown in this photograph is a new transmigrant community in the Indonesian part of the island of New Guinea.

Urban Settlement

Despite the relatively high level of economic development found in much of Southeast Asia, the region is not heavily urbanized. Even Thailand's population is roughly two-thirds rural, which is unusual for a country that has experienced so much industrialization. However, cities are growing rapidly throughout the region, increasing the rate of urbanization.

Many Southeast Asian countries have **primate cities**—single, large urban settlements that overshadow all others. Thailand's urban system, for example, is dominated by Bangkok, just as Manila far surpasses all other cities in the Philippines. Both have grown recently into megacities with more than 10 million residents (Figure 13.17). In Manila, Bangkok, Jakarta, and other massive cities, explosive growth has led to housing problems, congestion, and pollution. Bangkok suffers from some of the worst traffic in the world, although the Thai government has responded with large-scale highway and mass-transit

FIGURE 13.17 Bangkok Bangkok saw the development of an impressive skyline during its boom years from the late 1970s through the late 1990s. Unfortunately, transportation development did not keep pace with population and commercial growth, resulting in one of the most congested and polluted urban landscapes in the world.

construction programs. It is estimated that more than half of Manila's people live in squatter settlements, usually without basic water and electricity service. Most large Southeast Asian cities suffer from a lack of parks and other public spaces, which is one reason why massive shopping malls have become so popular. Bangkok's Paragon Mall has recently emerged as a major urban focus, complete with a conference center and a concert hall.

Urban primacy is less pronounced in other major Southeast Asian countries. Vietnam, for example, has two main cities: Ho Chi Minh City (formerly Saigon) in the south and the capital city of Hanoi in the north. Massive Jakarta is the largest urban area in Indonesia, and the 13th largest city in the world, but the country has several other large and growing cities, including Bandung and Surabaya. Yangon (formerly Rangoon) remains the primate city of Burma, with more than 4 million residents. Yangon is no longer the capital of Burma, however, as the government was moved to inaccessible Naypyidaw in 2006, due to security concerns.

Kuala Lumpur, the largest city in Malaysia, has received heavy investments from both the national government and the global business community. This has produced a modern, forward-looking city that is largely free of most of the traffic, water, and slum problems that plague most other Southeast Asian cities. Although the city of Kuala Lumpur has only around 1.5 million residents, its greater metropolitan area supports over 7 million, making it the primate city of Malaysia.

The independent republic of Singapore is essentially a city-state of 5 million people on an island of 274 square miles (710 square kilometers), about three times the size of Washington, DC (Figure 13.18).

FIGURE 13.18 Singapore   Singapore remains the economic and technological hub of Southeast Asia. It is famous for its clean, efficiently run, and very modern urban environment. Some residents complain, however, that Singapore has lost much of its charm as it has developed.

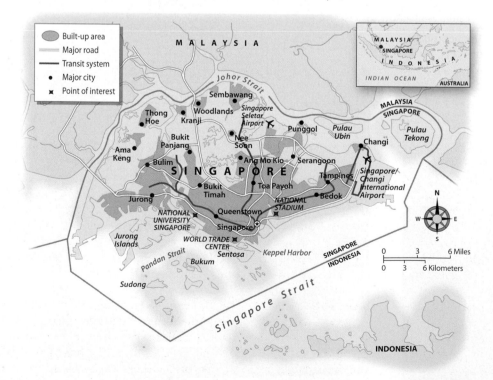

While space is at a premium, Singapore has been very successful at developing high-tech industries that have brought it great wealth. Unlike most other Southeast Asian cities, Singapore has no squatter settlements or slums.

> ### REVIEW QUESTIONS
>
> 1. Why do different parts of Southeast Asia vary so much in regard to both population density and population growth?
>
> 2. Why does Southeast Asia have such distinctive forms of agriculture?

CULTURAL COHERENCE AND DIVERSITY: A MEETING GROUND OF WORLD CULTURES

Unlike many other world regions, Southeast Asia lacks the historical dominance of a single civilization. Instead, the region has been a meeting ground for cultural influences from South Asia, China, the Middle East, Europe, and North America. Abundant natural resources, along with the region's strategic location on oceanic trading routes connecting major landmasses, have long made Southeast Asia attractive to outsiders.

The Introduction and Spread of Major Cultural Traditions

In Southeast Asia, contemporary cultural diversity is related to the historical influence of the major religions of the region: Hinduism, Buddhism, Islam, and Christianity (Figure 13.19).

South Asian Influences   The first major external religious influence arrived from South Asia some 2,000 years ago when small numbers of educated migrants from what is now India helped local leaders establish Hindu kingdoms in coastal locations in Burma, Thailand, Cambodia, Malaysia, and western Indonesia. Although Hinduism later faded away in most locations, it is still the dominant religion on the Indonesian island of Bali.

A second wave of South Asian religious influence reached mainland Southeast Asia in the 13th century in the form of Theravada Buddhism, which spread from Sri Lanka. Almost all of the people in lowland Burma, Thailand, Laos, and Cambodia converted to Buddhism at that time, and today this religion forms the foundation for their social institutions. Saffron-robed monks, for example, are a common sight and Buddhist temples abound.

Chinese Influences   Unlike most other mainland peoples, the Vietnamese were not heavily influenced by South Asian civilization. Instead, their early connections were to East Asia. Vietnam was

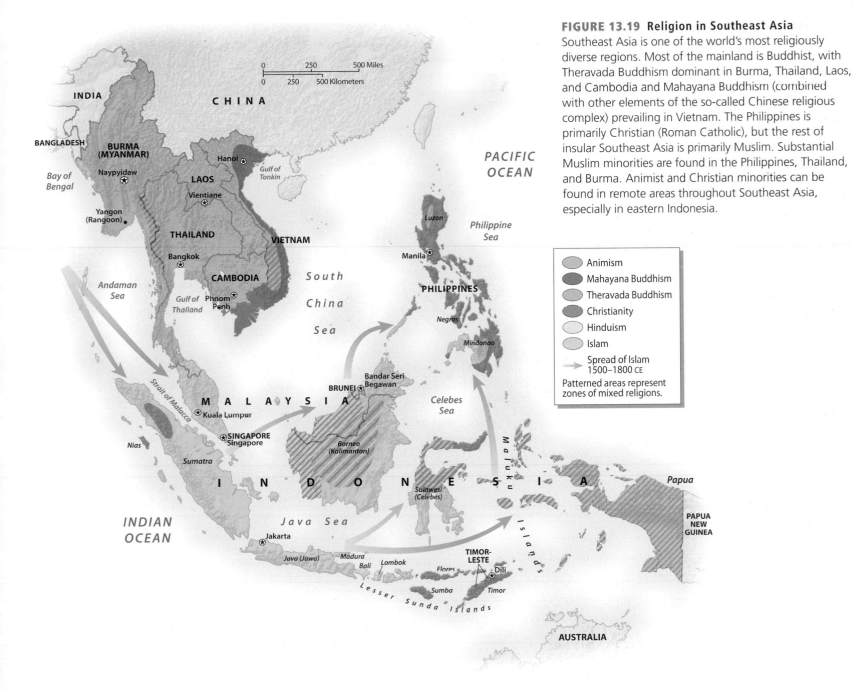

FIGURE 13.19 Religion in Southeast Asia
Southeast Asia is one of the world's most religiously diverse regions. Most of the mainland is Buddhist, with Theravada Buddhism dominant in Burma, Thailand, Laos, and Cambodia and Mahayana Buddhism (combined with other elements of the so-called Chinese religious complex) prevailing in Vietnam. The Philippines is primarily Christian (Roman Catholic), but the rest of insular Southeast Asia is primarily Muslim. Substantial Muslim minorities are found in the Philippines, Thailand, and Burma. Animist and Christian minorities can be found in remote areas throughout Southeast Asia, especially in eastern Indonesia.

a province of China until about 1000 CE, when the Vietnamese established a kingdom of their own. But while the Vietnamese rejected China's political rule, they retained many features of Chinese culture. The traditional religious and philosophical beliefs of Vietnam, for example, are centered on Mahayana Buddhism and Confucianism.

East Asian cultural influences in many other parts of Southeast Asia are directly linked to the more recent immigration of southern Chinese. This migration reached a peak in the 19th and early 20th centuries (Figure 13.20). China was then a poor and crowded country, which made sparsely populated Southeast Asia appear to be a place of opportunity. Eventually, distinct Chinese settlements were established in every Southeast Asian country, especially in urban areas. In Malaysia, the Chinese minority now constitutes roughly one-third of the population, whereas in Singapore some three-quarters of the people are of Chinese ancestry.

In many places in Southeast Asia, relationships between the Chinese minority and the native majority are strained. Even though their ancestors arrived generations ago, many Chinese are still considered resident aliens because they maintain their Chinese identities. A more significant source of tension is the fact that most Chinese communities in Southeast Asia are relatively wealthy. Many Chinese immigrants prospered as merchants, an occupation avoided by most local people. As a result, they have acquired substantial economic influence, which is often resented by others.

The Arrival of Islam Muslim merchants from South and Southwest Asia arrived in Southeast Asia hundreds of years ago and soon began converting many of their local trading partners. From an initial core established around 1200 CE in northern Sumatra, Islam spread through much of insular Southeast Asia. By 1650, it had

FIGURE 13.20 Chinese in Southeast Asia People from the southern coastal region of China have been migrating to Southeast Asia for hundreds of years, a process that reached a peak in the late 1800s and early 1900s. Most Chinese migrants settled in the major urban areas, but in peninsular Malaysia large numbers were drawn to the countryside to work in the mining industry and in plantation agriculture. Today Malaysia has the largest number of people of Chinese ancestry in the region. Singapore, however, is the only Southeast Asian country with a Chinese majority.

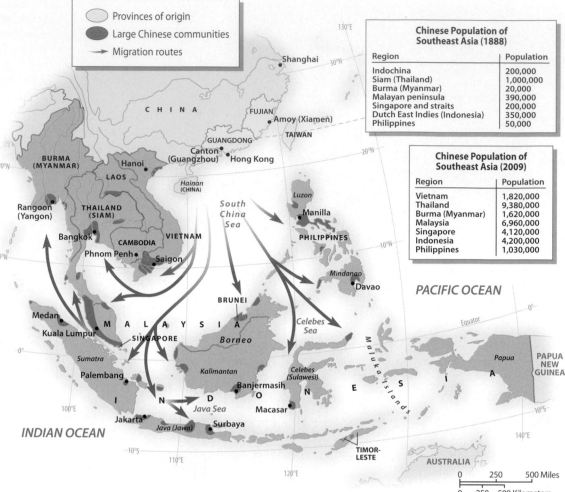

Chinese Population of Southeast Asia (1888)	
Region	Population
Indochina	200,000
Siam (Thailand)	1,000,000
Burma (Myanmar)	20,000
Malayan peninsula	390,000
Singapore and straits	200,000
Dutch East Indies (Indonesia)	350,000
Philippines	50,000

Chinese Population of Southeast Asia (2009)	
Region	Population
Vietnam	1,820,000
Thailand	9,380,000
Burma (Myanmar)	1,620,000
Malaysia	6,960,000
Singapore	4,120,000
Indonesia	4,200,000
Philippines	1,030,000

largely replaced Hinduism and Buddhism throughout Malaysia and Indonesia. The only significant holdout was the small, but fertile island of Bali.

Some 88 percent of Indonesia's inhabitants follow Islam, making it the world's most populous Muslim country (Figure 13.21). This figure, however, hides a significant amount of internal religious diversity. In some parts of Indonesia, such as northern Sumatra (Aceh), highly orthodox forms of Islam took root. In others, such as central and eastern Java, a more relaxed form of worship emerged that included certain Hindu and even animistic beliefs. Islamic reformers, however, have long tried to instill more mainstream forms of the faith among the Javanese. Recently, they have found much success, particularly among the young.

Christianity Islam was still spreading eastward through insular Southeast Asia when the Europeans arrived in the 16th century. When the Spanish claimed the Philippine Islands in the 1570s, they found the southwestern portion of the archipelago to be thoroughly Islamic. To this day, the southwest Philippines is still largely Muslim, although the rest of the country is mostly Roman Catholic. East Timor, long a Portuguese colony, is also a predominantly Roman Catholic country.

Christian missions spread through other parts of Southeast Asia in the late 19th and early 20th centuries, when European colonial powers controlled most of the region. While French priests converted many people in Vietnam to Catholicism, missionaries had

FIGURE 13.21 Tuban Mosque in Java Indonesia is often said to be the world's largest Muslim nation because more Muslims reside here than in any other country. Islamic architecture in Indonesia is often modern in style, especially when contrasted with the more traditional styles found in Southwest Asia and North Africa. The Tuban Mosque, shown here, is noted for its extravagant decorations.

Muslim sect, a group regarded as heretical by many mainstream Muslims. In 2011, the country was shocked when a mob of 1,500 people attacked a small group of Ahmaddiya adherents, killing 3.

Geography of Language and Ethnicity

The linguistic geography of Southeast Asia is complicated (Figure 13.23). The several hundred distinct languages of the region are all placed into five major linguistic families. Four of these families are discussed below, whereas the fifth, Papuan, is discussed in Chapter 14.

The Austronesian Languages One of the world's most widespread language families is Austronesian, which extends from Madagascar to Easter Island in the eastern Pacific. Today almost all insular Southeast Asian languages belong to the Austronesian family. But despite this common linguistic grouping, more than 50 distinct languages are spoken in Indonesia alone. In far eastern Indonesia, a variety of languages fall into the completely separate family of Papuan, closely associated with New Guinea.

The Malay language overshadows all others in insular Southeast Asia. Malay is native to the Malay Peninsula, eastern Sumatra, and coastal Borneo and was spread historically by merchants and seafarers. As a result, it became a common trade language, or **lingua franca**, throughout much of the insular realm. When Indonesia became an independent country in 1949, its leaders decided to use the lingua franca version of Malay as the basis for a new national language called *Bahasa Indonesia* (or simply *Indonesian*). Although Indonesian is slightly different from the Malaysian spoken in Malaysia, they form a single, mutually understandable language. Both are now written in the Roman script.

The goal of the new Indonesian government was to offer a common language that could overcome ethnic differences throughout the huge country. This policy has been generally successful, with the vast majority of Indonesians now using the language. Regionally-based languages, however, such as Javanese, Balinese, and Sundanese, continue to be the primary languages of most Indonesian homes.

The Philippines contains eight major languages and several dozen minor languages, all of which are closely related. Despite more than 300 years of colonialism by Spain, Spanish never became a unifying force for the islands. During the American period (1898–1946), English served as the language of government and education. After independence, Philippine nationalists selected Tagalog, the language of the Manila area, to replace English and help unify the new country. After Tagalog was standardized and modernized, it was renamed Filipino. Through its use in education, television, and films, Filipino has gradually emerged as the country's unifying national language.

Tibeto-Burman Languages Each country of mainland Southeast Asia is closely identified with the national language spoken in its core territory. This does not mean, however, that all the residents of these countries speak these official languages on a daily basis. In the mountains and other remote districts, other languages are commonly used. This linguistic diversity reinforces ethnic differences, often presenting challenges for programs designed to build national unity.

A good example of such linguistic challenges is Burma. Its national language is Burmese, a language that is closely related to Tibetan. Some 32 million people speak Burmese. Although the government

FIGURE 13.22 Protesting Burmese Monks In September 2007, more than 100,000 people took to the streets of Yangon (formerly Rangoon) to protest Burma's repressive government. The protest marches were led by robed monks chanting prayers for peace.

little influence in other lowland areas. They were more successful in highlands inhabited by tribal peoples worshiping nature spirits and their ancestors. The general name for such religions is **animism**. While many modern hill tribes remain animist today, others were converted to Christianity. As a result, significant Christian concentrations are found in several upland areas of Indonesia, Vietnam, and Burma.

Religious Persecution Religious persecution has recently become a serious issue in parts of Southeast Asia. Vietnam's communist government is struggling against a revival of faith among the country's Buddhist majority and its 8 million Christians. In 2011, Human Rights Watch accused the Vietnamese government of persecuting ethnic minority Christians. In Burma, the government has generally supported Buddhism, but when Buddhist monks led massive demonstrations against the government in late 2007, it cracked down hard, killing an estimated 30 to 40 monks (Figure 13.22). Burma has also severely repressed its Rohingya Muslim minority, driving hundreds of thousands out of the country. In late 2011, Burma agreed to let some Rohingyas return to the country, but allegations of persecution persist. In Indonesia, the government has sought to ban the Ahmaddiya

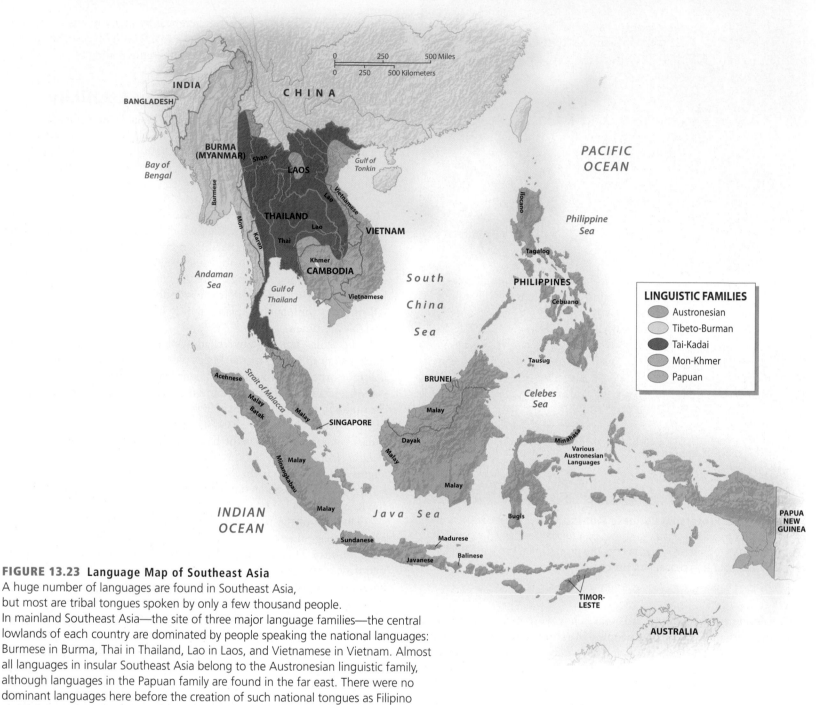

FIGURE 13.23 Language Map of Southeast Asia
A huge number of languages are found in Southeast Asia,
but most are tribal tongues spoken by only a few thousand people.
In mainland Southeast Asia—the site of three major language families—the central
lowlands of each country are dominated by people speaking the national languages:
Burmese in Burma, Thai in Thailand, Lao in Laos, and Vietnamese in Vietnam. Almost
all languages in insular Southeast Asia belong to the Austronesian linguistic family,
although languages in the Papuan family are found in the far east. There were no
dominant languages here before the creation of such national tongues as Filipino
and Bahasa Indonesia in the mid-20th century.

of Burma has sought to unify the population with one language, a major split has developed with the non-Burman peoples who live in the rough uplands on both sides of the Burmese-speaking Irrawaddy Valley. Although most of these tribal groups speak languages in the Tibeto-Burman family, they are quite distinctive from Burmese.

Tai-Kadai Languages The Tai-Kadai linguistic family probably originated in southern China and then spread into Southeast Asia starting around 1100. Today closely related languages within the Tai subfamily are found through most of Thailand and Laos, in the uplands of northern Vietnam, and in Burma's Shan Plateau. Most Tai

languages are spoken by members of small tribal groups. But two of them, Thai and Lao, are important national languages.

Historically, the main language of Thailand, called Siamese (just as the kingdom was called Siam), was restricted to the lower Chao Phraya Valley. In the 1930s, however, the country changed its name to Thailand to emphasize the unity of all the peoples speaking the closely related Tai languages within its territory. Siamese was similarly renamed Thai, and it has gradually become the country's unifying language. There is still much variation in dialect, however, with northern Thai sometimes considered a separate language. Somewhat more distinctive is Lao, the Tai language that became the national tongue of

Laos. In Thailand's Khorat Plateau, most people speak Isan, a dialect that is much closer to Lao than to standard Thai. It is notable that Isan food is also quite different from standard Thai food, as it is spicier and is based on glutinous ("sticky") rice.

Mon-Khmer Languages   The Mon-Khmer language family probably once covered virtually all of mainland Southeast Asia. It contains two major languages—Vietnamese and Khmer (the national language of Cambodia)—as well as a large group of minor languages spoken by hill peoples and a few lowland groups. Because of the historical Chinese influence in Vietnam, the Vietnamese language was written with Chinese characters until the French colonial government imposed the Roman alphabet, which remains in use today. Khmer, on the other hand, is—like Lao, Thai, and Burmese—written in its own Indian-derived script.

The most important aspect of linguistic geography in mainland Southeast Asia is the fact that in each country the national language is spoken mainly in the core lowlands, whereas the peripheral uplands are populated by tribal peoples speaking separate languages. In Vietnam, for example, Vietnamese speakers occupy less than half of the national territory, even though they constitute a sizable majority of the country's population. Ethnic tensions here have recently mounted, as Vietnamese speakers, aided by the country's major road-building program, have begun moving into the sparsely populated highlands.

Southeast Asian Culture in Global Context

Several Southeast Asian countries have been quite receptive to global cultural influences. This is particularly true in the case of the Philippines, where U.S. colonialism encouraged the country to embrace many forms of popular Western culture. As a result, Filipino musicians and other entertainers are in demand elsewhere in Asia (Figure 13.24). Thailand, which was never subjected to colonial rule, is also highly open to global culture.

Cultural globalization has also been challenged in Southeast Asian countries. The Malaysian government has been especially critical of American films and satellite television. Islamic revivalism in Indonesia and Malaysia also presents a challenge to cultural globalization. Islamic radicals have attacked several nightclubs and other tourist destinations, and anti-American sentiments spread rapidly after the U.S.-led invasion of Iraq in 2003.

The use of English as the global language also causes controversy. As the language of popular global culture, it is opposed by many conservatives, yet it must be mastered if citizens are to participate in global business and politics. In Malaysia, the widespread use of English grew increasingly controversial in the 1980s as nationalists stressed the importance of the native tongue. This worried the business community, which considers English vital to Malaysia's competitive position. It also troubled the influential Chinese community, for which Malaysian is not a native language.

In Singapore, the situation is more complex. Mandarin Chinese, English, Malay, and Tamil are all official languages. Furthermore, the languages of southern China are common in home environments, as 75 percent of Singapore's population is of southern Chinese ancestry. In recent years, the Singapore government has encouraged Mandarin Chinese and discouraged the southern Chinese dialects. It also launched a campaign against "Singlish," a popular form of speech

FIGURE 13.24 Filipino Entertainers   Filipino performers, who have long adopted global musical styles, are often in demand abroad. Charice (Charmaine Clarice Relucio Pempengco) is a young singer who rose to fame on the basis of her YouTube videos. She has performed extensively in other Asian countries and has appeared on the U.S. television show *Glee.*

based on English, but employing many words from Malay and Chinese (Figure 13.25).

In the Philippines, nationalists complain about the common use of English, but widespread fluency has proved beneficial for the millions of Filipinos who work abroad or for international businesses. Moreover, many people from other Asian countries come to the Philippines to study English. In 2011, more than 100,000 South Koreans were residing in the country for this purpose. Although the Philippine government has sought to gradually replace English with Filipino, English remains a widespread official language. At the same time, Filipino itself is increasingly incorporating words and phrases from English, giving rise to a hybrid dialect known as "Taglish."

REVIEW QUESTIONS

1. What major world religions have spread into Southeast Asia over the past 2,000 years, and in what parts of the region did they become established?

2. How have different Southeast Asian countries reacted to the challenges of cultural globalization?

FIGURE 13.25 Singlish Sign The Malay- and Chinese-influenced dialect of English known as "Singlish" is sometimes said to be the most important factor of cultural unity in Singapore. Singapore's government, however, discourages Singlish, viewing it as crude slang. The sign visible in this photograph ends with the common Singlish particle *la*, which is used to soften a sentence and to show agreement with listeners or readers.

GEOPOLITICAL FRAMEWORK: WAR, ETHNIC STRIFE, AND REGIONAL COOPERATION

Southeast Asia is sometimes defined as the geopolitical grouping of 10 different countries that have joined together as ASEAN (Figure 13.26). Although East Timor is not a member, its government applied for membership in 2011, and it will probably be admitted within a few years. ASEAN has significantly reduced geopolitical problems among its member states, while giving Southeast Asia as a whole a greater degree of regional coherence. But despite ASEAN's successes, many Southeast Asian countries still experience internal ethnic conflicts, as well as tensions with their neighbors.

Before European Colonialism

The modern countries of mainland Southeast Asia all existed in one form or another as kingdoms before European colonialism. Cambodia emerged first, over 1,000 years ago, and by the 1300s, independent states had been established by the Burmese, Siamese, Lao, and Vietnamese people.

The situation in insular Southeast Asia was different from that of the mainland, with the premodern map being completely different from that of the modern nation-states. Many kingdoms existed on the Malay Peninsula and on the islands of Sumatra, Java, and Sulawesi, but few were territorially stable. Indonesia, the Philippines, and Malaysia thus owe their statehood and their territorial shapes largely to European colonialism (Figure 13.27).

The Colonial Era

The Portuguese were the first Europeans to arrive (around 1500), lured mainly by the spices of the Maluku Islands in eastern Indonesia. In the late 1500s, the Spanish conquered most of the Philippines, which they used as a base for their silver trade between China and the Americas. By the 1600s, the Dutch began to establish trading bases, followed by the British. With superior naval weapons, the Europeans were able to conquer key ports and control strategic waterways. Yet for the first 200 years of colonialism, except in the Philippines, the Europeans made no major territorial gains.

By the 1700s, the Netherlands had become the most powerful naval force in the region. As a result, a Dutch Empire in the "East Indies" began appearing on world maps. This empire continued to grow into the early 20th century, when it defeated its last major enemy, the Islamic state of Aceh in northern Sumatra. Later, the Netherlands divided the island of New Guinea with Germany and Britain, rounding out its Indonesian colony.

The British, preoccupied with India, concentrated their attention on the sea-lanes linking South Asia to China. As a result, they established several fortified outposts along the Strait of Malacca, the most important of which was Singapore, founded in 1819. To avoid conflict, the British and Dutch agreed that the British would limit their attention to the Malay Peninsula and the northern portion of Borneo. The British allowed local rulers to retain limited powers, much as they had done in parts of India.

In the 1800s, European colonial power spread through most of mainland Southeast Asia. The British conquered the kingdom of Burma and extended their power into the nearby uplands. During the same period, the French moved into Vietnam's Mekong Delta, gradually expanding their territorial control into Cambodia and northward to China's border. Thailand, then called "Siam," was the only country to avoid colonial rule, although it did lose territories to the British and French. The final colonial power to enter the region was the United States, which took the Philippines first from Spain and then from Filipino nationalists between 1898 and 1900.

Organized resistance to European rule began in the 1920s, but it took the Japanese occupation of World War II to show that colonial power was vulnerable. Between 1942 and 1945, Japan occupied virtually the entire region. After Japan's surrender in 1945, pressure for independence intensified throughout Southeast Asia. Britain withdrew from Burma in 1948 and began to pull out of its colonies in insular Southeast Asia in the late 1950s, although Brunei did not gain independence until 1984. Singapore briefly joined Malaysia but then withdrew and became independent in 1965. In the Philippines, the United States granted long-promised independence on July 4, 1946, although it retained military bases for several decades. While the Dutch attempted to reestablish their colonial rule after World War II, they were forced to acknowledge Indonesia's independence in 1949.

The Vietnam War and Its Aftermath

After World War II, France was determined to regain control of its Southeast Asian colonies. Resistance to French rule was organized primarily by communist groups based mainly in northern Vietnam. Open warfare between French soldiers and the communist forces continued until 1954, when France agreed to withdraw after a major military defeat. An international peace council then divided Vietnam into a communist North Vietnam, allied with the Soviet Union and China, and a capitalist-oriented South Vietnam, with close ties to the United States.

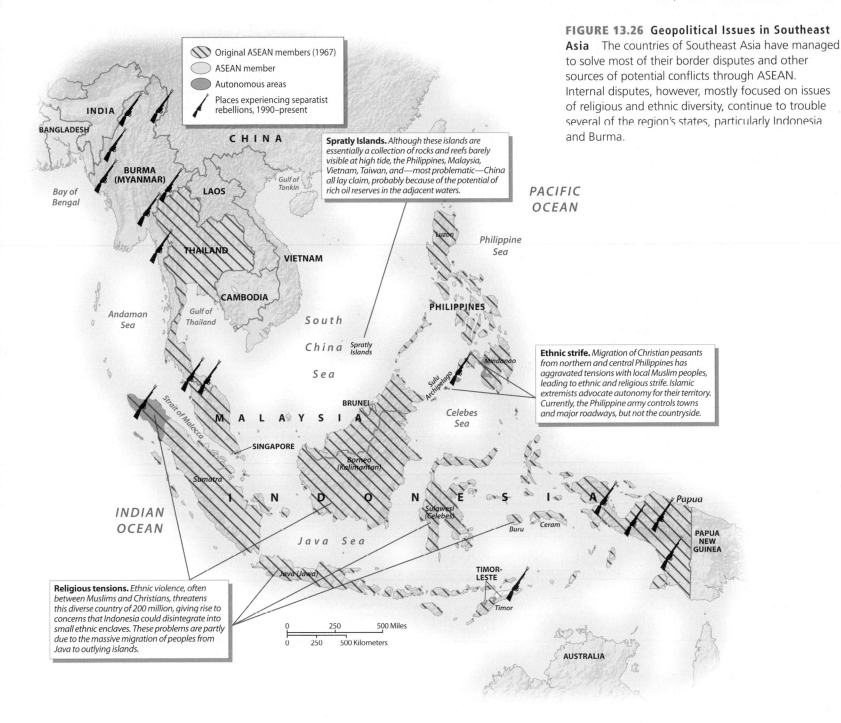

FIGURE 13.26 Geopolitical Issues in Southeast Asia The countries of Southeast Asia have managed to solve most of their border disputes and other sources of potential conflicts through ASEAN. Internal disputes, however, mostly focused on issues of religious and ethnic diversity, continue to trouble several of the region's states, particularly Indonesia and Burma.

Legend:

- Original ASEAN members (1967)
- ASEAN member
- Autonomous areas
- Places experiencing separatist rebellions, 1990–present

Spratly Islands. *Although these islands are essentially a collection of rocks and reefs barely visible at high tide, the Philippines, Malaysia, Vietnam, Taiwan, and—most problematic—China all lay claim, probably because of the potential of rich oil reserves in the adjacent waters.*

Ethnic strife. *Migration of Christian peasants from northern and central Philippines has aggravated tensions with local Muslim peoples, leading to ethnic and religious strife. Islamic extremists advocate autonomy for their territory. Currently, the Philippine army controls towns and major roadways, but not the countryside.*

Religious tensions. *Ethnic violence, often between Muslims and Christians, threatens this diverse country of 200 million, giving rise to concerns that Indonesia could disintegrate into small ethnic enclaves. These problems are partly due to the massive migration of peoples from Java to outlying islands.*

The peace accord did not, however, end the fighting. Communist guerrillas in South Vietnam fought to overthrow the new government and unite it with the north. North Vietnam sent troops and war materials across the border to aid the rebels. Most of these supplies reached the south over the Ho Chi Minh Trail, a confusing network of forest passages through Laos and Cambodia, thus steadily drawing these two countries into the conflict (Figure 13.28). In Laos, the communist Pathet Lao forces challenged the government, while in Cambodia the **Khmer Rouge** guerrillas gained considerable power.

In Washington, DC, the **domino theory** guided foreign policy. According to this notion, if Vietnam fell to the communists, then so would Laos and Cambodia. Once those countries were lost, Burma, Thailand, and perhaps Malaysia and Indonesia would also

become members of the communist bloc. Fearing such an outcome, the United States was drawn ever deeper into the war. By 1965, thousands of U.S. troops were fighting to support the government of South Vietnam. But despite superiority in arms and troops, U.S. forces gradually lost control over much of the countryside. As casualties mounted and the antiwar movement back home strengthened, the United States began secret talks in search of a negotiated settlement. Subsequent U.S. troop withdrawals began in earnest in the early 1970s.

With the withdrawal of U.S. forces, the noncommunist governments began to collapse. Saigon fell in 1975, and in the following year Vietnam was officially reunited under the government of the north. Reunification was a traumatic event in southern Vietnam. Hundreds

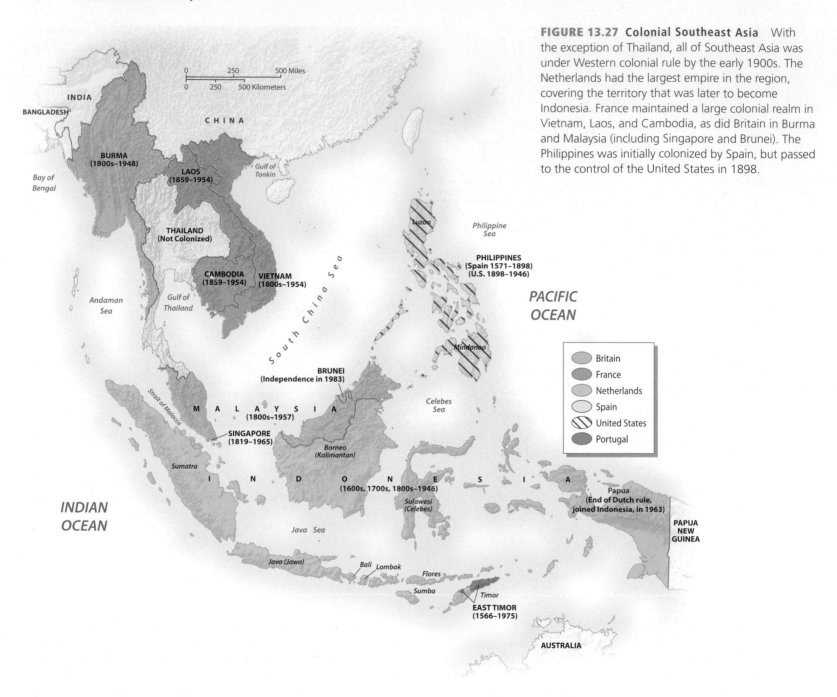

FIGURE 13.27 Colonial Southeast Asia With the exception of Thailand, all of Southeast Asia was under Western colonial rule by the early 1900s. The Netherlands had the largest empire in the region, covering the territory that was later to become Indonesia. France maintained a large colonial realm in Vietnam, Laos, and Cambodia, as did Britain in Burma and Malaysia (including Singapore and Brunei). The Philippines was initially colonized by Spain, but passed to the control of the United States in 1898.

of thousands of people fled from the new regime, with many settling in the United States.

Vietnam proved fortunate compared to Cambodia. There the Khmer Rouge installed one of the most brutal regimes the world has ever seen. Cities were largely evacuated as urban residents were forced into the countryside to become peasants, and most wealthy and educated people were executed. The Khmer Rouge's goal was to create an agriculturally self-sufficient society that would eventually provide the foundation for industrialization. After several years of horrific bloodshed, neighboring Vietnam invaded Cambodia and installed a far less brutal, but still repressive, regime. Fighting between different groups continued until 1991, when a comprehensive peace settlement was finally reached.

Geopolitical Tensions in Contemporary Southeast Asia

In several parts of Southeast Asia, local ethnic groups have been struggling against national governments that inherited their territory from former colonial powers. Tensions have also emerged where tribal groups attempt to preserve their homelands from logging, mining, or migrant settlers.

Conflicts in Indonesia When Indonesia gained independence in 1949, it included all of the former Dutch possessions in the region except western New Guinea (Papua). In 1962, the Netherlands

FIGURE 13.28 Ho Chi Minh Trail North Vietnam supplied communist insurgents in South Vietnam with weapons and other goods that they moved over the ill-marked Ho Chi Minh Trail, running through Laos and Cambodia. The United States extensively bombed the trail and sprayed chemical defoliants on the surrounding forests, but it was never able to shut down this supply route.

FIGURE 13.29 Grasberg Mine Grasberg, located in the Indonesian province of Papua, is the world's largest gold mine and third largest copper mine. Employing more than 19,000 people, Grasberg is of great economic importance to Indonesia. It is also highly polluting, generating 253,000 tons (230,000 metric tons) of tailings per day, and is thus opposed by many local inhabitants.

organized an election to see whether the people of this area wished to join Indonesia or form an independent country. The vote went for union, but many observers believe that the election was rigged by the Indonesian government. As a result, many of the local people began to rebel. Indonesia is determined to maintain control of the region, in part because it is the site of the country's largest source of tax revenue—the highly polluting Grasberg mine, run by Phoenix-based Freeport-McMoRan Corporation (Figure 13.29).

In 2000, the Indonesian government granted partial autonomy to Papua, which reduced support for the rebellion. A decade later, tensions again mounted. In October 2011, Indonesian security forces stormed a pro-independence rally, beating a number of protestors and killing three. One underlying source of conflict is the continuing immigration of people to Papua from central and western Indonesia.

The island of Timor has also experienced political bloodshed in recent years. The eastern half of this poor and rather dry island had been a Portuguese colony and had evolved into a largely Christian society. The East Timorese expected independence when Portugal finally withdrew in 1975. Indonesia, however, viewed the area as its own and immediately invaded. A brutal war followed, which the Indonesian army won in part by preventing food from reaching the province.

After a severe economic crisis in 1997, Indonesia's power in the region slipped. A new Indonesian government promised an election in 1999 to determine whether the East Timorese still wanted independence. At the same time, however, the Indonesian army began to organize militias to intimidate the people of East Timor into voting to remain within the country. When it was clear that the vote would be for independence, the militias began rioting, looting, and killing civilians. Under international pressure, Indonesia finally withdrew and the East Timorese began to build a new country. By 2012, East Timor had settled down enough for the UN peacekeeping force to finally withdraw from the country.

The Aceh region of northern Sumatra has also experienced prolonged political violence, as local rebels have sought to create an independent Islamic state. While the Indonesian government has given Aceh "special autonomy," it has been determined to prevent actual independence. Ironically, the devastation of the December 2004 tsunami seems to have generated a solid peace, as the needs of the province were so great that the separatist fighters finally agreed to lay down their weapons. Tensions persist in the area, however, as Aceh's autonomous government attempts to institute strict Islamic law. In a controversial move in December 2011, police rounded up large numbers of young fans of punk-rock music, forcibly shaved their heads, and then sent them off to a distant boarding school to be "re-educated."

Regional Tensions in the Philippines

The Philippines has also suffered from regional political violence. Its most persistent problem is the Islamic southwest, where rebels have long demanded independence. After successful negotiations with the main separatist group in 1989, the government created the Autonomous Region in Muslim Mindanao (ARMM), which it expanded in 2008. The more radical Islamist groups, however, rejected the settlement and continued to fight, setting off bombs and kidnapping civilians. In late 2012, however, the Philippine government signed a peace treaty with the country's largest Muslim insurgent group, the Moro Islamic Liberation Front."

The Muslim southwest does not present the Philippines' only political problem. A revolutionary communist group called the New People's Army operates in most parts of the country and still controls many rural districts. Furthermore, the country's national government, although democratic, is far from stable, suffering from periodic coup threats, corruption scandals, mass protests, and impeachment efforts.

Burma's Many Problems

Of all the countries of Southeast Asia, Burma has experienced the most extreme ethnic conflicts. Burma's simultaneous wars have pitted the central government, dominated by the Burmese-speaking ethnic group (the Burmans), against the country's varied non-Burman societies. Fighting intensified gradually after independence in 1948, and by the 1980s almost half of the country's territory had become a combat zone. In the 1990s, however, successful Burmese army offensives, combined with cease-fire agreements, resulted in a marked decrease in warfare. In early 2012, the Burmese government finally signed a cease-fire with the Karen ethnic rebels, although at the same time it intensified a military drive against the Kachin rebels in the far north (Figure 13.30).

Due to the repressive nature of its Burman-dominated government, Burma has long suffered from trade restrictions imposed by the United States and the European Union. In 2010, however, Burma held elections and began to reduce restrictions on its people. Many observers have been skeptical of such moves, fearing that they would not result in real change. But such concerns were reduced in 2012 when the Burmese government agreed to let noted opposition leader and 2001 Nobel Peace Prize winner Aung San Suu Kyi run for parliament. At that time, the United States reestablished diplomatic relations with Burma, while the European Union hinted that it might drop its trade sanctions on the country.

Thailand's Troubles

Unlike Burma, Thailand has enjoyed some basic human freedoms and a relatively free press for some time. Until 2006, most observers thought that it had emerged as a stable democracy. In that year, however, Thai Prime Minister Thaksin Shinawatra was overthrown by a military coup that apparently had the blessing of Thailand's revered king. Thaksin, a wealthy businessman, had gained the support of Thailand's poor in part by setting up a national health-care system, but he infuriated the middle and upper classes by taking too much power into his own hands. Huge protests and counterprotests followed for several years, causing major problems (Figure 13.31). In 2011, Thaksin's sister, Yingluck Shinawatra, became prime minister after winning a landslide victory. As she is opposed by the country's traditional elite, many observers foresee continuing instability.

FIGURE 13.30 Women Soldiers of the Kachin Liberation Army The Kachin Liberation Army has been fighting the Burmese government for decades. Like several other ethnic insurgent groups in Burma, it employs many female fighters. Although Burma has recently made peace with many insurgent groups, it stepped up its struggle against the Kachin in early 2012.

The recent political chaos in Thailand has made it difficult for the country to deal with its main political threat, the murky insurgency that plagues its southernmost provinces. Far southern Thailand is relatively poor, and its people are mostly Malay in language and Muslim in religion. They have long resented being ruled by Thailand, and periodic rebellions have flared up for decades. After 2004, violence sharply increased, resulting in almost 5,000 deaths by 2012. One

FIGURE 13.31 Anti-Government Thai Protests Thailand has recently been plagued by massive political protests that have challenged the country's democratic institutions. In this photograph, supporters of ousted Thai Prime Minister Thaksin Shinawatra protest outside the Finance Ministry in Bangkok on April 2, 2009.

odd feature of the insurgency in southern Thailand is the fact that no group has emerged to claim responsibility for the violence, and hence no political demands have been made.

International Dimensions of Southeast Asian Geopolitics

Despite ASEAN's general successes in encouraging peaceful relations among the countries of Southeast Asia, border disputes continue to flare up. In late 2008 and early 2009, a conflict between Thailand and Cambodia over the area around the 11th-century Preah Vihear Temple resulted in roughly 40 deaths. In December 2011, both countries agreed to withdraw their military forces from the area, but tensions remain pronounced.

A more complicated territorial dispute centers on the Spratly and Paracel islands in the South China Sea, two groups of tiny islands and reefs that might sit over substantial oil reserves (Figure 13.32). The Philippines, Malaysia, Brunei, and Vietnam have all claimed territory there, as have China and Taiwan. International tensions in the South China Sea began to heat up in 2010, as China increased its naval presence in the area. In 2011, Chinese ships opened fire on several Vietnamese fishing boats. As a result of such tensions, both Vietnam and the Philippines began to seek closer military cooperation with the United States.

One of the biggest problems faced by ASEAN leaders has been the establishment of radical Islamist networks in the region. The largest of these is Jemaah Islamiya (JI), a militant group dedicated to establishing an Islamic state that would contain all Muslim areas within Southeast Asia. JI agents are believed to have detonated bombs that killed 202 people in Bali in 2002 and to have set off major explosions in Jakarta in 2004 and Bali (again) in 2005. Indonesia responded by creating an elite counterterrorism squad (Detachment 88) and by establishing a "deradicalization program" aimed at convincing radical Islamists to change sides. By 2008, most observers had concluded that JI was no longer a serious threat, but in 2010 the Indonesian government accused the organization of running secret training camps in the autonomous region of Aceh.

FIGURE 13.32 The Spratly Islands The Spratly Islands are small and barely above water at high tide, but they are geopolitically important. Oil may exist in large quantities in the surrounding areas, heightening the competition over the islands. Southeast Asian countries are especially concerned about China's military activities in the Spratlys.

REVIEW QUESTIONS

1. How did European colonization influence the development of the modern countries of Southeast Asia, and how did that process differ in the insular and the mainland regions?

2. How did the emergence and spread of ASEAN reduce geopolitical tensions in Southeast Asia, and why were certain conflicts in the region not able to be solved by the ASEAN process?

ECONOMIC AND SOCIAL DEVELOPMENT: THE ROLLER COASTER RIDE OF DEVELOPING ECONOMIES

Over the past few decades, Southeast Asia has experienced major economic fluctuations (Table 13.2). An economic boom between 1980 and 1997 was followed by a major recession. Southeast Asian economies began to grow quickly after 2000, but the global economic crisis of 2008–2009 resulted in another blow. By 2010, however, most of the economies of the region had fully recovered.

Uneven Economic Development

While the region as a whole has experienced pronounced ups and downs, parts of Southeast Asia have done much better in the global economy than have others. Oil-rich Brunei and technologically sophisticated Singapore rank among the world's more prosperous countries, whereas Cambodia, Laos, Burma, and East Timor are among the poorest countries in Asia. Also, although Malaysia, Thailand, and Vietnam have seen tremendous economic gains over the past 50 years, the Philippines has experienced major disappointments during the same period.

The Philippine Decline and Recovery In the 1950s, the Philippines was the most highly developed Southeast Asian country. By the late 1960s, however, Philippine development had been derailed. Through the 1980s and early 1990s, the country's economy failed to outpace its population growth, resulting in declining living standards for both the poor and the middle class. The Philippine people are still well educated and reasonably healthy by world standards, but even the country's educational and health systems declined during this period.

Why did the Philippines fail despite its earlier promise? While there are no simple answers, it is clear that dictator Ferdinand Marcos (who ruled from 1968 to 1986) wasted—and perhaps even stole—billions of dollars, while failing to create conditions that would lead to genuine development. The Marcos regime instituted a kind of **crony capitalism** in which the president's friends were given huge economic favors, while those believed to be enemies had their properties taken.

By 2010, however, it appeared that the Philippines was finally recovering. In that year, its economy grew by more than 7.5 percent. A particularly bright spot in the Philippine economy has been the expansion of business outsourcing operations, attracted by the country's educated and English-speaking population. The recent development of modern communications systems in the region has been particularly beneficial. By 2011, almost 300,000 Filipinos worked in international call centers, handling telephone inquiries from customers in

Table 13.2   DEVELOPMENT INDICATORS

Country	GNI per capita, PPP 2010	GDP Average Annual %Growth 2000–10	Human Development Index (2011)[1]	Percent Population Living Below $2 a Day	Life Expectancy (2012)[2]	Under Age 5 Mortality Rate (1990)	Under Age 5 Mortality Rate (2010)	Adult Literacy (% ages 15 and older)	Gender Equity (2011)[3]
Burma (Myanmar)	1,950	—	.483	—	65	112	66	92	0.492
Brunei	50,180	—	.838	—	78	12	7	95	—
Cambodia	2,080	8.7	.523	53.3	62	121	51	78	0.500
East Timor	3,600	3.4	.495	72.8	62	169	81	51	—
Indonesia	4,200	5.3	.617	46.1	72	85	35	92	0.505
Laos	2,460	7.2	.524	66.0	65	145	54	73	0.513
Malaysia	14,220	5.0	.761	2.3	74	18	6	92	0.286
Philippines	3,980	4.9	.644	41.5	69	59	29	95	0.427
Singapore	55,790	6.0	.866	—	82	8	3	95	0.086
Thailand	8,190	4.5	.682	4.6	74	32	13	94	0.382
Vietnam	3,070	7.5	.593	43.4	73	51	23	93	0.305

[1]United Nations, *Human Development Report, 2011.*

[2]Population Reference Bureau, *World Population Data Sheet, 2012.*

[3]Gender Inequality Index—A composite measure reflecting inequality in achievements between women and men in three dimensions: reproductive health, empowerment and the labor market that ranges between 0 and 1. The higher the number, the greater the inequality.

Source: World Bank, *World Development Indicators, 2012.*

the United States and other wealthy countries (Figure 13.33). Filipino business leaders are understandably worried about moves by the U.S. Congress to place limits on outsourcing by American businesses.

The Regional Hub: Singapore   Singapore and Malaysia have been Southeast Asia's major developmental successes. Singapore has transformed itself from an **entrepôt** city—a place where goods are imported,

FIGURE 13.33 Philippine Call Center   The Philippines competes with India for the position as the world's leading international call center. Almost a million Filipinos now work at these centers, providing information and advice to English-speaking consumers in North America and Europe.

stored, and then transshipped—to one of the world's wealthiest and most modern states. Singapore is now the communications and financial hub of Southeast Asia, as well as a thriving high-tech manufacturing center. The Singaporean government has played an active role in the development process. Singapore has encouraged investment by multinational technology companies and has itself invested heavily in housing, education, and some social services (Figure 13.34). The Singaporean government, however, remains only partly democratic, as the ruling party maintains a firm grip on government and restricts free speech.

The Malaysian Boom   Although not nearly as well-off as Singapore, Malaysia has also experienced rapid economic growth. Development was initially concentrated in agriculture and natural resources, focused on tropical hardwoods, plantation products, and tin mining. More recently, manufacturing, especially in labor-intensive high-tech sectors, has become the main engine of growth.

The modern economy of Malaysia is not uniformly distributed across the country. One difference is geographical: Most industrial development has occurred on the west side of peninsular Malaysia. More important, however, are differences based on ethnicity. The industrial wealth generated in Malaysia has been concentrated in the Chinese community. Ethnic Malays remain less prosperous than Chinese-Malaysians, and those of South Asian descent are poorer still.

The unbalanced wealth of the local Chinese community is a feature of most Southeast Asian countries. The problem is particularly acute in Malaysia, however, because its Chinese minority is so large. The government's response was one of aggressive "affirmative action," by which economic power was transferred to the dominant Malay, or **Bumiputra** ("sons of the soil"), community through the so-called Malaysian New Economic Policy (NEP). This policy was reasonably successful. Since the economy as a whole expanded significantly, the Chinese community was equal extent. Most industrial development has occurred in the historical core, especially in the city of Bangkok and surrounding areas. The vulnerability of such industrial concentration was demonstrated in the summer of 2011, when extensive flooding in central Thailand disrupted not only the Thai national economy, but also global supply chains in the automobile and consumer electronics sectors (Figure 13.35).

Thailand's Lao-speaking northeast (the Khorat Plateau) and Malay-speaking far south remain the country's poorest regions. Because of the poverty of their homeland, northeasterners are often forced to seek employment in Bangkok: The men typically find work in the construction industry; the women not uncommonly make their living as prostitutes. In the far south, poverty and unemployment have contributed to the region's brutal insurgency.

Unstable Economic Expansion in Indonesia At the time of independence (1949), Indonesia was one of the poorest countries in the world. The Indonesian economy finally began to expand in the 1970s. Oil exports fueled the early growth, as did the logging of tropical forests. But unlike most other oil exporters, Indonesia continued to grow even after oil prices plummeted in the 1980s. Like Thailand and Malaysia, Indonesia attracted multinational companies looking for low wages and a relatively well-educated workforce. Large Indonesian firms, many of them owned by local Chinese families, have also capitalized on the country's human and natural resources. The national government has attempted to build technologically-oriented businesses, but their success remains uncertain.

Despite its recent economic growth, Indonesia remains a poor country. Its pace of economic expansion never matched those of Singapore and Malaysia, and it has remained much more dependent on the unsustainable exploitation of natural resources. The financial crisis of the late 1990s, moreover, hurt Indonesia more severely than any other country. The global economic crisis of 2008–2009 was not as severe in Indonesia as it was in the wealthier countries of Southeast Asia, and in 2010 and 2011 the country experienced solid economic growth.

FIGURE 13.34 Public Housing in Singapore Despite its free-market approach to economics, the government of Singapore has invested heavily in public housing. Most Singaporeans live in buildings similar to these. (*Dr. Pradeep Kumar, ProPhotoz*)

able to thrive even as its relative share of the country's wealth declined. The NEP was officially discontinued in 1990, but many of its features remain in place. Opposition to such policies is growing, particularly among the Chinese population, but many ethnic Malay professionals also want a more open and competitive society.

Thailand's Ups and Downs Thailand, like Malaysia, climbed rapidly during the 1980s and 1990s into the ranks of the world's newly industrialized countries. Japanese companies were leading players in the earlier Thai boom, attracted by Thailand's low-wage, yet reasonably well-educated, workforce. Thailand experienced a major downturn in the late 1990s, however, that undercut much of this development. Growth resumed by 2000, but Thailand's political crisis, coupled with the global economic recession, brought about another economic drop in 2008–2009. By 2010, the Thai economy was again expanding at a rapid rate, but widespread flooding in 2011 took a huge economic toll.

Thailand's economic growth over the past several decades has by no means benefited the entire country to an

FIGURE 13.35 Flooded Factories in Thailand Central Thailand is a highly global manufacturing center, with many factories owned by Japanese and other foreign companies. As a result, extensive flooding in the region in the summer of 2011 disrupted the international automobile and consumer electronics industries.

The Recent Rise of Vietnam and Cambodia

The former French colonial zone was long noted as one of the poorest and least globalized parts of Southeast Asia. From the time of the Vietnam War through the early 1990s, Vietnam experienced relatively little economic development. Frustrated with their country's economic performance, Vietnam's leaders began to follow China by embracing market economics, while retaining the political forms of a communist state. The Vietnamese economy subsequently began to expand at a very rapid rate, experiencing annual rates of growth of over 8 percent through much of the first decade of the new century. But Vietnam is still a poor country that has a long way to go to catch up with Thailand, let alone Malaysia.

Vietnam now welcomes multinational corporations, which are attracted by its extremely low wages and relatively well-educated workforce. Japanese and South Korean companies now often favor Vietnam over other Southeast Asian countries. Local businesses, however, complain of harassment by state officials, and development remains geographically uneven. Southern Vietnam is still much more entrepreneurial and capitalistic than the north, while deep and persistent poverty remains entrenched in many rural areas, particularly in the tribal highlands.

Cambodia's recent economic history is somewhat similar to that of Vietnam, only more extreme. Long burdened by war and corruption, Cambodia was one of Asia's poorest countries, its economy focused largely on subsistence agriculture. The discovery of oil and other mineral resources after 2000, however, combined with a thriving tourist economy and large-scale international investment, resulted in a major economic boom.

Cambodia's recent economic expansion has resulted in problems as well as opportunities. A property boom in Phnom Penh, for example, saw thousands of poor people being forced out of their homes to make way for development projects. Tourism has also proved to be a mixed blessing. Several Cambodian border towns, most notably Poipet, have set themselves up as gambling centers, attracting investment—and criminal activities—from Thai underworld figures (Figure 13.36). Many Cambodians fear that their country is coming under the economic domination of Thailand and Vietnam.

Persistent Poverty in Laos and East Timor

Like Cambodia, Laos has long been dominated by subsistence agriculture, which employs roughly three-quarters of its workforce. Laos has particular economic difficulties owing to its rough terrain and relative isolation; outside of its few cities, paved roads and reliable electricity are rare. As a result, it remains heavily dependent on foreign aid.

The Laotian government is pinning its economic hopes on hydropower development, mining, tourism, and investment from Thailand, Vietnam, and China. Hydropower is particularly important, as the country is mountainous, with many rivers, and could therefore generate and export large quantities of electricity. Laos has also benefited from the increasing volume of barge traffic going up the Mekong River to China. In 2011, China also began building a high-speed railroad system in Laos designed to reduce the country's isolation.

The weakest economy in Southeast Asia is undoubtedly that of East Timor—which counts as one of the world's poorest countries. East Timor has hardly begun to recover from the devastation that accompanied its independence, and it has been further weakened by the gradual withdrawal of international aid agencies. A recent agreement with Australia to share the revenues of substantial offshore natural gas and oil deposits, however, promises some hope for this impoverished country.

Burma's Troubled Economy

Burma also is near the bottom of the scale of Southeast Asian economic development. For all of its many problems, however, Burma remains a land of great potential. It has abundant natural resources (including oil and other minerals, water, and timber), as well as a large expanse of fertile farmland. Its population density is moderate, and its people are reasonably well educated. But despite these advantages, Burma's economy has remained relatively stagnant since independence in 1948. More recently, Burma has opened its economy to some degree, actively trading with China, India, and its Southeast Asian neighbors (see "Exploring Global Connections: Burma's Economic and Geopolitical Links with China—and the Rest of the World").

Many experts think that, if the Burmese government continues to pursue more open and globally engaged policies, it could begin to experience broad-based economic development. But major economic and political reforms are also needed. In 2011, Burma was ranked along with Afghanistan as the third most corrupt country in the world. Another indication of the country's economic failure is the fact that the official exchange rate, as of December 2011, was 6.51 Burmese kyat to the dollar, whereas the black-market rate was approximately 780 kyat per dollar!

FIGURE 13.36 Cambodian Casino   The Cambodian border city of Poipet has recently emerged as a major gambling center. Most of the investments, and most of the tourists, come from Thailand, angering many Cambodians. Organized crime is also a major problem in the city.

EXPLORING GLOBAL CONNECTIONS

Burma's Economic and Geopolitical Links with China—and the Rest of the World

In late 2011, Burma began opening up to the rest of the world, agreeing to reestablish diplomatic relations with the United States, for example, in January 2012. From the early 1960s until the 1980s, Burma had been one of the most economically isolated countries in the world, maintaining extremely high tariff barriers and limiting tourism. In the 1980s, however, the military government of Burma began to encourage both trade and tourism in order to improve its economy. Still, because of the repressive nature of the Burmese government, the United States, the European Union, and other states maintained sanctions on the country, restricting the extent of Burmese globalization.

Despite such opposition, Burma was not isolated from the rest of the world prior to opening up in 2011. Burma was already an ASEAN member, it maintained close diplomatic ties with Russia, and it developed significant economic links with India. But Burma's strongest ties by far have been with China, in regard to both economic and political issues. To a significant extent, China served as Burma's main partner during its long period of semi-isolation.

Chinese military ties with Burma strengthened after 1989. Such ties help China to project its own power on the global stage. Chinese investments have created a major deep-water port in the city of Kyaukpyu. The Kyaukpyu development is part of China's "String of Pearls" strategy, designed to give it naval access to the Indian Ocean region. China has also reportedly constructed a high-tech military surveillance facility on Burma's Great Coco Island, designed to spy on India's military activities, although both Burma and China deny the accusation.

Economic ties between Burma and China also heightened during the same period.

China has invested heavily in Burmese infrastructure, building roads, bridges, dams, and industrial facilities in many parts of the country. The Burmese oil and natural gas industries have been a particular focus of interest. The most important Chinese-financed project in Burma is a massive pipeline system, which was begun in 2009 (Figure 13.2.1). By importing fuel directly across Burma, China will be able to reduce its vulnerability to disruptions in the maritime trade.

As the connections between China and Burma deepened, some Burmese officials began to fear that their country was becoming too dependent on China. Anger was mounting among the Burmese people over dams and other environmentally damaging projects. Another cause of concern was the close relations that China maintained with several ethnic militias along the border between the two countries. In 2009, Burma angered China by sending its military in to reclaim Kokang, a Chinese-speaking "special region" in far northern Burma. Another insult followed when a government-run Burmese newspaper ran an article on the Dalai Lama of Tibet. China was again infuriated in 2011 when Burma canceled the Chinese-financed Myitsone Dam project.

Despite these instances, Burma and China will probably continue to maintain close economic and political relations. But in this age of globalization, it can be dangerous for a country to count too much on ties with a single other state, even one as powerful as China. As a result, the leaders of Burma have opted

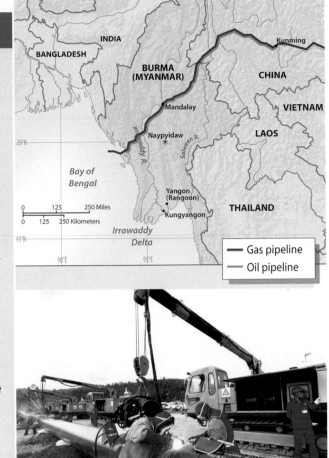

FIGURE 13.2.1 Chinese-Constructed Oil and Gas Pipelines in Burma   China has been investing massively in the infrastructure of Burma, Laos, Cambodia, and other Southeast Asian countries. New pipelines will soon extend across Burma into south-central China, as shown in this map. Chinese investments in oil and gas production in Burma are also prominent, as can be seen in this photo of a pipeline under construction.

to pursue a more balanced strategy, reducing dependence on China and seeking closer economic and diplomatic linkages with other countries.

Globalization and the Southeast Asian Economy

Southeast Asia as a whole has undergone rapid integration into the global economy (Figure 13.37). Singapore has thoroughly staked its future on the success of multinational capitalism, as have several other countries. Even communist Laos and once-isolationist Burma have opened their doors to the global system—although, in the case of Burma, with much hesitation.

Regardless of what happens in coming years, global economic integration has already brought significant development to Singapore, Malaysia, Thailand, and even Indonesia. Outside of Singapore and Malaysia, however, successful development has generally been based on labor-intensive manufacturing, in which workers are paid low wages and subjected to harsh discipline. Movements have thus begun in Europe, the United States, and elsewhere to pressure both multinational corporations and Southeast Asian governments to improve the working conditions of laborers in the export industries.

China figures prominently in the recent globalization of Southeast Asian economies. China has invested heavily in infrastructural projects, most notably in Laos, Burma, and Cambodia. Dams, highways, railroads, and ports in the region have all received large

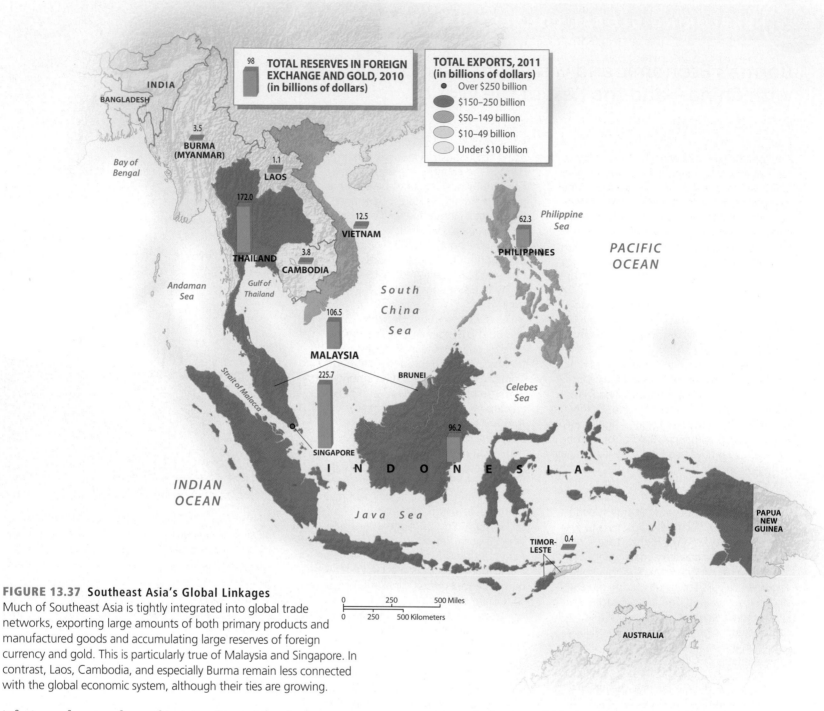

FIGURE 13.37 Southeast Asia's Global Linkages
Much of Southeast Asia is tightly integrated into global trade networks, exporting large amounts of both primary products and manufactured goods and accumulating large reserves of foreign currency and gold. This is particularly true of Malaysia and Singapore. In contrast, Laos, Cambodia, and especially Burma remain less connected with the global economic system, although their ties are growing.

infusions of money from China. Southeast Asian leaders are generally pleased with Chinese investments, but local residents are often angered over environmental degradation and the loss of land. Political strings are also sometimes tied to these projects. In 2010, Cambodia agreed to deport 20 ethnic Uighur asylum-seekers back to China; two days later, the Beijing government agreed to release $1.2 billion for infrastructure improvements in the country.

Issues of Social Development

As might be expected, several key indicators of social development in Southeast Asia are closely linked to levels of economic development. Singapore thus ranks among the world leaders in regard to health and education, as does Brunei. Laos and Cambodia, not surprisingly, come

out near the bottom of the chart. The people of Vietnam, however, are healthier and better educated than might be expected on the basis of their country's overall economic performance.

Life Expectancy and Education With the exceptions of Laos, Cambodia, East Timor, and Burma, Southeast Asia has achieved relatively high levels of social welfare. In Laos and Cambodia, life expectancy at birth hovers around 62 years (as compared to Thailand's 74 years), and the female literacy rate in Laos remains below 65 percent. But even the poorest countries of the region have made some improvements.

Most of the governments of Southeast Asia have placed a high priority on basic education. Literacy rates are relatively high in most countries of the region, even in the impoverished country of Burma. Much less success, however, has been realized in university and technical education. As Southeast Asian economies continue to grow, this educational gap is beginning to have negative consequences, forcing many students to study abroad for higher degrees. If Southeast Asian countries other than Singapore are to become fully developed, they will probably have to invest more money in their own human resources.

Gender Equity and the Sex Trade   In many ways, Southeast Asia exhibits contradictory tendencies in regard to the social position of women. Historically speaking, Southeast Asia has long been noted for its gender equity. Women in the region have played important economic roles as household managers and market venders. Early European visitors to the region were often shocked at how freely men and woman mixed and at how much authority women exercised. Such patterns have not completely disappeared. Southeast Asia has had a significant number of female leaders, including two recent presidents of the Philippines as well as Burma's opposition leader, Aung San Suu Kyi. Some anthropologists have gone so far as to describe the Minangkabau people of western Sumatra as a "modern matriarchy," since Minangkabau women have traditionally controlled their large households, which are based on descent from female ancestors.

Despite these positive tendencies, Southeast Asia is also the site of some of the world's most extensive sexual exploitation. Commercial sex is a huge business in Thailand. Despite its massive scope, prostitution is technically illegal in Thailand, which means that it is a major source of corruption. Other Southeast Asian countries, particularly the Philippines, Vietnam, and Cambodia, are also centers of a globally oriented commercial sex trade. Many workers in Southeast Asian brothels are underage, many have been coerced into the activity, and quite a few are held as virtual slaves. Young women, girls, and boys are frequently trafficked from the poorer parts of the region, often in connection with the drug trade. Sexually transmitted diseases, including HIV-AIDS, are associated with this activity, although the Thai government has engaged in a relatively successful public health campaign focused on condom use.

Two of the main centers of commercial sex in Southeast Asia developed around U.S. military bases during the Cold War: Pattaya in Thailand and Angeles City in the Philippines (Figure 13.38). Both cities have lost their military bases, but have expanded their economies by focusing on tourism, much of it sex-related. Pattaya now supports an estimated 20,000 sex workers. Here the high end of the business includes thousands of Russian and Ukrainian women. As a result, Pattaya now has a major Russian presence, attracting hundreds of thousands of Russian tourists every year in addition to wealthy Russian investors. Russian organized crime now plays a major role in the city, illustrating one of the seamier aspects of globalization in modern Southeast Asia.

FIGURE 13.38 Commercial Sex in Pattaya   The beach city of Pattaya was once a sleepy fishing village near an American military base. It is now a major center of the global sex trade, which employs an estimated 200,000 people. Russian interests figure prominently in the city's business community.

REVIEW QUESTIONS

1. Why have some Southeast Asian countries experienced sustained economic growth and social development, whereas other have more generally experienced stagnation in the same period?

2. Why have the major Southeast Asian economies experienced such shared booms and busts over the past several decades?

Aung San Suu Kyi is the internationally famous leader of Burma's National League for Democracy For most of the period between 1989 and 2010 she was held under house arrest by Burma's military leaders. After being released as Burma began its transition to a more democratic system, Suu Kyi was elected to the Burmese parliament. She received the Nobel Peace Prize in 1991."

Summary

- In many ways, Southeast Asia presents the prime example of diversity amid globalization. Globalization, as is true elsewhere, has created both challenges and opportunities. Some of the most serious problems that it has generated are environmental; for example, in most of the region, forests are now seriously depleted. In several Southeast Asian countries, however, major conservation efforts are now under way.

- Deforestation in Southeast Asia is also linked to domestic population growth and changes in settlement patterns. As people move from densely populated, fertile lowland areas into remote uplands, both environmental damage and cultural conflicts often follow. Population movements in Southeast Asia also have a global dimension. This is particularly true in regard to the Philippines, which has sent millions of workers to more prosperous parts of the world.

- Southeast Asia, unlike many other world regions, has never had a single, major cultural influence and is characterized today by tremendous cultural diversity. Some experts have argued, however, that globalization has helped Southeast Asia find a new sense of regional identity, as expressed through ASEAN. As a result, the historical and cultural unity that Southeast Asia has lacked may be forced on it by 21st-century globalization.

- The relative success of ASEAN has by no means solved all of Southeast Asia's political tensions. Many of its countries still argue about geographical, political, and economic issues, while ethnic and religious conflicts have generated major problems in Indonesia, the Philippines, and Thailand. Several of the region's countries, such as Cambodia, Laos, and especially Burma, have also been held back by repressive and corrupt governments. The current trend, however, is for more open governments and more global engagement.

- Although ASEAN has played an economic as well as a political role, its economic successes have been limited. Most of the region's trade is still directed outward toward the traditional centers of the global economy—North America, Europe, and East Asia. As is true in many parts of the world, China is becoming as key trading partner. This orientation is not surprising, considering the export-focused policies of most Southeast Asian countries. A significant question for Southeast Asia's future is whether the region will develop an integrated regional economy. A more important issue is whether social and economic development will be able to lift the entire region out of poverty instead of benefiting just the more fortunate areas.

Key Terms

animism 443
Association of Southeast Asian Nations
 (ASEAN) 426
Bumiputra 452
crony capitalism 451

domino theory 447
entrepôt 452
Golden Triangle 436
Khmer Rouge 447
lingua franca 443

primate city 439
swidden 436
transmigration 438
tsunami 432
typhoon 434

Thinking Geographically

1. What might be the fate of animism in the new millennium? Consider whether it is doomed to extinction before the forces of modern economics and national integration or whether it may persist as tribal peoples struggle to retain their cultural identities.

2. What should be the position of the English language in the education systems of Southeast Asia? What should be the position of each country's national language? What about local languages?

3. How might ethnic tensions and human rights abuses in Burma be reduced? What role should the international community play?

4. Can Singapore continue to experience economic growth and technological development while limiting freedom of expression? What role might the Internet play as Singapore struggles with these issues?

5. Is the Southeast Asian economic path of integration into the global economy, which is marked by an openness to multinational corporations and foreign investment, going to prove wise in the long run, or do its potential hazards outweigh its benefits?

Mastering Geography™

Looking for additional review and test prep materials? Visit the Study Area in MasteringGeography™ to enhance your geographic literacy, spatial reasoning skills, and understanding of this chapter's content by accessing a variety of resources, including MapMaster interactive maps, videos, RSS feeds, flashcards, web links, self-study quizzes, and an eText version of *Globalization and Diversity*.

Scan to visit the author's blog for chapter updates.

http://gad4blog.
wordpress.com/
category/southeast-
asia/

Authors' Blogs

Scan now to access the authors' blogs for up-to-date information on Southeast Asia.

Scan to Access Martin Lewis's blog.

http://geocurrents.
info/category/
place/southeast-
asia

Globalization and Diversity

Australia and Oceania were once isolated from the rest of the world; today, however, globalization has brought the region's various countries into the global community through a complex array of environmental, cultural, geopolitical, and economic arrangements. Traditionally oriented toward Europe because of shared colonial histories, the region is becoming increasingly linked to Asia.

ENVIRONMENTAL GEOGRAPHY

Extending across the Pacific Ocean from Hawaii to Australia and poleward to New Zealand's midlatitudes, this region is characterized by vastly different landscapes, from tropical islands to glacial mountains. Environmental issues are diverse as well, with drought due to global warming forecast for Australia, while flooding from sea-level rise threatens low-lying islands.

POPULATION AND SETTLEMENT

Australia and New Zealand, because of their higher living standards, continue to be targets for immigration from both within and outside of Oceania. Whereas earlier migrant streams to the region came primarily from Europe, today Asia is the major source for immigration.

CULTURAL COHERENCE AND DIVERSITY

Until 1973, Australia protected its European ethnic roots with the White Australia Policy, but more recently a new multicultural society has been created with immigration from Asia and Africa. Similar migration geographies are also affecting Oceania.

GEOPOLITICAL FRAMEWORK

From Hawaii to Australia, native peoples are demanding ownership of or, minimally, access to their ancestral lands. More often than not, however, these land claims are fraught with controversy and tension.

ECONOMIC AND SOCIAL DEVELOPMENT

Increased trade linkages with resource-hungry China have brought economic benefits to many countries in Oceania, particularly Australia, despite the geopolitical liabilities. In many Pacific islands, including Hawaii, tourism-based economies are slowly recovering from downturns suffered during the recent global recession.

➤ **This idyllic sight of tourists sunbathing in Kakadu National Park** masks the many tensions permeating this UNESCO World Heritage site. Because Aboriginals own most of the park's land (and claim the rest), there are considerable differences of opinion on matters of protection and development.

460

AUSTRALIA AND OCEANIA 14

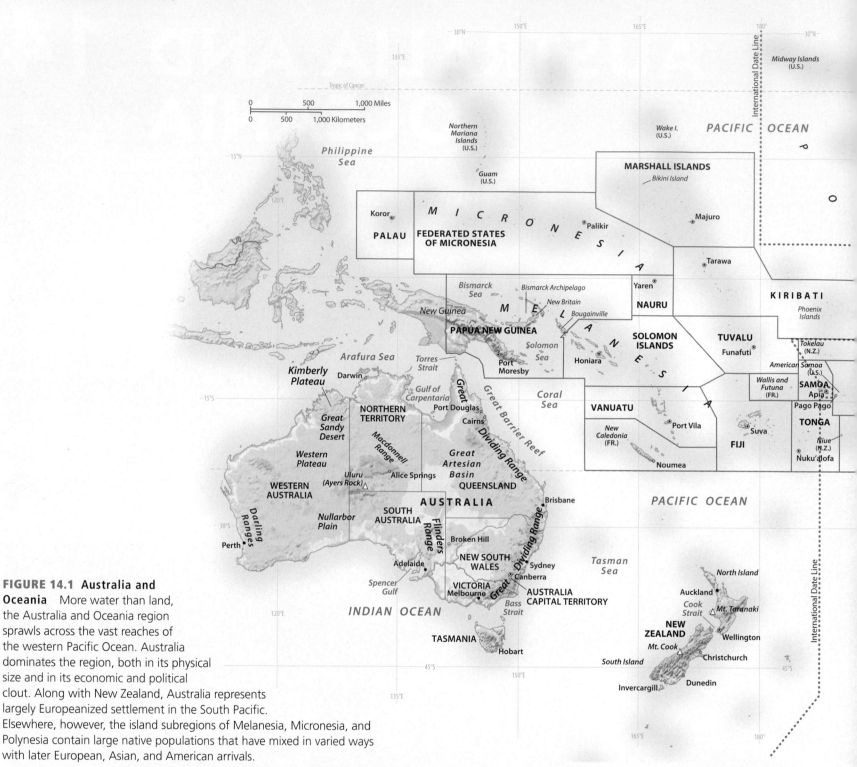

FIGURE 14.1 Australia and Oceania More water than land, the Australia and Oceania region sprawls across the vast reaches of the western Pacific Ocean. Australia dominates the region, both in its physical size and in its economic and political clout. Along with New Zealand, Australia represents largely Europeanized settlement in the South Pacific. Elsewhere, however, the island subregions of Melanesia, Micronesia, and Polynesia contain large native populations that have mixed in varied ways with later European, Asian, and American arrivals.

This vast world region includes the island continent of Australia, as well as **Oceania**, a collection of islands that reaches from New Guinea and New Zealand to the U.S. state of Hawaii in the mid-Pacific (Figure 14.1). Although native peoples settled the area long ago, more recent European and North American colonization began the process of globalization that is now producing new and sometimes unsettled environmental, cultural, and political geographies.

The microstate of Nauru, a tiny country of 14,000 people clustered on a small island of 8 square miles (21 square kilometers), offers insight into the mixed consequences of globalization as it has traveled a bumpy road from self-sufficient fishing society to bankrupt ward of the world community.

Nauru is one of several small islands in the Pacific where, over centuries, seabird droppings collected to form rich deposits of phosphate, a coveted mineral with many uses, including agricultural fertilizer and gunpowder. Although Nauru first allowed foreign mining companies access to its phosphate deposits as far back as the 1920s, it was only in the 1960s, when Nauru became an independent country, that it granted full mining rights to an Australian company. The result

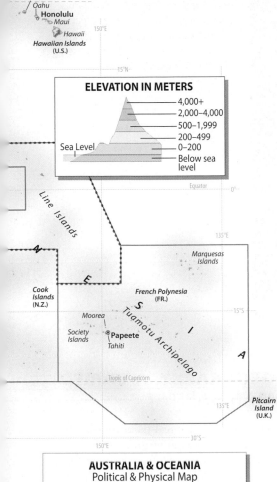

AUSTRALIA & OCEANIA
Political & Physical Map

⊛ ● Metropolitan areas more than 20 million

⊛ ● Metropolitan areas 10–20 million

⊛ ● Metropolitan areas 5–9.9 million

⊛ • Metropolitan areas 1–4.9 million

⊛ ○ Selected smaller metropolitan areas

⌐⌐ Plate boundaries

was an economic boom for the small island. As a result of the mining royalties, Nauru became known as the Kuwait of the Pacific, with its small population enjoying one of the world's highest per capita incomes.

With cash in their pockets, Naurans spent lavishly, building mansions, buying luxury cars, and traveling widely to faraway destinations for golf holidays. But with these new riches, Naurans also turned away from their traditional lifestyle. Many Naurans substituted imported food (reportedly mainly junk food) for local fish, shellfish, fruits, and vegetables. Unfortunately, this change of diet took an immediate toll. In 1975, half of the adult population was diagnosed with diabetes, mainly a result of obesity. Still today, Nauru has one of the highest rates of diabetes anywhere in the world, with its population suffering high rates of blindness and limb amputation.

Furthermore, the phosphate resources are now exhausted (Figure 14.2). Although Nauru's government had planned for a day when the mining royalties would end, its strategies for keeping the cash flowing through global connections have gone awry. Bad investments in Southeast Asian real estate and failed London musicals stand out, sapping the gains made from successful ventures in Oregon subdivisions and high-end Washington, DC condominiums. To compensate for these losses, Nauru developed an offshore Internet banking facility. Although this activity appeared to be successful, the international financial community shut it down when it was discovered that the major activity was laundering money for the Russian Mafia. Nauru then tried selling citizenship in its small nation, complete with Nauruan passports for those who bought into the scheme. Not surprisingly, global anti-terrorist organizations took a very dim view of this activity, and once again the international community applied pressure to stop operations.

Most recently, Nauru tried to capitalize on global mobility by building a detention center (a prison, basically) for people who entered Australia illegally and asked for political asylum. Although this plan was supported by the former conservative Australian Prime Minister, the current Australian government thought this a very bad idea and ended support for the Nauru detention center.

Today Nauru's future is grim—a poster child, some would say, for the complexities and mixed benefits of globalization.

LEARNING OBJECTIVES

After reading this chapter you should be able to:

- Describe the geographic characteristics of the region known as Oceania.
- Identify the major environmental issues problematic to Australia and Oceania, as well as the pathways toward solving those problems.
- Explain how the Pacific Rim of Fire is linked to the landforms of Oceania.
- Describe the array and location of climate types found in Australia and Oceania.
- Summarize the prehistoric peopling of the Pacific, as well as the colonial exploration and settlement of Australia and Oceania.
- Explain the changing migration patterns to and within postwar Australia and Oceania.

- Describe the historical and modern interactions between native peoples and Anglo-European migrants in Australia and Oceania.
- Identify and describe the different pathways to independence taken by countries in Oceania.
- List several geopolitical tensions that persist in Australia and Oceania.
- Describe the diverse economic geographies of Oceania.
- Explain the positive and negative interactions of Australia and Oceania with the global economy.

FIGURE 14.2 The Globalization of Nauru A Nauruan local points to the scarred landscape that remains after decades of phosphate mining by Australian companies. With no mineral riches left the islanders have tried several different schemes for finding their place in the global economy.

OVERVIEW OF THE REGION

The vast distances of the Pacific stretching from New Guinea to Hawaii help define the boundaries of this region, but many of the national boundaries were born from political convenience during an earlier period of colonial globalization. Australia (or "southern land"), which is often thought of as a continent with its population of 22 million people, forms a coherent political unit and subregion. To the east, New Zealand is a three-hour flight from Australia, has a much smaller population of 4.4 million people, and is linked to Australia by shared historical ties to Britain. However, New Zealand is considered part of **Polynesia** ("many islands") because of its native Maori people.

Hawaii, 4,400 miles (7,100 kilometers) northeast of New Zealand, shares the same Polynesian heritage as New Zealand. Hawaii is thought of as the northeastern boundary of Oceania, while the southeastern boundary of the region is usually delimited by the Polynesian islands of Tahiti, 3,000 miles (4,400 kilometers) to the southeast (Figure 14.3).

Four thousand miles (6,400 kilometers) west of French Polynesia, well across the International Date Line, lies the island of New Guinea, the accepted, yet sometimes confusing boundary between Oceania and Asia. Today an arbitrary boundary line bisects the island, dividing Papua New Guinea (the eastern half, which is usually considered part of Oceania) from neighboring Papua and West Papua (the western half, which, as part of Indonesia, is usually thought of as part of Southeast Asia). This western part of Oceania is sometimes called **Melanesia** (meaning "dark islands") because early explorers considered local peoples to be darker-skinned than those in Polynesia.

Finally, the more culturally diverse region of **Micronesia** (meaning "small islands") lies north of Melanesia and west of Polynesia. It includes microstates such as Nauru and the Marshall Islands, as well as the U.S. territory of Guam.

ENVIRONMENTAL GEOGRAPHY: A VARIED NATURAL AND HUMAN HABITAT

Behind the postcard images of Oceania as a tropical paradise of sandy beaches and verdant island forests lie serious environmental issues, caused by exploitation of the region's unique and fragile ecosystems (Figure 14.4; also see "Working Toward Sustainability: Easter Island on the Mind").

Global Resource Pressures

Globalization has exacted an environmental toll on Australia and Oceania. Specifically, the region's considerable base of natural resources has been opened to development, much of it by outside interests. While gaining from the benefits of global investment, the region has also paid a considerable price for encouraging development, and the result is an increasingly threatened environment.

Major mining operations have greatly affected Australia, Papua New Guinea, New Caledonia, and Nauru. Some of Australia's largest gold, silver, copper, and lead mines are located in sparsely settled portions of Queensland and New South Wales, polluting watersheds in these semiarid regions. In Western Australia, huge open-pit iron mines dot the landscape, unearthing ore that is bound for global markets, particularly China and Japan. To the northeast, gold mining is transforming the Solomon Islands, while an even larger gold mining venture has raised environmental concerns on the island of New Guinea (Figure 14.5). Elsewhere, as noted earlier, Micronesia's tiny Nauru has been virtually turned inside out as much of the island's jungle cover has been removed to get at some of the world's richest phosphate deposits.

Deforestation is another major environmental threat across the region. Vast stretches of Australia's eucalyptus woodlands, for example, have been destroyed to create better pastures. In addition, coastal rain forests in Queensland cover only a fraction of their original area, although a growing environmental movement in the region is fighting to save the remaining forest tracts. Tasmania has also been an environmental battleground, particularly given the biodiversity of its midlatitude forest landscapes. While the island's earlier European and Australian development featured many logging and pulp mill operations, more than 20 percent of the island (Figure 14.6) is now protected by national parks.

Many islands in Oceania are also threatened by deforestation. With limited land areas, islands are subject to rapid tree loss, which, in turn, often leads to soil erosion. Although rain forests still cover 70 percent of Papua New Guinea, more than 37 million acres (15 million hectares) have been identified as suitable for logging. Some of the world's most biologically diverse environments are being threatened in these operations, but landowners see the quick cash sales to loggers as attractive, even though this nonsustainable practice is contrary to their traditional lifestyles.

FIGURE 14.3 Tahiti, Outlier of Polynesia   The islands of Tahiti were first settled between 300 and 800 CE and form the southeastern corner of Polynesia some 3000 miles from New Zealand in the west and roughly the same distance from Hawaii to the north. This photo is of Cook's Bay on the island of Moorea.

Climate Change in Oceania

Even though Oceania contributes relatively little atmospheric pollution to the global atmosphere, the harbingers of climate change are already widespread and problematic. New Zealand mountain glaciers are melting, while Australia suffers from frequent droughts and devastating wildfires. Warmer ocean waters have caused widespread bleaching of the Great Barrier Reef off Australia's coast as microorganisms die, and rising sea levels are flooding several low-lying island nations, forcing residents to migrate to higher land (Figure 14.7). United Nations projections for the future are also highly disturbing: Stronger tropical cyclones could devastate Pacific islands, with widespread damage to land and life, and island inhabitants will suffer from reduced coastal resources as ocean waters warm. Increased wildfires will threaten population centers in southeast Australia, while agriculture will suffer from droughts and severe water shortages.

In response to these threats, the actions taken and policies implemented by Oceania's countries vary considerably. Until a 2007 change of government, Australia was the only industrial country besides the United States to not ratify the Kyoto Protocol. Perhaps the fact that Australia is the world's largest exporter of coal influenced that decision; not to be overlooked is that most of that country's emissions are produced by coal-fired power plants. While the former government rejected the Kyoto Protocol and downplayed the threat of global warming, the current government seems committed to a 5 percent reduction in greenhouse gas (GHG) emissions by 2020. In November 2011, the government approved an emission cap-and-trade program that will be fully operational by 2015.

Strong opposition to this plan, however, continues from three powerful components of the Australia economy: the coal industry, aluminum manufacturers (which use coal-based power plants to generate the energy needed to refine aluminum), and agriculture.

When New Zealand ratified the Kyoto agreement, it committed itself to a 5 percent reduction in GHG emissions over its 1990 baseline; however, a booming economy resulted in a 50 percent increase of emissions over the past decades. In response to this increase, the government is now proposing a 10 to 20 percent reduction in GHG emissions to be achieved by 2020. A major component of New Zealand's GHG pollution is methane emission from the country's large livestock population. These emissions, in fact, account for over half of the global warming pollution due to methane, all coming from that small country. As a result, New Zealand is discussing a much-publicized "flatulence tax," which would be levied on livestock operations (Figure 14.8). Livestock specialists are also experimenting with grass and grain fodder mixtures for sheep that might reduce their emissions.

Many Pacific nations—most notably, Tuvalu, Kiribati, and the Marshall Islands—maintain they are already experiencing problems from sea-level rise, coral bleaching, and degraded fishery resources because of warming ocean waters. These nations have banded together into a strident political union lobbying for a global solution to climate change. At the December 2011 Durban UN global warming conference, these small island nations were outspoken in their demands that developed nations such as the United States, Japan, and the countries of western Europe provide the island nations with financial aid to mitigate damage from global warming.

In a similar vein, Hawaii was one of the first U.S. states to formally address global warming, starting with an analysis of the sources of its

FIGURE 14.4 Environmental Issues in Australia and Oceania Modern environmental problems belie the myth that the region is an earthly paradise. Tropical deforestation, extensive mining, and a long record of nuclear testing by colonial powers have brought severe challenges to the region. Human settlements have also extensively modified the pattern of natural vegetation. Future environmental threats loom for low-lying Pacific islands as sea levels rise from global warming.

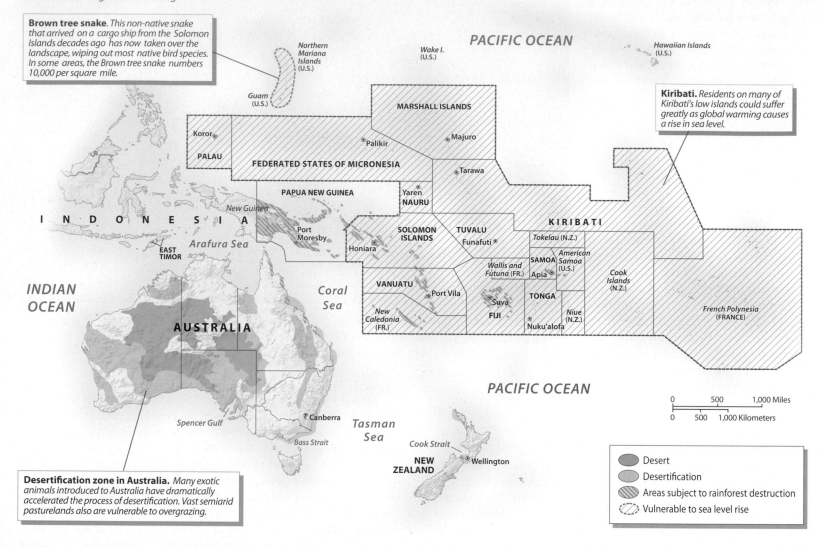

Brown tree snake. *This non-native snake that arrived on a cargo ship from the Solomon Islands decades ago has now taken over the landscape, wiping out most native bird species. In some areas, the Brown tree snake numbers 10,000 per square mile.*

Kiribati. *Residents on many of Kiribati's low islands could suffer greatly as global warming causes a rise in sea level.*

Desertification zone in Australia. *Many exotic animals introduced to Australia have dramatically accelerated the process of desertification. Vast semiarid pasturelands also are vulnerable to overgrazing.*

Legend:
- Desert
- Desertification
- Areas subject to rainforest destruction
- Vulnerable to sea level rise

FIGURE 14.5 Mining in Papua New Guinea Open-pit mining for gold, silver, copper, and lead marks the landscapes of Papua New Guinea, New Caledonia, and Nauru. Although bringing some economic benefit to local peoples, these activities also cause immense environmental damage to the region. In Papua New Guinea, for example, sediments from upland mines have severely damaged the Fly River ecosystem.

GHG emissions and then setting a target for reduction of those emissions. Since most of Hawaii's emissions result from oil- and coal-fired powered plants, the state, like Australia and other countries, is actively promoting sustainable energy generation through wind, tidal, and solar power (Figure 14.9).

Australian and New Zealand Environments

Curiously, Australia is one of the world's most urbanized societies, yet most people associate the country with its vast and arid **Outback**, a sparsely settled land of sweeping distances, scrubby vegetation, (Figure 14.10). In contrast, the two small islands that make up New Zealand are known for their humid landscapes of rolling foothills and rugged mountains.

Landform Regions Three major landform regions dominate Australia's physical geography (see Figure 14.2).

WORKING TOWARD SUSTAINABILITY:

Easter Island on the Mind

Pacific islands have long occupied a prominent place in conversations about ecology and sustainability because they are considered microcosms for our larger world. The notion of Earth Island, for example, is based on inferences of limited natural resources and fragile ecosystems, which must be carefully stewarded in a sustainable manner by its human inhabitants. Not that this imagery is unfounded; on the contrary, recent history is full of examples of how an island's ecological balance was demolished by the introduction of a plant or animal species that took over the island at the expense of native species. The brown gum snake has done this in Guam, and historically the rabbit flourished in Australia.

Energizing conversations about island sustainability is the specter of Easter Island, one of the remotest places on Earth. It is almost 2,000 miles (3219 kilometers) from the nearest populated area, but a prehistoric civilization may have vanished here because of an ecological disaster of its own making (Figure 14.1.1). The icons of Easter Island are the huge megalithic statues, hundreds of them, resembling haughty men gazing down on the island, keeping to themselves the secret of what went wrong.

Seafaring Polynesians came to Easter Island (which is also known as Rapa Nui) around 400 CE, forging a civilization that lasted until around 1600 CE. One mystery from those times is why and how they created those huge stone statues, but it is joined by the equally baffling riddle of what caused the society's downfall and destruction.

In his book *Collapse: How Societies Choose to Fall or Succeed,* scientist Jared Diamond posits that the island's civilization experienced a rapid decline because of an ecological imbalance brought about by deforestation, soil erosion, and overpopulation. Diamond also argues that these deteriorating conditions led to a civil war, during which warring factions toppled some of the stone statues. In short, Diamond maintains Easter Island society collapsed by committing "ecocide," the consequence of letting a formerly sustainable resource system deteriorate and run wild.

FIGURE 14.1.1 Monoliths of Easter Island Called "Moai" in Polynesian, these stone statues silently watch over Easter Island. Most were created between the period 1250–1500 CE, and may or may not be connected to the ecological disaster that the islands suffered.

Other scientists and prehistorians have added fuel to Diamond's argument by pointing to the possibility of a population explosion by an invasive species. An animal such as a rat could have easily stowed away on a Polynesian boat and then exploded in population to the point where it overran island food resources.

Regardless of the causes, the mystery of Easter Island, with its silent stone statues and vanished civilization, forms a cautionary tale that continues to influence conversations about Pacific island sustainability.

The Western Plateau occupies more than half of the continent. Most of the region is a vast, irregular plateau that averages only 1,000 to 1,800 feet (300 to 550 meters) in height. Farther east, the Interior Lowland Basins stretch north to south for more than 1,000 miles (1,600 kilometers) from the swampy coastlands of the Gulf of Carpentaria to the Murray and Darling valleys, Australia's largest river system. Finally, more forested and mountainous country exists near Australia's east coast. The Great Dividing Range extends over 2,300 miles (3,700 kilometers) from the Cape York Peninsula in northern Queensland to southern Victoria. Nearby, off the eastern coast of Queensland, the Great Barrier Reef offers a final dramatic subsurface feature: Over the past 10,000 years, one of the world's most spectacular examples of coral reef-building has produced a living legacy now protected by the Great Barrier Reef Marine Park (Figure 14.11).

Part of the Pacific Rim of Fire, New Zealand owes its geologic origins to volcanic mountain-building that produced two rugged and spectacular islands in the South Pacific. The North Island's active volcanic peaks, reaching heights of more than 9,100 feet (2,800 meters), and geothermal features reveal the country's fiery origins (Figure 14.12).

FIGURE 14.6 Grampian National Park This park, west of Melbourne, Australia, was listed on the country's national heritage list in 2006 because of a combination of natural beauty and rich indigenous rock art sites.

FIGURE 14.7 Sea-Level Rise and Pacific Islands   The world's oceans are forecast to rise at least 3.3 feet (1 meter) by the end of the century, flooding low-lying islands not just in Oceania, but also throughout the globe. This photo is of Funafuti Atoll, home to half of Tuvalu's population of just over 12,000 people. With its highest point 16 feet (5 meters) above sea level, Tuvalu is already flooded regularly by storm surf and seasonal high tides, events that will occur more often in the future because of global warming. Not surprisingly, Tuvalu is one of the most outspoken countries on the topic of international measures to alleviate global warming.

Even higher and more rugged mountains comprise the western spine of the South Island. Mantled by high mountain glaciers and surrounded by steeply sloping valleys, the Southern Alps are some of the world's most visually spectacular mountains, complete with narrow, fjord-like valleys that indent much of the South Island's isolated western coast.

FIGURE 14.8 New Zealand's Greenhouse Gas Problem   Worldwide, cattle and sheep emit the greenhouse gas methane, but only in New Zealand are these livestock emissions greater than those produced by human activities. While New Zealand has discussed a "flatulence tax" on sheep farms, the country's scientists are also working on an anti-flatulence inoculation that would reduce sheep emissions.

FIGURE 14.9 Wind Power in Australia   Currently, Australia produces most of its energy from coal-fired power plants, which is not surprising given the wealth of its coal resources. However, like many other countries attempting to cut down on their carbon emissions, Australia is building an increasing number of wind farms, which today supply about 2 percent of its energy needs. This array of turbines is on the coast of Western Australia.

Climate   Generally, zones of somewhat higher precipitation encircle Australia's arid center (Figure 14.13). In the tropical low-latitude north, seasonal changes are dramatic and unpredictable. For example, Darwin (see the climograph in Figure 14.13) can experience drenching monsoonal rains in the summer (December to March), followed by bone-dry winters (June to September). Indeed, life across the region is shaped by this annual rhythm of what is called locally "the

FIGURE 14.10 The Australian Outback   Arid and generally treeless, the vast lands of the Australian Outback resemble some of the dry landscapes of the U.S. West. In this photo, wildflowers blossom along a dirt road near Tom Price, in the Pilbara region of Western Australia.

FIGURE 14.11 The Great Barrier Reef Stretching along the eastern Queensland coast, the famed Great Barrier Reef is one of the world's most spectacular examples of coral reef-building. Threatened by varied forms of coastal pollution, much of the reef is now protected in a national marine park.

FIGURE 14.12 Mt. Taranaki New Zealand's North Island contains several volcanic peaks, including Mt. Taranaki. The 8,000-foot (2,400-meter) peak offers everything from subtropical forests to challenging ski slopes and attracts both local and international tourists.

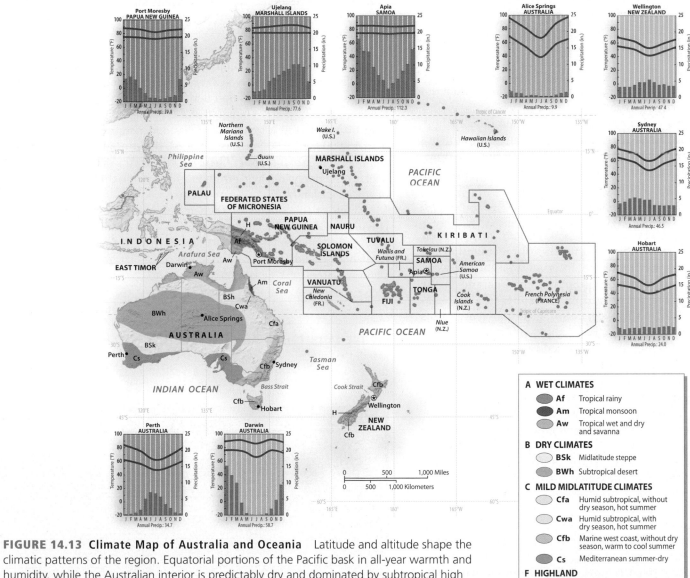

FIGURE 14.13 Climate Map of Australia and Oceania Latitude and altitude shape the climatic patterns of the region. Equatorial portions of the Pacific bask in all-year warmth and humidity, while the Australian interior is predictably dry and dominated by subtropical high pressure. Cool and moisture-bearing storms of the southern Pacific Ocean provide midlatitude conditions across New Zealand and portions of Australia. More locally, mountain ranges dramatically raise precipitation totals in many highland zones.

A WET CLIMATES
- **Af** Tropical rainy
- **Am** Tropical monsoon
- **Aw** Tropical wet and dry and savanna

B DRY CLIMATES
- **BSk** Midlatitude steppe
- **BWh** Subtropical desert

C MILD MIDLATITUDE CLIMATES
- **Cfa** Humid subtropical, without dry season, hot summer
- **Cwa** Humid subtropical, with dry season, hot summer
- **Cfb** Marine west coast, without dry season, warm to cool summer
- **Cs** Mediterranean summer-dry

F HIGHLAND
- **H** Complex mountain climates

FIGURE 14.14 Australian Wildfires Huge and savage dry-season wildfires (known as bushfires in Australia) threaten both rural settlements and sprawling city suburbs in the southeast. This fire in February 2009, just 70 miles from the heart of Melbourne, was the worst fire disaster in 25 years and may be a harbinger of even more damaging fires accompanying global warming.

wet" and "the dry." By the end of the dry season, wildfires often dot the landscape of northern Australia (Figure 14.14).

Along the east coast of Queensland, precipitation remains high (60 to 100 inches, or 150 to 250 centimeters), but it diminishes rapidly as you move into the interior. Rainfall at interior locations, such as the Northern Territory's Alice Springs, averages less than 10 inches (25 centimeters) annually. South of Brisbane, more mid-latitude influences dominate eastern Australia's climate. Coastal New South Wales, southeastern Victoria, and Tasmania experience the country's most dependable year-round rainfall, which averages 40 to 60 inches (100 to 150 centimeters) of precipitation per year. Nearby mountains see frequent winter snows. Farther west, summers are hot and dry in much of South Australia and in the southwest corner of Western Australia. These zones of Mediterranean climate produce the **mallee** vegetation, a scrubby eucalyptus woodland.

Climates in New Zealand are influenced by a combination of latitude, the moderating effects of the Pacific Ocean, and proximity to local mountains or mountain ranges. Most of the North Island is distinctly subtropical; the coastal lowlands near Auckland, for example, are mild and wet year-round. Still, local variations can be striking, since the area's volcanic peaks create their own microclimates. On the South Island, conditions become distinctly cooler as you move closer to the South Pole. Indeed, the island's southern edge feels the seasonal breath of Antarctic chill, as it lies more than 46° south of the equator. Mountain ranges on New Zealand's South Island also display incredible local variations in precipitation: West-facing slopes are drenched with more than 100 inches (250 centimeters) of precipitation annually, while lowlands to the east average only 25 inches (65 centimeters) per year. The Otago region, inland from Dunedin, sits partially in the rain shadow of the Southern Alps, and its rolling, open landscapes resemble the semiarid expanses of North America's West (Figure 14.15).

Environments of Oceania

The vast expanse of Pacific waters reveals another set of environmental settings that are as rich and complex as they are fragile. Oceanic currents and wind patterns have historically served as natural highways of movement between these island worlds. Those same forces define the rhythm of weather patterns across the region in a broad band that extends more than 20° north (Hawaiian Islands) and south (New Caledonia) of the equator.

Creating Island Landforms Much of Melanesia and Polynesia is part of the seismically active Pacific Basin. As a result, volcanic eruptions, major earthquakes, and **tsunamis**, or earthquake-induced sea waves, are not uncommon across the region, and they impose major environmental hazards on the population. For example, volcanic eruptions and earthquakes on the island of New Britain (Papua New Guinea) forced more than 100,000 people from their homes in 1994. Only four years later a massive tsunami triggered by an offshore earthquake swept across the north coast of New Guinea, killing 3,000 residents and destroying numerous villages. Such events are unfortunately a part of life in this geologically active part of the world.

Most of the islands of Polynesia and Micronesia are truly oceanic, having originated from volcanic activity on the ocean floor. The larger active and recently active volcanoes form **high islands**, which often rise to a considerable elevation and cover a large area. The island of Hawaii, the largest and youngest of the Pacific's high islands, is more than 80 miles (130 kilometers) across and rises to a height of more than 13,000 feet (4,000 meters). Indeed, the entire Hawaiian archipelago is a geological **hot spot**, where slowly moving oceanic crust passes over a vast supply of magma from Earth's interior, thus creating a chain of volcanic islands.

Many of the islands of French Polynesia, including Bora Bora (Figure 14.16), are smaller examples of high islands and are widely scattered throughout Micronesia and Polynesia. In tropical latitudes, most high islands are ringed by coral reefs that grow quickly in the shallow waters near the shore.

FIGURE 14.15 Central Otago, South Island On New Zealand's South Island, the Southern Alps capture rainfall on the west coast, but leave areas to the east in a drier rain shadow. As a result, the Central Otago region has a semiarid landscape resembling portions of the U.S. West.

FIGURE 14.16 Bora Bora   The jewel of French Polynesia, Bora Bora displays many of the classic features of Pacific high islands. As the island's central volcanic core retreats, surrounding coral reefs produce a mix of wave-washed sandy shores and shallow lagoons.

The combination of narrow sandy islands, barrier coral reefs, and shallow central lagoons is known as an **atoll**. The islands and reefs of an atoll characteristically form a circular or oval shape, although some are quite irregular (Figure 14.17). The world's largest atoll, Kwajalein in Micronesia's Marshall Islands, is 75 miles (120 kilometers) long and 15 miles (25 kilometers) wide. Polynesia and Micronesia are dotted with extensive atoll systems, as is Melanesia.

Island Climates   Many Pacific islands receive abundant precipitation, and high islands in particular are often noted for their heavy rainfall and dense tropical forests. In American Samoa, this environment has been protected in one of the nation's newest national parks. Much of the island zone is located in the rainy tropics or in a tropical wet-dry climate region where abundant summer rains and even tropical cyclones can bring heavy seasonal precipitation. In contrast to the high islands, low-lying atolls usually receive less precipitation than high islands and very often experience water shortages. During dry periods, the limited stores of water on these islands are quickly depleted.

REVIEW QUESTIONS

1. What was Nauru's major mineral resource, how was it formed, and what is it used for?

2. What is the major source of GHG emissions in New Zealand, and what steps are being taken to reduce these emissions?

3. Describe the different climate regions found in Australia and New Zealand. What climate controls produce those regions?

4. How are atolls and barrier reefs formed?

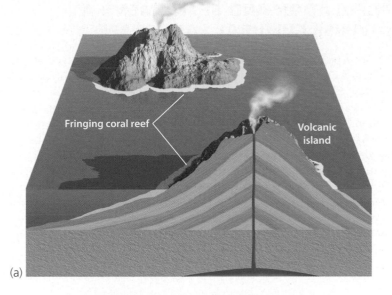

(a)

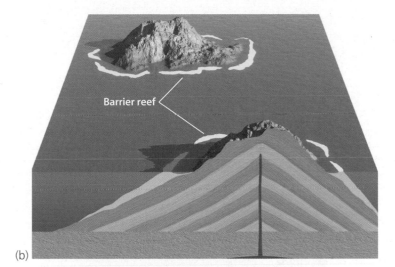

(b)

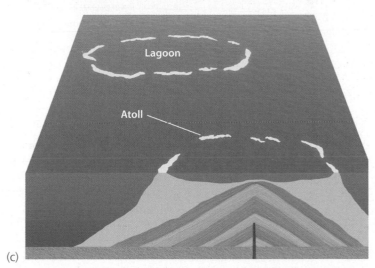

(c)

FIGURE 14.17 Evolution of an Atoll   Many Pacific islands begin as rugged volcanoes (a) with fringing coral reefs. However, as the extinct volcano subsides and erodes away, the coral reef expands, becoming a larger barrier reef (b). The term *barrier reef* comes from the hazards these features pose to navigation when approaching the island from the sea. Finally, all that remains (c) is a coral atoll surrounding a shallow lagoon.

POPULATION AND SETTLEMENT: A DIVERSE CULTURAL LANDSCAPE

Modern population patterns across the region reflect the combined influences of indigenous and European settlement. In countries such as New Zealand and Australia and in the Hawaiian Islands, Anglo-European migration has structured the distribution and concentration of contemporary populations. In contrast, on smaller islands elsewhere in Oceania, population geographies are determined by the needs of native peoples (Figure 14.18). More recently, however, migration patterns to both Australia and New Zealand have changed, with a greater number of migrants arriving from Asia. This pattern is coupled with an increase of intra-regional migration as people move about for complex reasons, including the push forces of unemployment, resource depletion, and even the threat of flooding associated with global warming.

Contemporary Population Patterns

Despite the stereotypes of life in the Outback, modern Australia has one of the most urbanized populations in the world (Table 14.1). About 90 percent of the country's residents live within either the Sydney or the Melbourne metropolitan area. Indeed, Australia's eastern and southern coasts are home to the majority of its 22.4 million people.

Inland, population densities decline as rapidly as the rainfall: Semiarid hills west of the Great Dividing Range still contain significant rural settlement, but the state's southwestern periphery remains sparsely populated. New South Wales is the country's most heavily populated state; its sprawling capital city of Sydney (over 4 million people), focused around one of the world's most magnificent natural harbors, is the largest metropolitan area in the entire South Pacific. In the nearby state of Victoria, Melbourne's 3.8 million residents have long competed with Sydney for status as Australia's premiere city, claiming cultural and architectural supremacy over their slightly

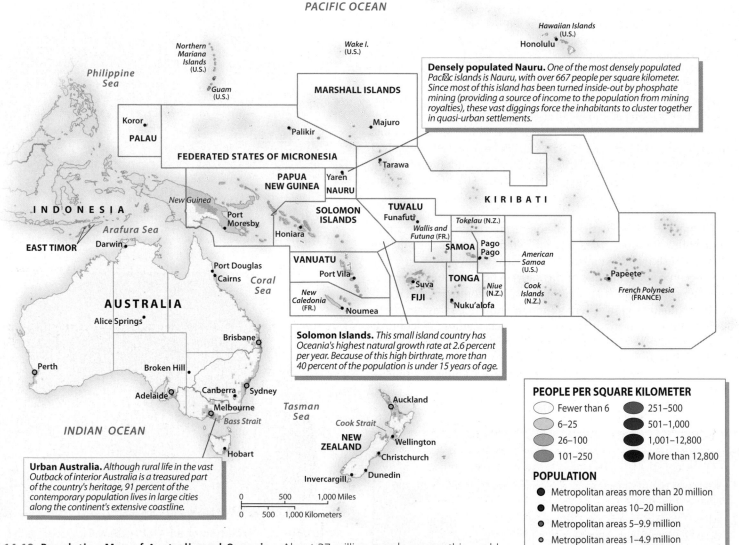

Densely populated Nauru. *One of the most densely populated Pacific islands is Nauru, with over 667 people per square kilometer. Since most of this island has been turned inside-out by phosphate mining (providing a source of income to the population from mining royalties), these vast diggings force the inhabitants to cluster together in quasi-urban settlements.*

Solomon Islands. *This small island country has Oceania's highest natural growth rate at 2.6 percent per year. Because of this high birthrate, more than 40 percent of the population is under 15 years of age.*

Urban Australia. *Although rural life in the vast Outback of interior Australia is a treasured part of the country's heritage, 91 percent of the contemporary population lives in large cities along the continent's extensive coastline.*

PEOPLE PER SQUARE KILOMETER

Fewer than 6	251–500
6–25	501–1,000
26–100	1,001–12,800
101–250	More than 12,800

POPULATION

- Metropolitan areas more than 20 million
- Metropolitan areas 10–20 million
- Metropolitan areas 5–9.9 million
- Metropolitan areas 1–4.9 million
- Selected smaller metropolitan areas

FIGURE 14.18 Population Map of Australia and Oceania   About 37 million people occupy this world region. While Papua New Guinea and many Pacific islands feature mainly rural settlements, most regional residents live in the large urban areas of Australia and New Zealand. Sydney and Melbourne account for almost half of Australia's population, and most New Zealand residents live on the North Island, home to the cities of Auckland and Wellington.

TABLE 14.1 POPULATION INDICATORS

Country	Population (millions) 2012	Population Density (per square kilometer)	Rate of Natural Increase (RNI)	Total Fertility Rate	Percent Urban	Percent <15	Percent >65	Net Migration (Rate per 1000) 2010–15[a]
Australia	22	3	0.7	1.9	82	19	14	6.5
Fed. States of Micronesia	0.1	152	1.9	3.5	22	31	6	−12.8
Fiji	0.8	46	1.4	2.7	51	29	5	−5.7
French Polynesia	0.3	69	1.1	2.1	51	25	6	−0.4
Guam	0.2	291	1.4	2.6	93	27	7	0.0
Kiribati	0.1	145	2.2	3.8	44	35	4	−2.8*
Marshall Islands	0.1	304	2.5	4.8	68	42	2	−5.1*
Nauru	0.01	485	2.1	3.3	100	35	1	
New Caledonia	0.3	14	1.2	2.2	58	26	7	4.9
New Zealand	4.4	16	0.7	2.1	86	20	14	3.2
Palau	0.02	45	0.6	2.0	77	20	6	
Papua New Guinea	7.0	15	2.1	4.1	13	38	3	0.0
Samoa	0.2	66	2.4	4.5	21	40	5	−13.5
Solomon Islands	0.6	19	2.6	4.2	20	40	3	0.0
Tonga	0.1	138	2.0	3.8	23	38	6	−15.4
Tuvalu	0.01	433	1.4	3.1	47	32	5	−6.97*
Vanuatu	0.3	21	2.5	4.0	24	37	3	0.0*

[a]Net Migration Rate from the United Nations, Population Division, *World Population Prospects: The 2010 Revision Population Database.*
* Additional data from the *CIA World Factbook, 2012*

Source: Population Reference Bureau, *World Population Data Sheet, 2012.*

larger neighbor (Figure 14.19). In between these two metropolitan giants, the location of the much smaller federal capital of Canberra (population 325,000) represents a classic geopolitical compromise, in the same spirit that created Washington, DC, located midway between the populous southern and northern portions of the eastern United States.

Aboriginal populations are widely, but thinly scattered inland across Western Australia and South Australia, as well as in the Northern Territory, creating smaller, but regionally important centers of settlement.

In New Zealand, more than 70 percent of the country's 4.4 million residents live on the North Island, with the Auckland region (over 1.1 million) dominating the metropolitan scene in the north and the capital city of Wellington (165,000) anchoring settlement along the Cook Strait in the south. Settlement on the South Island is mostly located in the somewhat drier lowlands and coastal districts east of the mountains, with Christchurch (340,000) serving as the largest urban center. Elsewhere, rugged and mountainous terrain on both the North and the South islands produces much lower densities.

In Papua New Guinea, only 13 percent of the country's population is urban, with most people living in the isolated interior highlands. The nation's largest city is the capital, Port

FIGURE 14.19 Downtown Melbourne Metropolitan Melbourne lies along the Yarra River. Capital of the Australian state of Victoria, Melbourne resembles many modern in likeness to North American cities, with its high-rise office buildings, entertainment districts, and downtown urban redevelopment.

Moresby (around 200,000), located along the narrow coastal lowland in the far southeastern corner of the country. In stark contrast to Papua New Guinea, the largest urban area on the northern margin of Oceania is Honolulu (1 million), on the island of Oahu. Here rapid metropolitan growth has occurred since World War II because of U.S. statehood and the scenic attractions of its mid-Pacific setting.

Historical Settlement

The region's remoteness from the world's early population centers meant that it lay beyond the dominant migratory paths of earlier peoples. Even so, prehistoric settlers eventually found their way to the isolated Australian interior and even the far reaches of the Pacific. Later, the pace of new in-migrations increased once Europeans explored the region and identified its resource potential.

Peopling the Pacific   The large islands of New Guinea and Australia, with their nearness to the Asian landmass, were settled much earlier than the more distant islands of the Pacific. By 60,000 years ago, the ancestors of today's native Australian, or **Aborigine**, population were making their way out of Southeast Asia and into Australia (Figure 14.20). The first Australians most likely arrived using some kind of watercraft, but because such boats were probably not capable of more lengthy voyages, the more distant islands remained inaccessible to settlement for tens of thousands of years. During the last glacial period, however, sea levels were much lower than they are now, which would have allowed easier movement to Australia across relatively narrow spans of water from what is now called Southeast Asia. It is not known whether the original Australians arrived in one wave of people or in many, but the available evidence suggests that they soon occupied large portions of the continent, including Tasmania, which was at that time connected to the mainland by a land bridge because of the lower sea level.

Eastern Melanesia was settled much later than Australia and New Guinea. By 3,500 years ago, some Pacific peoples had mastered long-distance sailing and navigation, which eventually opened the entire oceanic realm to human habitation. In that era,

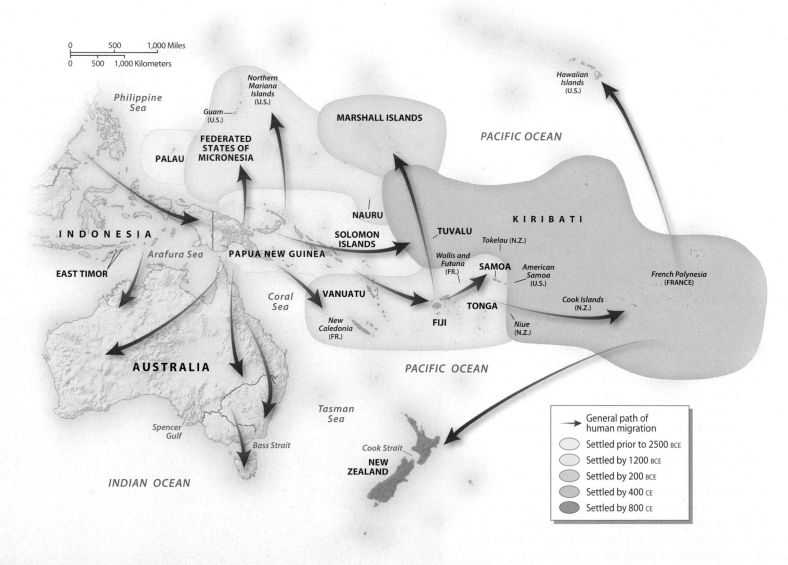

FIGURE 14.20 Peopling the Pacific   Recent settlement of Pacific islands by Austronesian peoples from Southeast Asia shaped cultural patterns across the oceanic portions of the realm. Eastward migrations through the Solomon Islands, Fiji, and the Cook Islands were followed by later movements to the north and south.

FIGURE 14.21 Polynesian Ocean-Voyaging Canoes   In 2011 seven ocean-voyaging canoes were built using a combination of traditional and modern materials (the hulls were fiberglass instead of wood) and then sailed between far-flung Polynesian islands to revive cultural traditions.

people gradually moved east to occupy New Caledonia, the Fiji Islands, and Samoa. From there, later movements took seafaring folk north into Micronesia, with the Marshall Islands occupied around 2,000 years ago.

Continuing movements from Asia further complicated the story of these migrating Melanesians. Some of the migrants mixed culturally and eventually reached western Polynesia, where they formed the core population of the Polynesian people. By 800 CE, they had reached such distant places as Tahiti, Hawaii, and Easter Island (Figure 14.21). Prehistorians hypothesize that population pressures may have quickly reached crisis stage on the relatively small islands, leading people to attempt dangerous voyages to colonize other Pacific islands. Equipped with sturdy outrigger sailing vessels and ample supplies of food, the Polynesians were quickly able to colonize most of the islands they discovered.

European Colonization   About six centuries after the Maori people brought New Zealand into the Polynesian realm, Dutch navigator Abel Tasman spotted the islands on his global exploration of 1642, marking the beginning of a new chapter in the human occupation of the South Pacific. British sea captain James Cook surveyed the shorelines of both New Zealand and Australia between 1768 and 1780, with the belief that these distant lands might be worthy of European development. By 1800, other European expeditions were also exploring the Pacific, placing most of Oceania's island groups on colonial maps.

European colonization of the region began in Australia when the British needed a remote penal colony to which convicts could be exiled. The southeastern coast of Australia was selected as an appropriate site, and in 1788 the First Fleet arrived with 750 prisoners in Botany Bay, near what is now Sydney. Other fleets and more convicts soon followed, as did boatloads of free settlers. Before long, free settlers outnumbered the convicts, who were gradually being released after serving their sentences. The growing population of English-speaking people soon moved inland and also settled other favorable coastal areas. British and Irish settlers were attracted by the agricultural and stock-raising potential of the distant colony, as well as by the lure of

gold and other minerals (a major gold rush occurred in Australia during the 1850s). The British government also encouraged the emigration of its own citizens, often paying the transportation fare of those too poor to afford it themselves.

These new settlers came into conflict with the Aborigines almost immediately after arriving. No treaties were signed, however, and in most cases Aborigines were simply displaced from their lands. In some places—most notably, Tasmania—they were hunted down and killed. In mainland Australia, the Aborigines were greatly reduced in numbers by disease, removal from their lands, and pure economic hardship. By the mid-19th century, Australia was primarily an English-speaking land as the native peoples were driven into submission.

The lush and fertile lands of New Zealand also attracted British settlers. European whalers and sealers arrived shortly before 1800, with more permanent agricultural settlement taking shape after 1840, when the British formally declared sovereignty over the region. As new arrivals grew in number and the scope of planned settlement colonies on the North and South islands expanded, tensions with the native Maori population increased. Organized in small kingdoms, or chiefdoms, the Maori were formidable fighters (Figure 14.22). Widespread

FIGURE 14.22 Maori Warrior   Body ornamentation, including tattoos, was common in traditional Maori culture, particularly among the high status warrior class. Before battle the Maori warriors would perform a ceremonial dance, the *Haka*, that included fierce facial contortions with the tongue and eyes as shown by this Maori dancer. *Haka* dances (along with the facial contortions) are now a common part of New Zealand culture, particularly among sports teams.

Maori wars began in 1845 and engulfed New Zealand until 1870. The British eventually prevailed, however, and as in Australia the native Maori lost most of their land.

The native Hawaiians also lost control of their lands to immigrants. Hawaii emerged as a united and powerful kingdom in the early 1800s, and for many years its native rulers limited U.S. and European claims to their islands. Increasing numbers of missionaries and settlers from the United States were allowed in, however, and by the late 19th century control of the Hawaiian economy had largely passed to foreign plantation owners. In 1893, U.S. interests were strong enough to overthrow the Hawaiian monarchy, resulting in formal political annexation to the United States in 1898. Hawaii became a state in 1959.

Settlement Landscapes

The settlement geography of Australia and Oceania offers an interesting mixture of local and global influences. The contemporary cultural landscape still reflects the imprint of indigenous peoples in those areas where native populations remain numerically dominant. Elsewhere, however, patterns of recent colonization have produced a modern scene mainly shaped by Europeans. The result includes everything from German-owned vineyards in South Australia to houses on New Zealand's South Island that appear to be plucked directly from the British Isles. In addition, processes of economic and cultural globalization have resulted in urban forms that make cities such as Perth or Auckland look strikingly similar to such places as San Diego or Seattle.

The Urban Transformation   Both Australia and New Zealand are highly urbanized, Westernized societies, and thus the vast majority of their populations live in city and suburban environments. As in Europe and North America, much of this urban transformation came during the 20th century as the rural economy became less labor intensive and as opportunities for urban manufacturing and service employment grew. As urban landscapes evolved, they took on many of the characteristics of their largely European populations, yet blended with a strong dose of North American influences. The result is an urban landscape in which many North Americans are quite comfortable, even though the varied local accents heard on the street and many features of the metropolitan scene are reminders of the strong and lasting attachments to British traditions (Figure 14.23).

The affluent Western-style urban environments in Australia and New Zealand offer a dramatic contrast with the urban landscapes found in less developed cities in the region. Walk the streets of Port Moresby in Papua New Guinea, and a very different urban landscape reveals evidence of the large gap between rich and poor within Oceania (Figure 14.24). Port Moresby is the country's political capital and largest commercial center. Rapid growth here has produced many of the classic problems of urban underdevelopment: There is a shortage of adequate housing, and the building of roads and schools lags far behind the need, while street crime and alcoholism rise. Elsewhere, urban centers such as Suva (Fiji), Noumea (New Caledonia), and Apia (Samoa) also reflect the economic and cultural tensions generated as local populations are exposed to Western influences. Rapid growth is a common problem in the smaller cities of Oceania because native people from rural areas and nearby islands

FIGURE 14.23 Downtown Sydney   In the foreground is the distinctive opera house that opened in 1973 and is now a World Heritage Site because of its unique architecture. The park on the left of the photo is the Royal Botanic Gardens

FIGURE 14.24 Port Moresby, Papua New Guinea Urban poverty and high crime afflict the city of Port Moresby, the capital of Papua New Guinea. The city's slums, many built out on the water, reflect stresses of recent urban growth as rural residents emigrate from the even poorer nearby highlands.

gravitate toward the job opportunities available. In the past 50 years, the huge global growth of tourism in places such as Fiji and Samoa has also transformed the urban scene, replacing traditional village life with a landscape of souvenir shops, honking taxicabs, and crowded seaside resorts.

The Rural Scene Rural landscapes across Australia and the Pacific region express a complex mosaic of cultural and economic influences. In some settings, Australian Aborigines or native Papua New Guinea Highlanders can still be found in their familiar homelands, their traditional way of life and settlements barely changed from pre-European times. Yet such settlement landscapes are becoming increasingly rare. Global influences penetrate the scene as the cash economy, foreign tourism and investment, and the currents of popular culture make their way from city to countryside.

Most of rural Australia is too dry for farming or serves as only marginally valuable agricultural land. Much of the remainder of the interior, however, features range-fed livestock, areas beyond the pale of any agricultural potential, and isolated areas where Aboriginal peoples still pursue their traditional forms of hunting and gathering. Sheep and cattle dominate rural Australia's livestock economy. Many rural landscapes in the interior of New South Wales, Western Australia, and Victoria, for example, are oriented around isolated sheep stations—ranch operations that move the flocks from one large pasture to the next. Cattle can sometimes be found in these same areas, although many of the more extensive range-fed cattle operations are concentrated farther north, in Queensland.

Croplands also vary across the region. A band of commercial wheat farming sometimes mingles with the sheep country across southern Queensland; the moister interiors of New South Wales, Victoria,

and South Australia; and a swath of favorable land east and north of Perth. Elsewhere, specialized sugarcane operations thrive along the narrow, warm, and humid coastal strip of Queensland. To the south and west, productive irrigated agriculture has developed in places such as the Murray River Basin, allowing for the production of orchard crops and vegetables. **Viticulture**, or grape cultivation, increasingly shapes the rural scene in places such as South Australia's Barossa Valley, the Riverina district in New South Wales, and Western Australia's Swan Valley. Indeed, the area under grape cultivation grew by 50 percent between 1991 and 1998 as the popular Chardonnay, Cabernet Sauvignon, and Shiraz varieties boosted revenues from wine production to more than $540 million per year (Figure 14.25).

New Zealand's Landscapes Although much smaller in area than Australia, New Zealand's rural settlement landscape includes a variety of agricultural activities. Ranching clearly dominates the New Zealand scene, with the vast majority of agricultural land devoted to livestock production, particularly sheep grazing and dairying. Commercial livestock outnumber people in New Zealand by a ratio of more than 20 to 1, and this is apparent everywhere in the countryside. Dairy operations are present mostly in the lowlands of the north, where they sometimes mingle with suburban landscapes in the vicinity of Auckland.

Rural Oceania Elsewhere in Oceania, varied influences shape the rural landscape. On high islands with more water, denser populations take advantage of diverse agricultural opportunities; on the more barren low islands, fishing is often more important. Several types of rural settlement can be identified across the island realm. In rural New Guinea, village-centered shifting cultivation dominates: Farmers

FIGURE 14.25 Vineyards in Australia The fourth largest wine exporter in the world, much of Australia's wine exports go to Asia, particularly to South Asia where Australian wines are second only to those from France. This photo is from the Hunter Valley wine region, close to Sydney in New South Wales, where the oldest vineyards were planted in the early 19th century.

FIGURE 14.26 Yam Harvest This farmer on the Melanesian island of Vakuta (Papua New Guinea) is harvesting yams. Traditional tropical agriculture features a mix of crops, often grown in the same field. Where possible, fields are periodically rotated to maintain productivity.

clear a patch of forest and then, after a few years, shift to another patch, thus practicing a form of land rotation. Subsistence foods such as sweet potatoes, taro (another starchy root crop), coconut palms, bananas, and other garden crops are often found in the same field. Increasing numbers of planters also include commercial crops such as coffee. In other parts of Oceania, traditional agricultural patterns are similar (Figure 14.26). Commercial plantation agriculture has also made its mark in many more accessible rural settings. In these places, settlements consist of worker housing near crops that are typically controlled by absentee landowners. For example, copra (coconut), cocoa, and coffee operations have transformed many agricultural settings in places such as the Solomon

Islands and Vanuatu. Sugarcane plantations have reshaped other island settings, particularly in Fiji and Hawaii.

Diverse Demographic Paths

A variety of population-related issues faces residents of the region today. In Australia and New Zealand, although populations grew rapidly (mostly from natural increases) in the 20th century, today's low birthrates parallel the pattern in North America, where population growth takes place from immigration (Figure 14.27).

Different demographic challenges grip many less-developed island nations of Oceania. High population growth rates of over 2 percent per year are not uncommon, as in Vanuatu and the Marshall Islands. The larger islands of Melanesia contain some room for settlement expansion, but competitive pressures from commercial mining and logging operations limit the amount of new agricultural land available. On some of the smaller island groups in Micronesia and Polynesia, population growth is a more pressing problem. Tuvalu (north of Fiji), for example, has just over 12,000 inhabitants, but they are crowded onto a land area of about 10 square miles (26 square kilometers), making it one of the world's most densely populated countries.

Out-migration from several island nations is very high. For example, in Tonga and Samoa the lack of employment is a considerable push force. In contrast, Australia and New Zealand remain attractive to migrants, as does New Caledonia because of its recent mining boom.

Too many people or not enough? According to politicians, talk radio pundits, and activists of different stripes, Australia is facing a severe population crisis. However, it is not clear if the country (a) already has too many people or (b) is not growing fast enough and has too small a population.

Does Australia have a population problem? It depends on whom you ask.

Groups like Sustainable Population Australia (SPA) call for an environmentally responsible population policy that restricts population size. They say that, because so much of the country is either arid or semiarid, it has a limited rural carrying capacity that demands vast quantities of water for people to survive and prosper. Global warming will

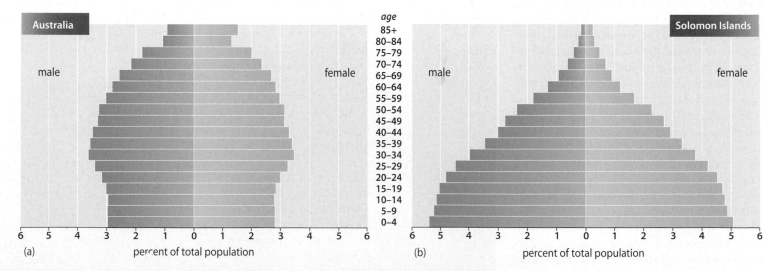

FIGURE 14.27 Comparative Population Pyramids, Australia and the Solomon Islands 2025 Like many developed countries, Australia has very low natural growth as is shown by the forecast for 2025. In contrast, like many developing countries, the Solomon Island forecast shows the classic pyramidal shape of a young and growing population.

make matters even worse, they argue, expanding Australia's dry interior and forcing even more people into the crowded coastal urban strips, where seasonal brushfires are already a serious problem. Tim Flannery—scientist, author, activist, and Australian of the Year in 2007—argues that the country is already overpopulated by some 8 million people; consequently, Australia should formulate a population policy that limits its immigration and is sensitive to the country's ecological constraints. Environmental groups advocating population limits often find their bandwagon crowded with an uneasy array of anti-immigration groups and their baggage of ethnic discrimination and racism (Figure 14.28).

On the other side of the controversy are those who say Australia's low birthrate will lead to a shrinking population unless immigration is increased. For example, the Australian Labor Party's minister for family and community services has said that current birthrates could halve the population by the end of the century. To combat this

so-called baby bust, the Labor Party and business coalitions such as the Australian Population Institute are calling for maternity leave, tax breaks, and, most of all, increased immigration to shore up Australia's population. Enhanced population growth, they argue, is absolutely necessary for the national economy, with a larger market, larger labor force, and broader base of taxpayers.

In a country basically the size of the mainland United States, yet with a total population only slightly larger than the greater Los Angeles metropolitan area, many find it difficult to believe Australia is already overpopulated. This is particularly true for those living on densely populated Pacific islands, casting a wary eye at rising sea levels and a longing eye at Australia's empty spaces.

> ### REVIEW QUESTIONS
>
> 1. Compare and contrast the populations of Australia and New Zealand as to size, density, and level of urbanization.
> 2. Describe the prehistoric peopling of the Pacific.
> 3. Is Australia over- or underpopulated? What are the arguments for each point of view?

CULTURAL COHERENCE AND DIVERSITY: A GLOBAL CROSSROADS

The Pacific world offers excellent examples of how culture is transformed as different groups migrate to a region, interact with one another, and evolve over time. As Europeans and other outsiders arrived in Oceania, colonization forced native peoples to resist or adjust. More recently worldwide processes of globalization have also redefined the region's cultural geography, provoking fears of homogenization that lead to cultural preservation efforts as native groups attempt to protect their heritage.

Multicultural Australia

Australia's cultural patterns illustrate many of the fundamental processes of globalization at work. Today, while still dominated by its colonial European roots, the country's multicultural character is becoming increasingly visible as the native people assert their cultural identity and as varied immigrant populations play larger roles in society, particularly in major metropolitan areas.

Aboriginal Imprints   For thousands of years, Australia's Aborigines—the indigenous people of Australia—dominated the cultural geography of the continent. They did not do any farming, opting instead for a hunting-and-gathering way of life that persisted up to the time of the European conquest. As the population consisted of foragers and hunters living in a relatively dry land, settlement densities remained low. Tribal groups were often isolated from one another, and the overall population probably never numbered more than 300,000 inhabitants. Because these people were clustered in many different areas, their language became fragmented. Although precise counts vary, there were probably 250 languages spoken at the time of European contact, and almost 50 indigenous languages can still be found today.

Radical cultural and geographic changes accompanied the arrival of Europeans, and Aboriginal populations were decimated in the process. The geographic results of colonization were striking: Aboriginals

FIGURE 14.28 Asians in Australia.   Although white Australians celebrate their immigrant past, controversy is growing about the large number of Asians inhabiting the country, most coming from China and India. Currently, about 10 percent of the Australian population was born in Asia, changing what was primarily a white Anglo society several decades ago into a multicultural nation.

FIGURE 14.29 Australian Aborigines The country's original inhabitants came to Australia somewhere around 70 thousand years ago, traveling primarily overland from southern Asia, then island-hopping through what is now Indonesia until reaching the subcontinent of Australia. At a national level Aborigines make up only 2 percent of the country's population, although in the Northern Territory population constitutes about 40 percent of the population. The Aborigines in this photo are in Alice Springs in the Northern Territory.

were relocated to the sparsely settled interior, particularly in northern and central Australia, where fewer Europeans competed for land. In most cases, the historical European attitude toward the Aboriginal population was even more discriminatory than it was toward the native peoples of North America.

Today Aboriginal culture perseveres in Australia, and a native people's movement is growing, similar to what is happening in the Americas (Figure 14.29). Indigenous people account for approximately 2 percent (or 430,000) of Australia's population, but their geographic distribution changed dramatically over the past century. Aborigines account for

almost 30 percent of the Northern Territory's population (many of these near Darwin), and other large native reserves are located in northern Queensland and Western Australia. Most native people, however, live in the same urban areas that dominate the country's overall population geography. Indeed, more than 70 percent of Aborigines live in cities, and very few of them still practice traditional hunting-and-gathering lifestyles. Processes of cultural assimilation are clearly at work: Urban Aborigines are frequently employed in service occupations, Christianity has often replaced traditional animist religions, and only 13 percent of the native population still speaks a native language.

Still, forces of diversity are also operative, as evidenced by a growing Aboriginal interest in preserving traditional cultural values. Particularly in the Outback, several Aboriginal languages remain strong and have growing numbers of speakers. In addition, cultural leaders are preserving some aspects of Aboriginal spiritualism, and these religious practices often link local populations to surrounding places and natural features that are considered sacred. In fact, a growing number of these sacred locations are at the center of land-use controversies (such as mining on sacred lands) between Aboriginal populations and Australia's European majority.

The future of Aboriginal cultures remains unclear: Pressures for cultural assimilation will be intense as many native people move to more Western-oriented urban settlements and lifestyles. At the same time, rapid rates of natural increase (almost twice the national average) and a growing cultural awareness of Aboriginal traditions will act to preserve elements of the country's indigenous cultures.

A Land of Immigrants Most Australians reflect the continent's more recent European-dominated migration history, but even these patterns have become more complex as a rising tide of Asian cultures becomes increasingly important (Figure 14.30). Overall, more than

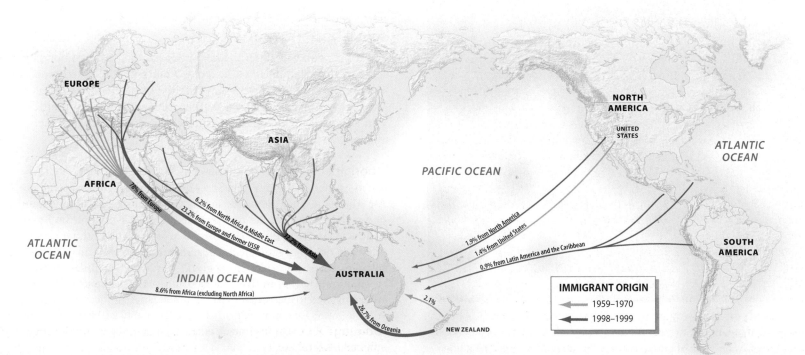

FIGURE 14.30 Changing Immigration to Australia This map shows the dramatic changes in the source areas for immigrants to Australia with the abolishing of the White Australia Policy in 1973.

70 percent of Australia's population continues to reflect a British or Irish cultural heritage. These groups dominated many of the 19th- and early 20th-century migrations into the country, and as a result, the close cultural ties to the British Isles remain strong.

A need for labor along the fertile Queensland coast caused European plantation owners to import inexpensive workers from the Solomons and New Hebrides in the late 19th century. These Pacific island laborers, known as **kanakas**, were spatially and socially segregated from their Anglo employers, but further diversified the cultural mix of Queensland's "sugar coast." Historically, however, nonwhite migrations to the country were strictly limited by what is often termed the **White Australia Policy**, in which governmental guidelines promoted European and North American immigration at the expense of other groups. This remained national policy until 1973.

Recent migration trends have reversed this historical bias, and more diverse inflows of new workers and residents are adding to the country's multicultural character. Since the 1970s, the government's Migration Program has been dominated by a variety of people chosen on the basis of their educational background and potential for succeeding economically in Australian society. Thus, a growing number of families have come from places such as China, India, Malaysia, and the Philippines. Smaller numbers have arrived as refugees from troubled parts of the world, such as Southeast Asia and the Balkans. The result is a more diverse foreign-born population. Indeed, today 25 percent of Australia's people are foreign-born, reflecting the country's global popularity as a migration destination. In the period 2000–2010, almost 40 percent of the settlers arriving in the country were from Asia. Major cities offer particularly attractive possibilities: Sydney's Asian population already exceeds 10 percent and is growing rapidly, while Perth's culture and economy are increasingly linked to Asian countries.

Cultural Patterns in New Zealand

New Zealand's cultural geography broadly reflects the patterns seen in Australia, although the precise cultural mix differs slightly. The native **Maori** people are more numerically important and culturally visible in New Zealand than their Aboriginal counterparts in Australia. While British colonization clearly mandated the dominance of Anglo cultural traditions by the late 19th century, the Maori survived, even though they lost most of their land in the process. After the initial decline, the native population began rebounding in the 20th century, and today the Maori account for more than 8 percent of the country's 4 million residents. Geographically, the Maori remain most numerous on the North Island, including a sizable concentration in metropolitan Auckland. While urban living is on the rise, many Maori, like their Aboriginal counterparts, are also committed to preserving their religion, traditional arts, and Polynesian culture (Figure 14.31). In addition, Maori joins English as an official language of New Zealand.

While many New Zealanders still identify with their largely British heritage, the country's cultural identity has increasingly separated from its British roots. Several processes have forged New Zealand's special cultural character. As Britain tightened its own links with the European continent after World War II, New Zealanders increasingly formed a more independent and diverse identity. In many ways, popular culture ties the country ever more closely to Australia, the United States, and continental Europe, a function of increasingly global mass media. Several major movies, for example, have been filmed

FIGURE 14.31 Maori Artwork Woodcarving is an active part of preserving Maori culture in New Zealand, as demonstrated by this carver at the Puia Cultural Center in Rotorua, North Island, New Zealand.

in New Zealand, including *The Adventures of Tintin, The Hobbit, Avatar, The Lord of the Rings* trilogy, and *Whale Rider.*

The Mosaic of Pacific Cultures

Native and colonial influences produce a variety of cultures across the islands of the South Pacific. In more isolated places, traditional cultures are largely insulated from outside influences. In most cases, however, modern life in the islands revolves around an intricate cultural and economic interplay of local and Western influences. The result illustrates the fact that the relative cultural insularity of the past is gone forever, and in its place is a Pacific realm rapidly adjusting to powerful forces of globalization.

Language Geography A modern language map reveals some significant cultural patterns that both unite and divide the region (Figure 14.32). Most of the native languages of Oceania belong to the **Austronesian** language family, which encompasses wide expanses of

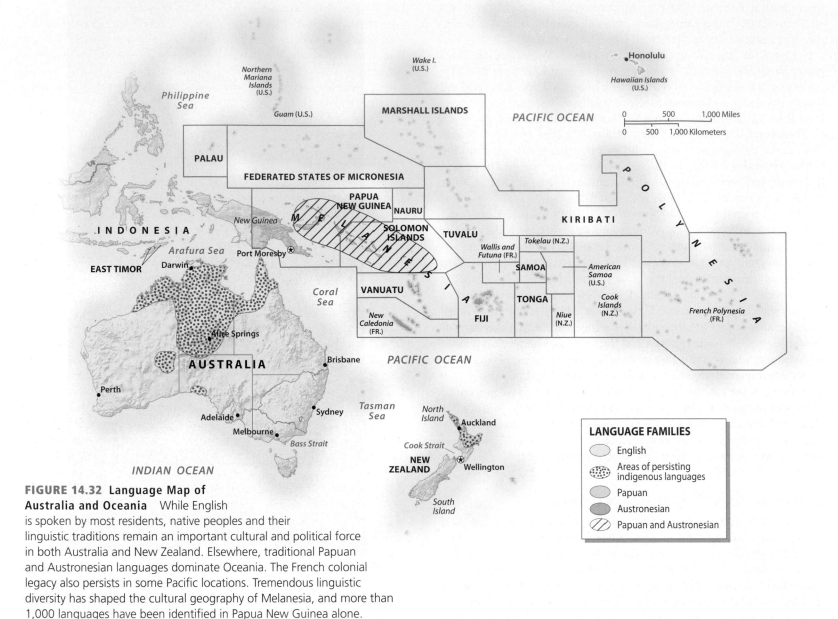

FIGURE 14.32 Language Map of Australia and Oceania While English is spoken by most residents, native peoples and their linguistic traditions remain an important cultural and political force in both Australia and New Zealand. Elsewhere, traditional Papuan and Austronesian languages dominate Oceania. The French colonial legacy also persists in some Pacific locations. Tremendous linguistic diversity has shaped the cultural geography of Melanesia, and more than 1,000 languages have been identified in Papua New Guinea alone.

LANGUAGE FAMILIES
- English
- Areas of persisting indigenous languages
- Papuan
- Austronesian
- Papuan and Austronesian

the Pacific, much of insular Southeast Asia, and, somewhat surprisingly, Madagascar. Linguists hypothesize that the first prehistoric wave of oceanic mariners spoke Austronesian languages and thus spread them throughout this vast realm of islands and oceans. Within this broad language family, the Malayo-Polynesian subfamily includes most of the related languages of Micronesia and Polynesia, suggesting a common cultural and migratory history for these widespread peoples.

Melanesia's language geography is more complex and still incompletely understood by outside experts. Although coastal peoples often speak languages brought to the region by the seafaring Austronesians, more isolated highland cultures, particularly on the island of New Guinea, speak varied Papuan languages. In fact, more than 1,000 languages have been identified, creating such linguistic complexity that many experts question whether they even constitute a unified "Papuan family" of related languages. Some scholars estimate that half of New Guinea's languages are spoken by fewer than 500 people, suggesting that the region's rugged topography plays a strong role

in isolating cultural groups. These New Guinea highlands may hold some of the world's few remaining **uncontacted peoples**—cultural groups that have yet to be "discovered" by the Western world.

Given the frequency of contact between different island cultures, it is no surprise that people have generated new forms of intercultural communication. For example, several forms of **Pidgin English** (also known simply as *Pijin*) are found in the Solomons, Vanuatu, and New Guinea, where it is the major language used between ethnic groups. In Pijin, a largely English vocabulary is reworked and blended with Melanesian grammar. Pijin's origin is commonly traced to 19th-century Chinese sandalwood traders ("pijin" is the Chinese pronunciation of the word for "business"). While of historical origin, Pijin has become a globalized language of sorts in Oceania as trade and political ties develop between different native island groups.

Village Life Traditional patterns of social life are as complex and varied as the language map. In many cases, life revolves around

free-range, emphasizing the concern women had with livestock trampling and consuming household subsistence gardens.

These gendered resource maps were invaluable to aid and development workers as they introduced conservation practices to the islanders. For example, when it was clear that a certain reef fish was threatened, men were the focus of informational campaigns. In contrast, meetings with women were held to solicit support for managing shellfish resources.

External Cultural Influences and Interactions

While traditional culture persists in some areas, most Pacific islands have witnessed tremendous cultural transformations in the past 150 years. Settlers from Europe, the United States, and Asia brought new values and technological innovations that have forever changed Oceania's cultural geography and its place in the larger world. The result is a modern setting where Pidgin English has mostly replaced native languages, Hinduism is practiced on remote Pacific islands, and people from traditional fishing communities now work at resort hotels and golf course complexes.

Anglo-European colonialism transformed the cultural geography of the Pacific world by introducing new political and economic systems. In addition, the region's cultural makeup was changed by new people migrating into the Pacific islands. Hawaii illustrates the pattern. By the mid-19th century, Hawaii's King Kamehameha was already entertaining a varied assortment of whalers, Christian missionaries, traders, and navy officers from Europe and the United States. A small elite group of **haoles**, or light-skinned European and American foreigners, was successfully profiting from commercial sugarcane plantations and Pacific shipping contracts. Labor shortages on the islands led to the importation of Chinese, Portuguese, and Japanese workers, who further complicated the region's cultural geography. By 1900, the Japanese had become a dominant part of the island workforce.

After the United States formally annexed the islands in 1898, the cultural mix revealed in the Hawaiian census of 1910 suggests the magnitude of culture change: More than 55 percent of the population was Asian (mostly Japanese and Chinese), native Hawaiians made up another 20 percent, and about 15 percent (mostly imported European workers) were white. By the end of the 20th century, however, the Asian population was less dominant, as about 40 percent of Hawaii's residents were white. In addition, the small number of remaining native Hawaiians had been joined by an increasingly diverse group of persons from other Pacific island. Ethnic mixing between these groups has produced a rich mosaic of Hawaiian cultures, showing a unique blend of North American, Asian, Pacific, and European influences.

Hawaii's story has also played out in many other Pacific island locations. In the Mariana Islands, Guam was absorbed into the United States' Pacific empire at the conclusion of the Spanish-American War in 1898. Thereafter, not only did native people feel the effects of Americanization (the island remains a self-governing U.S. territory today), but also thousands of Filipinos were moved there to supplement its modest labor force. To the southeast, the British-controlled Fiji Islands offered similar opportunities for redefining Oceania's cultural mix. The same sugar-plantation economy that spurred changes in Hawaii prompted the British to import thousands of South Asian Indian laborers to Fiji. The descendants of these Indians (most practicing Hinduism) now constitute almost half the island country's population and, as noted earlier, often come into sharp conflict with the native Fijians

FIGURE 14.33 Tonga Village Although the economic and technological effects of globalization have arrived in Tonga, village life remains important in many Polynesian settings. Most village housing in these tropical environments reflects the use of locally available construction materials.

predictable settings. For example, across much of Melanesia, including Papua New Guinea, most people live in small villages, often occupied by a single clan or family group. Many of these traditional villages contain fewer than 500 residents, although some larger communities may house more than 1,000 people. Life often revolves around the gathering and growing of food, annual rituals and festivals, and complex networks of kin-based social interactions.

Traditional Polynesian culture also focuses on village life (Figure 14.33), although strong class-based relationships often exist between local elites (who are often religious leaders) and ordinary residents. Polynesian villages are also more likely linked to other islands by wider cultural and political ties. Despite the Western stereotype of Polynesian communities as idyllic and peaceful, violent warfare was actually quite common across much of the region prior to European contact.

Gender and Island Resources One way for outsiders to better understand how island people use and manage their natural resources is to have them put down on paper their mental maps of local resources and important island features. One study done in the Fiji Islands provided important and useful information about gender roles and activities.

When the Fijian men drew resource maps, they emphasized agricultural activities such as the cultivation of *masi* (plants for mat weaving), yams, and coconuts, as well as pasture areas and water holes for cattle farming. Forest areas surrounding pastures were noted with detailed lists of specific tree species. Coastal resources were also noted, particularly those locales where men spearfished, including lists of the different fish caught in different areas.

In contrast, women's maps had different emphases, locating where water was found for household use, plots of subsistence gardens, and coastal resources that emphasized shellfish, rather than fish. Also prevalent on women's resource maps were places where local women were employed in service and handicraft jobs connected with tourism, underscoring the fact that it was common for women to work in these kinds of jobs to supplement household income. Cattle ranching, which was important in men's resource maps, was treated differently in those maps sketched by women; instead of noting pasture conditions, women noted whether cattle were contained by fences or

FIGURE 14.34 South Asians in Fiji British sugar-plantation owners imported thousands of South Asian workers to Fiji during the colonial era. Today almost half of Fiji's population is South Asian and is often in conflict with indigenous Fijians.

(Figure 14.34). In French-controlled portions of the realm, small groups of traders and plantation owners filtered into the Society Islands (Tahiti), but a larger group of French colonial settlers (many originally part of a penal colony) had a major impact on the cultural makeup of New Caledonia. Still a French colony, New Caledonia's population is more than one-third French, and its capital city of Noumea reveals a cultural setting forged from French and Melanesian traditions.

The American Football Connection European rugby is played widely throughout Oceania and ranks (along with cricket) as the national sport of Australia and New Zealand. However, in recent years, young men from the Pacific islands are looking to American football as their ticket to a better future, one that includes a U.S. college scholarship and perhaps even a chance to play in the National Football League (NFL).

At the other end of this island–mainland pipeline are American college football teams that increasingly recruit Pacific islanders on scholarship (Figure 14.35). In fact, rare is a top-ranked college team that doesn't feature at least one player from Oceania. American Samoans make up the majority because they are American citizens and need no visas, although football players from Tonga and Fiji are also recruited because of similar Polynesian cultural and physical traits. While a genetic heritage of large, strong males is important, coaches say that adding to the attraction is a Polynesian culture that emphasizes group activity and bonding, coupled with a strong work ethic. Some pundits add that a long history of tribal warfare doesn't hurt. Others say that the Polynesian tradition of male group dancing assures that even the largest men still have a remarkable agility that helps them on the football field.

Most credit Brigham Young University (BYU) in Utah as the first college to recruit Pacific islanders, back in the late 1950s, drawing on its ties to the Mormon church and its strong missionary presence in Oceania. But once BYU started the process, the floodgates opened, with other western teams—most notably, Washington State University, the University of Washington, UCLA, and the University of Arizona—filling their rosters with Pacific islanders. Not all players were 300-pound linemen, for Samoans have played at other positions. Jack Thompson (his Anglicized name), the "Throwin' Samoan," was a star Washington State quarterback and the National Collegiate Athletic Association's most prolific passer in 1978, before beginning an NFL career. More recently, his nephew, Tavita Pritchard, played the same position at Stanford University. Other notable

FIGURE 14.35 Samoan Football Players Even though rugby is the most popular sport in Oceania, there are a large number of Samoan football players in the United States, at both college and professional levels. Here, Joey Iosefa, running back for University of Hawaii moves the ball against San Jose State.

Samoan quarterbacks are Jeremiah Masoli (University of Oregon) and Marques Tavita Tuiasosopo (University of Washington).

REVIEW QUESTIONS

1. Compare the situations of the Australian Aborigines and the New Zealand Maori.
2. How has Australia's immigrant flow changed in the last 50 years?
3. Describe Hawaiian cultural changes over the last century.

GEOPOLITICAL FRAMEWORK: A REGION OF DYNAMIC POLITIES

Pacific geopolitics reflect a complex interplay of local, colonial-era, and global-scale forces (Figure 14.36). These complexities become apparent in the story of Micronesia's Marshall Islands. This sprinkling of islands and atolls (covering 70 square miles, or 180 square kilometers, of land) historically consisted of many ethnic groups that made up small political units. In 1914, the Japanese moved into the islands, and the area remained under their control until 1944, when U.S. troops occupied the region. Following World War II, a UN trust territory (administered by the United States) was created across a wide swath of Micronesia, including the Marshall group. Demands for local self-government grew during the 1960s and 1970s, resulting in a new constitution and independence for the Marshall Islands by the early 1990s. Today, still benefiting from U.S. aid, government officials in the modest capital city on Majuro Atoll struggle to unite island populations, protect large maritime sea claims, and resolve a generation of

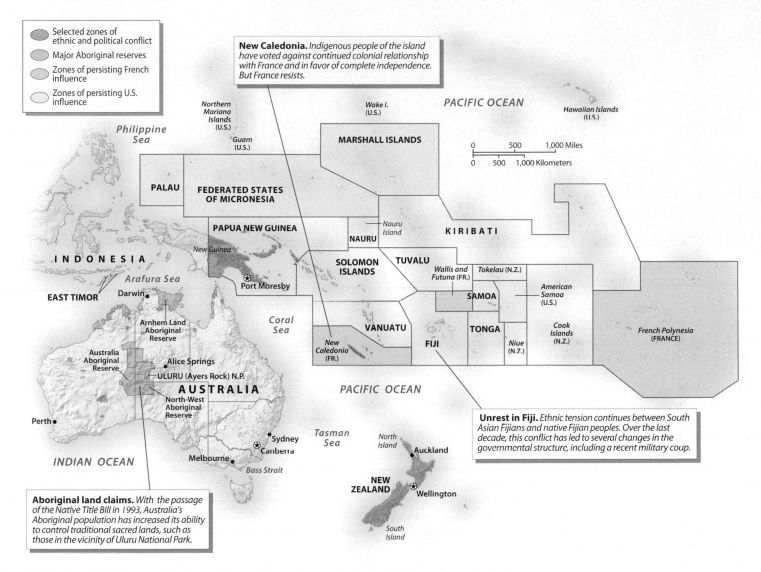

FIGURE 14.36 Geopolitical Issues in Australia and Oceania Native land claim issues increasingly shape domestic politics in Australia and New Zealand. Elsewhere, ethnic conflicts have raised political tensions in countries such as Fiji and Papua New Guinea. Colonialism's impact endures as well: American and French interests remain particularly visible in the region, including a legacy of nuclear testing that continues to affect certain Pacific island populations.

legal and medical problems that grew from U.S. nuclear bomb testing in the region. Similar stories are typical across the realm, suggesting a 21st-century political geography that is still very much in the making.

Roads to Independence

The newness and fluidity of the region's political boundaries are remarkable. The region's oldest independent states are Australia and New Zealand, and both were 20th-century creations that are still considering whether they want to complete their formal political separation from the British Crown. Elsewhere, political ties between colony and mother country are closer and, perhaps, more enduring. Even many of the newly independent Pacific **microstates**, with their tiny overall land areas, keep special political and economic ties to countries such as the United States.

Independent Australia (1901) and New Zealand (1907) gradually created their own political identities, yet both still struggle with the final shape of these identities. Although Australia became a commonwealth in 1901, it still acknowledges the British Crown as the symbolic head of its government. A national referendum in 1999 asked Australians to decide whether they would like to drop this remaining tie to Britain and instead have the country become a genuine republic with its own president replacing the British queen as head of state. Despite a strong movement toward complete independence from Britain, a slight majority (55 percent) voted to retain Australia's ties to the Crown; thus they remain today. In New Zealand, formal legislative links with Great Britain were not broken until 1947, and today New Zealand is discussing the same formal break with the British Crown being debated by the Australians.

Elsewhere in the Pacific, colonial ties were cut even more slowly, and the process has not yet been completed. In the 1970s, Britain and Australia began giving up their colonial empires in the Pacific. Fiji (Great Britain) gained independence in 1970, followed by Papua New Guinea (Australia) in 1975 and the Solomon Islands (Great Britain) in

FIGURE 14.37 United States Military Bases in the South Pacific This U.S. base is on Kwajalein Atoll in the Marshall Islands. During the European colonial period Germany used this as a copra trading center, but after their defeat in WW I the Japanese took over the island, which was subsequently taken over by the U.S. Army in 1944 where it served as a command center for post-war American nuclear testing at other Marshall Island atolls. Today the Kwajalein base is part of the Ronald Reagan Missile Defense Test Site.

Native Rights in Australia and New Zealand

Indigenous peoples in both Australia and New Zealand have used the political process to gain more control over land and resources in their two countries. The strategies these native groups have used parallel similar efforts in North America and elsewhere. In Australia, Aboriginal groups are discovering newfound political power from both more effective lobbying efforts by native groups and a more sympathetic federal government, interested in rectifying historical discrimination that left native peoples with no legal land rights. In recent years, the Australian government established several Aboriginal reserves, particularly in the Northern Territory, and expanded Aboriginal control over sacred national parklands such as Uluru (called Ayers Rock by the British settlers) (Figure 14.38). Further concessions to indigenous groups were made in 1993, when the government passed the **Native Title Bill**, which compensated Aborigines for lands already given up and gave them the right to gain title to unclaimed lands they still occupied. The bill also provided them with legal standing to deal with mining companies in native-settled areas.

Efforts to expand Aboriginal land rights, however, have met strong opposition. In 1996, an Australian court ruled that pastoral leases (the form of land interest held by the cattle and sheep ranchers who control most of the Outback) do not necessarily negate or replace Aboriginal land rights. Grazing interests were infuriated, which led the government to respond that Aboriginal claims allow them to visit

1978. The small island nations of Kiribati and Tuvalu (Great Britain) also became independent in the late 1970s.

The United States has recently turned over most of its Micronesian territories to local governments, while still maintaining a large influence in the area (Figure 14.37). After gaining these islands from Japan in the 1940s, the U.S. government provided large monetary subsidies to islanders and also utilized a number of islands for military purposes. Bikini Atoll was destroyed by nuclear tests, and the large lagoon of Kwajalein Atoll was used as a giant missile target. Moreover, a major naval base was established in Palau, the westernmost archipelago of Oceania.

By the early 1990s, both the Marshall Islands and the Federated States of Micronesia (including the Caroline Islands) had gained independence. Their ties to the United States, however, remain close. Several other Pacific islands remain under U.S. administration. Palau is a U.S. "trust territory," which gives Palauans some local autonomy. The people of the Northern Marianas chose to become a "self-governing commonwealth in association with the United States," a rather vague political position that allows them to become U.S. citizens. The residents of self-governing Guam and American Samoa are also U.S. citizens. Hawaii became a full-fledged U.S. state in 1959, yet debate continues regarding native land claims and sovereignty.

Other colonial powers appear less inclined to give up their oceanic possessions. New Zealand still controls substantial territories in Polynesia, including the Cook Islands, Tokelau, and the island of Niue. France has even more extensive holdings in the region. Its largest maritime possession is French Polynesia, which includes a large expanse of mid-Pacific territory. To the west, France still controls the much smaller territory of Wallis and Futuna in Polynesia and the larger island of New Caledonia in Melanesia.

FIGURE 14.38 Aborigines at Uluru National Park Uluru-Kata Tjuta National Park Ululu, also known as Ayers Rock, a large sandstone formation is of great spiritual significance to the Anangu Aboriginal people who live in this part of Australia's Northern Territory. Not only is the rock its surrounding area a national park but also a World Heritage site.

Persistent Geopolitical Tensions

Cultural diversity, colonial legacy, youthful states, and a rapidly changing political map contribute to ongoing geopolitical tensions in the

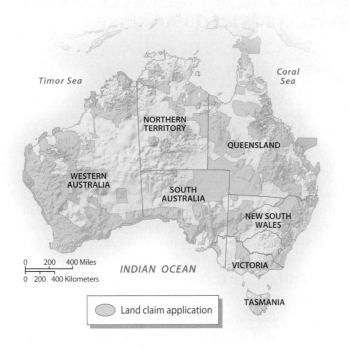

FIGURE 14.39 Applications for Native Land Claims in Australia This map shows the applications for native land claims in Australia filed by different Aboriginal groups as of 2010. Important to note is the fact that these are applications only—not government-approved claims. Nevertheless, the widespread extent of the claims shows why the topic is so contentious and controversial.

sacred sites and do some hunting and gathering, but do not give them complete economic control over the land (Figure 14.39).

In New Zealand, Maori land claims have generated similar controversies in recent years. Because the Maori constitute a far larger proportion of the overall population and because the lands they claim tend to be more valuable than rural Aborigine lands in Australia, the issues are even more complicated. Recent protests include civil disobedience, ever-increasing Maori land claims over much of North and South islands, and a call to return the country's name to the indigenous *Aotearoa,* "Land of the Long White Cloud." The government response, including a 1995 visit from Queen Elizabeth, has been to increasingly recognize Maori land and fishing rights. In 2009, for example, the Maori political party's support was crucial to formalizing New Zealand's global warming emissions trading scheme.

Native Rights in Hawaii
As is the case with indigenous peoples in Australia, New Zealand, and North America, native Hawaiians have issues with their government concerning human rights, access to ancestral land, and the political standing of native people. Attempts are being made to resolve these contentious issues in Hawaii, but the path forward is uncertain.

Native Hawaiians, who call themselves *Kanaka maoli,* are descendants of Polynesian people who arrived in the Hawaiian Islands a thousand years or so ago. They lived in an independent and sovereign state that was recognized by major foreign powers until the islands were annexed by the United States in 1898. The legality of this annexation, however, is still contested today and underlies native Hawaiian demands for a return of their historical sovereignty.

Giving credence to this demand is the U.S. admission on two occasions that the annexation process was illegal: initially in 1893 when the Hawaiian queen was displaced (President Cleveland opposed annexation of Hawaii at that time because of his concern about its legality) and more recently in 1993, on the 100th anniversary of that earlier event, when President Clinton signed the Apology Bill. This bill, which was passed without debate by Congress, once again referred to the illegal overthrow of Hawaiian sovereignty in 1893.

Today many Hawaiian nationalists support a return to Polynesian sovereignty. One pathway might be similar to that of the Navajo nation in Arizona, Utah, and New Mexico. Extremists, however, reject the idea of native Hawaiians having the same standing as U.S. Indian tribes and instead demand the United Nations invalidate the 1898 annexation, thereby granting native Hawaiians a return to complete international sovereignty (Figure 14.40). Were that to happen, native Hawaiian ancestral lands would be a separate country within the U.S. state of Hawaii.

Understandably, this extreme solution is resisted by private and corporate landowners in Hawaii, since it could call into question the legality of their ownership of, for example, residential properties in the Honolulu suburbs or hotel parcels along Waikiki Beach.

Another land issue is the disposition of those lands historically administered by the Hawaiian Monarch , lands that were ceded to the U.S. government in the 1898 annexation. These lands, which constitute almost half of the land area of the Hawaiian Islands, were transferred to the local government in 1959 upon statehood, with the stipulation that they be used for the benefit of native Hawaiians. While

FIGURE 14.40 Native Hawaiian Nationalism Thousands of red-shirted supporters of native Hawaiian sovereignty gather to march along Waikiki Beach in Honolulu in the annual Justice for Hawaiians rally.

many argue these lands should be turned over to the native Hawaiians to use as they please, this notion is resisted by the state government because these lands generate revenue through rental to public agencies (a portion of the Honolulu airport, for example, is built on ceded lands) and the U.S. military.

To resolve these issues, the U.S. Congress is pondering the Native Hawaiian Government Reorganization Act (commonly called the "Akaka Bill" after its author, Senator Daniel Akaka). If passed, this law would give native Hawaiians legal standing similar to that of Native Americans. Currently, unlike Native Americans, native Hawaiians have no legal standing or recognition in U.S. courts, a legality that prevents them from seeking redress for acts of discrimination or settling land claims. Equally important, the law would, for the first time, provide a legal mechanism for interaction Native Hawaiians and the United States government, a legal necessity that currently does not exist.

Influences from the Two Chinas Some political analysts foresee Asian influences in Oceania replacing the historical linkages to the United States, Australia, and New Zealand. Taiwan, also known as the Republic of China, or ROC, has long provided economic aid to the island nations of Nauru, Tuvalu, the Solomon Islands, the Marshall Islands, and Palau in return for their political allegiance. As a result, some experts think that ROC political influence has been behind recent unrest in the Solomon Islands and Fiji. However, with the recent emergence of the People's Republic of China as a regional player, matters could become even more complicated.

China appears to have two objectives as it expands its influence into Oceania: first, to draw on the region's natural resources to fuel its domestic economy and, second, to neutralize—and then replace—Taiwan's political and economic influence in the region. Since China's economic and political resources are far greater than Taiwan's, it seems inevitable that it will achieve these economic and political objectives sometime soon. Recent interaction between the current military leader of Fiji and China suggests that Fiji's former allegiance to ROC is negotiable.

Another factor is that Australia and New Zealand still play key political roles in the South Pacific. Although these two countries sometimes disagree on strategic and military matters, their size, wealth, and collective political influence in the region make them important forces for political stability. Also, special colonial relationships still connect these two nations with many Pacific islands. Australia maintains close political ties to its former colony of Papua New Guinea, and New Zealand's continuing control over Niue, Tokelau, and the Cook Islands in Polynesia confirms that its political influence extends well beyond its borders. How these two countries will respond to China's growing influence, however, is unclear.

REVIEW QUESTIONS

1. Describe the colonial history of the Pacific islands and contrast it to that of Australia and New Zealand.

2. What are the arguments for and against sovereignty for native Hawaiians?

3. Describe why and how China and Taiwan are causing geopolitical tensions in Oceania.

ECONOMIC AND SOCIAL DEVELOPMENT: A DIFFICULT PATH TO PARADISE

As with all other world regions, the Pacific realm contains a diversity of economic situations creating both wealth and poverty. Even within affluent Australia and New Zealand, pockets of pronounced poverty occur, and large economic disparities exist as well between those Pacific countries with global trade ties and small island nations lacking resources and external trade. While tourism offers some relief from abject poverty, the whims and economics of foreign tourists can be fickle, creating a Pacific version of boom and bust economies. Tourism in Hawaii, for example, suffered deeply during the global recession of 2008–2010. Overall, the economic future of the Pacific realm remains uncertain and highly variable because of its small domestic markets, its peripheral position in the global economy, and its diminishing resource base.

The Australian and New Zealand Economies

Much of Australia's past economic wealth has been built on the cheap extraction and export of abundant raw materials. Export-oriented agriculture, for example, has long been one of the key supports of Australia's economy. Australian agriculture is highly productive in terms of labor input, and it produces a wide variety of both temperate and tropical crops, as well as huge quantities of beef and wool for world markets. While farm exports are still important to the economy, the mining sector has grown more rapidly since 1970, making Australia one of the world's mining superpowers.

Mining has grown in the last two decades primarily due to increased trade with China, an activity that has made Australia the world's largest exporter of iron and coal (see "Exploring Global Connections: China Buys Australia"). Besides coal and iron ore, Australia produces an assortment of other materials—namely, bauxite (aluminium ore), copper, gold, nickel, lead, and zinc. As a result, the New South Wales–based Broken Hill Proprietary Company (BHP) is one of the world's largest mining corporations.

Growing numbers of Asian immigrants and economic links with potential Asian markets also offer promise for Australia's economic future. In addition, an expanding tourism industry is helping to diversify the economy. More than 7 percent of the nation's workforce is now devoted to serving the needs of more than 4 million tourists each year. Popular destinations include Melbourne and Sydney, as well as Queensland's resort-filled Gold Coast, the Great Barrier Reef, and the vast, arid Outback. Along the Gold Coast, most luxury hotels are owned by Japanese firms and provide a bilingual resort experience for their Asian clientele (Figure 14.41).

New Zealand is also a relatively wealthy country, although somewhat less well off than Australia. Before the 1970s, New Zealand relied heavily on exports to Great Britain, especially agricultural products such as wool and butter. Problems developed, however, with these colonial trade linkages in 1973 when Britain joined the European Union, with its strict agricultural protection policies. Unlike Australia, New Zealand lacked a rich base of mineral resources to export to global markets. As a result, the country had slipped into a serious recession by the 1980s. Eventually, the New Zealand government enacted drastic reforms, and the country that had previously been noted for its lofty taxes, high levels of social welfare, and state ownership of large companies changed dramatically. Privatization became the

EXPLORING GLOBAL CONNECTIONS

China Buys Australia

Around 600 miles (1,000 kilometers) northeast of Perth, Australia, in the Pilbara region of Western Australia, where road signs warn that the nearest gas station is hundreds of miles away, lies Mt. Whaleback, the world's largest iron ore mine. Mount Whaleback, once 1,500 feet (450 meters) high, however, is no longer a mountain (Figure 14.2.1). Instead, it's a very large hole in the ground because the mountain, rich in iron ore, was recently shipped piecemeal to China.

Ton by ton, Australia's coal, iron ore, and natural gas are being bought up by resource-hungry China, resulting in one of the world's largest transfers of natural resources. While this trade has enriched many Australians, it has also alarmed others.

Hardly affected by the global recession of 2008–2009, China's appetite for natural resources is huge—equaled, apparently, only by Australia's willingness to sell its resources.

Even during the recession, China pumped $40 billion into the Australian economy. Besides natural resources, half a million Chinese tourists visited Australia and added to the local economy by keeping busy Australian lifeguards, blackjack dealers, and real estate brokers. In addition, 70,000 Chinese students are currently attending Australian universities.

"China is remaking the social and political fabric of this country," said Chen Jie, a senior lecturer in international relations at the University of Western Australia, who immigrated to Australia from China over 20 years ago.

No city better illustrates the China boom than Perth, the capital of Western Australia, a state five times as big as Texas that holds the bulk of Australia's mineral wealth. The state's average income has jumped $10,000 in five years to more than $70,000, thanks to China's purchases of iron ore, natural gas, and other resources. In Perth, the median price for a

home just broke $500,000. The city's unemployment rate is a measly 2.6 percent, while the national rate is 5.5 percent. However, that figure is predicted to drop in the near future thanks to even more trade with China.

Beyond the numbers, Perth even looks like booming China, with a skyline filled with skyscrapers and cranes. The city is bristling with more than $1 billion in new construction, including a hospital, a museum, highways, offices, and a vast indoor entertainment and sports complex (Figure 14.2.2). More than 110 people move into Western Australia each day, making it the fastest-growing state in Australia. Statewide, projects worth more than $95 billion are transforming the economy of the region and contributing to a historic shift in power from "old Australia" in Sydney and Melbourne to the resource-rich Australia of the west.

FIGURE 14.2.1   Mt. Whaleback   This iron ore mine in Western Australia is the largest in the world and has been a major supplier of iron ore to China. Mt. Whaleback itself was once 1500 feet (457 meters) high, but as can be seen in this photo, is now primarily a deep pit.

FIGURE 14.2.2   Booming Perth   Largely because of the rich iron and coal resources in Western Australia, commodities sold mainly to China, Perth has turned into a boom town of plentiful jobs and ubiquitous constructions projects. Because of this economic vitality many opine that the political and economic geography of Australia will shift to the west away from the tradition centers of Sydney and Melbourne in the southeast.

watchword, and most state industries were sold off to private parties. As a result, New Zealand has been transformed into one of the most market-oriented countries in the world and has largely recovered from its earlier recession.

Oceania's Economic Diversity

Varied economic activities shape the Pacific island nations. One way of life is oriented around subsistence-based economies, such as

shifting cultivation or fishing. In other places, a commercial extractive economy dominates, with large-scale plantations, mines, and timber activities competing for land and labor with the traditional subsistence sector. Elsewhere, the huge growth in global tourism has transformed the economic geographies of many island settings, forever changing the way that people make a living. In addition, many island nations still benefit from direct subsidies and economic assistance from present and former colonial powers, designed to promote development and stimulate employment.

FIGURE 14.41 Queensland's Gold Coast Many of these luxury hotels in the Surfer's Paradise section of the Gold Coast are owned by Japanese firms specializing in accommodations for Asian tourists.

Melanesia is the least developed and poorest part of Oceania because these countries have benefited the least from tourism and from subsidies from wealthy colonial and ex-colonial powers. Today most Melanesians still live in remote villages that remain somewhat isolated from the modern economy. The Solomon Islands, for example, with few industries other than fish canning and coconut processing, has a per capita gross national income (GNI) of only $2,220 per year. Similarly, Papua New Guinea's economy produces a per capita GNI for PING is $2.420. Although traditional exports such as coconut products and coffee have increasingly been supplemented with the rapid development of tropical hardwoods, the economic returns remain low. Further, gold and copper mining has dramatically transformed the landscape, although political instability has often interfered with mineral production in places such as Bougainville. In New Guinea's interior highlands, much village life is focused on subsistence activities. In contrast, Fiji is the most prosperous Melanesian country, with a per capita GNI of $4,510., largely because of its tourist economy, popular with North Americans and Japanese.

The Economic Impact of Mining Among the smaller islands of Melanesia and Micronesia, mining economies dominated New Caledonia and Nauru. New Caledonia's nickel reserves, the world's second largest, are both a blessing and a curse (Figure 14.42). Although they currently sustain much of the island's export economy, income from nickel mining will lessen in the near future as reserves dwindle. Dramatic price fluctuations for the industrial economy also hamper economic planning for the French colony. Other activities include coffee growing, cattle grazing, and tourism. To the north, the tiny, phosphate-rich island of Nauru also depended on mining; however, as noted earlier, that day is past, with deposits exhausted and the future uncertain.

FIGURE 14.42 Nickel Mine in New Caledonia New Caledonia has a quarter of the world's nickel resources, most of which are located in the area of the North Star Nickel Mine, the second largest nickel mine in the world. While mining is a mainstay of the New Caledonia economy, fluctuations in the demand and price for nickel cause sporadic problems for the country. Even more worrisome is the fact that the fiscal future is not particularly bright because of dwindling reserves in the North Star mine.

FIGURE 14.43 Trouble in the Tuna Industry These Samoan fishermen return to traditional subsistence fishing in their outriggers after the economic collapse of the Samoan tuna fishing and canning industry because of competition from lower-wage paying countries from East Asia.

The South Pacific Tuna Fishery The South Pacific is home to the world's largest tuna fishery, and this resource contributes significantly to local island economies, both directly and indirectly. Tuna fishing and processing accounts for about 10 percent of all wage employment in southern Oceania, although 90 percent of the working tuna boats come from far-flung Pacific nations—namely, China, Japan, South Korea, and the United States. These boats are charged access fees for fishing in island territory (Figure 14.43).

More specifically, international law allows the extension of sovereign rights 200 miles (320 kilometers) beyond a country's coastline in an **Exclusive Economic Zone (EEZ)**. Within the EEZ, that country has economic control over resources such as fisheries, as well as retaining ocean floor mineral rights. In Oceania, because island countries are often composed of a series of islands and because each country's EEZ is delimited from its outermost points, the South Pacific tuna fishery is composed of a patchwork of intersecting EEZs, with different fishing regulations and fees. Complicating the matter is the fact that, as tuna become more scarce (and more valuable as a resource), many island nations have closed their EEZs to outside countries, reserving the tuna resource for local boats and processing. Whether this fragmentation will aid or injure the sustainability of the tuna fishery is not clear.

Micronesian and Polynesian Economies Throughout Micronesia and Polynesia, economic conditions depend on both local subsistence economies and economic linkages to the wider world beyond. Many archipelagos export a few food products, but native populations survive mainly on fish, coconuts, bananas, and yams. Some island groups enjoy large subsidies from either France or the United States, even though such support often comes with a political price. China is also increasingly involved with economic development plans that often imply political support.

Other Polynesian island groups have been completely transformed by tourism. In Hawaii, more than one-third of the state's economy flows directly from tourist dollars. With almost 7 million visitors annually (including more than 1.5 million from Japan), Hawaii represents all the classic benefits and risks of the tourist economy. While job creation and economic growth have reshaped the island realm, congested highways, high prices, and the unpredictable spending habits of tourists have put the region at risk for future problems. Elsewhere, French Polynesia has long been a favored destination of the international jet set. More than 20 percent of French Polynesia's GNI is derived from tourism, making it one of the wealthiest areas of the Pacific. More recently, Guam has emerged as a favorite destination of Japanese and Korean tourists, especially those on honeymoons (Figure 14.44).

The Global Economic Setting

Many international trade flows link the area to the far reaches of the Pacific and beyond. Australia and New Zealand dominate global trade patterns in the region. In the past 30 years, ties to Great Britain, the British Commonwealth, and Europe have weakened in comparison with growing trade links to Japan, East Asia, the Middle East, and the United States. Australia, for example, now imports more manufactured goods from China, Japan, and the United States than it does from Britain and Europe. Other global economic ties have come in the form of capital investment in the region. Today U.S. and Japanese banks and other financial institutions dot the South Pacific landscape from Sydney to Suva. Both Australia and New Zealand also participate in the Asia-Pacific Economic Cooperation Group (APEC), an organization designed to encourage economic development in Southeast Asia and the Pacific Basin. The region's economic ties to Asia also carry risks, however; the Asian downturn in the late 1990s slowed the

FIGURE 14.44 Fewer Asian Honeymooners in Guam During the 1970s Guam and other Pacific islands built a lucrative tourist business catering to Japanese honeymooners. However, today that industry is suffering because of the decreasing rate of marriage in Japan, China, and South Korea.

TABLE 14.2   DEVELOPMENT INDICATORS

Country	GNI per Capita, PPP, 2010	GDP Average Annual % Growth 2000–10	Human Development Index (2011)[1]	Percent Population Living Below $2 a Day	Life Expectancy (2012)[2]	Under Age 5 Mortality Rate (1990)	Under Age 5 Mortality Rate (2010)	Adult Literacy (% ages 15 and older)	Gender Inequality Index (2011)[3]
Australia	36,910	3.2	.929	—	82	9	5	—	0.136
Fed. States of Micronesia	3,490	—	.636	44.7	68	56	42	—	—
Fiji	4,510	—	.688	22.9	69	30	17	—	—
French Polynesia	18,000*	—	—	—	75			98*	—
Guam	15,000*	—	—	—	79			99*	—
Kiribati	3,520	—	.624	—	61	87	49	—	—
Marshall Islands	2,500*	—	—	—	68	51	26	94*	—
Nauru	5,000*	—	—	—	56			—	—
New Caledonia	15,000*	—	—	—	77			96	—
New Zealand	28,100	3.3	.908	—	81	11	6	—	0.195
Palau	11,000	—	.782	—	69	33	19	—	—
Papua New Guinea	2,420	3.8	.466	57.4	62	90	61	60	0.674
Samoa	4,250	—	.688	—	73	27	20	99	—
Solomon Islands	2,220	—	.510	—	67	45	27	—	—
Tonga	4,580	—	.704	—	70	25	16	99	—
Tuvalu	3,400*	—	—	—	64	57	33	—	—
Vanuatu	4,310	—	.617	—	71	39	14	82	—

[1]United Nations, *Human Development Report, 2011.*
[2]Population Reference Bureau, *World Population Data Sheet, 2012.*
[3]Gender Inequality Index—A composite measure reflecting inequality in achievements between women and men in three dimensions: reproductive health, empowerment and the labor market that ranges between 0 and 1. The higher the number, the greater the inequality.
*Additional data from the *CIA World Factbook*

Source: World Bank, *World Development Indicators, 2012.*

popular Korean tourist trade to New Zealand and also lowered China's demand for a variety of Australian raw material exports.

To promote more economic integration within the region, Australia and New Zealand signed the **Closer Economic Relations (CER) Agreement** in 1982, which successfully slashed trade barriers between the two countries. As a result, New Zealand benefited from the opening of larger Australian markets to New Zealand exports, and Australian corporate and financial interests gained new access to New Zealand business opportunities. Since the CER Agreement's signing, trade between the two countries has expanded almost 10 percent per year. Today more than 20 percent of New Zealand's imports and exports come from Australia, and this pattern of regional free trade is likely to strengthen in the future.

Smaller nations of Oceania, while often closely tied to countries such as Japan, the United States, and France, also benefit from their proximity to Australia and New Zealand. More than half of Fiji's imports come from those two nearby nations, and countries such as Papua New Guinea, Vanuatu, and the Solomon Islands enjoy a similarly close trading relationship with their more developed Pacific neighbors.

Continuing Social Challenges

Australians and New Zealanders enjoy high levels of social welfare, but face some of the same challenges evident elsewhere in the developed world (Table 14.2). Life spans average about 80 years in both

PERCENT POPULATION IN POVERTY BY CENSUS TRACT

- Greater than 20
- 7–20
- Less than 7
- No data

PACIFIC OCEAN

CITY AND COUNTY OF HONOLULU

0   5   10 Miles

0   5   10 Kilometers

FIGURE 14.45 Poverty in Hawaii This map of the distribution of poverty on the Hawaiian Island of Oahu shows the highest levels of poverty are in the urban pockets around Honolulu in the southeast, as well as in the western census tracts, which have the highest number of Native Hawaiians.

countries, and rates of child mortality have fallen greatly since 1960. However, echoing patterns in North America and Europe, cancer and heart disease are leading causes of death, and alcoholism is a continuing social problem, particularly in Australia. Furthermore, Australia's rate of skin cancer is among the world's highest, the result of having a largely fair-skinned, outdoors-oriented population from northwest Europe in a sunny, low-latitude setting. Overall, Australia's Medicare program (initiated in 1984) and New Zealand's system of social services provide high-quality health care to their populations.

Not surprisingly, the social conditions of the Aborigines and Maoris are much less favorable than those of the population overall. Schooling is irregular for many native peoples, and levels of postsecondary education for Aborigines (12 percent) and Maoris (14 percent) remain far below the national averages (32 to 34 percent). Many other social measures reflect this same pattern. Less than one-third of Aboriginal households own their own homes, while more than 70 percent of white Australian households are homeowners. Furthermore, considerable discrimination against native peoples continues in both countries, a situation that has been aggravated and publicized with the recent assertion of indigenous political rights and land claims. As with African-American, Hispanic, and Native American populations in North America, simple social policies do not yet exist as solutions to these lasting problems.

Even in Hawaii, the social welfare of the native people is problematic, as the proportion of native Hawaiians living below the poverty level is much higher than that of any other ethnic group (Figure 14.45). This group also has the shortest life expectancy and the highest infant mortality rate; rates of death from cancer and heart disease are almost 50 percent higher than for other groups in the United States; and native Hawaiian women have the highest rate of

breast cancer in the world. Further, 55 percent of native Hawaiians do not complete high school, and only 7 percent have college degrees.

In other parts of Oceania, levels of social welfare are higher than you might expect based on the region's economic situation. Many of its countries and colonies have invested heavily in health and education services and have achieved considerable success. For example, the average life expectancy in the Solomon Islands, one of the world's poorer countries as measured by per capita GNI figures, is a respectable 67 years. By other social measures as well, the Solomon Islands and several other Oceania states have reached higher levels of human well-being than exist in most Asian and African countries with similar levels of economic output. This is partly a result of successful policies, but it also reflects the relatively healthy natural environment of Oceania because many of the tropical diseases that are so troublesome in Africa simply do not exist in the Pacific islands.

REVIEW QUESTIONS

1. Explain what an Exclusive Economic Zone (EEZ) is and why it is important to Pacific island economies.

2. What is the Closer Economic Relations (CER) Agreement, and why is it important to understand the economics of Oceania?

3. What are the major social development challenges in Oceania?

4. How has Australia's foreign trade changed over the last 20 years in terms of its trading partners and the goods being exported and imported?

Summary

- Globalization has brought new interconnections to Oceania as Japanese-financed golf courses pop up along tropical shores, Chinese mining companies invest in Australian iron ore and coal, and Korean newlyweds honeymoon on tropical islands. How native cultures will respond and what new cultural hybrids will emerge as Pacific, European, North American, and Asian peoples mingle and interact are uncertain. Nonetheless, whatever the expression, these issues will give Australia and Oceania a new globalized face.

- The natural environment, which has been transformed over millennia by indigenous and colonial settlement, has witnessed accelerated change in the past 50 years. Urbanization, tourism, extractive economic activity, exotic species, and climate change from global warming are altering the landscape and increasing the vulnerability of island environments. As the country with the highest GHG emissions in the region, Australia is implementing a carbon cap-and-trade arrangement, while New Zealand is experimenting with ways to limit methane emissions from its large livestock population. On low-lying Pacific islands threatened by sea-level rise, grim scenarios for evacuation and migration are discussed.

- Migration from both outside and within the region is forcing countries to address multiculturalism in its many forms. Also, the native peoples of the region—particularly the Aborigines of Australia, the Maori of New Zealand, and native Hawaiians—are increasingly strident about regaining their land and civil rights.

- The region's contemporary political geography reveals a fluid and changing character as countries struggle to disentangle themselves from colonial ties by asserting their own political identities. Globalization complicates the process, with new economic linkages and alliances replacing historical colonial ties.

- If Australia sits on the edge of Asia, New Zealand is the edge of Polynesia as its Maori population increases due to migration from other parts of Oceania. In addition, New Zealand is taking an active role in the affairs of the entire Pacific Basin, thus becoming in many ways Oceania's leading state, economically and politically, as Australia becomes increasingly oriented toward Asia.

Key Terms

Aborigine 474
atoll 471
Austronesian 481
Closer Economic Relations (CER) Agreement 492
Exclusive Economic Zone (EEZ) 491
haole 483
high island 470

hot spot 470
kanaka 481
mallee 470
Maori 481
Melanesia 464
Micronesia 464
microstate 485
Native Title Bill 486

Oceania 462
Outback 466
Pidgin English 482
Polynesia 464
tsunami 470
uncontacted people 482
viticulture 477
White Australia Policy 481

Thinking Geographically

1. Working in a small group, describe what you see as the three principal social and economic challenges facing a Pacific island of your choice. Discuss how the country's physical and human geography may affect its future prospects.

2. Should New Zealand welcome or be wary of increased economic and political ties with Australia? Include in your response perspectives from different segments of New Zealand's population, such as sheep ranchers, Maoris, dairy farmers, fishermen, and middle-class urbanites.

3. In a group of three, have one person represent the issues facing Australian Aborigines, New Zealand Maoris, and native Hawaiians in their native countries. Then, as a group, discuss the similarities and differences, noting also the public legal and social programs helping each group, as you discuss where each native group might be in 10 years.

4. Working with several other students, each of whom will represent a specific country in Oceania (including Australia), become better acquainted with the geopolitical and economic ties and tensions between that country and China. Then think about how those might change in the next five years.

5. You are part of a team responsible for a comprehensive economic and social development plan for a small South Pacific island. Choose a specific island, become conversant with its needs and resources, define your goals, and describe a plan for achieving them in three or four pages, complete with multimedia support.

MasteringGeography™

Looking for additional review and test prep materials? Visit the Study Area in MasteringGeography™ to enhance your geographic literacy, spatial reasoning skills, and understanding of this chapter's content by accessing a variety of resources, including **MapMaster** interactive maps, videos, RSS feeds, flashcards, web links, self-study quizzes, and an eText version of *Globalization and Diversity*.

Scan to visit the author's blog for chapter updates.

http://gad4blog.
wordpress.com/
category/australia-
and-oceania/

Authors' Blogs

Scan now to access the authors' blogs for up-to-date information on Australia and Oceana.

Scan to visit the GeoCurrents blog.

http://geocurrents.
info/category/place/
australia-and-pacific

GLOSSARY

Aborigine   An indigenous inhabitant of Australia.

acid rain   A harmful form of precipitation high in sulfur and nitrogen oxides. Caused by industrial and auto emissions, acid rain damages aquatic and forest ecosystems in regions such as eastern North America and Europe.

adiabatic lapse rate   The rate a moving air mass cools or warms with changes in elevation, which is usually around 5.5 degrees F per 1000 feet or 1 degree C per 100 meters. Contrast with environmental lapse rate.

African diaspora   The forced removal of Africans from their native area and their resettlement throughout the world, especially in the Americas.

African Union (AU)   A mostly political body that has tried to resolve regional conflicts. Founded in 1963, the organization grew to include all the states of the continent except South Africa, which finally was asked to join in 1994. In 2004, the body changed its name from the Organization of African Unity to the African Union.

agrarian reform   A popular but controversial strategy to redistribute land to peasant farmers. Throughout the 20th century, various states redistributed land from large estates or granted title from vast public lands in order to reallocate resources to the poor and stimulate development. Agrarian reform occurred in various forms, from awarding individual plots or communally held land to creating state-run collective farms.

alluvial fan   A fan-shaped deposit of sediments dropped by a river or stream flowing out of a mountain range.

Altiplano   The largest intermontane plateau in the Andes, which straddles Peru and Bolivia and ranges in elevation from 10,000 to 13,000 feet (3,000 to 4,000 meters).

altitudinal zonation   The relationship between elevation, temperature, and changes in vegetation that result from the environmental lapse rate (average 3.5°F for every 1,000 feet [6.5°C for every 1,000 meters]). In Latin America, four general altitudinal zones exist: tierra caliente, tierra templada, tierra fria, and tierra helada.

animism   A wide variety of tribal religions based on the worship of nature's spirits and human ancestors.

anthropogenic   An adjective for human-caused change to a natural system, such as the atmospheric emissions from cars, industry, and agriculture that are causing global warming.

anthropogenic landscape   A landscape heavily transformed by human agency.

apartheid   The policy of racial separateness that directed the separate residential and work spaces for white, blacks, coloureds, and Indians in South Africa for nearly 50 years. It was abolished when the African National Congress came to power in 1994.

Arab League   A regional political and economic organization focused on Arab unity and development.

Arab Spring   A series of public protests, strikes, and rebellions in the Arab countries, often facilitated by social media, that have called for fundamental government and economic reforms.

archipelago   An island group. Many archipelagos are oriented in an elongated pattern.

areal differentiation   The geographic term for description and analysis of how physical or human traits differ within a spatial unit or area on the surface of Earth.

areal integration   The geographic term for description and analysis of how different places or points on Earth interact with each other.

Association of Southeast Asian Nations (ASEAN)   A supranational geopolitical group linking together the 10 different states of Southeast Asia.

atoll   A low, sandy island made from coral. Atolls are often oriented around a central lagoon.

Austronesian   A language family that encompasses wide expanses of the Pacific, insular Southeast Asia, and Madagascar.

autonomous areas   Minor political subunits created in the former Soviet Union and designed to recognize the special status of minority groups within existing republics.

autonomous region   In the context of China, provinces that have been granted a certain degree of political and cultural autonomy, or freedom from centralized authority, due to the fact that they contain large numbers of non-Han Chinese people. Critics contend that they have little true autonomy.

balkanization   A geopolitical term and concept to describe the breaking up of large political units into smaller ones, the type example being the replacement of the former Yugoslavia with smaller independent states such as Bosnia, Macedonia, Kosovo, and so on.

Berlin Conference   A 1884 conference that divided Africa into European colonial territories. The boundaries created in Berlin satisfied European ambition but ignored indigenous cultural affiliations. Many of Africa's civil conflicts can be traced to ill-conceived territorial divisions crafted in 1884.

biofuels   Energy sources derived from plants or animals. Throughout the developing world, wood, charcoal, and dung are primary energy sources for cooking and heating.

biome   Ecologically interactive flora and fauna adapted to a specific environment. Examples are deserts and tropical rainforests.

bioregion   A spatial unit or region of local plants and animals adapted to a specific environment, such as a tropical savanna.

Bolsa Familia   This is a Brazilian conditional cash transfer program created to reduce extreme poverty. Families who qualify receive a monthly check from the government as long as they keep their children in school and take them for regular health checkups.

Bolshevik   A member of the Russian Communist movement led by Lenin that successfully took control of the country in 1917.

boreal forest   A coniferous forest found in a high-latitude or mountainous environment in the Northern Hemisphere.

brain drain   Migration of the best-educated people from developing countries to developed nations where economic opportunities are greater.

brain gain   The potential of return migrants to contribute to the social and economic development of a home country with the experiences they have gained abroad.

British East India Company A private trade organization that acted as an arm of colonial Britain in ruling most of South Asia until 1857, when it was abolished and replaced by full governmental control.

bubble economy A highly inflated economy that cannot be sustained. Bubble economies usually result from rapid influx of international capital into a developing country.

buffer zone An array of nonaligned or friendly states that "buffer" a larger country from invasion. In Europe, keeping a buffer zone has been a long-term policy of Russia (and also of the former Soviet Union) to protect its western borders from European invasion.

Bumiputra The name given to native Malay (literally, "sons of the soil"), who are given preference for jobs and schooling by the Malaysian government.

bustees Settlements of temporary and often illegal housing in Indian cities, caused by rapid urban migration of poorer rural people and the inability of the cities to provide housing for this rapidly expanding population.

capital leakage The gap between the gross receipts an industry (such as tourism) brings into a developing area and the amount of capital retained.

carbon inequity The position taken by developing countries such as China and India, which argue that, because Western industrial countries in North America and Europe have been burning large amounts of fossil fuels since the mid-19th century and because CO_2 stays in the atmosphere for hundreds of years, these countries caused the global warming problem and therefore should fix it.

Caribbean Community and Common Market (CARICOM) A regional trade organization established in 1972 that includes former English colonies in the Caribbean Basin as its members.

Caribbean diaspora The economic flight of Caribbean peoples across the globe.

caste system The complex division of South Asian society into different hierarchically ranked hereditary groups. The caste system is most explicit in Hindu society but is also found in other cultures to a lesser degree.

Central American Free Trade Association (CAFTA) A trade agreement between the United States and Guatemala, El Salvador, Nicaragua, Honduras, Costa Rica, and the Dominican Republic to reduce tariffs and increase trade between member countries.

centralized economic planning An economic system in which the state sets production targets and controls the means of production.

centrifugal forces Cultural and political forces—such as linguistic minorities, separatists, and fringe groups—that pull away from and weaken an existing nation-state.

centripetal forces Cultural and political forces—such as a shared sense of history, a centralized economic structure, and the need for military security—that promote political unity in a nation-state.

chernozem soils A Russian term for dark, fertile soil, often associated with grassland settings in southern Russia and Ukraine.

China proper The eastern half of the country of China, where the Han Chinese form the dominant ethnic group. The vast majority of China's population is located in China proper.

choke point Strategic setting where narrow waterways or other narrow passages are vulnerable to military blockade disruption.

choropleth map A thematic map in which areas are colored or shaded to depict differences in whatever is being mapped.

clan A social unit that is typically smaller than a tribe or an ethnic group but larger than a family, based on supposed descent from a common ancestor.

climate region A region of similar climatic conditions. An example is the marine west coast climate regions found on the west coasts of North America and Europe.

climograph A graph of average annual temperature and precipitation data by month and season.

Closer Economic Relationship (CER) Agreement An agreement signed in 1982 between Australia and New Zealand, designed to eliminate all economic and trade barriers between the two countries.

Cold War An ideological struggle between the United States and the Soviet Union that was conducted between 1946 and 1991.

Collective Security Treaty Organization (CSTO) A Russian-led military association that includes Belarus, Armenia, Kazakhstan, Kyrgyzstan, Tajikistan, and Uzbekistan. The CSTO and SCO work together to address military threats, crime, and drug smuggling.

colliding plate boundary A tectonic boundary where two different tectonic plates are moving against each other.

colonialism Formal, established (mainly historical) rule over local peoples by a larger imperialist government for the expansion of political and economic empire.

coloured A racial category used throughout South Africa to define people of mixed European and African ancestry.

Commonwealth of Independent States (CIS) A loose political union of former Soviet republics (without the Baltic states) established in 1992 after the dissolution of the Soviet Union.

concentric zone model A simplified description of urban land use in which a well-defined central business district (CBD) is surrounded by concentric zones of residential activity, with higher-income groups living on the urban periphery.

Confucianism A philosophical system based on the ideas of Confucius, a Chinese philosopher who lived in the 6th century BCE. Confucianism stresses education and the importance of respecting authority figures, as well as the importance of authority figures acting in a responsible manner. Confucianism is historically significant throughout East Asia.

connectivity The degree to which different locations are linked with one another through transportation and communication infrastructure.

continental climate A climate region in a continental interior, removed from moderating oceanic influences, characterized by hot summers and cold winters. In such a climate, at least one month must average below freezing.

continentality Tendency of land to experience more thermal variation than water. Continental climates exhibit temperature extremes of warmth during the summer and cold during the winter.

core–periphery model A conceptualization of the world into two economic spheres. The developed countries of western Europe, North America, and Japan form the dominant core, with less-developed countries making up the periphery. Implicit in this model is that the core gained its wealth at the expense of peripheral countries.

Cossacks Highly mobile Slavic-speaking Christians of the southern Russian steppe who were pivotal in expanding Russian influence in 16th- and 17th-century Siberia.

counterinsurgency The suppression of a rebellion or insurgency by both military and political means, which includes not just armed warfare but also winning the support of local peoples by improving local infrastructure (schools, roads, etc.).

counterurbanization The movement of people out of metropolitan areas toward smaller towns and rural areas.

creolization The blending of African, European, and some Amerindian cultural elements into the unique sociocultural systems found in the Caribbean.

crony capitalism A system in which close friends of a political leader are either legally or illegally given business advantages in return for their political support.

cultural assimilation A process in which immigrants are culturally absorbed into the larger host society.

cultural homeland A culturally distinctive settlement in a well-defined geographic area, whose ethnicity has survived over time, stamping the landscape with an enduring personality.

cultural imperialism The active promotion of one cultural system over another, such as the implantation of a new language, school system, or bureaucracy. Historically, cultural imperialism has been primarily associated with European colonialism.

cultural landscape A physical or natural landscape that has been changed considerably by the influences of human settlement.

cultural nationalism A process of protecting, either formally (with laws) or informally (with social values), the primacy of a certain cultural system against influences (real or imagined) from another culture.

cultural syncretism or hybridization The blending of two or more cultures, which produces a synergistic third culture that exhibits traits from all cultural parents. Also called *cultural hybridization*.

culture Learned and shared behavior by a group of people that gives them a distinct "way of life." Culture is made up of both material (technology, tools, etc.) and abstract (speech, religion, values, etc.) components.

culture hearth An area of historical cultural innovation.

Cyrillic alphabet An alphabet based on the Greek alphabet and used by Slavic languages heavily influenced by the Eastern Orthodox Church. It is attributed to the missionary work of St. Cyril in the 9th century.

Dalit The currently preferred term used to denote the members of India's most discriminated against ("lowest") caste groups, those people previously referred to as "untouchables."

deciduous tree A tree that looses its leaves to inhibit or halt photosynthesis during the harsh season. While most commonly leaves fall before the cold winter, deciduousness can also take place as plants prepare for a hot and dry summer.

decolonialization The process of a former colony's gaining (or regaining) independence over its territory and establishing (or reestablishing) an independent government.

demographic transition model A five-stage model of population change derived from the historical decline of the natural rate of increase as a population becomes increasingly urbanized through industrialization and economic development.

denuclearization The process whereby nuclear weapons are removed from an area and dismantled or taken elsewhere.

dependency theory A popular theory to explain patterns of economic development in Latin America. Its central premise is that underdevelopment was created by the expansion of European capitalism into the region that served to develop "core" countries in Europe and to impoverish and make dependent peripheral areas such as Latin America.

desertification The spread of desert conditions into semiarid areas due to improper management of the land.

diaspora The scattering of a particular group of people over a vast geographic area. Originally, the term referred to the migration of Jews out of their homeland, but now it has been generalized to refer to any ethnic dispersion.

divergent plate boundary A geologic boundary where tectonic plates move away from each other, in opposite diections, thereby creating either a rift zone, which is a depression, or, in other places, a ridge built of volcanic material.

diversity Difference among cultures and ethnicities.

dollarization An economic strategy in which a country adopts the U.S. dollar as its official currency. A country can be partially dollarized, using U.S. dollars alongside its national currency, or fully dollarized, in which case the U.S. dollar becomes the only medium of exchange and the country gives up its own national currency. Panama fully dollarized in 1904; more recently, Ecuador fully dollarized in 2000.

domestication The purposeful selection and breeding of wild plants and animals for cultural purposes.

domino theory A U.S. geopolitical policy of the 1970s that stemmed from the assumption that if Vietnam fell to the Communists, the rest of Southeast Asia would soon follow.

Dravidian language A strictly South Asian language family that includes such important languages as Tamil and Telugu. Once spoken through most of the region, Dravidian languages are now largely limited to southern South Asia.

Eastern Orthodox Christianity A loose confederation of self-governing churches in eastern Europe and Russia that are historically linked to Byzantine traditions and to the primacy of the patriarch of Constantinople (Istanbul).

economic convergence The notion that globalization will result in the world's poorer countries gradually catching up with more advanced economies.

edge city Suburban node of activity that features a mix of peripheral retailing, industrial parks, office complexes, and entertainment facilities.

El Niño An abnormally large warm current that appears off the coast of Ecuador and Peru in December. During an El Niño year, torrential rains can bring devastating floods along the Pacific coast and drought conditions in the interior continents of the Americas.

entrepôt A city and port that specializes in transshipment of goods.

environmental lapse rate The decline in temperature as one ascends higher in the atmosphere. On average, the temperature declines 3.5°F for every 1,000 feet of elevation, or 6.5°C for every 1,000 meters. Not to be confused with the adiabatic lapse rate.

ethnic religion A religion closely identified with a specific ethnic or tribal group, often to the point of assuming the role of the major defining characteristic of that group. Normally, ethnic religions do not actively seek new converts.

ethnicity A shared cultural identity held by a group of people with a common background or history, often as a minority group within a larger society.

Euroland The 17 states that form the European Monetary Union, with its common currency, the euro. The euro completely replaced national currencies in July 2002.

European Union (EU) The current association of 27 European countries that are joined together in an agenda of economic, political, and cultural integration.

exclave A portion of a country's territory that lies outside its contiguous land area.

Exclusive Economic Zone (EEZ) An area of the ocean decreed by international law where one local country has more rights to fishing and mineral rights than do other countries.

exotic river A river that issues from a humid area and flows into a dry area otherwise lacking streams.

federal state Nations that allocate considerable political power to units of government beneath the national level.

Fertile Crescent An ecologically diverse zone of lands in Southwest Asia that extends from Lebanon eastward to Iraq and that is often associated with early forms of agricultural domestication.

fjord Flooded, glacially carved valley. In Europe, fjords are found primarily along Norway's western coast.

formal region A geographic concept used to describe an area where a static and specific trait (such as a language or a climate) has been mapped and described. A formal region contrasts with a functional region.

fossil water Water supplies that were stored underground during wetter climatic periods.

free trade zone (FTZ) A duty-free and tax-exempt industrial park created to attract foreign corporations and create industrial jobs.

functional region A geographic concept used to describe the spatial extent dominated by a specific activity. The circulation area of a newspaper is an example, as is the trade area of a large city.

gender The social and cultural expressions of male- and femaleness, which contrasts with sex, which is the biological distinction between male and female.

gender equity A social condition, state, or goal of complete parity and equality between males and females.

gender gap The difference in parity or equity between males and females in a specific social or cultural context. A term often used to describe gender differences in salary, working conditions, or political power.

gender roles How female and male behavior differs in a specific cultural context.

genocide The deliberate and systematic killing of a racial, political, or cultural group by a state.

gentrification A process of urban revitalization in which higher-income residents displace lower-income residents in central city neighborhoods.

geographic information system (GIS) A computerized mapping and information system that analyzes vast amounts of data that may include many layers of specific kinds of information, such as microclimates, hydrology, vegetation, or land-use zoning regulations.

geomancy The traditional Chinese and Korean practice of designing buildings in accordance with the principles of cosmic harmony and discord that are thought to course through the local topography.

geopolitics The relationship between politics and space and territory.

glasnost A policy of greater political openness initiated during the 1980s by then Soviet President Mikhail Gorbachev.

global positioning system (GPS) Originally used to describe a very accurate satellite-based location system, but now also used in a general sense to describe smartphone location systems that may use cell phone towers as a subsitute for satellites.

globalization The increasing interconnectedness of people and places throughout the world through converging processes of economic, political, and cultural change.

global warming An increase in the temperature of Earth's atmosphere.

Golden Triangle The world's second largest opium and heroin producing area, located in northern Laos, Thailand, and Burma.

graphic or linear scale A ruler-like symbol on a map that translates the map's cartographic scale into visual terms.

grassification The conversion of tropical forest into pasture for cattle ranching. Typically, this process involves introducing species of grasses and cattle, mostly from Africa.

Great Escarpment A landform that rims southern Africa from Angola to South Africa. It forms where the narrow coastal plains meet the elevated plateaus in an abrupt break in elevation.

Greater Antilles The four large Caribbean islands of Cuba, Jamaica, Hispaniola, and Puerto Rico.

Green Revolution Highly productive agricultural techniques developed since the 1960s that entail the use of new hybrid plant varieties combined with large applications of chemical fertilizers and pesticides. The term is generally applied to agricultural changes in developing countries, particularly India.

greenhouse effect The natural process of lower atmospheric heating that results from the trapping of incoming and reradiated solar energy by water moisture, clouds, and other atmospheric gases.

gross domestic product (GDP) The total value of goods and services produced within a given country (or other geographical unit) in a single year.

gross national income (GNI) The value of all final goods and services produced within a country's borders (gross domestic product) plus the net income from abroad (formerly referred to as gross national product).

gross national income (GNI) per capita The figure that results from dividing a country's GNI by the total population.

Group of Eight (G8) A collection of powerful countries—United States, Canada, Japan, Great Britain, Germany, France, Italy, and Russia—that confers regularly on key global economic and political issues.

guest workers Workers from Europe's agricultural periphery—primarily Greece, Turkey, southern Italy, and the former Yugoslavia—solicited to work in Germany, France, Sweden, and Switzerland during chronic labor shortages in Europe's boom years (1950s to 1970s).

Gulag Archipelago A collection of Soviet-era labor camps for political prisoners, made famous by writer Aleksandr Solzhenitsyn.

Hajj An Islamic religious pilgrimage to Makkah. One of the five essential pillars of the Muslim creed to be undertaken once in life, if an individual is physically and financially able to do it.

haoles Light-skinned Europeans or U.S. citizens in the Hawaiian Islands.

high islands Large, elevated islands, often focused around recent volcanic activity.

Hindi An Indo-European language with more than 480 million speakers, making it the second-largest language group in the world. In India, it is the dominant language of the heavily populated north, specifically the core area of the Ganges Plain.

Hindu nationalism A contemporary "fundamental" religious and political movement that promotes Hindu values as the essential—and exclusive—fabric of Indian society. As a political movement, Hindu nationalism appears to be less tolerant of India's large Muslim minority than do other political movements.

hiragana   The main Japanese syllabary, used for writing indigenous words. Each symbol stands for a particular vowel–consonant combination.

Horn of Africa   The northeastern corner of Sub-Saharan Africa that includes the states of Somalia, Ethiopia, Eritrea, and Djibouti. Drought, famine, and ethnic warfare in the 1980s and 1990s resulted in political turmoil in this area.

hot spot   A supply of magma that produces a chain of mid-ocean volcanoes atop a zone of moving oceanic crust.

Human Development Index (HDI)   For the past three decades, the United Nations has tracked social development in the world's countries through the Human Development Index (HDI), which combines data on life expectancy, literacy, educational attainment, gender equity, and income.

human trafficking   A practice in which women are lured or abducted into prostitution.

hurricane   A storm system with an abnormally low-pressure center, sustaining winds of 75 miles per hour (121 km/hour) or higher. Each year during hurricane season (July–October), a half dozen to a dozen hurricanes form in the warm waters of the Atlantic and Caribbean, bringing destructive winds and heavy rain.

hydropolitics   The interplay of water resource issues and politics.

ideographic writing   A writing system in which each symbol represents not a sound but a concept.

indentured labor   Foreign workers (generally South Asians) contracted to labor on Caribbean agricultural estates for a set period of time, often several years. Usually the contract stipulated paying off the travel debt incurred by the laborers. Similar indentured labor arrangements have existed in most world regions.

Indian diaspora   The historical and contemporary propensity of Indians to migrate to other countries in search of better opportunities. This has led to large Indian populations in South Africa, the Caribbean, and the Pacific islands, along with western Europe and North America.

Indian subcontinent   The name frequently given to South Asia in reference to its largest country. It forms a distinct landmass separated from the rest of the Eurasian continent by a series of sweeping mountain ranges, including the Himalayas—the highest mountains in the world.

informal sector   A much-debated concept that presupposes a dual economic system consisting of formal and informal sectors. The informal sector includes self-employed, low-wage jobs that are usually unregulated and untaxed. Street vending, shoe shining, artisan manufacturing, and self-built housing are considered part of the informal sector. Some scholars include illegal activities such as drug smuggling and prostitution in the informal economy.

insolation   A measure of solar radiation often expressed in units of solar energy received over a specific area (square foot or square meter) over a specific period of time.

insurgency   A political rebellion or uprising.

internally displaced persons (IDPs)   Groups and individuals who flee an area due to conflict or famine but still remain in their country of origin. These populations often live in refugee-like conditions but are difficult to assist because they technically do not qualify as refugees.

Iron Curtain   A term coined by British leader Winston Churchill during the Cold War to define the western border of Soviet power in Europe. The notorious Berlin Wall was a concrete manifestation of the Iron Curtain.

irredentism   A state or national policy of reclaiming lost lands or those inhabited by people of the same ethnicity in another nation-state.

Islamic fundamentalism   A movement within both the Shiite and Sunni Muslim traditions to return to a more conservative, religious-based society and state. Often associated with a rejection of Western culture and with a political aim to merge civic and religious authority.

Islamism   A political movement within the religion of Islam that challenges the encroachment of global popular culture and blames colonial, imperial, and Western elements for many of the region's problems. Adherents of Islamism advocate merging civil and religious authority.

isolated proximity   A concept that explores the contradictory position of the Caribbean states, which are physically close to North America and economically dependent upon that region but also have strong loyalties to locality and limited economic opportunity.

Jainism   A religious group in South Asia that emerged as a protest against orthodox Hinduism around the 6th century BCE. Its ethical core is the doctrine of noninjury to all living creatures. Today, Jains are noted for their nonviolence, which prohibits them from taking the life of any animal.

kanakas   Melanesian agricultural workers imported to Australia, historically concentrated along Queensland's "Sugar Coast."

kanji   The Chinese characters, or ideographs, used in Japanese writing.

Khmer Rouge   Literally, "Red (or Communist) Cambodians," the left-wing insurgent group that overthrew the royal Cambodian government in 1975 and subsequently created one of the most brutal political systems the world has ever seen.

kleptocracy   A state where corruption is so institutionalized that politicians and bureaucrats siphon off a huge percentage of a country's wealth.

laissez-faire   An economic system in which the state has minimal involvement and in which market forces largely guide economic activity.

latifundia   A large estate or landholding in Latin America.

Lesser Antilles   The arc of small Caribbean islands from St. Maarten to Trinidad.

Levant   The eastern Mediterranean region.

lingua franca   An agreed-upon common language to facilitate communication on specific topics such as international business, politics, sports, or entertainment.

linguistic nationalism   The promotion of one language over others that is, in turn, linked to shared notions of nationalism. In India, some Hindu nationalists promote Hindi as the national language, yet this is resisted by many other groups in which that language is either not spoken or does not have the same central cultural role as in the Ganges Valley. The lack of a national language in India remains problematic.

location factors   The various influences that explain why an economic activity takes place where it does.

loess   A fine, wind-deposited sediment that makes fertile soil but is very vulnerable to water erosion.

Maghreb   A region in northwestern Africa that includes portions of Morocco, Algeria, and Tunisia.

maharaja   Regional Hindu royalty, usually a king or prince, who ruled specific areas of South Asia before independence but who was usually subject to overrule by British colonial advisers.

mallee   A tough and scrubby eucalyptus woodland of limited economic value that is common across portions of interior Australia.

Maori   Indigenous Polynesian people of New Zealand.

map projections   The cartographic and mathematical solution to translating the surface of a rounded globe (usually Earth) to a flat surface (usually a piece of paper) with a minimum of distortion.

map scale   The relationship between distances on a mapped object such as Earth and depection of that space on a map. Large scale maps cover small areas in great detail, whereas small scale maps depict less detail but over large areas.

maquiladora   Assembly plants on the Mexican border built by foreign capital. Most of their products are exported to the United States.

marine west coast climate   A moderate climate with cool summers and mild winters that is heavily influenced by maritime conditions. Such climates are usually found on the west coasts of continents between the latitudes from 45 to 50 degrees.

maritime climate   A climate moderated by proximity to oceans or large seas. It is usually cool, cloudy, and wet and lacks the temperature extremes of continental climates.

maroons   Runaway slaves who established communities rich in African traditions throughout the Caribbean and Brazil.

Marxism   A philosophy developed by Karl Marx, the most important historical proponent of communism. Marxism, which has many variants, presumes the desirability and, indeed, the necessity of a socialist economic system run through a central planning agency.

medieval landscape   An urban landscape from 900 to 1500 CE, characterized by narrow, winding streets, and three- or four-story structures (usually in stone, but sometimes wooden), with little open space except for the market square. These landscapes are still found in the centers of many European cities.

medina   The original urban core of a traditional Islamic city.

Mediterranean climate   A unique climate, found in only five locations in the world, that is characterized by hot, dry summers with very little rainfall. These climates are located on the west side of continents, between 30 and 40 degrees latitude.

megacity   An urban conglomeration of more than 10 million people.

Megalopolis   A large urban region formed as multiple cities grow and merge with one another. The term is often applied to the string of cities in eastern North America that includes Washington, DC; Baltimore; Philadelphia; New York City; and Boston.

Melanesia   A Pacific Ocean region that includes the culturally complex, generally darker-skinned peoples of New Guinea, the Solomon Islands, Vanuatu, New Caledonia, and Fiji.

Mercosur   The Southern Common Market, established in 1991, which calls for free trade among member states and common external tariffs for nonmember states. Argentina, Paraguay, Brazil, Uruguay, and Venezuela are members; Chile, Peru, Bolivia, Ecuador and Colombia are associate members.

meridians (lines of longitude)   Meridians run north-south, from pole to pole, and measure distance east or west from the Prime Meridian (0 degrees) located near London, England.

mestizo   A person of mixed European and Indian ancestry.

Micronesia   A Pacific Ocean region that includes the culturally diverse, generally small islands north of Melanesia. Micronesia includes the Mariana Islands, Marshall Islands, and Federated States of Micronesia.

microstates   Usually independent states that are small in both area and population.

mikrorayon   A large, state-constructed urban housing project built during the Soviet period in the 1970s and 1980s.

Millennium Development Goals   A program of the United Nations, in collaboration with the World Bank, that aims to reduce extreme poverty by focusing resources on improving basic education, health care, and access to clean water in developing countries. The targeted goals are based on 1990 baselines and are supposed to be reached by 2015. Many countries in the developing world will reach their targets; it appears that many Sub-Saharan African countries will not.

minifundia   A small landholding farmed by peasants or tenants who produce food for subsistence and the market.

mono crop production   Agriculture based on a single crop.

monotheism   A religious belief in a single God.

Monroe Doctrine   A proclamation issued by U.S. President James Monroe in 1823 that the United States would not tolerate European military action in the Western Hemisphere. Focused on the Caribbean as a strategic area, the doctrine was repeatedly invoked to justify U.S. political and military intervention in the region.

monsoon   The seasonal pattern of changes in winds, heat, and moisture in South Asia and other regions of the world that is a product of larger meteorological forces of land and water heating, the resultant pressure gradients, and jet-stream dynamics. The monsoon produces distinct wet and dry seasons.

moraines   Hilly topographic features that mark the path of Pleistocene glaciers. They are composed of material eroded and carried by glaciers and ice sheets.

Mughal (or Mogul) Empire   The powerful Muslim state that ruled most of northern South Asia in the 1500s and 1600s. The last vestiges of the Mughal dynasty were dissolved by the British following the rebellion of 1857.

multinational   Strictly speaking, "multinational" simply means "occurring in many different nations." A multinational company is one that establishes significant production operations outside of the country in which it is based.

nation-state   A relatively homogeneous cultural group (a nation) with its own political territory (the state).

Native Title Bill   A bill by the Australian legislation signed in 1993 that provides Aborigines with enhanced legal rights regarding land and resources within the country.

neocolonialism   Economic and political strategies by which powerful states indirectly (and sometimes directly) extend their influence over other, weaker states.

neoliberalism   Economic policies widely adopted in the 1990s that stress privatization, export production, and few restrictions on imports.

neotropics   Tropical ecosystems of the Americas that evolved in relative isolation and support diverse and unique flora and fauna.

net migration rate   A statistic that depicts whether more people are entering or leaving a country.

new urbanism   An urban design movement stressing higher density, mixed-use, pedestrian-scaled neighborhoods where residents might be able to walk to work, school, and local entertainment.

North American Free Trade Agreement (NAFTA)   An agreement made in 1994 between Canada, the United States, and Mexico that established a 15-year plan for reducing all barriers to trade among the three countries.

northern sea route   An ice-free channel along Siberia's northern coast that will grow in importance given sustained global warming.

Oceania   A major world subregion that is usually considered to include New Zealand and the major island regions of Melanesia, Micronesia, and Polynesia.

offshore banking   Financial services offered by islands or micro-states that are typically confidential and tax exempt. As part of a global financial system, offshore banks have developed a unique niche, offering their services to individual and corporate clients for set fees. The Bahamas and the Cayman Islands are leaders in this sector.

Organization of American States (OAS)   Founded in 1948 and headquartered in Washington, DC, an organization that advocates hemispheric cooperation and dialogue. Most states in the Americas, except Cuba, belong to the OAS.

Organization of the Petroleum Exporting Countries (OPEC)   An international organization (formed in 1960) of 12 oil-producing nations that attempts to influence global prices and supplies of oil. Algeria, Gabon, Indonesia, Iran, Iraq, Kuwait, Libya, Nigeria, Qatar, Saudi Arabia, the United Arab Emirates, and Venezuela are members.

orographic effect   The influence of mountains on weather and climate, usually referring to the increase of precipitation on the windward side of mountains, and a drier zone (or rain shadow) on the leeward or downwind side of the mountain.

orographic rainfall   Enhanced precipitation over uplands that results from lifting and cooling of air masses as they are forced over mountains.

Ottoman Empire   A large, Turkish-based empire (named for Osman, one of its founders) that dominated large portions of southeastern Europe, North Africa, and Southwest Asia between the 16th and 19th centuries.

Outback   Australia's large, generally dry, and thinly settled interior.

outsourcing   A business practice that transfers portions of a company's production and service activities to lower-cost settings, often located overseas.

Palestinian Authority (PA)   A quasi-governmental body that represents Palestinian interests in the West Bank and Gaza.

parallels (lines of latitude)   Parallels circle the mathematical globe in an east-west direction, and measure a location north or south of the Equator, which is 0 degrees. When coupled with longitude, latitude provides a precise and unique mathematical address for any point on Earth.

pastoral nomadism   A traditional subsistence agricultural system in which practitioners depend on the seasonal movements of livestock within marginal natural environments.

pastoralists   Nomadic and sedentary peoples who rely on livestock (especially cattle, camels, sheep, and goats) for sustenance and livelihood.

perestroika   A program of partially implemented, planned economic reforms (or restructuring) undertaken during the Gorbachev years in the Soviet Union and designed to make the Soviet economy more efficient and responsive to consumer needs.

permafrost   A cold-climate condition in which the ground remains permanently frozen.

physiological density   A population statistic that relates the number of people in a country to the amount of arable land.

Pidgin English   A version of English that also incorporates elements of other local languages, often utilized to foster trade and basic communication between different culture groups.

plantation America   A cultural region that extends from midway up the coast of Brazil, through the Guianas and the Caribbean, and into the southeastern United States. In this coastal zone, European-owned plantations, worked by African laborers, produced agricultural products for export.

plate tectonics   A geophysical theory that postulates Earth's surface is made up of numerous large segments (tectonic plates) that move very slowly about. The tensions between these tectonic plates, through long periods of time, have shaped Earth's surface and topography. In the shorter term plate pressures and movement cause some areas of Earth to be more prone to earthquakes and volcanoes.

podzol soil   A Russian term for an acidic soil of limited fertility, typically found in northern forest environments.

pollution exporting   The process of exporting industrial pollution and other waste material to other countries. Pollution exporting can be direct, as when waste is simply shipped abroad for disposal, or indirect, as when highly polluting factories are constructed abroad.

Polynesia   A Pacific Ocean region, broadly unified by language and cultural traditions, that includes the Hawaiian Islands, Marquesas Islands, Society Islands, Tuamotu Archipelago, Cook Islands, American Samoa, Samoa, Tonga, and Kiribati.

population density   The population of an area as measured by people per spatial unit, usually people per sqare mile or square kilometer.

population pyramid   The structure of a population measuring the percentage of young and old, presented graphically as a pyramid-shaped graph. This graph plots the percentage of all different age groups along a vertical axis that divides the population into male and female.

postindustrial economy   An economy in which the tertiary and quaternary sectors dominate employment and expansion.

prairie   An extensive area of grassland in North America. In the more humid eastern portions, grasses are usually longer than in the drier western areas, which are in the rain shadow of the Rocky Mountain range.

primate city   The largest urban settlement in a country that dominates all other urban places, economically and politically. Often, but not always, the primate city is also the country's capital. Primate cities are usually three to four times larger than the next largest city in a country.

prime meridian   Zero degrees longitude, from which locations east and west are measured in a system of latitude and longitude. Currently, the most used Prime Merdian is that established in 1851 at the Naval Observatory in Greenwich, England (in southeastern London). Before that other countries and cultures established their own prime meridians upon which to base their maps and navigation systems.

privatization   The process of moving formerly state-owned firms into the contemporary capitalist private sector.

purchasing power parity (PPP)   An important qualification to these GNI per capita data is the concept of adjustment through PPP, an adjustment that takes into account the strength or weakness of local currencies.

Quran (or Koran)   A book of divine revelations received by the prophet Muhammad that serves as a holy text in the religion of Islam.

rain shadow   A drier area of precipitation, usually on the leeward or downwind side of a mountain range, that receives less rain and snowfall than the windward or upwind side. A rain shadow is caused by the warming of air as it descends down a mountain range; this warming increases the ability of an air mass to hold moisture.

rate of natural increase (RNI) The standard statistic used to express natural population growth per year for a country, a region, or the world, based on the difference between birthrates and death rates. RNI does not consider population change from migration. Though most often a positive figure (such as 1.7 percent), RNI can also be expressed as a negative (such as −.08 percent figure) for no-growth countries.

refugee A person who flees his or her country because of a well-founded fear of persecution based on race, ethnicity, religion, ideology, or political affiliation.

region A geographic concept of areal or spatial similarity, large or small.

remittances Monies sent by immigrants working abroad to family members and communities in their countries of origin. For many countries in the developing world, remittances often amount to billions of dollars each year. For small countries, remittances can equal 5 to 10 percent of a country's gross domestic product.

remote sensing A method of digitally photographing Earth's surface from satellites or high altitude aircraft so that the information captured can be manipulated by computers to translate information into certain electromagnetic bandwidths, which, in turn, emphasizes certain features and patterns on Earth's surface.

Renaissance–Baroque period A historical period, dated roughly from the 16th to the 19th century, characterized by certain urban planning designs and architectural styles that are still found today in many European cities: wide, ceremonial boulevards; large monumental structures (palaces, public squares, churches, and so on); and ostentatious housing for the urban elite.

renewable energy source An energy source such as hydroelectric, solar, wind, and geothermal that is known for its enduring availability and its potentially lower environmental costs.

representative fraction The cartographic and mathematical expression of the relationship between distance on a map and that on Earth's surface which is expressed as a fraction depicting the scale of the map. For example, a common scale is that of one inch on a map depicting one mile on the surface, which is then rounded off to the representative fraction of 1/62,500.

rimland The mainland coastal zone of the Caribbean, beginning with Belize and extending along the coast of Central America to northern South America.

Russification A policy of the Soviet Union designed to spread Russian settlers and influences to non-Russian areas of the country.

rust belt Regions of heavy industry that experience marked economic decline after their factories cease to be competitive.

Sahel The semidesert region at the southern fringe of the Sahara, and the countries that fall within this region, which extends from Senegal to Sudan. Droughts in the 1970s and early 1980s caused widespread famine and dislocation of population.

salinization The accumulation of salts in the upper layers of soil, often causing a reduction in crop yields, resulting from irrigation using water with high natural salt content and/or irrigation of soils that contain a high level of mineral salts.

Sanskrit The original Indo-European language of South Asia, introduced into northwestern India perhaps 4,000 years ago, from which modern Indo-Aryan languages evolved. Over the centuries, Sanskrit has become the classical literary language of the Hindus and is widely used as a scholarly second language, much like Latin in medieval Europe.

Schengen Agreement The 1985 agreement between some—but not all—European Union member countries to reduce border formalities in order to facilitate free movement of citizens between member countries of this new "Schengenland." For example, today there are no border controls between France and Germany or between France and Italy.

sectoral transformation The evolution of a labor force from being highly dependent on the primary sector to being oriented around more employment in the secondary, tertiary, and quaternary sectors.

secularism The term describes both the separation of politics and religion, as well as the non-religious segment of a population. An example of the first usage is the secularism of the United States constitution, which clearly separates State from Church; whereas in the second usage it is common to refer to the growing secularism of Europe, referring to the disinterest in religion of a large part of the population.

secularization The widespread movement in western Europe away from regular participation and engagement with traditional organized religions such as Protestantism or Catholicism.

sediment load The amount of sand, silt, and clay carried by a river.

Shanghai Cooperation Organization (SCO) Formed in 2001, a geopolitical group composed of China, Russia, Kazakhstan, Kyrgyzstan, Uzbekistan, and Tajikistan that focuses on common security threats and works to enhance economic cooperation and cultural exchange in Central Asia.

Shi'a Islam The second largest denomination of Islam, whose adherents are called Shi'ites or Shias. While difficult to assess accurately, generally speaking 75–90 percent of the world's Muslims are Sunni, with Shias making up the remainder of 10–25 percent.

shield A large upland area of very old exposed rocks. Shields range in elevation from 600 to 5,000 feet (200 to 1,500 meters). The three major shields in South America are the Guiana, Brazilian, and Patagonian.

Shiite A Muslim who practices one of the two main branches of Islam. Shiites are especially dominant in Iran and nearby southern Iraq.

shogun, shogunate The true ruler of Japan before 1868. In contrast, the emperor's power was merely symbolic.

Sikhism An Indian religion combining Islamic and Hindu elements, founded in the Punjab region in the late 15th century. Most of the people of the Indian state of Punjab currently follow this religion.

siloviki Members of military and security forces in the Russian domain.

Slavic peoples A group of peoples in eastern Europe and Russia who speak Slavic languages, a distinctive branch of the Indo-European language family.

social and regional differentiation A process by which certain classes of people, or regions of a country, grow richer when others grow poorer.

socialist realism An artistic style once popular in the Soviet Union that was associated with realistic depictions of workers in their patriotic struggles against capitalism.

Spanglish A hybrid combination of English and Spanish spoken by Hispanic Americans.

special administrative region In China, a region of the country that temporarily maintains its own laws and own system of government. When Hong Kong was rejoined with China in 1997, it became a special administrative region, a position that it is scheduled to keep until 2047. In 1999 Macao passed from Portuguese rule to become China's second special administrative region.

Special Economic Zones (SEZs)   Relatively small districts in China that have been fully opened to global capitalism.

spheres of influence   In countries not formally colonized in the 19th and early 20th centuries (particularly China and Iran), limited areas gained by particular European countries for trade purposes and more generally for economic exploitation and political manipulation.

squatter settlement   Makeshift housing on land not legally owned or rented by urban migrants, usually in unoccupied open spaces within or on the outskirts of a rapidly growing city.

steppe   Semiarid grasslands found in many parts of the world. Grasses are usually shorter and less dense in steppes than in prairies.

structural adjustment programs   Controversial yet widely implemented programs used to reduce government spending, encourage the private sector, and refinance foreign debt. Typically, these International Monetary Fund and World Bank policies trigger drastic cutbacks in government-supported services and food subsidies, which disproportionately affect the poor.

subduction zone   A tectonic boundary where one colliding plate slips under another.

subnational organizations   Groups that form along ethnic, ideological, or territorial lines that can induce serious internal divisions within a state.

Suez Canal   A pivotal waterway connecting the Red Sea and the Mediterranean opened by the French in 1869.

Sunni   A Muslim who practices the dominant branch of Islam.

Sunni Islam   The major denomination of Islam, with its adherents comprising from 75 to 90 percent of the world's Muslims.

superconurbation   A massive urban agglomeration that results from the coalescing of two or more formerly separate metropolitan areas.

supranational organizations   Governing bodies that include several states, such as trade organizations, and often involve a loss of some state powers to achieve organizational goals.

sustainable agriculture   A system of agriculture where organic farming principles, a limited use of chemicals, and an integrated plan of crop and livestock management combine to offer both producers and consumers environmentally friendly alternatives.

sweatshop   Crude factories in developing countries in which workers perform labor-intensive tasks for extremely low wages.

swidden   Also called slash-and-burn agriculture, a form of cultivation in which forested or brushy plots are cleared of vegetation, burned, and then planted to crops, only to be abandoned a few years later as soil fertility declines. Also often called *shifting cultivation*.

syncretic religions   Religions that feature a blending of different belief systems. In Latin America, for example, many animist practices were folded into Christian worship.

taiga   The vast coniferous forest of Russia that stretches from the Urals to the Pacific Ocean. The main forest species are fir, spruce, and larch.

terrorism   The systematic use of terror to achieve political or cultural goals.

theocracy   A political state led by religious authorities. Also called a theocratic state.

theocratic state   A political state led by religious authorities. Also called a *theocracy*.

tonal   A language in which the same set of phonemes (or basic sounds) may have very different meanings, depending on the pitch in which they are uttered.

total fertility rate (TFR)   The average number of children who will be borne by women of a hypothetical, yet statistically valid, population, such as that of a specific cultural group or within a particular country. Demographers consider TFR a more reliable indicator of population change than the crude birthrate.

transhumance   A form of pastoralism in which animals are taken to high-altitude pastures during the summer months and returned to low-altitude pastures during the winter.

transmigration   The planned, government-sponsored relocation of people from one area to another within a state territory.

transnational firm   A firm or corporation that, although it may be chartered and have headquarters in one specific country, does international business through an array of global subsidiaries.

Trans-Siberian Railroad   A key southern Siberian railroad connection completed during the Russian empire (1904) that links European Russia with the Russian Far East terminus of Vladivostok.

Treaty of Tordesillas   A treaty signed in 1494 between Spain and Portugal that drew a north–south line some 300 leagues west of the Azores and Cape Verde islands. Spain received the land to the west of the line and Portugal the land to the east.

tribal peoples   Peoples who were traditionally organized at the village or clan level, without broader-scale political organization.

tribalism   Allegiance to a particular tribe or ethnic group rather than to the nation-state. Tribalism is often blamed for internal conflict in Sub-Saharan states.

tribe   A group of families or clans with a common kinship, language, and definable territory but not an organized state.

tsar   A Russian term (also spelled *czar*) for "Caesar," or ruler. Tsars were the authoritarian rulers of the Russian empire before its collapse in the 1917 revolution.

tsetse fly   A fly that is a vector for a parasite that causes sleeping sickness (typanosomiasis), a disease that especially affects humans and livestock. Livestock is rarely found in areas of Sub-Saharan Africa where the tsetse fly is common.

tsunami   A very large sea wave induced by earthquakes.

tundra   Arctic region with a short growing season in which vegetation is limited to low shrubs, grasses, and flowering herbs.

typhoon   A large tropical storm, similar to a hurricane, that forms in the western Pacific Ocean in tropical latitudes and can cause widespread damage to the Philippines and coastal Southeast and East Asia.

UNASUR (Union of South American Nations)   A supranational organization that seeks to integrate trade and population movements within South America. Created in 2008, it is modeled after the European Union.

uncontacted peoples   Cultures that have yet to be contacted and influenced by the Western world.

unitary state   A political system in which power is centralized at the national level.

universalizing religion   A religion, usually with an active missionary program, that appeals to a large group of people, regardless of local culture and conditions. This contrasts with ethnic religions. Christianity and Islam both have strong universalizing components.

urban decentralization   A process in which cities spread out over a larger geographic area.

urban heat island   An effect in built-up areas in which development associated with cities often produces nighttime temperatures some 9 to 14°F (5 to 8°C) warmer than nearby rural areas.

urban primacy   A state in which a disproportionately large city (for example, London, New York or Bangkok) dominates the urban system and is the center of economic, political, and cultural life.

urban realms model   A simplified description of urban land use, especially descriptive of the modern North American city. It features a number of dispersed, peripheral centers of dynamic commercial and industrial activity linked by sophisticated urban transportation networks.

urbanized population   The percentage of a country's population living in settlements characterized as cities. Usually, high rates of urbanization are associated with higher levels of industrialization and economic development, since these activities are usually found in and around cities. Conversely, lower urbanized populations (less than 50 percent) are characteristic of developing countries.

Urdu   One of Pakistan's official languages (along with English), Urdu is very similar to the Indian language of Hindi, although it includes more words derived from Persian and Arabic and is written in a modified form of the Persian Arabic alphabet. Although most Pakistanis do not speak Urdu at home, it is widely used as a second language and is extensively employed in education, the media, and government, thus giving Pakistan a kind of cultural unity.

vernacular region   An areal unit of similarity that is neither formal nor functional, but, instead, is more informal and cognitive, thus a region of the mind having imprecise boundaries. Vernacular regions can be thought of as spatial stereotypes used by humans to characterize an area generally. Examples abound: "The Village" in Manhattan; "The Hill" in Washington, DC; "Watts" in Los Angeles, and so on.

viticulture   Grape cultivation.

water stress   A condition where water availability is less than water demand, either currently or projected for the future.

White Australia Policy   Before 1973, a set of stringent Australian limitations on nonwhite immigration to the country. It has been largely replaced by a more flexible policy today.

World Trade Organization (WTO)   Formed as an outgrowth of the General Agreement on Tariffs and Trade (GATT) in 1995, a large collection of member states dedicated to reducing global barriers to trade.

CREDITS

Chapter 1 opening photo   Sculpies / Shutterstock.com   Figure 1.1.1 Roger de la Harpe/R de la Harpe/Africaimagery/Newscom   Figure 1.1 Rob Crandall   Figure 1.2 Sean Sprague/The Image Works   Figure 1.3 Rob Crandall   Figure 1.4 Stephanie Kuykendal / Corbis Images   Figure 1.6 Interfoto / Alamy   Figure 1.7 Heritage Imagestate / Glow Images   Figure 1.8 Tim Boyle / Getty Images   Figure 1.9 Thorvaldur Kristmundsson / AP Images   Figure 1.11 Eye Ubiquitous / SuperStock   Figure 1.12 Jay Directo / Getty Images   Figure 1.13 Earth Satellite Corporation/Science Photo Library / Photo Researchers, Inc.   Figure 1.14 Bill Bachmann / PhotoEdit, Inc.   Figure 1.16 Maurice Joseph / Robert Harding World Imagery   Figure 1.21 NASA   Figure 1.25 Adrian Bradshaw/EPA/Newscom   Figure 1.2.1 Adafir / Alamy   Figure 1.26 Hemis / Alamy   Figure 1.29 James Steidl / SuperFusion / SuperStock   1.30 a Craig Aurness/Corbis   1.30 b Tony Waltham/Robert Harding   1.31 Ajit Solanski/AP Photo   Figure 1.33 Arko Datta/Reuters   Figure 1.34 AP Photo/Khalil Hamra   Figure 1.35 Danita Delimont / Getty Images   Figure 1.36 Rob Crandall   Figure 1.38 Thavornc / Shutterstock.com   Figure 1.39 Gaizka Iroz / Getty Images   Figure 1.42 Hubert Boesl / Corbis   Figure 1.44 Ajay Verma/Reuters   Figure 1.46 Rob Crandall   Figure 1.48 Map adapted from *Where the Poor Are: An Atlas of Poverty*, Center for International Earth Science Information Network.   Figure 1.49 SUPRI/Reuters   Figure 1.50 Photononstop /SuperStock

Chapter 2 opening photo   The Africa Image Library / Alamy   Figure 2.5 Reuters/Kyodo   Figure 2.6 Arctic-Images / SuperStock   Figure 2.4 NHPA / SuperStock   Figure 2.7 Andrey Podkorytov / Alamy   Figure 2.8 Eddie Gerald / Alamy   Figure 2.11 Erlend Kvalsvik/iStockphoto.com   Figure 2.10 Nomad / SuperStock   Figure 2.14 NASA   Figure 2.20 Tony Cunningham / Alamy   Figure 2.22 Ashley Cooper pics / Alamy   Figure 2.24 Ajay Verma/Reuters   Figure 2.26 Michael Doolittle / Alamy   Figure 2.27 Louise Murray / Alamy   Figure 2.28 Fabio Pili / Alamy   Figure 2.29 James P. Blair/Getty Images, Inc.   Figure 2.30 Eye Ubiquitous / SuperStock   Figure 2.31 Debra Behr / Alamy   Figure 2.1.1 Gualberto Becerra / Shutterstock.com   Figure 2.33 H. Mark Weidman Photography / Alamy   Figure 2.32 cappi thompson / Shutterstock.com   Figure 2.34 Calvin Larsen/Photo Researchers, Inc.   Figure 2.35 Julien Menghini / Photo Researchers, Inc.   Figure 2.2.1 Lynnette Peizer / Alamy

Chapter 3 opening photo   Inga Spence, Getty Images   Figure 3.2 William Wyckoff   Figure 3.3 Danny E. Hooks / Shutterstock.com   Figure 3.1.1 Bruce Coleman Inc. / Alamy   Figure 3.5 William Wyckoff   Figure 3.6 Rubenstein, James M., *Contemporary Human Geography*, 1st Ed., © 2010. Reprinted and Electronically reproduced by permission of Pearson Education, Inc., Upper Saddle River, New Jersey   Figure 3.7 Worldsat International Inc. / Photo Researchers, Inc.   Figure 3.8 Joe Sohm / The Image Works   Figure 3.10a Karen Holzer / Biological Resources Division, U.S. Geological Survey   Figure 3.10b Glacier Bay National Park and Preserve   Figure 3.12 Songquan Deng / Shutterstock.com   Figure 3.15 William Wyckoff   Figure 3.17 Rob Crandall / The Image Works   Figure 3.18 Soffer Organization   Figure 3.19 Craig Aurness/CORBIS / Glow Images   Figure 3.20 Modified from James M. Rubenstein, *An Introduction to Human Geography*, Upper Saddle River, NJ: Prentice Hall, 2008   Figure 3.2.1 AFP/Getty Images/Newscom   Figure 3.2.2 Richard Green / Alamy   Figure 3.21 William Wyckoff   Figure 3.22 U.S. Census, 2010   Figure 3.23 Modified from U.S. Census, 2010.   Figure 3.24 Modified from Terry G. Jordan, Mona Domosh, and Lester Rowntree, *The Human Mosaic*, Upper Saddle River, NJ: Prentice Hall, 1998   Figure 3.25

William Wyckoff   Figure 3.26 Jim Noelker / The Image Works   Figure 3.27 Rubenstein, James M., *The Cultural Landscape: An Introduction to Human Geography*, 9th Ed., © 2008. Reprinted and Electronically reproduced by permission of Pearson Education, Inc., Upper Saddle River, New Jersey.   Figure 3.28 Data from Beverage Industry Magazine, 2011; and *Modern Brewery Age*, 2003   Figure 3.30 Aurora Photos / Alamy   Figure 3.31 Reuters/Erik de Castro   Figure 3.32 Modified from David L. Clawson and James S. Fisher, *World Regional Geography*, Upper Saddle River, NJ: Prentice Hall, 2004; and Howard Veregin, ed., *Goode's World Atlas*, 22nd ed., Upper Saddle River, NJ: Prentice Hall, 2010   Figure 3.33 Scott Berner / age fotostock   Figure 3.34 Craig Ruttle / Alamy   Figure 3.35 Dennis MacDonald / PhotoEdit, Inc.   Figure 3.36 David South / Alamy   Figure 3.37 U.S. Department of Homeland Security, 2011   Figure 3.38 Comstock / Photolibrary

Chapter 4 opening photo   AP Photo/Marco Ugarte   Figure 4.2 Rob Crandall   Figure 4.3 Adapted from DK World Atlas, London: DK Publishing, 1997, pp. 7, 55   Figure 4.4a NASA   Figure 4.4b NASA   Figure 4.5 Rob Crandall   Figure 4.6 Martin Bernetti/AFP/Getty Images/Newscom   Figure 4.7 AP Photo/Fernando Vergara   Figure 4.8 Hubert Stadler / Corbis   Figure 4.9 Stephanie Maze / National Geographic Stock Collection   Figure 4.10 Rob Crandall   Figure 4.11 Sue Cunningham/Worldwide Picture Library / Alamy   Figure 4.12 Temperature and precipitation data from E. A. Pearce and Charles Gordon Smith, *The World Weather Guide*, London: Hutchinson, 1984   Figure 4.14 Rob Crandall   Figure 4.16a Rob Crandall   Figure 4.16b Rob Crandall   Figure 4.16c Rob Crandall   Figure 4.16d Rob Crandall   Figure 4.16 line art Courtesy American Geographical Society and the The Geographic Review   Figure 4.17 Adapted from David L. Clawson, *Latin America and the Caribbean: Lands and Peoples*   Figure 4.19 Orlando Kissner / Getty Images   Figure 4.20 Rob Crandall   Figure 4.21 Rob Crandall   Figure 4.1.2 Rob Crandall   Figure 4.22 Adapted from Atlas of the World's Languages, New York: Routledge, 1994   Figure 4.23 Rob Crandall   Figure 4.24 Chris Brunskill/UPI/Newscom   Figure 4.25 Lombardi, Cathryn L., and John V. Lombardi, and K. Lynn Stoner. *Latin American History: A Teaching Atlas*, © 1993 by the Regents of the University of Wisconsin System. Reprinted by permission of the University of Wisconsin Press   Figure 4.28 Bob Daemmrich / Corbis   Figure 4.29 Rob Crandall   Figure 4.30 Paulo Fridman / Corbis   Figure 4.31 Rob Crandall   Figure 4.32 Data from International Development Bank, Remittance Map, 2006; Population Reference Bureau, World Population Data Sheet, 2006; The World Bank, World Development Indicators, 2009   Figure 4.2.1 Yao Dawei/Xinhua/Photoshot/Newscom   Figure 4.33 Rob Crandall   Figure 4.34 The World Bank, World Bank Development Indicators, 2010

Chapter 5 opening photo   Oscar Elias / Iberfoto / The Image Works   Figure 5.2 Rob Crandall   Figure 5.3 Rob Crandall   Figure 5.4 NASA   Figure 5.5 European Space Agency   Figure 5.6 Temperature and precipitation data from E. A. Pearce and Charles Gordon Smith, *The World Weather Guide,* London: Hutchinson, 1984   Figure 5.7 Alejandro Ernesto/EPA/Newscom   Figure 5.8 Adapted from *DK World Atlas,* London: DK Publishing, 1997, pp. 7, 55   Figure 5.9 Gideon Mendel / Corbis   Figure 5.10 Rob Crandall   Figure 5.11 Rob Crandall   Figure 5.14 Data from Barry Levin, *Caribbean Exodus,* New York: Praeger, 1987   Figure 5.15 British Library / HIP / Art Resource, NY

Figure 5.16 Rob Crandall Figure 5.17 Rob Crandall Figure 5.18 Data based on Philip Curtin, *The Atlantic Slave Trade, A Census*, Madison: University of Wisconsin Press, 1969, p. 268 Figure 5.1.1 AP Photo/Javier Galeano Figure 5.19 Frans Lemmens/Robert Harding Figure 5.20 Reuters/Andrea De Silva Figure 5.22 Rob Crandall Figure 5.2.1 Rob Crandall Figure 5.2.2 Paivi Vaisanen / Association of Samba Schools Finland Figure 5.23 Rob Crandall Figure 5.24 Data from Barbara Tenenbaum, ed., *Encyclopedia of Latin American History and Culture*, 1996, vol. 5, p. 296, with permission of Charles Scribner's Sons; and John Allcock, *Border and Territorial Disputes*, 3rd ed., Harlow, Essex, England: Longman Group, 1992 Figure 5.25 Reuters/David Mercado Figure 5.26 Lynne Sladky / AP Images Figure 5.28 Laif/Frank Heuer / Redux Pictures Figure 5.29 UN World Tourism Organization, *Yearbook of Tourism Statistics*, 2009 ed. Figure 5.30 photoshot/ Newscom Figure 5.31 *United Nations, Human Development Report*, 2009 and 2011

Chapter 6 opening photo John Warburton-Lee Photography / Alamy Figure 6.2 Dung Vo Trung / Corbis Figure 6.3 Rob Crandall Figure 6.4 AfriPics.com / Alamy Figure 6.5 Mint Images Limited / Alamy Figure 6.6 Temperature and precipitation data from E. A. Pearce and Charles Gordon Smith, *The World Weather Guide*, London: Hutchinson, 1984 Figure 6.7 Rob Crandall Figure 6.8 Chris Howes/Wild Places Photography / Alamy Figure 6.9 Adapted *from DK World Atlas*, London: DK Publishing, 1997, p. 75 Figure 6.10 Daniel Berehulak / Getty Images Figure 6.11a USGS Center for Earth Resources Observation and Science Figure 6.11b USGS Center for Earth Resources Observation and Science Figure 6.12 Roberto Schmidt / Getty Images Figure 6.13 Rob Crandall Figure 6.14 From *Famine Early Warning System (FEWS) Networks*, April to September 2010 Report, U.S. Agency for International Development Figure 6.15 Data from U.S. Census Bureau, International Programs Figure 6.16 Jake Lyell / Alamy Figure 6.17 Data from Population Reference Bureau, World Population Data Sheet, 2010 Figure 6.18 Gideon Mendel / Corbis Figure 6.20 Jon Hrusa / Corbis Figure 6.21 Picture Contact BV / Alamy Figure 6.22 Google Figure 6.23 UIG via Getty Images Figure 6.25 Rob Crandall Figure 6.26 Ed Kashi / Corbis Figure 6.27 From Claude S. Phillips, The African Political Dictionary, Santa Barbara, CA: Clio Press, 1984, p. 196 Figure 6.28 Greenshoots Communications / Alamy Figure 6.1.1 Rob Crandall Figure 6.30 Amar Grover / Getty Images Figure 6.31 Streeter Lecka / Getty Images Figure 6.32 Aryeety-Attoh, Samuel; Mcdade, Barbara Elizabeth; Chintuwa Obia, Godson; Oppong, Joseph Ransford; Osei, William Yaw; Yeboah, Ian; Johnston-Anumonwo, Ibipo, *Geography of Sub-Saharan Africa*, 3rd Ed., © 2010, p. 62. Reprinted and Electronically reproduced by permission of Pearson Education, Inc., Upper Saddle River, New Jersey Figure 6.33 From *The Times Atlas of World History*, Maplewood, NJ: Hammond, 1989 Figure 6.34 Data from UNHCR Global Trends, 2011 Figure 6.35 Liba Taylor / Robert Harding/Newscom Figure 6.37 Simon Maina/ Stringer / Getty Images Figure 6.38 G P Bowater / Alamy Figure 6.39 Africa Media Online / Alamy Figure 6.40 From World Development Indicators, Washington, DC: World Bank, 2012 Figure 6.41 Reuters / Siphiwe Sibeko Figure 6.43 imagebroker / Alamy Figure 6.2.1 MCT via Getty Images

Chapter 7 opening photo Rex Features via AP Images Figure 7.2 Motty Levy / Getty Images Figure 7.4 Egmont Strigl / age Fotostock Figure 7.5 David Buimovitch/ AFP/Getty Images/Newscom Figure 7.8 Ccharleson / Dreamstime Figure 7.7 Worldspec/NASA / Alamy Figure 7.9 NASA/John F. Kennedy Space Center Figure 7.10 Alan Carey / Corbis Figure 7.12 Buiten Beeld / Alamy Figure 7.13 Dott. Giorgio Gualco / Photoshot Holdings, Ltd. Figure 7.14 Patrick Syder / age fotostock Figure 7.15 Imagebroker / Alamy Figure 7.18 Glen Allison / Getty Images Figure 7.1.2 Juan Carlos Muñoz / age Fotostock Figure 7.19 AP Photo/Ben Curtis Figure 7.20 Stuart Franklin / Magnum Photos Figure 7.21 Dave Bartruff / Corbis Figure 7.22a Gavin Hellier / Robert Harding World Imagery Figure 7.22b Rob Crandall Figure 7.24 Kuwait Programme on Development, Governance and

Globalisation in the Gulf States, Labour Immigration and Labour Markets in the GCC Countries: National Patterns and Trends, March 2011 Figure 7.26 Awad Awad / Newscom Figure 7.30 Marwan Naamani/AFP/Getty Images/ Newscom Figure 7.2.1 Chris Hondros / Getty Images Figure 7.32 Modified from James M. Rubenstein, *An Introduction to Human Geography*, 10th ed. Upper Saddle River, NJ: Prentice Hall, 2011 Figure 7.33 West Bank Modified from James M. Rubenstein, *An Introduction to Human Geography*, 10th ed. Upper Saddle River, NJ: Prentice Hall, 2011 Figure 7.34 Reuters/ Nir Elias Figure 7.35 Kevin Unger-KPA/Zuma Press/Newscom Figure 7.36 Reuters/Handout Figure 7.38 Modified from James M. Rubenstein, *An Introduction to Human Geography*, 10th ed. Upper Saddle River, NJ: Prentice Hall, 2011 Figure 7.39 Mehmet Biber/dpa /Landov Figure 7.40 Ricki Rosen / Corbis Figure 7.41 Shawn Baldwin/New York Times / Corbis Figure 7.42 Reuters/Louafi Larbi Figure 7.43 Patstock / age Fotostock

Chapter 8 opening photo Martin Froyda / Shutterstock.com Figure 8.2 Sean Gallup / Getty Images Figure 8.4 P. Vauthey / Corbis Figure 8.5 Christophe Boisvieux / Corbis Figure 8.7 smart.art / Shutterstock.com Figure 8.9 Richard Packwood / Getty Images Figure 8.10 Sarah Leen / National Geographic Figure 8.1.1 Les Rowntree Figure 8.13 Reuters/ Kacper Pempel Figure 8.2.1 AFP/Getty Images Figure 8.15 Les Rowntree Figure 8.16 Les Rowntree Figure 8.17 Les Rowntree Figure 8.19 David Brauchli / AP Images Figure 8.21 Les Rowntree Figure 8.22 Embassy of the Federal Republic of Germany Figure 8.23 Olivier Hoslet/Epa/Newscom Figure 8.26 Les Rowntree Figure 8.27a Bettmann / Corbis Figure 8.27b Stephen Ferry/Getty Images Figure 8.28 Data from CIA World Factbook, 2010 Figure 8.29 Fehim Demir/AFP/Newscom Figure 8.30 James Marshall / Corbis Figure 8.32 Aeroview Figure 8.34 Rob Crandall Figure 8.35 Bloomberg via Getty Images Figure 8.37 Simela Pan Tzartzi/ EPA/Newscom

Chapter 9 opening photo Sailorr / Fotolia Figure 9.2 Sisse Brimberg / National Geographic Stock Figure 9.3 WoodyStock / Alamy Figure 9.4 Masami Goto / Glow Images Figure 9.1.1 LOOK Die Bildagentur der Fotografen GmbH / Alamy Figure 9.1.2 AP Photo/Alexander Zemlianichenko Figure 9.6 YuI / Fotolia Figure 9.8 Clawson, David L; Fisher, James; Aryeetey-Attoh, Samuel A; Theide, Roger; Williams, Jack F; Johnson, Merrell L; Johnson, Douglas L; Airriess, Christopher A; Jordan-Bychkov, Terry G; Jordan, Bell, *World Regional Geography: A Development Approach*, 8th Ed., © 2004. Reprinted and Electronically reproduced by permission of Pearson Education, Inc., Upper Saddle River, New Jersey. p. 302 Figure 9.9 ZUMA Press/Newscom Figure 9.10 sergbob / Fotolia Figure 9.11 NASA Figure 9.12 Sovfoto/Eastfoto Figure 9.13 Modified from Howard Veregin, ed., *Goode's World Atlas*, 22nd ed., Upper Saddle River, NJ: Prentice Hall, 2010 Figure 9.15 CNES/Spot Image / Photo Researchers, Inc. Figure 9.16 Jon Arnold Images Ltd / Alamy Figure 9.17 Jon Arnold Images Ltd / Alamy Figure 9.18 Reuters/Sergei Karpukhin Figure 9.20 Lain Masterton / Alamy Figure 9.21 Grigory Dukor / Corbis Figure 9.22 Ivan Vdovin / Alamy Figure 9.23 U.S. Census Bureau, International Database Figure 9.24 Bergman, Edward F.; Renwick, William H., *Introduction to Geography: People, Places, and Environment*, 1st Ed., © 1999. Reprinted and Electronically reproduced by permission of Pearson Education, Inc., Upper Saddle River, New Jersey Figure 9.26 Dean Conger / Corbis Figure 9.27 U.S. Central Intelligence Agency Figure 9.28 Jochem D Wijnands / Getty Images Figure 9.29 Mikhail Japaridze / AP Images Figure 9.30 Sergie IInitsky / Corbis Figure 9.31 Rubenstein, James M., *The Cultural Landscape: An Introduction To Human Geography*, 6th Ed., © 1999. Reprinted and Electronically reproduced by permission of Pearson Education, Inc., Upper Saddle River, New Jersey Figure 9.32 Rubenstein, James M., *The Cultural Landscape: An Introduction to Human Geography*, 10th Ed., © 2011. Reprinted and Electronically reproduced by permission of Pearson Education, Inc., Upper Saddle River, New Jersey Figure 9.33 Gleb Garanich / Reuters

INDEX

World – Physical

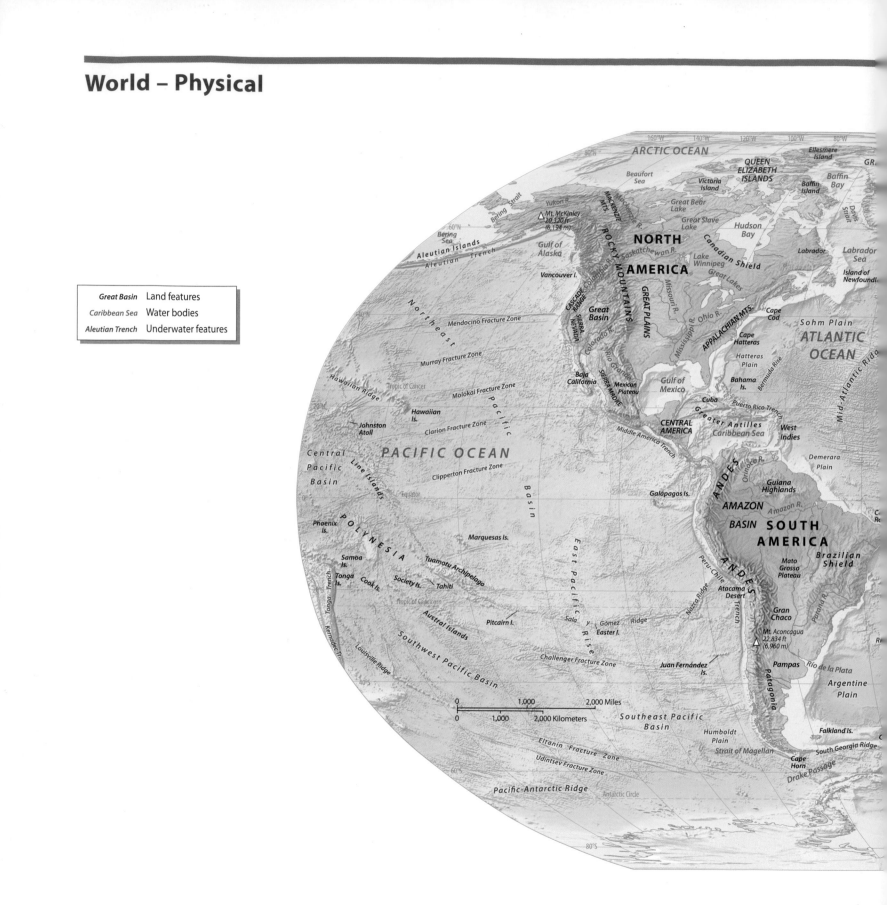

Great Basin	Land features
Caribbean Sea	Water bodies
Aleutian Trench	Underwater features

ARCTIC OCEAN

Beaufort
Sea

QUEEN
ELIZABETH
ISLANDS

Ellesmere
Island

GR.

Victoria
Island

Baffin
Island

Baffin
Bay

Davis
Strait

Bering
Strait

Great Bear
Lake

MACKENZIE MTS.

Mackenzie R.

Yukon R.

△ Mt. McKinley
20,320 ft
(6,194 m)

Great Slave
Lake

Hudson
Bay

Labrador

Labrador
Sea

Bering
Sea

Gulf of
Alaska

Saskatchewan R.

ROCKY MOUNTAINS

NORTH
AMERICA

Canadian Shield

Island of
Newfoundl.

Aleutian Islands

Aleutian Trench

Vancouver I.

Columbia R.

Lake
Winnipeg

Great Lakes

Northeast

Mendocino Fracture Zone

CASCADE RANGE

Great
Basin

GREAT PLAINS

Missouri R.

Mississippi R.

Ohio R.

APPALACHIAN MTS.

Cape
Cod

Sohm Plain

ATLANTIC
OCEAN

SIERRA NEVADA

Colorado R.

Cape
Hatteras

Murray Fracture Zone

Pacific

Hatteras
Plain

Bermuda Rise

Tropic of Cancer

Molokai Fracture Zone

Baja
California

SIERRA MADRE

Rio Grande

Mexican
Plateau

Gulf of
Mexico

Bahama
Is.

Mid-Atlantic Ridge

Hawaiian Ridge

Hawaiian
Is.

Cuba

Puerto Rico Trench

Johnston
Atoll

Clarion Fracture Zone

CENTRAL
AMERICA

Greater Antilles

Caribbean Sea

West
Indies

Central
Pacific
Basin

PACIFIC OCEAN

Middle America Trench

Line Islands

Clipperton Fracture Zone

Demerara
Plain

Equator

Galápagos Is.

ANDES

Orinoco R.

Guiana
Highlands

Co.
Ro.

Phoenix
Is.

POLYNESIA

AMAZON

Amazon R.

BASIN

SOUTH
AMERICA

Brazilian
Shield

Marquesas Is.

East

Basin

Samoa
Is.

Tonga Trench

Cook Is.

Society Is.

Tahiti

Tuamotu Archipelago

Pacific

Mato
Grosso
Plateau

Paraná R.

Tropic of Capricorn

ANDES

Nazca Ridge

Peru-Chile Trench

Atacama
Desert

Gran
Chaco

Kermadec Tr.

Austral Islands

Pitcairn I.

Sala y Gómez Ridge

Easter I.

Rise

△ Mt. Aconcagua
22,834 ft
(6,960 m)

Pampas

Río de la Plata

Louisville Ridge

Southwest Pacific Basin

Challenger Fracture Zone

Juan Fernández
Is.

Patagonia

Argentine
Plain

0   1,000   2,000 Miles
0   1,000   2,000 Kilometers

Southeast Pacific
Basin

Humboldt
Plain

Falkland Is.

Eltanin Fracture Zone

Strait of Magellan

South Georgia Ridge

Udintsev Fracture Zone

Cape
Horn

Drake Passage

Pacific-Antarctic Ridge

Antarctic Circle